AN INTRODUCTION TO STATISTICAL METHODS

An Introduction to
STATISTICAL METHODS

C.B. GUPTA
Adviser, Birla Institute of Management Technology, Delhi
Formerly-Principal, Shri Ram College of Commerce,
University of Delhi, and
Director, Institute of Management Technology, Ghaziabad

&

VIJAY GUPTA
Professor, Indian Institute of Technology,
Kanpur

Twenty-Third Revised Edition

VIKAS® PUBLISHING HOUSE PVT LTD

VIKAS® PUBLISHING HOUSE PVT LTD
A-22, Sector-4, **Noida**-201301 (UP)
Phone: 0120-4078900 • Fax: 4078999
VIKAS® Regd. Office: 576, Masjid Road, Jangpura, **New Delhi**-110 014
E-mail: ***helpline@vikaspublishing.com*** • ***www.vikaspublishing.com***
- First Floor, N.S. Bhawan, 4th Cross, 4th Main, Gandhi Nagar, **Bangalore**-560 009 • Phone: 080-22204639, 22281254
- Damodhar Centre, New No. 62, Old No. 59, Nelson Manickam Road, Aminjikarai, **Chennai**-600 029 • Phone: 044-23744547, 23746090
- P-51/1, CIT Road, Scheme - 52, **Kolkata**-700014 • Ph. 033-22866995, 22866996
- 67/68, 3rd Floor, Aditya Industrial Estate, Chincholi Bunder, Malad (West), **Mumbai**-400 064 • Ph. 022-28772545, 28768301

Distributors:

UBS PUBLISHERS' DISTRIBUTORS PVT LTD
5, Ansari Road, **New Delhi**-110 002 • Ph. 011-23273601, 23266646 • Fax: 23276593, 23274261
- 10, First Main Road, Gandhi Nagar, **Bangalore**-560 009 • Ph. 080-22253903
- Z-18, M P Nagar, Zone-1, **Bhopal**-462 011 • Ph. 0755-4203183, 4203193
- Ist Floor 145, Cuttack Road, **Bhubaneshwar**-751 006 • Ph. 0674-2314446
- 60, Nelson Manickam Road, Aminjikarai, **Chennai**-600 029 • Ph. 044-23746222
- 2nd & 3rd Floor, Sri Guru Towers, No. 1-7, Sathy Road, Cross III, Gandhipuram, **Coimbatore**-641 012
- 40/7940, Convent Road, **Ernakulam**-682 035 • Ph. 0484-2353901, 2363905
- 3rd Floor, Alekhya Jagadish Chambers, H. No. 4-1-1058, Boggulkunta, Tilak Road, **Hyderabad**-500 001 • Ph. 040-24754472 / 73 / 74
- 8/1-B, Chowringhee Lane, **Kolkata**-700 016 • Ph. 033-22521821, 22522910
- 9 Ashok Nagar, Near Pratibha Press, Gautam Buddha Marg, Latush Road, **Lucknow**-226 001 • Ph. 0522-2294134, 3014010
- 2nd Floor, Apeejay Chambers, 5 Wallace Street, Fort, **Mumbai**-400 001 • Ph. 022-66376922-3, 66102069 • Fax: 66376921
- GF, Western Side, Annapoorna Complex, Naya Tola, **Patna**-800 004 • Ph. 0612-2672856, 2673973
- 680 Budhwar Peth, 2nd Floor, Appa Balwant Chowk, **Pune**-411 002 • Ph. 020-24461653, 24433976

Twenty Third Revised Edition 2004
Reprinted in 2005, 2006, 2007, 2008
Second Reprint 2009

Printed at City Printers, Delhi-110095

Preface to the Twenty-third Edition

An Introduction to Statistical Methods was first published in 1957. It was written to serve as a text for students of management, commerce, economics and other social sciences. One of the challenges in teaching statistical methods to the students of social sciences is the fact that many of them have not been exposed to much mathematics beyond elementary algebra. I have, therefore, consciously kept the exposition of the basic principles to such a level that a knowledge of elementary algebra is all that is a pre-requisite for understanding the statistical concepts and methods. The encouraging reception to the book accorded by my colleagues teaching statistical methods and the continuing demand for it make me believe that I have succeeded in my endeavour.

This also provides me, every time that a new edition is prepared, the necessary stimulus to expand its coverage and improve the presentation of the fundamental statistical concepts. This helps in keeping pace with the rapidly changing courses in statistical methods being taught in our universities, management schools and other professional institutes.

This revised edition has a little more orientation towards the applications of statistical methods to various types of problems in commerce, business and other social sciences. As in the previous few editions, I have been joined by Prof. Vijay Gupta of Indian Institute of Technology, Kanpur, to go through the entire text in the light of the feedback received from the users of the book. In addition, a new chapter on *Theory of Games* has been added in view of its increased application in business management.

The large number of illustrations contained in the book serve a definite purpose—to demonstrate the applications of statistical techniques. These are, however, intended only as supplement to the main text and not as a substitute thereof.

I hope the book continues to be useful for the students.

C. B. Gupta

CONTENTS

Chapter 1

INTRODUCTION

"The fundamentalgospel of statistics is to push back the domain of ignorance, prejudice, rule of thumb, arbitrary or premature decisions, traditions and dogmatism, and to increase the domain in which decisions are made and principles are formulated on the basis of analysed quantitative facts."

ROBERT W. BUGESS

1.1. WHAT IS STATISTICS?

The term statistics is used to mean either statistical data or statistical method.

When it is used in the sense of statistical data it refers to *quantitative* aspects of things, and is a numerical description. Thus, the distribution of family incomes is a quantitative description, as also the annual production figures of various industries. These quantities are numerical to begin with. But there are also some quantities which are not in themselves numerical, but can be made so by *counting*. The sex of a baby is not a number, but by counting the number of boys, we can associate a numerical description to the sex of all new born babies, for example, when saying that 54 per cent of all live-born babies are boys. This information, then, comes within the realm of statistics. Likewise, the statistics of students of a college include count of the number of students, and separate counts of numbers of various kinds, as males and females, married and unmarried, or postgraduates and undergraduates. They may also include such measurements as their heights and weights. In addition, there may also be numbers computed on the basis of these measurements or counts, *e.g.*, the proportion of female students, their average height or average weight. An Example of statistical data is given in table 1.1.

The other aspect of statistics is as a body of theories and techniques employed in analysing the numerical information and using it to make wise decisions. It is a branch of the scientific method, used in dealing with those phenomena which can be described numerically, either by measurements or by counts. For example, if a preliminary test using a new vaccine shows that in a sample of 100 cases the incidence of the disease is reduced to 8 while that in the unvaccinated

TABLE 1.1. STATISTICS OF STUDENTS OF A COLLEGE

Total number of students : 1,000

Their sex-wise distribution			*Their class-wise distribution*		
Sex	*Number*	*Percentage*	*Course*	*Number*	*Percentage*
Male	900	90	Undergraduate	800	80
Female	100	10	Postgraduate	200	20
Total	1,000	100	Total	1,000	100

Distribution According to Height

Height	*Number*
From 160 cm to 170	600
From 170 cm to 180	390
From 180 cm to 190	10
Total	1,000

population it is 10, can we say if the vaccine is effective or not ? For surely we can select 100 individuals in the unvaccinated population in which the incidence is 8 or even less. Is the observed reduction due to chance causes or does it show the effect of the vaccine? The statistical method provides with theories and techniques for checking this out. In this text, we will be primarily concerned with the statistical method,its theories and its techniques.

Consider the following situations typical of what are faced by the decision-makers. In each of these the method at arriving at policy decision consists of first understanding the parameters of the problem, for which statistical data is called for.

1. A large technical university experiences a fall in the number of persons seeking admission. Is this fall due to factors which are peculior to that particular university, or is it a country-wide trend? The exact course of policy action will depend upon the answer to this question.

2. At what age should one retire airline pilots? Does the increasing age affect the safety record ? How is it balanced by the increasing experience, if at all?

3. In certain circles it is taken as axiomatic that more the people

travel in the country, the more emotional integration is achieved. Recently there has been some doubt cast on this promise. The implications of this are wide ranging, affecting the government attitude towards promotion of tourism. How does one verify the truth of the contention?

4. More than 50 per cent of heart-transplant patient die within a year. Is heart-transplant surgery beneficial? Wouldn't a heart-patient be better off without a transplant than with it? Wouldn't he live longer?

5. Is advertising on television more cost effective than advertising in print? Are street-corner hoardings effective at all?

Anybody would see immediately that we need data to answer any of these questions. But mere data would not help. The data will have to be systematically collected and analysed so that our answers are not affected by other factors. For example, how does one separate the effect of age and experience when one knows that experience increases with age? Also in each of these there are a whole lot of chance factors interacting with the outcome. Therefore, we have to isolate the effect that we want to study, and for this specialized statistical methods are called for.

The statistical method, when used properly, helps in understanding phenomena using numerical evidence. As a further example, suppose we want to understand the factors that affect the yield of farms. We may note that various factors such as rainfall, soil fertility, quality of seed, soil nutrients used, method of cultivation, etc. are all more or less important. One can never for sure predict the influence of one parameter, because we cannot control all of these independently. But it is possible to design experiments and collect data so that one is able to, more or less, isolate each effect to a predetermined level of certainity. The procedures for doing so are provided by statistical method.

As a further example, let us suppose we are interested in studying the level of income of the people living in a certain village.

For this purpose the following procedure may be adopted with advantages:

1. *Collect data.* Information should be collected regarding
   (*a*) the number of persons living in the village;
   (*b*) the number of persons who are getting income; and
   (*c*) the daily income of each earning member.

2. Organise the data obtained above so as to show the number of persons within different income groups, and in that way reduce its bulk.

3. Present this information by means of diagrams or other visual aids.

4. Analyse the data above to determine the 'average' income of the people and the extent of disparities that exist.

5. On the basis of the above it would be possible to have an understanding of the phenomenon (income of people), and one would know (*i*) the 'average income of people', and (*ii*) the extent of disparity in the level of incomes.

6. All this may lead to a policy decision for improvement of the existing situation.

1.2. FUNCTIONS OF STATISTICS

The proper function of statistics is to enlarge our knowledge of complex phenomena, and to lend precision to our ideas that would otherwise remain vague and indeterminate. Our knowledge of such things as 'national income,' 'population,' 'national resources,' etc., would not have been so definite and precise, if there were no reliable statistics pertaining to each one of these. To say that the per capita income in India is low, is a vague statement. 'Low' to one individual may mean one thing while to another it might mean something altogether different. I may take it to be near about Rs. 100 while someone else may think it to be in the neighbourhood of Rs 5,000. But the moment we say that our per capita income is Rs 750 we make a statement which is precise and convincing. Again a statement, viz., the per capita income in agricultural sector is lower than in the industrial sector, is vague and indefinite. But if the per capita incomes for both these sectors are ascertained, the comparison would be easier and even a layman would be able to appreciate the difference in the productivity of these two occupations. It can thus be said that 'statistics increases the field of mental vision as an opera glass or telescope increases the field of physical vision.' Statistics is able to widen our knowledge because of the following services that it renders.

It presents facts in a definite form. It is the quality of definiteness which is responsible for the growing universal application of statistical methods. The conclusions stated numerically are definite and hence more convincing than conclusions stated qualitatively. This fact can be readily understood by a simple example. In an advertisement, statements expressed numerically have greater attraction and are more appealing than those expressed in a qualitative manner. The caption, 'we have sold *more* cars this year,' is certainly less attractive than Record Sale of 10,000 cars in 1985 as compared to 6,000 in 1984.' The latter statement emphasises in a much better manner the growing popularity of the advertiser's cars.

Statistics simplifies unwieldy and complex mass of data and presents them in such a manner that they at once become intelligible. The complex data may be reduced to totals, averages, percentages, etc., and presented either graphically or diagrammatically. These devices help us to understand quickly the significant characteristics of the numerical data, and consequently save us from a lot of mental strain. Single figures in the form of averages and percentages can be grasped more easily than a mass of statistical data comprising thousands of facts. Similarly, diagrams and graphs, because of their greater appeal to the eye and imagination tender valuable assistance in the proper understanding of numerical data. Time and energy of business executives are thus economised, if the statistician supplies them with the results of production, sales and finances in a condensed form.

Statistics classifies numerical facts. The procedure of classification brings into relief the salient features of the variable that is under investigation. This can be clearly illustrated by an example. If we are given the marks in mathematics of each individual student of a class and if it is desired to judge the performance of the class on the basis of these data it will not be an easy matter. Human mind has its limitations and cannot easily grasp a multitude of figures. But if the students are classified *i.e.*, if we put into one group all those boys who get more than second division marks, in still another group those who get third division marks, and have a separate group of those who fail to get pass marks, it will be easier for us to form a more precise idea about the performance of the class.

Statistics furnishes a technique of comparison. The facts, having been classified, are now in a shape when they can be used for purposes of comparisons and contrasts. Certain facts, by themselves, may be meaningless unless they are capable of being compared with similar facts at other places or at other periods in time. We estimate the national income of India not essentially for the value of that fact itself, but mainly in order that we may compare the income of today with that of the past and thus draw conclusions as to whether the standard of living of the people is on the increase, decrease or is stationary. Statistics affords suitable technique for comparison. It is with the help of statistics that the cost accountant is able to compare the actual accomplishment (in terms of cost) with programmes laid out (in terms of standard cost). Some of the modes of comparison provided by statistics are : Totals, ratios, averages or measure of central tendencies, graphs and diagrams, and coefficients. Statistics thus 'serves as a scale in which facts in various combinations are weighed and valued'.

Statistics endeavours to interpret conditions. Like an artist statistics renders useful service in presenting an attractive picture of the phenomenon under investigation. But it frequently does far more than this by enabling the interpretation of condition, by developing possible causes for the results described. If the production manager discovers that a certain machine is turning out some articles which are not up to the standard specifications, he will be able to find statistically if this condition is due to some defect in the machine or whether such a condition is normal.

1.3. STATISTICAL METHOD

Statistical approach to a problem may broadly be summarised as: (*i*) collection of facts: (*ii*) organisation of facts; (*iii*) analysis of facts; and (*iv*) interpretation of facts.

A detailed discussion of the various methods of collection, presentation, analysis and interpretation of facts is given later. Here the intention is to give only a bird's eye-view of the entire statistical procedure.

1. Collection of facts is the first step in the statistical treatment of a problem. Numerical facts are the raw materials upon which the statistician is to work and just as in a manufacturing concern the quality of a finished product depends, *inter alia*, upon the quality of the raw material, in the same manner, the validity of statistical conclusions will be governed, among other considerations, by the quality of data used. Assembling of the facts is thus a very important process and no pains should be spared to see that the data collected are accurate, reliable and thorough.

One thing that should be noted here is that the work of collecting facts should be undertaken in a planned manner. Without proper planning the facts collected may not be suitable for the purpose and a lot of time and money may be wasted.

2. The data so collected will more often than not be a huge mass of facts running into hundreds and thousands of figures. Human mind has its limitations. No one can appreciate at a glance or even after a careful study hold in mind the information contained in a hundred or a thousand schedules. For a proper understanding of the data their irregularities must be brushed off and their bulk be reduced, *i.e.*, some

process of condensation must take place. Condensation implies the organisation, classification, tabulation and presentation of the data in a suitable form.

3. The process of statistical analysis is a method of abstracting significant facts from the collected mass of numerical data. This process includes such things as 'measures of central tendency'—the determination of Mean, Median and Mode—'measures of dispersion' and the determination of trends and tendencies, etc. This is more or less a mechanical process involving the use of elementary mathematics.

4. The interpretation of the various statistical constants obtained through a process of statistical analysis is the final phase or the finishing process of the statistical technique. It involves those methods by which judgements are formed and inferences obtained. To make estimates of the population parameters on the basis of sample statistics in an example of the problem of interpretation. For the interpretation of results a knowledge of advanced mathematics is essential.

1.4. CHARACTERISTICS OF STATISTICAL DATA

Even a casual look at Table 1.2 would lead us to the conclusion that statistical data always denotes 'figures', *i.e.*, numerical descriptions. Whereas this is true, it must be remembered that all numerical descriptions are not statistical data. In order that numerical descriptions may be called statistics they must possess the following characteristics:

(*i*) They must be in aggregates.
(*ii*) They must be affected to a marked extent by a multiplicity of causes.
(*iii*) They must be enumerated or estimated according to reasonable standard of accuracy.
(*iv*) They must have been collected in a systematic manner for a predetermined purpose.
(*v*) They must be placed in relation to each other.

Let us explain these characteristics:

Statistics are aggregates of facts. This means that statistics are a 'number of facts.' A single fact, even though numerically stated, cannot be called statistics. 'A single death, an accident, a sale, a shipment does not constitute statistics. Yet numbers of deaths, accidents, sales and shipments are statistics.' Observe carefully Table 1.2 containing information about the population of India. Table 1.2*a* states the population only for one year whereas Table 1.2*b* gives population figures for seven different years. The data given in Table 1.2*b* are statistics whereas the figure given in Table 1.2*a* is not so, for the simple reason that it is a single solitary figure.

TABLE 1.2

1.2(a)		1.2(b)	
Year	*Population (in lakhs)*	*Year*	*Population (in lakhs)*
1951	3,569	1911	2.490
		1921	2,481
		1931	2,755
		1941	3,128*
		1951	3,569
		1961	4,390
		1971	5,470

*After deducting estimated amount of inflation of returns in West Bengal and Punjab (20 lakhs).

They must be affected to a marked extent by a multiplicity of causes. The term statistical data can be used only when we cannot predict exactly the values of the various physical quantities. This means that the numerical value of any quantity at any particular moment is the result of the action and interaction of a number of forces, differing amongst themselves and it is not possible to say as to how much of it is due to any one particular cause. Thus, the volume of wheat production is attributable to a number of factors, viz., rainfall, soil, fertility, quality of seed, methods of cultivation, etc. All these factors acting jointly determine the amount of the yield and it is not possible for any one to assess the individual contribution of any one of these factors.

Statistics must be enumerated or estimated according to reasonable standards of accuracy. This means that if aggregates of numerical facts are to be called' statistics' they must be reasonably accurate. This is necessary because statistical data are to serve as a basis for statistical investigations. If the basis happens to be incorrect the results are bound to be misleading. It must, however, be clearly stated that it is not 'mathematical accuracy, but only reasonable accuracy' that is necessary in statistical work. What standard of accuracy is to be regarded as reasonable will depend upon the aims and objects of inquiry. Where precision is required accuracy is necessary; where generel impressions are sufficient, appreciable errors may be tolerated. Again, whatever standard of accuracy is once adopted, it should be uniformly maintained throughout the inquiry.

Statistics are collected in a systematic manner for a predetermined purpose. Numerical data can be called statistics only if they have been compiled in a properly planned manner and for a purpose about which the enumerator had a definite idea. So long as the compiler is not clear about the object for which facts are to be collected, he will not be able to distinguish between facts that are relevant and those that are unnecessary; and as such the data collected will, in all probability, be a hecterogeneous mass of unconnected facts. Again, the procedure of data collection must be properly planned, *i.e.*, it must be decided beforehand as to what kind of information is to be collected and the method that is to be applied in obtaining it. This involves decisions on matters like 'statistical unit,' 'standard of accuracy,' 'list of questions,' etc. Facts collected in an unsystematic manner, and without a complete awareness of the object, will be confusing and cannot be made the basis of valid conclusions.

Statistics should be placed in relation to each other. Numerical facts may be placed in relation to each other either in point of time, space or condition. The phrase 'placed in relation to each other' suggests that the facts should be comparable. Facts are comparable in point of time when we have measurements of the same object, obtained in an identical manner, for different periods. They are said to be related in point of space or condition when we have the measurements of the same phenomenon at different places or in different conditions, but at the same time. Numerical facts will be comparable, if they pertain to the same inquiry and have been compiled in a systematic manner for a predetermined purpose.

Putting all these characteristics together, Secrist has defined statistics (numreical descriptions) as: "Aggregates of facts, affected to a marked extent by multiplicity of causes, numerically expressed, enumerated or estimated, according to reasonable standard of accuracy, collected in a systematic manner, for a predetermined purpose, and placed in relation to each other."

1.5. SOME OTHER* DEFINITIONS OF STATISTICS

As Numerical Data

Wester has defined statistics as "classified facts respecting the con-

*These definitions are out of date, but are included here to provide a glimpse of the historical evolution of the subject.

dition of the people in a state . . . especially those facts which can be stated in numbers or in tables or in any other tabular or classified arrangement." No doubt, this definition was correct at a time when statistics were collected only for purposes of internal administration or for knowing, for purposes of war, the wealth of the State. The scope of statistics is now considerably wider and it has almost a universal application. Obviously, therefore, the definition is inadequate.

Bowley defines statistics as 'numerical statements of facts in any department of inquiry placed in relation to each other.' This is somewhat more accurate. It means that if numerical facts do not pertain to a department of inquiry or if such facts are not related to each other they cannot be called statistics. The leads us to the conclusion that 'all statistics are numerical facts but all numerical facts are not statistics.' This definition is certainly better than the previous one. But it is not comprehensive enough in as much as it does not give any importance either to the nature of facts or the standard of accuracy.

As Statistical Methods

Bowley has called it 'the science of measurement of the social organism, regarded as a whole, in all its manifestations.' This definition is too narrow as it confines the scope of statistics only to human activities. Statistics in fact has a much wider application and is not confined only to the social organism. Besides, statistics is not only the technique of measuring but also of analysing and interpreting. Again, statistics, strictly speaking, is not a science but a scientific method. It is a device of inferring knowledge and not knowledge itself.

Bowley has also called statistics 'the science of counting,' and 'the science of average.' These definitions are again incomplete in the sense that they pertain to only a limited field. True, statistical work includes counting and averaging, but it also includes many other processes of treating quantitative data. In fact, while dealing with large numbers, actual count becomes illusory and only estimates are made. Thus these definitions can also be discarded on the ground of inadequacy.

1.6. ORIGIN OF STATISTICS

Statistics originated from two quite dissimilar fields, viz., games of chance and political states. These two different fields are also termed as two distinct disciplines—one primarily analytical and the other essentially descriptive. The former is associated with the concept of chance and probability and the latter is concerned with the collection of data.

The theoretical development of the subject has its origin in the mid-seventeenth century and many mathematicians and gamblers of France, Germany and England are credited for its development. Notable amongst them are Pascal (1623-1662), who investigated the properties of the coefficients of binomial expansion and James Bernouilli (1654-1705). who wrote the first treatise on the theory of probability.

As regards the descriptive side of statistics it may be stated that statistics is as old as statecraft. Since time immemorial men must have been compiling information about wealth and manpower for purpose of peace and war. This activity considerably expanded at each upsurge of social and political development and received added impetus in periods of war.

The development of statistics can be divided into the following three stages :

The empirical stage (down to 1600). During this, the primitive stage of the subject, numerical facts were utilized by the rulers, principally as an aid in the administration of Government. Information was gathered about the number of people and the amount of property held by them—the former serving the ruler as an index of human fighting strength and the latter as an indication of actual and potential taxes.

The Comparative stage (1600-1800). During this period statisticians frequently made comparisons between nations with a view to judging their relative strength and prosperity. In some countries enquiries were instituted to judge the economic and social conditions of their people. Colbert introduced in France a 'mercantile' theory of Government whose basis was essentially statistical in character. In 1719, Frederick William I began gathering information about population occupation, house-taxes, city finance, etc., which helped to study the condition of the people.

The modern stage (1800 up to date). During this period statistics is viewed as a way of handling numerical facts rather than a mere device of collecting numerical data. Besides, there has been a considerable extension of the field of its applicability. It has now become a useful tool and statistical methods of analysis are now being increasingly used in biology, psychology, education, economics and business.

PROBLEMS

1. What is statistics?
2. Discuss the meaning and scope of statistics.
3. 'Statistics are numerical statements of facts in any department of inquiry, placed in relation to each other.' Explain.
4. Comment on the following:
   "Statistics is always concerned with mass phenomena and never with a single observation."
5. "Statistics is a science of counting." Comment and give a comprehensive definition of statistics.
6. "Statistics is the science of measurement of social organism regarded as a whole in all its manifestations." Comment.
7. Write short notes on:
   (*i*) multiplicity of causes,
   (*ii*) statistical methods,
   (*iii*) origin of statistics.
8. Examine critically the important definitions of statistics, pointing out the one you consider to be the best.
9. What is statistical enquiry? Describe the main stages in a statistical enquiry.
10. Describe the main steps that must be considered in planning a statistical survey.

Chapter 2

THE FIELD OF STATISTICS

2.1. IMPORTANCE OF STATISTICS

Statistical methods have become useful tools in the world of affairs. A man who opens and slams the door of a car before he buys it is using the techniques of statistical analysis as much as the scientist who is trying to assess the impact of increased solar activity on the onset of monsoon. In both cases the person concerned is trying to draw inferences from the limited (numerical) information that is available to him. At times he can seek more information. But obtaining information costs money and time, and one may be forced to make a decision in the absence of complete information. At other times more information may be impossible to obtain. If one wants to estimate the life of a lot of electrical bulbs, one can take each bulb and test it to burn-out to obtain complete information. But the information so obtain is useless, besides being expensive, for then one has no bulb left.

It is, therefore, essential that one make decisions on the basis of incomplete information. As stated in Chapter 1, the science of statistics provides techniques for doing this. The economy afforded and the high degree of flexibility provided by Statistical methods have rendered them specially useful to economists, businessmen and scientists. We give below some of the major applications of the statistical method.

2.2. STATISTICS IN BUSINESS

The need for statistical information in the smooth functioning of an undertaking increases along with its size. The bigger the concern the greater is the need for statistics. In the era preceding the Industrial Revolution the master craftsman was in intimate touch with the sources of the supply of raw materials. He worked in his own home with the help of the members of his family and a few other employees whom he knew rather well. His customers were few and he knew them all personally. Thus, he had almost all information about his business and obviously no technique for the supply of this information was necessary.

Today also, in an era of mass production technology, the business executive needs all such information for the successful conduct of affairs. But he cannot, even if he were to try, get this information in the same manner as the master craftsman did. Naturally, therefore, he has to resort to the statistical technique and statistics takes the place of personal observation. 'For better or wrose, the modern business executive is largely dependent on statistical data and methods of analysis for essential information'.

No business, large or small, public or private, can flourish in these days of large-scale production and cut-throat competition without the help of statistics. Statistical information is needed from the time the business is launched till the time of its exit. At the time of the floatation of the concern facts are required for the purposes of drawing up the financial plan of the proposed unit. All the factors that are likely to affect judgement on these matters are quantitatively weighed and statistically analysed before taking any decisions.

A shrewd manufacturer must know in advance 'how much is to be produced,' 'how many workers and how much raw material is needed to produce that estimated quantity' and what quality, type size, colour or grade of the product is to be manufactured.' In short, he must have a production plan. Now such a plan—requiring all the details given above—cannot be framed without quantitative facts. Statistics thus help in planning and formulation of future policies.

Quantitative data will have to be collected and analysed, if a workable personnel plan is to be carried out. The only route for a personnel officer or a labour officer to get acquainted with the labour force numbering hundreds, or thousands, or even lakhs, is to know its members through statistical analysis of information, largely quantitative. Wage levels and wage standards also require the statistical study of different jobs within the same organisation and the study of wages in like business undertakings.

In a labour dispute it is the official of the union that generally represents the workers. It is through statistical data that a man representing the workers knows about the working conditions, rates of wages, frequency of lock-outs, monthly earnings and other matters in the industry where the dispute may arise. Again, in negotiation conferences, proper data, competently collected and honestly analysed, may lead to an early and just solution of the differences.

Statistical methods of analysis are helpful in the marketing function of an enterprise through its enormous help in market research, advertisement campaigns and in comparing the sales performances. Statistics also directs attention towards the effective use of the advertising funds.

Above all, statistical methods of analysis provide an important tool to the management for cost and budgetary control. The most elementary use to the management is in the balancing of the activities of one part of a system against those of another, to secure that supplies equal requirements and that there are no 'bottlenecks' or parts that are not employed to the full.

Various statistical techniques viz., index numbers and analysis of time series help in the study of price behaviour; correlation and regression help in the estimation of relationships between dependent and one or more independent variables, *e.g.*, relationships are established between market demand and per capita income, inputs and outputs etc.

The theory and technique of sampling can be used in connection with various business surveys with a considerable saving in time and money. Likewise these techniques are now being extensively used in test checking of accounts.

Statistical quality control is now being used in industry for establishing quality standards for products, for maintaining the requisite quality, and for assuring that the individual lots sold are of a given standard of acceptance.

Statistics is thus a useful tool in the hands of the management. But it must be remembered that no volume of statistics can replace the knowledge and experience of the executives. Statistics supplements their knowledge with more precise facts than were hitherto available.

We give below some typical situations in business which can be analysed using statistical techniques:

(*a*) **Survey of consumer tastes.** To predict the acceptability of a synthetic soft-drink concentrate, a manufacturer distributed code-marked samples along with the samples of a leading brand. By analysing consumer preferences, he could select the sales territory in which to concentrate his effort, and obtained very good results.

(*b*) **Quality control.** An electric lamp manufacturer wanted to control the average life of his product. Since he obviously could not test every bulb to burnout, he devised a sampling plan wherein 25 out of every lot of 1,000 bulbs were tested. The average life of these 25 bulbs provided a check on the quality of the whole lot.

(*c*) **Optimum inventory size**. Large dealerships require one to maintain a stock large enough to service all customers but not so large that money is tied up unnecessarily in idle stock. An auto spare-parts manufacturer solved this problem by collecting statistics over a year and determining the probable distribution of demand, and calculated the optimum inventory levels.

(*d*) **Overbooking**. Because of the passengers dropping out at last moment, all airlines overbook their flights in the hope that some passengers will not show up. This is a risky proposition because if all do show up, one has a lot of irate passengers on one's hand.By proper statistical analysis one can determine the optimum overbooking so as to minimize this risk and yet maximize the load factor.

2.3. STATISTICS IN ECONOMICS

Statistical data and methods of statistical analysis render valuable assistance in the proper understanding of the economic problems and the formulation of economic policy. Economic problems almost always involve facts that are capable of being expressed numerically, *e.g.*, volume of trade, output of industries—manufacturing, mining and agriculture—wages, prices, bank deposits, Clearing House returns, etc. These numerical magnitudes are the outcome of a multiplicity of causes and are consequently subjects to varriations from time to time, or between places or among particular cases. Accordingly the study of economic problems is specially suited to statistical treatment.

Let us take an example to clarify this point. A proper appreciation of the nature and magnitude of the problems of unemployment would necessitate a knowledge about the following: Is unemployment increasing or decreasing? It is widespread or largely confined to certain area? Does it affect the educated and uneducated alike or is more pronounced in any particular class? which industries are expanding and which are contracting? Has there been any remarkable increase in the population? All these questions can be answered statistically, and the resultant data will enable us to form a correct estimate of the problem. It is natural, therefore, that there is a growing emphasis on the importance of collecting systematic and regular statistic bearing on every aspect of our economy. Mere collection of data, however, is not enough. The complexity of all such problems make it imperative that the collected data be condensed and analysed—condensed in order that it may be possible for limited human faculties to handle, and analysed in order that the elements in the problem may be distinguished and their significance appreciated.

A statistical approach to an economic problem not only leads to its correct description but also indicates lines along which it is to be tackled. The great emphasis that was laid upon the development of agriculture by the Planning Commission in the First Five Year Plan can be defended with the help of factual data bearing on our economy. In a country where agriculture contributes nearly 50 per cent of its total income, and where the net output (in agriculture) per engaged person is the lowest, being only Rs 500 per year, it is in the fitness of things that the planners have laid so much emphasis on the improvement of our agriculture.

Apart from economic policy, the development of economic theory has also been facilitated by the use of statistics. The complexity of modern economic organisation has rendered deductive reasoning inadequate and difficult. Statistics is now being used increasingly not only to develop new economic concepts but also to test the old ones. The increasing importance of statistics in the study of economic problems has resulted in a new branch of study called Econometrics.

2.4. STATISTICS IN THE BIOLOGICAL SCIENCES

Statistics is being used more and more in biological sciences as an aid to the intelligent planning of experiments, and as a means of assuring the significance of the results of such experiments. Experiments about the growth of animals under different diets and environments, or the crop yields with different seeds, fertilizers and types of soil are frequently designed and analysed according to statistical principles. The entire theory of heredity rests on statistical basis, and its development has been intimately related to the development of statistics. The following are some specific examples:

(*a*) **Fish populations**. To estimate the population of fish in a lake, biologists catch a sample, count them, mark them (say, with a metal tag, and release them back into the lake. They then catch another sample. If the percentage of marked fish is 5 in this second catch then one fairly accurate estimate of the total fish population is 20 times the size of the original catch. To control the accuracy of this estimate one uses statistical theories and determines the size of the catch required for a given accuracy.

(*b*) **Mendelian heredity**. Gregor Mendel, the famous biologist developed his theory of heredity using statistical techniques on the distribution of various characteristics in the various generation of common garden peas.

2.5. STATISTICS IN THE PHYSICAL SCIENCES

In the beginning of the nineteenth century some of the statistical methods were not only applied but also developed in the fields of astronomy, geology and physics. But for a considerable time afterwards these sciences did not share later developments in statistics to the same extent as biological sciences. This was due mainly to their relatively high precision of measurements. In fact, for a long time statistics did not make any progress in physical sciences beyond the calculation of standard error, and fitting of curves. Currently, however, the physical sciences are making increasing use of statistics in the treatment of the complex problems of molecular, atomic and nuclear structure, The following are some typical examples:

(*a*) **Statistical model of gases.** According to the current concepts a gas consists of a large number of molecules moving randomly and at various velocities. It is possible to construct a probability distribution for the speed of particles and from it to obtain estimates of the various properties of the gases.

(*b*) **Radiocarbon dating**. A certain form of the element carbon emits radioactive particles. The content of this radiocarbon decays with the age of non-living things. Therefore, by measuring the amount of radiocarbon we can fix the age of the object. This process is called radiocarbon dating. But the particle emmission are a matter of chance and, therefore, statistical techniques are required to relate the emission with the amount of radiocarbon present in a sample.

2.6. LIMITATION OF STATISTICS

That statistical technique, because of its flexibility and economy, is growing in popularity and is being successfully employed by the seekers of truth in numerous fields of learning is a fact that cannot be denied. But it is not without limitations. It cannot be applied to all kinds of phenomena and cannot be made to answer all our queries.

Statistics deals with only those subjects of inquiry which are capable of being quantitatively measured and numerically expressed. This is an essential condition for the application of statistical methods.

Now all subjects cannot be expressed in numbers. Health, poverty, intelligence (to name only a few) are instances of the objects that defy the measuring rod, and hence are not suitable for statistical analysis. It is true that efforts are being made to accord statistical treatment to subjects of this nature also. Health of the people is judged by a study of its death rate, longevity of life and the prevalence of any disease or diseases. Similarly intelligence of the students may be compared on the basis of marks obtained by them in a class test. But these are only indirect methods of approaching the problem and subsidiary to quite a number of other considerations which cannot be statistically dealt with.

Statistics deals only with aggregates of facts and no importance is attached to individual items. It is, therefore, suited only to those problems where group characteristic are desired to be studied. But where the knowledge about individual cases is necessary statistical technique proves inadequate. The per capita consumption of food-grains in a state will camouflage cases of starvation, if any. The scarcity felt by the poorer section may be more than made up by the extravagance of the rich. In such cases, therefore, statistics, will fail to reveal the real position.

Statistical data is only approximately and not mathematically correct. Greater and greater emphasis is being laid on sampling technique of collecting data. This means that by observing only a limited number of items we make an estimate of the characteristic of the entire population. This system works well so long as the mathematical accuracy is not essential. But when exactness is essential statistics will fail to do the job.

Statistics can be used to establish wrong conclusions and, therefore, can be used only by experts. Since many of the statistical conclusions are based on sample studies, it is very common to come to wrong conclusions if one is not very careful about the techniques of analysis. In fact, one is so often deceived by "correct" facts that there is a general distrust of things "proved statistically". Usually, most of these can be traced to incorrect application of methods. The next few sections illustrate some of the common statistical fallacies.

2.7. DISTRUST OF STATISTICS

In spite of the very valuable service that statistics renders to business community and to scientists, both social and natural, there is some amount of misgiving in the minds of a few people with regard to their reliability and usefulness. This feeling has been given expression in a number of ways of which the following are the often quoted examples:

'There are three kinds of lies—namely, lies, demand lies, and statistics—wicked in the order of their naming.'

'With statistics anything can be proved.'

Such misgivings can be attributed mainly to two clauses: (*i*) figures carry conviction and are capable of being easily manipulated; (*ii*) the presence in this world of persons who are selfish and unscrupulous.

If an argument is supported by facts, stated numerically, it has a much greater appeal than the one without them. It is because of this that in a discussion the winner is almost invariably the one who is able to substantiate his point with facts and figures. Naturally, therefore, if a lie is to be pushed through, the best way is to give figures in support of it. Statistical data do not carry the label of their quality and as such can be manipulated in any desired manner without causing the least suspicion. Statistical data and conclusions may be manipulated in many ways, some of which are given in the following sections.

2.8. MISUSE BY SHIFTING DEFINITION

A slight alteration in the definition of a key term might provide a basis for conclusions which are not warranted by facts. Thus, while making comparison of the number of workers employed in two industrial units misleading conclusions may be obtained, if the meaning assigned to the term 'workers' is not identical while conducting the census of workers in the two units. In one case 'workers' may include casual workers also while in other, they may be excluded. The following are some other examples:

1. A recent press release showed the development cost of the Indian SLV-3 rocket to be very low compared to the costs of similar systems developed elsewhere. This was due to the fact that the establishment cost of the space research organization whose principal task was the development of this vehicle was not charged to this head, contrary to the practise elsewhere.

2. In a labour dispute the union claimed that average hourly rate has decreased while the management claimed that it has increased. On closer scrutiny it was found that the management has taken a straight average of income over number or hours worked, while the union has accounted for the increased rate during overtime. Over the period in question, the amount of overtime had increased resulting in the descrepancy observed.

2.9. MISUSE BY METHOD OF SELECTING CASES

If it is desired to ascertain the number of school age children per family it may be done in either of the two ways: (*i*) by questioning a number of children in schools, (*ii*) by conducting a survey of a number of families. It is quite possible that the results obtained by the former method may be different from those obtained by the latter method. If we take a simple case of two families, one with a single school age child, and the other with six, the number of school age children per family would be 3.5 under the second method. But if we apply the first method the result would be higher *i.e.*, if each of the seven children were asked the number of school age children in his family, the total of seven replies would be 37 and the average 5.286 children per family. The following are some more examples:

1. A market research company claimed that in over 50 per cent of the cases handled by them they could improve the decisions taken by the company executives and went on to claim that the typical company executive was proved to err about half the time. This is a mis-statement because obviously a company consulted the researcher only when in doubt. The cases referred to him do not reflect a typical sample of cases handled by the marketing executive.

2. A survey of those attending the showing of an off-beat film which had difficulty finding a distributor showed that 80 per cent liked it. The report went on to claim that the distributors are poor judges of public taste. This conclusion seems to be based on the assumption that those who attended the survey represent a typical sample of movie-going public. This is highly doubtful since only those who like such movies are likely to attend it.

2.10. MISUSE BY INAPPROPRIATE COMPARISON

Wrong conclusion may be obtained by comparing 'statistics' which are not essentially comparable. Thus, if the items included in the construction of a 'price index' have changed over a period of time the comparison of the price index as it was in the beginning of the period with that at the end of the period would lead to misleading results. Some more examples are given below:

1. One newspaper report claimed that it is far more dangerous to drive on streets of New Delhi than fight in a war. While only just over 1,000 persons lost their lives in the last war, well over 2,000 persons lose their lives annually in road accidents in New Delhi. The report is clearly erroneous because the number of persons using

Delhi's streets is far larger. The appropriate comparison is between proportions and not numbers.

2. It was claimed sometimes ago that there has been no advance made in treatment of cancer because the death rate due to it has remained constant over the years. The comparison is inappropriate because while the persons contracting cancer may still be dying of it, the advancement of treatment should be measured not by decrease in death rate but by the increase in expected life-span of cancer patients due to improved treatment.

3. A recent headline claimed that there has been no increase in the incidence of Malaria, noting that it is "still one half of what it was in 1952. The comparison here is again inappropriate, if we compare the present incidence with that a few years ago, a different conclusion may emerge.

2.11. MISUSE BY MISINTERPRETATION OF ASSOCIATION OR CORRELATION

Sometimes a certain degree of association or correlation may be apparent from a set of figures when none actually exists. A statement—"those who drink die before reaching the age of 100 years and hence drinking is harmful for longevity"—is a case in point. Unless it is shown that those who do not drink live up to 100 years or a higher age than those who drink, the conclusion contained in the statement is fallacious.

Some other examples are:

1. A social scientist claimed that the more children you have, the less likely it is that you will ever be divorced. He based his claim on the statistics of divorce applications which clearly showed that couple with more children are less likely to file for divorce. But this is misrepresentation of correlation because more children also means that a couple has been longer together and thus a lower divorce rate may mean that divorces occur early in the marriage.

2. In a recent election it was observed that a particular party polled generally lesser votes in those constituencies where a very senior leader campaigned personally. This led some to doubt the popularity of the leader. The conclusion may be erroneous because it is possible that the senior leader was asked to campaign in only the doubtful constituencies and not in the safer ones where the election was presumed to be 'won'.

3. A survey of the alumuni of a leading technical institute showed that the average salary of those who graduated near the top of the

class was less than that of the rest. This can give a misleading impression unless it is noted that the top of the class might have decided to opt for academic or research jobs which generally pay less than other jobs.

3. "Three men set about investigating what caused intoxication drank whiskey and water one day, rum and water the second day, and vodka and water the third day, and concluded that since water was the common constituent, that is what caused intoxication."

2.12. MISUSE DUE TO INADEQUATE SAMPLE SIZE

Statistical data may lead to misleading results, if the investigator jumps to a conclusion on the basis of too small a sample or one which does not represent the whole population adequately. Thus, if a coin is tossed four times and the head turns up thrice the conclusion that the probability of getting a head is 0.75 is wrong because it is based on inadequate sample. The size of the sample should be sufficiently large and it must be representative of the population.

Now, if such manipulated figures are quoted in support of one's point of view they are likely to mislead people. And since there is not a great scarcity of such selfish and unscrupulous people who do not hesitate in distorting facts, statistics are used to prove the worst lies.

But if statistics are made to prove anything, it is certainly not the fault of statistics as such but of the person who makes them the tool of his personal aggrandizement.

It must, therefore, be clearly stated that the useful application and coordination of statistics is not exclusively dependent upon the degree of skill, ability and special experience of the statistician. No matter how best the statistical techniques are used in analysing, representing or interpreting the information, the conclusion so derived may become unreliable and even useless, if the enumerators or investigators are biased and prejudiced.

PROBLEMS

1. Clearly explain what do you understand by statistics. Discuss its scope and limitations.
2. Write a short essay on the uses of statistics in commerce and industry.
3. What are the uses and limitations of statistics? Discuss the importance of statistics as an aid to commerce.

4. Explain clearly what you understand by the science of statistics. Discuss its scope and limitations.
5. "Statistics should not be used as a blind man uses a lamp post for support rather than for illumination." Comment.
6. Write a note on the importance of statistics to the businessman, the economist, and government.
7. Explain and illustrate the uses of statistics in commerce and business.
8. 'Statistics can prove anything'
   'Figures cannot lie'
   Comment on the above two statements, indicating reasons for the existence of such divergent views regarding the nature and functions of statistics.
9. 'Statistics widens the field of knowledge.'
   Elucidate the above treatment.
10. Discuss how in modern times statistics is the science of human welfare.
11. "Statistics only furnishes a tool though imperfect which is dangerous in the hands of those who do not know its use and deficiencies.'. Comment.
12. Explain elearly the characteristics and limitations of statistics.
13. Write a critical note on the limitation and distrust of statistics. Discuss the important causes of distrust, and show how statistics could be made more reliable,
14. Define 'statistics'. Give three illustrations to show how statistics may be misused.
15. "Sciences without statistics bear no fruit; Statistics without sciences have no root." Explain the above statement showing the relationship of statistics with some other sciences.
16. Statistics are like clay of which you can make a God or Devil, as you please." In the light of the above statement, discuss the uses and limitations of statistics.
17. Bring out clearly the scope of statistics and discuss its limitations.
18. What is statistics? Bring out clearly the importance of statistics methods to business.
19. Discuss the various definitions of statistics. Describe the important uses and limitations of statistics.
20. In what way statistics is useful to the state, the economist, the industrialist and trader?
21. Explain with the help of suitable illustrations, the functions of statistics. Give two imaginary examples to show how statistics may be misused.
22. Define statistics and show how it can help the extension of scientific knowledge, the establishment of a sound business and the formulation of a plan for national economic development.
23. "Without an adequate understanding of statistics the investigator in the social sciences may frequently be like the blind man groping in a dark closet for a black cat that isn't there." Comment.

Can the statement be extended to the field of natural sciencсs also?

24. The following represent some or the other misuse of statistics. Find out the flaws in the arguments:
    (*a*) The number of men in the prisons today is less than that twenty years ago. That means we are doing a better job at controlling crimes than we did earlier.
    (*b*) The dental profession is understaffed. Only 12 per cent of the population receives dental care in our country compared to an average of 22 per cent for western countries.
    (*c*) Since most of the persons get involved in giving or receiving illegal favours at one time or the other, therefore, only a small percentage persons who are prosecuted for graft have been involved in behaviour which is materially different from the behaviour of most of the populations.
    (*d*) "It is three times as dangerous to be a pedestrian while intoxicated as to be a driver. This is shown by the fact that last year 13,000 intoxicated pedestrians were injured while only 4,300 intoxicated drivers were injured."
    (*e*) Inspite of higher average per day charges of the hospital, it is cheaper now to go to hospital than it was 30 years ago. This is because the improved diagnostic and treatment techniques have made hospital stays shorter, and as the statistics bear out, the charge per hospital stay has gone down.
    (*f*) In an attempt to minimize the seriousness of the problem of torture of minority community at police stations, it was claimed that more men belonging to majority community were allegedly tortured than those belonging to the minority community.
    (*g*) In a study of unemployment it was noticed that more males than females were on the employment exchange register. This showed that contrary to popular belief, it is men who find it difficult to get jobs.
    (*h*) The latest census report shows that the death rate among widowers and widows were almost twice that among married persons showing that demise of one spouse hastens the death of the other.
    (*i*) The 1981 census report shows that the average Indian's age come down by 0.9 years in the ten years since the last census.
    (*j*) The termination of service as a punishment for indiscipline has little value since the High Court invalidates most of those that are disputed before it.

Chapter 3

STATISTICAL INQUIRIES AND SAMPLING

3.1. INTRODUCTION

By 'inquiry' we mean 'a search for knowledge.' Statistical inquiry, therefore implies a search conducted according to the statistical technique. The technique of statistics, however, cannot be applied to all kinds of phenomena. It application is restricted to only those subjects which can be measured quantitatively. It is possible to know statistically as to whether the distribution of wealth in a particular country is equitable or not, for a simple reason that both wealth and the concept of equity are capable of being expressed numerically. But if it is desired to know whether such a distribution is justifiable, statistical methods will not be of much assistance as 'justifiability' cannot be quantitatively measured.

It should be obvious that before we conduct an inquiry its purpose and scope should be clear to us and the inquiry should be carefully planned to ensure that what we measure has a relationship with what we want to study. Since it is mere numbers that a statistician is dealing with, he should ensure before hand that the numbers are relevant to the phenomenon he wants to measure. If we want to measure the cost of living, an obvious question that comes up is the cost of what? Since it is not the cost of only one thing that controls the cost of living, and since the cost of various items vary at different dates, we have to take a kind of an average cost. But does one average the cost of an apple with that of a house or an automobile? Obviously different weighting factors have to be used.

Similarly if we want to measure the spread of literacy in a community, it will be worthwhile to spend some time deliberating on what is literacy and what measure of it would have a meaning. Should a person who can sign his name with difficulty, or a person who could read a child's reader be treated as literate? Similar questions can arise in almost any field of statistical inquiry. A meterologist who wants to study the change of weather reads his thermometers, barometers, anemometers, psychrometers, hydrometers, precipitation gauges, sunshine recorders and comes up with a conclusion regarding weather. But at best he is measuring only a few parameters. Surely, these few parameters donot 'define' the weather completely.

The problem of the relationship between a number and what is intended to be measured can be tricky even in situations which appear straightforward. A museum took pride in its attendance figures thinking that it represented the interest the public had in its displays. One year a small building went up next to the museum and the attendance figure dropped down to 10 per cent of the original level. What was the new building? A public toilet.

3.2. POPULATION

In statistics the term *population* is used to mean the totality of cases (items) in a investigation. For instance, if the inquiry is intended to determine the yield of wheat per acre in a state in a particular year, the population will include all farms on which wheat was grown in the state in that year. Again, if we wish to study the family incomes in a certain region, the population will comprise of all the households in that region. Likewise, if it is intended to study the average life of a lot of 1,000 electric bulbs the population will include all these 1,000 bulbs.

3.3. CENSUS ENUMERATION

A straightforward method which an investigator may adopt for his inquiry is to observe each item constituting the population and obtain the information relevant to his purpose. Thus, he may prepare a list of all farms growing wheat and obtain information from all of them about their areas and yields. An investigator studying the national income may ask every individual of the country his personal income and add it all up. This method of inquiry is termed as census enumeration.

In most cases, the data obtained by such a procedure is too cumbersome to be directly useful. It is usually described by such characteristics as mean and spread. Some of these characteristics are discussed in the following chapters. The discriptive characteristics so obtained from the entire population through census enumeration are termed as *population parameters*. These are most often all that we can handle and also all that we are interested in determining. Statistical inquiries are usually aimed at determining these parameters.

3.4. SAMPLING

As the population in most inquiries are quite large the cost of such census enumerations in time and money will be substantial. But since the parameters are. in a sense, only the indicators of the quantitative nature of the population data, we in almost all situations do not lose much if certain amount of accuracy is sacrificed for savings in census costs. If we can estimate the parameters without measuring every item of the population, and if we can determine the reasonable bounds of errors of such estimates from the true values of the population parameters, that is about all we need in most cases. Such a procedure of estimating population parameters from only a few items is called *sampling*.

When we look at a handful of grain to evaluate the quality of wheat in a sack, we are conducting a sampling enquiry. Again, when we take out a few screws from a production run and test them to pass judgement on the quality of all screws, it is a sampling procedure. In both these cases, we estimate the parameters from the study of samples. In both these we could collect data on every single item and then evaluate the parameters. But the additional accuracy will be achieved at considerable cost and may not be worth it

In all the above cases, the populations are very large, but still finite. It is possible at least in theory to conduct a complete census. There are some cases when the census is impossible because of the infinite size of the population. Consider, for example, the problem of determining the proportion of times the head will come up in tossing of a bent coin. The population in this case consists of all possible tosses of this coin. But, that is an infinite number. We will have to make do with a finite number of tosses and from that estimate the proportion of times the coin will turn up heads in all possible tosses.

Again, consider the problem where a large number of springs are to be tested for strength. To find the strength of a spring, one may load the spring, and keep on increasing the load till the spring fails. Due to factors beyond human control, no two springs have exactly the same strength, but the strengths are distributed about an average. To find this average strength we could test every spring and calculate the mean of the strengths so obtained. But in the end we would not have a single spring left for use. It would be of great value if we could draw a sample from the population of springs, test them for strength, and estimate the population mean from the sample data so obtained.

Sampling is thus used for a variety of reasons which include: (1) Savings in time and money; (2) When the population is infinitely large; and (3) Where measurement technique destroys or in someways, alters irreversibly the items which are measured.

A single population can give rise to many different samples. The population, therefore, is stable but samples and their characteristics vary. The central problem in statistics is to determine the characteristics of the population (which is fixed but unknown) from the known statistics of the sample (which may vary from sample to sample and thus, there is no one-to-one correspondence between population parameters and sample statistics).

The fact that the characteristics of the sample (*i.e.*, sample statistics) are able to provide an approximately correct idea about the population parameters is borne out by the theory of probability. We give here two laws which form the basis of the sampling procedure.

3.5. LAW OF STATISTICAL REGULARITY

The law of statistical regularity enunciates that a group of objects selected at random from a population tends io possess the characteristics of the larger group. This law operates only if the samples are randomly chosen, which means that they are chosen by a method in which every item of the population has an equal chance of selection, *i.e.*, the selection process does not favour or disfavour any item or group of items. Even with randomly chosen samples the statistical regularity is only a tendency. Put in other words, the law states that *the pattern of variations within the sample depends upon that of the population*. Thus, if the population is skewed to right, the sample, more often than not, will be skewed to right. The characteristics of the sample will not be identical to, but only in the near neighbourhood of the characteristics of the population. This should be obvious since, if a number of samples are taken from a particular population, and the items are not all identical, the results will generally vary from sample to sample, and thus may differ from the population characteristics. But the over all pattern of variation within the population will be reflected in the sample. The size of the spread around the population value within which the sample value may lie, is governed by the law of large numbers.

3.6. LAW OF LARGE NUMBERS

This law lays down that *large samples are more stable in their characteristics than the small ones.* This means that as we increase the sample size the characteristics like mean and standard deviation tend to become more steady and converge on to the population parameter. To take an example of the operation of this law, consider a 'fair' coin which is as likely to turn up 'heads' as 'tails.'

If we take all possible tosses of this coin, the population proportion of heads will be 0.5. If we toss it only a few times, the proportion will almost certainly be different from 0.5. As we increase the number of tosses (increase the sample size), the proportion will fluctuate. These fluctuations will tend to die out as the sample size becomes large, and the proportion will converge on to 0.5. Thus, the larger the sample, more representative is it of the population from which it is drawn.

There is one aspect of this law of large numbers on which some confusion prevails. It is sometimes wrongly assumed that the tosses of a fair coin approach 50 per cent heads because if there is an unusual runs of heads, it will more likely be followed by tails. In other words, it is thought that law operates *by compensation* that is, the movement in one direction due to one part will be offset by the movement in the other direction by some other part. This is actually not so. Even after an usually long runs of head, the next toss is as likely to be heads as tails. The law works not on compensation but on a sort of '*swamping*.' An unusual run is not compensented, but the effect of that run on the total proportion is 'swamped' by the bulk of data which is not that unusual. The basis of the law is that the probabilities of long runs are very low, but the probability of a sequence of long runs is still lower, so that in a very large sample, the chances that the net proportion will be severely affected by an unusual occurrence will be very much lower than what they would be for a small sample.

To illustrate the above, suppose 20 tosses of a fair coin show 14, or 70 per cent, heads. This is an excess of 20 per cent or 4 heads over the 50 per cent population heads. If we toss the coin 80 more times and there are 44 heads in that, we have a total excess of 8 heads in 100 tosses, or only 8 per cent. Thus, though the subsequent tosses did not compensate for the unusually high proportion of heads in the first 20 tosses, the sheer bulk swamped that excess, and the sample proportion came down to 0.58 from 0.70. In this manner the law of large number assures that larger samples generally give better estimates than the smaller samples. It is of course possible that we could get more than 56 heads in the next 80 tosses to get the overall proportion more than 7.0, but the chances of that are very low compared to the chances of 14 in first 20 tosses.

As was stated earlier, *the sample statistic is only an estimate of the population parameter* and there is always an associated *quantum of uncertainty* no matter how large the sample. The magnitude of this uncertainty, or the spread of the region about the sample statistic in which the population parameter may lie can be determined by the laws of probability, and is of concern to us in the theory of hypothesis.

3.7. CENSUS VS SAMPLE ENUMERATION

1. In the method of complete enumeration (census) since all items constituting the population have been observed, information is available for each separate part of the universe. It is not so in the sample method. Here (as has been already explained) only a part of the universe is observed and from the results the estimates for the population are inferred. It, therefore, follows that whenever overall results of the whole population are required sample method would admirably suit the purpose. But if detailed results of the different parts of the population are needed, sampling would fail to give the required information and census method will have to be adopted.

2. Another factor that should be taken into account in choosing the method of enumeration is the relative difficulty and cost of organising a sample and a complete enumeration. Sample method needs better organisation than the census method. In fact, in some cases of census enumeration the work may be done by an already existing government machine and no organisation of any importance may be necessary. The *cost per unit*, therefore, is greater for a sample than for a complete census. But if the size of the sample represents only a small fraction of the whole population the expense in money and effort will be much less in a sample than in the census method.

3. Sample method has another advantage over the census method. If the information is collected from only a small proportion of the population, its completeness and accuracy can be easily ensured. When the numbers involved are few, more attention can be given to each such member and by persistent effort complete information can be obtained. This, obviously, is not possible when numbers are large.

Again, the information supplied by each individual can be carefully scrutinized if the numbers are few and in case of any doubt inquiries can be undertaken for verification.

5. In a sample method it is possible to collect more detailed information as an individual would be more willing to supply information in detail, if he knows that he represents a sample of the whole population.

6. Sample method requires comparatively a much less time both for the collection of data and their analysis. Time, as has been already pointed out elsewhere, is an important factor in statistical investigation and as such sample methods have a superiority over the method of complete enumeration.

7. Again, in investigation of a sociological type where information is to be collected from persons many of whom are illiterate, or in investigation which requires skilled physical observation and measurements, really qualified enumerators are needed. In a sample method few such enumerators would be required and hence it would be possible to undertake a sample enumeration by a qualified staff rather than a complete enumeration by untrained and less capable enumerators.

3.8. RANDOM SAMPLING

The method of selection of a sample is termed *random sampling* if it is such that each item wtthin the population has equal chance of being selected. A sample selected randomly is sometimes referred to as a random sample, but it must be understood that the random here is not a characteristic of the sample but that of the sampling procedure. It is the same as in a fair hand in a card game where fair does not describe the hand, but the fact that in dealing it out, the dealer was not prejudiced. Thus, if a person holds all four aces, it may not mean that the hand is unfair. We call a sample of size n a random sample, if it was obtained by a process which gave each possible combination of n items in the population an equal chance of being actually selected. Randomness in sample is important because only then the law of statistical regularity, operates. *Only if the sample is random, can we presume that the pattern of distribution of sample value follows the pattern of distribution in the population.* Only under such conditions do the laws of the theory of probability operate and the magnitude of the uncertainties in the estimates of population parameters computed.

It is not very easy to ensure that the procedure of sampling one uses in practice is really random. If we were to write the digits to 9 randomly on a piece of paper and we started writing them without any apparent thought it may appear to us there is no pattern, but the digits still would not be random. This is because our mind subconsciously imposes subtle patterns. You can try it and detect one common pattern very easily; no digit is followed by itself.

A random sample can generally be selected, by a kind of a lottery method in which all items of the population are numbered. Then the numbers are selected randomly. This selection. of random numbers may be done in a number of ways. Perhaps the most straight forward is write on identical pieces of paper all numbers. mix them up in a bowl and start selecting the items corresponding to the number selected from the sample. We could simulate an infinite population by simply selecting the numbers after replacing the previously selected numbers. In this way the chance of the selection of a number on first attempt is the same as that on the subsequent attempts. Under certain circumstances. sampling with replacement is to be preferred.

The lottery can also be drawn using a random number table. This is a table which simplifies the generation of random numbers. In this table digits from 0 to 9 have been pre-arranged scientifically in a random manner. We determine whether we want two digit, a three digit or a large number. and keep on writing the numbers of required digits by following the reference in the table.

3.9. NON-RANDOM SAMPLING

Although the principle of random sampling is the surest, there are circumstances in which non-random sampling may be called for. One of the reason is that the random sampling is often more costly than non-random sampling, and there are circumstances in which the non-random sampling if properly done gives fairly good results. This saving is primarily because of the fact that representative non-random samples can be obtained which are much smaller than random samples which give the same order of accuracy. Some the major non-random sampling techniques are:

Systematic sampling or quasi-random sampling. When a complete list of the population is available a common method of selecting a sample is to take every *n*th item from this list. The method is called systematic or quasi-random sampling. Thus, if the list of the population contains 20,000 items and a sample of 50 items is to be taken, the selection of every 40th item will give the required sample. The first entry is determined by selecting a number at random between 1 and 40. Thus, if the first item obtained in this manner is fifteenth, fifty-fifth and ninety-fifth will be the second and third in the sample and in this manner all items will be picked up. This method is based on the assumption that a complete list of the population under study is available. In order to obtain good results from this method it is necessary that the list is arranged wholly at random.

Stratified sampling. The simple random sampling technique is based upon a fundamental assumption that the population to be sampled is homogenous. Often the population is not homogenous. There may be clear cut groups within the population which are radically different as regards the characteristics under study. The economic conditions of the poor and the rich are bound to be dissimilar. The views of adolescents and grown-ups may be divergent as regards a social problem. When the population is markedly heterogeneous it is first sub-divided into groups or 'stratas' in such a manner that all items in any particular group are similar with regard to the characteristic under consideration. From each such 'strata' items are chosen at random. The number of items taken from each group may be in proportion to its relative strength. The sample so formed is called stratified.

Multi-stage sampling. In multi-stage sampling, the population is distributed into a number of first-stage sampling units and a sample is taken of these first stage units by some suitable method. This is the first stage of sampling process. Each of these (selected) first sample units is further sub-divided into second-stage units, and from these again a sample is taken by some suitable method. Further stages may be added if required.

Thus, if a socio-economic survey is to be organised in a State, the fist stage of the sampling process would be to divide the area into regions and select in a suitable manner a sample of these regions. The selécted regions will again be divided into smaller sampling units (say villages) and from out of these will be selected the sample units that are to serve as sample for the purpose of the investigation.

PROBLEMS

1. Discuss what went wrong in the following examples: Ruber life-rafts for navy were tested individually for impact resistance by dropping them into water from a height equal to that of a ship's deck. Only those that survived were passed. Users, however, reported that a large percentage of those that passed in this test failed in actual use.
2. If you are a cricket fan, try carefully to device a numerical, *i.e.*, statistical measure of "all-round performance". You would hopefully, notice the truth of the statement that "Statistics deals with numbers which are mere shadows of the phenomenon being studied."
3. "Based on the responses to a mailed questionnaire, a company was dis-

turbed to find that about 30 per cent of its customers were dissatisfied by the service they receive from the field agents. This was in sharp constrast to the field agents' response which indicated that only about 2 per cer customers were having problems." Discuss.

4. 'The characteristics of tbe sample are able to provide an approximatel correct idea about the characteristics of the population.'
   Do you agree with the above statement? Give arguments in support of youi answer.
5. Explain, giving suitable illustrations, the Law of Statistical Regularity.
6. Explain the law of large numbers. In what sense is this law relevant in a statistical investigation.
7. Distinguish between the 'census' and 'sampling methods' of collecting data and compare their merits.
8. Point out the significance of sampling. Distinguish between random and deliberate sampling.
9. Compare the various methods of selecting a sample.
10. Bring out clearly the important features of (*i*) systematic sampling, (*ii*) stratified sampling, and (*iii*) multi-stage sampling.
11. "Sampling is necessary under certain conditions." Explain with illustrative examples.

Chapter 4

COLLECTION OF DATA - PRELIMINARY CONSIDERATIONS

4.1. STATEMENT OF PURPOSE

The first and the most essential thing which the statistical investigator must do is the preparation of a statement of the purpose of the statistical inquiry in hand. Failure to work out a statement of purpoes, clearly and carefully, can lead only to misunderstanding and confusion. It will result in diffusion of effort, gradual over-expansion of the field covered and aimlessness.

Objects of statistical inquiry. The purpose of a statistical inquiry may be either:

(*i*) to supplement, disprove or simply to test some theory or hypothesis, which is current, or

(*ii*) to discover a new theory or hypothesis, or

(*iii*) to know the existing state of affairs, or

(*iv*) to solve a problem involving the inter-relations of several groups of facts.

A statistical inquiry may, for example, be concerned with the problem of unemployment. Now, there are various aspects of this problem—urban or rural unemployment, educated unemployment or unemployment of the un-educated, partial or full unemployment. So long as the investigator does not know the objects of his inquiry he will not be sure as to what facts are to be collected.

It is of imperative necessity that statement of purpose must be precise. A general statement is inadequate. Even a slight variation in the purpose of inquiry may require entirely or partially different statistical data. Thus, if an economist desires to study and compare the scale of wages in different localities, he must be very clear about the following points:

(*a*) Whether the requirements of his problem demand information on wage rates or actual earning?

(*b*) If on wage rates, should it refer to rates approved by collective negotiations or to rates actually paid? Should adjustments be made for overtime, undertime, bonuses and fines?

(*c*) Whether supervisory and clerical workers are to be included?

(*d*) Whether receipts in kind be added to wage rates?

Having ascertained the purpose of inquiry, it is desirable for the investigator to acquire some general information about the thing he is investigating. If the purpose of the inquiry, for example, is to determine the best site and location of a proposed plant in a particular area, the investigator must have some knowledge of the technical requirements of that plant.

4.2. SCOPE OF INQUIRY

The scope of a particular statistical inquiry will be decided with reference to either (*a*) the space, (*b*) the time, or (*c*) the number of items to be covered. As regards space, the statistician would fix such limits as might serve his purpose in the best possible manner. In practice, however, the following limits are generally used:

Political and administrative divisions, such as the country, the state, the district, the city, the municipality, the ward, the circle, the block, etc.

Economic divisions, such as agriculture and animal husbandary, mining, manufacturing, trade, transport, communications, banking, professional, liberal arts, etc.

Natural or climatic divisions, such as the plains, the mountains, the plateaus, the forests, etc.

As regards time, it may be noted that the work of collection of the data must be finished within a reasonable time. For, if more than reasonable time is taken in the collection of the numerical facts, conditions might change and the data collected may be rendered useless for the purpose of the inquiry in hand. 'Reasonable time' depends upon the nature of the phenomenon under investigation. If the phenomenon is such where the conditions change quickly and frequently, the duration of process of investigation should be narrowed to such an extent that there is no possibility of a change affecting the data. Certain problems are of topical interest only and in such cases the plan of statistical investigation should be so devised that the entire work is finished before the inquiry ceases to have an interest.

The decision with regard to the number of items that are to constitute the data is, in fact, a very important one. In other words, it means the question of choice between the census and the sampling technique of data collection. By census method we mean a method where each item constituting the population or the universe is enumerated. Sample method, on the other hand means a system where only a limited number of items is taken into account. This limited number of items is regarded as the sample of the population. The way or ways in which sample should be chosen, and the question of deciding between these two alternative methods of data collection, are discussed in detail in Chapter 3.

4.3. CHOICE OF STATISTICAL UNITS

For the correct solution of any statistical problem, it is not enough that the facts are collected with the maximum of accuracy, but it is essential too that the unit employed for expressing the numerical information is appropriate. The mistake in the selection of the unit is more harmful than mistakes in the collection of data. As an illustration. suppose that it is desired to restrict the output of cotton textiles and as such production quotas are to be fixed on the basis of the *size of mills*. Now the size of a manufacturing concern may be measured in terms of either the 'volume of output,' 'capital investment,' 'number of wage-earners employed', 'the volume of input,' or 'the number of looms installed.' For our purpose, however, it is the volume of output that is the best criterion of determining the size. If a unit other than the 'volume output' is chosen in this particular case the results are bound to be incorrect and misleading. The importance of the unit is still greater in the interpretation of data for the simple reason that more people use statistics than compile them. If we collect data regarding strikes and lockouts in cotton textile industry from the different states of India during a particular year, the figures so collected will show total number of manhours lost due to work-stoppages. Now, if work-stoppages are not properly defined some states might give figures only for those work-stoppages which lasted for more than a month, some might give figures only for those in which the work-stoppages lasted for more than a week, and some states might give figures which include those which lasted for only one or even less than one day. The obvious result would be that the data so obtained would not be comparable and would fail to give an idea about the relative condition of industrial relations in different states.

The appropriateness of the unit depends on the purpose of statistical investigation. The purpose of the statistical inquiry and definition of the unit are reciprocal. The latter is determined by the former and the former is governed by the latter. 'Statistical units cannot be defined without regard to the object in view and the purpose of inquiry cannot be stated with sufficient accuracy without a clear-cut notion of the units.'

Characteristics of statistical unit. 1. The unit of measurement must be definite and specific.

2. The unit must be of such a nature that it may be correctly ascertained.

3. Care must be taken to ensure homogeneity and uniformity. A given choice of unit is unsatisfactory, if it implies different properties on different occasions or by different persons.

4. The unit should be stable.

5. The unit must be appropriate for the purpose. Thus the strength of an army from the fighting standpoint depends upon the number of combatants, whilst from a commissariat standpoint it depends upon combatants, plus auxiliary services, plus sick and wounded.

Classification of units. The classification may be made either on the basis of (*i*) the origination of the unit, or (*ii*) the function performed by the unit.

Classification of statistical units as regards their origin. Statistics has been defined as the technique of counting and measurement. In the counting operations the unit employed is invariably a unit which arises spontaneously such as a man, a house, an accident, a city, a town, etc. In the process of measurement the unit is artificially designed solely for the object of measurement, such as a kilogram, a rupee, a kilometre, etc. In other words, individuals or objects. separated by distinctive marks, may be counted, whilst other objects must be measured by reference to arbitrary or conventional units.

Very often, in the fields of commerce and economics the units used are the pecuniary value units which measure the financial importance in terms of rupee, dollar, pound sterling etc. Thus, the gross national product in rupee term measures the product in pecuniary units.

The advantage of pecuniary value unit is its wide applicability as a common denominator and the ease of utilising it. In discussing the foreign trade of India or the balance of payment problem, the pecuniary value unit is employed. Had there been no pecuniary value unit, it would have been a problem 'how to add tons, pieces, yards, bales and bushels, etc.' For purposes of comparison between diverse articles, pecuniary value unit serves as a common denominator. Thus, we can compare the production of cereals with that of minerals. The principal defect inherent in such a unit is that statistics stated in rupee can only yield indices of value and not of quantity. As an example in the field of business, the amount of turnover (on an average) of a grocery and provision merchant is much less than the amount of turnover of a jeweller; whereas in reality the grocer might have sold more quantities as regards weight and number than the jeweller.

Classification according to functions of units. As regards their use in statistics, units may be classified as:

Units of enumeration,

Units of analysis and interpretation, and

Units of presentation.

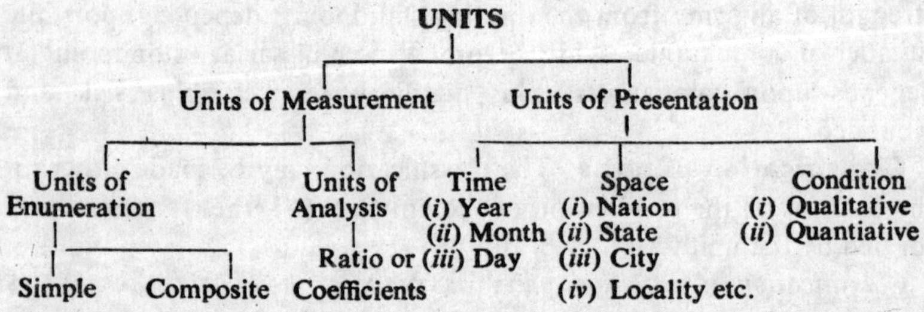

Untis of enumeration or estimation. This units may be divided into (*i*) simple, and (*ii*) composite. A simple unit is one in which 'one determining consideration is prescribed,' *i.e.*, a unit in which the ideas expressed are general and only class differences are distinguished. The examples are: an animal, a farm, a mile, a ton, an accident, a house, a man, a dispute, a room, a city, etc. Such units have no limiting qualifications, and are easily defined.

If we add a qualifying limit to the simple unit the unit becomes composite. The effect of adding a qualifying phrase may be (*i*) to dercribe more correctly the general concept, (*ii*) to limit the class which it names, (*iii*) to add to the difficulty of defining it. Examples of composite units are a farm animal, a ton-mile, an industrial accident, a rent-free house, an industrial dispute, a vacant room, credit sale, chain-store, etc. The chance of error and bias increases as we add, limiting conditions to simple units.

Units of analysis and interpretation. Such units are those in which things or attributes of things are not only named but also compared. To compare things rates, ratios and coefficients are used. The ratios and coefficient, may relate in time, space, or the conditions in time or space. Road accidents may be expressed by number or by severity, but related to months or year. The production of food crop may be expressed in quintals, but related to the area under food crops. Deaths during a given period, or for a particular region, may be expressed as a number, but they may be related to the entire population of the same age, caste and sex characteristics.

Units of presentation. Units of presentotion are of three types: (*i*) time, (*ii*) space, and (*iii*) condition. For instance, the manufacturing expenses of a group of concerns may be measured and presented by years, by location, by size, by nature of industry, or by the efficiency and nature of management. The turnover may be shown by months or by zones of territory.

4.4. TECHNIQUE OF DATA COLLECTION

Having determined the scope of inquiry and units of measurement, the next step in drawing up the plan of investigation is to determine the best method of data collection.

In order to obtain statistical material the investigator may either (*i*) go to the records of some institution, whether public or private, that collects and publishes data as a routine or (*ii*) make a special survey, *i.e.*, conduct a field inquiry. Thus, there are two sources of information available to the statistician. The former is usually termed as *secondary source* and the latter as *primary source.* Each source of information has its own merits and demerits. The selection of a particular source is dependent upon a variety of factors such as (*a*) the purpose of inquiry, (*b*) time required, (*c*) the accuracy desired, (*d*) funds available, (*e*) other facilities available, and (*f*) nature of the person conducting investigation. However, this subject is so broad that a subsequent chapter is devoted to a discussion of the techniques of data collection.

4.5. STANDARD OF ACCURACY

Another important step in planning a statistical investigation is the determination of the standard of accuracy that is to be observed in the collection of statistical material. While determining the standard it is necessary to bear in mind two things: (*i*) the degree of accuracy that is necessary, and (*ii*) the accuracy which is actually attainable. Absolute accuracy is neither possible nor very necessary. When the number of informants is very large and there is a great divergence in the standard of their education and intelligence it is not possible to get correct information about each one of them. Some may unconsciously be left out of consideration altogether and a few may fail to supply information of a requisite standard. Thus some inaccuracy is bound to creep in even in spite of our best efforts. The results of statistical analysis, however, will not be significantly affected by this unavoidable inaccuracy. Statistics is concerned with an overall picture of the universe and minor discrepancies will not alter the ultimate conclusions. Mathematical exactness in statistical investigation, therefore, is neither possible nor very necessary. But statistical-data must possess a reasonable degree of accuracy, and every effort should be made to attain it. What is 'reasonable' accuracy will depend upon the circumstances of each case. For instance, in measuring the height of an individual even centimetres will be considered, while in measuring the distance between two towns even hectometres may not be given any importance.

4.6. APPROXIMATION

In some cases, the results of the measurement, estimation (or calculation) of the statistical phenomena are not given in all the digits that are made available. Thus, if the census of population of a certain district gives a total of 85,39,421, the last three digits are dropped by approximation and the population is given as 85,39,000. This is called the method of approximation and is adopted because it facilitates calculation and simplifies comparisons. Besides if the results are put to the last digit they are likely to give an impression of the accuracy.

Methods of approximation. The first thing that is necessary in this connection is to determine the degree to which accuracy has to be maintained. Thus, if the data are to be correct to the nearest thousand it means that the last three digits are to be rounded. In this case amounts less than 1,000 will be regarded as fractions of a thousand—more than 500 being counted, more than one-half and less than 500 being taken as less than one-half. This being understood, the rules of approximation may be given as follows:

A method of approximation that is commonly adopted is to ignore fractions less than half and count fractions more than half as full, Fractions equal to exactly half may be counted as full or dropped at the discretion of the investigator. Thus, if accuracy is to be maintained up to the nearest thousand:

48,27,681 will become 48,28,000
48,27,481 will become 48,27,000
48,27 500 will become either 48,27,000 or 48,28,000

If accuracy is to be maintained up to the nearest hunderd:

48,27,681 will become 48,27,700
48,27,481 will become 48,27,500.

If accuracy is to be maintained up to the nearest tenth of a unit (nearest to first decimal place):

15.213 will read as 15.2
15.486 will read as 15.5.

If accuracy is to be maintained to the nearest hundredth of a unit (nearest to two decimal places):

15.213 will become 15.21
15.486 wiil become 15.49.

Note. (1) Approximation is sometimes done by discarding the number of digits not required and no consideration is made of the fraction which is more than half or less than half. Thus,

28,97,835 will become 28,97,000
89,97,349 will become 21,97,000.

(2) Approximation can also be done by counting the fraction, whether more or less than half, as full. Thus,

28,97,835 will become 28,98,000
28,97,349 will become 28,98,000,

PROBLEMS

1. Explain the main steps that must be considered in planning a statistical survey.
2. Describe the process of planning a statistical enquiry with special reference to its scope and purpose.
3. What is a 'statistical unit'? What are the characteristics that a statistical unit must possess? Give reasons in support of your answer.
4. Bring out clearly the importance of statistical unit in a statistical investigation Give example in support of your answer.
5. "Statistical unit is necessary not only for the collection of data, but also for their interpretation and presentation." Explain the above statement.
6. Write short notes on:
   (*i*) Natural Units. (*ii*) Produced Units. (*ili*) Artificial Units.
7. Give a method of classifying units on the basis of their functions.
8. What steps would you take in collecting the necessary data for introducing primary education in the state.
9. Giving appropriate reasons, state what units can be used for the following:
   (*i*) Production of cotton textile industry.
   (*ii*) Labour employed in an industry.
   (*iii*) Cansumption of electricity.
10. What considerations should determine the degree of accuracy in planning a statistical Investigation?
11. Explain the significance of approximation in a statistical enquiry and illustrate the various methods used for this purpose.

Chapter 5

COLLECTION OF DATA-TECHNIQUES

5.1. PRIMARY AND SECONDARY DATA

The term *primary data* refers to the statistical material which the investigator originates for the purpose of the inquiry in hand. Thus, if it is desired to conduct an inquiry into the cost of living of the workers in a certain mill, and if the facts pretaining to this inquiry are collected by the investigator or his representatives from the workers themselves such data would be termed as primary data.

The term *secondary data* on the other hand refers to that statistical material which is not originated by the investigator himself, but which he obtains from someone else's records. Thus, if instead of obtaining our data from the workers themselves we get them from the records of the Trade Union, or from some other source, the data will be called secondary data.

The difference between primary and secondary data is largely one of degree. Data which are primary in the hands of one person may be secondary in the hands of another. Thus, the data collected during census operations are primary to the census department of the Government of India, but to a person who makes use of these data for further research they will be termed 'secondary.'

5.2. PRIMARY METHODS OF DATA COLLECTION

Those methods that aim at collecting primary data are termed as primary methods. These consist of direct personal observations (as in physical sciences) or of direct communication (as in social services) with people either orally or in writing. While the details of such methods are beyond the scope of this book, we can, however, give some general principles. One thing that should always be borne in mind while collecting data is that the information so obtained is always susceptible to conscious or unconscious error or abuse. These errors may creep in because of carelessness on the part of the person collecting data, or it may be inherent in the nature of data to be obtained. Even the seasoned researchers do not always know the right question to ask. One should always start an inquiry with determining what the information being sought is and how to obtain it.

For obtaining information by communication an investigator may adopt any one of the following methods:

(*i*) Personal interviews

(*ii*) Mailed questionnaires

(*iii*) Questionnaires in charge of enumerators.

All these methods have advantages and disadvantages. The direc personal interview technique is perhaps best suited to situations where the problems are not completely understood and the investigator presents himself personally before the informant and questions him carefully. This technique is used very successfully in social anthropological research where the questions cannot be formulated before hand and one question leads to another. This technique is also useful in situations where great depth in study is required. Obviously this time consuming method is not suited to large groups of informants.

In those cases one drafts a detailed questionnaire. These questionnaires can either be mailed to the respondents for filling in and returning, or can be put in charge of enumerators who go around and fill them after obtaining the desired information.

In the mailed questionnaire method the costs are relatively less, but it suffers from some serious drawbacks. One of these is in the nature of questionnaires itself. It is very difficult to design a questionnaire that all can understand and fill. Often the questions, or at least their intent is misunderstood leading to inaccurate or even irrelevant answers. This problem can be taken care of to some entent by redesigning the questionnaire, but there still will be some gray areas. Again, one might have to give detailed instructions for filling, that increases the effort a respondent has to put in and this reduces the percentage of response

The other fault with mailed questionnaire method lies in the fact that there is no way of ensuring that questionnaires are filled in and returned. There may not be enough motivation with the targetted respondents. This often results in very poor percentages of returns which in turn may cast serious doubts on the validity of responses. This is because those responding may not form a fair sample of the population. A weekly magazine once conducted a mailed questionnaire survey of the reading public's dissatisfaction with the civic services in their town. This gave absurd results because only those who were seriously dissatisfied responded.

Another serious fault with the mailed questionnaire lies in the possibility that the respondent may lack the knowledge of the facts wanted. But even with these faults, one often goes for mailed questionnaires because of the convenience and the lower cost afforded.

In the second method, one delegates the task of interviewing informants to select agents. These agents are termed as 'enumerators'. They may be employed either on a pay basis or they may offer their services free. This method should be adopted in those cases where the informants are illiterate. It ensures great reliability as the accuracy can be checked by supplementary questions.

The enumerators who are incharge of our population census questionnaires are honorary workers. These agents are provided with standardised questionnaires and specific training and instructions are given to them regarding the way in which the schedules are to be filled and the information elicited.

For private investigation, this procedure is too expensive to be undertaken. This method has an advantage over the previous one inasmuch as schedules will be complete and comparatively accurate. The enumerator will see that only relevant answers are received from the informants. By cross-questioning, the enumerator will be able to get correct answers. The advantages of the agency method are, however, very largely nullified by careless drafting of schedules, inadequate instructions to agents and selection of incompetent enumerators.

The enumerators must possess certain qualities, viz., (*i*) Sufficient knowledge to comprehend and follow the instructions; (*ii*) Ability to distinguish between good and bad answers; (*iii*) Tact in extracting information desired; (*iv*) Open-mindedness to record the facts as they appear; (*v*) Unprejudiced and unbiased attitude; and (*vi*) Courtesy.

A high degree of uniformity in the information collected will be possible only if the enumerators are competent and well trained. Again, it is only competent enumerators who will be able to get correct answers to otherwise complex questions.

The way in which the schedules are to be collected will also have an important bearing on the number, nature and form of the questions asked. If the schedules are to be filled by the enumerators, the questions may be more complex and more numerous than in the case when the schedules are directly mailed to the informants. The enumerators may explain to the informants the nature and purpose of the inquiry, and also the questions asked. He may also clear up doubtful points and ask corroborative questions In the case when information is obtained by mailing the schedules, the questions themselves must carry conviction, be self-explanatory, consistent and persuasive. Personal appeal for information, best made by human contact, is then made through the printed rather than the spoken word.

5.3. DRAFTING QUESTIONS AND QUESTIONNAIRES

The following are some of the points that should be kept in mind while drafting questionnaires:

1. **Clarity**. The individual question should be as simple and as clear as possible. This is particularly necessary when these questions are to be answered by the informants, without any assistance from the investigator or his representative. Thus a question on marital status, if put in the following manner is simple and quite clear for anyone to answer.

Marital Status (Check against appropriate answers)
Newly married—
Married—
Unmarried—
Separated—
Divorced—
Widowed—

2. **Avoid certain type of question**. Those questions should be avoided which are likely either

(*a*) to offend or frighten the informants, or
(*b*) to arouse the resentment of the informant, or
(*c*) to allow evasive answers, or
(*d*) to be answered in prejudice.

For example, questions like "Have you stopped drinking"? or "Do you take your job seriously," or "Is your married life happy" should better be avoided.

3. **Objectivity of answers**. When the enquiry is aimed at making a factual study the questions should be such as result in objective answers. Thus a question aimed at ascertaining the condition of class room furniture may better be asked in the following manner:

Is furniture in good condition?
Does it need repairs?
Is it unfit for use?

4. **Definiteness of answers**. The questions should be so framed that the answers to them are perfectly definite.

5. **Number of questions**. The number should be consistent with the scope of the investigation.

6. **Units**. The unit in terms of which information is to be given must be clearly mentioned.

7. **Arrangements of questions**. In planning the arrangement of the questions the following points should be observed:

(*a*) *Space*. There should be proper space for answers.

(*b*) *Order*. The order in which questions are arranged should be in a logical sequence so as to avoid skipping back and forth from one topic to another.

8. **Instructions and definitions**. The forms must contain necessary instructions so that the enumerator and the informant are in no doubt as to what information is desired.

5.4. SOURCES OF SECONDARY DATA

It is not always necessary to conduct special surveys for the purpose of obtaining statistical material. Such material may be obtained from the records of institutions that collect and publish statistics as a part of their routine duties. The most important routine compilers and suppliers of statistics are governments. Next in importance come the semi-official institutions like municipalities, railways and state-owned undertakings. In addition to these, there are trade associations that publish statistical data. Statistical material also appears in trade journals, maket reports, magazines and other periodicals. With reference to the manner of its publication, the secondary data may be divided into three groups:

1. **Continuous or regular data**. Statistical data published regularly at short, known intervals are called continuous or regular data. The examples are—the weekly index number of wholesale prices, monthly figures of exports and imports, monthly figures of workers involved in work-stoppages, monthly statistics of the production of certain selected industries in India, and so on.

2. **Periodical data**, which are regularly published at long intervals such as Indian Census, Statistical Abstract for India, Agricultural Statistics of India, Trade Statistics, and Statistical Tables relating to Banks in India.

3. **Irregular data**, consisting of a special studies of statistical phenomenon, with no regular dates of publication, *e.g.*, the reports of the National Income Committee, Tariff Commission, reports and the reports of the various committees and commissions appointed by the Government from time to time.

5.5. EDITING PRIMARY DATA

The returned questionnaires (filled by the informants or by the enumerators) should be scrutinized at an early stage with a view to detect errors, omissions and inconsistencies. Before the collected information is tabulated, each individual schedule must be checked in detail to ascertain whether or not it has been answered in full and accurately. Defective schedules should be returned for amendment. But if the figures are manifestly erroneous, they may be rejected or corrected. In order that the schedules may be corrected by the investigator himself he must have reasonable grounds upon which to act. The work of editing requires skill and scientific impartiality of a high degree. Bailey and Cummings name four types of editing: editing for consistency, uniformity, completeness and accuracy.*

In order to judge whether the returned schedules are consistent, we must compare the answers of those questions which were designed to be mutually confirmatory. If the answers of such two questions appear to be mutually contradictory, it is essential to determine which, if either, is correct. To obviate this difficulty, the most obvious solution is to send back the schedule, though this procedure will not always yield a decisive result and will in any case consume considerable time and effort. An alternative method is to discard such schedules entirely, but again this is not desirable as we would have to sacrifice some important items of information bearing upon the problem in hand.

The returns may be edited to avoid lack of uniformity. Even in primary collections, there will be an occasional return in which the answers are submitted in the wrong manner. Perhaps the most general mistake of this sort is in the statement of units: time may be expressed in years rather than month. Such mistakes can be remedied easily.

On the other hand, there are some such mistakes which can only be remedied by further correspondence. Instances of the mistakes of this type are the stating of figures for wholesale trade when information is sought for retail trade, the reporting of figures for a calendar year when it is sought for the fiscal year, etc.

Editing for completeness is the main and a straightforward operation. Certain space often appears on the schedule for such things as totals, ratios, etc. which the informant was not asked to fill in. These spaces are to be filled by the investigator himself. If, however, the return is incomplete, in the sense that the informant has not

*Quoted by Crum, Pattcu and Tebbutt, *Introduction to Economic Statistics*, p. 57.

answered a particular question, then further correspondence is essential.

Editing for accuracy is the difficult task of the editor. Inaccuracies, other than inconsistencies are seldom apparent from internal evidence in the return. Only experience can help an investigator to detect such errors.

The editor should bear in mind that he must never destroy, by clipping or erasure, the original reply and the corrected items should be entered separately. The process of editing is by no means an unimportant and routine operation; rather it requires marked ability, scrupulous care and a rigid adherence to scientific objectivity.

5.6. EDITING SECONDARY DATA

Statistical material obtained from secondary sources is not always as reliable as that from the primary source. Statistics, especially other people's statistics, are full of pitfalls for the user. Terms may be used in peculiar senses; meanings may have imperceptibly changed; external factors may operate to produce discrepancies. As such it is necessary to scrutinize the secondary data in the light of the following points:

1. The type and purpose of the institution that publishes statistics as a routine.
2. The purpose for which the data are issued and the consumers to whom they are addressed. The purpose may be (*i*) general or specific, (*ii*) restrictive or inclusive, (*iii*) transient or permanent, and (*iv*) scientific or unscientific.
3. The nature of the data themselves. Are the data biased? Bias may be due to (*i*) wilfully eliminating parts of the facts, (*ii*) basing comparisons on inadequate data, and (*iii*) relating them to unrepresentative periods or conditions. The next question that should be asked is: 'Are the data samples only or complete enumeration?'
4. In what types of units are the data expressed? Are they the same at different times, at different places, and for all cases at the same time or place?
5. Are the data accurate?
6. Do the data refer to homogeneous condition?
7. Are the data germane to the problem under study?

PROBLEMS

1. Distinguish between primary and secondary data.
2. What are the various methods of collecting primary data? Give examples of each with relative merits and demerits.
3. Define secondary data. What are their sources and the precautions necessary for using them?
4. Describe the primary and secondary methods of data collection. In what special circumstances are the two methods suitable.
5. (*a*) Explain the advantages of direct personal [investigation as compared to other methods generally used in collecting data.
   (*b*) Describe clearly 'Indirect personal interviews' as a method of collecting statistical data. Under what conditions is it most suitable? Give reasons.
6. Distinguish clearly between 'mailed questionnaire method' and 'questionnaires in charge of enumerators' as methods of data collections. What are the respective area of applicability.
7. (*a*) How would you ensure the success of a 'mailed questionnaire method'?
   (*b*) Explain what precautions must be taken while drafting a questionnaire in order that it may be really useful. Illustrate your points.
   (*c*) What is a questionnaire? Discuss the main points that you will take into account while designing a questionnaire.
   (*d*) Describe briefly the chief features of a good questionnaire. Draft a suitable questionnaire for studying the Five-Year Plans consciousness in your area.
8. Define 'secondary data.' State their chief sources and point cut the dangers involved in their use and what precautions are necessary before using them.
9. What do you understand by 'Editing' the data? Explain clearly the procedure that you would adopt in editing 'Primary data' and 'Secondary data.'
10. In collection of data "common sense is the chief requisite and experience the chief teacher." Discuss.
11. (*a*) Design a questionnaire to ascertain the view of the students concerning the measures to be undertaken for improving the bus services in Delhi.
    (*b*) Make up a questionnaire from which you would hope to obtain information you needed in an investigation of housing conditions in a small town.
12. Describe the different methods of collecting data indicating the merits and demerits of each of them. Which method is suitable to the following type, of inquiries?
    (*a*) Inquiry into the food situation by a committee appointed by the government.
    (*b*) Enquiry by a research organisation into the living conditions of the cotton textile workers.
    (*c*) Study of the socio-cultural aspects of the life in Arunachal.

Chapter 6

CLASSIFICATION AND TABULATION

6.1. INTRODUCTION

The data collected for the purpose of a statistical inquiry sometimes consist of a few fairly simple figures which can be easily understood without any kind of special treatment. But more often there is an overwhelming mass of raw material and detail without any form or structure. Data obtained from primary sources, obviously enough, are in a raw state for they have not gone through any statistical treatment. But the secondary data too are no better in this respect inasmuch as the form in which they are obtained is more often than not unsuited for the purpose of the inquiry.

This unwieldy, unorganised and shapeless mass of collected data is not capable of being rapidly or easily assimilated or interpreted. At best only a hazy impression and that too of a doubtful reliability may be obtained by its *perusal.* Consider the marks listed in Table 6.1, obtained by the students of a class. Without reorganising these data, a cursory survey yields a general impression that most of the students obtained marks in between 45 and 55. A more laborious examination shows the lowest marks to be 14 and the highest 74, so that the range may be stated as 74 – 14 = 60. Beyond this it is difficult to go with the data in their present form.

In order to make the data easily understandable, the first task of the statistician is to condense and simplify them in such a manner that irrelevant details are eliminated and their significant features stand out prominently. The procedure that is adopted for this purpose is known as the method of classification and tabulation,

6.2. CLASSIFICATION

Classification is the process of arranging things (either actually or notionally) in groups or classes according to their resemblances and affinities, and gives expression to the unity of attributes that may

TABLE 6.1. DATA AS ORIGINALLY COLLECTED

Marks in 'Statistics' obtained by 105 students of a college						
40	37	61	67	59	70	39
46	68	41	60	38	39	47
51	37	40	72	39	50	44
41	25	42	38	40	50	30
33	54	58	14	71	55	33
65	55	66	40	62	54	39
48	55	38	40	20	69	41
43	49	59	73	28	47	59
30	38	52	68	38	48	45
71	44	52	45	56	64	
51	59	47	46	57	65	
39	73	28	30	46	54	
56	44	14	62	47	52	
50	50	50	23	41	74	
58	47	52	61	42	28	
40	56	63	56	66	30	

subsist amongst a diversity of individuals. It serves the following purposes:

1. eliminates unnecessary details,
2. brings out clearly the points of similarity and dissimilarity,
3. allows comparisons and drawing of inferences.

The first step in the process of classification is to select the basis of classification. Statistical facts are classified according to their characteristics. Thus, the students of a college may be classified according to their marital status, height, religion. etc. When a particular characteristic has been chosen for this purpose the next step in the process of classification would be to note the similarity and dissimilarity as regards this chosen characteristic in the various items. Items that would be alike in respect of this characteristic will be grouped together. Thus, if the students are to be classified according to their marital staus all married students would be put into one group and all unmarried in another. If the students are classified on the basis of religion, there will be different groups for Hindus, Muslims, Sikhs, etc. When the clssification is made according to heights, each group will include only those students whose height lies within a certain range. It will be noted that the three different characteristics (marital status, height and religion) give us groups that are significantly different from one another. Thus, in the first case we have groups where the characteristic is either present or absent, *e.g.*, married or unmarried; in the second we have groups where the characteristic is of differing quality *e.g.*, students may be Hindus, Muslims or Sikhs. In the third illustration we have groups where the characteristic is present in different degrees, *e.g.*, there will be a group of students, whose height is between 150 em and 158 cm, another group whose height is between 158 cm and 165 cm, and so on.

The set of characteristics one choses as the basis of classifiction depends on what we want to study. Thus, if we wanted to study the religous mix of the students, we will classify on the basis of religion. If we wanted to study the dependence of the marital status of students on their religion we will have to classify the students on the two sets of the characteristics, and form classes that look like:

Hindu married,
Hindu unmarried,
Muslim married,
Muslim unmarried, etc.

The characteristics of a population may be broadly divided into two categories: *attributes and variables*. Attributes are qualitative characteristics which are not capable of being described numerically, *e.g.*, sex, nationality, colour of eye, etc. These characteristics are called 'attributes' or 'attributive variates' or 'descriptive characteristics.' When classification is to be made on the basis or attributes, groups are differentiated either by the presence and absence of the attribute (*e.g.*, Muslims and Non-Muslims) or by its differing qualities. The qualities of an attribute can easily be differentiated by means of some natural or physical line of demarcation, and their natural differences determine the group into which a particular item is to be placed. Thus, if we select 'colour of eye' as the basis of classification there will be a group of 'brown eyed' people, another of 'blue eyed' people, and so on. If the data are classified on the basis of one attribute only, the process is termed as simple classification. In cases, where more than one attribute is studied, resulting in a subdivision of classes, the classification is known as manifold. Thus the population of a city may be divided into literate and illiterate. Literate persons may again be divided into literate males and literate females. The following illustration depicts an example of manifold classification:

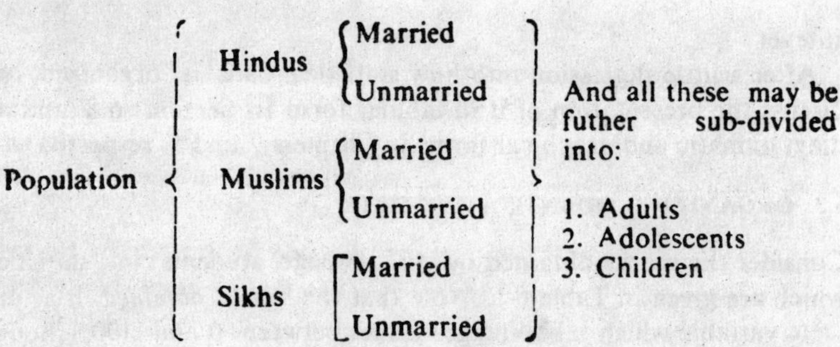

Variables are quantitative characteristics which are numerically described. Thus, height, weight, turnover, age, export, death rate, inventory volume, test scores, etc. are quantitative variables, or simply, *variables*.

A variable may be (*i*) continuous, or (*ii*) discrete. If a variable can take any numerical value within a certain range it is called a continuous variable. Such, for example, are heights, weights, rainfall records, barometric readings, etc. In measuring the height of university students it is possible to come across any measure (say) between 167 cm and 170 cm; there may, for example, be a student whose height is 167.12034 cm. The point is that any conceivable height within this range can occur. If we prepare a list of all persons in India whose height is between 167 cm and 170 cm, it is quite possible that for every height (say 167.001 cm, 167.002 cm and so on) there will be a person available who has it. Thus a variable is said to be continuous when it may 'pass from one value to the next by infinitely small gradation.' If a man says he has driven 50 kilometers, we understand that the distance is approximately 50 kilometers. In order to increase his distance from 50 to 51, he must pass through all the infinitely small gradations of distance between the two. Distance cannot increase at a leap by a whole unit discretely and must increase continuously.

Variables which take only discrete or exact values are called discrete, *e.g.*, test scores, family members, etc. There cannot be test scores in fraction and the members in a family will always be integers, *i.e.*, 1, 2, 3, 4. . .and so on. Thus, a variable is said to be discrete 'when there are gaps between one value and the next possible one.' A count of the number of the students in a class room may yield, say, 60. If one more student enters, the count leaps from 60 to 61 without passing through any intermediate fractional values. At no time it is 'slightly under 60' or 'a little more than 60.' The count is exact, and proceeds by a series of leaps from integer to integer.

After a little digression into how statistical data is organized, we discuss the presentation of it in tabular form in Section 6.8 and in diagrammatic and graphical forms in Chapters 7 and 8, respectively.

6.3. ORGANIZING QUANTITATIVE DATA

Consider the marks obtained by 105 college students in 'statistics' which are given in Table 6.1. Note that the *marks obtained* is a discrete variable which takes integer values between 0 and 100, though in the present case there are no values between 0 to 14 and 75 to 100. As was maintained earlier, the data of Table 6.1 is a shapeless mass not capable of being readily assimilated or interpreted. It is, therefore, to be organised properly.

The first thing that is to be done in the matter of arranging the collected data is to prepare an *'array'*. The array is prepared by arranging the values of the variable in an ascending or descending order. This will enable us to know the range over which the items are spread and we will also get an idea of their general distribution. The data of Table 6.1 when rearranged takes the form of Table 6.2.

An inspection of Table 6.2 provides a much clearer impression of the distribution of marks than could be obtained from the unorganised marks in the original compilation. The highest and lowest

TABLE 6.2. DATA ARRANGED IN ORDER OF SIZE

Marks in 'statistics' obtained by 105 students of a college					
14	38	42	50	56	64
14	38	43	50	56	65
20	39	43	50	56	65
23	39	44	50	56	66
25	39	44	50	57	66
28	39	45	51	58	67
28	29	45	51	58	68
28	40	46	52	59	68
30	40	46	52	59	69
30	40	46	52	59	70
30	40	47	52	59	71
33	40	47	54	60	71
33	40	47	54	61	72
33	40	47	54	61	73
37	41	48	55	62	73
37	41	48	55	62	74
38	41	49	55	63	
38	41				
38	42				

marks are immediately seen and the middle marks, or mid-marks can be readily obtained by counting from either end. The marks which occur most frequently are readily identified as 40 (occurring 7 times), and 38, 39 and 50 (appearing 5 times each). No other marks are as common as these.

The tabulation shown in Table 6.2 however, is rather difficult to

TABLE 6.3. FREQUENCY DISTRIBUTION OF THE MARKS OBTAINED BY 105 STUDENTS

Marks	Tallies	Frequency	Marks	Tallies	Frequency
14	II	2	45	II	2
15	—	-	46	III	3
16	—	-	47	IIII	4
17	—	-	48	II	2
18	—	-	49	I	1
19	—	-	50	~~IIII~~	5
20	I	1	51	II	2
21	—	-	52	IIII	4
22	—	-	53	—	-
23	I	1	54	III	3
24	—	-	55	III	3
25	I	1	56	IIII	4
26	—	-	57	I	1
27	—	-	58	II	2
28	III	3	59	IIII	4
29	—	-	60	I	1
30	III	3	61	II	2
31	—	-	62	II	2
32	—	-	63	I	1
33	III	3	64	I	1
34	—	-	65	II	2
35	—	-	66	II	2
36	—	-	67	I	1
37	II	2	68	II	2
38	~~IIII~~	5	69	I	1
39	~~IIII~~	5	70	I	1
40	~~IIII~~ II	7	71	II	2
41	IIII	4	72	I	1
42	II	2	73	II	2
43	II	2	74	I	1
44	II	2			

prepare, cumbersome to print, and not particularly effective in its presentation. After arranging the data as in Table 6.2 their bulk must be reduced so that the 'eye can take them in easily, the mind comprehend them, and computational method deal with them efficiently.' A first step in such a condensation would be achieved by representing the repetitions of a particular mark by tallies instead of rewriting the 'marks' itself, as in Table 6.3. The number of tallies corresponding to any given marks is the *frequency* .of that 'marks,' usually represented by the letter *f*. The traditional method of tallying is to record the frequencies by marks until four have been made, then to make a cross mark for fifth score. This procedure makes up the preliminary sheet. (*See Table 6.3*).

By breaking down the data into the form of Table 6.3 much more information becomes apparent. A superficial scrutiny of Table 6.3 reveals the number of students getting a certain number of marks. Thus we see that 50 marks are obtained by 5 students; 51 marks by 2 students; and 54 marks by 3 students. The figures 5, 2 and 3 are the frequencies of 50, 51 and 54 respectively. *Frequency thus means the number of times a certain value of the variables is repeated in the given data.* A table so formed is known as *frequency distribution.*

Even after so much simplification, the data contain too many figures. In order that it may be more readily comprehensible the bulk of this data is further reduced by preparing a *grouped frequency distribution*.

A grouped frequency distribution is one where the total number of items possessing a certain number of value of the variable under study are put together and are stated as the frequency of those values.

TABLE 6.4. FREQUENCY DISTRIBUTION SHOWING THE MARKS OBTAINED BY 105 STUDENTS

Marks	*Frequeucy*
14—23	4
24—33	10
34—43	27
44—53	25
54—63	23
64—73	15
74—83	1
Total	105

In our data given in Table 6.1 the variable under study is 'Marks,' and its different values are the marks which may be (or are) obtained by 105 students individually. *The objects in relation to which the values of the variable are obtained are called 'items.'* In our data of Table 6.1 'students' are the items.

Now, if we add up the number of items (students) obtaining marks equal to any 10 consecutive values beginning with 14, the grouped frequency distribution will take the following form:

Each group of 10 consecutive values of Marks, viz. 14—23, 24—33 34—43, etc., is called a *class*. Since each class includes ten values, it is the magnitude or width of the class. The first figure of each class is known as the *lower limit* of the class, and the last figure of each class is called its *upper limit*. Thus, 14, 24, 34 are respectively the lower limits of the first three classes and 23, 33 and are their upper limits.

A look at this table makes it evident that the process of grouping the frequencies makes possible a considerable reduction in the size of the data. As a result of this, it is made more comprehensible by highlighting the salient features of variable. Thus we find that a very large number of students have marks varying between 34 and 63, very few having less than 34 and only one having more than 73.

It should be noted that there is a price that we have paid for the simplification obtained. Thus, while the frequency distribution of Table 6.3 had all the details, the grouped frequency distribution has masked certain details. We no longer know how the 27 'items' within the class 34—43 are distributed. Are they all close to 34 or to 43 or more or less uniformly distributed across the class interval. This 'smudging' of the details is the price we pay for easier comprehension of the essential pattern of the distribution. Clearly the larger the class interval we select the more would be this 'loss of information'. There is thus an optimum class size for each problem at which the best compromise is made between the demands of details and simplification.

6.4. SELECTING CLASS INTERVALS

The main problem in preparing a frequency distribution as detailed in Sec. 6.3 is that of selecting class intervals. The first question that has to be settled is the number of classes to be used. No hard and fast rules can be given for determining the number of classes. This is decided by balancing the two extremes, one, where the number is so large that enough simplification is not obtained, and two, when the number is so small that much useful details are lost. A good rule of thumb is to choose between 5 and 15 classes, the exact number being determined by other considerations as detailed below.

The class width used should be some convenient number like 1, 5, 10, 25, 100 etc. and ordinarily not 7, 11, 13, 26, 29 etc. Thus, in the distribution of marks of Table 6.4 the class width is 10 giving 7 intervals. We could easily have choosen 9 or 11, both leading to convenient number of classes, but both would have been awkward.

After the width and number of class intervals are fixed, there is the problem of locating them, *i.e.*, of fixing their limits or centres. Since for many of the subsequent analysis, we assume that all the items within an interval are concentrated at the mid-point, the intervals should be so located as to make this assumption as nearly true as possible, and to make the computations simple. For example, if we inquire about the salaries of individual workers, they will report generally, (but not always) in multiple of 50, giving numbers which are close to, say, 400, 450, 500 etc. Thus, the boundaries of the classes should be located at 425, 475, 525 etc., so that the reported values are close to the mid-values of the classes.

For discrete variable data, *i.e.*, where there are finite jumps between the consecutive possible values of the variable, the limits can be specified in a number of ways. One way is that shown in Table 6.4 for the 'marks obtained'. Here the limits are *both inclusive*, *i.e.*, items having values equal to the lower and the upper limits of a class are included in that class. In the other method, we use what are called *exclusive classes* wherein the items equal to the size of either the upper limit or the lower limit are *excluded* from the frequency of that class. These may be either of the 'lower limit excluded' type as given in Table 6.5, or the 'upper limit excluded' type as given in Table 6.6.

TABLE 6.5

Marks (*Lower limit excluded*)	*Frequency*
13—23	4
23—33	10
33—43	27
43—53	25
53—63	23
63—73	15
73—83	1
Total	105

TABLE 6.6

Marks (*Upper limit excluded*)	*Frequency*
13—23	3
23—33	8
33—43	28
43—53	27
53—63	22
63—73	14
73—83	3
Total	105

When the variable involved is a continuous one, *i.e.*, when the variable takes values on a continuous scale, the meaning of the class limit becomes slightly different. Now, we cannot have the limits as in Table 6.4. The intervals must be such that they cover the entire range and therefore there cannot be any gaps. When classifying weights of individuals, we cannot use classes like 33—37, 38—42, 43—47, etc. (all in kg). This is so, because the weights are not restricted to whole numbers. A weight of 42.6 kg will not fit any of these classes. A little thought will reveal that only such classes as 33—38, 38—42, 42—47, etc. will do. Here any weight less than 38 will be included in the first class, and any greater than 38 in the second class. How about a weight of exactly 38 kg? Theoretically, a weight of exactly 38 kg has very little chance of occuring. If it was measured accurately enough, may be we can find that it is 1 gm (or even less) on one side or the other. But usually these weights are not measured to such accuracy, and if we get one that is *reported* at 38 kg, where do we place it? there are at least three courses that are available to a statistician: (1) include half an item in the class 33—38 and the other half in 38—42. (2) toss a coin to decide, and (3) make a thumb rule to include in the higher class the first time such an item occurs, and in the lower class when it occurs the second time. The first option is perhaps the best statistically, but the idea of half a boy in 38—42 class is diverting and is best avoided. The second option suffers from the fact that it will not permit rechecking the results. The third or some variant of it is usually employed.

Illustration 6.1. From the following observations prepare a frequency distribution in ascending order starting with 5—10 (Exclusive method).

TABLE 6.7

When upper limit is excluded			When lower limit is excluded		
Marks	Tally bars	Frquency	Marks	Tally bars	Frequency
5—10	III	3	0—5	I	1
10—15	IIII	4	5—10	IIII	4
15—20	~~IIII~~ I	6	10—15	III	3
20—25	~~IIII~~	5	15—20	~~IIII~~ IIII	9
25—30	IIII	4	20—25	I	1
30—35	III	3	25—30	~~IIII~~ I	6
35—40	III	3	30—35	I	1
40—45	I	1	35—40	IIII	4
45—50	I	1	40—45	I	1
			45—50		

Marks in English:

12	36	40	30	28	20	19	10	10	16
19	27	15	26	20	19	7	45	33	21
56	37	6	20	11	17	37	30	20	5

Solution. Note the variable involved is a discrete one. We chose the 'exclusive' type of limits. The resulting distribution is as given in Table 6.7.

Illustration 6.2. Prepare a frequency table for the following reported weights (in kg) 40 individuals.

47	50	79	45	46	80	82	72
75	74	57	69	65	52	55	60
64	73	61	60	71	70	68	68
65	55	59	61	60	66	54	70
62	53	65	56	52	72	67	58

Solution. The variable is a continuous one with values ranging from 45 to 82. If we select an interval of 5 kg, it gives 8 classes, and hence is quite appropriate. We note the prepondarance of figures ending in 5 and 0 (15 out of 40 compared to a 'normal' of about 8... denoting gross rounding off.) Therefore, the mid-points may be placed at 0's and 5's. Thus, we select the class intervals 42.5—47.5, 47.5—52.5, ... 77.5—82.5. The resulting distribution is given in Table 6.8.

TABLE 6.8

Class	*Frequency*
42.5—47.5	3
47.5—52.5	3
52.5—57.5	6
57.5—62,5	8
62.5—67.5	6
67 5—72 5	8
72.5 77.5	3
77.5–82.5	3

6.5. TWO-WAY FREQUENCY DISTRIBUTION

In the preceding sections we have explained how to construct frequency tables where only one variable was observed for the given number of items. We now proceed to describe the construction of frequency tables when two variables are involved. As an example we start with the data in Table 6.9 regarding the marks in 'Statistics' and in 'Accountancy'. A frequency table should be able to show the performance of students in the two subjects simultaneously.

TABLE 6.9

Roll number of Students	1	2	3	4	5	6	7	8	9	10	11	12
Marks in Statistics	15	1	1	3	16	2	18	5	4	17	6	19
Marks in Accountancy	13	1	2	7	8	9	12	9	17	16	6	18
Roll number of Students	13	14	15	16	17	18	19	20	21	22	23	24
Marks in Statistics	14	9	8	13	10	13	11	11	12	18	9	7
Marks in Accountancy	11	3	5	4	10	11	14	17	18	15	15	3

A two-way frequency table has class intervals for one variable as columns and for the other variable as rows, as in Table 6.10. The boxes formed at the intersection of rows and columns thus represent a 'joint-class'. The frequency of this joint class is the number of items that has the value of the first variable in the class given by the *column* heading and the value of the second variable in the class given by the *row* heading.

The method of construction of the two-way table consists of the following steps.

1. Determine the class intervals for each of the variables. For data of Table 6.9. Since the variables are discrete we use inclusive-type intervals for both variables.Use of 4 classes of width 5 for each variable gives 16 *joint-classes*, which is a 'reasonable' number.

2. Place one of the variables at the top of the table (here, 'marks in statistics') and the other on the left-hand side.

TABLE 6.10. TWO-WAY FREQUENCY TABLE FOR MARKS IN STATISTICS AND ACCOUNTANCY

Statistics / *Accounts*	1—5	5—10	11—15	16—20	Total
1—5	II (2)	III (3)	I (1)		6
6—10	III (3)	II (2)		I (1)	6
11—15		I (1)	IIII (4)	II (2)	7
16—20	I (1)		II (2)	II (2)	5
Total	6	6	7	5	24

3. Place each item in the approximate box. Thus, Roll no. 1 with 15 in statistics and 13 in Accountancy belongs to box at the intersection of *column 11—15* and *row 11—15*.

4. Total the tallies in each box and in each row and colunm. The grand total of rows and columns should check with the total number of items.

The two-way table for marks in Statistics and Accountancy constructed in this manner is shown as Table 6.10.

6.6. CUMULATIVE FREQUENCIES

Consider the distribution of the weekly earnings of handloom employees given in Table 6.11.

TABLE 6.11. WEEKLY EARNINGS OF HANDLOOM EMPLOYEES

Earnings	*Number of employees*
30 and under 40	20
40 ,, ,, 50	50
50 ,, ,, 60	100
60 ,, ,, 70	40
70 ,, ,, 80	35
80 ,, ,, 90	10
90 ,, ,, 100	5

It is at times desirable to know the number of employees earning less than or more than certain amounts. This information is useful if we are planning some social security measures based on their earnings.

In order to provide this information, it is necessary to change the form of the frequency distribution from a simple distribution as shown above to a *cumulative frequency distribution* as shown below. When frequencies of two or more classes are added up, such totals are called *cumulative frequencies*. There are, in general, two types of cumulative frequencies used. The 'less than' and 'more than' cumulative frequencies.

If we are interested in preparing 'Less than' frequency table the cumulation is started from the lowest size of the variable to the highest size. Thus the frequency of 'less than 40' is frequency of class-interval '30 and under 40'; of 'less than 50' is the total of the frequencies of first two classes; of 'less than 60' the total of the frequencies of the first three classes, and so on.

When it is desired to construct a 'more than' cumulative frequency distribution,' the cumulation proceeds from the greatest to the least. Thus in order to determine the workers whose earnings are Rs. 70 or more the frequencies of the last three classes are added up. Tables 6.12 and 6.10 shows the two types of cumulative frequency distributions.

TABLE 6.12. LESS THAN CUMULATIVE DISTRIBUTION

Weekly earnings	*Number of workers who fell below the upper limit of each class*
Less than 40	20
,, ,, 50	70
,, ,, 60	170
,, ,, 70	210
,, ,, 80	245
,, ,, 90	255
,; ,, 100	260

TABLE 6.13. MORE THAN CUMULATIVE FREQUENCY DISTRIBUTION

Weekly earnings	*Number af workers who equaled or exceeded the lower limit of the class*
30 or more	260
40 or more	240
50 or more	190
60 or more	90
70 or more	50
80 or more	15
90 or more	5

6.7. STATISTICAL SERIES

A series is a systematic arrangement of items. When statistical data are arranged in a systematic manner, the arranged data is called a 'Statistical series.' These are of three types: (*a*) Categorical, (*b*) Time, and (*c*) Frequency.

Categorical series. If the data are arranged on the basis of a qualitative characteristics, such as geographical location, or some other attribute, the series so obtained are called *Categorical*. Tables given on next page are examples of categorical series. In Table 6.14

the arrangement is on the basis of geographical location while in Table 6.15 the outlay for village and small industries in public sector is classified on the basis of the type of industry.

TABLE 6.14. EMPLOYMENT IN FACTORIES (*in thousands*)

State	*Average daily number of workers employed*
Assam	79
Bihar	258
Gujarat	405
Haryana	77
Kerala	204
Tamil Nadu	411
Uttar Pradesh	384
West Bengal	850

TABLE 6.15. OUTLAY FOR VILLAGE AND SMALL INDUSTRIES IN PUBLIC SECTOR, 1966-69 (*in crores of rupees*)

Industry	*Expenditure*
Small Scale Industry	39.35
Industrial Estates	7.58
Handloom Industry	14.05
Village Industry	89.33
Coir Industry	1.79

There may be a wide variety of categorical series. For example, a series where new buildings are classified according to the purpose for which they are erected: series where wage rates are classified according to occupation; series where population is classified according to religion, etc.

Time series. If the different values that a variable has taken in a period of time are arranged in a chronological order, the series so obtained is called a time series. Below is given an example of a time series:

TABLE 6.16. EXPORT TRADE OF INDIA (*in crores of rupees*)

Year	*Export*
1964–65	816.30
1965—66	805.64
1966—67	1,156.56
1967—68	1,198.69
1968—69	1,357,87
1969—70	1,413.21

Frequency series. If the various items wherein a certain variable has been studied are classified according to the size of this variable the series that we obtain in called a frequency series. The following is an example:

TABLE 6.17. AGE STRUCTURE, 1961

Age group	*Percentage of total population*
Up to 4	15.0
5—14	26.0
15—24	16.7
25—34	15.4
35—44	11.0
45—54	8.0
55—64	4.8
65—74	2.1
75 and above	1.0
Total	100

6.8. STATISTICAL TABLES

A statistical table is an orderly and systematic presentation of numerical data in columns and rows. Columns are vertical arrangements; rows are horizontal. The main object of a statistical table is to so arrange the physical presentation of numerical facts that the attention of the reader is automatically directed to the relevant information. Some of the main advantages of tabular presentation over descriptive statements are;

1. Tabulated data can be easily understood than facts stated in the form of descriptions.
2. They leave a lasting impression.
3. They facilitate quick comparison.
4. Statistical tables make easier the summation of items and detection of errors and omissions.
5. When data are tabulated all unnecessary details and repetitions are avoided.
6. A tabular arrangement makes it unnecessary to repeat explanations, phrases and headings.

Parts of Tables

The following parts must be present in all tables;

(1) Title, (2) Caption, (3) Stubs, (4) Body. There are, however,

other parts whose presence depends upon the specific purpose. They are (5) Head note (or Prefatory note). (6) Footnote, and (7) Source note.

1. **Title**. A complete title explains in brief and concise language (*a*) what the data are, (*b*) where the data are, (*c*) time period of data, and (*d*) how the data are classified.

2. **Captions**. The title of the columns are given in captions. In case there is a sub-division of any column, there would be sub-caption headings also.

3. **Stubs**. The titles of the rows are called *stubs*. The box over the stub on the left of the table gives description of the stub contents, and each stub labels the data found in its row of the table.

4. **Body**. The body of the table contains the numerical information.

5. **Headnote**. (or prefatory note) is a statement, given below the title, which clarifies the contents of the table.

6. **Footnote**. It is a statement which clarifies some specific items given in the table, or explains the omission thereof. Thus, if we look into a table, giving yearly figures of wheat production in India, the sudden fall in the figure for 1947 would be misleading unless there is a foonote to point out that the figures for 1947 relate to India after partition.

7. **Source**. The source from where the data contained in the table has been obtained should be stated. This would permit the reader to check the figures and gather, if necessary, additional information.

TABLE 6.18. TITLE

(Description of Units and Year, Place, etc.) HEAD NOTE

(Stub box) *(D)*	*(A) Caption*		*(B) Caption*	
	(1)	*(2)*	*(3)*	*(4)*
Stub X				
Y	B	O	D	Y
Z				
Total				

***Notes*:** Any definition.
Any explanation.
Source from which derived.

6.9. TYPES OF TABLES

Tables may be classified according to the number of characteristics used for tabulation. A *simple* or a *one-way* table uses only one characteristic against which the frequency distribution is given, as in Table 6.19 where the characteristic used is the age of student.

TABLE 6.19. AGE-WISE DISTRIBUTION OF THE STUDENTS OF A COLLEGE

Age in year	*Students*
16—17	—
17—18	—
18—19	—

In a two-way table, on the other hand, two characteristics are used. In this case one characteristic is taken as column headings, and the other as row stubs. Example of a two-way table showing a two-way frequency distribution is shown in Table 6.20.

TABLE 6.20. AGE AND SEX-WISE DISTRIBUTION OF THE STUDENTS OF A COLLEGE

Age in years	*Student*		*Total*
	Male	*Female*	
16—17 17—18 18 and on			

When it is desired to represent three or more characteristics in a single table, such a table is called Higher order table. Thus, if it is desired to represent the 'age,' 'sex' and 'course,' of the students, the table would take the form as shown on page. 70 and would be called a higher order table.

TABLE 6.21. TABLE SHOWING DISTRIBUTION OF THE STUDENTS OF A COLLEGE ACCORDING TO 'AGE', 'SEX' AND 'COURSE'

Age in years	*Course*						*Total*
	Arts		*Science*		*Commerce*		
	Male	*Female*	*Male*	*Female*	*Male*	*Female*	
16—17							
17—18							
18 and over							
Total							

Illustration 6.3. Draft a form of tabulation to show:

(*a*) Sex,

(*b*) three ranks—supervisors, assistants and clerks,

(*c*) years--1976 and 1979,

(*d*) age group—18 years and under, over 18 but less than 55 years, over 55 years.

Solution. In the above question we have to prepare a table to

TABLE 6.22. TABLE SHOWING THE DIVISION OF THREE RANKS OF EMPLOYEES ACCORDIMG TO SEX AND AGE GROUP FOR 1976 AND 1979

	Age groups	*1976*				*1979*			
		Supervisors	*Assistants*	*Clerks*	*Total*	*Supervisors*	*Assistants*	*Clerks*	*Total*
Males	0—18 18—55 55 and above								
	Total								
Females	0--18 18—55 55 and above								
	Total								

show four characteristics, *i.e.*, sex, three ranks of the employees, as given, for two different years and the data is to be divided according to age groups already given here. We can prepare a blank table to incorporate all these characteristics. (Table 6.22)

Illustration 6.4. The city of Timbaktoo was divided into three areas: the administrative district, other urban districts, and rural districts. A survey of housing conditions was carried out and the following information was gathered:

There were 6,77,100 buildings of which 1,76,100 were in rural districts. Of the buildings in other urban districts 4,06,400 were inhabited and 4,500 were under construction In the administative district 4,000 buildings were inhabited and 500 were under construction of the total of 61,600.

The total buildings in the city that are under construction are 6.200 and those uninhabited are 44,900.

Tabulate the above information so as to give the maximum possible information. How many buildings are under construction in rural areas?

Solution.

TABLE 6.23. DISTRIBUTION OF BUILDINGS IN THE THREE DISTRICTS OF TIMBAKTOO ACCORDING TO INHABITATION

(in hundreds)

District	*Inhabited*	*Uninhabited*	*Under Construction*	*Total*
Administrative	571	40	5	616
Other Urban	4,064	285	45	4,394
Rural	1,625	124	12	1,761
Total	6,260	449	62	6,771

The table clearly shows that there are 1,200 buildings under construction in rural areas.

Illustration 6.5. An investigation conducted by the education department in a public library revealed the following facts. You are required to tabulate the information as neatly and clearly as you can.

"In 1960, the total number of readers was 46,000 and they borrowed some 16,000 volumes. In 1965 the number of books borrowed increased by 4,000 and the borrowers by 50 per cent."

The classification was on the basis of three sections: Literature, Fiction and Illustrated News. There were 10,000 and 30,000 readers in the section Literature and Fiction, respectively, in the year 1960. In the same year 2,000 and 10,000 books were lent in the sections--Illustrated News and Fiction, respectively. Marked changes were seen in 1965. There were 7,000 and 42,000 readers in the Literature and Fiction section respectively. So also 4,000 and 13,000 books were lent in the sections Illustrated News and Fiction respectively.

Solution.

TABLE 6.24. SHOWING THE CHANGES IN THE NUMBER OF READERS AND TYPE OF BOOKS IN THE YEAR 1975 AS COMPARED TO 1970

Type of books	*1970*		*1975*		*Changes in 1975 over 1970*	
	Number of readers	*Number of baoks borrowed*	*Number of readers*	*Number of books borrowed*		
Fiction	30,000	10,000	42,000	13,000	+12,000	+3,000
Literature	10,000	4,000	7,000	3,000	−3,000	−1,000
Illustrated News	6,000	2,000	20,000	4,000	+18,000	+2,000
Total	46,000	16,000	69,000	20,000	27,000	4,000

Illustration 6.6. Prepare a two-way frequency table and marginal frequency tables for 25 values of the two variables x and y given below. Take class interval of x as 10—20, 20—30, etc., and that of y as 100—200, 200—300, etc.

x	12	24	33	22	44	37	26	36	
y	140	256	360	470	470	380	280	315	
x	55	48	27	57	21	51	27	42	
y	420	390	440	390	590	250	550	360	
x	43	52	57	44	48	48	52	41	69
y	570	290	416	380	452	370	312	330	590

[*C.A., May 1980; B.A. Eco., Bombay*]

Solution.

TABLE 6.24. BIVARIATE FREQUENCY TABLE

Y \ X	10- 20	20—30	30—40	40—50	50—60	60—70	Total
100—200	1	—	—	—	—	—	1
200- 300	—	2	—	—	2	—	4
300 · 400	—	—	3	5	2	—	10
400 –500	—	2	—	2	2	—	6
500—600	—	2	—	1	—	1	4
Total	1	6	3	8	6	1	25

TABLE 6.25. MARGINAL DISTRIBUTION OF X

x	*f*
10—20	1
20—30	6
30—40	3
40—50	8
50—60	6
60—70	1
Total	25

TABLE 6.26. MARGINAL DISTRIBUTION OF Y

y	*f*
100 200	1
200—300	4
300 - 400	10
400—500	6
500—600	4
Total	25

Illustration 6.7. In a trip organised by a college, there were 80 persons, each of whom paid Rs 15.50 on an average. There were 60 students, each of whom paid Rs 16. Members of teaching staff were charged at a higher rate, The number of servants (all males) was six, and they were not charged anything. The number of ladies was 20 per cent of the total, and there was only one lady staff member. Tabulate this information.

[*B. Com., Bombay, 1981*; *B. Com., Poona*]

Solution.

Total contribution $=80\times 15.50=$ Rs. 1240.00

TABLE 6.27. SHOWING PARTICIPANTS, SEX AND CLASSWISE

Class	*Sex*		*Totals*	*Contribution rate*	*Contribution Total*
	Males	*Females*			
Students	45	15	60	16	960
Teaching staff	13	1	14	20	280
Servants	6	—	6	—	—
Totals	64	16	80	15.50	1240

Illustration 6.8. Prepare a bivariate frequency distribution for the following data:

Marks in Law:	10	11	10	11	11	14	12	12	13	10	13
Marks in Statistics:	20	21	22	21	23	23	22	21	24	23	24

Marks in Law:	12	11	12	10	14	12	13	10	14
Marks in Statistics:	23	22	23	22	22	20	24	23	24

[*B. Com., Delhi, 1982*]

Solution.

TABLE 6.28. BIVARIATE FREQUENCY DISTRIBUTION

Marks in Statistics / *Law*	20	21	22	23	24	Total
10	1	—	2	2	—	5
11	—	2	1	1	—	4
12	1	1	1	2	—	5
13	—	—	—	—	3	3
14	—	—	1	1	1	3
Totals	2	3	5	6	4	20

PROBLEMS

1. Describe the considerations which are to guide you in fixing the range, the class interval and upper and lower limits of class intervals for a frequency distribution.
2. Differentiate between the data which are spread over a time and the data which exist at a point of time.
3. Discuss the objects of classification of a raw mass of collected data.
4. Distinguish between:
   (*a*) Continuous series and discrete series.
   (*b*) Exclusive and inclusive class intervals.
   (*c*) Ordinary and cumulative frequencies.
   (*d*) More than and less than frequency tables.
   (*e*) Historical and non-historical series.
5. What is a statistical table? What are its objects and importance in a general scheme of any statistical inquiry?
6. Describe what considerations are to guide you in constructing a statistical table.
7. Prepare a blank table to give as much information as possible of the summary results of the distribution of population according to sex and four

religion, in five groups of different states of India.

8. Following are the marks abtained by a group of 35 students in class test carrying maximum marks 5. Tabulate the data in the form of frequency distribution.

3, 2, 0, 1, 3, 4, 2, 5, 3, 3, 1, 3, 2, 3, 1, 3, 3, 0.
4, 3, 5, 2, 2, 5, 3, 1, 4, 2, 1, 2, 3, 4, 1, 3, 2.

9. Compile a table showing the number of letters in each word of the extract given below, treating the number of letters in a word as variable.
"Sample method has another advantage over the census method. If the information is collected from only a small proportion of the population, its completeness and accuracy ean be easily ensured."

10. The following is a record of sales of a shop on 70 days. Tabulate the data in the form of frequency distribution taking the lowest class as 60—69 rupees.
61, 73, 93, 107, 112, 76, 78, 69, 96, 72, 80, 88, 95, 109, 103, 84, 84, 106, 91, 75, 91, 92, 102, 91, 101, 90, 77, 105, 90, 86, 113, 101, 114, 72, 77, 118, 95, 63, 99, 82, 100, 106, 87, 89, 92, 107, 111, 75, 83, 86, 106, 107, 62, 94, 73, 108, 115, 85, 98, 93, 109, 97, 74, 98, 67, 82, 104, 88, 88, 92.

11. Following are the marks (out of 100) obtained by 50 students in statistics.
70, 45, 33, 64, 50, 25, 65, 75, 30, 20, 55, 60, 65, 58, 52, 36, 45, 42, 35, 40, 51, 47, 39, 61, 53, 59, 49, 41, 15, 55, 42, 63, 82, 65, 45, 63, 54, 52, 48, 46, 57, 53, 55, 42, 45, 32, 64, 35, 26, 18.
Make a frequency distribution taking a class-interval of 10 marks (Take the first class interval as 0—10). *(B. Com., Delhi, 1968)*

12. Following are the wages in rupees of 70 workers. Tabulate them by taking the class interval of size 10.
32, 47, 57, 67, 62, 92, 117, 87, 27, 102, 93, 63, 73, 83, 123, 108, 63, 98, 113, 68, 63, 78, 98, 133, 98, 128, 118, 68, 73, 92, 82, 62, 57, 82, 72, 92, 52, 42, 36, 46, 41, 86, 136, 146, 96, 66, 46, 26, 114, 89, 79, 129, 24, 89, 99, 94, 84, 85, 102, 115, 40, 35, 125, 105, 35, 75, 45, 76, 84, 125.

13. Following figures gtve the height in centimetres of 80 plants. Represent the data by a frequency distribution with suitable class-interval.
62.1, 65.5, 63.0, 62.2, 64.7, 63.1, 65.8, 62.3, 60.7, 63.2, 64.1, 59.6, 64.5, 61.1, 65.7, 60.2, 64.6, 67.3, 64.5, 66.4, 64.2, 62.4, 63.3, 64.0, 62.5, 63.4, 66.3, 59.9, 63.5, 61.8, 65.4, 67.3, 60.4, 65.6, 59.1, 64.8, 61.9, 62.6, 67.0, 68.1, 59.4, 63.6, 64.4, 62.0, 63.7, 66.3, 63.8, 66.7, 63.9, 60.8, 63.0, 64.3, 61.2, 62.7, 64.6, 64.9, 60.5, 64.4, 61.7, 66.5, 65.3, 63.5, 65.2, 66,2, 59.7, 67.6, 63.5, 67.4, 63.5, 68.6, 60.0, 61.3, 63.6, 61.5, 65.1, 63.8, 61.6, 64.0, 68.7, 66.6.

14. Prepare a statistical table from the following:

Weekly wages of workers (in Rupees) in factory A

88	23	27	28	88	96	94	93	86	99
82	24	24	55	88	99	55	86	82	36
96	39	26	54	87	100	56	84	84	46
102	48	27	26	29	100	59	83	84	48
104	46	30	29	40	101	60	89	45	49
106	33	36	30	40	103	70	90	49	50
104	36	37	40	40	106	72	94	50	60

24	39	49	46	66	107	76	96	46	67
26	78	50	44	43	46	79	99	36	68
29	67	56	99	93	48	80	102	32	51

15. The following is a record of marks obtained by students in two sections, A and B, out of maximum of 150 marks. Tabulate the data in the form of a frequency distribution in such a way that a visual look at them would easily indicate the comparative performance of students in the two sections without the use of other sophisticated tools of statistical analysis.

Section A; 61, 73, 93, 107, 112, 80, 88, 96, 109, 103, 91, 92, 102, 91, 101, 113, 101, 114, 72, 77, 100, 106, 87, 89, 92, 106, 107, 62, 94, 73, 109, 97, 74, 98, 67.

Section B: 76, 78, 69, 96, 72, 84, 84, 106, 91, 75, 90, 77, 105, 90, 86, 118, 95, 63, 99, 82, 107, 111, 76, 83, 86, 108, 115, 85, 98, 93, 82, 104, 88, 88, 92.

16. Following are the marks obtained by students of a class in a certain test of statistics and law. Represent the data by one frequeney table.

No. of students	1	2	3	4	5	6	7	8	9	10	11	12
Marks in statistics	15	0	1	3	16	2	18	5	4	17	6	19
Marks in Law	13	1	2	7	8	9	12	9	17	16	5	18

No. of students	13	14	15	16	17	18	19	20	21	22	23	24
Matks in statistics	14	9	8	13	10	13	11	11	12	18	9	7
Marks in Law	11	3	5	4	10	11	14	7	18	15	15	3

17. The ages of 20 husbands and wives are given below. Form a two way frequency table showing the relationship between the ages of husbands and wives with the class-intervals 20-25, 25-30, etc.

S. No.	*Age of husband*	*Age of wife*	*S. No.*	*Age of husband*	*Age of wife*
1	28	23	11	27	24
2	37	30	12	39	34
3	42	40	13	23	20
4	25	26	14	33	31
5	29	25	15	36	29
6	47	31	16	32	35
7	37	35	17	22	23
8	35	25	18	29	27
9	23	21	19	38	34
10	41	38	20	48	47

18. A class of 32 students obtained the following marks in 1978 and 1979.

S. No. of students	Marks in 1978	Marks in 1979	S. No. of students	Marks in 1978	Marks in 1979
1	41	42	17	38	43
2	53	54	18	78	64
3	46	31	19	10	16
4	49	56	20	36	33
5	40	30	21	38	31
6	22	24	22	48	53
7	57	50	23	21	11
8	75	62	24	24	26
9	48	51	25	18	24
10	31	35	26	61	68
11	78	63	27	29	14
12	23	26	28	13	20
13	31	37	29	30	36
14	19	25	30	36	30
15	45	31	31	38	30
16	36	45	32	37	42

Classify these marks into the following form (your answer should clearly show the actual process of classification)

Marks group	Numbers of students who obtained: More marks in 1978 than in 1979	More marks in 1979 than in 1978	Total
Less than 30			
30-47			
48-59			
60 and above			

19. Point out the mistakes made in the following blank table drawn to show the disiribution of population according to sex, age, civil conditions and literacy.

	0-25				25-50				50-75				75 and above			
	Married		Un-married		Married		Un-married		Married		Un-married		Married		Un-married	
	M	F	M	F	M	F	M	F	M	F	M	F	M	F	M	F
Literates																
Illiterates																

Reconstruct the above table.

23. Tabulate the following information:

In a trip organized by a college, there were 80 persons each of whom paid

Rs 15.50 on an average. There were 60 students each of whom paid Rs 16. Members of the teaching staff were charged at a higher rate. The number of servants was 6 (all males) and they were not charged anything. The number of ladies was 20% of the total out of which 1 was a lady staff member.

21. Out of the total number of 1807 women who were interviewed for employment in a textile factory of Bombay, 512 were from textile areas and the rest from non-textile areas. Amongst the married women who belonged to textile areas, 247 were experienced and 73, inexperienced, while for non-textile areas, the corresponding figures were 49 and 550. The total number of inexperienced women was 134 of whom 111 resided in textile areas. Of the total number of women, 918 were unmarried, and of these; the number of experienced women in the textile and non-textile areas was 154 and 16 respectively. Tabulate.

22. In a newspaper account, describing the incidence of influenza among tubercular persons living in the same family, the following paragraph appeared. "Exactly a fifth of the 100,000 inhabitants showed signs of tuberculosis and no fewer than 5,000 among them had an attack of influenza, but among them only 1,000 lived in uninfected houses. In constrast with this 1/15th of the tubercular persons who did not have influenza were still exposed to infection. Altogether 21,000 were attacked by influenza and 41,000 were exposed to risk of infection, but the number, who having influenza but not tuberculur lived in houses where no other cases of influenza occured, was only 2,000."

Redraft the information in a conscise and elegant rabular form.

23. Present the following information in a suitable tabular form.

In 1975-76 the total production in India (in thousand tons) of the principal oilseeds was as follows: Groundnuts 3702; linseed 434; rape and mustard 1193; castor 105; sesamum 433. Next year the production of each of the first three items fell by 36% and the remaining items fell by 10% each.

In 1977-78 there was an increase compared to the preceding year of 8% in ground nuts, 12% in linseed, 1% in rape and mustard, 50% in castor, and 10% in sesamum. In the next year the figures were respectively 3823, 395, 955, 140 and 447.

24. The total number of accidents in Southern Railway in 1970 was 3500 and it decreased by 300 in 1971 and by 700 in 1972. The total number of accidents in Metre Gauge section showed a progressive increase from 1970 to 1972. It was 245 in 1970, 346 in 1971 and 428 in 1972. In the Metre Gauge 'not compensated' cases were 49 in 1970, 76 in 1971 and 108 in 1972. Compensated cases in the Broad Gauge section were 2867, 2587 and 2152 in these three years respectively.

From the above report, you are required to prepare a neat table as per rule of tabulation.

Chapter 7

DIAGRAMMATIC REPRESENTATION

As the eye is the best judge of proportion, being able to estimate it with more quickness and accuracy than any other of our organs, it follows that wherever relative qualities are in question, a gradual increase or decrease of any value is to be stated, this mode (Diagrammatic) of representing is peculiarly applicable; it gives a simple, accurate and permanent idea, by giving form and shape to a number of separate ideas, which are otherwise abstract and unconnected.

—WILLIAM PLAYFAIR

7.1. INTRODUCTION

Figures are not always interesting, and as their size and number increase they become confusing and uninteresting to such an extent that no one (unless he is specially interested) would care to study them. Their study is a great strain upon the mind without, in most cases, any scientific result. The aim of statistical methods, *inter alia*, is to reduce the size of statistical data and to render them easily intelligible. To attain this object the methods of classification, tabulation, averages, percentages, and index numbers are generally used. But the method of diagrammatic representation (visual aids) is probably simpler and more easily understandable. It consists in presenting statistical material in geometric figures, pictures or maps and lines or curves. In this chapter we shall discuss the first two, *i.e.*, geometric forms and pictures. The 'line' (or curve) forms will be discussed in Chapter 8.

The essential advantage of diagrams lies in the fact that they facilitate comparisons, and therefore, should be used only where comparisons are called for since they give only an approximate idea, they are not suitable where greater accuracy is desired. They fail to bring out small differences. Also, they take more time for construction and should, therefore, not be used indiscriminately.

7.2. BAR DIAGRAMS

Various kinds of devices have been developed for presenting statistical data—each suitable for a particular kind of situation. *One-way* or *simple* frequency distribution is usually presented through what are known as one-dimensional diagrams. The most common form of one-dimensional diagram is the *bar diagram*. In this diagram the length of a bar associated with a value or a class of values represents the cor-

TABLE 7.1. BIRTH RATE OF A FEW COUNTRIES OF THE WORLD DURING THE YEAR 1967

Country	*Birth rate*
Iran	33
Kenya	16
Libya	20
Malaysia	40
Mexico	30
Sweden	15

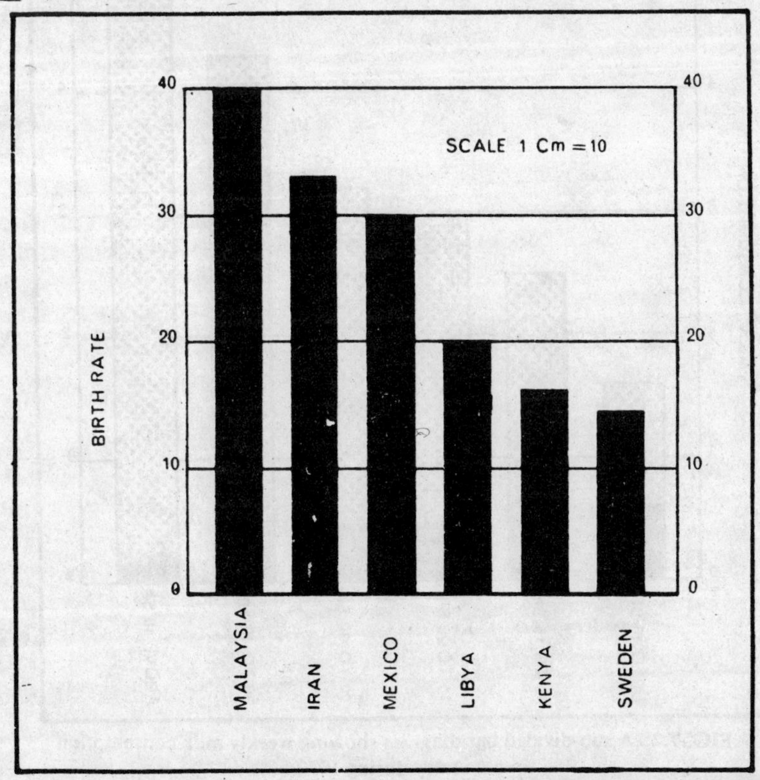

FIG. 7.1. A simple bar diagram showing birth rate of a few countries of the world during 1934.

responding frequency. Figure 7.1 is an example of a small bar diagram which presents graphically the data of Table 7.1. Note that the vertical axis here has the frequency. The bars should be equally spaced and laid out so that the diagram is pleasing to look at. The scale caption, or designation, etc. must be clearly marked.

Variations of this simple bar diagram can be used to present more complicated data, *i.e.*, when sub-divisions within the frequency exist, or when corresponding to one statistical unit, various series of data are to be presented.

Figure 7.2 shows what is termed as a *sub-divided bar diagram* and

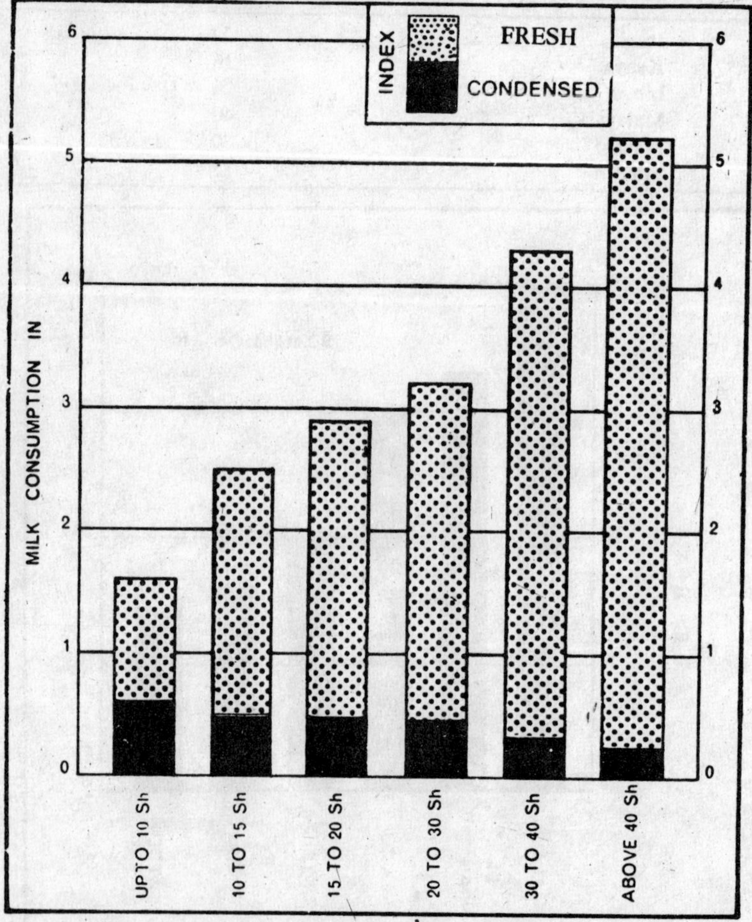

FIG. 7.2. A sub-divided bar diagram showing weekly milk consumption in a country during 1970.

presents the data of Table 7.2. In this the given magnitude (or frequency) is split up into sub-divisions. Thus, this diagram not only shows at a glance the trend of increasing milk consumption, but also the fact that the proportion of fresh milk is also increasing with income.

TABLE 7.2. WEEKLY MILK CONSUMPTION PER HEAD IN A COUNTRY DURING THE YEAR 1970

Income per head per week	*Fresh milk (in lbs.)*	*Condensed milk (in lbs.)*
Up to 10 shillings	1.0	0.6
10 shillings to 15	2.0	0.5
15 ,, ,, 20	2.4	0.5
20 ,, ,, 30	2.8	0.4
30 ,, ,, 40	4.0	0.3
Above 40 shillings	5.0	0.2

Another variation is the *percentage bar-diagram* as shown in Fig. 7.3 which illustrates the data of Table 7.3. Such diagrams are used when relative differences or changes in the size of components (*i.e.*, the constitution of the mix) are to be highlighted. The calculation of percentages and cumulative percentages as shown in Table 7.3 is an important step in construction of these diagrams.

TABLE 7.3. MONTHLY EXPENDITURE OF TWO FAMILIES

	Family A			*Family B*		
	Rs.	*%age*	*Cumulative %age*	*Rs.*	*%age*	*Cumulative %age*
Food	140	28	28	240	30	30
Clothing	80	16	44	160	20	50
House Rent	100	20	64	120	15	65
Education	30	6	70	80	10	75
Fuel and Lighting	40	8	78	40	5	80
Miscellaneous	40	8	86	80	10	90
Saving	70	14	100	80	10	100

In a multiple bar diagram two or more sets of inter-related data are represented. The technique of drawing such a diagram is that of a simple bar diagram. The different bars are differentiated using different shades, colours, dots or crossings.

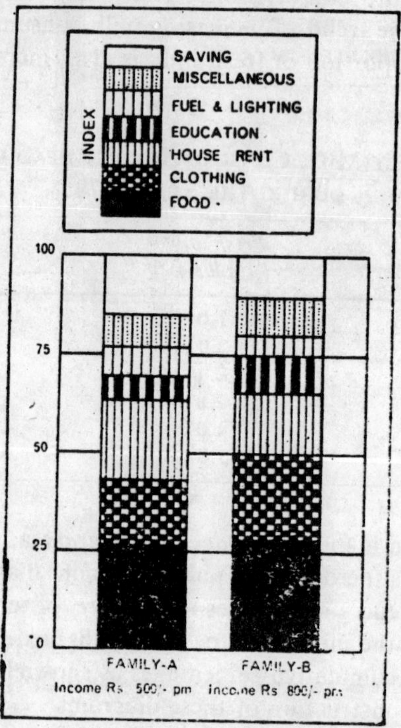

FIG. 7.3. A percentage-bar diagram showing details for monthly expenditure of two families, *A* and *B*.

The data of Table 7.2 may be represented by multiple bars. This is shown in Fig. 7.4. In this two bars are shown side by side, one is (shaded) representing fresh milk, and the other representing conden-sed milk.

Another important type of bar-diagrams in the *bilateral bar-diagram*. also called 'gain or loss chart'. It employs bars to show plus and minus directions from the point of reference as is done in Fig. 7.5 which shows cost, sales proceeds and loss or profit per ton for the data of Table 7.4.

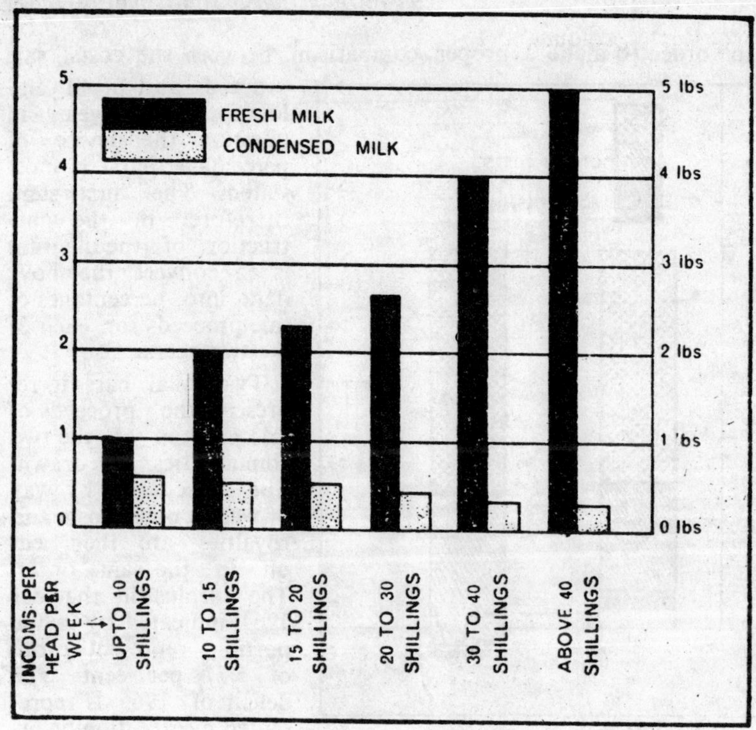

FIG. 7.4. A multiple-bar diagram showing weekly milk consumption in a country during 1970.

TABLE 7.4. COST AND SALES PROCEEDS PER TON OF COAL

	1964	*1965*
Cost per ton:	Rs	Rs
Wages	12.74	7.95
Other costs	5.46	4.51
Royalties	0.56	0.50
Total Rs	18.76	12.96
Proceeds of sale per ton	19.91	12.16
Profit (+) or Loss (−) per ton	+1.15	−0.80
Percentage (of sales proceeds):		
Cost per ton:		
Wages	63.99	65.38
Other costs	27.43	37.09
Royalties	2.81	4.11
Total Rs	94.23	106.58
Proceeds of sale per ton	100	100
Profit (+) or Loss (—) per ton	+5.77	—6.58

In order to make a proper comparison between the costs, sale proceeds and profit and loss for the two years in question the device of percentage bars is most suited. The first step, therefore, in the construction of the diagram is to convert the above data into percentages of sale proceeds for each of the two years.

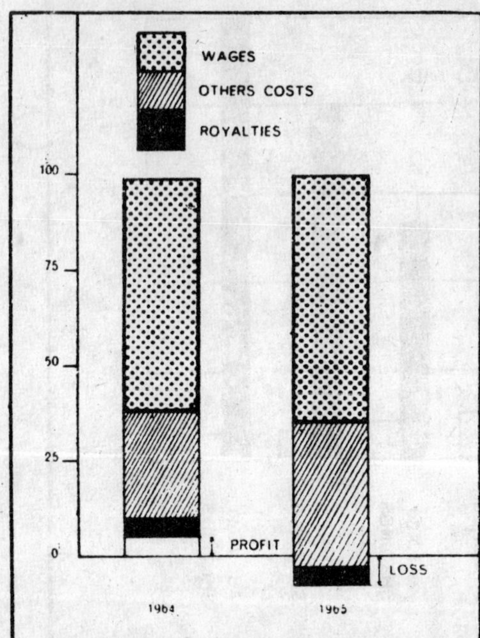

FIG. 7.5. A bilateral bar diagram showing costs, sales proceeds, profits or loss.

Two equal bars to represent the proceeds of sale per ton of the two commodities are drawn. The percentage by way of wages, other costs and royalties are then cut off in the same order. The surplus in the year 1964, indicated by white portion, represents profit of 5.77 per cent. The deficit of 1965 is represented by extention below the base portion of the bar.

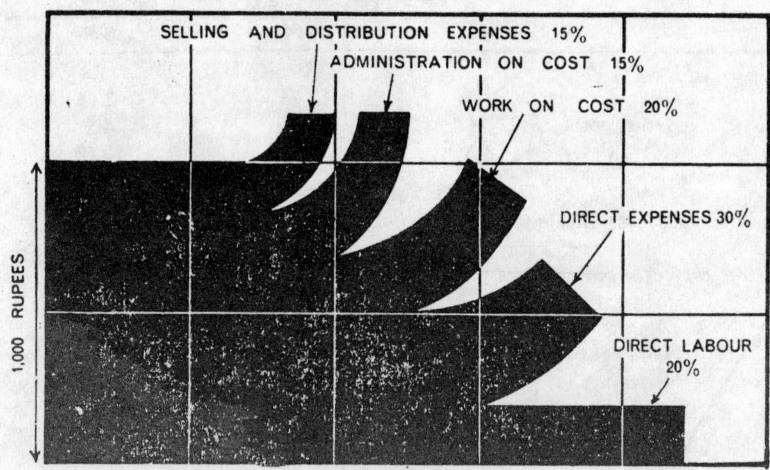

FIG. 7.6. A split-bar diagram.

For the purpose of showing diagrammatically component parts of a total, a split bar can also be used. Fig. 7.6 illustrates its use to show the proportion that the constituent divisions of cost bear to the total selling cost.

Illustration 7.1. Table 7.5 gives the birth rate and death rate of different countries during the year 1967. Construct a suitable diagram to represent the data.

TABLE 7.5. BIRTH RATE AND DEATH RATE OF A FEW COUNTRIES OF THE WORLD DURING THE YEAR 1967

Country	*Birth rate*	*Death rate*
India	33	24
Germany	16	11
Irish Free State	20	14
Soviet Russia	40	18
New Zealand	18	8
Sweden	15	12

Solution. Since the data contains two sets of values corresponding to each 'class', a sub-divided bar or a multiple-bar diagram can be used. Fig. 7.7 shows the former while Fig. 7.8 shows the latter.

Illustration 7.2. Table 7.6 gives the assets of the issue department of the Reserve Bank of India. Show this by means of a suitable diagram.

TABLE 7.6.

	Assets of the Issue Department in lakhs of rupee			
	Gold Coin & bullion	*Foreign securities*	*Rupee coin*	*Rupee coin*
1971-72	44,42	3,19,11	22,23	1,39,33
1972-73	44,42	6,43,52	14,28	85,45
1973-74	44,42	8,63,73	13,52	57,95
1974-75	44,42	10,61,26	15,53	57,84
1975-76	44,42	11,33,88	19,43	57,84

Solution. A split-bar diagram is the most suitable device for this data and is drawn as Fig. 7.9.

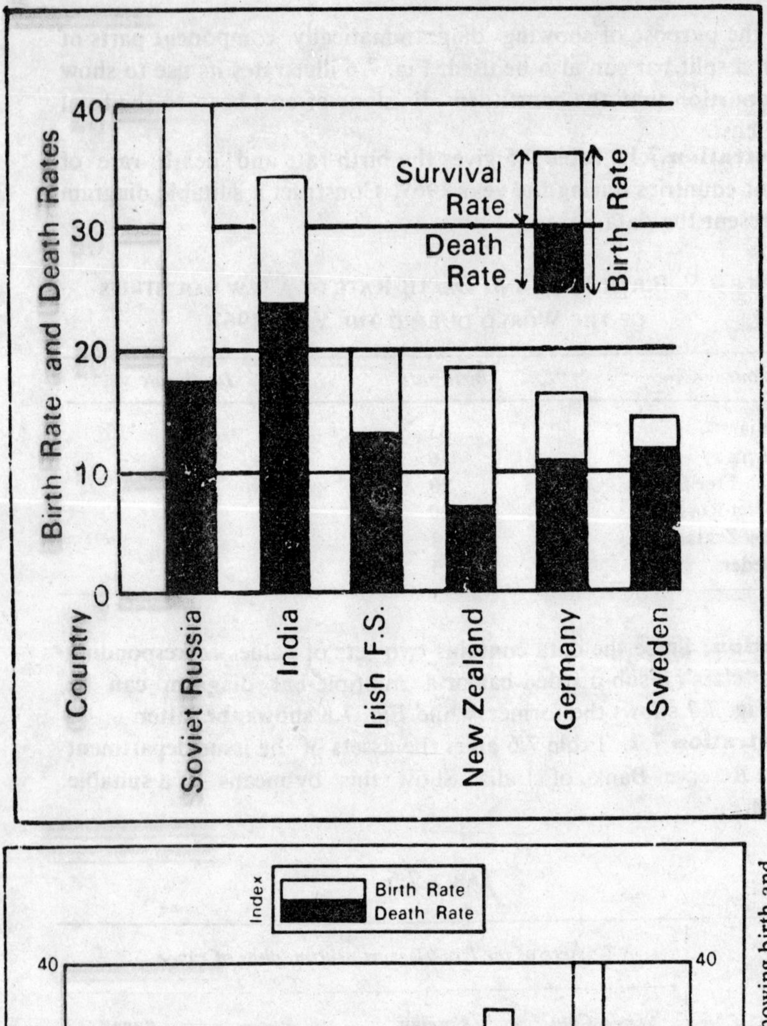

FIG. 7.7. A sub-divided bar diagram showing birth and death rate.

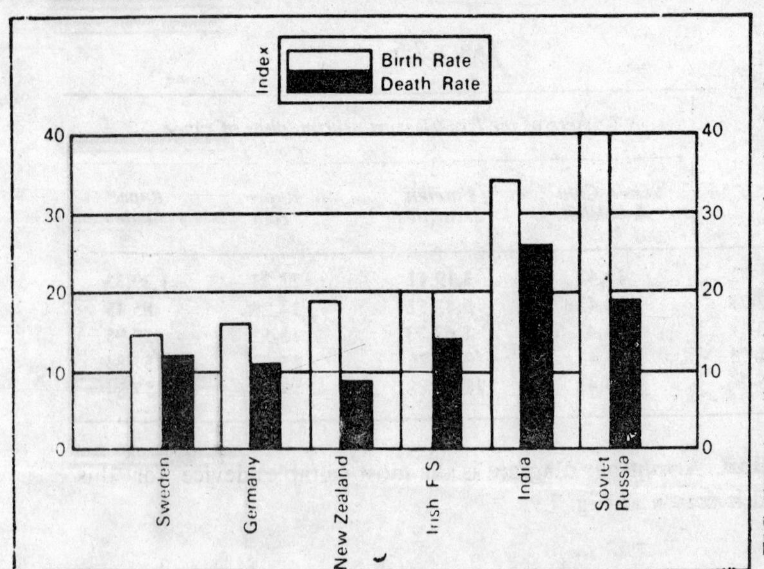

FIG. 7.8. A multiple-bar diagram showing birth and death rates.

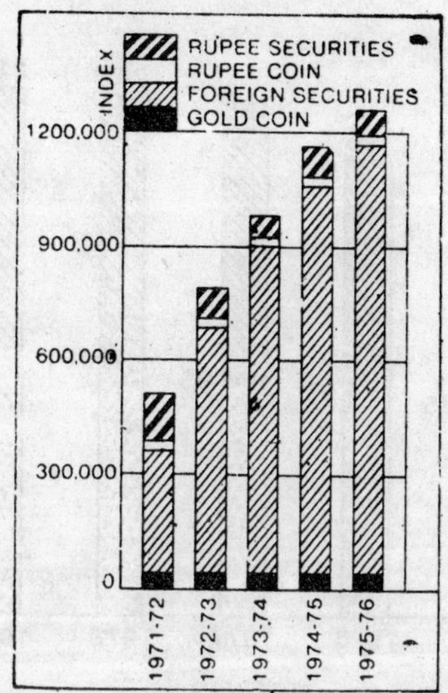

FIG. 7.9. A split-bar diagram showing assets of the Issue Department.

Illustration 7.3. Construct an appropriate diagram to present the data in Table 7.7.

TABLE 7.7. YIELD OF PADDY (*in thousands of tons*)

Years	*Karnatka*	*Kerala*	*Total*
1975	85	87	172
1976	91	95	186
1977	95	103	198
1978	101	106	207
1979	103	107	210

Solution. Figs. 7.10 and 7.11 show two diagrams, a sub-divided bar and a multiple bar to present the above data.

7.3. TWO-DIMENSIONAL DIAGRAMS

Though the 'bar' is the most common method of representing statistical data, there are occasions when it does not serve the purpose and as such other forms, viz., rectangles and circles are employed. A rectange is a two-dimensional diagram, *i.e.*, its height as well as width are taken into consideration for purposes of representation. (It may

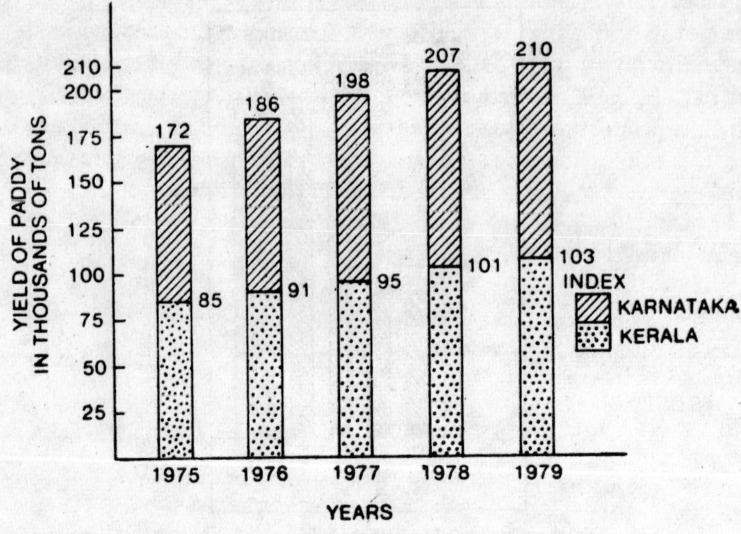

FIG. 7.10

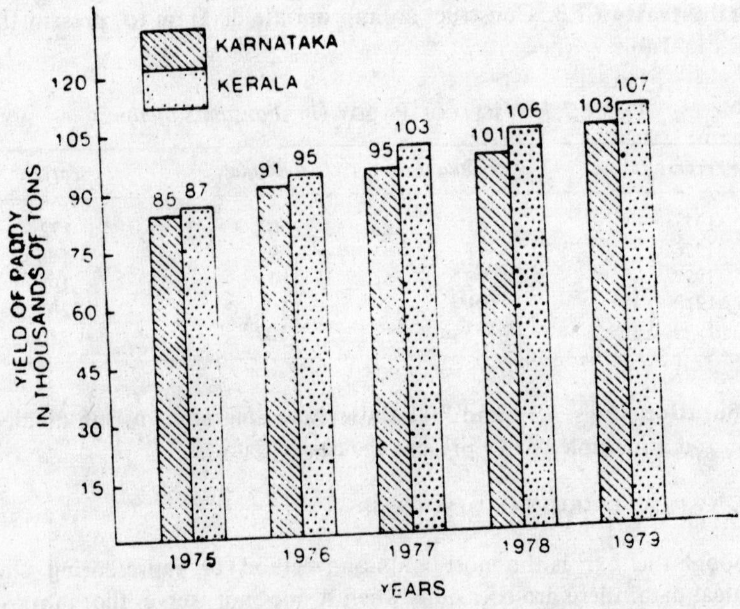

FIG. 7.11

be repeated here that in a bar diagram it is only the height of the bar and not its width that is significant.) Rectangles are used when it is desired to give a more detailed information than can be conveyed by the bar. In these (as well as in circles) it is the *area* of the diagram which is proportional to the frequency, unlike in the bars where the *height* is proportional to the frequency. The following illustrations demonstrate the use and construction of the rectangle.

Illustration 7.4. Represent the following data by a suitable diagram. Also indicate the 'profit' or 'loss' per unit.

TABLE 7.8

Factory	*Wages* *Rs*	*Materials* *Rs*	*Other costs* *Rs*	*Profit* *Rs*	*No. of units produced* *Rs*
A	3,000	5,000	1,000	1,000	1,000
B	2,000	3,000	800	590	700

Solution. Since there are two types of data involved, costs and no. of units produced, a one-dimensional bar diagram does not suffice and we use a *rectangle*. Since it is the area which represents the 'value' in a rectangle, we reduce costs and profits to per unit basis, and construct a rectangle where no. of units produced are taken horizontally while the *per unit* costs/profits are taken vertically as in Fig. 7.12. Thus, the area of each shade in each rectangle is proportional to the corresponding 'total' cost or profit.

The products of factory *A* realise Rs 10 per unit.

$$\frac{3000+5000+1000+1000}{1000} = \frac{10000}{1000} = \text{Rs } 10$$

The products of factory *B* realise Rs 9 per unit.

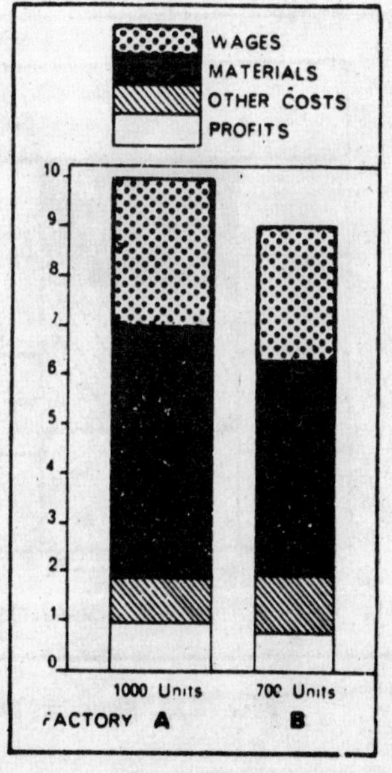

FIG. 7.12. Showing cost, profits, total number of units produced, etc,

$$\frac{2000+3000+800+500}{700} = \frac{6300}{700} = \text{Rs } 9$$

Thus, the height of the two rectangles will be in the ratio of 10 : 9 and their width in the ratio of units produced by the two factories, *i.e.*, 10 : 7.

Illustration 7.5. Represent diagrammatically the following:

TABLE 7..9 DETAILS OF THE COST OF TWO COMMODITIES

	A *Rs*	*B* *Rs*
Price per unit	4	5
Quantity sold	40	30
Value of raw material used	52	50
Other production expenses	64	60
Profit	44	40

Solution. Since two types of values are to be presented in this problem too, a rectangle is to be used as shown in Fig. 7.13.

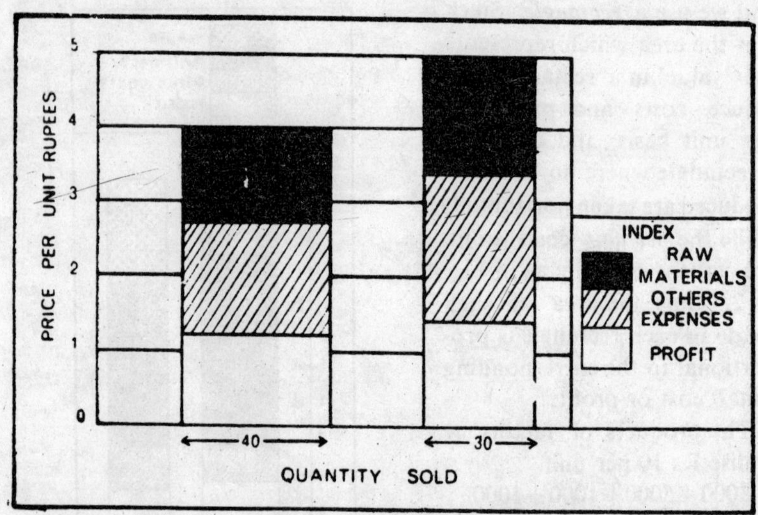

FIG. 7.13. Showing the cost of two commodities.

Illustration 7.6. Table 7.10 gives the comparative budgets of three families *A*, *B* and *C*. Present this data diagrammatically.

TABLE 7.10. FAMILY BUDGET OF THREE FAMILIES

Family Items of expenditure	*A* Actual expenditure (1)	% (2)	*B* Actual expenditure (1)	% (2)	*C* Actual expenditure (1)	% (2)
	Rs		Rs		Rs	
Food	12	60	30	50	90	30
Clothing	2	10	7	11-2/3	35	11-2/3
House rent	2	10	8	13-1/3	40	13-1/3
Education	1-50	7-1/2	3	5	12	4
Litigation	1	5	5	8-1/3	40	13-1/3
Conventional necessaries	0-50	2-1/2	3	5	60	20
Miscellaneous	1	5	4	6-2/3	23	7-2/3
Total	20	100	60	100	300	100

Solution. Since we want a comparative statement of the budgets, a

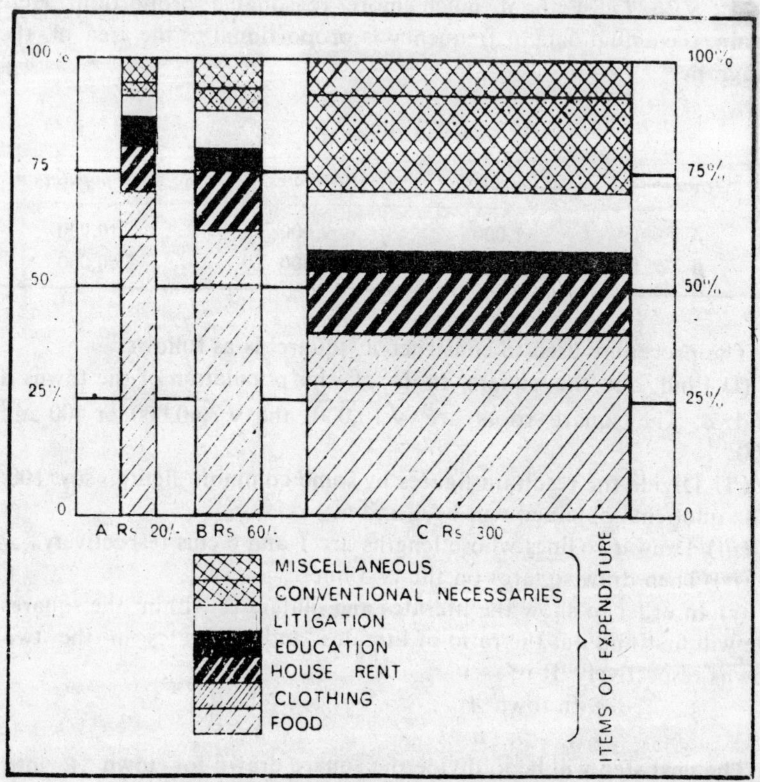

FIG. 7.14. Showing expenditure of three families *A*, *B* and *C*.

percentage bar diagram is suitable. But since, in addition, we want to present the total expenditure as well, we employ rectangles as shown in Fig. 7.14. Here the total expenses are taken horizontally and the percentage expenses (data of columns 2 in Table 7.10) is taken vertically. Thus, area of each shade of each column is proportional to the rupee expense.

Another common two-dimensional diagram used is the so called *square*. These are used when it is desired to compare quantities that differ widely in magnitude. If we are to present diagrammatically the population of two towns as it is given in Table 7.11, bar diagrams will fail to do the job. The difference in the two populations is so great that the height of one bar shall be 36 times as great as that of the other so that one bar would become too big and the other too small. To overcome this difficulty squares are used as in Fig. 7.15. The side of a square varies as the square root of its area. If the areas of two squares are in the ratio of 1 : 36, the ratio of its sides would be 1 : $\sqrt{36}$, *i.e.*, 1 : 6, a much more reasonable proportion. Here again, the actual data or frequency is proportional to the area of the diagram.

TABLE 7.11

Town	*Litrerates*	*Illiterates*	*Total population*
A	1,000	9,000	10,000
B	1,60,000	2,00,000	3,60,000

The procedure used to construct a 'square' is as follows:

(*i*) Find out the square roots of the population of the towns *A* and *B*. The square roots are $\sqrt{10{,}000}$ and $\sqrt{3{,}60{,}000}$ or 100 and 600.

(*ii*) Divide the resultant figures by some common figure, say 100. The quotients obtained will be 1 and 6.

(*iii*) Draw two lines whose lengths are 1 and 6 cms respectively.

(*iv*) Then draw squares on the two lines.

(*v*) In order to show the literates and illiterates within the squares we will first find out the ratio of literates and illiterates in the two towns respectively. It is:

In town *A* 1 : 9
" " *B* 4 : 5

The next step will be io divide the square drawn for town *A*, into 10 parts, the portion showing literates will be shown as in Fig. 7.14.

Similarly, the literates will also be shown in the square representing population of town *B*.

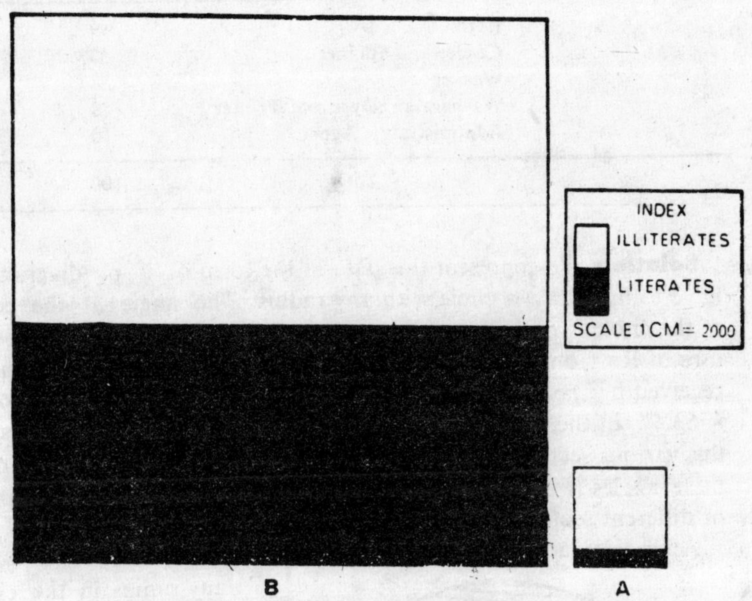

FIG. 7,15. Showing literates and illiterates in towns *A* and *B*.

Perhaps the most commonly used two-dimensional diagram is a *circle* or a *pie diagram*. Circular diagrams are alternative to square diagrams. Just as the areas of squares vary in the same proportion as the squares of their sides, likewise the areas of circles vary as the squares of their radii. It follows that if the radii of two circles are in the same proportion as the sides of two squares, the areas of the circles would also be in the same proportion as the areas of the squares. Hencc the lengths, which are used as the sides of squares, may also be used as the radii of circles. In Fig. 7.15 the sides of the squares are in the proportion of 1 : 6. By keeping the radii of the two circles in the proportion of 1:6 we can as well draw the circles representing the data of Table 7.11. Just as bars, rectangles and squares may be sub-divided in order to represent component parts, similarly circles may be sub-divided into various sectors. Such a sub-divided circle is known as a *pie diagram*.

Illustration 7.7. A rupee spent on 'khadi' is distributed according to data in Table 7.12. Present the data in the form of a pie-diagram.

TABLE 7.12. DISTRIBUTION OF A RUPEE SPENT ON KHADI

Head	*Paise*
Former	19
Carder and Spinner	35
Weaver	28
Washerman, Dyer and Printer	8
Administrative Agency	10
Total	100

Solution. To represent this data in the form of a pie-diagram, we have to first draw a circle with any radius. The angle at the centre of the circle is of 360° and it represents 100 paise, *i.e.*, the expenditure of Re 1 on Khadi. Thus, the sector to represent the amount received by the farmer, 19 paise, would have an angle of $\frac{19}{100} \times 360° = 68.4°$. In the same way we can calculate the degrees of angles for the various sectors of the circle and the total of the degree of all these angles would be 360°. Table 7.13 gives the degrees of angles of different sectors relating to the data in the above illustration.

After calculating the degrees of angles, the next step is to draw any radius in the circle. From the base line, an angle of 68.4° would be marked and a line would be drawn from the centre to touch the circumference. This sector would represent the amount received by the farmer. From the second line, an angle of 126° would be drawn representing the amount received by carder and spinner. In the same way other sectors would be completed. (*see* Fig. 7.16).

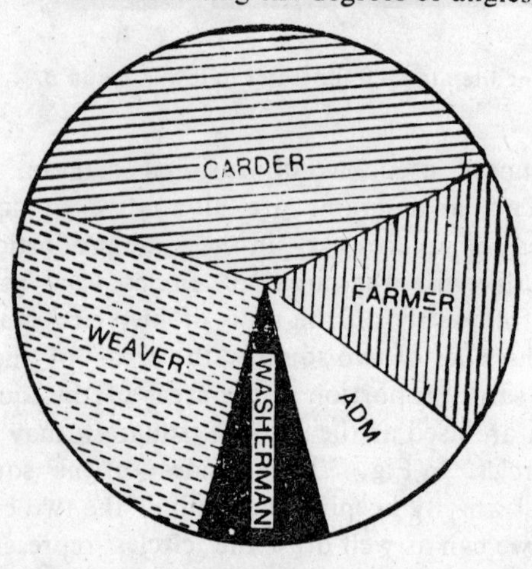

FIG. 7.16.

Pie diagrams show the changes in total and component parts. In Fig. 7.17 the two circles show (*i*) the changes in the total of clearing house statistics, and (*ii*) the share of individual towns in such totals. Interpretations of angular diagrams require a visual comparison of

TABLE 7.13. DISTRIBUIING OF A RUPEE SPENT ON 'KHADI'

	Paise	*Degree of angles*
Farmer	19	68.4
Carder and Spinner	35	126
Weaver	28	100.8
Washerman, Dyer and Printer	8	28.8
Administrative Agency	10	36
Total	100	360.0

the area of the two circles which is quite difficult, if not impossible, to make. That is why pie charts are not generally used.

TABLE 7.14. CLEARING HOUSE STATISTICS IN 1947-48 AND 1958-59

Town	*Total amount Rs. Lakhs*	*1947-48*	*Degrees*	*Total amounts Rs. Lakhs*	*1958-59*	*Degrees*
Bombay	80,232	$\frac{80232 \times 360}{207260}$	139.4	2,55,264	$\frac{255264 \times 360}{658014}$	139.7
Calcutta	100,853	$\frac{100853 \times 360}{207260}$	175.2	2,59,996	$\frac{259996 \times 360}{658014}$	142.3
Delhi	2,853	$\frac{2853 \times 360}{207260}$	5.0	12,646	$\frac{12646 \times 360}{658014}$	6.9
Kanpur	1,920	$\frac{1920 \times 360}{207260}$	3.3	10,983	$\frac{10983 \times 360}{658014}$	6.0
Bangalore	4,676	$\frac{4676 \times 360}{207260}$	8.1	27,481	$\frac{27481 \times 360}{658014}$	15.0
Poona	1,633	$\frac{1633 \times 360}{207260}$	2.8	4,954	$\frac{4954 \times 360}{658014}$	2.7
Madras	10,865	$\frac{10865 \times 360}{207260}$	18.9	34,794	$\frac{34794 \times 360}{658014}$	19.0
Others	4,228	$\frac{4228 \times 360}{207260}$	7.3	51,896	$\frac{51896 \times 360}{658014}$	28.4
Total	2,07,260		360°	6,58,014		360°

7.4. PICTURES

The main object of representing the data in the shape of pictures is to help in quick visualisation of comparison of magnitudes: for example, a pictogram might represent by a picture of men the population of India accompanied by a picture of proportionately fewer or more men representing, respectively, the populations of Britain, the USA, China, Japan, etc. More often pictures are adopted to help in visuali-

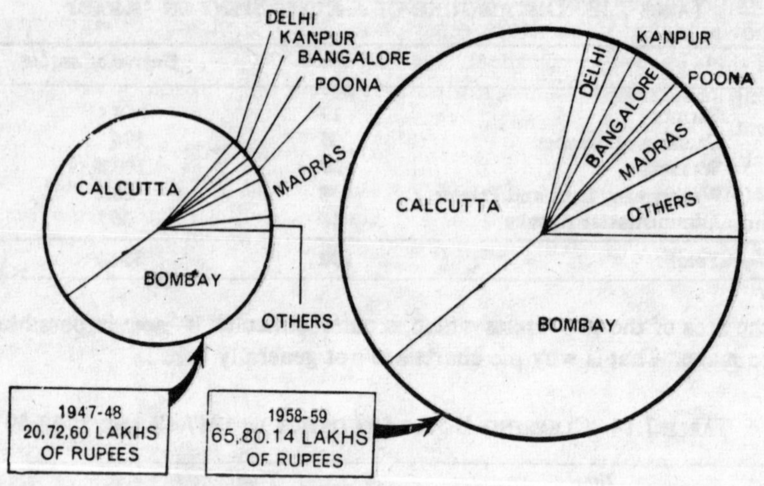

FIG, 7.17. Showing Clearing House Statistics,

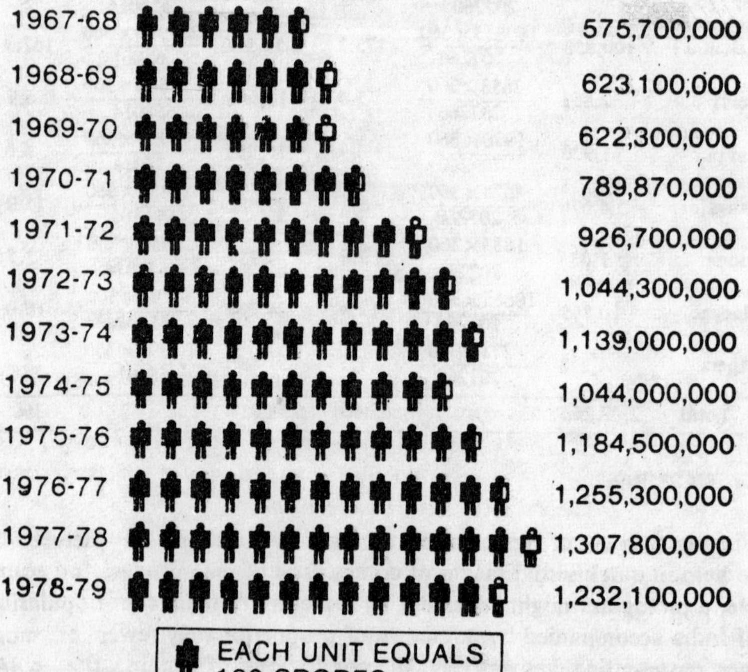

FIG. 7.18

sing the proportional parts of a whole magnitude, *e.g.*, a rupee is shown divided into sectors, representing the way in which the revenue of the Government is spent. Fig. 7.18 illustrates the number of passengers carried by Indian Railways. The picture of each man represents 100 crores passengers.

Figure 7.19 is much more explanatory and illustrates more than one variable. It shows the area under main oil-seeds and their yield in the Indian Union in the year 1947-48. The area is illustrated by the use of squares, whereas the yield is represented by the picture of a bag within the square.

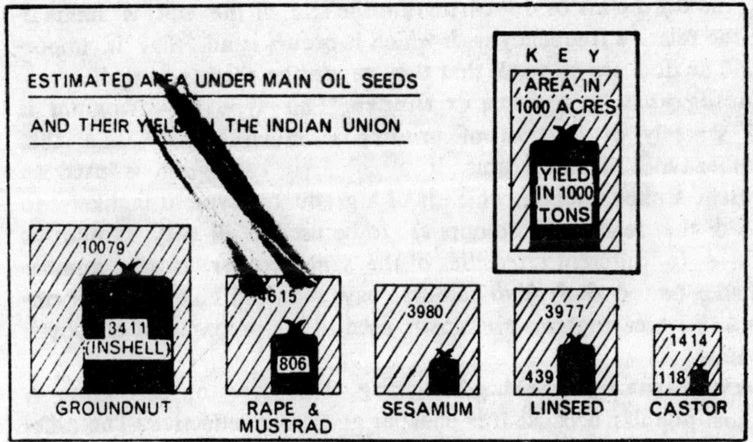

FIG. 7.19

7.5. STATISTICAL MAPS OR CARTOGRAMS

The statistical maps or cartograms represent frequency distributions associated with geographical areas. These are of three general types: (*i*) Cartograms by dots or points, (*ii*) Cartograms by colours or shades, and (*iii*) Cartograms by cross-hatching.

Cartograms by dots. In such types of maps the frequency of the data is represented by various types of dots. This type of cartogram may be further sub-divided into three classes.

In the first class, dots varying in size for different quantities or frequencies are used. This type of cartogram is not considered satisfactory, because of the necsssity of using varying sizes of circles as dots.

In the second class of dot cartogram, dots of uniform size are used. They can be counted to figure out the total. Dots may be shaded to indicate different values. Generally greatest quantity is shown by a solid dot, three-quarter, one-quarter, and other shading indicating less quantities. This type of cartogram is also unpopular because of the mechanical difficulty of arriving at the proper magnitude to allot to each dot of uniform size. If the magnitude assigned to each dot is too large, it becomes difficult to show graphically the small quantities relating to geographical locations where the characteristic is scarce. On the other hand, if the magnitude assigned to each dot is too small, this results in too great a crowding of the dots in areas where the characteristic is very plentiful.

In the third class of dot cartogram the size of the dot is immaterial; the relative frequency with which it occurs is all that is important. The dots are so small that they cannot be easily counted.

Cartograms by colours or shades. This type of cartogram is used scarcely as its cost of printing is much higher than that of other types of cartograms. If a coloured cartogram is used to represent a phenomenon, which is of a graduated type, it is likely to mislead the reader. If colours are to be used at all they should be confined to different intensities of the same colour. If the number of shades be too many, two colours, say red and blue, may be employed. Such cartograms are mostly used in the preparation of physical maps.

Cartograms by crossing-hatching. This type of cartogram is the most popular because it is cheaper and more effective. The different types of phenomena under this method are represented by lines, dots, circles and such other symbols.

PROBLEMS

1. Explain the usefulness of diagrams in presenting statistical data.
2. What, in your opinion, are the tests of a good diagram?
3. The marits of diagrammatic representation are classified under three main heading, attraction, effective impression and comparison. Explain and illustrate these points.
4. Discuss the merits and limitations of a diagrammatic presentation.
5. What points should be taken into consideration while presenting a statistical table diagrammatically?
6. Diagrams help us to visualise the whole meaning of a numerical complex at a single glance. Comment.
7. 'Diagrams are handy tools in the hands of a sales executive.' Discuss.
8. State briefly, giving reasons, the kind of diagram you consider more appropriate for use with each of the following classes of statistical data:
   (*a*) Number of children per family in a large town.

(*b*) Monthly rainfall for a period of three years.

(*c*) Monthly output of steel for one year according to the principal grades of quality.

9. Draw a sjmple diagram to represent the following statistics relating to the earning of a business house:

Years	*Rs (in lakhs)*
1971-72	244
1972-73	222
1973-74	184
1974-75	234
1975-76	258
1976-77	265
1977-78	294

10. Represent the following data relating to the values of imports and exports during four years ending 1976-77 by a suitable diagram:

VALUES OF EXPORTS AND IMPORTS (1973-74 TO 1976-77)

(*in crores of rupees*)

Years	*Exports*	*Imports*
1973-74	610.36	624.65
1974-75	955.39	742.78
1975-76	660.65	578.36
1976-77	565,25	527.98

11. Represent the following data diagrammatically:

PERCENTAGE SHARES IN INDIA'S EXPORTS

Countries	*1975*	*1976*	*1977*	*1978*
UK	34.1	29.2	30.0	29.3
Total other empire countries	18.6	37.5	35.4	30.4
US	8.3	17.6	22.0	23.2
Other non-empire countries	39.0	15.7	13.6	17.1

12. Represent the following figures by a suitable diagram:

AMOUNT TO BE SPENT ON VARIOUS HEADS UNDER GAHDHIAN PLAN

Various Heads	*Rupee (in crores)*
Agriculture	1,175
Rural Industries	350
Large Scale and Key Industries	1,000
Public Utilities	10
Transport	400
Public Health	26
Education	295
Research	20
Total	3,276

13. A rupee on 'Khadi' is distributed as follows:

	Paise
Farmer	19
Cardor and Spinner	35
Weaver	28
Washerman, dyer and Printer	8
Administrative Agency	10
Total	100

Present the data in the form of a pictogram.

14. Represent the followidg data pertaining to Indian Railways by suitable bar diagram:

(in crores of rupees)

	1976-77	*1977-78*	*1978-79*
1. Gross Income	390	422	468
2. Gross Expenditure	331	353	389
3. Net Income	59	69	79

15. Illustrate by a suitable diagram the following data of expenditure of an average working class family:

Item of expenditure	*Per cent of total expenditure*
Food	65
Clothing	10
Housing	12
Fuel and lighting	5
Miscellaneous	8

16. The following table gives the details of monthly expenditure of three families:

Items of expenditure	*Family A*	*Family B*	*FamilyC*
	Rs	*Rs*	*Rs*
Food	12,00	30	90
Clothing	2.00	7	35
House Rent	2.00	8	40
Education	1.50	3	12
Litigation	[illegible]	5	40
Conventional necessity	0.50	3	60
Miscellaneous	1.00	4	23

Represent the above figures by a suitable diagram. Which family is spending the money most wisely? Give reasons.

17. The following table gives the details of the cost and construction of a house in Allahabad:

	Rs
Land	4,500
Labour	2,500
Bricks	2,000
Iron	1,800
Timber	1,500
Cement	800
Ltme	800
Stone	600
Sand	200
Other things	1,300

Represent the above figures by a suitable diagram.

18. Draw a suitable diagram to represent the following informations:

(a) Town	[illegible]	*Illiterate*	*Total Population*
A	1,000	9,000	10,000
B	1,60,000	2,00,000	3,60,000

(b) Factory	*Wages*	*Material*	*Profits*	*Units*
A	2,000	3,000	1,000	1,000
B	1,400	2,400	1,000	800

19. Represent the following data by means of a suitable diagram:

Year	*Number Employed*			*Total*
	Men	*Women*	*Children*	
1968	1,80,000	1,10,000	70 000	3,60,000
1978	3,50,000	2,10,000	1,60,000	7,20,000

20. Represent the followihg data by a "Pie Diagram".
Cheques cleared in India in clearing houses in the year 1974 and 1979:

	Amount in crores of Rs	
	1974	*1979*
Bombay	829	2,670
Calcutta	1,070	2,443

Madras	108	274
Other Centres	313	515
	2,320	5,902

21. Represent the following by a suitable diagram:

(in lakhs of Rs)

Principal heads of revenue	*1978*	*1979*
Customs	4,050	4,588
Central Exeise Duty	868	652
Corporation Tax	204	238
Taxes on Income	1,374	1,420
Salt	812	1,080
Opium	50	46
Other Heads	112	130

22. The following data gives the birth-rate and death rate of a few countries of the world during the year 1978: Present it by a suitable diagram.

Country	*Birth rate*	*Death rate*
Egypt	44	27
Canada	24	11
USA	19	12
India	33	24
Japan	32	19
Germany	16	11
France	18	16
Norway	17	11

23. Represent the following by sub-divided bars drawn on a percentage basis:

COST, PROCEEDS, PROFIT OR LOSS PER CHAIR DURING 1977 AND 1979

Particulars	*1977*	*1978*	*1979*
Cost per Chair		(in Rupee)	
Wages	4.5	7.5	10.5
Other Cost	3.0	5.1	7.0
Polishing	1.5	2.4	3.5
Total Cost	9.0	15.0	21.0
Proceed per Chair	10.0	14.0	20.0
Profit or Loss per Chair	+1.0	-- 1.0	-- 1.0

24. Show the details of monthly expenditure of two families given on p. 105 by means of two dimensional diagrams.

Items of expenditure	*Family A* *Income Rs 500 p.m.*	*Family B* *Income Rs 403 p.m.*
Food	140	120
Clothing	80	80
House Rent	100	60
Education	30	40
Fuel and Lighting	40	20
Miscellaneous	40	40

25. Details the prices, cost and quantity sold of three commodities are given below. Represent this data by a suitable diagram:

Price of a commodity	*I* *Rs 3 per unit*	*II* *Rs 4 per unit*	*III* *Rs 5 per unit*
Quantity Sold	100	80	70
Value of Raw Materials	Rs 115	Rs 120	Rs 130
Expenses of Management	Rs 30	Rs 30	Rs 35
Expenses on Labour	Rs 40	Rs 45	Rs 55
Other Expenses	Rs 15	Rs 25	Rs 30
Profits	Rs 100	Rs 100	Rs 100

26. Draw a rectangular diagram to represent the following information:

	Factory A	*Factory B*
Price per unit of a Commodity	Rs 6	Rs 6
Quantity Produced	1,000 units	800 units
Value of Raw Materials used	Rs 3,000	Rs 2,400
Other Expenses of Production	Rs 2,000	Rs 1,400
Profits	Rs 1,000	Rs 1,000

27. Represent the following data by a rectangular diagram:

	Commodities	
	A *Rs*	*B* *Rs*
Price per Unit of Commodity	10	12
Quantity Sold	20	24
Cost of Raw Materials used	100	120
Other Costs	60	96
Profit	40	72

28. On page 106 is given average expenditure of some families in a year:

Item	*Average expenditure per annum Rs*
Food	945
Clothing	325
Rent	520
Medical care	210
Other items	400
Total	2,400

Draw a bar diagram and pie diagram to the above data.

29. Present the following data of area and production of rice in India graphically:

Year	*Area (Million area)*	*Production (Million tonnes)*
1972—73	74.1	25.5
1973—74	77.3	27.8
1974—75	76.1	24.8
1975—76	77.9	27.2
1976—77	79.3	28.3
1977—78	79.1	24.8

30. Represent by suitable diagram the following data of patients who died in hospitals and dispensaries in India:

Disease	*1976*	*1977*	*1978*
Cholera	1445	4595	5172
Dysentery	2001	2302	2629
Tubercle of Lungs	3455	3234	4043
Disease of Nervous System	7100	1974	1655
Disease of Respiratory System	1540	1796	1741

31. The following figure give the net change in Business Inventories in billions of dollars in various years in America. Represent them by suitable diagrams:

Year	*Change*
1971	1.6
1972	—0.3
1973	—1.4
1974	—2.6
1975	—1.6
1976	—1.1
1977	0.9
1978	1.0

32. The following table gives the value of Imports and Exports of country A for two years 1976-77 and 1977-78 in crores of Rs;

Months	*1976-77*		*1978-79*	
	Imports	*Exports*	*Imports*	*Exports*
April	22	28	26	18
May	24	28	21	20
June	26	23	19	17
July	28	21	18	17
August	31	20	21	27
September	29	22	20	20
October	32	21	23	18
November	30	20	26	20
December	32	20	23	22
January	31	20	28	25
February	25	18	20	25
March	24	20	20	30

Plot the above figures on a graph paper and show also the Balances of Trade.

Chapter 8

GRAPHIC REPRESENTATION

8.1 INTRODUCTION

Graphs are perhaps the most commonly used devices for presenting statistical data. Instead of showing individual bars for each class, a graph uses continuous scales in rectangular coordinates. In presenting time series, *i.e.*, the variations in the value of a variable with time, it is convenient to take the time (the independent variable) along the horizontal axis and the value (the dependent variable) along the vertical axis, thus, the variation of annual food grain production or the daily movement of share-price index is best shown by a line graph with time taken horizontally.

In presenting frequency distributions, the values of the variables are treated as independent and taken along the horizontal axis, and the frequencies attached with the different classes are taken along the vertical axis.

We will first present the line graphs for time series.

8.2. LINE GRAPHS FOR TIME SERIES

In a time series, there are two variables, viz., (*i*) time, and (*ii*) the variable under study *e.g.*, quantity of crops or sales. 'Time' is always the independent variable in a time series. The other variable 'output' or 'sales' as the case may be is called a dependent variable.

1. The independent variable is measured along the *X*-axis, and dependent variable along the *Y*-axis.

2. Equal distances on the *Y*-axis should mean equal absolute amounts. Thus, if one centimetre represents Rs 1,000 sales revenue, two cm would represent Rs 2,000, and three cm would indicate Rs 3,000.

Likewise on the *X*-axis equal distances should mean equal time duration.

3. The scale along the *Y*-axis must begin at zero as origin. This would facilitate a correct understanding of the information presented by the line graphs.

4. The scale should be so chosen as to give a clear picture of the data. It should be remembered if the *X*-axis is stretched the fluctuations would be de emphasised. If on the other hand, the *Y*-axis is stretched the fluctuations would be over-emphasised. By 'stretching' here we mean showing comparatively smaller quantity on a given magnitude (say one cm).

5. For actual plotting of the data, it should be remembered that for every value of the independent variable (*i.e.*, time) there is corresponding value of the dependent variable (imports in Table 8.1). It is these matched values (pair of values) that are to be plotted. Each pair of value is represented on the graph by a point which corresponds to the value of the independent variable on the *X*-axis and the value of the dependent variable on the *Y*-axis. If this procedure is repeated for each pair in the data given in Table 8.1 we would have as many points on the grid as are the pairs of values. In Fig, 8.1 these points bear the numbers of the corresponding pairs in the table.

6. Now, the points (as obtained) are connected by straight lines. The succession of connecting lines gives a line (or curve) graph.

TABLE 8.1. IMPORTS TRADE OF INDIA

(in crores of Rs)

Year	*Value of Imports*	*Pair*
1961	1,139	(1)
1962	1,107	(2)
1963	1,135	(3)
1964	1,222	(4)
1965	1,349	(5)
1966	1,408	(6)
1967	1,078	(7)
1968	2,007	(8)

7. The title, headnote and footnote are to be shown in the same manner as in a diagram.

8. The scale caption for the *X*-axis is placed under the centre of the horizontal axis. The scale caption for the *Y*-axis is placed at the top of the *Y* scale.

9. If more than one graph is plotted on the same graph paper, then a different type of line should be used for each curve, *e.g.*. a full line, dotted line, dashed line, dot cum dash line, etc.

If two or more variables pertaining to the same phenomenon and expressed in similar units are shown on a graph paper, a comparison in their movements as also in their absolute amounts in facilitated.

Illustration 8.1 is an example of this. The method of plotting the points is exactly similar to the one explained with reference to Fig.

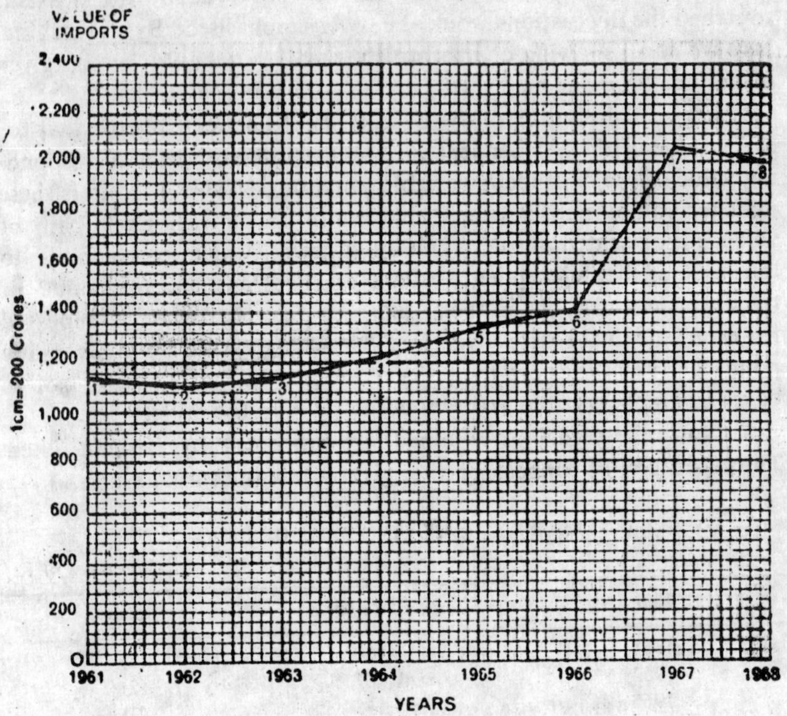

FIG. 8.1

8.1. It should, however, be remembered that the lines depicting these variables are easily distinguishable. This is done by using different types of lines (*see* Fig. 8.2). The line showing "steam' is full; the one showing 'Hydro' is 'dotted and dashed', and the third for 'Diesel' is 'dashed'.

When the fluctuations in a variable are small relative to its size and it is desired to visualize these fluctuation properly, the vertical scale may be amplified (or stretched). This can be done if, instead of showing the entire scale from zero to the highest value involved, only as much is shown as is necessary for the purpose. The portion which lies between zero and the lowest value of the variable is left out. This method is termed as *false base line* approach of showing line graphs. Illustration 8.2 is an example of false base line graph. One has to gaurd very carefully in interpretation of such graphs, because the false impression about the magnitudes of changes may be created by such devices. Fig. 8.3 illustrates this quite well. A cursory look at Fig. 8.3 may suggest a far spectacular rise of import trade than it actually is.

It is at times desirable to present two variables on the same graph even when they are measured in different units. For example presenting the volume and value of trade (as in Illustration 8.3) on one graph can be illustrative. In such cases the following method may be adopted.

1. Represent one of these variables on 'y' axis on the left side of the grid, and the other variable on the 'y' axis on the right side of the grid.

2. Plot the points of one variable with respect to X-axis, and connect these points.

3. Plot the points of the second variable, represented on the right side of the grid. with respect to X-axis, and connect these points.

The impropriety of showing on a single graph two variables measured in different units can be overcome by converting the data in percentages. The process automatically eliminates units and the data is reduced to pure numbers. This can be seen from Illustration 8.4 where the data of Table 8.3 is expressed in the form of percentages:

Illustration 8.1. Represent the following data by line charts:

TABLE 8.2. PROGRESS OF ELECTRICITY SUPPLY

Year	*Installed capacity (MW)*		
	Steam	*Diesel*	*Hydro*
1971	2,436	300	1,917
1972	2,471	329	2,419
1973	2,538	327	2,936
1974	3,008	401	3,167
1975	3,605	403	3,389
1976	4,417	486	4,124
1977	4,887	448	4 757
1978	5,975	421	5,487

Solution. Here the three different variables are plotted as three different lines as shown in Fig. 8.2.

Illustration 8.2. The value of imports as given in Table 8.1 and plotted in Fig. 8.1 have the lowest value of Rs 1.107 crores and vary to about twice the value. As such most of the space in Fig. 8.1 is unused. Plot the same data using a false base line.

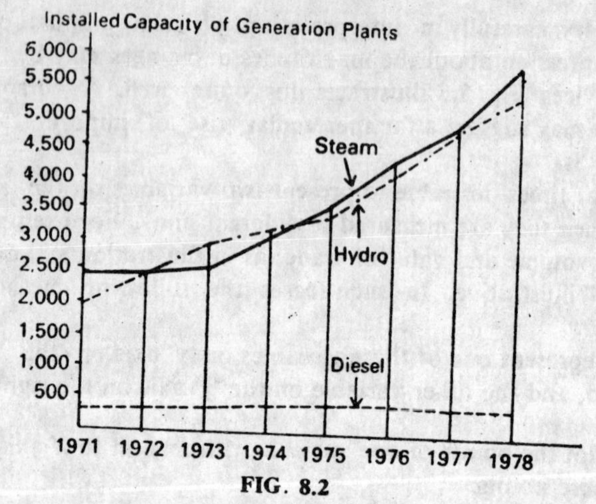

FIG. 8.2

Solution. We start the dependent variable scale at 1,100 crore. The fact that the scale does not start at zero is shown as a break in the *y*-axis and the squiggly lines across the graph. Fig. 8.3 is the resulting line graph.

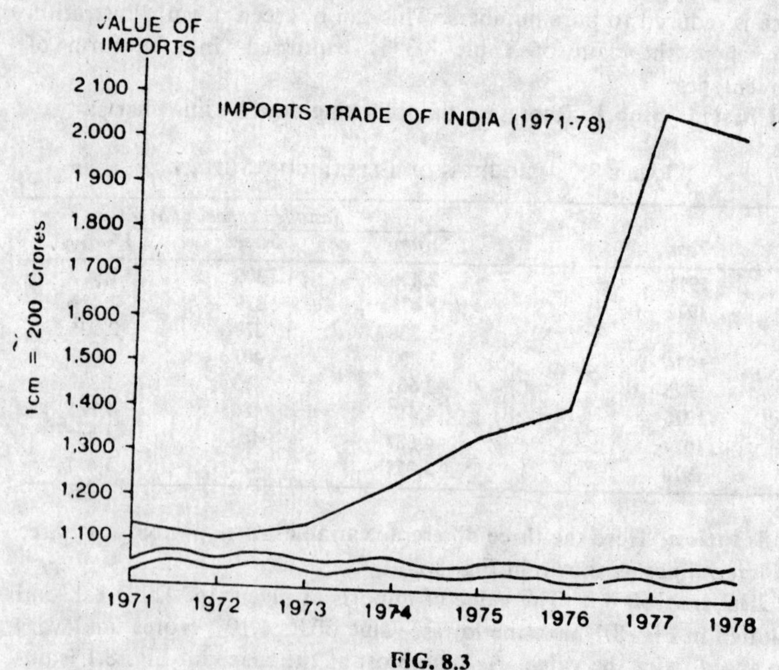

FIG. 8.3

Illustration 8.3. Show graphically the following:

TABLE 8.3. EXPORT OF IRON ORE

Year	*Quantity (million tons)*	*Value (Rs crores)*
1973	9.3	35.4
1974	10.0	35.4
1975	11 3	39.4
1976	13 3	58.9
1977	13.5	75.6
1978	**15.7**	**87.5**

Solution. In this case we take one variable along the left, and the other along the right. The resulting graph is shown as Fig. 8.4.

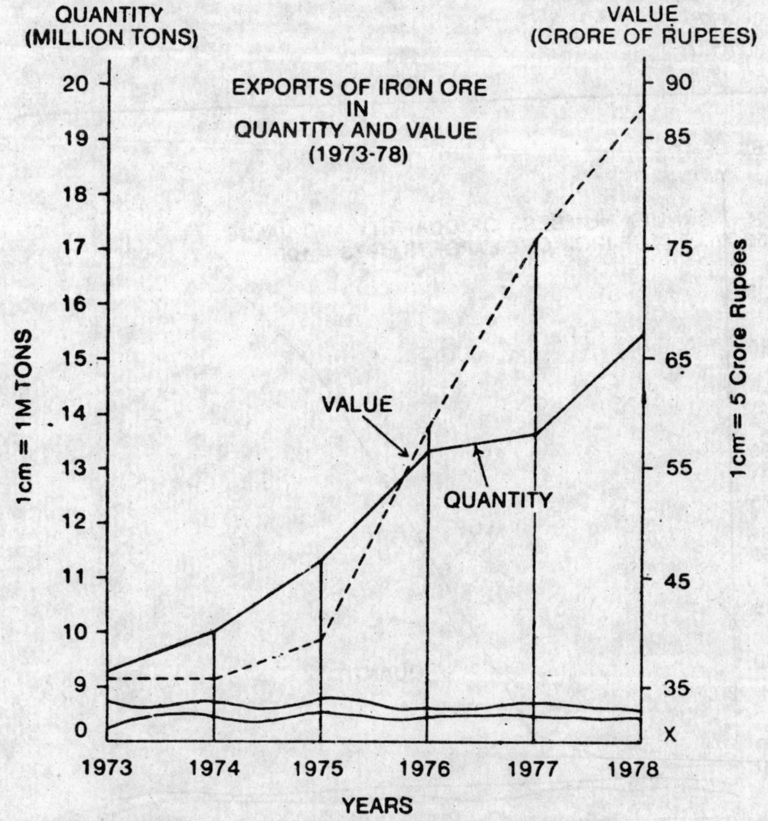

FIG. 8.4

Illustration 8.4. Show the data of Table 8.3 eliminating the different units of measurements.

Solution. This is done by converting the data into percentages. These percentages are termed as *index numbers* and are calculated in Table 8.4 and plotted in Fig. 8.5.

TABLE 8.4. INDEX NUMBERS OF QUALITY AND VALUE OF IRON ORE EXPORTS
(*1971 = 100*)

Year	*Index of quantity*	*Index of value*
1973	100	100
1974	108	100
1975	122	111
1976	144	166
1977	146	215
1978	169	245

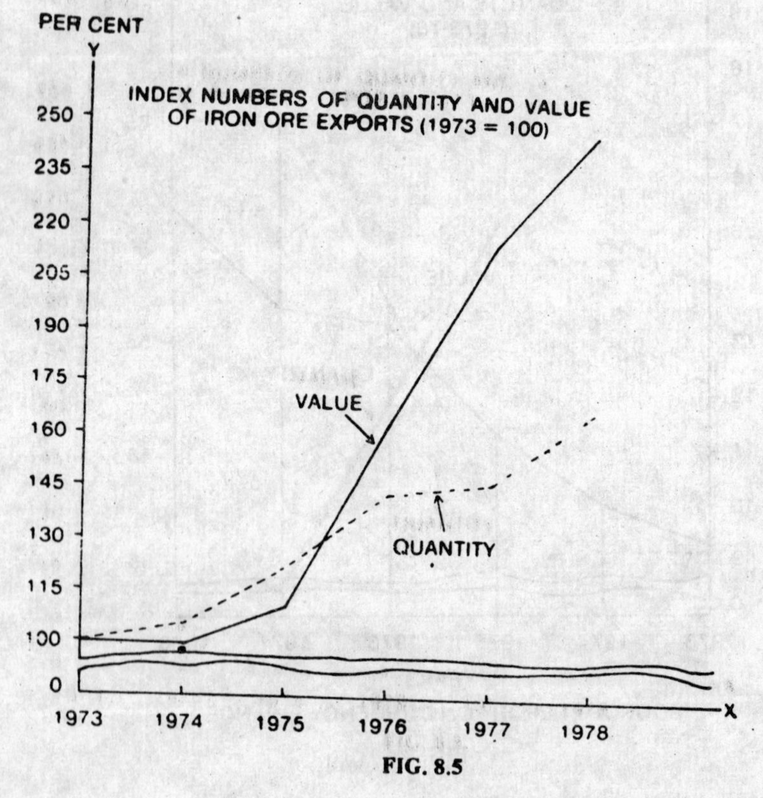

FIG. 8.5

This figure brings out clearly that the value of exports is increasing much faster than the quantity.

Illustration 8.5. Table 8.5 gives the highest and lowest prices of silver during the period 1962-69. Plot these as a graph.

TABLE 8.5. PRICES OF SILVER—BOMBAY (*Rs per kilogram*)

	Highest	*Lowest*
1962-63	242	187
1963-64	269	222
1964-65	281	244
1965-66	406	281
1966-67	385	336
1967-68	565	346
1968-69	633	508

Solution. Fig. 8.6 shows the highest and the lowest prices yearwise. The distance between the two curves represent the range of prices in a given year. Such graphs are known as *range charts.*

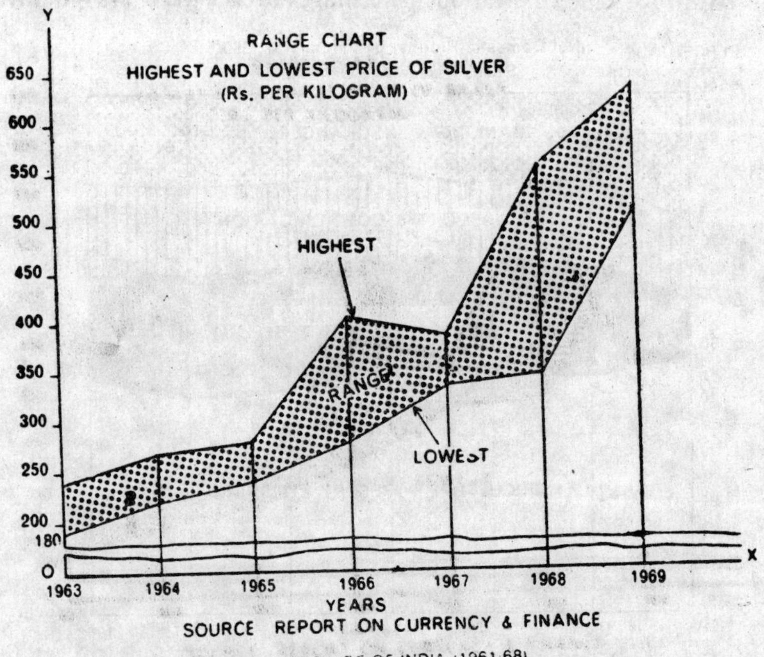

FIG. 8.6

Illustration 8.6. Table 8.6 gives the net domestic product by industry of origin as a percentage distribution. Presents the data by a suitable graph.

TABLE 8.6. NET DOMESTIC PRODUCT BY INDUSTRY OF ORIGIN AT 1960-61 PRICES

(*Percentage distribution*)

Industry	*1961-62*	*1962-63*	*1963-64*	*1964-65*	*1965-66*	*1966-67*	*1967-68*
1. Agriculture Foresting and Fishing	49.6	47.3	46.0	48.8	42.3	41.8	45.3
2. Mining, Quarrying, Large and Small Scale Manufacture	20,8	21.9	22.6	22.3	24.0	23.9	22.1
3. Transport and Communication	14.5	15.0	15.3	15.1	16.3	16.5	15 6
4. Banking and Insurance	15.1	15.8	16.1	15.8	17.4	17.8	17.0
Total	100.0	100.0	100.0	100.0	100.0	100 0	100.0

Solution. Since the various percentages over a year are additive,

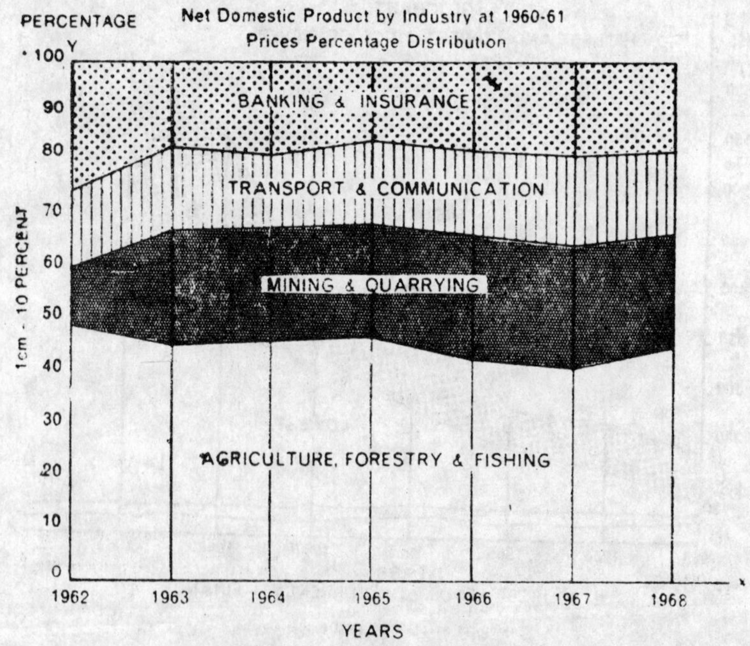

FIG. 8.7

the data is best presented by a *component-part line chart,* which serves the same function as a subdivided bar diagram. The graph is shown in Fig. 8.7.

8.3. CHARTING FREQUENCY SERIES—HISTOGRAMS

For charting the frequencies associated with different classes of a continuous variable, one commonly used technique is the construction of *histogram.* A histogram consists of bars erected upon the class interval columns taken along the horizontal or the x-axis.

A histogram is essentially an area chart in which the area of a bar represents the frequency associated with the corresponding interval. The height of the bar, then, is essentially a measure of *frequency density i.e.*, of the concentration of frequency. The scale of y-axis, therefore, is frequency per unit width of class interval. We illustrate the construction of a histogram by plotting the data of Table 8.7 which gives the weight of 1,515 students of a college.

Note that since the variable is a continuous one, the classes are effectively 90.5-100.5, 100.51-110.5, etc. The class width is 10 pounds in each case, and therefore, the vertical axis can conveniently be taken as frequency per 10 pounds.

TABLE 8.7. WEIGHTS OF COLLEGE STUDENTS IN POUNDS.

Size	*Frequency*
91 to 100	5
101 to 110	34
111 to 120	139
121 to 130	300
131 to 140	367
141 to 150	319
151 to 160	205
161 to 170	76
171 to 180	43
181 to 190	16
191 to 200	3
201 to 210	4
211 to 220	3
221 to 230	1
Total	1,515

To prepare a histogrom from the above data, the following steps should be taken:

1. The fourteen class intervals of 10 pounds each are laid off to a convenient scale on a horizontal axis; the value increasing from left to right. The end point of these intervals should correspond to the *real limits* of the class.

2. The next step would be to erect a scale of frequency density at right angles to the scale of size on *X*-axis. It is not essential to show the zero point on the horizontal axis, but it is necessary to show it on the vertical axis or on the scale of frequency.

Note. With a view of reduce the distance between 'zero' and the 'minimum value' on the horizontal scale the method of 'Kinked Line' may be adopted as shown in Fig. 8.8.

3. The third step would be to erect the bars. The length of the bar would represent the frequency density of that particular class interval.

Since the class intervals are equal in this case, the heights of the bars is also proportional to the frequency of each class. Fig. 8.8 shows the resulting histogram.

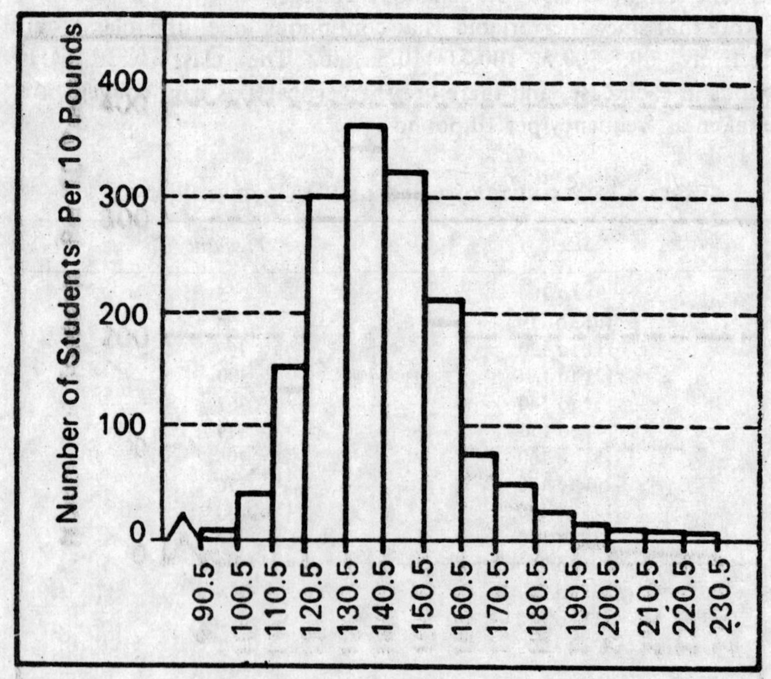

FIG. 8.8

When the class intervals are unequal, we have first to calculate the frequency density on a convenient scale. It is usually advisable to calculate frequency on the basis of smallest class interval. The procedure is explained through Illustration 8.7.

Illustration 8.7. Present the data of Table 8.8 by a suitable histogram.

TABLE 8.8. AVERAGE MONTHLY EARNING OF 1035 EMPLOYEES IN CONSTRUCTION INDUSTRY

Monthly earnings	*Number of workers*	*Frequency density of Workers (per Rs 10 of earning)*
60-70	25	25
70-80	100	100
80-90	150	150
90-100	200	200
100-120	240	120
120-140	160	80
140-150	50	50
150-180	90	30
180 or more	20	—

Solution. This data has two complications: one, the class intervals are unequal, and, two, the last class is open ended. The first is taken

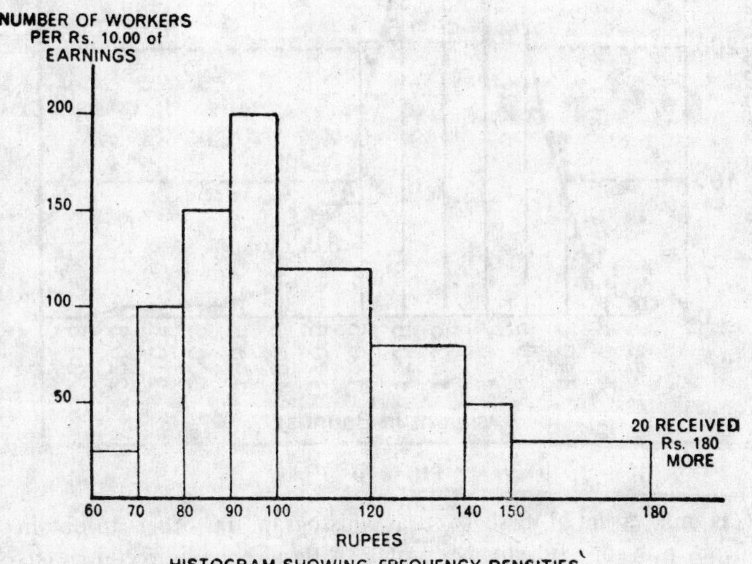

HISTOGRAM SHOWING FREQUENCY DENSITIES

FIG. 8.9

care of by calculating the density on the basis of Rs. 10 of earning, and plotting the histogram as in Fig. 8.9.

It is not possible to make an adjustment for the open-ended class, and the frequency associated with it is given just as a note.

8.4 FREQUENCY POLYGON

If we join the middle points of the tops of the adjacent rectangles of a histogram (Fig. 8.8) with line segments, as indicated in Fig. 8.10, a frequency polygon is obtained. When the polygon is continued to the X-axis just outside the range of lengths, as in the figure, the total area under polygon will be equal to the total area under histogram. Note that in Fig. 8.10 the triangles 1, 2, 3, . . . 14 are congruent respectively to triangles 1′, 2′, 3′, . . . 14′.

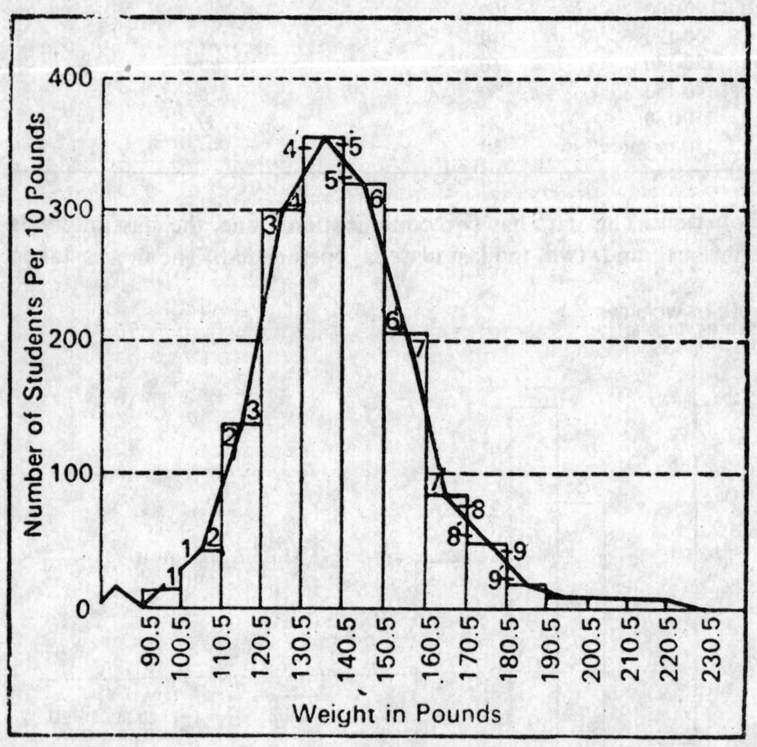

FIG. 8.10

It is not essential first to draw histogram in order to obtain frequency polygon. It can be drawn without erecting rectangles as in Fig. 8.11.

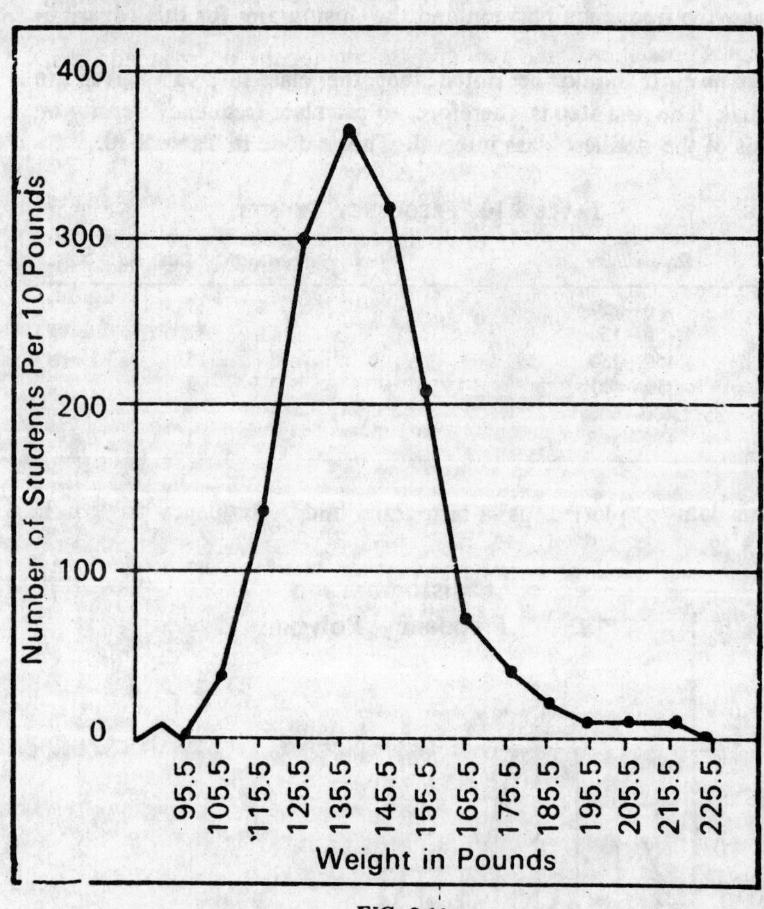

FIG. 8.11

Illustration 8.8. The monthly expenditure on food incurred by a sample of 100 families is given in Table 8.9.

TABLE 8.9. MONTHLY EXPENDITURE ON FOOD

Expenditure	*Frequency*
0 — 50	7
50 —150	24
150—250	30
250—400	27
400—600	8
600—800	4
	100

Draw the frequency polygon and the histogram for this distribution.

Solution. It should be noted that the class-intervals given are unequal. The first step is, therefore, to calculate frequency density on a basis of the smallest class interval. This is done in Table 8.10.

TABLE 8.10. FREQUENCY DENSITY

Expenditure	*Frequency* per Rs. 50
0— 50	7
50—150	12
150—250	15
250—400	9
400—600	2
600—800	1

This data is plotted as a histogram and a frequency polygon in Fig. 8.12.

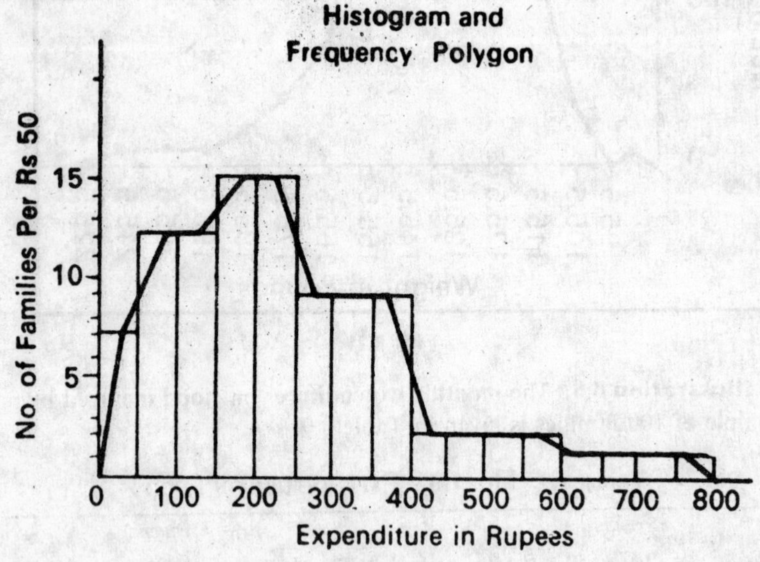

FIG. 8.12.

8.5. SMOOTH FREQUENCY CURVE

The frequency polygon described above consists of a number of points connected by straight lines. If we subdivide class intervals further the number of points becomes larger and the straight lines segments consequently shorter. In the limit as the class interval becomes very small, the straight line segments are so small that the frequency polygon can be rounded into a smooth curve which is termed as the frequency curve. In fact it plots the frequency density against the variable. *It is conventional to calculate frequency density on the basis of the class of unit width.* We will have occasion to deal with smooth curves extensively in the following chapters.

8.6. OGIVE OR CUMULATIVE FREQUENCY CURVE

So far we have been discussing the charting of simple frequency distributions where each frequency refers to the measurement or the class interval against which it is placed. Sometimes it may become necessary to know the number of items whose values are more or less than a certain amount. We may, for example, be interested in knowing the number of students whose weight is less than 140.5 lbs. or more than (say) 150.5 lbs. (Table 8.7)

To get this information it is necessary to change the form of the frequency distribution from a 'simple' to a 'cumulative' distribution. In a cumulative frequency distribution the frequency of each class is made to include the frequencies of all the lower or all the upper classes depending upon the manner in which cumulation is done. The manner of cumulation, in its turn, depends upon our purpose. If we are interested to know the number of items that are 'less than' a certain size, the cumulation will proceed from the least to the greatest size, and the series so obtained will be called 'less than' cumulative frequency distribution. When it is desired to know the number of items whose size is 'more than' a certain size, cumulation will proceed from the greatest to the least, and the series so obtained will be called 'more than' cumulative frequency distribution.

Construction of a 'Less Than' cumulative frequency table. If we look carefully at the lower end of Table 8.7 we will find that 5 students have a weight of 100.5 lbs. or less (considering the real limits of the class interval) and 34 have between 100.5 lbs. and 110.5 lbs. The total number of students whose weight is 110.5 lbs. or less is, therefore, 34+5, or 39. The number of students between 110.5 lbs. and 120.5 lbs. is 139. Adding this to be accumulated frequency of the two classes (139+34+5=178), we find that 178 students have a weight of 120.5 lbs. or less. When the frequencies of each age group are added in this way distribution given in column 2 of Table 8,11 is secured. With the help of this distribution it is possible to find out the number of students whose weight is equal to or less than a certain amount. Thus, there are 1,507 students whose weight is 200.5 lbs. or less.

Construction of a 'More Than' cumulative frequency table. When a 'more than' cumulative frequency table is to be constructed we start at the upper end of the distribution. In Table 8.7 we find that there is one student whose weight is 220.5 lbs. or more and 3

TABLE 8.11. DISTRIBUTION OF WEIGHT OF 1,515 STUDENTS OF A COLLEGE (IN LBS.)

Class Interval	*Cumulative Frequency (less than)*
110.5 pounds or less	5
110.5 ,, ,,	39
120.5 ,, ,,	178
130.5 ,, ,,	478
140.5 ,, ,,	845
150.5 ,, ,,	1164
160.5 ,, ,,	1369
170.5 ,, ,,	1445
180.5 ,, ,,	1488
190.5 ,, ,,	1504
200.5 ,, ,,	1507
210.5 ,, ,,	1511
220.5 ,, ,,	1514
230.5 ,, ,,	1515

whose weight is between 210.5 lbs. and 220.5 lbs. Thus, the total, number of students whose weight is 210.5 lbs. or more is 3+1=4. The frequency of the next lower class is 4. Adding this to the cumulated frequencies of the two higher classes, we get 4+3+1=8, the number of students whose weights are 200.5 lbs. or more.

When the frequencies of each group are cumulated in this way we get Table 8.12. From this table we can find the number of students whose weight is more than 160.5 lbs. as 146.

It must always be remembered that the frequencies of a 'less than' cumulative frequency table refer to the upper limit of the class intervals and those of the 'more than' cumulative frequency table refer to the lower limit.

When cumulative frequency distributions are charted on a graph paper the curve so obtained is called 'ogive' or a cumulative frequency curve.' In charting cumulative frequency distributions, the following steps are necessary:

TABLE 8.12. DISTRIBUTION OF WEIGHT OF 1,515 STUDENTS OF A COLLEGE (IN LBS.)

Class Width	*Cumulative Frequency (more than)*
90.5 pounds or more	1515
100 5 ,, ,,	1510
110.5 ,, ,,	1476
120.5 ,, ,,	1337
130.5 ,, ,,	1037
140 5 ,, ,,	670
150.5 ,, ,,	351
160.5 ,, ,,	146
170.5 ,, ,,	70
180.5 ,, ,,	27
190.5 ,, ,,	11
200 5 ,, ,,	8
210.5 ,, ,,	4
220.5 ,, ,,	1
230.5 ,, ,,	0

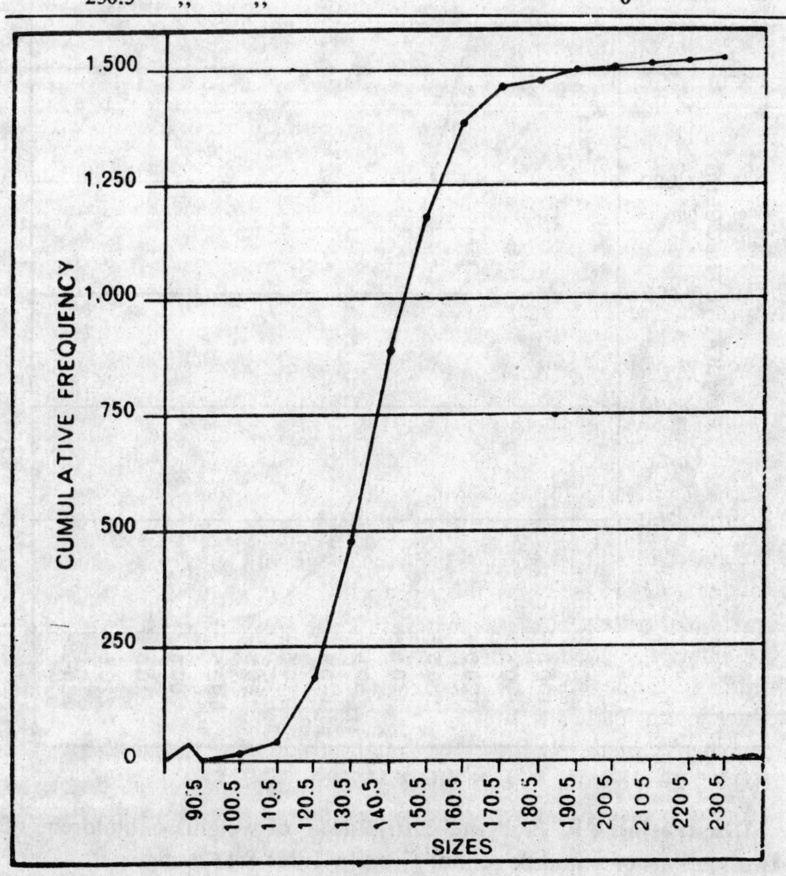

FIG. 8.13

1. Scale the cumulative frequencies along the *Y*-axis, and class intervals along the *X*-axis.

2. The scale along *Y*-axis should be such that it may accommodate the total frequency.

3. The accumulated frequencies for each class are plotted:

(*a*) against the upper limit of the class in the case of 'less than' cumulative frequency distributions (see Fig. 8.13), and

(*b*) against the lower limit of the class in the case of 'more than' cumulative frequency distributions (see Fig. 8.14).

The data of Tables 8.11 and 8.12 are used in Figs. 8.13 and 8.14.

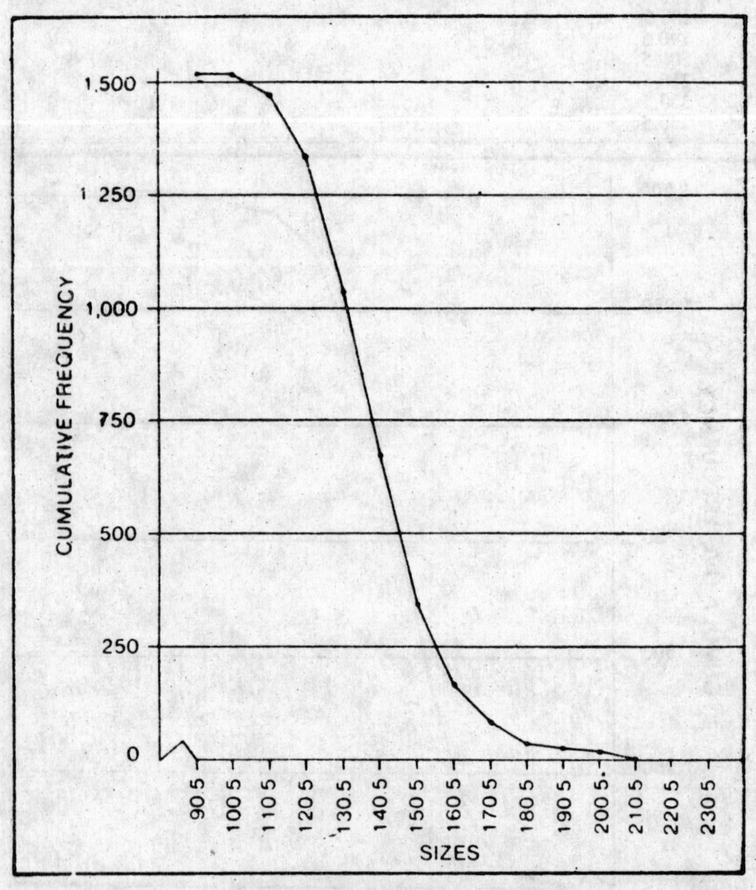

FIG. 8.14.

Illustration 8.9. Plot the distribution of weights of children of two sections of a middle school given in Table 8.13 as ogive.

TABLE 8.13. FREQUENCY DISTRIBUTION SHOWING WEIGHTS OF THE CHILDREN OF TWO SECTIONS OF A MIDDLE SCHOOL CLASS

Weight (in Kg.)	Section A (50 students)			Section B (78 students)		
	Fre.	Cum. Fre.	Cum. %	Fre.	Cum. Fre.	Cum. %
30 – 33	2	2	4	10	10	12.8
34—37	2	4	8	10	20	25.6
38—41	14	18	36	16	36	46.6
42—45	6	24	48	10	46	59.0
46—49	2	26	52	10	56	71.8
50—53	8	34	68	8	64	82 1
54—57	10	44	88	10	74	94.9
58—61	4	48	96	2	76	97.4
62—65	—	48	96	—	76	97.4
66—69	2	50	100	2	78	100.0
	50			78		

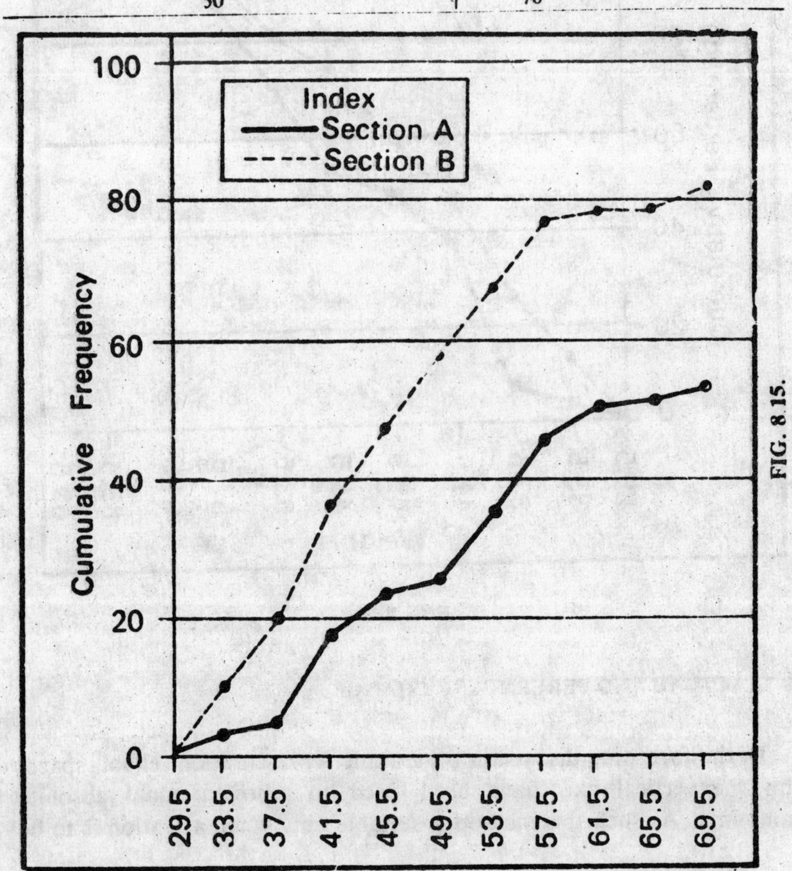

FIG. 8.15.

Solution. The calculation of "less then" cumulative frequencies are given in Table 8.13. There are plotted as two ogives in Fig. 8.15.

A better comparison of the two section is however, obtained when the cumulative frequencies are converted into cumulative percentages (as is done in Table 8.13). The percentage ogives as plotted in Fig. 8.16 give a better comparison of the distribution of weights in the two sections.

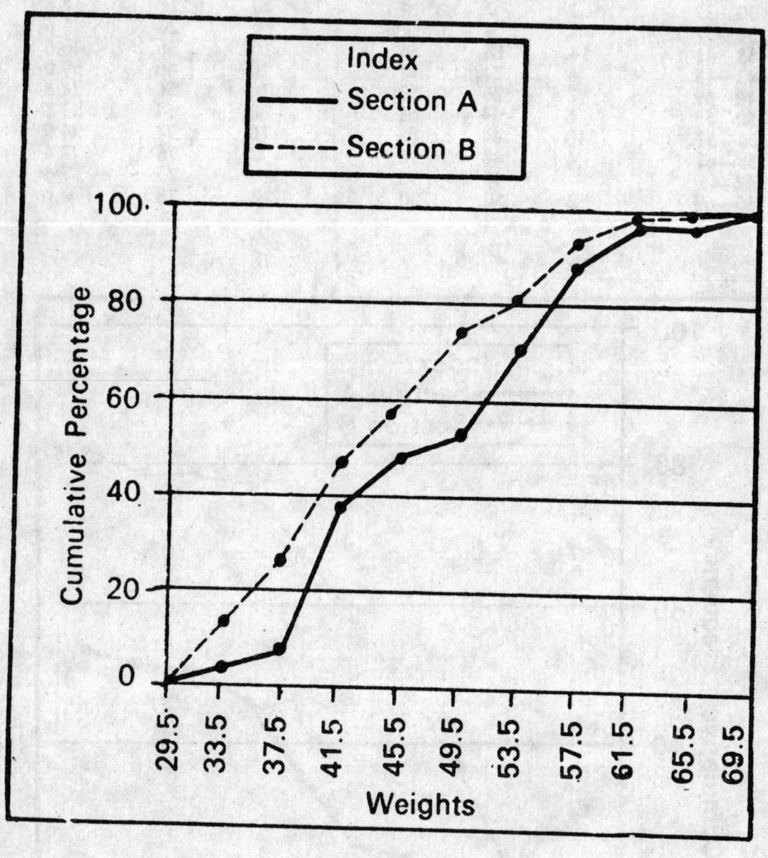

FIG 8.16

8.7. ACTUAL AND PERCENTAGE CHANGE

In the foregoing discussion on graphic representation equal spaces on the vertical axis have been used to represent equal absolute amounts. As such this method is suitable only when attention is to be focussed on the amount of change in the variable or variables under study. But when it is desired to emphasise the rate of change in a variable, or when a comparison is to be made between the rates of change of two or more variables natural (arithmetic) scale will not serve the purpose.

TABLE 8.14. SALE OF A LTD., FROM JANUARY TO DECEMBER 1971

Year & months	*Amount Rs*	*Actual increase or decrease over the preceeding month*	*Actual percentage change over the preeding month*	*Logarithm of amouut*
1	*2*	*3*	*4*	*5*
1971				
January	287	–	–	2.46
February	195	– 92	— 32	2.29
March	638	+443	+226	2.80
April	826	+188	+ 30	2.90
May	308	– 518	— 63	2.50
June	515	+207	+ 67	2,70
July	235	—280	— 54	2.38
August	80	—155	— 66	1.90
September	231	+151	+188	2.36
October	277	+ 46	— 20	2.44
November	222	— 55	— 20	2,35
December	459	+237	+106	2.66

Observe Table 8.14 giving monthly sales of a company for a year. Column 3 of this table shows the actual increases or decreases in sales each month as compared to the previous month. Arithmetic line curve emphasises these changes as is evident from Fig. 8.17. The ascents and descents of the moving line from one point of time to the other indicates actual change that has taken place. Thus if we move from April to May the decline of 518 stands out clearly.

This arithmetic line graph, however, will not reflect the percentage increase or decrease in the variable during any month as compared to the previous month. In other words, it is not easy to determine from this chart whether the rate of increase in sales, in any one month was greater or less than that in another month. Thus an examination of this chart shows that the increase in sales in April over March was more than the increase in September sales over those of August. But it fails to emphasise that the comparative rate of change was precisely opposite to the amount of change. The rate of increase in April was only 30 per cent as compared to 188 per cent in the month of September.

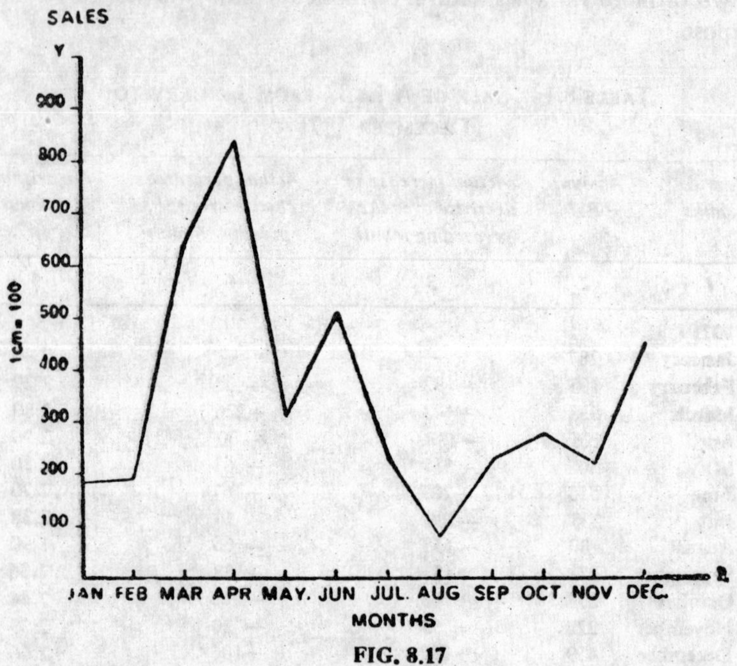

FIG. 8.17

8.8. SEMI-LOGARITHMIC LINE GRAPHS OR RATIO SCALE CHARTS

If it is desired to emphasise the rate of change the scale along the vertical axis is to be such in which equal intervals present *equal percents of change* in the variable (the horizontal axis, showing time, would continue to be on natural scale). This can be done in either of the two way:

(*i*) Convert the values of the variable into their logarithms, and the scale along the vertical axis should be so arranged that equal distances represent equal amount of change in the logarithm (see Fig. 8.18). The logarithms of the given values are then plotted on this grid in the usual manner.

Column 5 of Table 8.14 gives the logarithms of the scales in different months. The logarithms are plotted in Fig. 8.18.

If we look at this curve it would be easily noticed that the percentage increase of sales in September over August is much more than that of April over March. As such the curve of Fig. 8.18 gives a true picture of the percentage changes.

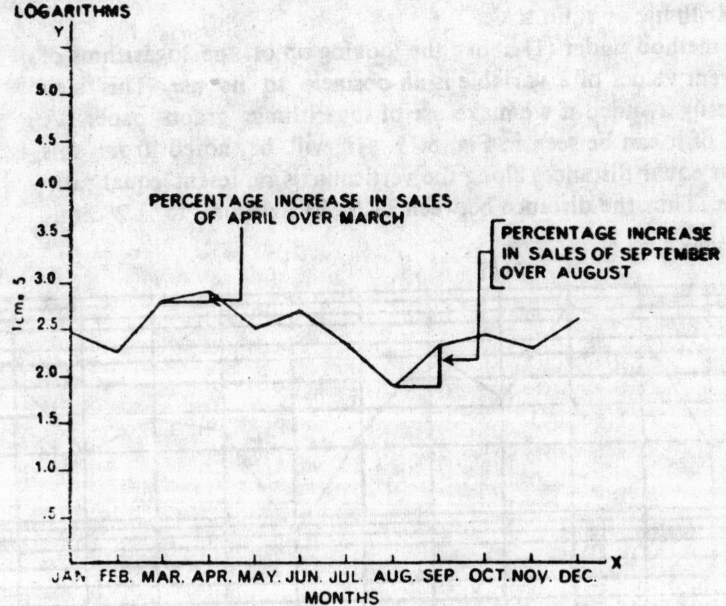

FIG. 8.18

The reason for this is obvious and can be easily understood from a principle of logarithm. According to this principle, the difference between the logarithms of a pair of numbers is greater than the difference between the logarithms of another pair of numbers if the ratio between the numbers of the first pair is greater than the corresponding ratio between the numbers of the second pair.

Let us take two pairs of months to illustrate this principle:

		Actual amount	*Actual increase*	*Percentage charge*	*Log. of amount*	*Diff. in logs.*
I Pair	March	638	188	+30	2.80	
	April	824			2.90	.10
II Pair	August	80	151	+188	1.90	
	September	231			2.36	.46

From the above data it is clear that the difference in the logarithms of the second pair is greater than that between the first pair, and reflects greater percentage increase in sales during September over August as compared to increase in April over March.

(*ii*) Plotting the given data on paper on which vertical axis is ruled on a logarithmic or ratio scale.

In the method under (1) above the looking up of the logarithms of the different values of a variable is an obstacle to its use. This step can be easily avoided if we make use of logarithmic graph paper. A specimen of it can be seen in Fig. 8.19. It will be noted from this figure that equal distances along the vertical axis represent equal ratio of change. Thus the distance between 10 to 20 is equal to 1.2 centi-

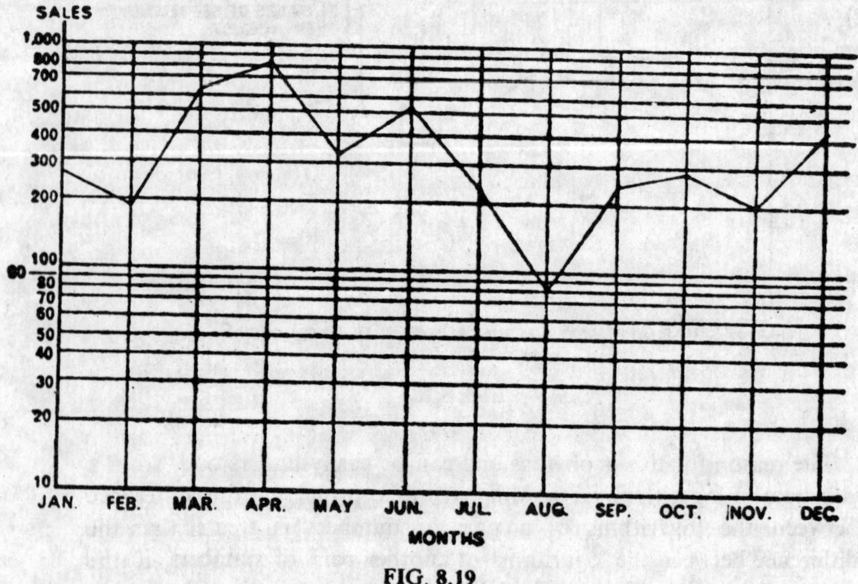

FIG. 8.19

metres. Likewise between 20 and 40 and between 30 and 60, it is 1.2 centimetres. These measurements reveal that any two numbers of ratio 1 : 2 are separated by 1.2 centimetres on the vertical axis. Similarly any two numbers of ratio 1 : 4 are separated by 2.4 centimetres. This explains the reason why the semi-logarithmic chart is called ratio chart.

The vertical axis shown in Fig. 8.19 is divided into two parts—one from 10 to 100 and the other from 100 to 1000. These parts are generally called cycles. Thus the paper on which Fig. 8.19 is shown is referred to as 'two cycle, semi-logarithmic paper'.

It should be remembered that the logarithmic scale must begin with a positive value. If *it* were to begin with zero all values]on scale would also be zeros.

When the data of Table 8.14 is plotted on this semi-logarithmic graph the relatives changes get due emphasis.

Distinction between a Ratio Scale and a Natural Seale

From the foregoing discussion a ratio scale can be distinguished from a natural scale in the following respects:

(1) In the natural scale equal distances on the vertical axis represent equal absolute amounts, whereas in a ratio scale equal distances on the vertical axis represent equal ratio of change.

(2) A line chart drawn on natural scale shows absolute changes from one point of time (or period) to another point of time whereas the ratio chart shows the rate of change between any two points of time.

(3) The scale on the vertical axis in a ratio scale must begin with a positive number, whereas the scale in the natural scale chart should begin with a zero.

(4) On a ratio scale chart variables expressed in different units can be charted for purposes of comparisons whereas on a natural scale such comparison may more often be misleading.

(5) When the variable under study has a wide range of values the ratio scale would be more helpful in its graphic representation than the natural scale.

Shape of the curve on ratio and natural scales. When a variable increases by a constant amount the line on the arithmetic scale would be a straight line sloping upward as curve *A* in Fig. 8.20. When plotted on ratio scale, however, the line would be concave to the base and if it were increased far enough it would run almost parallel to the *X*-axis as curve *B*. This happens because a series which increases by equal magnitudes is increasing at a diminishing rate.

When the time series increases at a constant proportional rate, the curve on the ordinary scale would be a line ascending rapidly, *i.e.*, convex to the base as curve *C* in Fig. 8.20. If the same data are plotted on the semi-logarithmic scale, the curve will be a straight line like curve *D* in Fig. 8.20. In case where series diminish by a constant amount their shape would be like curves *E* and *F*. In curve *F* (drawn on a ratio scale) the line is rapidly declining. This is so because as the base value becomes smaller, the constant decrease is a progressively greater percentage.

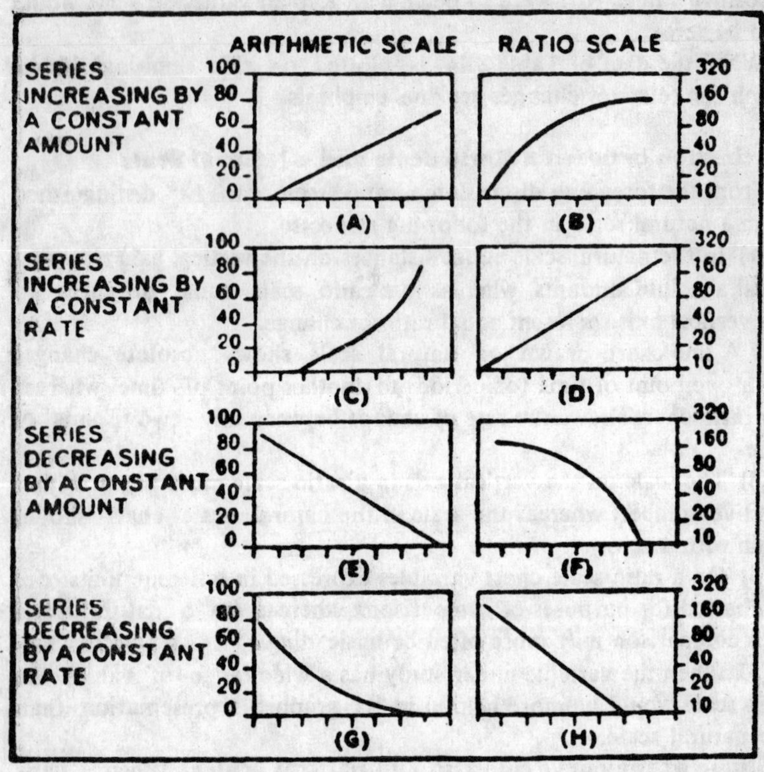

FIG. 8.20

Uses of Semi-logarithmic Scale

The semi-logarithmic scale is employed usefully under the following circumstances:

(1) When comparison between series of widely different magnitudes is desired. About this James A. Field wrote 'it is far superior to the natural scale for effecting comparison when very small and very large quantities must be taken into account concurrently. Whenever a historical curve records extreme growth the same advantage is found. It is not necessary to dwarf the small beginnings in order to keep the later development within manageable dimensions.'

(2) When comparison between series of different units is desired.

(3) When the data are to be examined to see whether they are characterised by a constant rate of change.

(4) When proportionate variations are more important than abso-

lute variations.

The merits of the 'semi-logarithmic' chart have been summarised by Prof. Irving Fisher as: 'The eye reads a ratio chart more rapidly than a difference chart or a table of figures.'

Rules for Interpretation of Semi-log Curves

(1) If we see a curve ascending, and nearly straight, we know that the statistical magnitude it represents is increasing at a nearly constant rate.

(2) If the curve is descending, and nearly straight, the statistical magnitude is decreasing at a nearly constant rate.

(3) If the curve bends upward, the rate of growth is increasing.

(4) If the curve bends downward, the rate of change is decreasing.

(5) If the direction of the curve in one portion is the same as in some other portion, it indicates the same percentage rate of change in both.

(6) If the curve is steeper in one portion than in another, it indicates a more rapid rate of change in the former than in the latter.

(7) If two curves on the same ratio chart run parallel they represent equal percentage rates of changes.

(8) If one is steeper than another the first is changing at a faster rate than the second.

PROBLEMS

1. Discuss the procedure that you would adopt in constructing the graphs.
2. What is false base line? Under what conditions would its use be desirable?
3. What are the uses of graphic presentation of the statistical data? Discuss its limitations.
4. What is histogram? Explain the method of representing a frequency distribution by a histogram when the class-intervals are (1) of equal size are (2) of unequal size.
5. Describe the method of constructing 'cumulative frequency curve'. How would you derermine the median and quartiles from such a curve.
6. Write short notes on:
   (*a*) Frequency polygon,
   (*b*) Ogive curve, and
   (*c*) Histogram.
7. Represent the following data graphically:

INDEX NUMBERS OF WHOLESALE PRICES

	Cereals	*Pulses*	*Fibres*	*Oilseeds*
1973 (Average)	443	424	432	499
1974	465	438	446	593
1975	471	449	476	665
1976	483	506	622	679
1977	450	483	454	484
1978	451	494	420	473

8. The following table gives the proportions of married women in 1977 and 1978 from women of different ages. Show graphically that the increase was most marked for the women of younger years.

Percentage of Married Women (imaginary data)

Age	*1977*	*1978*
18	17.0	19.2
20	36.2	38.4
22	50.7	52.9
24	62.0	64.2
26	55.7	67.8

9. Plot the following figures of population from one period to another:

Year	*Population (000,000's omitted)*
1872	210
1881	250
1891	290
1901	295
1911	315
1921	320
1931	350
1941	399

10. Represent the following data by a suitable diagram showing the difference between proceeds and cost;

Proceed and Costs of a Farm

Year	*Total Proceeds*	*Total Costs*
1960	22.0	19.5
1961	27.3	21.7
1962	28.2	30.0
1963	30.3	25.6
1964	32.7	26.1
1965	33.3	34 2

11. Draw histogram of the following distribution:

Life of Electric lamps (in hours) Mid values	1010	1030	1050	1070	1090
Firm A	13	130	482	360	18
Firm B	287	105	26	230	352

12. The following table gives the cumulative frequency distribution of the ages of a group of 199 teachers. Draw the 'Less than' curve and hence or otherwise find the following:

(*i*) The number of teachers of the age group 38-58. (*ii*) The number of teachers that will have to retire if the retiring age is fixed at 58 years.

Age in Year	*Cumulative Frequency*
20–25	21
25–30	40
30–35	90
35–40	130
40–45	146
45–50	166
50–55	176
55–60	186
60–65	195
65–70	199

13. The following table gives the distribution of monthly income of 600 middle class families in a certain city:

Monthly income (in Rs)	*Frequeney*	*Monthly income (in Rs)*	*Frequency*
Below 75	69	300 – 375	58
76 – 150	167	375—450	25
150 – 225	207	450 and over	9
225 – 300	65		

Draw an ogive for the above data,

14. The following table gives the distribution of the wages of 65 employees in a factory:

Wages in Rs (equal to or more than)	50	60	70	80	90	100	110	120
Number of employees	65	57	47	31	17	7	2	0

Draw a 'less than' curve from the above data. Hence or otherwise estimate the number of employees earning at least Rs 63 but less than Rs 75.

15. The following table gives publicity expenses and sales (in different) units over a twelve month period of a business house:

	Jan.	*Feb.*	*Mar.*	*Apr.*	*May*	*June*
Publicity expenses Rs	53	68	80	68	80	90
Sales (Number of units)	58	65	72	83	70	80
	July	*Aug.*	*Sep.*	*Oct.*	*Nov.*	*Dec.*
Publicy expenses Rs	80	92	102	90	79	92
Sales (Number of units)	100	90	100	110	97	92

Draw graphs for comparison

16. Represent **the following data by means of a histogram:**

Weekly Wages in Rs	*No. of Workers*
10—15	7
15—20	19
20—25	27
25—30	15
30—40	12
40—60	12
60—80	8

17. **Draw histogram, frequency polygon and ogive for the following data:**

Marks	*Frequency*	*Marks*	*Frequency*
0—10	4	40—50	20
10—20	10	50--60	18
20—30	16	60—70	8
30—40	22	70—80	2

18. **The following table relates to rupee loans and small savings in India during 1951-60:**

Year	*Loans (Rs Lakhs)*	*Savings (Rs Lakhs)*	*Year*	*Loans (Rs Lakhs)*	*Savings (Rs Lakhs)*
1969	200	100	1974	220	95
1970	222	95	1975	226	100
1971	240	105	1976	240	80
1972	220	110	1977	236	75
1973	212	90	1978	244	70

Represent the above data by means of an index histogram.

Chapter 9

MEASURES OF CENTRAL TENDENCY

9.1. DESCRIPTIVE STATISTICS

It has been pointed out earlier that for a proper understanding of the quantitative data they should be classified and converted into a frequency distribution. This process of condensation reduces their bulk and gives prominence to the underlying structure of the data. But classification is only the first step in statistical analysis. If the characteristics of given data are to be properly revealed or if one distribution is to be compared with another, it is necessary that the frequency distribution itself must be summarised and condensed in such a manner that its essence is expressed in as few figures as possible.

A single number describing some feature of a frequency distribution is called a *descriptive statistic*. The main thrust of a statistician presenting a mass of data is to evolve a few such descriptive statistics which describe the essential nature of the frequency distribution.

For a proper appreciation of the various descriptive statistics involved, it is necessary to note that most of the statistical distributions have some common features. Though the size of the variables vary from item to item, most of the items are distributed in such a manner that if we move from the lowest value to the highest value of the variable the numbers of items at each successive stage increase with a certain amount of regularity till we reach a maximum; and then as we proceed further, they decrease with the similar regularity. If we plot the percentage frequency density, i.e., the percentage of cases in an interval of *unit variable width* we get frequency curves of the type shown in Fig. 9.1. (Note that the area under each curve should be equal to 100, the total percentage points.)

There are various 'gross' ways in which frequency curves can differ from one another. Even when the 'general' shapes of the curves are the same (the area under them already made equal by the strategy of plotting the per cent density), the details of the shape may change. Thus the curve B has a smaller spread than A, the curve C is more peaky and curve E is less symmetrical. Even when the curves

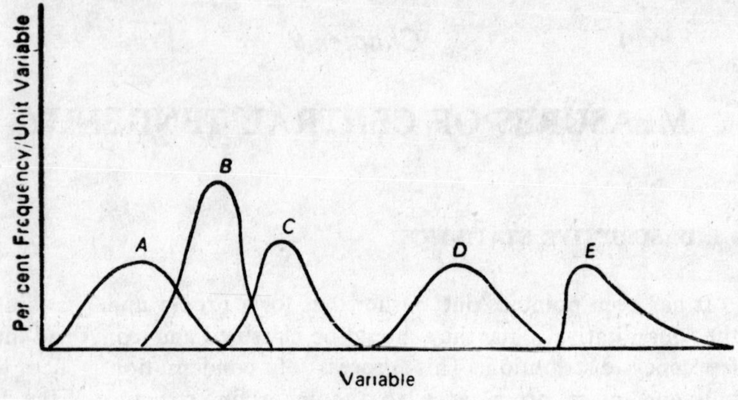

FIG. 9.1.

have almost the same shape (i.e., same spread, peakiness, symmetry etc.) as in curves A and D, the two may differ in location along the variable axis. Thus the items of distribution D are generally larger than those of A. So also are those of B compared to A. Thus, a kind of an 'average' location of the distribution along the variable axis is an important descriptive statistics. These statistics are collectively known as measures of location or of central tendency.

9.2. MEASURES OF CENTRAL TENDENCY

As mentioned above, these statistics indicate the location of the frequency curve along the x-axis and ignore all other features of the distribution. There are various possible measures that can be used to 'locate' a frequency distribution, such as in Fig. 9.2.

A, the minimum value

B, the value of maximum concentration

C, the value which divides the distribution into half, such that one half of the items have value less than this and the other half more.

D, the average value of all items

E, the 95th percentile, *i.e.*, the value below which 95% items lie, and

F, the maximum value.

If the shape of the frequency distributions were fixed, then all these measures are equally descriptive, and fix the location of the curve. But the practical distributions that we deal with always have some change in shape depending on the samples we take, even

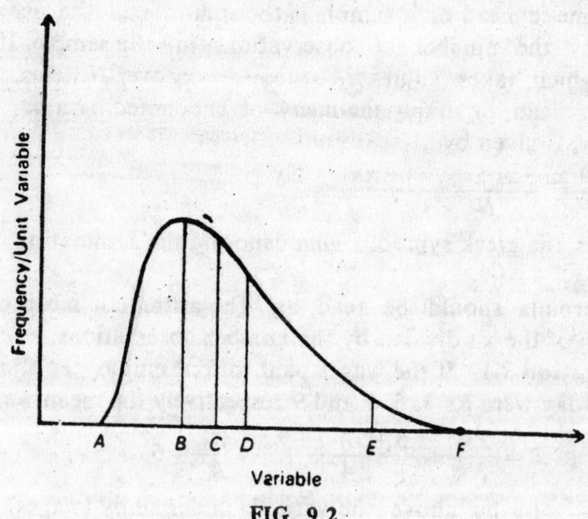

FIG. 9.2.

though the general shapes are quite similar. It is, therefore, necessary that we choose those measures of location which are not very sensitive to the specific values of items, in particular the extreme values. Thus measures *A* and *F* are generally meaningless because they depend on the values of the lowest and the highest items respectively. The other measures, on the contrary, are less susceptible to extreme values because they are somehow related to the entire distributions. Thus, we treat *B*, *C*, *D* and *E* as the most common measures of location. There are some more of such measures which we will consider shortly.

The most important object of calculating and measuring central tendency is to determine a 'single figure' which may be used to represent a whole series involving magnitudes of the same variable. In that sense it is an even more compact description of the statistical data than the frequency distribution.

Since an 'average' represents the entire data, it facilitates comparison within one group or between groups of data. Thus the performance of the members of a group can be compared by relating it to the average performance of the group. Likewise the achievements of groups can be compared by a comparison of their respective averages.

9.3. THE AIRTHMETIC MEAN

The arithmetic mean of a sample is the sum of all the observations divided by the number of observations in the sample. If x is the variable which takes values $x_1, x_2, x_3, \ldots x_n$ over N items, then the arithmetic mean, or simply the *mean* of x, denoted by a bar over the variable x, is given by

$$\bar{x}=\frac{x_1+x_2+x_3+\ldots+x_N}{N}=\frac{\Sigma x}{N} \tag{9.1}$$

where Σ is the greek symbol *sigma* denoting the summation over all values of x.

The formula should be read as: 'The arithmetic mean of the x's is the sum of the x's divided by the number observations.

Illustration 9.1. If the wages paid to four employees of a firm on a certain day were Rs 3, 5, 7 and 9 respectively the mean wage, i.e.,

$$\bar{x}=\frac{\Sigma x}{N}=\frac{3+5+7+9}{4}=\frac{24}{4}=6.$$

In the above the variable under study (wages) is represented by x's: $\Sigma\bar{x}$ denotes the sum of different values of x (*i.e.* 3, 5, 7. and 9 are x_1, x_2, x_3 and x_4 respectively); N represents the number of values of x's,*i.e.*, 4 in the above case: and $\bar{x}$ (read x bar) represents the arithmetic mean (6 in our illustration).

Illustration 9.2. The following table gives the daily income of ten operators in a machine tool factory. Find the Arithmetic Mean.

Name of operator	*A*	*B*	*C*	*D*	*E*	*F*	*G*	*H*	*I*	*J*
Income	12	15	18	20	25	30	22	35	37	26

Solution. If income is represented by x, then,

$$\bar{x}=\frac{x_1+x_2+x_3+x_4+x_5+x_6+x_7+x_8+x_9+x_{10}}{N}$$

or

$$\bar{x}=\frac{12+15+18+20+25+30+22+35+37+26}{10}$$

$$=\frac{240}{10}=24.$$

If we have a discrete frequency distribution with frequencies f_1, $f_2, \ldots, f_n$ associated with the values $x_1, x_2, \ldots x_n$ of the variable, it can be seen that the sum of all the items equals $f_1\, x_1+f_2\, x_2 ..+f_n x_n$ since there are now f_1 items with value x_1, f_2 with value x_2, and so on. The total number of items is obviously $N=\Sigma f$, and the arithmetic mean is, therefore, given by

$$\bar{x}=\frac{\Sigma fx}{\Sigma x} \tag{9.2}$$

where Σfx now stands for the summation of all products of f's and their *respective* values of x.

Illustration 9.3. Calculate the mean of the marks of 46 students given in Table 9.1.

TABLE 9.1. FREQUENCY OF MARKS OF 46 STUDCNTS

Marks (*X*)	*Frequency* (f)
9	1
10	2
11	3
12	6
13	10
14	11
15	7
16	3
17	2
18	1
Total	46

Solution. This is a discrete frequency distribution, and to calculate its mean by formula (9.2) we need to obtain Σfx, which is done in Table 9.2.

TABLE 9.2

x	f	fx
9	1	9
10	2	20
11	3	33
12	6	72
13	10	130
14	11	154
15	7	105
16	3	48
17	2	34
18	1	18
Σ	46	623

By Eq. (9.2)

$$\bar{x} = \frac{\Sigma fx}{\Sigma f} = \frac{623}{46} = 13.54.$$

9.4. ARITHMETIC MEAN OF GROUPED DATA

In all of the cases considered in Sec. 9.3, the exact value of each item is known. If however the data is grouped such that we are given frequency of finite-sized class intervals we do not know the value of every item. The calculation of arithmetic mean in such a case is then necessarily, a process of estimation, based on some assumption. The standard assumption for this purpose is that all the items within a particular class are concentrated at the mid-value of the class and thus jx corresponding to the f items of a class equals fm where m is the mid-point of the class interval, and the arithmetic mean is then given by

$$\bar{x} = \frac{\Sigma fm}{\Sigma f} \tag{9.3}$$

The determination of the mid-point of a class-interval requires some consideration. The position of the mid-point is determined by *real* as distinguished from *apparent* class limits.

If a frequency table records the distribution of a discrete variable the real and the apparent class limits are the same (unless the class-interval is exclusive). This is due to the fact that discrete data are always expressed in whole numbers and are always characterised by gaps at which no measure may ever be found. Thus, if the class intervals of discrete variable are:

6—10
11—15
16—20

the apparent limits 6 and 10, 11 and 15, 16 and 20 are real limits also. The class 6—10 includes only those items whose sizes are 6, 7, 8. 9 or 10. Any item whose size is more than 10, *i.e.*, 11, 12, etc., or less then 6, *i.e.*, 5 is not included in this class, but in the next higher or the next lower class. There are, of course, no values between 10 and 11 or 15 and 16. In such a case, therefore the mid-point is the middle of the five values included in a class, viz., 8 in 5—10 class, 13 in 11—15 class, and 18 in 16—20 class.

If, however. the class-interval is exclusive, the apparent limits are not real and before finding the mid-point real limits should be determined. If the class interval in given in the following manner, it is said to be exclusive class interval:

(*i*) 5—10
(*ii*) 10—15
(*iii*) 15—20

This means that an item having a value 15 is to be included either in class (*ii*) or class (*iii*). If it is included in class (*ii*) it means value 10 is included in class (*i*). Hence the real limits of class (*ii*) are 11—15 and their mid-point is 13. If 15 is not included in class (*ii*) but in class (*iii*) the real limits of class (*ii*) are 10—14 and the mid-point is 12. It, therefore, follows that, whenever we have an exclusive class interval we must decide as to which limit of the class is excluded and it is only then that the mid-point should be ascertained.

If the frequency table records the distribution of a continuous variable, the real limits are not the same as the apparent limits. This is because theoretically such variables can be measured to an infinitesimal fraction of a unit, the measures that are obtained are only approximations to absolute accuracy. While measuring the weight of boys, for example, we seldom go to a unit smaller than the pound. Thus, when we say that the weight of an individual is 140 pounds, what we really mean is that his weight is nearer 140 pounds than 139 or 141 pounds. Thus means that it is somewhere between 139.5 and 140.5 pounds.

From this it follows that if in any frequency distribution of weights we find a class interval identified by the interval limit (say, 140—144) we must conclude (*i*) that weights have been measured correct to the nearest pound, and (*ii*) hence the real limits of the interval extend by .5 pounds on either side and the class-interval strictly speaking, is 139.5—144.5. The mid-point of this class is to be determined from these limits. The method of finding the mid-value in this case is as follows:

$$\text{Lower limit of the class} + \frac{\text{Upper limit—Lower limit}}{2}$$

$$= 139.5 + \frac{5}{2} = 139.5 + 2.5 = 142.0$$

or $$\frac{\text{Upper limit} + \text{Lower limit}}{2} = \frac{139.5 + 144.5}{2} = \frac{284}{2} = 142.0.$$

If the weight has been measured correct to the nearest tenth of a pound we will have class intervals like the following:

140 – 144.9
145 – 149.9

On the basis of what has been said earier the real limits are:

139.95—144.95
144.95—149·95

Here the mid-point will be $\frac{139.95 + 144.95}{2} = 142.45$, *i.e.*, **142.5.**

Illustration 9.4. The first and third columns of Table 9.3 give the frequency distribution of the average monthly earnings of male workers. Calculate the mean earnings.

TABLE 9.3. DISTRIBUTION OF MALE WORKERS BY AVERAGE MONTHLY EXPENSES

(Computation of Arithmetic Mean . . . Long Method)

Monthly earnings Rs	*Mid-point* Rs *m*	*No. of workers* *f*	Rs *fm*
27.5 - 32.5	30	120	3,600
32.5—37.5	35	152	5,320
37.5—42.5	40	170	6,800
42.5 47.5	45	214	9,630
47.5 - 52.5	50	410	20,500
52.5 57.5	55	429	23,595
57.5 62.5	60	568	34,080
62.5 – 67.5	65	650	42,250
67.5 – 72.5	70	795	55,650
72.5 – 77.5	75	915	68,625
77.5—82.5	80	745	59,600
82.5—87.5	85	530	45,050
87.5—92.5	90	259	23,310
92.5—97.5	95	152	14,440
97.5 - 102.5	100	107	10,700
102.5—107.5	105	50	5,250
107.5 – 112.5	110	25	2,750
Σ		6,291	4,31,150

Solution. Since the variable is a continuous one, the mid-points are calculated simply as (lower limit + upper limit)/2, and are shown in second column as *m*. The *fm* values are calculated in column 4. The mean is calculated as

$$\bar{x} = \frac{\Sigma fm}{\Sigma f} = \frac{4,31,150}{6,291} = \text{Rs } 68.5$$

If we compute the arithmetic mean from unclassified data it may differ slightly from Rs 68.5. This lack of agreement is due to the inadequacy of the mid-value assumption. It is almost always true that none of the mid-value is actually the true concentration point of this class. But in the case of symmetrical distributions there is greater possibility of errors compensating, some of the mid-points erring by being too low and others erring by being too high. However, if the frequency tails off towards either the high or low values, *i.e.*, if it departs seriously from a symmetrical distribution, the arithmetic average computed* will be somewhat in error because of the failure of the known errors in the mid-point assumption to compensate.

9.5. PROPERTIES OF THE MEAN

From the foregoing discussion we are in a position to understand the properties inherent in an arithmetic mean.

1. The product of the arithmetic mean and the number of values on which the mean is based is equal to the sum of all given values. In other words if we replace each item in series by the mean, then the sum of these substitutions will equal the sum of individual items. Thus, in the figures 3, 5, 7, 9, if we substitute the mean for each item, 6, 6, 6, 6 the total is 24 both in the original series and in the substitution series.

This can be shown like this,

Since $\bar{x} = \frac{\Sigma x}{N}$

$\therefore \quad N\bar{x} = \Sigma x.$ **(9.4)**

This property provides a test to check if the computed value is the correct arithmetic mean.

2. A second property of the mean is that the algebraic sum of the deviations of the various values from the arithmetic mean is equal to zero, *i.e.* $\Sigma(x - \bar{x}) = 0$. As such the mean may be characterised as a point of balance. On this property is based the short method of computing mean given in sec. 9.6.

3. Another property of the arithmetic mean is that the sum of the squares of deviations from arithmetic mean is the least. For example, if we have a series of values 3, 5, 7, 9, the mean is 6. The squared deviations are:

x	$x - X$	=	x'	x'^2	
3	3—6	=	—3	9	
5	5—6	=	—1	1	
7	7—6	=	1	1	
9	9—6	=	3	9	$\Sigma x'^2 = 20.$

If the deviations from some other value are squared their sum would be larger than 20. For example, we may find the sum of the squared deviations of the values from 5;

x	$x-5$	x'	x'^2
3	3--5	--2	4
5	5--5	0	0
7	7--5	2	4
9	9--5	4	16 $\Sigma x'^2 = 24$.

The importance of this property will be seen when we discuss standard deviation.

4. The mean of all the sums (or, differences) of corresponding observations in two series, number of observations being equal in the two, is equal to the sum (or, difference) of the means of the two series.

Let $x=x_1 \pm x_2$,

$$\Sigma x = \Sigma x_1 \pm \Sigma x_2.$$

Divide by N (the total number of observations in each series),

$$\bar{x}=\bar{x}_1 \pm \bar{x}_2. \tag{9.4}$$

Illustration 9.5. The mean age of a group of 100 persons (grouped in intervals 10—, 12— . . .etc.) was found to be 32.02. Later, it was discovered that age 57 was misread as 27. Find the corrected mean.

Solution. Let the items be denoted by x. So putting the given values in the formula of A.M., we have

$$32.02=\frac{\Sigma x}{100} \quad i.e., \ \Sigma x=3202$$

correct $\Sigma x=3202-27+57=3232$

$\therefore$ Corrrect $\text{A.M.}=\frac{3232}{100}=32.32.$

Illustration 9.6. Out of the total population of certain town in South Africa, 60% belonged to the Black Race and the rest belonged to the White Race. It was estimated that their mean incomes were respectively 2,000 and 5,000 pounds. Find the average income of the entire town.

Solution. Total population $N=100$

Black Race $(N_1)=60$

White Race $(N_2)=40$

Average Income Black $(\bar{x}_1)=2{,}000$

Average Income White $(\bar{x}_2)=5{,}000$

$$\bar{x}_{1+2}=\frac{\Sigma x}{N}=\frac{N_1\bar{x}_1+N_2\bar{x}_2}{N_1+N_2}$$

$$=\frac{2000\times 60+5000\times 40}{60+40}$$

$$=\frac{320000}{100}=3{,}200 \text{ pounds.}$$

Illustration 9.7. The mean weight of 150 students in certain class is 60 kilograms. The mean weight of boys in the class is 70 **kg** and that of the girls is 55 kg. Find the number of boys and the number of girls in the class.

Solution. Let N_1 be the number of boys and N_2 be the number of girls in the class. Also let $\bar{x}_1$, $\bar{x}_2$, and $\bar{x}_{1+2}$ be the mean of boys, girls and of all the students respectively.

Then $\bar{x}_1=70,\ \bar{x}_2=55,\ \bar{x}_{12}=60,\ N_1+N_2=150$

Now $$\bar{x}_{1+2}=\frac{\Sigma x}{N}=\frac{N_1\bar{x}_1+N_2\bar{x}_2}{N_1+N_2}$$

or $$60=\frac{N_1\times 70+N_2\times 55}{150}$$

also $$9000=70N_1+55N_2 \quad \ldots (i)$$
$$150=N_1+N_2 \quad \ldots (ii)$$

solving (i) and (ii), we get

$$N_1=50 \text{ and } N_2=100.$$

Illustration 9.8. The mean monthly salary paid to all employees in a company was Rs 500. The monthly salaries paid to male and female employees average Rs 520 and Rs 420 respectively. Determine the percentage of males and females employed by the company.

Solution. Let N_1 be the number of males and N_2 be the number of females employed by the company. Also let x_1 and x_2 be the monthly average salaries paid to male and female employees and x_{12} be the mean monthly salary paid to all the employees:

$$\bar{x}_{12}=\frac{N_1\bar{x}_1+N_2\bar{x}_2}{N_1+N_2}$$

or $$500=\frac{520N_1+420N_2}{N_1+N_2}$$

or $$20N_1=80\ N_2$$

or $$\frac{N_1}{N_2}=\frac{80}{20}=\frac{4}{1}$$

Hence the males and females are in the ratio of 4 : 1 or 80 per cent are males and 20 per cent are females in those employed by the company.

9.6. SHORT-CUT METHOD FOR CALCULATING MEAN

We can simplyfy the calculations of mean by noticing that if we subtract a constant amount A from each item x to define a new

variable $x'=x-A$, the mean $\bar{x}'$ of x' differs from $\bar{x}$ by A. This generally simplifies the calculations and we can then add back the constant A, termed as the *assumed mean*:

$$\bar{x}=A+\bar{x}'=A+\frac{\Sigma fx'}{\Sigma f} \tag{9.5}$$

Table 9.5 illustrates the procedure of calculation for data of Table 9.1. The choice of A is made in such a manner as to simplify calculation the most, and is generally in the region of the concentration of data.

TABLE 9.4

x	f	*Deviation from assumed mean (13)* x'	fx'
9	1	−4	−4
10	2	−3	−6
11	3	−2	−6
12	6	−1	−6
13	10	0	−22*
14	11	+1	+11
15	7	+2	+14
16	3	+3	+ 9
17	2	+4	+ 8
18	1	+5	+ 5
			47
			−22
Σ	46		25

*Since there will be no entry in the fx' column corressponding to $x'=0$, this is a convenient place to write the sum (here—22) of the negative entries in the fx' column. The sum of the positive products in the fx' column, namely 47, is written in the same line as the total N. The final sum 25 it then easily obtained.

The mean

$$\bar{x}=A+\frac{\Sigma fx'}{\Sigma f}=13+\frac{25}{46}=13.54$$

the same as calculated in Illustration 9.3.

In the case of grouped frequency data, the variable x is replaced by mid-value m, and in the short-cut technique we subtract a constant value A from each m, so that the formula becomes.

$$\bar{x}=A+\frac{\Sigma f(m-A)}{\Sigma f} \tag{9.6}$$

In cases where the *class intervals are equal*, we may further simplify calculation by taking the factor i from the variable $m-A$, defining

$$x'=\frac{m-A}{i} \tag{9.7}$$

where i is the class width. It can be verified, then, that with x' defined as in Eq. (9.7), the mean of the distribution in given by

$$\bar{x}=A+\frac{\Sigma fx'}{\Sigma f}\times i \tag{9.8}$$

We illustrate below the use of short-cut method.

Illustration 9.9. Table 9.5 gives the distribution of 500 firms in an industry according to the sales volume. Calculate the arithmetic mean of the sales.

TABLE 9.5

Sales in ·000 rupees	*No. of firms*
0– 500	3
500–1000	24
1000—1500	55
1500—2000	98
2000—2500	120
2500—3000	95
3000—3500	51
3500 4000	39
4000—4500	15
Total	500

Solution. We have a grouped frequency distribution with $i=500$. Since the data is concentrated around the cass 2000—2500, its mid-value 2250 is selected as A. The calculations are then as shown in Table 9·6.

The mean sales is

$$\bar{x}=A+\frac{\Sigma fx'}{\Sigma f}\times i=2250+\frac{82}{500}\times 500$$
$$=2332$$

Thus, the mean sales is Rs 23,32,000.

Illustration 9.10. The distribution of male workers by average monthly earnings is given in Table 9.7. Calculate the arithmetic mean.

Solution. For the short-cut method we set the mid-value of the class with the highest frequency as A. The variable x' is then defined

TABLE 9.6

Sales in '000 rupees	*Mid-point* x	*No. of firms* f	*Deviation in interval units* $x'=\frac{x-2250}{500}$	fx'
0— 500	250	3	—4	— 12
500 – 1000	750	24	—3	— 72
1000—1500	1250	55	—2	—110
1500—2000	1750	98	– 1	— 98
				—292
2000—2500	2250	120	0	
2500—3000	2750	95	1	95
3000—3500	3250	51	2	102
3500—4000	3750	39	3	117
4000—4500	4250	15	4	60
				374
		500		82

TABLE 9.7

Monthly earnings Rs	*Mid-point*	*No. of workers*	*Deviation tn inierval units* $x'=\frac{x-70}{5}$	$f x'$
27.5—32.5	30	120	—8	– 960
32.5—37.5	35	152	—7	—1064
37.5—42.5	40	170	—6	– 1020
42.5—47.5	45	214	—5	—1070
47.5—52.5	50	410	– 4	—1640
52.5 – 57.5	55	429	—3	—1287
57.5 – 62.5	60	568	—2	—1136
62.5—67.5	65	650	—1	— 650
				—8827
67.5—72.5	70	795	0	
72.5—77.5	75	915	1	915
77.5—82.5	80	745	2	1490
82.5—87.5	85	530	3	1590
87.5 – 92.5	90	259	4	1036
92.5 – 97.5	95	152	5	760
97.5—102.5	100	107	6	642
102.5 – 107-5	105	50	7	350
107.5—112.5	110	25	8	200
				+6983
Σ		6291		—1844

as $(x-70)/5$. The calculation of $\Sigma fx'$ is shown in Table 9.7. Thus.

$$\bar{x}=A+\frac{\Sigma fx'}{\Sigma f}\times i=70+\frac{(-1844)}{6291}\times 5$$
$$=68.53$$

Illustration 9.10. The ages of twenty husbands and wives are given in Table 9.8. Form a two-way frequency table showing the relationship between the ages of husbands and wives with class-intervals 20--24; 25--29, etc.

Calculate the Arithmetic Mean of the two groups after the classification.

TABLE 9.8

S. No.	*Age of husband*	*Age of wife*
1	28	23
2	37	30
3	42	40
4	25	26
5	29	25
6	47	41
7	37	35
8	35	25
9	23	21
10	41	38
11	27	24
12	39	34
13	23	20
14	33	31
15	36	29
16	32	35
17	22	23
18	29	27
19	38	34
20	48	47

Solution.

TABLE 9.9. FREQUENCY DISTRIBUTION OF AGE OF HUSBANDS AND WIVES

Age of Husband	*Age of Wife* 20—24	25—29	30—34	35—39.	40—44	45—49	*Total*
20–24	III						3
25—29	II	III					5
30—34			I	I			2
35—39		II	III	I			6
40 – 44				I	I		2
45—49					I	I	2
Total	5	5	4	3	2	1	20

TABLE 9.10. CALCULATION OF ARITHMETIC MEAN, HUSBANDS

Class-intervals	*Mid-values* m	*Husband frequency* (f_1)	$x_1' = \frac{m-37}{5}$	f_1x_1'
20—24	22	3	—3	— 9
25—29	27	5	—2	—10
30—34	32	2	—1	— 2
35—39	37	6	0	—21
40—44	42	2	1	2
45—49	47	2	2	4
				6
Σ		20		—15

Husband Age, Arithmetic Mean

$$\bar{x} = \frac{\Sigma f_1 x_1'}{N} \times i + A = \frac{-15}{20} \times 5 + 37 = 33.25.$$

TABLE 9.11. CALCULATION OF MEAN, WIVES

Class-intervals	*Mid-values* m	*Wife frequency* (f_2)	$x_2' = \frac{m-37}{5}$	f_2x_2'
20—24	22	5	—3	—15
25—29	27	5	—2	—10
30—34	32	4	—1	— 4
35—39	37	3	0	0
40—44	42	2	1	2
45—49	47	1	2	2
Σ		25		—25

Wife Age, Arithmetic Mean

$$\bar{x} = \frac{\Sigma f_2 x_2'}{N} \times i + A = \frac{-25}{20} \times 5 + 37 = 30.75.$$

Illustration 9.11. Calculate the number of shops corresponding to class-interval 30—40 of the following distribution:

Profit per shop:	0—10	10—20	20—30	30—40	40—50	50—60
No. of shops:	12	18	27	?	17	6

The mean profit per shop is 28.

Solution.

Let f_1 be the missing frequency.

TABLE 9.12. CALCULATION OF MISSING FREQUENCY

Profit	*No. of shops* (f)	*Mid-point* m	$x' = \frac{m-25}{10}$	fx'
0—10	12	5	−2	−24
10—20	18	15	−1	−18
20—30	27	25	0	0
30—40	f_1	35	1	f_1
40—50	17	45	2	34
50—60	6	55	3	18
Σ	$80+f_1$			f_1+10

Now Arithmetic Mean $= \bar{x} = \frac{\Sigma fx'}{N} \times i + A$

or $$28 = \frac{f_1+10}{80+f_1} \times 10 + 25$$

or $$3(80+f_1) = (f_1+10) \times 10$$

or $$7f_1 = 140$$

or $$f_1 = 20$$

∴ Number of shops corresponding to profit 30—40 is 20.

9.7. THE WEIGHTED ARITHMETIC MEAN

In the computation of arithmetic mean we had given equal importance to each observation in the series. This equal importance may be misleading if the individual values constituting the series have different importance as in the following illustration:

The Raja Toy shop sells

Toy Cars at	Rs 3 each
Toy Locomotive at	Rs 5 each
Toy Aeroplanes at	Rs 7 each
Toy Double Decker at	Rs 9 each

What shall be the average price of the toys sold. If the shop sells 4 toys one of each kind,

Mean Price *i.e.* $$\bar{x} = \frac{\Sigma x}{4}$$

$$= \text{Rs } \frac{24}{4} = \text{Rs } 6.$$

In this case the importance of each observation (Price quotation) is equal in as mach as one toy of each variety has been sold. In the above computation of the arithmetic mean this fact has been taken care of by including 'once only' the price of each toy.

But if the shop sells 100 toys: 50 cars, 25 locomotives, 15 aeroplanes and 10 double deckers, the importance of the four Price quotations to the dealer **is not equal** as a source of earning revenue. In fact their respective importance is equal to the number of units of each toy sold, *i.e.*,

the importance of		Toy Car is	50
,,	,,	Locomotive	25
,,	,,	Aeroplane	15
,,	,,	Double Decker	10

It may be noted that 50, 25, 15, 10 are the quantities of the various classes of toys sold. It is for these quantities that the term 'weights' is used in statistical language. Weight is represented by symbol 'w'; and Σw represents the sum of weights.

While determining the 'average price of toy sold' these weights are of very great importance and are taken into account in the manner illustrated below: $\bar{x}=\dfrac{(w_1x_1)+(w_2x_2)+(w_3x_3)+w_4x_4)}{w_1+w_2+w_3+w_4}$.

$$\bar{x}=\frac{\Sigma wx}{\Sigma w}$$

when w_1, w_2, w_3, w_4 are the respective weights of x_1, x_2, x_3, x_4 which in their turn represent the price of 4 varieties of toys,viz., car, locomotive, aeroplane and double decker respectively.

$$\bar{x}=\frac{(50\times3)+(25\times5)+(15\times7)+(10\times9)}{50+25+15+10}$$

$$=\frac{(150)+(125)+(105)+(90)}{100}$$

$$=\frac{470}{100}=\text{Rs } 4.70$$

The table given on next page summarises the steps taken in the computation of the weighted Arithmetic Mean.

$$\Sigma w=100; \qquad \Sigma wx=470$$

$$\bar{x}=\frac{\Sigma wx}{\Sigma w}=\frac{470}{100}=4.70.$$

The weighted arithmetic mean is particularly useful where we have to compute the *mean of means*. If we are given two arithmetic means, one for each of two different series, in respect of the *same variable*, and are required to find the arithmetic mean of the combined series, the weighted arithmetic mean is the only suitable method of its determination.

WEIGHTED ARITHMETIC MEAN OF TOYS SOLD BY THE RAJA TOY SHOP

Toy	*Price per toy Rs x*	*Number sold w*	*Price 0 weight wx*
Car	3	50	150
Locomotive	5	25	125
Aeroplane	7	15	105
Double Decker	9	10	90
Σ		100	470

Illustration 9.12. The arithmetic mean of daily wages of two manufacturing concerns *A* Ltd. and *B* Ltd. is Rs 5 and Rs 7 respectively. Determine the average daily wages of both concerns if the number of workers employed were 2,000 and 4,000 respectively.

Solution. (*i*) Multiply each average (viz. 5 and 7) by the number of workers in the concern it represents.

(*ii*) Add up the two products obtained in (*i*) above and

(*iii*) Divide the total obtained in (*ii*) by the total number of workers.

TABLE 9.13. WEIGHTED MEAN OF MEAN WAGES OF A LTD. AND B LTD.

Manufacturing concern	*Mean wages x*	*Workers employed w*	*Mean wages × Workers employed wx*
A Ltd.	5	2,000	10,000
B Ltd.	7	4,000	28,000
		$\Sigma w = 6{,}000$	$\Sigma wx = 38{,}000$

$$\bar{\bar{x}} = \frac{\Sigma wx}{\Sigma w}$$

$$= \frac{38{,}000}{6{,}000} = \text{Rs } 6.33.$$

Illustration 9.13. Admissions during 1971 to 'Vidya Mandir,' and 'Children's Home' increased by 20% and 40% respectively over those during 1970. If the admission in 1970 were 800 and 200 respectively find the mean increase in the admission of the two schools.

Solution. To each per cent increase a weight equal to the admission of the previous year is assigned.

	1970 admission w	*Per cent increase* x	wx
Vidya Mandir	800	20	16,000
Children's Home	200	40	8,000
	1;000		24,000

Thus the mean per cent increase is

$$\bar{X}=\frac{24000}{1000}=24$$

One can check that the answer is correct by calculating the number of admission in 1971 as $800\times 120/100+\frac{200\times 140}{100}=1{,}240$ representing a 24 per cent increase over 1,000 admissions of 1970.

From the foregoing discussion it should be clearly understood that 'Arithmetic means and percentage' are not original data. They are derived figures and their importance are relative to the original data from which they are obtained. This relative importance must be taken into account by weighting while averaging them (means and percentage).

9.8. THE MEDIAN

The second measure of central tendency that has a wide usage in statistical works, is the median. Median is that *value* of a variable which divides the series in such a manner that the number of items below it is equal to the number of items above it. Half the total number of observations lie below the median, and half above it. The median is thus a positional average.

The median of ungrouped data is found easily if the items are first arranged in order of magnitude. The median may then be located simply by counting, and its value can be obtained by reading the value of the middle observations. If we have five observations whose values are 8, 10, 1, 3 and 5, the values are first arrayed: 1, 3, 5, 8 and 10. It is now apparent that the value of the median is 5, since two observations are below that value and and two observations are above it. When there is an even number of cases, there is no actual middle item and the median is taken to be the average of the values of the items lying on either side of $(N+1)/2$, where N is the total number of items. Thus if the values of six items of a series are 1, 2, 3, 5, 8 and 10. The median is the value of item number $(6+1)/2=3.5$, which is approximated as the average of the third and the fourth items, *i.e.*, $(3+5)/2=4$.

Thus the steps required for obtaining median are;

1. Arrange the data as an array of increasing magnitude
2. Obtain the value of the $(N+1)/2$th item.

Even in the case of grouped data, the procedure for obtaining median is straight forward as long as the variable is discrete or non-continuous as is clear from the following illustration.

Illustration 9.14. Obtain the median size of shoes sold from the following data.

TABLE 9.14. NUMBER OF SHOES SOLD BY SIZE IN ONE YEAR

Size	*Number of pairs*	*Cumulative total*
5	30	30
$5\frac{1}{2}$	40	70
6	50	120
$6\frac{1}{2}$	150	270
7	300	570
$7\frac{1}{2}$	600	1170
8	950	2120
$8\frac{1}{2}$	820	2940
9	750	3690
$9\frac{1}{2}$	440	4130
10	250	4380
$10\frac{1}{2}$	150	4530
11	40	4570
$11\frac{1}{2}$	39	4609
	Total 4,609	

Solution. Median, is the value of $\frac{(N+1)}{2}$ th $=\frac{4609+1}{2}$ th $=$ 2305th item. Since the items are already arranged in ascending order (size-wise), the size of 2305th item is easily determined by constructing the cumulative frequency. Thus, the median size of shoes sold is $8\frac{1}{2}$, the size of 2305th item.

In the case of grouped data with continuous variable, the determination of median is a bit more involved. Consider as example the data relating to the distribution of male workers by average monthly earnings given as Table 9.15. Clearly the median of 6,291 cases is the earnings of $(6{,}291+1)/2=3{,}146$th worker arranged in ascending order of earnings.

TABLE 9.15. DISTRIBUTION OF MALE WORKERS BY AVERAGE MONTHLY EARNINGS

Group No.	Monthly earnings Rs.	No. of workers	Cumulative No. of workers
1	27.5—32.5	120	120
2	32.5—37.5	152	272
3	37.5—42.5	170	442
4	42.5—47.5	214	656
5	47.5—52.5	410	1066
6	52.5—57.5	429	1495
7	57.5—62.5	568	2063
8	62.5—67.5	650	2713
9	67.5—72.5	795	3508
10	72,5—77.5	915	4423
11	77.5—82.5	745	5168
12	82.5—87.5	530	5698
13	87.5—92.5	259	5957
14	92.5—97.5	152	6109
15	97.5—102.5	107	6216
16	102.5—107.5	50	6266
17	107.5—112.5	25	6291
	Total	6,291	

From the cumulative frequency, it is clear that this worker has his income in the class 67.5—72.5. But it is impossible to determine his exact income. We, therefore, resort to approximation by assuming that the 795 workers of this class are distributed *uniformly* across the interval 67.5 to 72.5. The median worker is (3146—2713)=433rd of these 795, and hence, the value corresponding to him can be approximated as

$$67.5+\frac{433}{795}\times(72.5-67.5)=67.5+2.73=70.23.$$

The value of the median can thus be put in the form of the formula

$$Me=l+\frac{\frac{N+1}{2}-C}{f}\times i \qquad (9.9)$$

where l is the lower limit of the median class, i its width, f its frequency, C the cumulative frequency upto (but not including) the median class, and N is the total number of cases.

Illustration 9.15. Recast the following cumulative table into the form of an ordinary frequency distribution and determine value of the median:

No. of days absent		*No. of students*
less than	5	29
,,	10	224
,,	15	465
,,	20	582
,,	25	634
,,	30	644
,,	35	650
,,	40	653
,,	45	655

Solution.

Class-intervals	*Cumulative frequency*	*Frequency*
0—5	29	29—0 =29
5—10	224	224—29 =195
10—15	465	465—224=241
15—20	582	582—465=117
20—25	634	634—582= 52
25—30	644	644-634= 10
30—35	650	650—644= 6
35—40	653	653—650= 3
40—45	655	655—653= 2
	Total	655

Median is the value of $(655+1)/2=328$th item. The median group is the class in which 328th item lies, *i.e.*, 10—15. With the reference to formula (9.9)

$$l=10,\ \frac{N+1}{2}=328,\ C=224,\ f=241,\ \text{and}\ i=5$$

Thus,

$$Me=10+\frac{328-224}{241}\times 5=12.2$$

9.9. LOCATION OF MEDIAN BY GRAPHICAL ANALYSIS

The median can quite conveniently be determined by reference to the ogive which plots the cumulative frequency against the variable. The value of the item below which half the items lie can easily be read from the ogive as is shown in Illustration 9.16.

Illustration 9.16. Obtain the median of data given as Table 9.16.

TABLE 9.16

Monthly earnings	*Frequency*	*Less than*	*More than*
27.5	—	0	6291
32.5	120	120	6171
37.5	152	272	6019
42.5	170	442	5849
47.5	214	656	5635
52.5	410	1066	5225
57.5	429	1495	4796
62.5	568	2063	4228
67.5	650	2713	3578
72.5	795	3508	2783
77.5	915	4423	1868
82.5	745	5168	1123
87.5	530	5698	593
92.5	259	5957	334
97.5	152	6109	182
102.5	107	6216	65
107.5	50	6266	25
112.5	25	6291	0

Solution. It is clear that this is grouped data. The first class is 27.5—32.5 whose frequency is 120, and the last class is 107.5—112.5 whose frequency is 25. Figure 9.2 shows the ogive of less than cumulatiue frequency. The median is the value below which $N/2$ items lie, is $6291/2=3145.5$ items lie, which is read of from Fig. 9.2. as about 70. More accuracy than this is unobtainable because of the space limitation on the earning scale.

The median can also be determined by plotting both less than and 'more than' cumulative frequency as in Fig. 9.3. It should be obvious that the two curves should intersect at the median of the data.

9.10 QUARTILES. DECILES AND PERCENTILES

We have defined the median as the value of the item which is located at the centre of the array, we can define other measures which are located at other specified points. Thus, the *N*th *percentile* of an array is the value of the item such that N per cent items lie *below* it. Clearly then the N_{th} percentile P_n of grouped data is given by

$$P_n = l + \frac{\frac{nN}{100} - C}{f} \times i \qquad (9.10)$$

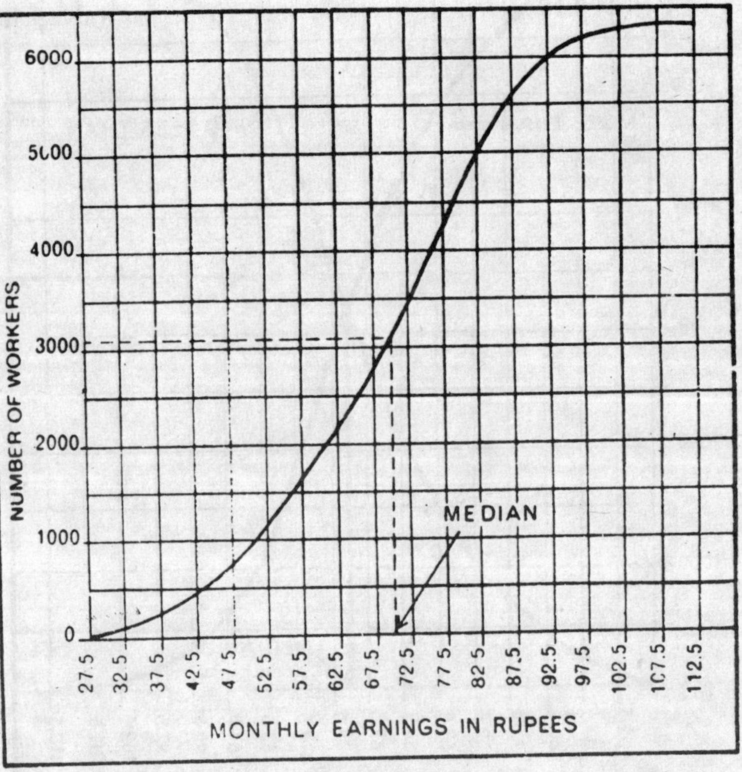

FIG. 9.2.

where l is the lower limit of the class in which $nN/100$th item lies, i its width, f its frequency, C the cumulative frequency upto (but not including) this class, and N is the total number of items.

We similarly define the N_{th} *decile* as the value of the item below which $(nN/10)$ items of the array lie. Clearly,

$$D_n = P_{10n} = l + \frac{\frac{nN}{10} - C}{f} \times i \qquad (9.11)$$

where the symbols have the obvious meanings.

The other most commonly referred to measures of location are the quartiles. Thus, n_{th} quartile is the value of the item which lie at the $n(N/4)_{th}$ item. Clearly Q_2, the second quartile is the median. For grouped data,

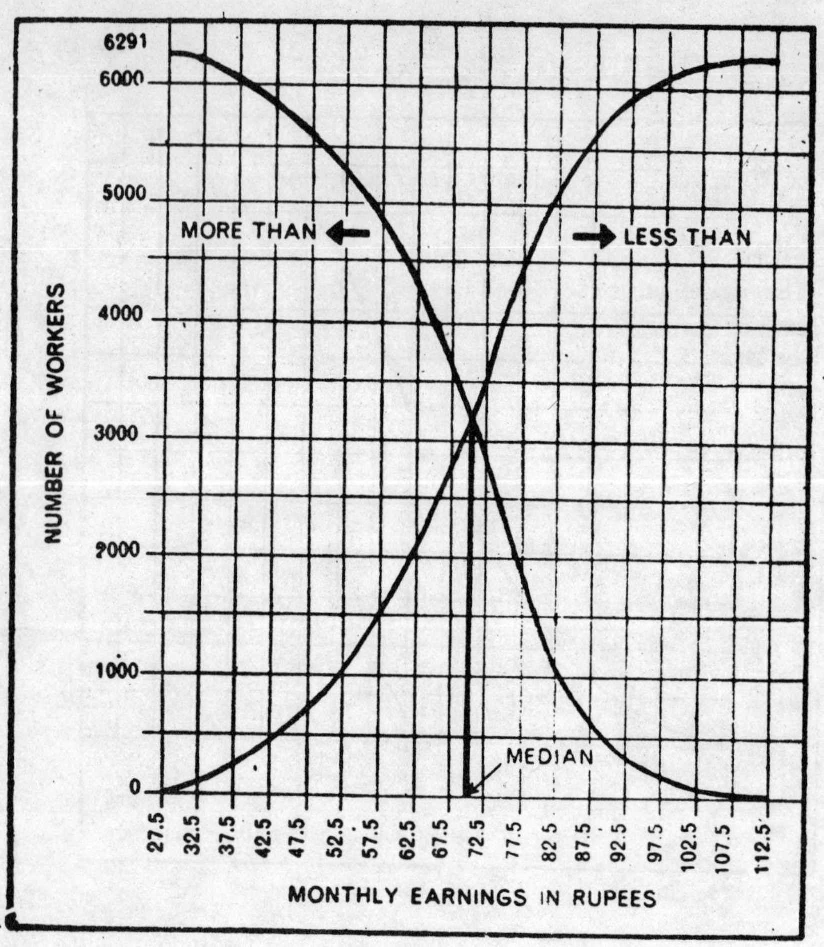

FIG. 9.3.

$$Q_n = P_{25n} = l + \frac{\frac{nN}{4} - C}{f} \times i \qquad (9.12)$$

Illustration 9.17. Find the first and the third quartiles and the 90th percentile of the data given in Table 9.15.

Solution. The first quartile Q_1 is the value of the N/4 =6291/4 = 1572.75th item. Thus the appropriate class is 57.5--62.5, and by Eq. 9.12

$$Q_1 = 57.5 + \frac{(1572.75 - 1495)}{568} \times 5 = 58.18$$

The third quartile Q_3 is the value of the $3N/4 = 3 \times 6291/4$ $= 4718.25$th item, or Q_3 class is 77.5—82.5, and

$$Q_3 = 77.5 + \frac{(4718.25-4423)}{745} \times 5 = 79.5$$

Similarly P_{90} lies in 82.5—87.5 class, and

$$P_{90} = 82.5 + \frac{(5661.9-5168)}{530} \times 5 = 87.16$$

or 90 per cent workers earn less than Rs 87.16.

9.11. THE MODE

The mode, strictly defined, is that value of the variable, which occurs or repeats itself the greatest number of times. The mode is the most 'fashionable' size in the sense that it is the most common and typical, and is defined by Zizek as "the value occuring most frequently in a series (or group) of items and around which the other items are distributed most densely."

The mode of a distribution is the value at the point around which the items tend to be most heavily concentrated. It is the most frequent or the most common value, provided that a sufficiently large number of items are available to give a smooth distribution. It will correspond to the value of the maximum point (ordinate) of a frequency distribution if it is an 'ideal' or smooth distribution. It may be regarded as the most typical of a series of values. The modal wage, for example, is the wage received by more individuals than any other wage. The modal hat size is that which is worn by more persons than any other single size.

It may be noted that the occurrence of one or a few extremely high or low values has no effect upon the mode. If a series of data are unclassified, not having been either arrayed or put into a frequency distribution, the mode cannot be readily located.

Taking first an extremely simple example, if seven men are receiving daily wages of Rs 5, 6, 7, 7, 7, 8 and 10, it is clear that the modal wage is Rs 7 per day. If we have a series such as 2, 3, 5, 6, 7, 10 and 11, it is apparent that there is no mode.

There are several methods of estimating the value of the mode. But it is seldom that the different methods of ascertaining the mode give us identical results. Consequently it becomes necessary to decide as to which method would be most suitable for the purpose in hand. In order that a choice of the method may be made intelligently we should understand carefully each of the methods and the differences that exist among them.

The four important methods of estimating mode of a series are:

(*i*) Locating the most frequently repeated value in the ar

(*ii*) Estimating the mode by interpolation; (*iii*) Locating the mode by graphic method; and (*iv*) Estimating the mode from the mean and the median.

(*i*) **Locating the most frequently repeated value in the array**. The data are first arrayed, and not infrequently the mode is at once apparent—the value coming most frequently being the mode of that distribution. If we look at Table 6.2 we will find that 40 is repeated more often (7 times) than any other value. Thus, '40', being the most frequently value, is the mode of this series. This is evident from Table 6.3 which gives the frequency distribution of marks.

TABLE 9.17. NUMBER OF ARTICLES SOLD BY SIZES

	f					
Size	*1*	*2*	*3*	*4*	*5*	*6*
1	2					
		9				
2	7			22		
			20			
3	13	—			35	
		28				
4	15		—	—		48
			35			
5	20	—		60	—	
		45				
6	25		—	—		
			48		68	
7	23	—		—		72
		47				
8	24		—	—	—	
9	20	—	44	67		
		43				
10	23		38	—	67	
11	15		—	—	—	58
		29				
12	14		—	52		
13	26	—	40	—	55	
		45				
14	19		--	—	—	59

ANALYSIS TABLES

(showing the Occurence in the maximum frequency class for each column)

Column No.	*Sizes*					
	5	*6*	*7*	*8*	*9*	*13*
1						1
2			1	1		
3		1	1			
4			1	1	1	
5	1	1	1			
6		1	1	1		
No. of items	1	3	5	3	1	1

When there is no apparent mode clearly revealed by the frequency distribution, it is necessary to re-group the figures. This is done by widening the classes; a procedure which smooths out irregularities. This is shown in Table 9.17.

The procedure is as follows: the frequencies are grouped by two's to obtain column 2; and again by two's, starting with the second item, to obtain column 3. Then they are grouped by three items, starting with the first item, then with the second, and lastly with the third item to obtain columns 4, 5 and 6. If necessary, grouping can be done in fours. As each of the groupings is completed the maximum grouped frequency is underlined or indicated in bolder figures. It will be seen that the 6th size is included in 3 maximum of grouped frequencies, the 7th in 5 and 8th in 3, hence the mode is the size the 7th group, *i.e.*, 7; and the other most popular sizes are 6 and 8. (See Analysis Table above).

(*ii*) **Estimating the mode by interpolation**. In the case of continuous frequency distributions, the problem of determining the value of the mode is not so simple as it might have appeared from the forgoing description. Having located the modal class of the data, the next problem in the case of continuous series is to interpolate the value of the mode within this 'modal' class.'

The interpolation is made by the use of any one of the formula;

$$(i)\ Mo = l_1 + \frac{f_2}{f_0+f_2} \times i; \quad (ii)\ Mo = l_2 - \frac{f_0}{f_0+f_2} \times i$$

or

$$(iii)\ Mo = l_1 + \frac{f_1 - f_0}{(f_1 - f_0) + (f_1 - f_2)} \times i$$

where l_1 is the lower limit of the modal class, l_2 is the upper limit of the model class, f_0 equals the frequency of the class next below modal class in value, f_1 equals the frequency of the modal class in value, f_2 equals the frequency of the following class (class next above modal class) in value and i equals the interval of the modal class.

Illustration 9.18. Determine the mode for the data of Table 9.18.

TABLE 9.18

Wage-group		*Frequency*
Exceeding	but not exceeding	
14	 18	6
18	 22	18
22	 26	19
26	 30	12
30	 34	5
34	 38	4

38	 42	3
42	 46	2
46	 50	1
50	 54	0
54	 58	1

Solution. In the given data 22—26 is the modal class, since it has the largest frequency, the lower limit of the modal class is 22, its upper limit 26, its frequency 19, the frequency of the preceding class is 18, and of the following one 12. The class interval is 4. Using the various methods of determining mode, we have

(*i*) $Mo = 22 + \frac{12 \times 4}{18+12}$ $= 22 + \frac{8}{5}$ $= 23.6.$

(*ii*) $Mo = 26 - \frac{18 \times 4}{18+12}$ $= 26 - \frac{12}{5}$ $= 23.6.$

(*iii*) $Mo = 22 + \frac{19-18}{2 \times 19 - 18 - 12} \times 4$

$22 + \frac{4}{8} = 22{\cdot}5.$

The formulae (*i*) and (*ii*), it will be noticed, use the frequency of the classes adjoining the modal class to pull the estimate of the mode away from the mid-point towards either the upper or lower class limit. In this particular case the frequency of the class preceding the modal class is more than the frequency of the class following and, therefore, the estimated mode is less than the mid-value of the modal class. This seems quite logical. If the frequencies are more on one side of the modal class than on the other, it can be reasonably concluded that the items in the modal class are concentrated more towards the class limit of the adjoining class with the larger frequency.

The formula (*iii*) is also based on a logic similar to that of (*i*) and (*ii*). In this case, to interpolate the value of the mode within the modal class, the differences between the frequency of the modal class, and the respective frequencies of the classes adjoining it are used. This formula usually gives results better than the values obtained by the other and exactly equal to the results obtained by graphic method. The formulae (*i*) and (*ii*) give values which are different from the value obtained by formula (*iii*) and are more close to the central point of modal class. If the frequencies of the class adjoining the modal are equal, the mode is expected to be located at the mid-value of the modal class, but if the frequency on one of the sides is greater the mode will be pulled away from the central point. It will be pulled more and more if the difference between the frequencies of the classes adjoining the modal class is higher and higher. In the example given above the frequency of the modal class is 19 and that of preceding class is 18. So the mode should be quite close to the lower limit of the modal class. The mid-point of the modal class is 24 and lower limit of the modal class is 22.

(*ii*) **Locating the mode by the graphic method.** The method of graphic interpolation is illustrated in Fig. 9.4. The upper corners of the rectangle over the modal class have been joined by straight lines to those of the adjoining rectangles as shown in the diagram;

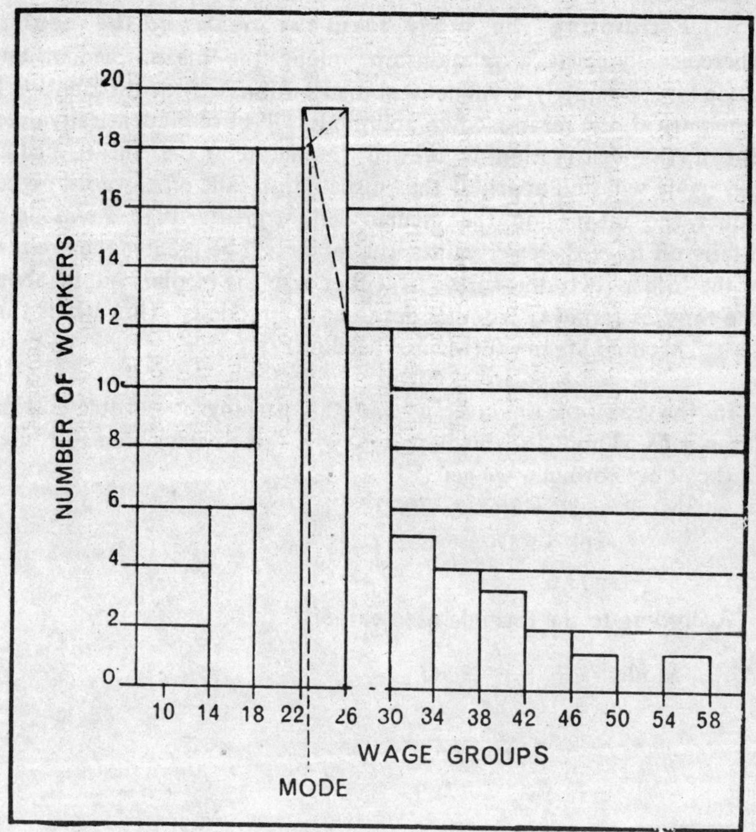

FIG. 9.4.

the right corner to the corresponding one of the adjoining rectangle on the left etc. If a perpendicular is drawn from the point of intersection of these lines, we have a value for the mode indicated on the base line. The graphic approach is in principle similar to the arithmetic interpolation explained above.

The mode may also be determined graphically from an ogive or cumulative frequency curve. It is found by drawing a perpendicular to the base from that point on the curve where the curve is most nearly vertically, *i.e.*, steepest (in other words, where it passes through the greatest distance vertically and smallest distance horizontally). The point where it cuts the base gives us the value of the mode. How accurately this method determines the mode is governed by (1) the shape of the ogive, (2) the scale on which the curve is drawn.

(*vi*) **Estimating the mode from the mean and the median.** There usually exists a relationship among the mean, median and mode for moderately asymmetrical distributions. If the distribution is symmetrical, the mean median and mode will have indentical values. But if the distribution is skewed (moderately) the mean, median and mode will pull apart. If the distribution tails off towards higher values, the mean and the median will be greater than the mode: if it tails off towards lower values, the mode will be greater than either of the other two measures. In either case the median will be about one-third as far away from the mean as the mode is. This means that

$$\text{Mode} = \text{Mean} - 3\,(\text{Mean} - \text{Median})$$
$$= 3\ \text{Median} - 2\ \text{Mean}$$

In the case of the average monthly earnings (see Table 9.2) the mean is 68.53 and the median is 70.2 If these values are substituted in the above formula, we get

$$\text{Mode} = 68.5 - 3(68.5 - 70.2)$$
$$= 68.5 + 5.1$$
$$= 73.6$$

According to the formula used earlier,

$$\text{Mode} = l_1 + \frac{f_2}{f_0 + f_2} \times i$$
$$= 72.5 + \frac{745}{795 + 745} \times 5$$
$$= 72.5 + 2.4 = 74.9.$$

OR

$$\text{Mode} = l_1 + \frac{f_1 - f_0}{2f_1 - f_0 - f_2} \times i$$
$$72.5 + \frac{915 - 795}{2 \times 915 - 795 - 745} \times 5$$

$$72.5+\frac{120}{290}\times 5$$
$$=74.57.$$

The difference between the two estimates is due to the fact that the assumption of relationship between the mean, median and mode may not always be true, It is obviously not valid in this case.

Illustration 9.19. (a) In a moderetely asymmetrical distribution. the mode and mean are 32.1 and 35.4 respectively. Calculate the median.

(*b*) If the mode and median of a moderately asymmetrical series are respectively 16″ and 15.7″, what would be its most probable median?

(*c*) In a moderately skewed distribution, the mean and the median are respectively 25.6 and 26.1 inches. What is the mode of the distribution?

Sotution. (*a*) We know,

Mean – Mode = 3 (Mean—Median)

or 3 Median = Mode + 2 Mean

or Median $=\frac{1}{3}(32.1+2\times 35.4)=\frac{102.9}{3}=34.3$

(*b*) 2 Mean = 3 Median—Mode

or Mean $=\frac{1}{2}(3\times 15.7-16.0)=\frac{31.1}{2}=15.55$

(*c*) Mode = 3 Median – 2 Mean

$=3\times 26.1-2\times 25.6=78.3-51.2=27.1.$

Illustration 9.20. The following Table shows the distribution of 105 families according to their expenditure per week. Number of families corresponding to the expenditure groups Rs (10-20) and Rs. (30-40) are missing from the Table. The median and mode for the distribution are Rs 25 and Rs 24 respectively. Calculate the missing frequencies and then calculate the arithmetic mean of the data.

Expenditure:	0—10	10—20	20—30	30—40	40—50
No. of families:	14	?	27	?	15

Solution. Let us suppose that the frequencies of the groups Rs (10–20) and Rs (30–40) are f_1 and f_2 respectively. We can write the Table in the following manner:

Expenditure (*x*)	*No. of families* (*f*)	*Cumulative frequencies*
0—10	14	14
10—20	f_1	$14+f_1$
20—30	27	$41+f_1$
30—40	f_2	$41+f_1+f_2$
40—50	15	$56+f_1+f_2$

Since the median and mode are Rs 25 and 24 respectively the group Rs (20—30) is the median as well as the modal class. Hence we have

$$25=20+\frac{\frac{56+f_1+f_2}{2}-(14+f_1)}{27}\times 10 \qquad ...(i)$$

and
$$24=20+\frac{27-f_1}{2\times 27-f_1-f_2}\times 10 \qquad (ii)$$

Equations (*i*) and (*ii*) can be written as

$$f_1-f_2=1 \qquad ...(iii)$$

and
$$3f_1-2f_2=27 \qquad ...(iv)$$

solving (*iii*) and (*iv*), we get the values of f_1 and f_2 as

$$f_1=25 \text{ and } f_2=24.$$

Hence the number of families in the groups Rs (10–20) and Rs (30–40) are 25 and 24 respectively.

CALCULATION OF ARITHMETIC MEAN

Expenditure	*No. of families*	*Mid-value*	$x'=\frac{m-25}{10}$	fx'
0–10	14	5	–1	–14
10—20	25	15	–2	–50
20—30	27	25	0	0
30—40	24	35	+1	+24
40—50	15	45	+2	+30
	105			–10

$$\bar{x}=A+\frac{\Sigma fx'}{N}\times i$$

$$=25+\frac{(-10)}{105}\times 10=24.05.$$

9.12. THE GEOMETRIG MEAN

The geometric mean of *n* positive values is defined as the *n*th root of their product. Thus, it is obtained by multyplying together all the values and then extracting the relevant root of the product. It can be represented as;

$$\text{Geometric Mean}=\sqrt[n]{x_1 \cdot x_2 \cdot x_3 \ldots x_n}$$

where n stands for the number of items and $x_1, x_2, x_3 \ldots x_n$ are the various values. For instance the geometric mean of 4, 8, 16 is

$$GM=\sqrt[3]{4\times 8\times 16}$$

$$=\sqrt[3]{512}=8.$$

The above method of calculating geometric mean is satisfactory only if there are two or three items. But if n is a large number the problem of computing the nth root of the product of these values by simple arithmetic is a tedious work. To facilitate the computation of geometric mean we make use of logarithms. The above formula when reduced to its logarithmic form will be:

$$\log GM = \frac{\log x_1 + \log x_2 + \log x_3 + \ldots \log x_n}{n}$$

The logarithm of the geometric mean is equal to the arithmetic mean of the logarithms of individual values.

Illustration 9.21. Find the GM of 2, 4, 8, 12, 16, 24.

	log
2	0·3010
4	0.6021
8	0.9031
12	1.0792
16	1.2041
24	1.3802
	5.4697

Solution.

$$\text{Geometric Mean} = \text{antilog}\,\frac{5.4697}{6}$$
$$= \text{antilog}\ .9116$$
$$= 8.158\ \textbf{Ans.}$$

It is easily verified that the geometric mean (GM) of a frequency distribution is given by

$$\log GM = \frac{f_1 \log x_1 + f_2 \log x_2 + f_3 \log x_3 \ldots f_n \log x_n}{N} \qquad (9.17)$$

Similarly for grouped data,

$$\log GM = \frac{\Sigma f \log m}{N} \qquad (9.18)$$

where m is the mid-value of a particular class.

Illustration 9.22. Calculate the GM of the data given in the first two columns Table 9.19.

Solution. For the grouped frequency data, we complete the table as shown and calculate GM using Formula (9.18),

$$GM = \text{antilog}\ \frac{146.7497}{105}$$
$$= \text{antilog}\ 1.3976 = 24.98$$

TABLE 9.19

Closs intervals	Mid-value x	Frequency f	log x	$f \times (\log x)$ x
9.5—14.5	12	10	1.0792	10.7920
14.5—19.5	17	15	1.2304	18.4560
19.5—24.5	22	17	1.3424	22.8208
24.5—29.5	27	25	1.4314	35.7850
29.5—34.5	32	18	1.5051	27.0918
34.5—39.5	37	12	1.5682	18.8184
39.5- 44.5	42	8	1.6232	12.9856
		105		146.7496

The geometric means are used extensively where we are interested in averaging the ratios, for example, in construction of index numbers.

9.13. THE HARMONIC MEAN

Another important mean is the harmonic mean which is used for averaging the *rates*. It is defined by

$$\frac{1}{HM} = \left(\frac{1}{x_1} + \frac{1}{x_2} + \frac{1}{x_3} + \ldots + \frac{1}{x_N} \right) / N \qquad (9.19)$$

where N is the number of items in the series $x_1, x_2, x_3, \ldots, x_N$.

Thus, if a man travels 200 km. each on three days at speeds of 60, 50 and 40 km ph respectively, his average speed is given by the HM of the three speeds, namely

$$\text{HM} = \frac{3}{\frac{1}{60} + \frac{1}{50} + \frac{1}{40}} = 48{\cdot}65 \text{ kmph}$$

(Note that HM gives the correct average speed because the man travelled equal distances on three speeds. If, however, he had travelled for equal times, the AM would have been the correct average).

9.14. THE CHOICE OF AVERAGE

The choice of a particular measure of central tendency or location depends on the purpose of investigation. It should be noted that the arithmetic mean (AM) is quite precisely defined and is therefore more amenable to further mathematical manipulation. On the other hand the mode can be located by inspection, but can not be manipulated easily.

There are certain other properties of these which also determine there use. It has been noted that sum of deviations of individual items from AM is zero and, therefore, AM is useful in those situations where the effect of positive deviations cancels out the effect of negative deviations. Thus, the average height of boys or the average marks of students is generally the AM. On the other hand mode is the most typical item and thus more representative of the series. Thus, if we want to determine the height of chair seats most suitable to a class room, we should use most typical value, *i.e.*, the mode. That value will satisfy the most number of students. If we had chosen the arithmetic mean it will ensure that the 'total discomfort' of those who find it too high balances the 'total discomfort' of those who find it too low which obviously has nothing to recommend for itself. The median on the other hand is a value, the total of *absolute values of* deviations from which is the minimum. This properly makes it useful in those situations where we want to minimize the total 'discomfort'.

It should further be noted that AM takes all the items into consideration, while median and mode are unaffected by extreme values In many situation this makes AM less attractive. Thus if we calculate average income for the purposes of determining the general well being of a population, the astronomical incomes of a few individuals give a wrong tilt to the AM income. In such a situation the median income will serve the purpose a lot better.

9.15. MISUSE OF AVERAGES

A lot of mischief can be done by wrong interpretation of the meaning of averages. It was reported in 1943 that the death rate in the state of Arizona came down to 8.4 per thousand from 16.8 per thousand in a matter of 5 years making it a much healthier place to live in. A closer investigation revealed that the decrease was essentially because of the influx of a large number of young recruits to army posts swamping a relatively small base of population. Similarly, it was reported that in a village in Rajasthan, a cloud burst produced a rainfall in one night which was 20 times the average annual rainfall of last 60 years. This seems like quite a cloud burst till it is mentioned that in those 60 years there was only one other recorded rainfall of a 3 mm, so that a mere 1 mm rainfal gave that dramatic headline.

9.16. MISCELLANEOUS ILLUSTRATIONS

Illustration 9.23. The table on the next page shows the number of

Summary of the Properties of the Various Measures of Location.

Property	*Arithmetic Mean AM*	*Median Me*	*Mode Mo*	*Geometric Mean GM*	*Harmonic Mean HM*
1. Rigidity defined	yes	yes	not very	yes	yes
2. Based on all values of series	yes	no	no	yes	yes
3. Ease in calculation	moderate	very easy	very easy	very difficult	difficult
4. Understanding by common man	most easily	quite	quite	not easy	not easily
5. Amenable to algebraic treatment	most	no	no	yes	yes
6. Effect of sample variations (Small effect is desirable)	least	moderate	moderate	moderate	moderate
7. Effect of extreme values (Small effect in desirable)	large	none	none	very low	very low
8. Difficulties with open-ended classes	yes	no difficulty	no difficulty	yes	yes
9. Use as typical value of series	no	more	most typical	no	no
10. Most useful in	general purpose problems	'least net discomfort problems	typifying series	averaging ratios	averaging rates

and unskilled workers in two small communities, together with their average hourly wages.

	Ram Nagar		Shyam Nagar	
Worker Category	*Number*	*Wage per hour*	*Number*	*Wage per hour*
Skilled	150	Rs 1.80	350	Rs 1.75
Unskilled	850	Rs 1.30	650	Rs 1.25

Determine the hourly wage for each community. Also give reasons why the result show that the average hourly wage in Shyam Nagar exceeds the average hourly wage in Ram Nagar, even though in Shyam Nagar the ayerage hourly wage of both categories of workers is lower.

Solution.

	Ram Nagar			Shyam Nagar		
Worker Category	*Number*	*Wage per hour*	*Total wage per hour*	*Number*	*Woge per hour*	*Total wage per hour*
Skilled	150	Rs 1.80	270	350	Rs 1.75	612.50
Unskilled	850	Rs 1.30	1105	650	Rs 1.25	812.50
Total	1,000		1,375	1,000		1425.00

The hourly rate of wages for Ram Nagar is Rs 1.375.

The hourly rate of wages for Shyam Nagar is Rs 1.425.

The reason for the wage rate in Shyam Nagar being higher than in Ram Nagar is the fact the number of skilled workers getting a higher wage rate is proportionately more than in Ram Nagar. In Shyam Nagar their number is 350 out of 1,000 as compared to 150 out of 1,000 in Ram Nagar.

Illustration 9.24. Calculate the Arithmetic Mean and apply Charlier's check to test accuracy of your calculations:

Age under (years)	10	20	30	40	50
Frequency	5	13	25	31	35

Solution. Charlier's check is applied to check the accuracy of calculations while finding out mean by short-cut method. The formula for checking the accuracy is

$$\Sigma fx' = \Sigma\{f(x'+1)\} - \Sigma f$$

CALCULATION OF THE AVERAGE AGE BY SHORT-CUT METHOD

Age in years	Mid-value (x)	Frequency	Deviation from assumed mean $x'=x-25$	fx'
0—10	5	5	—20	—100
10—20	15	8	—10	— 80
20—30	25	12	0	0
30—40	35	6	10	60
40—50	45	4	20	80
Total		$N=35$		$\Sigma fx'=-40$

$$\bar{X}=A+\frac{\Sigma fx'}{N}=25+\left(\frac{-40}{35}\right)$$
$$=23.86 \text{ years.}$$

APPLICATION OF THE CHARLIER'S CHECK

f	x'	$x'+1$	$f(x'+1)$	fx'
5	—20	—20+1=—19	—95	—100
8	—10	—10+1=— 9	—72	— 80
12	0	0+1=1	12	0
6	10	10+1=11	66	60
4	20	20+1=21	84	80
Total $\Sigma f=35$			$\Sigma f(x'+1)=-5$	$\Sigma fx'=-40$

Charlier's check formula is

$$\Sigma fx'=\Sigma\{f(x'+1)\}-\Sigma f$$

$\therefore$ $$-40=-5-35$$

or $$-40=-40$$

Hence the calculations are correct.

Illustration 9.25. From the following data calculate the missing frequency.

No. of tablets	4-8	8-12	12-16	16-20	20-24	24-28	28-32	32-36	36-40
No. of persons cured	11	13	16	14	?	9	17	6	4

The average number of tablets to cure fever was 19.9.

Solution. Let the missing frequency be f_1.

Now $$A.M.=\bar{X}=\frac{\Sigma fx}{N}$$

or $$19.9=\frac{1772+22f_1}{90+f_1}$$

Class-interval	*Frequency* (f)	*Mid-point* (x)	fx
4—8	11	6	66
8—12	13	10	130
12—16	16	14	224
16—20	14	18	252
20—24	f_1	22	$22f_1$
24—28	9	26	234
28—32	17	30	510
32—36	6	34	204
36—40	4	38	152
Total	$90+f_1$		$1772+22f_1$

or $\quad 1791+19.9f_1=1772+22f_1$

or $\quad f_1=9.0$

Hence the missing frequency is 9.

Illustration 9.26. The age distribution of the members of a certain children's club is as follows:

Age on last birthday (in years)	*Frequency*
4	5
5	9
6	18
7	35
8	42
9	32
10	15
11	7
12	3

There is a member A such that there are twice as many members older than A as there are younger than A. Estimate his age (in years up to two places of decimals).

Solution. Since age on last birthday is given

No. of persons who are in 4—5 group=5

No. of persons who are in 5—6 group=9

No. of persons who are in 6—7 group=18 and so on.

Now $\quad$ Age of A=size of $\frac{166}{3}$ th item

which clearly lies in 7--8 group whose frequency is 35.

$$\text{Age of } A=7+\frac{\left(\frac{166}{3}-32\right)}{35}(8-7)$$

$$=7.67.$$

Age	*Frequency*	*Cumulative Frequency*
4 5	5	5
5—6	9	14
6—7	18	32
7—8	35	67
8—9	42	109
9—10	32	141
10—11	15	156
11—12	7	163
12—13	3	166

Illustration 9.27. For a group of 100 saree weavers of Varanasi, the median and quartile earnings per week are Rs 88.6, Rs 86.0 and 91.8 respectively. The earnings for the group range between Rs 80 and Rs 100. Ten per cent of the group earn under Rs 84 week, 13 per cent earn Rs 94 and over and 6 per cent Rs 96 and over.

Put these data into the form of a frequency distribution and obtain an estimate of the mean wage.

Solution.

Earnings (Rs)	*No. of Weavers (f)*	*C.f.*	*Details*	*Mid-point*	*fx*
80—84	10	10	10 per cent 100=10	82.00	820.00
84—86	25−10=15	25	Q_1=86.0, ¼ of 100=25	85.00	1275.00
86—88.6	50—25=25	50	Med.=86.6.½ of 100=50	87.30	2182.50
88.6—91.8	75—50=25	75	Q_3=91.8,3/4 of 100=75	90.20	2255.00
91.8—94	12	87	13 per cent of 100=13,	92.90	1114.80
94—96	7	94	13−6=7	95.00	665.00
96—100	6	100	6 per cent of 100=6	98.00	588.00
Total	N=100				Σfx =8900.30

$$\text{Mean wage}=\frac{\Sigma fx}{N}=\frac{8900.30}{100}=\text{Rs } 89.00 \text{ approx.}$$

Illustration 9.28. Comment on the performance of the students

University Course of study	*Bombay*		*Calcutta*		*Madras*	
	% of pass	*No. of students in hundreds*	*% of pass*	*No. of students in hundreds*	*% of pass*	*No. of students in hundreds*
M.A.	71	3	82	2	81	2
M.Com.	83	4	76	3	76	3.5
B.A.	73	5	,3	6	74	4.5
B.Com.	74	2	76	7	58	2
B. Sc.	65	3	65	3	70	7
M. Sc.	66	3	60	7	73	2

in the three universities given below using simple and weighted averages.

Solution.

BOMBAY

Course of study	*% of pass* (x_1)	*No. of students in hundreds* (w_1)	$x_1' = x_1 - 75$	$w_1 x_1'$
M.A,	71	3	−4	−12
M. Com.	83	4	8	32
B.A.	73	5	−2	−10
B. Com.	74	2	−1	− 2
B. Sc.	65	3	−10	−30
M. Sc.	66	3	−9	−27
		20	−18	−49

Simple Arithmetic Average: $\bar{X}_1 = \frac{\Sigma x_1'}{N} + A = \frac{-18}{6} + 75 = 72.0$

Weighted Average; $\bar{X}_{1w} = \frac{\Sigma x_1' w_1}{\Sigma w} + A = \frac{-49}{20} + 75 = 72.55.$

CALCUTTA

Course of study	*% of pass* (x_2)	*No. of students in hundreds* (w_2)	$x_2' = x_2 - 75$	$w_2 x_2'$
M.A.	82	2	7	14
M. Com.	76	3	1	3
B.A.	73	6	−2	−12
B. Com.	76	7	1	7
B. Sc.	65	3	−10	−30
M. Sc.	60	7	−15	−105
		28	−18	−123

Simple Arithmetic Average $= \bar{X} = \frac{-18}{6} + 75 = 72.0$

Weighted Average $= \bar{X}_{2w} = \frac{-123}{28} + 75 = 70.61.$

MADRAS

Course of study	*% of pass* (x_3)	*No. of students in hundred* (w_3)	$x_3' = x_3 - 75$	$w_3 x_3'$
M.A.	81	2	6	12
M. Com.	76	3.5	1	3.5
B.A.	74	4.5	−1	4.5
B. Com.	58	2	−17	−34
B. Sc.	70	7	−5	−35
M. Sc.	73	2	−2	−4
		21	−18	−62

Simple Arithmetic Average $= \bar{X}_3 = \frac{-18}{6} + 75 = 72.0$

Weighted Average $= \bar{X}_{3w} = \frac{-62}{21} + 75 = 72..05.$

Comments 1. Since the simple arithmetic mean gives equal importance to all examinations and ignores their importance in terms of the number of students taking each examination, it is a misleading figure.

2. The weighted arithmetic mean, on the other hand, gives the correct percentage of passes of each university.

Illustration 9.29. The mean wage of 100 labourers working in a factory running two shifts of 60 and 40 workers respectively, is Rs 38. The mean wage of 60 labourers working in the morning shift is Rs 40. Find the mean wage of 40 labourers working in the evening shift.

Solution. We know

$$\bar{Y} = \frac{(N_1\bar{X}_1 + N_2\bar{X}_2)}{N_1 + N_2}$$

or $$38 = \frac{(60 \times 40) + (40 \times \bar{X}_2)}{100}$$

or $$40\bar{X}_2 = 3800 - 2400$$

$$\bar{X}_2 = 35$$

Hence the mean wage of 40 workers in the evening shift is Rs 35.

Illustration 9.30. Fifty students took up a test. The result of those who have passed the test is given below:

Marks	*No. of students*
4	8
5	10
6	9
7	6
8	4
9	3

If the avernge marks for all the fifty students were 5.16, find out the average marks of the students who failed.

Solution.

Total marks for all the 50 students $= 50 \times 5.16 = 258$

Total for 40 students who passed $= 237$

$\therefore$ Total for 10 students who failed $= 258 - 237 = 21$

Hence average of the students who failed $= \frac{21}{10} = 2.1$

Marks x	*f*	*f x*
4	8	32
5	10	50
6	9	54
7	6	42
8	4	32
9	3	27
	$N=40$	$\Sigma fx=237$

Illustration 9.31. Twenty per cent of the workers in a firm employing a total of 2,000 earn less than Rs 2 per hour, 440 earn from Rs 2 to 2.24 per hour, 24% earn from 2.25 to 2.49 per hour, 370 earn from Rs 2.50 to 2.74 per hour, 12% earn from Rs 2.75 to 2.99 per hour and the rest earn Rs 3 or more per hour. Set up a frequency table and calculate the modal wage.

Solution

Earnings per hour	*No. of workers*
Less than Rs 2.00	400
Rs 2.00 to Rs 2.24	440
Rs 2.25 to Rs 2.49	480
Rs 2.50 to Rs 2.74	370
Rs 2.75 to Rs 2.99	240
Rs 3.00 and more	70
Total	2,000

In the given data 2.25 to 2.49 is the modal class. The real limits of this class are 2.245 to 2.495.

Using the mode formula we get

$$\text{Mode}=l_1+\frac{f_1-f_0}{2f_1-f_0-f_2}\times i$$

$$=2.245+\frac{480-440}{960-440-370}\times 0.25$$

$$=2.245+\frac{40}{40+110}\times 0.25$$

$$=2.245+.07=\text{Rs } 2.315.$$

Illustration 9.32. Following is the distribution of marks obtained by 50 students in Mercantile Law.

Calculate the median marks. If 60 per cent of the students pass this test, find the minimum marks obtained by a pass candidate.

Marks	Number of Students
(More than)	
0	50
10	46
20	40
30	20
40	10
50	3

Solution.

Marks	f	c. f. (less than)
0—10	4	4
10—20	6	10
20—30	20	30
30—40	10	40
40—50	7	47
50—60	3	50

Median has $\frac{N}{2}$ *i.e.* $\frac{50}{2}$ or 25 items below it.

It lies in the group 20−30

Using formula

$$\text{Median} = 20 + \frac{25-10}{20} \times 10 = 27.5.$$

In order to find the minimum marks obtained by a pass candidate if 60% students pass, it is sufficient to obtain the value 40_{th} percentile of the above distribution, since 40 per cent students have obtained less marks.

Now

P_{40} has $40\left(\frac{N}{100}\right)$ *i.e.*, $\frac{40 \times 50}{100}$ or 20 items below it.

∴ P_{40} lies in the class 20−30. So by Formula (12),

$$P_{40} = l_1 + \frac{\frac{40N}{100} - c.f.}{f} \times i$$

$$= 20 + \frac{20-10}{20} \times 10 = 25$$

Hence minimum marks obtained by a pass candidate should be 25.

Illustration 9.33. The following data represent travel expenses (other than transportation) for 7 trips made during November by a salesman of a small firm.

Trip	*Days*	*Expense* *Rs*	*Expense per day* *Rs*
1	0.5	13.50	27.00
2	2.0	12.00	6.00
3	3.5	17.50	5.00
4	1.0	9.00	9.00
5	9.0	27.00	3.00
6	0.5	9.00	18.00
7	8.5	17.00	2.00
Total	25.0	105.00	70.00

An auditor criticized these expenses as excessive, asserting that the average expense per day is Rs 10 (Rs 70 divided by 7). The salesman replied that the average is only Rs 4.20 (Rs 105 divided by 25) and that in any event the median is the appropriate measure and is only Rs 3.00. The auditor rejoined that the arithmetic mean is the appropriate measure, but the median is Rs 6.00.

1. Explain the proper interpretation of each of the four averages mentioned.

2. Which average seems to you appropriate?

Solution 1.

(*i*) Rs 10 is the simple arithmetic mean of the 'Expenses per day' for seven trips. Here it has been (wrongly) assumed that each trip is of one day's duration.

(*ii*) Rs 4.20 is the weighted arithmetic mean of the 'Expenses per day.' In this the Expenses per day for each trip have been weighted by the duration of each trip.

Of the two Rs 4.20 is the correct arithmetic mean.

(*iii*) In computing Rs 3 as the median, the salesman has taken into account the time duration of each trip *i.e.* he has computed median by regarding 'Expenses per day' as the 'Value' and 'Days of each trip' as its frequency. This is a correct way of determining 'Median'.

(*iv*) The auditor, in the computation of the median also has committed the same mistake as that in the computation of the arithmetic mean. As such his calculation giving Rs 6 as the median is not correct.

2. From the point of business mean is more appropriate because it enables one to make an estimate of the cost involved in the 'trips' that the salesman has to undertake for purpose of business.

Illustration 9.34. A manufacturer of hand shovels is deciding what length handles to use. Studies of user preference reveal that the average, the median and the modal preferred length are all different. What are the implications of using each of these values? Which value would you decide?

Solution. If the average length is used it is quite possible that it may suit none. Those who prefer shorter than the average length, as also those who prefer longer than the average length will be put to discomfort. That is to say, a very large number (perhaps all) users should not take it.

In case median is used a similar situation would arise except that *half* the users would be uncomfortable because the handle is too short and another half would be uncomfortable because the handle is too long.

If the *mode* is used the largest number of persons would be satisfied.

Hence the modal length handles should be used.

Illustration 9.35. Twenty boats make 6 transatlantic trips each per year, and 10 boats make 4 trips per year. What is the average number of days for a 'turn around' (that is the time between consecutive departures from the same part? Take the year as 360 days for convenience.

Solution.

Trip	*Boats*	*Total trips*
6	20	120
4	10	40

$$\text{Average trips per boat} = \frac{160}{30} = 5.33$$

$$\text{Average number of days for a trip} = \frac{360}{16/3} = 67.5 \text{ days.}$$

or

This average may also be determined in another manner also *viz* the use of harmonic mean, since we are, in effect, averaging speeds.

Number of days needed for a trip are 60 and 90 days respectively for the groups consisting of 20 and 10 boats.

$$\text{Harmonic mean} = \frac{20+10}{\frac{1}{60}\times 20 + \frac{1}{90}\times 10}$$

$$=\frac{30}{\frac{1}{3}+\frac{1}{9}}$$

$$=\frac{30\times 9}{4}=67.5 \text{ days.}$$

Illustration 9.36. A given machine is assumed to depreciate 40 per cent in value in the first year, 25 per cent in the second year and 10 per cent per annum for the next three years, each percentage being calculated on the diminishing value. What is the average percentage depreciation, reckoned on the diminishing value for the five years?

Solution.

Average diminishing value

$$\sqrt[5]{(0.60\times 0.75\times 0.90\times 0.90\times 0.90)}$$

$$= \text{Antilog}\left[\frac{1}{5}\left(-1.7782-1.8751-1.9542-1.9542-1.9542\right)\right]$$

$$= \text{Antilog}\left[-1.9032\right] = 0.80$$

Thus depreciation is 20 per cent.

PROBLEMS

1. Define the various M's of central tendency. What purposes do their measurement serve?
2. Define geometric and harmonic mean and clearly explain their uses.
3. Show the relative positions of different averages in a moderately symmetrical series.
4. What do you mean by:
   (*a*) Quartiles,
   (*b*) Deciles.
   (*c*) Percentiles?
5. What are the qualities which an average must possess? Which of the averages you know, possess most of these qualities?
6. What do you mean by 'weights'? Why are they assigned? Point out a few cases in which weighted average should be used.
7. Differentiate between crude and corrected death rates.
8. The following are the monthly salaries, in rupees, of the employees in a branch bank. Calculate the arithmetic mean.

   10, 17, 29, 95, 95, 100, 100, 175, 250 and 750.

   (**Ans.** Arithmetic Mean of Salaries = Rs 202.1)

9. The expenditure of ten families in rupees are given below:

Family	A	B	C	D	E	F	G	H	I	J
Expenditure	30	70	10	75	500	8	42	250	40	36

Calculate the Arithmetic Average by (*a*) Direct Method and (*b*) Short-cut Method.
(Ans. A.M. of expenditure = Rs 106.1)

10. Eight coins were tossed together and the number of heads resulting was observed. The operation was performed 256 times and the frequencies that were obtained for the different values of x, the number of heads, are shown in the following table. Calculate mean, median and quartiles of the distribution of x.

x	0	1	2	3	4	5	6	7	8
Frequency	1	9	26	59	72	52	29	7	1

(Ans. A.M. of x = 3.97)

11. Find the mean of the following distribution:

Breadth in m.m.	19—21	22—24	25—27	28—30	31—33	34—36	37—39
	6	13	19	23	18	12	9

(Ans. A.M. = 29.11 m.m.)

12. The following table shows the number of persons employed in certain units of an industry. Find the average number of persons employed.

No. of persons below	20	20—30	30—50	50—100	100—200	200 and above
	2	5	6	3	2	2

(Ans. Average size is 71 men per unit.)

13. Calculate arithmetic mean for the following data:

Class interval	5—10	10—15	15—20	20—25	25—30	30—35	35—40	40—45
Frequency	6	5	15	10	5	4	3	2

(Ans. A.M. = 21.2)

14. From the following table calculate mean and median.

Crop cutting experimental data on plot yields of wheat

Yield (in lbs.)	*No. of Plots*	*Yield (in lbs.)*	*No. of Plots*
Over 0	216	Over 300	31
,, 60	210	,, 360	13
,, 120	156	,, 420	7
,, 180	98	,, 480	2
,, 240	57	Up to 540	216

(Ans. Mean yield = 189.4 lbs. per plot)

15. Find arithmetic mean, median and mode from the following:

Marks below	10	20	30	40	50	60	70	80
No. of students	15	35	60	84	96	127	198	250

(Ans. A.M. = 50.4)

16. The wages of 1060 employees range from Rs 300 to Rs 450. They are grouped in 15 classes with a common class interval of Rs. 10 Class frequencies from lowest to the highest are 6, 17, 35, 48, 65, 90, 131, 173, 155, 177, 75, 52, 21, 9, 6. Tabulate the data and calculate the mean wage.
(Ans. Mean wage = Rs 378.33. approx)

17. From the table given below, find the mean.

Salary per day	*No. of persons*	*Salary per day*	*No. of persous*
1— 5	7	26—30	18
6—10	10	31—35	10
11—15	16	36—40	5
16—20	32	41—45	1
21—25	24		

(Ans. Mean Salary=20.4 Rupees per day.)

18. Calculate arithmetic mean from the following data:

Temp. °C	—40 to —30	to —20	to —10	to 0	to 10	to 20	to 3o
No. of days	10	28	30	42	65	180	10

(Ans. A.M.=4.29°C)

19. The following figures represent the number of books issued at the counter of a commerce library in 11 different days. Calculate the median.

96, 180, 98, 75, 270, 20, 102, 100, 94, 75, 200.

(Ans. Median=98).

20. According to the census of 1941 the following are the population figures in 000' of the 1st 36 cities in India. 2488, 591, 437, 20, 213, 143, 1490. 407, 284, 176, 169, 181, 777, 387, 302, 213, 204, 153, 733, 391, 263, 176, 178, 142, 522, 360, 260, 193, 131, 92, 672, 258, 239, 160, 147, 151.

Find the median and quartiles.

(Ans. Median=216; Q_3=403.)

21. Find the mean, median, 20th percentile and the 8th decile of heights from the following dala.

Height	*Frequency*	*Height*	*Frequency*
5′—6″	1	5′—11″	1
5′—7″	2	6′—0″	2
5′—8″	4	6′—1″	1
5′—9″	3	6′—2″	1
5′—10″	2	6′—3″	1

(Ans. A.M.=5′—9.8″, Median=5′—9″. P_{20}=5′—7.8″)

22. The following distribution represents the number of minutes spent by a group of teenagers in going to movies. What is median?

Minutes/Week	*No. of teenagers*	*Minutes/Week*	*No. of teenagers*
0—99	27	400—499	58
100—199	42	500—599	32
200—299	65	600 or more	9
300—399	78		

(Ans. Median is 333.5 minutes per week)

23. The following table shows the age distributions of persons in a particular region:

Age (years)	Below 10	20	30	40	50	60	70	70 & over
No. persons (in 000's)	2	5	9	12	14	15	15.5	15.6

(*i*) Find the median age.

(*ii*) Why is the median a more suitable measure of a central tendency than the mean in this case?

(Ans. Median=27 years)

24. Calculate the median, 3rd decile and 20th percentile from the following data:

Central size	2.5	7.5	12.5	17.5	22.5
Frequency	7	18	25	30	20

(Ans. Median=15, D_3=13 and P_{20}=8.6)

25. Calculate median, mode, quartiles, 7th decile and 87th percentile from the following data:

Variate value	*Freq.*	*Variate value*	*Freq.*
7—10.99	5	31—34.99	12
11—14.79	9	35—38.99	7
15—18.99	13	39—42.99	5
19—22.99	21	43—46.99	3
23—26.99	17	47—50.99	2
27—30.99	15	51—54.99	1

(Ans. Median=24.6, Q_1=19.1, Q_3=318, D_7=30.2 and P_{87}=37.1)

26. In the frequency distribution of 100 families given below, the number of families corresponding to expenditure groups 20—40 and 60—80 are missing from the table. However the median is known to be Rs 50. Find the missing frequencies:

Expenditure: (Rs)	0—20	20—40	40—60	60—80	80—100
No. of families:	14	?	27	?	15

(Ans. 23 and 21).

27. The following data relate to the thickness of 25 bomb bases (inches)

0.134	0.144	0.135
0.141	0.114	0.155
0.140	0.147	0.138
0.168	0.159	
0.150	0.143	
0.150	0.123	
0.145	0.156	
0.167	0.128	
0.130	0.150	
0.130	0.132	
0.171	0.180	

Classify into groups 0.1095—0.1195, 0.1195–0.1295, etc. sketch the histogram and calculate the mean and median from the frequency distribution.

28. The following table gives the weekly wages in rupees in a certain commercial

organisation:

Weekly wages (Rs)	30—	32—	34—	36—	38—	40—
Frequency	2	9	25	30	49	62
Weekly wages (Rs)	42—	44—	46—	48—50		
Frequeney	39	20	11	3		

Calculate from the above data:

(*i*) the median and the third quartile wages,

(*ii*) the number of wage-earners receiving between Rs 37 and Rs 45 per week.

29. Find the mode of the following series:

Size	4	5	6	7	8	9	10	11	12	13	14	15	16	17	18	19
Frequency	40	48	52	56	60	63	57	55	50	52	41	57	63	52	48	40

(Ans. Mode=9)

Amend the following table and locate the median from the amended table:

Size	*Frequency*
10—15	10
16—17.5	15
17.5—20	17
20—30	25
30—35	28
35—40	30
40 and onward	40

30. Obtain the mode of the following data:

Monthly rent in (Rs)	*No. of families paying the rent*	*Monthly rent in (Rs)*	*No. of families paying the rent*
20—40	6	100—120	20
40—60	9	120—140	15
60—80	11	140—160	10
80—100	14	160—180	8
		180—200	7

(Ans. Modal rent=Rs 110.9)

31. Find out the median and mode from the following table:

No. of days absent	*No. of students*	*No. of days absent*	*No. of students*
less than 5	29	less than 30	644
,, ,, 10	244	,, ,, 35	650
,, ,, 15	465	,, ,, 40	653
,, ,, 20	582	,, ,, 45	655
,, ,, 25	635		

(Ans. Median=12.1 days
Mode=11.5 days.)

32. (*a*) From the data given below find the mode.

Ages	20-25	25-30	30-35	35-40	40-45	45-50	50-55	55-60
No. of persons	50	70	80	180	150	120	70	50

(*b*) If the mode and mean of a moderately asymmetrical series are respectively 16 inches and 20.2 inches, compute most probable median.
(Ans. Mode=42 years, Median=18.8 inches).

33. Following is the distribution of the size of certain farms selected at random from a district, Calculate the mode of distribution:

Central size of the farm in acres	10	20	30	40	50	60	70
No. of farms	7	12	17	29	31	5	3

(Ans. Mode=41.5 acres.)

34. Draw a histogram from the following data and measure the modal value:

Size class	*Frequency*	*Size class*	*Frequency*
0—10	5	50—60	10
10—20	11	60—70	8
20—30	19	70—80	6
30—40	21	80—90	3
40—50	16	90—100	1

35. Monthly incomes of the families are given below in rupees:
2000, 35, 400, 15, 40, 1500, 300, 6, 90, 250, 20, 12, 450, 10, 150, 8, 25, 30, 1200, 60.
Calculate the geometric mean and harmonic mean of the above series.
(Ans. G.M.=78.11, H.M.=26.07 rupees)

36. The following table gives weights of 31 persons in a sample inquiry. Calculate mean weight using (*i*) Geometric mean and (*ii*) Harmonic mean;

Weight in lbs.	130	135	140	145	146	148	149	150	157
No. of persons	3	4	6	6	3	5	2	1	1

(Ans. G.M.=142.5 lbs. H.M.=142.4 lbs.)

37. An investor buys Rs 1200 worth of shares in a company each month. During first five months, he bought the shares at a price of Rs 10, Rs 12, Rs 15, Rs 20 and Rs 24 per share. After 5 months what is the average price paid for the shares by him?
(Ans. Rs 14.63 per share)

38. Peter travelled by car for 4 days. He drove 10 hours each day. He drove: first day at the rate of 45 km per hour and fourth day at the rate of 37 km per hour. What was his average speed?

39. The price of a commodity was four times higher in 1970 than what it was a decade back. Find the average rate of growth of price of the commodity.
(Ans. Average percentage rise=17.4% per year.)

40. The price of certain articles becomes $1\frac{1}{2}$ time in first years, $1\frac{5}{8}$ times in the second year and $\frac{7}{9}$ times in the third year. What is the average change per year?
(Ans. Average change in price is 1.238 times.)

41. You take a trip which entails travelling 900 miles by train at an average speed of 60 m.p.h., 3000 miles by boat at an average of 25 m.p.h., 400 miles by plane at 350 m.p.h., and finally 15 miles by taxi at 25 m.p.h. What is

your average speed for the entire distance (4315 miles)?
(**Ans.** Average speed = 31.6 approx.)

42. Calculate the simple average and weighted average of the following items:

Items	68	85	101	102	108	110	112	113	124	128	143	146	151	153	172
Weights	1	46	31	1	11	7	23	17	9	14	2	4	6	5	2

Account for the difference in the two averages.
(**Ans.** Simple average = 121.1 approx.)

43. The following table gives the results of certain examinations of three universities in the year 1971. Which is the best university? Give reasons for any answer.

University examinations	*Percentage results in the university*		
	A	*B*	*C*
M.A.	80	75	70
M.Sc.	70	76	60
B.A.	65	80	60
B.Sc.	60	70	80
B. Com.	75	65	65

44. Arithmetic mean of a group of 100 items is 50 and of another group of 150 items is 100. What will be the mean of all the items?
(**Ans.** A.M. = 80)

45. Arithmetic mean of 98 items is 50. Two items 60 and 70 were left out at the time of calculations. What the correct mean of all the items?
(**Ans.** Correct A.M. = 50.3)

46. Arithmetic mean of 50 items were 100 and its median 95. At the time of calculations two items 180 and 90 were wrongly taken as 100 and 10. What is the correct mean and median?
(**Ans.** Correct A.M. = 103.2)

47. The following marks have been obtained in three papers of statistics in examination by 12 students. In which paper is the general level of the knowledge of the students highest? Give reasons.

A—36, 56, 41, 46, 54, 59, 55, 51, 52, 44, 37, 59
B—58, 54, 21, 51, 59, 46, 65, 31, 68, 41, 70, 36
C—65, 55, 26, 40, 30, 74, 45, 29, 85, 32, 80, 39.

(**Ans.** A.M. of paper *A* = 49.2 marks approx.
A.M. of paper *B* = 50 marks.
A.M of paper *C* = 50 marks.)

48. The following is the distribution of 136 individuals by 10 years age groups. Calculate the measure of central tendency, whieh will appropriately describe the distribution.

Age group	0-9	10-19	20-29	30-39	40-49	50-59	60-69	70 and over
No. of persons	48	26	27	11	13	4	3	4

(**Ans.** Average age, represented by median, is 17.2 years.)

49. In a shooting competition a person shoots 200 times a running object. If he shoots in front of the target it is taken as positive distance and if he shoots behind the target it is taken as negative distance. The distribution of distance missed (in cm.) by different shots are as follow:

Distance in cm	−30 to −20	−20 to −10	−10 to 0	0	0 to 10	10 to 20	20 to 30
No. of shots	5	15	30	90	43	9	8

Find the average distance by which (*i*) shot is ahead of the target (*ii*) shot is behind the target, (*iii*) shot missing the target misses it and (*iv*) a shot misses the target.

(Ans. (*i*) Average distance by which shot ahead of target is in front of the target = 9.17 cm.
(*ii*) Average distance by which a shot behind the target is behind the target = 10.0 cm.
(*iii*) Average distance by which shot which misses the target misses the target = 9.45 cm.
(*iv*) The average distance by which 'a shot' misses the target is 1050/200 = 5.25 cm.

50. From the following data of calculation of arithmetic mean find the missing item.

Housing rent in Rs	110	112	113	117	—	125	128	130
No. of houses	25	17	13	15	14	8	6	2

Mean rent = 115.86.

(Ans. The missing figure = Rs 120.)

51. Monthly expenditure for a group of families is as below:

Expenditure in Rs	100-200	200-300	300-350	350-400	400-500
No. of families	8	—	20	12	5

Median of expenditure is known to be as 317.5. Determine the number of families having expenditure between Rs 200 to Rs 300 per month.

(Ans. The number of families with expenditure Rs 200-300 is 15.)

52. Modal marks for a group of 47 students is 27. 5 students got marks between 0 to 10 and 15 students got marks between 20-30. Maximum marks in the test were 50, and 7 students got marks between 40-50. Tabulate the data in class interval of size 10 and calculate the missing frequencies.

(Ans. The missing frequency of the group 10—20 = 8 and the missing frequency of the group 30—40 = 12)

53. Weekly wages for a group of 100 persons are given below:

Wages in Rs	0-5	5-10	10-15	15-20	20-25
No. of persons	7	—	25	30	—

The 3rd decile for this group is Rs 11. Calculate the missing frequency.

(Ans Missing frequency of the group 20-25 = 20).

54. The following table gives the distribution of monthly income of 600 families in a certain city.

Monthly income (Rs)	*Number of families*
Below 75	60
75—150	170
150—225	200
225—300	60

300—375	50
375—450	40
450 and over	20

Draw a 'less than' and a 'more than' ogive curve for above data on the same graph. Read the median income and obtain the limits of income of central 50% of the observed frequencies.

55. Calculate the crude and standardized death rates from the following data:

Age group	*Population*	*Death rate per 1000*	*Standard age distribution*
0-10	400	40	600
10-20	1500	4	1000
20-60	2400	10	3000
60 and over	700	30	400

(Ans. Crude death rate = 13.4‰.
Standardized death rate = 14.0‰.)

56. Do you agree with the following?

(i) Rate for a certain commodity in the first week is 4 kilos for a rupee and in the second week is 8 kilos for a rupee. So the average price is (4+8)/2 = 6 kilos for a rupee.

(*ii*) Usually the attendance of B. Com. 1st year class in a college is 40 students per day. Therefore the total attendance for 100 working days is 4000.

(*iii*) An ordinary person consumes 30 g of salt per week. So 32 crores of persons living in India will consume 19.2 crore kg of salt in 5 months.
(1 month = 4 weeks)

(*iv*) The increase in the price of commodity *x* was 20%. Then the price decreased 25% and again increased 15%. So the resultant increase is (*i*) 15% and (*ii*) 10%.

(*v*) The rate of increase in the number of cows in India is greater than that of population. So the people of India are now getting more milk per head.

57. The table below shows the distribution of the number of processed articles per day per person and the rate of payment:

Daily No. of articles processed per person	*No. of persons processing*	*Rate of payment per article processed*
80—100	7	3.1
100—120	50	3.2
120—140	80	3.3
140—160	60	3.4
160—180	3	3.5

Calculate the rate of payment per person per article produced.

58. Draw the histogram and obtain graphically the mode from the following data:

Profit in Rs	5-9	10-14	15-9	20-24	25-29	30-34	35-39
No. of shops	8	18	27	21	10	28	8

59. Calculate graphically the value of median, quartiles and 35th percentile from the data given below:

Profit per shop	0-10	10-20	20-30	30-40	40-50	50-60
No. of shops	12	18	27	20	17	7

(**Ans.** Median $Q_1=17.5$ approx., $P_{35}=22$ approx., $Q_3=39$ approx.)

60. (*a*) The average rainfall for a week excluding Sunday was 0,50 inch. Due to heavy rainfall on Sunday the average for the week rose to 1.5 inches. How much rainfall was there on Sunday?

(*b*) A train runs 25 miles at a speed of 30 m.p.h. another 50 miles at a speed of 40 m.p.h., then due to repairs of the track, travels for 6 minutes at a speed of 10 m.p.h. and finally covers the remaining distance of 24 miles at a speed of 24 m.p.h. What is the average speed in miles per hour?

(*c*) The annual rates of growth of output of a factory in 5 years are 5.0, 7.5, 2.5, 5.0 and 10.0 per cent respectively. What is the compound rate growth of output per annum for the period?

(**Ans.** (*a*) 7.5 inches.
(*b*) Average speed = 31.41 m.p.h.
(*c*) Annual average growth = 5.9% per annum).

61. The population of a country was 300 million in 1951. It became 520 million in 1969. Calculate the percentage compound rate of growth per annum.

(**Ans.** Percentage compound rate of growth = 3.1%).

62. Calculate median and mode from the following data:

Class	0-7	7-14	14-21	21-28	28-35	35-42
Frequency	7	11	24	19	12	9

(**Ans.** Median = 40, Mode = 19.1).

63. An aeroplane covered a distance of 800 miles with four different speeds of 100, 200, 300 and 400 m p.h. for the first, second, third and fourth quarter of the distance respectively. Find the average speed in miles per hour.

(**Ans.** 192 m.p.h.)

64. Given below is the distribution of profits (in 000's rupees) earned by 94 book depots in a certain territory. Find the modal value.

Profits below	20	30	40	50	60	70	80	90
No. of Book Depots	5	14	27	48	68	83	91	94

(**Ans.** Mode = 47.3).

65. The following figures represent the number of books issued at the counter of a commerce college library on 12 different days, 96, 180, 98, 75, 270, 80 102, 100, 94. 75, 200, 600. Calculate the arithmetic mean, median and the mode of this data. Which of these would represent the above data best?

(**Ans** A M. = 114.2, Median = 99, Mode = 75).

66. Find the mean, median and mode from the following frequency distribution:

Output in units	*No. of workers*
300—309	9
310—319	20
320—329	24
330—339	38
340—349	48
350—359	27
360—369	17
370—379	6

(Ans. $\bar{X}$**=339.05,** Mode = 342.73, **Median=** 340.23**).**

67. From the following frequency table calculate mean, median and mode·

Monthly rent (in Rs)	*No. of families paying the rent*	*Monthly rent (in Rs)*	*No. of families paying the rent*
20—40	6	120—140	15
40—60	9	140—160	10
60—80	11	160—180	8
80—100	14	180—200	7
100—120	20		

(Ans. $\bar{X}$**=110, Median=110, Mode=110.9)**

Chapter 10

MEASURES OF DISPERSION

10.1. INTRODUCTION

The measures of central tendency or the 'averages' described in the last chapter give a number which is the typical value of the distribution. It is computed to see through the variability or *dispersion* of the individual values. But the dispersion is in itself a very important property of a distribution and needs to be measured by an appropriate statistics. The distributions *A* and *B* in Fig. 10.1 have the same mean, mode and median, but are quite different in their scatter about the mean. This scatter is brought out by the various measures of dispersion described in this chapter.

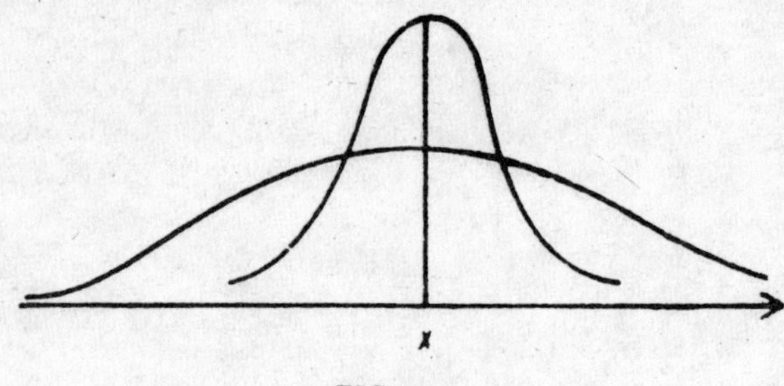

FIG. 10.1

The measures of dispersion and scatter are useful in many situations. The average income in a community is not an adequate index of the well being of a community because it glosses over the inequalities of the distribution of income. The measures of dispersion bring out this inequality. In engineering problems too the variability is an important concern. The amount of variability in dimensions of nominally identical components is critical in determining whether or not the components of a mass-produced item will be really interchangeable. The scatter in the life of light bulb is at times more important than the average life. Thus, the measures of dispersion are useful in:

(*a*) determining how representative the average is as a description of the data, as in the example of the average income of a community.

(*b*) Comparing two or more series with regard to their scatter, and

(*c*) designing a production control system which is based on the premise that if a process is under control, the variability it produces is same over a period of time. If the scatter produced by a process changes over time, it invariably means that something has gone wrong and needs to be corrected.

10.2 MEASURES OF DISPERSION

A measure of dispersion, or simply *dispersion* may be defined as a statistics signifying the extent of the scatteredness of items around a measure of central tendency.

A measure of dispersion may be expressed in an 'absolute form'; or in a 'relative form'. It is said to be in an absolute form when it states the actual amount by which the value of an item *on an average* deviates from a measure of central tendency. Absolute measures are expressed in concrete units *i.e.*, units in terms of which the data have been expressed *e.g.*, rupees, centimetres, kilograms etc., and are used to describe frequency distribution.

A relative measure of dispersion is a quotient by dividing the absolute measures by a quantity in respect to which absolute deviation has been computed. It is as such a pure number and is usually expressed in a percentage form. Relative measures are used for making comparisons between two or more distributions.

A measure of dispersion should possess all those characteristics which are considered essential for a measure of central tendency, *viz*.

(1) It should be based on all observations.

(2) Should be readily comprehensible.

(3) Should be fairly easily calculated,

(4) It should be affected as little as possible by fluctuations of sampling, and

(5) Be amenable to algebraic treatment.

The following are some more common measures of dispersion:

(*i*) the range, (*ii*) the semi-interquartile range or the quartile deviation, (*iii*) the mean deviation, and (*iv*) the standard deviation. Of these, the standard deviation is the best measure. We describe these measures in the following sections.

10.3. THE RANGE

The crudest measure of dispersion is the range of the distribution. The range of any series is the difference between the highest and the lowest values in the series. If the marks received in an examination taken by 248 students are arranged in ascending order, then the range will be equal to the difference between the highest and the lowest marks.

In a frequency distribution, the range is taken to be the difference between the lower limit of the class at the lower extreme of the distribution and the upper limit of the class at the upper extreme.

TABLE 10.1. WEEKLY EARNINGS OF LABOURERS IN FOUR WORKSHOPS OF THE SAME TYPE

Weekly earnings Rs	*No. of workers*			
	Workshop A	*Workshop B*	*Workshop C*	*Workshop D*
15—16	...	...	2	...
17—18	...	2	4	...
19 - 20	...	4	4	4
21 - 22	10	10	10	14
23—24	22	14	16	16
25—26	20	18	14	16
27—28	14	16	12	12
29—30	14	10	6	12
31—32	...	6	6	4
33—34	...	...	2	2
35—36	...	...	...	...
37—38	...	...	4	...
Total	80	80	80	80
Mean	25.5	25.5	25.5	25.5

Consider the data on weekly earnings of worker in four workshops given in Table 10.1. We note the following.

Workshop	*Range*
A	9
B	15
C	23
D	15

From the above figures it is clear that the greater the range, the greater is the variation of the values in the group.

The range is a measure of absolute dispersion and as such cannot be usefully employed for comparing the variability of two distributions expressed in different units. The amount of dispersion measured, say, in pounds, is not comparable with dispersion measured in inches. So the need of measuring relative dispersion arises. An absolute measure can be converted into relative measure if we divide it by some other value regarded as standard for the purpose. We may use the mean of the distribution or any other positional average as the standard.

For Table 10.1, the relative dispersion would be:

Workshop $A = \frac{9}{25.5}$ Workshop $C = \frac{23}{25.5}$

Workshop $B = \frac{15}{25.5}$ Workshop $D = \frac{15}{25.5}$

An alternate method of converting an absolute variation into a relative one would be to use the total of the extremes as the standard. This will be equal to dividing the difference of the extreme items by the total of the extreme items. Thus,

$$\text{Relative Dispersion} = \frac{\text{Difference of extreme items, } i.e. \text{ Range}}{\text{Sum of extreme items}}$$

The relative dispersion of the series is called the coefficient or ratio of dispersion. In our example of weekly earnings of workers taken above the coefficients would be:

$$\text{Workshop } A = \frac{9}{21+30} = \frac{9}{51}$$

$$\text{,, } \quad B = \frac{15}{17+32} = \frac{15}{49}$$

$$\text{,, } \quad C = \frac{23}{15+38} = \frac{23}{53}$$

$$\text{,, } \quad D = \frac{15}{19+34} = \frac{15}{53}$$

Merits and Limitations of Range

Merits. Of the various characteristics that a good measure of dispersion should possess, the range has only two *viz.*, (1) It is easy to understand, and (2) its computation is simple.

Limitations. Besides the aforesaid two qualities the range does not satisfy the other test of a good measure and hence it is often termed as a crude measure of dispersion.

The following are the limitations that are inherent in the range as a concept of variability:

(1) Since it is based upon two extreme cases in the entire distribution, the range may be considerably changed if either of the extreme cases happens to drop out, while the removal of any other case would not effect it at all.

(2) It does not tell anything about the distribution of values in the series relative to a measure of central tendency.

(3) It cannot be computed when distribution has open-end classes.

(4) It does not take into account the entire data. These can be illustrated by the following illustration. Consider the data given in Table 10.2.

TABLE 10.2. DISTRIBUTION HAVING THE SAME NUMBER OF CASES, BUT DIFFERENT VARIABILITY

Class	*No. of students*		
	Section A	*Section B*	*Section C*
0–10	...	...	...
10–20	1	...	...
20–30	12	12	19
30–40	17	20	18
40–50	29	35	16
50–60	18	25	18
60–70	16	10	18
70–80	6	8	21
80–90	11	...	...
90–100	...	...	...
Total	110	110	110
Range	80	60	60

The table is designed to illustrate three distributions with the same number of cases but different variability. The removal of two extreme students from section *A* would make its range equal to that of *B* or *C*. The greater range of *A* is not a description of the entire group of 110 students but of the two most extreme students only. Further, though sections *B* and *C* have the same range the students in section *B* cluster more closely around the central tendency of the group than they do in section *C*. Thus, the range fails, to reveal the greater homogeneity of *B* or the greater dispersion of *C*. Due to this defect, it is seldom used as a measure of dispersion.

Specific Uses of Range

In spite of the numerous limitations of the range as a measure of dispersion, there are circumstances when it is the most appropriate one:

1. In situations where the extremes involve some hazard for which preparation should be made, it may be more important to know the most extreme cases to be encountered than to know anything else about the distribution. An explorer, *e.g.* would like to know the lowest and highest temperatures on record in the region he is about to enter; or an engineer would like to know the maximum rainfall during 24 hours for the construction of a storm water drain.

2. In the study of prices of securities, range has a special field of activity. Thus to highlight fluctuations in the prices of shares or bullion it is a common practice to indicate the range over which the prices have moved during a certain period of time. This information besides being of use to the operators gives an indication of the stability of the bullion market, or that of the investment climate.

3. In statistical quality control the range is used as a measure of variation. We, *e.g.* determine the range over which variations in quality are due to random causes, which is made the basis for the fixation of control limits.

10.4. QUARTILE DEVIATION

Another measure of dispersion, much better than the range, is the semi-interquartile range, usually termed as 'Quartile Deviation.' As stated earlier, quartiles are the points which divide the array in four equal parts. More precisely, Q_1 gives the value of the item $\frac{1}{4}$th the way up the distribution and Q_3 the value of the item $\frac{3}{4}$th the way up the distribution. Between Q_1 and Q_3 are included half the total number of items. The difference between Q_1 and Q_3 includes only the central items but excludes the extremes. Since under most circumstances, the central half of the series tends to be fairly typical of all the items, the interquartile range (Q_3-Q_1) affords a convenient and often a good indicator of the absolute variability. The larger the interquartile range, the larger the variability.

Usually, one-half of the difference between Q_3 and Q_1 is used and to it is given the name of *Quartile Deviation* or semi-interquartile range. The interquartile range is divided by two for the reason that half of the interquartile range will, in a normal distribution, be equal to the difference between the median and any quartile. This means that 50% items of a normal distribution will lie within the interval defined by the median plus and minus the semi-interquartile range. Symbolically:

$$Q.D. = \frac{Q_3 - Q_1}{2} \tag{10.1}$$

Let us find quartile deviations for the weekly earnings of labour in the four workshop whose data is given in Table 10.1. The computations are as shown in Table 10.3.

As shown in the table Q.D. of workshop *A* is Rs 2.12 and Median Value in 25.3. This means that if the distribution is symmetrical the number of workers, whose wages vary between (25.3 – 2.1) Rs 23.2 and (25.3+2.1) Rs 27.4, shall be just half of the total cases. The other half of the workers will be more than Rs 2.1 removed from the median wage. As this distribution is not symmetrical the distance between Q_1 and the median Q_2 is not the same as between Q_3 and the median, hence the interval defined by median plus and minus semi-inter-quartile rage will not be exactly the same as given by the value of the two quartiles. Under such conditions the range between Rs 23.2 and Rs 27.4 will not include precisely 50% of the workers.

If quartile deviation is to be used for comparing the variability of any two series it is necessary to convert the absolute measure to a coefficient of quartile deviation. To do this the absolute measure is divided by the average size of the two quartile.

Symbolically:

Coefficient of quartile deviation

$$= \frac{Q_3 - Q_1}{Q_3 + Q_1} \tag{10.2}$$

Applying this to our illustration of four workshops the coefficients of Q.D. are as given below.

	A	*B*	*C*	*D*
Coefficient of quartile deviation=	$\frac{27.64-23.41}{27.64+23.41}$	$\frac{28-23.07}{28+23.07}$	$\frac{28.17-22.5}{28.17+22.5}$	$\frac{28.17-22.75}{28.17+22.75}$
$\frac{Q_3-Q_1}{Q_3+Q_1}$	=.083	=.097	=.112	=.106

Characteristics of quartile neviation. (1) The size of the Quar-

TABLE 10.3. CALCULATION OF QUARTILE DEVIATION

		Workshop A	*Workshop B*	*Workshop C*	*Workshop D*
Location of Q_2	$\frac{N}{2}$	$\frac{80}{2}=40$	$\frac{80}{2}=40$	$\frac{80}{2}=40$	$\frac{80}{2}=40$
	Q_2	$24.5+\frac{40-32}{20}\times 2$ $=24.5+.8$ $=25.3$	$24.5+\frac{40-30}{18}\times 2$ $=24.5+1.1$ $=25.61$	$24.5+\frac{40-36}{14}\times 2$ $=24.5+.57=25.07$	$24.5+\frac{40-34}{16}\times 2$ $=24.5+.75=25.25$
Location of Q_1	$\frac{N}{4}$	$\frac{80}{4}=20$	$\frac{80}{4}=20$	$\frac{80}{4}=20$	$\frac{80}{4}=20$
	Q_1	$22.5+\frac{20-10}{22}\times 2$ $=22.5+.91$ $=23.41$	$22.5+\frac{20-16}{14}\times 2$ $=22.5+.57$ $=23.07$	$20.5+\frac{20-10}{10}\times 2$ $=20.5+2$ $=22.5$	$22.5+\frac{20-18}{16}\times 2$ $=22.5+.25=22.75$
Location of Q_3	$\frac{3N}{4}$	$3\times\frac{80}{4}=60$	60	60	60
	Q_3	$26.5+\frac{60-52}{14}\times 2$ $=26.5+1.14$ $=27.64$	$26.5+\frac{60-48}{16}\times 2$ $=26.5+1.5$ $=28.0$	$26.5+\frac{60-50}{12}\times 2$ $=26.5+1.67=28.17$	$26.5+\frac{60-50}{12}\times 2$ $=26.5+1.67=28.17$
Quartile Deviation $\frac{Q_3-Q_1}{2}$		$\frac{27.64-23.41}{2}$ $\frac{4.23}{2}=$ Rs 2.12	$\frac{28-23.07}{2}=$ $\frac{4.93}{2}=$ Rs 2.46	$\frac{28.17-22.5}{2}=$ $\frac{5.67}{2}=$ Rs 2.83	$\frac{28.17-22.75}{2}$ $\frac{5.42}{2}=$ Rs 2.71

tile Deviation gives an indication about the uniformity or otherwise of the size of the items of a distribution. If the Q.D. is small it denotes large uniformity. Thus a coefficient of quartile deviation may be used for comparing uniformity or variation in different distributions.

(2) Quartile Deviation is not a measure of dispersion in the sense that it does not show the scatter around an average, but only a distance on scale. Consequently quartile deviation is regarded as a measure of partition.

(3) It can be computed when the distribution has open-end classes.

Limitations of quartile deviation. Except for the fact that its computation is simple and it is easy to understand a quartile deviation does not satisfy any other test of a good measure of variation

10.5. MEAN DEVIATION

A weakness of the measures of dispersion discussed above, based upon the range or a portion thereof, is that the precise size of most of the variants has no effect on the result. As an illustration, the quartile deviation will be the same whether the variates between Q_1 and Q_3 are concentrated just above Q_1 or they are spread uniformly from Q_1 to Q_3. This is an important defect from the viewpoint of measuring the divergence of the distribution from its typical value. The mean deviation is employed to answer the objection.

Mean deviation, also called average deviation, of a frequency distribution is the *mean of the absolute values of the deviation from some measure of central tendency*. In other words, mean deviation is the arithmetic average of the variations (deviations) of the individual items of the series from a measure of their central tendency.

We can measure the deviations from any measure of central tendency, but the most commonly employed ones are the median and the mean. The median is preferred because it has the important property that the *average deviation from it is the least*.

Calculation of the mean deviation then involves the following steps:

(*a*) Calculate the median (or the mean) Me (or $\bar{X}$).

(*b*) Record the deviations $|d| = |x - Me|$ of each of the items, ignoring the sign.

(*c*) Find the average value of deviations

$$\text{Mean Deviation} = \frac{\Sigma |d|}{N} \qquad (10.3)$$

Illustration 10.1. Calculate the mean deviation from the following data giving marks obtained by 11 students in a class test.

14, 15, 23, 20, 10, 30, 19, 18, 16, 25, 12.

$$\text{Median}=\text{Size of } \frac{11+1}{2} \text{ th item}$$

$$=\text{size of 6th item}=18.$$

Serial No.	*Marks*	$\mid x-\text{Median} \mid$ $\mid d \mid$
1	10	8
2	12	6
3	14	4
4	15	3
5	16	2
6	18	0
7	19	1
8	20	2
9	23	5
10	25	7
11	30	12
		$\Sigma \mid d \mid=50$

Mean deviation from Median

$$=\frac{\Sigma \mid d \mid}{N}$$

$$=\frac{50}{11}=4.54 \text{ marks.}$$

For grouped data, it is easy to see that the mean deviation is given by

$$\text{Mean Deviation, M.D.}=\frac{\Sigma f \mid d \mid}{\Sigma f} \qquad (10.4)$$

where $\mid d \mid = \mid x-\text{median} \mid$ for grouped discrete data, and $\mid d \mid = \mid M-\text{median} \mid$ for grouped continuous data with M as the mid-value of a particular group. The following examples illustrate the use of this formula.

Illustration 10.2. Calculate the mean deviation from the following data

Size of item	6	7	8	9	10	11	12
Frequency	3	6	9	13	8	5	4

Solution.

Size	*Frequency* f	*Cumulative frequency*	*Deviations from median* (9) $\lvert d \rvert$	$f \lvert d \rvert$
6	3	3	3	9
7	6	9	2	12
8	9	18	1	9
9	13	31	0	0
10	8	39	1	8
11	5	44	2	10
12	4	48	3	12
	48			60

Median = the size of $\frac{48+1}{2}$ = 24.5th item which is 9.

Therefore, deviations d are calculated from 9, *i.e.*, $\lvert d \rvert = \lvert x-9 \rvert$.

$$\text{Mean Deviation} = \frac{\Sigma f \lvert d \rvert}{\Sigma f}$$

$$= \frac{60}{48} = 1.25$$

Illustration 10.3. Calculate the mean deviation from the following data:

x	f
0—10	18
10–20	16
20—30	15
30—40	12
40—50	10
50–60	5
60—70	2
70–80	2

Solution.

This is a frequency distribution with continuous variable. Thus deviations are calculated from mid-values.

x	mid-value	f	Less than c.f.	Deviation from median | d |	f | d |
0—10	5	18	18	19	342
10–20	15	16	34	9	144
20–30	25	15	49	1	15
30—40	35	12	61	11	132
40—50	45	10	71	21	210
50—60	55	5	76	31	155
60—70	65	2	78	41	82
70—80	75	2	80	51	102
		80			1,182

$$\text{Median} = \text{the size of } \frac{80}{2} \text{ th item}$$

$$= 20 + \frac{6}{15} \times 10$$

$$= 24$$

and then, $\text{Mean deviation} = \frac{\Sigma f \mid d \mid}{\Sigma f}$

$$= \frac{1{,}182}{80} = 14.775.$$

Merits and Demerits of the Mean Deviation

Merits. 1. It is easy to understand.

2. As compared to standard deviation (discussed later) its computation is simple,

3. As compared to standard deviation it is less affected by extreme values,

4. Since it is based on all values in the distribution it is better than range or quartile deviation.

Demerits. 1. It lacks those algebraic properties which would facilitate its computation and establish its relation to other measures,

2. and, as such, it is not suitable for further mathematical processing.

10 6. COEFFICIENT OF MEAN DEVIATION

The coefficient or relative dispersion is found by dividing the mean deviation by that measure of central tendency about which deviations were recorded. Thus,

$$\text{Coefficient of M.D.} = \frac{\text{Mean Deviation}}{\text{Mean}} \quad (10.5)$$

(when deviations were recorded from the mean)

or
$$= \frac{\text{M.D.}}{\text{Median}} \quad (10.6)$$

(when deviations were recorded from the median)

Applying the above formula to Illustration 10.3.

$$\text{Coefficient of Mean Deviation} = \frac{14.775}{24}$$

$$= .616$$

10.7. STANDARD DEVIATION

By far the most universally used and the most useful measure of dispersion is the standard deviation or *root mean-square deviation* about the mean. We have seen that all the methods of measuring dispersion so far discussed are not universally adopted for want of adequacy and accuracy. The range is not satisfactory as its magnitude is determined by most extreme cases in the entire group. Further, the range is notable because it is dependent on the item whose size is largely a matter of chance. Mean deviation method is also an unsatisfactory measure of scatter, as it ignores the algebraic signs of deviation. We desire a measure of scatter which is free from these shortcomings. To some extent standard deviation is one such measure.

The calculation of standard deviation differs in the following respects from that of mean deviation. *First*, in calculating standard deviation, the deviations are squared. This is done so as to get rid of negative signs without committing algebraic violence. Further, the squaring of deviations provides added weight to the extreme items, a desirable feature for certain types of series.

Secondly, the deviations are always recorded from the arithmetic mean. because, although the sum of deviations is the minimum from the median, the sum of squares of deviations is minimum when deviations are measured from the arithmetic average. The deviation from $\bar{x}$ are represented by d.

Thus, standard deviation, σ (sigma) is defined as the square root of the mean of the squares of the deviations of individual items from their arithmetic mean.

$$\sigma = \sqrt{\frac{\Sigma(x - \bar{x})^2}{N}} \quad (10.7)$$

For grouped data (discrete variables)

$$\sigma=\sqrt{\frac{\Sigma f(x-\bar{x})^2}{\Sigma f}} \tag{10.8}$$

and, for grouped data (continuous variables)

$$\sigma=\sqrt{\frac{\Sigma f(M-\bar{x})}{\Sigma f}} \tag{10.9}$$

where M is the mid-value of the group.

The use of these formulae is illustrated by the following examples.

Illustration 10.4. Compute the standard deviation for following data:

11, 12, 13, 14, 15, 16, 17, 18, 19, 20, 21.

Solution. Here formula (10.7) is appropriate. We first calculate the mean as $\bar{x}=\Sigma x/N=176/11=16$, and then calculate the deviations as below:

x	$(x-\bar{x})$	$(x-\bar{x})^2$
11	−5	25
12	−4	16
13	−3	9
14	−2	4
15	−1	1
16	0	0
17	+1	1
18	+2	4
19	+3	9
20	+4	16
21	+5	25
176		110

Thus by formula (10.7),

$$\sigma=\sqrt{\frac{110}{11}}=\sqrt{10}=3.16$$

Illustration 10.5. Find the standard deviation of the data in the following distribution:

x	f
12	4
13	11
14	32
15	21
16	15
17	8
18	5
20	4

Solution. For this discrete variable grouped data, we use formula 8. Since for calculation of $\bar{x}$, we need Σfx and then for σ we need $\Sigma f(x-\bar{x})^2$, the calculations are conveniently made in the following format

x	f	fx	$d=x-\bar{x}$	d^2	fd^2
12	4	48	−3	9	36
13	11	143	−2	4	44
14	32	448	−1	1	32
15	21	315	0	0	0
16	15	240	1	1	15
17	8	136	2	4	32
18	5	90	3	9	45
20	4	80	5	25	100
	100	1500			304

Here $\bar{x}=\Sigma fx/\Sigma f=1500/100=15$,

and $\sigma=\sqrt{\dfrac{\Sigma fd^2}{\Sigma f}}=\sqrt{\dfrac{304}{100}}=\sqrt{3.04}=1.74$

Illustration 10.6. Calculate the standard deviation of the following data

Class	*Frequency*
1—3	1
3—5	9
5—7	25
7—9	35
9—11	17
11—13	10
13—15	3

Solution. This is an example of continuous frequency series and formula 10.9 appropriate.

Class	*Mid-point* x	*Frequency* f	fx	*Deviation of mid-point x from Mean 8*	*Squared deviation* d^2	*Squared deviation times frequency*
1−3	2	1	2	−6	36	36
3−5	4	9	36	−4	16	144
5−7	6	25	150	−2	4	100
7−9	8	35	280	0	0	0
9—11	10	17	170	2	4	68
11—13	12	10	120	4	16	160
13−15	14	3	42	6	36	108
		100	800			616

First the mean is calculated as

$$\bar{x}=\Sigma fx/\Sigma f=800/100=8.0$$

Then the deviations are obtained from 8.0. The standard deviation

$$\sigma=\sqrt{\frac{\Sigma f(M-\bar{x})^2}{\Sigma f}}$$

$$\sigma=\sqrt{\frac{\Sigma fd^2}{\Sigma f}}=\sqrt{\frac{616}{100}}=2.48$$

10.8. CALCULATION OF STANDARD DEVIATION BY SHORT-CUT METHOD

The three illustrations worked out above have one common simplifying feature, namely $\bar{x}$ in each turned out to be an integer, thus, simplifying calculations. In most cases, it is very unlikely that it will turn out to be so. In such cases, the calculation of d and d^2 becomes quite involved. Short-cut methods have consequently been developed. These are on the same lines as those for calculation of mean itself.

In the short-cut method, we calculate deviations x' from an assumed mean A. Then,

For ungrouped data

$$\sigma=\sqrt{\frac{\Sigma x'^2}{N}-\left(\frac{\Sigma x'}{N}\right)^2} \quad (10.10)$$

and for grouped data

$$\sigma=\sqrt{\frac{\Sigma fx'^2}{\Sigma f}-\left(\frac{fx'}{\Sigma f}\right)^2} \quad (10.11)$$

This formula is valid for both discrete and continuous variables. In case of continuous variables x in the equation $x'=x-A$ stands for the mid-value of the class in question.

Note that the second term in each of the formulae is a correction term because of the difference in the values of A and $\bar{x}$. When A is taken as $\bar{x}$ itself, this correction is automatically reduced to zero. Illustrations 10.7 to 10.11 explain the use of these formulae.

Illustration 10.7. Compute the standard deviation by the short method for the following data:

11, 12, 13, 14. 15, 16, 17, 18, 19, 20, 21.

Solution. Let us assume that $A=15$

x	x' $(x-15)$	x'^2
11	−4	16
12	−3	9
13	−2	4

14	−1	1
15	0	0
16	1	1
17	2	4
18	3	9
19	4	16
20	5	25
21	6	36
$N=11$	$\Sigma x'=11$	$\Sigma x'^2=121$

$$\sigma=\sqrt{\frac{\Sigma x'^2}{N}-\left(\frac{\Sigma x'}{N}\right)^2}$$

$$=\sqrt{\frac{121}{11}-\left(\frac{11}{11}\right)^2}$$

$$=\sqrt{11-1}$$

$$=\sqrt{10}$$

$$=3.16.$$

Another method. If we assume A as zero, then the deviation of each item from the assumed mean is the same as the value of item itself. Thus 11 deviates from assumed mean of zero by 11, 12 deviates by 12, and so on. As such, we work with deviations without having to compute them, and the formula takes the following shape:

x	x^2
11	121
12	144
13	169
14	196
15	225
16	256
17	289
18	324
19	361
20	400
21	441
176	2.926

$$\sigma=\sqrt{\frac{\Sigma x^2}{N}-\left(\frac{\Sigma x}{N}\right)^2}$$

$$=\sqrt{\frac{2926}{11}-\left(\frac{176}{11}\right)^2}$$

$$=\sqrt{266-256}$$

$$=3.16$$

Illustration 10.8. Calculate the standard deviation of the following data by short method.

Person	1	2	3	4	5	6	7
Monthly income (Rupees)	300	400	420	440	460	480	580

Solution. In this data, the values of the variable are very large making calculations cumbersome. It is advantageous to take a common factor out. Thus, we use $x'=\frac{x-A}{20}$. The standard deviation is calculated using x' and then the true value of σ is obtained by multiplying back by 20. The effective formula then is

$$\sigma=C\times\sqrt{\frac{\Sigma x'^2}{N}-\left(\frac{\Sigma x'}{N}\right)^2}$$

where C represents common factor.

Using $x'=(x-420)/20$

x	*Deviation from Assumed mean* $(x-420)$	x'	x'^2
300	−120	−6	36
400	−20	−1	1
420	0	−7	—
440	20	1	1
460	40	2	4
480	60	3	9
580	160	8	64
		+14	
$N=7$		7	115

$$\sigma=20\times\sqrt{\frac{\Sigma x'^2}{N}-\left(\frac{\Sigma x'}{N}\right)^2}$$

$$=20\sqrt{\frac{115}{7}-\left(\frac{7}{7}\right)^2}$$

$$=78.56$$

Illustration 10.9. Calculate standard deviation from the following data:

Size	6	9	12	15	18
Frequency	7	12	19	10	2

Solution.

x	Frequency f	Deviation from assumed Mean 12	Deviation divided by common factor 3 x'	x' times frequency fx'	x'^2 times frequency fx'^2
6	7	−6	−2	−14	28
9	12	−3	−1	−12	12
12	19	0	0		
15	10	3	1	10	10
18	2	6	2	4	8
	$N=50$			$\Sigma fx'=-12$	$\Sigma fx'^2=58$

Since deviations have been divided by a common factor, we use

$$\sigma=C\sqrt{\frac{\Sigma fx'^2}{N}-\left(\frac{\Sigma fx'}{N}\right)^2}$$

$$=3\sqrt{\frac{58}{50}-\left(\frac{-12}{50}\right)^2}$$

$$=3\sqrt{1.1600-.0576}$$

$$=3\times 1.05=3.15.$$

Illustration 10.10. Obtain the mean and standard deviation of the first N natural numbers, *i.e.* of 1, 2, 3, ..., $N-1$, N.

Solution. Let x denote the variable which assumes the values of the first N natural numbers.

Then

$$\bar{x}=\frac{\sum\limits_{1}^{N} x}{N}=\frac{\frac{N(N+1)}{2}}{N}=\frac{N+1}{2}$$

because $\sum\limits_{1}^{N} x=1+2+3+......+(N-1)+N=\dfrac{N(N+1)}{2}$

To calculate the standard deviation σ, we use 0 as the assumed mean A. Then

$$\sigma=\sqrt{\frac{\Sigma x^2}{N}-\left(\frac{\Sigma x}{N}\right)^2}$$

But $\Sigma x^2=1^2+2^2+3^2+......(N-1)^2+N^2=\dfrac{N(N+1)(2N+1)}{6}$

Therefore,

$$\sigma=\sqrt{\frac{N(N+1)(2N+1)}{6N}-\frac{N^2(N+1)^2}{4N^2}}$$

$$=\sqrt{\frac{(N+1)}{2}\left[\frac{2N+1}{3}-\frac{N+1}{2}\right]}$$

$$=\sqrt{\frac{(N+1)\,(N-1)}{12}}$$

Thus for first 11 natural numbers

$$\bar{x}=\frac{11+1}{2}=6,$$

and

$$\sigma=\sqrt{\frac{(11+1)\,(11-1)}{12}}$$

$$=\sqrt{10}=3.16$$

Illustration 10.11.

	Mid-point x	*Frequency*	*Step deviation from class of assumed mean* x'	*Step deviation time frequency* fx'	*Squared step deviation times frequency* fx'^2
0—10	5	18	−2	−36	72
10—20	15	16	−1	−16	16
				——	
20—30	25	15	0	−52	
				——	
30—40	35	12	1	12	12
40—50	45	10	2	20	40
50—60	55	5	3	15	45
60—70	65	2	4	8	32
70—80	75	1	5	5	25
		79		60	242
				−52	
				——	
				$\Sigma fx'=8$	

Solution.

Since the deviations are from assumed mean and expressed in terms of class-interval units.

$$\sigma=i\times\sqrt{\frac{\Sigma fx'^2}{N}-\left(\frac{\Sigma fx'}{N}\right)^2}$$

$$=10\times\sqrt{\frac{242}{79}-\left(\frac{8}{79}\right)^2}$$

$$=10\times1.75=17.5.$$

10.9. COMBINING STANDARD DEVIATIONS OF TWO DISTRIBUTIONS

If we were given two sets of data of N_1 and N_2 items with means $\bar{x}_1$ and $\bar{x}_2$ and standard deviations σ_1 and σ_2 respectively, we can obtain the mean and standard deviation $\bar{x}$ and σ of the combined distribution by the following formulae:

$$\bar{x}=\frac{N_1\bar{x}_1+N_2\bar{x}_2}{N_1+N_2} \tag{10.12}$$

and

$$\sigma=\sqrt{\frac{N_1\sigma_1^2+N_2\sigma_2^2+N_1(\bar{x}-\bar{x}_1)^2+N_2(\bar{x}-\bar{x}_2)^2}{N_1+N_2}} \tag{10.13}$$

Illustration 10.12. Mean and standard deviations of two distributions of 100 and 150 items are 50, 5 and 40, 6 respectively. Find the standard deviation of all the 250 items taken together:

Solution. Combined mean

$$\bar{x}=\frac{N_1\bar{x}_1+N_2\bar{x}_2}{N_1+N_2}$$

$$=\frac{100\times 50+150\times 40}{100+150}=44$$

Combined standard deviation

$$\sigma=\sqrt{\frac{N_1\sigma_1^2+N_2\sigma_2^2+N_1(\bar{x}-\bar{x}_1)^2+N_2(\bar{x}-\bar{x}_2)^2}{N_1+N_2}}$$

$$=\sqrt{\frac{100\times(5)^2+150(6)^2+100(44-50)^2+150(44-40)^2}{100+150}}$$

$=7.46.$

Illustration 10.12. A distribution consists of three components with 200, 250, 300 items having mean 25, 10 and 15 and standard deviation 3, 4 and 5. Find the standard deviation of the combined distribution.

Solution. In the usual notations, we are given here

$N_1=200,\ N_2=250,\ N_3=300,$

$\bar{x}_1=25,\ \bar{x}_2=10,\ \bar{x}_3=15$

The formulae (12) and (13) can easily be extended for combination of three series as

$$\bar{x}=\frac{N_1\bar{x}_1+N_2\bar{x}_2+N_3\bar{x}_3}{N_1+N_2+N_3}$$

$$=\frac{200\times 25+250\times 10+300\times 15}{200+250+300}$$

$$=\frac{12000}{750}=16$$

and $\sigma=\sqrt{\dfrac{N_1\sigma_1^2+N_2\sigma_2^2+N_3\sigma_3^2+N_1(\bar{x}-\bar{x}_1)^2+N_2(\bar{x}-\bar{x}_2)+N_3(\bar{x}-\bar{x})^2}{N_1+N_2+N_3}}$

$$=\sqrt{\frac{200\times 9+250\times 16+300\times 25+200\times 81+250\times 36+300\times 1}{200+250+300}}$$

$$=\sqrt{51.73}=7.19$$

10.10. COMPARISON OF VARIOUS MEASURES OF DISPERSION

The range is the easiest to calculate measure of dispersion, but since it depends on extreme values, it is extremely sensitive to the size of the sample, and to the sample variability. Infact, as the sample size increases the range increases dramatically, because more the items one considers more likely it is that some item will turn up which is larger than the previous maximum or smaller than the previous minimum. So, it is, in general, impossible to interpret properly the significance of a given range unless the sample size is constant. It is for this reason that there appears to be only one valid application of the range, namely in statistical quality control where the same sample size is repeatedly used, so that comparison of ranges are not distorted by differences in sample size.

The quartile deviations and other such positional measures of dispersions are also easy to calculate but suffer from the disadvantage that they are not amenable to algebraic treatment. Similarly the mean deviation is not suitable bacause we cannot obtain the mean deviation of a combined series from the deviations of component series. But it is easy to interpret and easier to calculate than the standard deviation.

The standard deviation of a set of data, on the other hand, is one of the most important statistics describing it. It lends itself to rigorous algebraic treatment, is rigidly defined and is based on all observations. It is, therefore, quite insensitive to sample size (provided the size is 'large enough') and is least affected by sampling variations.

We will see in later chapters that it is used extensively in testing of hypothesis about population parameters based on sampling statistics.

Infact, the standard deviations has such stable mathematical properties that it is used as a standard scale for measuring deviations from the mean. If we are told that the performance of an individual is 10 points better than the mean, it really does not tell us enough, for 10 points may or may not be a large enough difference to be of significance. But if we know that the σ for the score is only 4 points, so that on this scale, the performance is 2.5σ better than the mean, the statement becomes meaningful. As we shall see later, this indicates an extremely good performance. This sigma scale is a very commonly used scale for measuring and specifying deviations which immediately suggest the significance of the deviation.

The only disadvantages of the standard deviation lies in the amount of work involved in its calculation, and the large weight it attaches to extreme values because of the process of squaring involved in its calculations.

10.11. VARIANCE AND COEFFICIENT OF VARIATION

The square of standard deviation, namely σ^2, is termed as *variance* and is more often specified than the standard deviation. Clearly, it has the same properties as standard deviation.

As is clear, the standard deviation σ or its square, the variance, can not be very useful in comparing two series where either the units are different or the mean values are different. Thus, a σ of 5 on an exam where the mean score is 30 has an altogether different meaning than on an exam where the mean score is 90. Clearly the variability in the second exam is much less. To take care of this problem, we define and use a *coefficient of variation*, *V*,

$$V = \frac{\sigma}{\bar{x}} \times 100$$

expressed as percentage.

Illustration 10.14. The following are the scores of two batsmen *A* and *B* in a series of innings:

A	12	115	6	73	7	19	119	36	84	29
B	47	12	76	42	4	51	37	48	13	0

Who is the better run-getter? Who is more consistent?

Solution. In order to decide as to which of the two batsman, *A* and *B*, is the better run-getter, we should find their batting averages. The one whose average is higher will be considered as a better batsman.

To determine the consistency in batting we should determine the coefficient of variation. The less this coefficient the more consistent will be the player.

A			*B*		
Score x	*x*	x^2	*Scores x*	*x*	
12	—38	1,444	47	14	196
115	+65	4,225	12	--21	441
6	—44	1.936	76	43	1,849
73	+23	529	42	9	81
7	- 43	1,849	4	—29	841
19	—31	961	51	18	324
119	+69	4,761	37	4	16
36	—14	196	48	15	225
84	+34	1,156	13	—20	400
29	—21	441	0	—33	1,089
$\Sigma x=500$		17,498	$\Sigma x=330$		5,462

Batsman *A*:

$$\bar{x}=\frac{500}{10}=50.$$

$$\sigma=\sqrt{\frac{17{,}498}{10}}$$

$$=41.83.$$

$$V=\frac{41.83\times 100}{50}$$

$$=83.66 \text{ per cent.}$$

Batsman *B*

$$\bar{x}=\frac{330}{10}=33$$

$$\sigma=\sqrt{\frac{5{,}462}{10}}$$

$$=23.37.$$

$$V=\frac{23.37}{33}\ 100$$

$$=70.8 \text{ per cent.}$$

A is a better batsman since his average is 50 as compared to 33 of *B*. But *B* is more consistent since the variation in his case is 70.8 as compared to 83.66 of *A*.

Illustration 10.15. The following table gives the age distribution of students admitted to a college in the years 1914 and 1918. Find which of the two groups is more variable in age.

Age	*Number of students in*	
	1914	*1918*
15	—	1
16	1	6
17	3	34
18	8	22
19	12	35
20	14	20
21	13	7
22	5	19
23	2	3

24	3	—
25	1	—
26	—	—
27	1	—

Solution.

Age	Assumed Mean-21 1914				Assumed Mean-19 1918			
	f	x'	fx'	fx'^2	f	x'	fx	fx'^2
15	0	−6	0	0	1	−4	−4	16
16	1	−5	−5	25	6	−3	−18	54
17	3	−4	−12	48	34	−2	−68	136
18	8	−3	−24	72	22	−1	−22	22
19	12	−2	−24	48	35	0	−112	
20	14	−1	−14	14	20	1	20	20
21	13	0	−79		7	2	14	28
22	5	1	5	5	19	3	57	171
23	2	2	4	8	3	4	12	48
24	3	3	9	27				
25	1	4	4	16				
26	0	5	0	0				
27	1	6	6	36				
	63		+28	299	147		+103	495
			−51				−9	

1914 Group $\sigma=\sqrt{\frac{\Sigma fx'^2}{N}-\left[\frac{\Sigma(fx')}{N}\right]^2}$

$$=\sqrt{\frac{299}{63}-\left(\frac{-51}{63}\right)^2}$$

$$=\sqrt{4.476-.655}$$

$$=\sqrt{4.091}$$

$$=2.02.$$

$$\bar{x}=21+\left(\frac{-51}{63}\right)$$

$$=21-.8$$

$$=20.2.$$

$$V=\frac{2.02}{20.2}\times 100$$

$$=\frac{202}{20.2}=10$$

1918 Group:

$$\sigma=\sqrt{\frac{495}{147}-\left(\frac{-9}{147}\right)^2}$$

$$=\sqrt{3.3673-.0037}$$

$$=\sqrt{3.3636}$$

$$=1.834.$$

$$\bar{x}=19+\left(\frac{-9}{147}\right)$$

$$=19-.06=18.94.$$

$$V=\frac{1.834}{18.94}\times 100.$$

$$=9.68.$$

The coefficient of variation of the 1914 group is 10 and that of the 1918 group 9.68. This means that the 1914 group is more variable, but only barely so. We will later learn to evaluate the significance of these differences.

Illustration 10.16. You are supplied the following data about height of boys and girls studying in a college.

	Boys	Girls
Number	72	38
Average height (inches)	68	61
Variance of distribution	9	4

You are required to find out:

(*a*) In which sex, boys or girls, is there greater variability in individual heights.

(*b*) Common average height in boys and girls.

(*c*) Standard deviation of the height of boys and girls taken together.

(*d*) Combined variability.

Solution.

(*a*) C.V. of boys height $=\frac{\sigma_1}{\bar{x}_1}\times 100=\frac{\sqrt{9}}{68}\times 100$

$$=4.41\%$$

C.V. of girls height $=\frac{\sigma_2}{\bar{x}_2}\times 100=\frac{\sqrt{4}}{61}\times 100=3.28\%.$

Thus there is a greater variability in the height of boys than that of the girls, (*b*) Height of boys and girls combined is

$$\bar{x}_{12}=\frac{N_1\bar{x}_1+N_2\bar{x}_2}{N_1+N_2}=\frac{72\times 68+38\times 61}{72+38}=\frac{7214}{110}$$

$$=65.58 \text{ inches approx.}$$

(*c*) The combined standard deviation may be calculated by applying the following formula:

$$\sigma_{12}^2=\frac{N_1\sigma_1^2+N_2\sigma_2^2}{N_1+N_2}+\frac{N_1(\bar{x}-\bar{x}_1)^2+N_2(\bar{x}-\bar{x}_2)^2}{N_1+N_2}$$

$$=\frac{72\times 9+38\times 4}{72+38}+\frac{72(65.58-68)^2+38(65.58-61)^2}{72+38}$$

$$=\frac{2018.794}{110}=18.35$$

$$\sigma_{12}=4.28 \text{ inches.}$$

(*d*) Combined variability $=\frac{\sigma}{\bar{x}}\times 100=\frac{4.28}{65.58}\times 100=6.53$

Illustration 10.17. The following table gives weight in pounds of fat bullocks and fat sheep.

Fat bullocks (weight in lbs.)	*Number*	*Fat sheep (weight in lbs.)*	*Number*
850–900	2	150–175	8
900–950	24	175–200	30
950–1000	45	200–225	59
1000–1050	120	225–250	70
1050–1100	110	250–275	98
1100–1150	140	275–300	60
1150–1200	66	300–325	37
1200–1250	42	325–350	23
1250–1300	20	350–375	15
1300–1350	15	375–400	5

Determine if the bullocks or the sheep are more variable in weight. Let the assumed averages be 1075 and 262.5 respectively.

BULLOCKS

Class	*Mid-point* (x)	$x'=\frac{x-1075}{50}$	f	fx'	fx'^2
850—900	875	—4	2	—8	32
900—950	925	—3	24	—72	216
950—1000	975	—2	45	—90	180
1000—1050	1025	—1	120	—120	120
				—290	
1050—1100	1075	0	110		0
1100—1150	1125	1	140	140	140
1150—1200	1175	2	66	132	264
1200—1250	1225	3	42	126	378
1250—1300	1275	4	20	80	320
1300 1350	1325	5	15	75	375
			584	+263	2025

$$\bar{x}=1075+\frac{263}{584}\times 50=1075+22.5171$$

$$=1097.5171$$

$$\sigma=\sqrt{\frac{2025}{584}-\left(\frac{263}{584}\right)^2}\times 50$$

$$=\sqrt{3.4675-0.2028}\times 50$$

$$=\sqrt{3.2647}\times 50=1.807\times 50=90.34$$

$$V=\frac{90.34}{1097.52}\times 100=8.23$$

SHEEP

Class	*Mid-point* (x)	$x'=\frac{x-262.5}{25}$	f	fx'	fx'^2
150—175	162.5	—4	8	—32	128
175—200	187.5	—3	30	—90	270
200—225	212.5	—2	59	—118	236
225—250	237.5	—1	70	—70	70
				—310	
250—275	262.5	0	98		0
275—300	287.5	1	60	60	60
300—325	312.5	2	37	74	148
325—350	337.5	3	23	69	207
350—375	362.5	4	15	60	240
375—400	387.5	5	5	25	125
			405	—22	1484

$$\bar{x} = 262.5 + \frac{-22}{405} \times 25 = 262.5 - 1.35$$
$$= 261.15$$
$$\sigma = \sqrt{\frac{1484}{405} - \left(\frac{-22}{405}\right)^2} \times 25$$
$$= \sqrt{3.66 - .0029} \times 25$$
$$= 1.91 \times 25 = 47.75$$
$$V = \frac{47.75}{261.15} \times 100$$
$$= 18.25.$$

Thus, even though the σ of sheep is much smaller, the variability is much larger.

Illustration 10.18. In a co-educational college boys and girls formed separate groups on the foundation day when every one had to put in physical labour. Compute standard deviation for boys and girls separately and for the combined group. Did the separation by sex make each work-group more homogeneous.

Minutes of labour given by each individual	*No. of girls*	*No. of boys*
60	20	120
55	60	100
50	100	200
45	450	355
40	450	350
35	300	500
30	250	350
25	100	20

Solution.

Minutes of labour given by each individaul (x)	$x' = \frac{x-45}{5}$	*No. of girls* f_1	fx'	$f_1 x'^2$	*No. of boys* f_2	$f_2 x'$	$f_2 x'^2$
60	+3	20	60	180	120	360	1080
55	+2	60	120	240	100	200	400
50	+1	100	100	100	200	200	200
45	0	450	0	0	355	0	0
40	−1	450	−450	450	350	−350	350
35	−2	300	−600	1200	500	−1000	2000
30	−3	250	−750	2250	350	−1050	3150
25	−4	100	−400	1600	20	−80	320
Total		1730	−1920	6020	1995	−1720	7500

Girls. $\bar{x}_1=45-\frac{1920}{1730}\times 5=45-5.55=39.45$

$$\sigma_1=\sqrt{\frac{6020}{1730}-\left(\frac{-1920}{1730}\right)^2}\times 5=\sqrt{3.4798-1.2317)}\times 5$$

$$=1.5\times 5=7.5$$

$$V_1=\frac{7.5}{39.45}\times 100=19.0\%.$$

Boys. $\bar{x}_2=45-\frac{1720}{1995}\times 5=45-4.31=40.69$

$$\sigma_2=\sqrt{\frac{7500}{1995}-\left(\frac{-1720}{1995}\right)^2}\times 5=\sqrt{(3.7594-0.7434)}\times 5$$

$$=1.7366\times 5=8.68$$

$$V_2=\frac{8.68}{40.69}\times 100=21.34\%$$

Combined Mean. $\bar{x}_{12}=\frac{1730\times 39.45+1995\times 40.69}{1730+1995}$

$$=\frac{149424.78}{3725}=40.11$$

$$d_1=0.66 \text{ and } d_2=-0.58$$

$$\sigma_{12}{}^2=\frac{1730(7.5)^2+1995(8.68)^2+1730(0.66)^2+1995(-0.58)^2}{1730+1995}$$

$$= 66.86$$

$$\sigma_{12} = 8.18.$$

$$\text{C.V.} = \frac{8.18}{40.11}\times 100 = 20.38\%$$

Therefore, there doesn't seem to be any evidence that each work-group was more homogeneous than the total population.

Illustration 10.19. The daily temperature recorded in a city in Russia in a year is given below:

Temperature °C	*No. af days*
—40 to —30	10
—30 to —20	28
—20 to —10	30
—10 to 0	42
0 to 10	65
10 to 20	180
20 to 30	10
	365

Calculate the mean and standard deviation.

Solution.

CALCULATION OF MEAN STANDARD DEVIATION

Temperature °C	*Mid-value* (*x*)	*No. of days* *f*	$x' = \frac{x-0}{5}$	*fx'*	*fx'*²
−40 to −30	−35	10	−7	−70	490
−30 to −20	−25	28	−5	−140	700
−20 to −10	−15	30	−3	−90	270
−10 to 0	−5	42	−1	−42	42
0 to 10	+5	65	1	+65	65
10 to 20	+15	180	3	+540	1620
20 to 30	+25	10	5	+50	250
Total		365		313	3437

$$\bar{x}=0+\frac{313}{365}\times 5=4.3°\text{C}$$

Standard Deviation $\sigma=\sqrt{\frac{3437}{365}-\left(\frac{313}{365}\right)^2}\times 5$

$=\sqrt{9.41-(0.86)^2}\times 5=\sqrt{8.68}\times 5$

$=14.7°$C.

Illustration 10.20. An Association doing charity work decided to give old age pensions to people over sixty years of age.

The scale of pensions were fixed as follows:

Age group 60 to 65 ... Rs 25 per month
Age group 65 to 70 ... Rs 30 per month
Age group 70 to 75 ... Rs 35 per month
Age group 75 to 80 ... Rs 40 per month
Age group 80 to 85 ... Rs 45 per month

The age of 25 persons who secured the pension right are given below:

74 62 84 72 61 83 72 81 65 71 63 61
60 67 74 66 64 79 73 75 76 69 68 78 61

Calculate the monthly average pension payable and the standard deviation.

Solution. The above data can be summarised into a frequency distribution so as to find out the number of persons in different classes.

Age group	*Tally Marks*	*Frequencp*
60—65	~~IIII~~ II	7
65—70	~~IIII~~	5
70—75	~~IIII~~ I	6
77—80	IIII	4
80—85	III	3

CALCULATION OF ARITHMETIC MEAN AND ($\bar{X}$) STANDARD DEVIATION (σ)

Pension in Rs (*x*)	*No. of People* (*f*)	$x'=\frac{x-35}{5}$	fx'	fx'^2
25	7	−2	−14	28
30	5	−1	−5	5
35	6	0	0	0
40	4	+1	+4	4
45	3	+2	+6	12
	$N=25$		$\Sigma fx'=-9$	$\Sigma fx'^2=49$

$$\text{Arithmetic Mean}=\bar{x}=35+\frac{-9}{25}\times 5=33.20$$

$$\text{Standard deviation}=\sigma=5\times\sqrt{\frac{49}{25}-\left(\frac{-9}{25}\right)^2}$$

$$=6.76$$

∴ Monthly average pension, is Rs 33.20 with a standard deviation of Rs 6.76,

Illustration 10.21. The values of the arithmetic mean and the standard deviation of the following frequency distribution of a continuous variable derived from short-cut method are 135.3 lbs. and 9.6 lb. respectively.

x	−4	−3	−2	−1	0	1	2	3	Total
Freq.	2	5	8	18	22	13	8	4	80

Determine the actual class interval.

Solution. Calculation of standard deviation

x	−4	−3	−2	−1	0	1	2	3	Total
Freq. (*f*)	2	5	8	18	22	13	8	4	80
fx	−8	−15	−16	−18	0	13	16	12	−16
fx^2	32	45	32	18	0	13	32	36	208

$$\text{Standard deviation}=i\times\sqrt{\frac{\Sigma fx}{n}-\left(\frac{\Sigma fx}{n}\right)^2}$$

∴ Putting the known values, we have

$$9.6 = i \times \sqrt{\frac{208}{80} - \left(\frac{-16}{80}\right)^2}$$

$$= i \times \sqrt{2.6 - .04}$$

or $$9.6 = i \times \sqrt{2.56} = i \times 1.6$$

∴ $$i = \frac{9.6}{1.6} = 6.$$

Arithmetic mean $= A + \frac{\Sigma fx}{n} \times i$

∴ Putting the known values, we have

$$135.3 = A + \frac{-16}{80} \times 6 = A - 1.2$$

or $$A = 135.3 + 1.2 = 136.5.$$

A or assumed mean is the mid-point corresponding to the class having *x* value 0. As the class interval is of 6 and the variable under studying is a continuons one, the class for which $x=0$ will be 136.5 —3 to 136.5+3, *i.e.*, 133.5−139.5. A class next lower than this is 133.5−6 to 133.5, *i.e.*, 127.5 to 133.5.

Similarly other class can be calculated. So all the class intervals are:

109.5—115.5. 115.5—121.5, 121.5—127.5, 127.5—133.5,
133.5—139.5, 139.5—145.5, 145.5—151.5, 151.5—157.5.

Adjusting Values of Mean and Standard Deviations for Mistakes

It may sometimes happen that while making computations of mean or that of standard deviation a correct value in the original data is replaced by an incorrect one. When such mistakes are discovered it becomes necessary to correct the values of the mean and standard deviation. Instead of going through the entire process of calculation these corrections can be brought about with comparative ease as is evident from the following illustrations:

Illustration 10.22. The mean and standard deviation of a set of 100 observations were worked out as 40 and 5 respectively by a computer who by mistake took the value 50 in place of 40 for one observation. Recalculate the correct mean and variance.

Solution.

Calculated Mean

$$\bar{X} = \frac{\Sigma X}{N}$$

$$40 = \frac{\Sigma X}{100}$$

$$4{,}000 = \Sigma X$$

Correct $\Sigma X = 4{,}000 + 40 - 50 = 3{,}990$

Correct mean $\frac{3{,}990}{100} = 39.90.$

Computed variance

$$\sigma^2 = \frac{\Sigma X^2}{N} - \bar{X}^2$$

$$25 = \frac{\Sigma X^2}{100} - 40^2$$

$$2{,}500 = \Sigma X^2 - 1{,}60{,}000$$

or $\Sigma X^2 = 1{,}62{,}500$

Correct Value $\Sigma X^2 = 1{,}62{,}500 - 50^2 + 40^2$

$$= 1{,}62.500 - 2{,}500 + 1{,}600$$

$$= 1{,}61{,}600$$

Correct variance $= \frac{1{,}61{,}600}{100} - (39.9)^2$

$$= \frac{1{,}61{,}600 - 1{,}59{,}201}{100}$$

$$= \frac{2399}{100} = 23.99$$

Correct Mean = 39.90.

Correct variance = 23.99.

Illustration 10.23. The number of employees and the variance of the wages per employee for two factories are given below.

	Factory A	Factory B
No. of employees	50	100
Average wages per employee	120	85
Variance of the wages per employee per month (Rs)	9	16

(*a*) In which factory is there greater variance in the distribution of wage per employee?

(*b*) Suppose in factory *B*, the wages of an employee were wrongly noted as Rs 120 instead of Rs 100, what would be the corrected variance for factory *B*.

Solution. The test of greater variation is higher coefficient of variation. Coefficient of variation:

(*a*) Factory $A = \frac{\sqrt{9}}{120} \times 100$

$$= 2.5\%$$

(*b*) Factory $B=\frac{\sqrt{16}}{85}\times 100$

$=4.7\%$

Hence greater variation is in the distribution of wages in Factory *B*.

(*b*) Total wage bill for factory $B=100\times 85=8{,}500$

Correct wage bill should be Rs $8{,}500+100-120=8{,}480$

Correct arithmetic average$=\frac{8{,}480}{100}=84.80$

$$\text{Variance}=\frac{\Sigma X^2}{N}-\bar{X}^2$$

Substituting the values

$$16=\frac{\Sigma X^2}{100}-85^2$$

or $\quad 1600=\Sigma X^2-722500$

or $\quad \Sigma X^2=724100 \quad$ (calculated)

Correct Value $\quad \Sigma X^2=724100-120^2+100^2$

$=724100-14400+10{,}000$

$=719700.$

Substituting the correct values of ΣX^2 and $\bar{X}$ in the formula for variance, we get

Correct variance $=\frac{719700}{100}-(84.8)^2$

$=7197-7191.04$

$=5.96$

Illustration 10.24. A sample of 10 numbers gave a mean of 13 and a variance of 4. Later it was discovered that the number 12 included in the sample should have been 21. Find the corrected mean and variance.

Solution. In the usual notation, we are given

$$N=10,\ \bar{X}=13 \text{ and } \sigma^2=4.$$

$$\Sigma X=N\bar{X}=10\times 13=130.$$

Corrected $\quad \Sigma X=130-12+21=139$

Corrected $\quad \Sigma\bar{X}=\frac{\text{corrected } \Sigma X}{N}=\frac{139}{10}=13.9$

Calculated $\quad \Sigma X^2 = n\ (\sigma^2+x^2) = 10\ (4+169) = 1730$

Corrected $\quad \Sigma X^2 = 1730 - (12)^2+(21)^2 = 2027$

Corrected $\quad \sigma^2=\frac{\text{corrected } \Sigma X^2}{N}-\left(\frac{\text{corrected } \Sigma X}{N}\right)^2$

$$= \frac{2027}{10} - \left(\frac{139}{10}\right)^2 = 202.7 - 193.21$$

$$= 9.49$$

Corrected $\sigma = 3.08$

Determining Overall Performance

If a number of candidates each write papers in several subjects, then the total marks obtained by candidates will not be a correct basis for determining their merit. This is also because the marks in different subjects are likely to have different amounts of spread and means. The extent of the spread of marks in a paper introduces 'weights' which will be to the advantage of those getting high marks in subjects where the spread is great, and to the disadvantage of those gaining high marks in subjects where range is small.

A method to correct for the errors so introduced consists of converting the marks into 'standard scores' defined as $\frac{x - \bar{x}}{\sigma}$. This measures deviation of marks of a student from the mean of that subject in the units of standard deviation, and are termed as z-scores as well. This corrects both for variations of $\bar{x}$ and of σ in different subjects.

The z-scores of a student in different subjects are then added to give a true measure of relative performance.

Illustration 10.25.

Candidate	Marks in .conomics	Commerce	Total
A	84	75	159
B	74	85	159

Average for Economics is 60 with standard deviation 13
Average for Commerce is 50 with standard deviation 11
Whose performance is better—*A*'s or *B*'s?

Solution.

Z Scores *A*: Economics $\frac{84-60}{13} = 1.85$, Commerce $\frac{75-50}{11} = 2.27$ → 4.12

B: Economics $\frac{74-60}{13} = 1.08$, Commerce $\frac{85-50}{11} = 3.18$ → 4.26

Since *B*'s *Z* score is higher his performance is better.

Illustration 10.26. Three candidates in a certain examination had the same aggregate marks and were bracketed equal. Use the following data to determine whether or not this placing was equitable.

Marks awarded out of 100 for

Candidate	English	Science	Mathematics	Total
A	95	70	61	226
B	69	83	74	226
C	70	74	82	226

The mean marks were 55 for English, 53 for Science and 50 for Mathematics, and the standard deviation, in the same order were, 16, 12 and 11.

Solution.

Z Score: *A*: English $\frac{95-55}{16}=\frac{40}{16}=2.50$

Science $\frac{70-53}{12}=\frac{17}{12}=1.42$ } 4.92

Maths. $\frac{61-50}{11}=\frac{11}{11}=1.00$

B: English $\frac{69-55}{16}=\frac{14}{16}=.87$

Science $\frac{83-53}{12}=\frac{30}{12}=2.50$ } 5.55

Maths. $\frac{74-50}{11}=\frac{24}{11}=2.18$

C: English $\frac{70-55}{16}=\frac{15}{16}=0.94$

Science $\frac{74-53}{12}=\frac{21}{12}=1.75$ } 5.60

Maths. $\frac{82-50}{11}=\frac{32}{11}=2.91$

Equitable order: *C*, *B* and *A*.

10.12 LORENZ CURVE

The Lorenz curve is a graphic method of measuring deviations from the average. It was devised by Dr Lorenz for measuring the inequalities in the distribution of wealth. But it can be applied with equal advantage for comparing the distribution of profits amongst different groups of business and such other things. It is a *cumulative percentage curve.* In it the percentages of items are combined with the percentages of such other things as wealth, profits or turn-over, etc.

In drawing a Lorenz curve the following steps are necessary:

1. The various groups of each variable should be reduced to percentages. Thus, if it is desired to show the distribution of income amongst the various groups of population of a country the various groups of population should be reduced in the form of percentages of total population; so also the incomes derived by these groups in terms of the total income of the country.

2. The two sets of the percentages obtained by step 1 should then the cumulated and cumulative percentages thus determined.

3. The cumulative percentages of these two variables should then be plotted along the axis of Y and axis of X. The scale along the axis of Y begins from zero at the point of intersection and goes upward up to 100, while the scale along the axis of X begins with 100 at the point of intersection and goes up to zero towards the right.

4. The points 100, 100 along the axis of Y and the points 0, 0 along the axis of X should be joined by a straight line. The line so obtained is called the line of equal distribution, and serves as the basis for the determination of the extent to which the actual distribution deviates from the ideal distribution given by this line.

5. The actual data map now be plotted on this graph in the ordinary manner and the plotted points may be connected by means of a curve.

The farther the curve obtained under step 5 is from the line of equal distribution, the greater is the deviation.

Illustration 10.27. The following table gives the population and earnings of two town *A* and *B*. Represent the data graphically so as to bring out the inequlity of the distribution of earnings.

Town A		*Town B*	
Persons	*Earning (Daily)*	*Persons*	*Earning (Daily)*
100	75	50	80
100	100	70	120
100	150	30	60
100	225	25	140
100	325	100	200
100	375	45	200
100	450	30	140
100	600	80	460
100	850	20	120
100	1,850	50	480
1,000	5,000	500	2,000

Solution.

POPULATION AND DAILY EARNINGS OF TWO TOWNS A AND B

Town A				*Town B*			
Persons		*Earnings*		*Persons*		*Earnings*	
Cumulative Total	*Cumulative Percentage*	*Cumulative Total*	*Cumulative Percentage*	*Cumulative Total*	*Cumulative Percentage*	*Cumulative Total*	*Cumulative Percentage*
100	10	75	1.5	50	10	80	4
200	20	175	3.5	120	24	200	10
300	30	325	6.5	150	30	260	13
400	40	550	11	175	35	400	20
500	50	875	17.5	275	55	600	30
600	60	1,250	25	320	64	800	40
700	70	1,700	34	350	70	940	47
800	80	2,300	46	430	86	1,400	70
900	90	3,150	63	450	90	1,520	76
1,000	100	5,000	100	500	100	2,000	100

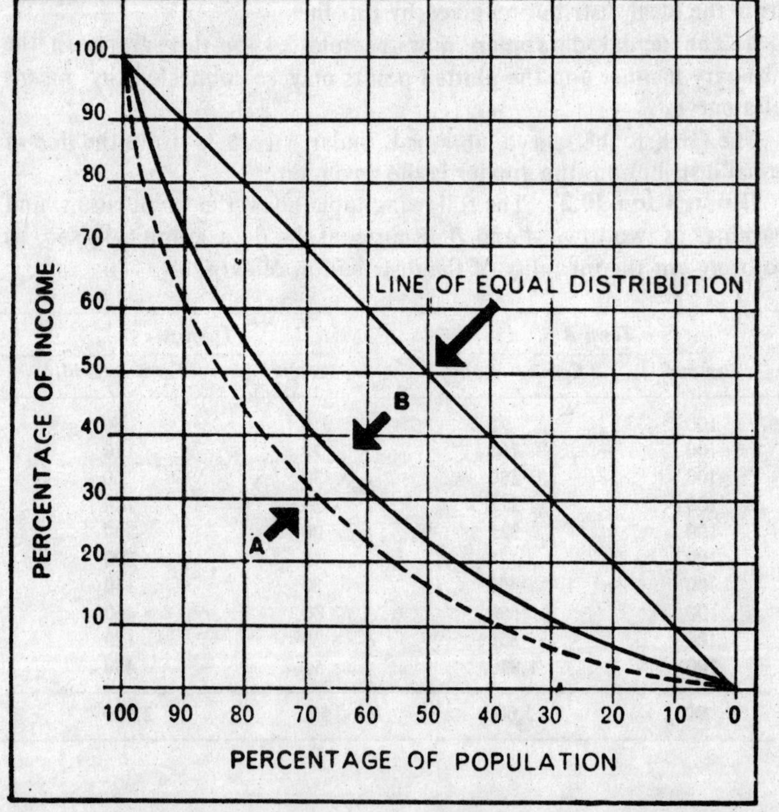

FIG. 10.2.

10.13. MISCELLENEOUS ILLUSTRATIONS

Illustration 10.28. Compile a table, showing the frequencies with which words of different numbers of letters occur in the extract reproduced below (omitting punctuation marks) treating as the variable the number of letters in each word, and obtain the mean, median and coefficient of variation of the distribution:

"Success in the examination confers no absolute right to appointment, unless Government is satisfied, after such equiry as may be considered neçessary, that the candidate is suitable in all respect for appointment to the public service."

Solution.

No. of letters	*Frequency*	*c.f.*	$x'=x-6$	fx'	fx'^2
2	9	9	—4	—36	**144**
3	6	15	—3	—18	**54**
4	2	17	—2	—4	**8**
5	2	19	—1	—2	**2**
6	2	21	0	0	**0**
7	4	25	+1	+4	**4**
8	3	28	+2	+6	**12**
9	3	31	+3	+9	**27**
10	2	33	+4	+8	**32**
11	3	36	+5	+15	**75**
	36 words			—18	**358**

$$\text{Median}=\text{the value of } \frac{N+1}{2} \text{ th item}=\frac{36+1}{2}=18.5\text{th item}$$

$$=5 \text{ letters}$$

$$\bar{x}=A+\frac{\Sigma fx'}{N}=6-\frac{18}{36}=6—0.5=5.5 \text{ letters}$$

$$\sigma=\sqrt{\frac{\Sigma fx'^2}{N}-\left(\frac{\Sigma fx'}{N}\right)^2}$$

$$=\sqrt{\frac{358}{36}-\left(\frac{-18}{36}\right)^2}$$

$$=3.12$$

Coefficient of variation

$$=\frac{\sigma}{\bar{x}}\times 100=\frac{3.12}{5.5}\times 100=56.7\%$$

Illustration 10.29. Find the standard deviation and the coefficient of variation of the following data:

Interval	Frequency	Interval	Frequency
3.00 – 3.25	6	4.00—4.25	47
3.25 – 3.50	19	4.25—4.50	29
3.50 – 3.75	35	4.50—4.75	15
3.75—4.00	44	4.75—5.00	5

Solution.

Interval	f	Mid-value (x)	$x'=\frac{x-3.875}{0.25}$	fx'	fx'²
3.00—3.25	6	3.125	—3	−18	54
3.25—3.50	19	3.375	—2	—38	76
3.50—3.75	35	3.625	—1	—35	35
3.75—4.00	44	3.875	0	0	0
4.00—4.25	47	4.125	+1	+47	47
4.25—4.50	29	4.375	+2	+58	116
4.50—4.75	15	4.625	+3	+45	135
4.75—5.00	5	4.875	+4	+20	80
	N=200			Σfx' = +79	Σfx'² = 543

$$\bar{x}=A+\frac{\Sigma fx'}{N}\times i=3.875+\frac{79}{200}\times 0.25$$

$$=3.974$$

Standard deviation

$$\sigma=\sqrt{\frac{\Sigma fx'^2}{N}-\left(\frac{\Sigma fx'}{N}\right)^2}\times i$$

$$=\sqrt{\frac{543}{200}-\left(\frac{79}{200}\right)^2}\times 0.25$$

$$=0.40$$

Coefficient of variation (C.V.)

$$=\frac{\sigma}{\bar{x}}\times 100$$

$$=\frac{0.40}{3.974}\times 100=10.06\%$$

Illustration 10.30. A collar manufacturer is considering the production of a new style of collar to attract young men. The following statistics of neck circumference are available based upon measurements of a typical group of college students:

Mid-value (inches)	No. of students
12.5	4
13.0	19
13.5	30
14.0	63

14.5	66
15.0	29
15.5	18
16.0	1
16.5	1

Compute the standard deviation and use the criterion ($M \pm 3\sigma$), where σ is standard deviation, M is arithmetic average, to determine the largest and the smallest sizes of collars he should make in order to meet the needs of practically all his customers, bearing in mind that collars are worn on average 3/4 inches larger than the neck size.

Solution.

Mid value (m)	*f*	*x′=m—14.5*	*fx′*	*fx′²*
12.5	4	—2.0	—8.0	16.00
13.0	19	—1.5	—28.5	42.75
13.5	30	—1.0	—30.0	30.00
14.0	63	—0.5	—31.5	15.75
14.5	66	0	0	0
15.0	29	+0.5	+14.5	7.25
15.5	18	+1.0	+18,0	18.00
16.0	1	+1.5	+ 1.5	2.25
16.5	1	+2.0	+ 2.0	4.00
Total	231		—62.0	136.00

$$\bar{x}=A+\frac{\Sigma fx'}{N}=14.5-\frac{62}{231}=14.5-0.27$$

$$=14.23 \text{ inches}$$

$$\sigma=\sqrt{\frac{\Sigma fx'^2}{N}-\left(\frac{\Sigma fx'}{N}\right)^2}$$

$$=\sqrt{\frac{136}{231}-\left(\frac{-62}{231}\right)^2}$$

$$=0.72 \text{ inches}$$

Using the criterian $M \pm 3\sigma = \left(14.23 \pm 0.72 \times 3 + \frac{3}{4}\right)$

The largest size of the collar$=14.23+2.16+0.75=17.14$ inches

The smallest size of the collar$=14.23-2.16+0.75=12.82$ inches.

Illustration 10.31. Find the missing information from the following:

	Group I	*Group II*	*Group III*	*Combined*
Number	50	?	90	200
Standard Deviation	6	7	?	7.746
Mean	113	?	115	116

Solution. In usuual notations, here we are given

$$N_1=50,\ N_2=?,\ N_3=90$$

$$\bar{X}=116,\ \bar{X}_1=113,\ \bar{X}_2=?,\ \bar{X}_3=115$$

and $$\sigma=7.746,\ \sigma_1=6,\ \sigma_2=7 \text{ and } \sigma_3=?$$

(*i*) Since $N_1+N_2+N_3=200$

or $$N_2=200-50-90=60.$$

(*ii*) The mean of the combined distribution is given by

$$\bar{X}=\frac{N_1\bar{X}_1+N_2\bar{X}_2+N_3\bar{X}_3}{N_1+N_2+N_3}$$

or $$116=\frac{50(113)+60(\bar{X}_2)+90(115)}{200}$$

or $$60\ \bar{X}_2=23200-16000=7200$$

or $$\bar{X}_2=120$$

(*iii*) Standard deviation of the combined distribution is given by

$$\sigma^2=\frac{N_1\sigma_1^2+N_2\sigma_2^2+N_3\sigma_3^2}{N_1+N_2+N_3}+\frac{N_1(\bar{X}-\bar{X}_1)^2+N_2(\bar{X}-\bar{X}_2)^2+N_3(\bar{X}-\bar{X}_3)^2}{N_1+N_2+N_3}$$

or $$(7.746)^2=\frac{50(6)^2+60(7)^2+90\sigma_3^2}{200}+\frac{50(113-116)^2+60(120-116)^2+90(115-116)^2}{200}$$

or $$60=\frac{6240+90\sigma_3^2}{200} \text{ or } 90\sigma_3^2=12000-6240$$

or $$\sigma_3=\sqrt{64}=8.$$

Illustration 10.32. The arithmetic mean and standard deviation of a series of 20 items were calculated by a student as 20 cm and 5 cm respectively. But while calculating them an item 13 was misread as 30. Find the correct arithmetic mean and standard deviation.

Solution. In the usual notations, we are given

$$N=20,\ \bar{X}=20 \text{ and } \sigma=5$$

$$\Sigma X=N\bar{X}=20\times 20=400$$

Corrected $$\Sigma X=400-30+13=383$$

Corrected $$\bar{X}=\frac{\text{Corrected }\Sigma X}{N}=\frac{383}{20}=19.15$$

Also we know

$$\sigma^2=\frac{\Sigma X^2}{N}-(\bar{X})^2$$

or $$\Sigma X^2=N((\sigma^2+\bar{X}^2)$$

$$=20(25+400)=8500.$$

Corrected $\Sigma X^2 = 8500 - (30)^2 + (13)^2$
$= 8500 - 900 + 169 = 7769$

Corrected $\sigma^2 = \frac{\text{Corrected } \Sigma X^2}{N} - \left(\frac{\text{Corrected } \Sigma X}{N}\right)^2$

$$= \frac{7769}{20} - \left(\frac{383}{20}\right)^2 = 388.45 - 366.72$$

$\sigma = 4.66.$

Hence the correct mean is 19.15 and correct standard deviation 4.66.

Illustration 10.33. Mean, and standard deviation of the following continuous series are 31 and 15.9 respectively. The distribution after taking step deviation is as follows:

x' :	−3	−2	−1	0	1	2	3
f :	10	15	25	25	10	10	5

Determine the actual class intervals.

Solution.

x' :	−3	−2	−1	0	1	2	3	Total
f ;	10	15	25	25	10	10	5	100
fx' :	−30	−30	−25	0	10	20	15	−40
fx'^2:	90	60	25	0	10	40	45	270

Standard deviation $= \sqrt{\frac{\Sigma fx'^2}{N} - \left(\frac{\Sigma fx'}{N}\right)^2} \times i$

Putting the known values, we have

$$15.9 = \sqrt{\frac{270}{100} - \left(\frac{-40}{100}\right)^2} \times i$$

$$= \sqrt{2.70 - 0.16} \times i = 1.59 \times i$$

$$\therefore \quad i = \frac{15.9}{1.59} = 10$$

Arithmetic mean $= A + \frac{\Sigma fx'}{N} \times i$

$\therefore$ Putting the known values, we have

$$31 = A + \frac{-40}{100} \times 10$$

or $A = 31 + 4 = 35.$

A or assumed mean is the mid-point corresponding to the class having x value 0. As the class interval is of 10 and the variable under study is a continuous one, the class for which $x=0$ will be $35-5$ to $35+5$, *i.e.* 30 to 40. A class next lower than this is $30-10$ to 30 *i.e.* 20 to 30.

Similarly other classes can be calculated. So all the class-intervals are:

0—10 10—20 20—30 30—40 40—50 50—60 60—70

Illustration 10.34. The mean of 50 readings of a variable was 7.43 and their S.D. was 0.28. The following ten additional readings became available: 6.80, 7.81, 7.58, 7.70, 8.05, 6.98, 7.78, 7.85, 7.21 and 7.40. If these are included with original 50 readings find (*i*) the mean, (*ii*) the standard deviation of the whole set of 60 readings.

Solution. Mean of 50 readings=7.43

Mean of 10 additional readings$=\frac{\Sigma x}{N}$

$$=\frac{6.80+7.81+7.58+7.70+8.05+6.98+7.78+7.85+7.21+7.40}{10}$$

$=7.516.$

Mean of 60 readings $=\frac{7.43\times 50+7.516\times 10}{50+10}$

$=\frac{371.5+75.16}{60}=7.44$

Standard Deviation

$$.28=\sqrt{\frac{\Sigma X^2}{50}-(7.43)^2}$$

$$0.0784=\frac{\Sigma X^2}{50}-55.2$$

$$\Sigma X^2=(.0784+55.2)50$$

$$=2764.165$$

Sum of square of 10 additional readings

46.24+61.00+57.46+59.29+64.80+48.72+60.53+61.62+51.99+54.76

$=566.55$

Sum of the square of 60 readings=2764.165+566.55

$=3330.71$

$\therefore$ S.D. of 60 readings $=\sqrt{\frac{3330.71}{60}-(7.44)^2}$

$=\sqrt{55.52-55.35}=\sqrt{0.17}=0.41$

Illustration 10.35. Two workers on the same job show the following results over a long period of time

		A	*B*
Mean time of completing the job (minutes)		30	25
S.D.	(″)	6	4

(*i*) Which worker appears to be more consistent in the time he requires to complete the job?

(*ii*) Which worker appears to be faster in completing the job?

Solution. (*i*) The coefficient of variation

$$= \frac{\sigma}{\bar{x}} \times 100$$

$$A = \frac{6}{30} \times 100 = 20\%$$

$$B = \frac{4}{25} \times 100 = 16\%$$

The coefficient of variation is more for worker *A*, hence worker *B* appears to be more consistent in the time requires to complete the job.

(*ii*) Worker *B* appears to be faster in completing the job as he takes on an average 25 minutes as against 30 minutes taken by worker *A*.

Illustration 10.36. The first of two sub-groups has 100 items with mean 15 and S.D. 3. If the whole group has 250 items with mean 15.6 and S.D. $\sqrt{13.44}$ find the S.D. of the second group.

Solution. Combined A.M. $= \bar{X} = \frac{N_1\bar{X}_1 + N_2\bar{X}_2}{N_1 + N_2}$

$$15.6 = \frac{(100 \times 15) + (150 \times \bar{X}_2)}{250}$$

$$3900 = 1500 + 150\bar{X}_2$$

$$\bar{X}_2 = 16.$$

Therefore the A.M. of the second group is 16

Combined S.D.

$$\sqrt{13.44} = \sqrt{\frac{100 \times 9 + 100(15 - 15.6)^2 + 150\sigma_2^2 + 150\,(16 - 15.6)^2}{250}}$$

$$13.44 = \frac{900 + 36 + 150\sigma_2^2 + 24}{250} \text{ or } 3360 = 960 + 150\sigma_2^2$$

$$150\,\sigma_2^2 = 2400$$

$$\sigma_2^2 = 16.\ \sigma_2 = 4$$

Therefore the S.D. of the second group is 4.

Illustration 10.37. Particulars regarding income of two villages are given below:

	Village	
	A	*B*
No. of people	600	500
Average Income	175	186
Variance of Income	100	81

(*a*) In which village is the variation in income greater?

(*b*) What is the total income of both the villages put together?

(*c*) What is the average income of the people *A* and *B* villages put together?

(*d*) What is the combined standard deviation?

Solution. (*a*) *Variation in distribution of Income*

$$A \qquad\qquad\qquad\qquad B$$

$$\text{C.V.} = \frac{\sigma}{\bar{x}} \times 100 \qquad\qquad \text{C.V.} = \frac{\sigma}{\bar{x}} \times 100$$

$$= \frac{\sqrt{100}}{175} \times 100 = 5.7 \qquad \frac{\sqrt{81}}{186} \times 100 = 4.8$$

The coefficient of variation is greater for village *A*, hence there is greater variation in the distribution of income in village *A*.

(*b*) Total income = No. of people (*A*) × Av. Income (*A*)

+ No. of people (*B*) × Av. Income (*B*)

= 175 × 600 + 186 × 500 = Rs 1,98,000.

(*c*) Combined Average Income

$$\frac{175 \times 600 + 186 \times 500}{1100} = \text{Rs } 180$$

(*d*) *Combined Standard deviation.*

$$= \sqrt{\frac{600[100+(180-175)^2]+500[81+(186-180)^2]}{1100}}$$

$$= \sqrt{\frac{600(100+25)+500(81+36)}{1100}}$$

= Rs 11.02

Illustration 10.38. The A.M. of 5 observations is 4.4 and the variance is 8.24. If 3 of the 5 observations are 1, 2 and 6, find the other two.

Solution. Let the two observations be x and y.

$$\bar{x} = \frac{\text{Sum of all the observations}}{\text{no. of items}}$$

Sum of all the observations = 5 × 4.4 = 22.

$$1+2+6+x+y=22$$

$$\therefore \qquad x+y=13 \qquad \ldots(1)$$

$$\sigma^2 = \frac{\Sigma X^2}{N} - (\bar{X})^2$$

$$8.24 = \frac{x^2+y^2+1^2+2^2+6^2}{5} - (4.4)^2$$

or $\qquad x^2+y^2=97$

$$(x-y)^2=(x+y)^2-4xy=(x+y)^2-2[(x+y)^2-(x^2+y^2)]$$
$$=169-2(13^2-97)=25$$
$$x-y=5$$

Subtracting (2) from (1), we get

$$2y=8 \text{ or } y=4.$$

Putting this an equation (2)

$$x=5+4=9.$$

∴ Other two observations are 4 and 9.

Illustration 10.39. Suppose that a prospective buyer tests bursting pressure of samples of polythene bags received from two manufactures *A* and *B* The tests reveal the following results:

Bursting pressure (lbs)	5—10	10—15	15—20	20—25	25—30
No. of Bags *A*	2	9	29	54	6
,, ,, *B*	9	15	30	32	14

Which manufacture's bags judging from these two samples have the higher average bursting pressure? Which of them are more uniform in bursting pressure?

Solution. *Calculation of coefficient of variation*

Assumed Mean $(A)=17.5$ Common factor $i=5$

Bursting Pressure	*Mid-point x*	$x'=\frac{x-A}{5}$	*Variable A* f	fx'	fx'^2	*Variable B* f	fx'	fx'^2
5—10	7.5	—2	2	—4	8	9	—18	36
10—15	12.5	—1	9	—9	9	15	—15	15
15—20	17.5	0	29	0	0	30	0	0
20—25	22.5	1	54	54	54	32	32	32
25—30	27.5	2	6	12	24	14	28	56
			100	53	95	100	27	139

Coefficient of Variation $\frac{\sigma}{\bar{x}}\times 100$.

Manufacturer A

$$\bar{x}=17.5+\frac{53}{100}\times 5=20.15 \text{ lbs.}$$

$$\sigma=\sqrt{\frac{95}{100}-\left(\frac{53}{100}\right)^2}\times 5$$

$$=\sqrt{.95-.2809}\times 5=\sqrt{.6691}\times 5=4.09 \text{ lbs.}$$

$$\therefore \quad \text{C.V.}=\frac{4.09}{20.15}\times 100=20.3\%.$$

Manufacturer B

$$\bar{x}=17.5+\frac{27}{100}\times 5=17.5+1.35=18.85 \text{ lbs.}$$

$$\sigma=\sqrt{\frac{139}{100}-\left(\frac{27}{100}\right)^2}\times 5$$

$$=\sqrt{1.39-.0729}\times 5=1.148\times 5=5.74 \text{ lbs.}$$

$$\therefore \quad \text{C.V.}=\frac{5.74}{18.85}\times 100=30.4\%$$

As the $\bar{x}$ of A is higher and C.V. lower than that of B, bags of A give higher bursting pressure and are more uniform in bursting pressure.

Illustration 10.40. You are given the following distribution of monthly income per family. Calculate the mean ($\bar{X}$) and S.D. (S) for this distribution. What % of families fall in the interval $\bar{X}-S$ and $\bar{X}+S$?

Monthly income	*No. of families*
100 - 120	30
120—160	25
160 - 200	20
200—240	15
240—280	10

Solution. Assumed Mean=180; $i=10$

Monthly income (Rs)	*Mid-value x*	*No. of families f*	$x'=\frac{x-180}{10}$	fx'	fx'^2
100—120	110	30	—7	—210	1470
120—160	140	25	—4	—100	400
160—200	180	20	0	0	0
200—240	220	15	4	60	240
240—280	260	10	8	80	640
		100		—170	2750

$$\bar{X}=180+\frac{(-170)}{100}\times 10=180-17=163.$$

$$S=\sqrt{\frac{2750}{100}-\left(\frac{-170}{100}\right)^2}\times 10$$

$$=\sqrt{27.5-2.89}\times 10=\sqrt{24.61}\times 10=\text{Rs } 49.6.$$

Families falling in the interval $X \pm S$ *i.e.* 113.4 to 212.6 by interpolation:

(Assuming that class frequencies are evenly distributed)

$$= \frac{120-113.4}{120-100} \times 30+25+20+ \frac{212.6-200}{240-200} \times 15$$

$$=59.62\%.$$

Illustration 10.41. You are given the two variables *A* and *B*. Using quartile deviations, state which of the two is more dispersed?

A		*B*	
Mid-point	*Frequency*	*Mid-point*	*Frequency*
15	15	100	340
20	33	150	492
25	56	200	890
30	103	250	1420
35	40	300	620
40	32	350	360
45	10	400	187
		450	140

Solution. To compare the variability comparison of coefficient of quartile deviations is required.

Coefficient of quartile deviation is$=\dfrac{Q_3-Q_1}{Q_3+Q_1}$

Variable A

Mid-point	*Class interval*	*f*	*c.f.*
15	12.5—17.5	15	15
20	17.5—22.5	33	48
25	22.5—27.5	56	104
30	27.5—32.5	103	207
35	32.5—37.5	40	247
40	37.5—42.5	32	279
45	42.5—47.5	10	289

Q_1 has $\frac{N}{4}$ *i.e.* $\frac{289}{4}$ or 72.25 items below it.

∴ It lies in the group 22.5—27.5

$$Q_1 = 22.5 + \frac{72.25-48}{56} \times 5 = 24.67$$

Q_3 has $\frac{3N}{4}$ or 216.75 items below it.

∴ Q_3 lies in the group 32.5—37.5

$$Q_3 = 32.5 + \frac{216.75-207}{40} \times 5 = 33.72$$

$$\text{Coefficient of } Q.D. = \frac{33.72-24.67}{33.72+26.67} = 0.15$$

Variable B

Mid-point	*Class interval*	*f*	*c.f.*
100	75—125	340	340
150	125—175	492	832
200	175—225	890	1722
250	225—275	1420	3142
300	275—325	620	3762
350	325—375	360	4122
400	375—425	187	4309
450	425—475	140	4449

Q_1 has $\frac{N}{4}$ *i.e.* 1112.25 items below it.

∴ It lies in the group 175—225

$$Q_1 = 175 + \frac{1112.25-832}{890} \times 50 = 190.7$$

Q_3 has $\frac{3N}{4}$ item below it

Q_3 lies in the group 275—325

$$Q_3 = 275 + \frac{3336.75-3142}{620} \times 50 = 290.7$$

$$\text{Coefficient of } Q.D. = \frac{290.7-190.7}{290.7+190.7} = 0.21$$

As coefficient of quartile deviation for *B* is higher, it is more variable.

Illustration 10.42. From the data given below about four sub-groups, calculate the average and the standard deviation of the whole group.

Sub-group	*No. of men*	*Average wage (Rs)*	*Standard deviation wage (Rs)*
A	50	61.0	8
B	100	70.0	9
C	120	80.5	10
D	30	83.0	11
	300		

Solution.

Sub-group	*Men* N	*Average wage* $\bar{X}$	$N\bar{X}$	σ	$N\sigma^2$	$\bar{X}-\bar{X}_c$	$N(\bar{X}-\bar{X}_c)^2$
A	50	61	3050	8	3200	—13	8,450
B	100	70	7000	9	8100	—4	1,600
C	120	80.5	9660	10	12000	6.5	5,070
D	30	83	2490	11	3630	9.0	2,430
	300		22,200		26930		17,550

Combined mean $(\bar{X})_c = \frac{\Sigma N\bar{X}}{N} = \frac{22,200}{300} = \text{Rs } 74$

(Combined standard deviation)^2

$$= \frac{\Sigma N\sigma^2}{\Sigma N} + \frac{\Sigma N(\bar{X}-\bar{X}_c)^2}{\Sigma N}$$

$$= \frac{26930}{300} + \frac{17,550}{300}$$

$$= \frac{44480}{300} = 148.27$$

$$\sigma = \sqrt{148.27} = \text{Rs } 12.18$$

Illustration 10.43. For a certain group of wage-earners, the median and quartile wages per week were Rs 44.3, Rs 43.0 and Rs 45.9 respectively. Wages for the group ranged between Rs 40 and Rs 50. 10% of the group had under Rs 42 per week, 13% had Rs 47 and over and 6% Rs 48 and over. Put these data into the form of a frequency distribution, and hence obtain an estimate of the mean wage and the standard deviation.

Solution. Assuming that the group has 100 workers the frequency distribution will take the following shape.

Earnings Rs	*No, of wage-earners (f)*	*MID Vaiue (x)*	*fx*	*d*	*fd*	*fd²*
40–42	10	41.00	410	−3.50	−35	122.50
42–43	15	42.50	637.50	−2 00	−30	60.00
43–44.3	25	43.65	1091.25	- 0.85	−21.25	18.06
44.3–45.9	25	45.10	1127.50	+0.60	15.00	9.00
45.9–47	12	46.45	557.40	1.95	23.40	43.63
47–48	7	47.50	332.50	3.00	21.00	63.00
48–50	6	49.00	294 00	4.50	27.00	121.50
	$\Sigma f=100$		4450.15			437.69

$$\bar{X}=\frac{\Sigma fx}{N}=\frac{4450.15}{100}=\text{Rs } 44.50$$

$$\sigma=\sqrt{\frac{437.69}{100}}=\text{Rs } 2.1 \text{ (approx.)}$$

Illustration 10.44. A calculating machine while calculating mean and standard deviation of 25 readings misread one observation as 36 instead of 26. The following results were given by a machine

$$\sigma=5.0 \text{ and } \bar{X}=30$$

What are the correct values of $\bar{X}$ and σ ?

Solution.

The error in sum of values $=36-26=10$

Hence the corrected $\bar{X}=\dfrac{(30\times 25)-10}{25}=29.6$

To find the corrected standard deviation, note that the sum of the squares of deviation as calculated from 30 is $25\times(5)^2=625$. Subtracting from it $(36-30)^2$ and adding $(26-30)^2$ we get

$$625-36+16=605.$$

Sum of squares of deviation from 30 is 605 and the sum of deviation is 10 hence the corrected deviation from the mean 29.6 is

$$=\sqrt{\frac{\Sigma x^2}{N}-\left(\frac{\Sigma x}{N}\right)^2}$$

$$=\sqrt{\frac{605}{25}-\left(\frac{10}{25}\right)^2}=\sqrt{24.2-.16}$$

$$=\sqrt{24.04}=4.09$$

Illustration 10.45. In a surprise check on passengers in local bus 20 ticketless passengers were caught. The sum of squares and the standard deviation of the amount found in their pockets were Rs 2,000.00 and Rs 6.00 respectively. If the total fine imposed is equal to the amount discovered on them, and fine imposed is uniform. What is the amount of that each one of them will have to pay as fine?

Solution.
$$\sigma=\sqrt{\frac{\Sigma X^2}{N}-\bar{X}^2}$$
$$\sigma^2=\frac{\Sigma X^2}{N}-\bar{X}^2$$
$$36=\frac{2,000}{20}-\bar{X}^2$$
$$\bar{X}^2=100-36$$
$$\bar{X}^2=64$$
$\bar{X}=8$; Total amount recovered $8\times 20=$ Rs 160

Each of the passangers shall have to pay Rs 8 as fine.

PROBLEMS

1. You are given two variables *A* and *B*. Using quartile deviation, state which is more variable.

A		B	
Mid-point	*Frequency*	*Mid-point*	*Frequency*
15	15	100	340
20	33	150	492
25	56	200	890
30	103	250	1420
35	40	300	620
40	32	350	360
45	10	400	187
		450	140

(Ans. *B* is more variable)

2. Calculate mean deviation and its coefficient about median, arithmetic mean and mode for the following figures, and show that mean deviation about median is least.

(103, 50, 68, 110, 108, 105, 174, 103, 150, 200, 225, 350, 103)

(Ans. Mean Deviation about madian=52.1, about mean=59.7, and about mode=52.8.
Coeff. of M.D. about median=0.482, about mean=0.420, and about mode=0.512.)

3. Compute mean deviations of the two series and point out which is more variable

Month	*Index No. Calcutta*	*Index No. Delhi*	*Manth*	*Index No. Calcutta*	*Index No. Delhi*
1970 April	93	107	1970 Oct.	97	107
,, May	97	108	,, Nov.	97	105
,, June	95	102	., Dec.	92	101
,, July	95	102	1971 Jan.	93	100
,, August	95	102	,, Feb.	89	97
,, Sept.	95	104	,, March	89	96

(Ans. Coeff. of *M.D.* of Calcutta Index No. 0.022 and Coeff. of *M.D.* of Delhi Index No. 0.027.)

4. Calculate (*a*) median coefficient of dispersion and (*b*) Mean coefficient of dispersion from the following data:

Size of items	14	16	18	20	22	24	26
Frequency	2	4	5	3	2	1	4

(Ans. (*a*) Median Coeff. of dispersion = 0.405 (*b*) Mean Coeff. of dispersion = 0.342)

5. Compute the mean deviation from the median and from the mean the following distribution of the scores of 50 college students.

Scores	140—	150—	160—	170—180—	190—200	
Frequency	4	6	10	18	9	3

(Ans. Mean deviation about Median = 10.24
Mean deviation about Mean = 10.56)

6. Find the mean deviation about mean of the following data of ages of married men in a certain town.

Ages	15—24	25—34	35—44	45—54	55—64	65—74
No. of men	33	264	303	214	128	58

(Ans. Mean deviation = 10.368.)

7. Calculate the mean deviation from the following data. What light does it throw on the social conditions of the community?
Difference in age between husband and wife :

Diff. in yrs.	0—5	5—10	10—15	15—20	20—25	25—30	30—35	35—40
Frequency	449	705	507	281	109	52	16	4

(Ans. 5.24 yrs.)

8. Following figures give the income of 10 persons in rupees. Find the standard deviation.
114, 115, 123, 120, 110, 130, 119, 118, 116, 115.
(Ans. 5.53 rupees.)

9. Calculate the mean and standard deviation of the following values of the world's annual gold output (in millions of pound) for 20 different years. 94, 95, 96, 93, 87, 79, 73, 69, 68, 67, 78, 82, 83, 89, 95, 103, 108, 117, 130, 97. Also calculate the percentage of cases lying outside the mean at distances $\pm\sigma$, $\pm 2\sigma$, $\pm 3\sigma$ where σ denotes standard deviation.
(Ans. Mean = 90.15 million pounds, St. Dev. = 15.99 million pounds and percentage of cases outside mean $\pm 3\sigma = 0\%$.)

10. Calculate standard deviation for the series 1, 2, 3, 5, 7.
(Ans. 2.15.)

11. Calculate standard deviation from the following data:

Size of item	6	7	8	9	10	11	12
Frequency	3	6	9	13	8	5	4

(Ans. 1.61 approx.)

12. From the following information about the accidents on a road in 200 days,) calculate the mean number of accidents and the variance of accidents:

No. of accidents per day	0	1	2	3	4	5
No. of day	46	76	38	25	10	5

(Ans. Mean = 1.46 accidents per day, and variance = 1.56 accidents per day.

13. Calculate the arithmetic mean and the standard deviation for the following data:

Class interval	5-10	10-15	15-20	20-25	25-30	30-35	35-40	40-45
Frequency	6	5	15	10	5	4	3	2

(Ans. Mean = 9.1.)

14. Calculate the mean and the standard deviations from the following:

Age group below	20-25	25-30	30-35	35-40	40-45	45-50	50-55	55 and above
No. of Employees	26	44	60	101	109	84	66	10

(Ans. Median = 39.51 yrs., St. Dev. 9.57 yrs.)

15. Calculate the appropriate measure of dispersion from the following data:

Wages in Rs per week	*No. of wage earners*
Less than 35	14
35—37	62
58—40	99
41—43	18
over 43	7

16. Given the following frequency distribution of the hailstorms in a mountain village, choose an appropriate measure of dispersions. If you reject any well known measure, justify your rejection.

Hailstorm	*No. of days in a year*
0 —2″	150
2″—4″	150
4″—6″	60
6″—8″	5

17. In the following data, two class frequencies are missing:

C.I-	*Frequency*	*C.I.*	*Frequency*
100—110	4	150—160	—
110—120	7	160—170	16
120—130	15	170—180	10
130—140	—	180—190	6
140—150	40	190—200	3

However, it was possible to ascertain that the total number of frequencies was 150 and that the median has been correctly found out as 146.25. You are required to find with the help of information given:

(*i*) The two missing frequencies.

(*ii*) Having found the missing frequencies calculate Arithmetic mean and Standard deviation.

18. A purchasing agent obtained samples of incandescent lamps from two suppliers. He had the samples tested in his own laboratory for length of life with the following results:

Length of life in hours	*Samples from*	
	Comp. A	*Comp. B*
700 and under 900	10	3
900 and under 1,100	16	42
1,100 and under 1,300	26	12
1,300 and under 1,500	8	3

Which company's lamps are more uniform.
(**Ans.** Sample *A*, C.V.=16.7%, Sample *B*, C.V.=11.9%. As the coefficient of variation of company *B* is smaller, its lamps are more uniform.)

19. Calculate mean and standard deviation for the following data:

Age under	10	20	30	40	50	60	70	80
No. of persons dying	15	30	53	75	100	110	115	125

20. The following table gives the area under cultivation, total production and cost per unit of production. Compare the variability of productivity (production per unit area) and the total cost of production.

Years	*1964*	*1965*	*1966*	*1967*	*1968*
Area under cultivation	10	12	9	14	15
Production per unit	60	72	63	91	90
Cost of production	4.0	4.5	5.0	5.5	5.8

(**Ans.** C V. of $x=7.7\%$. approx. C.V. of $y=31\%$ approx., ∴ Variation in total cost is much higher than the variation in productivity.)

21. The following gives the distribution by size of 40 different farms selected at random from 800 farms. Obtain a rough estimate of the (*i*) total acreage of 800 farms (*ii*) standard deviation of the acreage of 800 farms

Farm acreage to nearest acre	1-10	11-50	51-100	101-200	201-300	301-400	401-500
No. farms	13	9	0	7	4	5	2

(**Ans.** $\sigma=141$ acres.)

22. The number of employees, wages per employee and the variance of the wages per employee for two factories are given below:

	Factory A	*Factory B*
No. of employees	50	100
Average wages per employee per month (Rs)	120	85
Variance of the wages per employee per month (Rs)	9	16

(*a*) In which factory is there greater variation in the distribution of wages per employee?

(*b*) Suppose in factory *B*, the wages of an employee were wrongly noted as Rs 120 instead of Rs 100. What would be the correct variance for factory *B* ?

(Ans. (*a*) C.V. for factory *B*=4.7%, (*b*) Correct Variance=5.96.)

23. Particulars relating to the wage distribution of two manufacturing firms are given below:

	Firm A	*Firm B*
Mean	Rs 75	Rs 80
Median	Rs 72	Rs 70
Mode	Rs 67	Rs 62
Quartiles	Rs 62 and 78	Rs 65 and 85
Standard Deviation	Rs 13	Rs 17

Compare the features of the two distributions.

24. The marks obtained by the students of class *A* and *B* are given below:

Marks	5—10	10—15	15—20	20—25	25—30	30—35	35—40	40—45
Class *A*	1	10	20	8	6	3	1	—
Class *B*	5	6	15	10	5	4	2	2

Calculate mean, median, mode and standard deviation for the distributions. Explain your results regarding composition of the class in respect to intelligence.

25. Explain clearly the ideas implied in using arbitrary working origin and scale for the calculation of the arithmetic mean and standard deviation of frequency distribution.

The values of arithmetic mean and standard deviation of the following frequency distribution of a continuous variable derived from analysis are Rs 135.33 and Rs 9.6 respectively. Find the upper and lower limits of the various classes:

x'	−4	−3	−2	−1	0	+1	+2	+3	Total
f	2	5	8	18	22	13	8	4	

(Ans. $A=147.3$, $i=6$)

26. From the data given below state which series is more variable:

Variable	*Series A*	*Series B*
10—20	10	18
20—30	18	22
30—40	32	40
40—50	40	32
50—60	22	29
60—70	18	10

(Ans,	*A.M.*	*Median*	*Mode*	*C.V.*
Class *A*	19.6	18.4	17.3	32.6%
Class *B*	21.0	19.5	18.2	41.7%

The above figures clearly show that the average intelligence and variability of the class *B* is more than class A.)

27. (*a*) Coefficients of variation of two series are 58% and 69%. Their standard deviations are 21.2 and 15.6. What are their arithmetic means?

(*b*) When can coefficient of variations be greater than 100%? What can you say about the items of the given data in such a case?

(**Ans.** (*a*) For the first series, A.M.=36.6 approx., for the second series A.M.=22.6 approx.)

28. Two brands of tyres are tested with the following results:

Life thousands of miles	*Brand X*	*Brand B*
20—25	1	0
25—27.5	6	4
27.5—30	15	20
30—31	10	32
31—32	15	30
32—33	17	12
33—34	13	2
34—35	9	0
35—37.5	8	0
37.5—40	2	2
40—45	3	0
	100	180

(*a*) Draw a histogram for each frequency distribution.

(*b*) Which brand of tyre would you use on your fleet of trucks, and why?

(*c*) If the law forbids truck tyres to be used for more than 30,000 miles, how does that change your answer, if at all?

29. The following is a record of the number of bricks each day for 20 days by two brick layers *A* and *B*.

A: 725, 700, 750, 650, 675, 725, 675, 725, 625, 675, 700, 725, 675, 800, 650, 675, 625, 700, 650.

B: 575, 625, 600, 575, 675, 625, 575, 550, 650, 625, 550, 700, 625, 600, 625, 650, 575, 675, 625, 600.

Calculate the coefficient of variation in each case and discuss the relative consistency of the two brick layers. If the figures for *A* were in every case 10 more and those of *B* in every case 20 more than the figures given above, how would the answer be affected?

(**Ans.** Coefficient of variation A=6.0% and coefficient of variation B= 6.4%. So Brick layer *A* will continue to be more consistent than brick layer *B*).

30. (*a*) Mean of 100 items is 50 and their standard deviation is 4. Find the sum and sum of squares of all the items.

(*b*) The mean and the standard deviation of a sample of 100 observations were calculated as 40 and 5.1 respectively by a student, who took by mistake 50 instead of 40 for one observation. Calculate the correct mean and standard deviation.

(**Ans.** (*a*) $\Sigma X^2=251600$, (*b*) Correct standard deviation= $\sqrt{(25)}=5.0$.)

31. For a group of 50 male workers the mean and standard deviation of their weekly wages ars Rs 63 and Rs 9 respectively. For a group of 40 female workers these are Rs 54 and 6 respectively. Find the standard deviation of the combined group of 90 workers.

32. (*a*) Mean and standard deviations of two distributions of 100 and 150 items are 50, 5 and 40, 6 respectively. Find the mean and standard deviations of all the 250 items taken together.
(*b*) Mean and standard deviations of 100 items are found by a student as 50 and 5. If at the time of calculations two items are wrongly taken as 40 and 50 instead of 60 and 30, find the correct mean and standard deviations.
[**Ans.** (*a*) Standard deviation of 250 items = 7.46 (*b*) Correct standard deviation = 5.4 approx.]

33. The following data gives the arithmetic averages and standard deviations of the three sub-groups, calculate the arithmetic average and standard deviation of the whole group.

Sub-group	*No. of men*	*Average wage (Rs)*	*St. Dev. of wage (Rs)*
A	50	61.0	8.0
B	100	70.0	9.0
C	120	80.5	10.0

(**Ans.** Standard deviation = 11.9 rupees)

34. For a group containing 100 observations, the arithmetic mean and standard deviation are 8 and $\sqrt{10.5}$. For 50 observations selected from these 100 observations the mean and the standard deviation are 10 and 2 respectively. Find the arithmetic mean and the standard deviation of the other half.
(**Ans.** Standard deviation = 3 rupees)

35. A group has $\sigma = 10$, $N = 60$, $\sigma^2 = 4$. A sub-group of this has $\bar{X}_1 = 11$, $N_1 = 40$, $\sigma_1^2 = 2.25$. Find the mean and standard deviation of the other sub-group.

36. The mean of 5 observations is 4.4, and the variance is 8.24. If three of the five observations are 1, 2 and 6, find the other two.
(**Ans.** Two missing values are 4 and 9)

37. Two cricketers scored the following runs in the several innings. Find who is a better run-getter and who is more consistent player.

A 42, 17, 83, 59, 72, 76, 64, 45, 40, 32
B 28, 70, 31, 0, 59, 108, 82, 14, 3, 95

(**Ans.** *A* is better and more consistent)

38. The following are some of the particulars of the distribution of weights of boys and girls in a class.

	Boys	*Girls*
Number	100	50
Mean weight	60 kg	45 kg
Variance	9	4

(*i*) Find standard deviation of the combined data.
(*ii*) Which of the two distributions is more variable.

39. The ages of twenty husbands and wives are given below. Form a two-way frequency table showing the relationship between the ages of husbands and

wives with class intervals 20-25; 25-30; etc. Calculate the arithmetic mean and standard deviation of the ages after the classification:

S.No.	*Age of husband*	*Age of wife*	*S.No.*	*Age of husband*	*Age of wife*
1	28	23	11	27	24
2	37	30	12	39	34
3	42	40	13	23	30
4	25	26	14	33	31
5	29	25	15	36	29
6	47	41	16	32	35
7	37	35	17	22	23
8	35	25	18	29	27
9	23	21	19	38	34
10	41	38	20	48	4

Chapter 11

MOMENTS, SKEWNESS AND KURTOSIS

A quantity of data which by its mere bulk may be incapable of entering the mind is to be replaced by relatively few quantities which shall adequately represent the whole, or which, in other words, shall contain as much as possible, ideally the whole, of the relevant information contained in the original data."

—R. A. Fisher

11.1. INTRODUCTION

We have already introduced two parameters which describe the frequency distribution. These are mean $\bar{x}$ which locates the distribution, and the standard deviation σ which measures the scatter of the items about that mean. These two important parameters go a long way in describing the distribution, but there are many features of it which are not brought out. How symmetrical the distribution is about the mean, or how 'peaky' is the distribution are some other features that specify the distribution.

It can be shown that we can define a whole series of measures known as *moments* which when properly interpreted, give a wealth of information about the 'shape' of the distribution. It will be seen that the arithmetic mean $\bar{x}$ and the standard deviation σ are but the first two members of the series. The symmetry and the peakedness of the distribution can also be obtained from the higher members of this series.

Though the concept of moments is a highly mathematical one, an elementary introduction which brings out the physical significance is given here.

11.2. MOMENTS

The term *moment* is obtained from mechanics where the 'moment of a force' describes the tendency or capacity of a force to turn a pivoted lever (Fig. 11.1). Force F_1 has a counter-clockwise moment $F_1.x_1$ and force F_2 has a clockwise moment $F_2.x_2$.

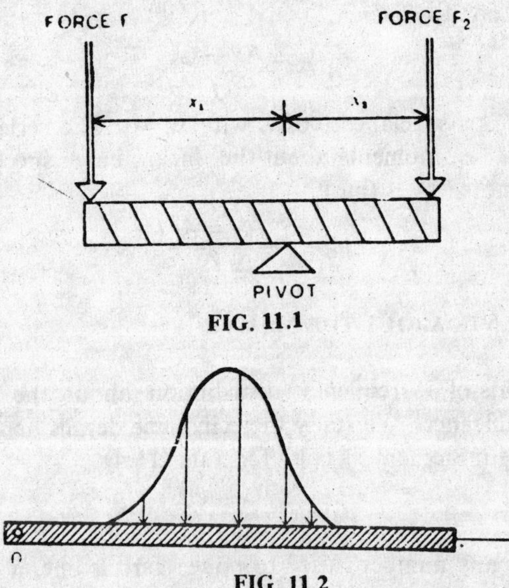

FIG. 11.1

FIG. 11.2

For a frequency distribution we imagine that at a distance x from the origin 0, a force equal to the frequency f associated with x acts and thus the moment about 0 is equal to fx. Taking the contributions of the whole distribution, the moment then is Σfx. This then, is, the moment on the lever produced if the whole distribution was sitting on a lever pivoted at the origin. To correct for the number of items involved (since we are interested in specifying only the 'shape' of the distribution) we divide by Σf, *i.e.* the total number of items. This 'first' moment about the origin then is *nu one prime*,

$$\nu_1' = \frac{\Sigma fx}{\Sigma f} \qquad (11.1)$$

Just like in mechanics, we can define higher order moments as well by multiplying f by higher powers of x. Thus

$$\nu_2' = \frac{\Sigma fx^2}{\Sigma f}$$

$$\nu_3' = \frac{\Sigma fx^3}{\Sigma f}, \text{ and so on.}$$

In general

$$\nu_r' = \frac{\Sigma fx^r}{\Sigma f} \qquad (11.2)$$

Instead of taking the moments about the origin we may also take them about any other point x_o (equivalent to pivoting the lever bar about that point). Then

$$\nu_r = \frac{\Sigma f(x-x_o)^r}{\Sigma f} \tag{11.3}$$

Thus, ν_r' is a special case of ν_r with $x_o=0$. The series of moments with $x_o=\bar{x}$, *i.e.*, moments about the mean have special significance and are denoted by μ (mu)

$$\mu_r = \frac{\Sigma f(x-\bar{x})^r}{\Sigma f} \tag{11.4}$$

11.3. MOMENTS ABOUT THE MEAN

The moments of a frequency distribution about the mean $\bar{x}$ have special significance. We study these in some details here.

The zeroeth moment μ_0 is by formula (11.4)

$$\mu_0 = \frac{\Sigma f(x-\bar{x})^0}{\Sigma f}$$

But since any number raised to power zero is one, it is clear that

$$\mu_0 = \frac{\Sigma f}{\Sigma f} = 1$$

For all distributions

Similarly,

$$\mu_1 = \frac{\Sigma f(x-\bar{x})}{\Sigma f}$$

But, by definition of the mean $\bar{x}$, the algebraic sum of the deviations about it, *i.e.*, $\Sigma f(x-\bar{x})$ is zero, so that $\mu_1=0$ *for all distributions.*

Next $\mu_2=\Sigma f(x-\bar{x})^2/\Sigma f$ is by definition the variance of the distribution. Thus,

$$\mu_0=1$$
$$\mu_1=0$$
$$\mu_2=\sigma^2 \tag{11.5}$$

for all distributions.

There is yet another point which can be deduced about the moments. We have already seen, while discussing the properties of the arithmetic mean that the sum of deviations below the mean equals the sum of deviations above the mean. This shows that the negative and positive deviations cancel out. This would be so in all distributions whether symmetrical or asymmetrical, when the deviations are raised to the first power.

When the deviations are raised to any even power their signs will all be positive and will no longer cancel out.

When the deviations are raised to any odd power (other than 1) and the sum of the negative deviations equals the sum of the positive deviations the distribution is symmetrical. Thus in *symmetrical distribution only*

$$\mu_3=0$$
$$\mu_5=0$$
$$\mu_7=0, \text{ etc.}$$

For this reason we can use these moments as *measures of asymmetry.*

Relation between μ and ν.* If we are given the moments about any arbitrary origin (including 0), then we can compute moments about mean by the following formula:

$$^{**}\mu_1=\nu_1-\nu_1=0$$
$$\mu_2=\nu_2-(\nu_1)^2 \qquad (11.6)$$
$$\mu_3=\nu_3-3\nu_2\times\nu_1+2(\nu_1)^3$$
$$\mu_4=\nu_4-4\nu_3\times\nu_1+6\nu_2\times(\nu_1)^2-3(\nu_1)^4$$

An important corollary follows from the above, *i.e.*, the mean square deviation about the mean of the observations is less than the mean square deviation about any arbitrary origin. In other words, the mean square deviation or variance (σ^2) about the mean is minimum— smaller than it would be if computed from any other average. So from the equation; since ν_2 is positive quantity, being a square μ_2 must be less than ν_2.

Illustration 11.1. Find the first four moments about the mean from the following data.

*The method of computing moment about the mean from moments about the arbitrary origin can be easily remembered by understanding the following.

$$\mu_1=(\nu-\nu_1) \text{ or } (\nu_1-d), \text{ where } d=(\bar{X}-A)=\frac{\Sigma(fx')}{N}=\nu_1$$
$$\mu_2=(\nu-d)^2=\nu_2-2\mu_1 d+d^2$$
$$=\nu_2-d^2=\nu_2-\nu_1^2$$
$$\mu_3=(\nu-d)^3=\nu_3-3\nu_2 d+3\nu_1 d^2-d^2$$
$$=\upsilon_3-3\nu_2\nu_1+2\nu_1^3$$
$$\mu_4=(\nu-d)^4=\nu_4-4\nu_3 d+6\upsilon_2 d^2-4\nu_1 d^3+d^4$$
$$=\nu_4-4\nu_3\nu_1+6\nu_2\nu_1^2-3\nu_1^4$$

**These measures may be in terms of class interval units or units of one. In the former case we will have to multiply μ_1, μ_2, μ_3 and μ_4 and i, i^2, i^3 and i^4 respectively, where i represents the class interval.

$\mu_2=\sigma^2$ if both are expressed either in class interval units or in units of one.

Class-interval	0—10	10—20	20—30	30—40
Frequency	1	3	4	2

Solution.

Size	x	f	$x' = \frac{x-25}{10}$	fx'	fx'^2	fx'^3	fx'^4
0—10	5	1	—2	—2	4	—8	+16
10—20	15	3	—1	—3	3	—3	+3
20—30	25	4	0	0	0	0	0
30—40	35	2	1	2	2	2	+2
		10		—3	9	—9	21

Moments about Arbitrary Mean

$$\nu_1 = \frac{-3 \times 10}{10} = -3$$

$$\nu_2 = \frac{9}{10} \times 10^2 = 90$$

$$\nu_3 = \frac{-9}{10} \times 10^3 = -900$$

$$\nu_4 = \frac{21}{10} \times 10^4 = 21{,}000.$$

Moments about Mean

$$\mu_1 = (\nu_1 - \nu_1) = 0$$

$$\mu_2 = \nu_2 - (\nu_1)^2 = 90 - 9 = 81$$

$$\mu_3 = \nu_3 - 3\nu_2 \times \nu_1 + 2(\nu_1)^3$$

$$= -900 - 270 \times (-3) + 2 \times (-27)$$

$$= -900 + 810 - 54 = -144$$

$$\mu_4 = \nu_4 - 4\nu_3 \times \nu_1 + 6\nu_2 \times (\nu_1)^2 - 3(v_1)^4$$

$$= 21{,}000 - 4 \times (-900) \times (-3) + (6 \times 90 \times 9) - 3 \times 81$$

$$= 21{,}000 - 10{,}800 + 4860 - 243 = 14{,}817.$$

11.5. SKEWNESS

When a frequency distribution is not symmetrical it is said to be asymmetrical or skewed. The nature of symmetry and the various types of asymmetry are illustrated in the example given below.

The following table shows the heights of the students of a college:

TABLE 11.1

Class	*A* f	*B* f	*C* f	*D* f
56.5—58.5	5	3	0	4
58.5—60.5	25	5	4	8
60.5—62.5	15	20	40	20
62.5—64.5	10	44	24	24
64.5—66.5	15	20	20	40
66.5—68.5	25	5	8	4
68.5—70.5	5	3	4	0
N	100	100	100	100
Mean	63.5	63.5	63.5	63.5
Median	63.5	63.5	63	64
Mode	—	63.5	61.9	65.1

The histograms and the corresponding curves are drawn in Figs. 11.3 and 11.4.

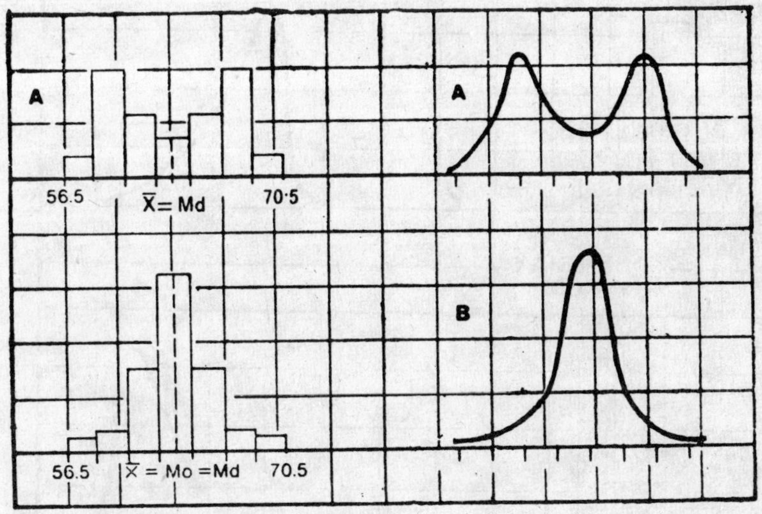

FIG. 11.3

A glance at the data of each of the four classes given above makes a very interesting study.

The shape of the curves, histograms and placement of equal items at equal distances on either side of the median clearly show that distributions *A* and *B* are symmetrical. If we fold these curves, or histograms on the ordinate at the mean, the two halves of the curve or histograms will coincide. In distribution *B* all the three measures of central tendency are identical. In *A*, which is a bimodal distribution mean and median have the same value.

Distributions *C* and *D* are asymmetrical. This is evident from the shape of the histograms and curves, and also from the fact that items at equal distances from the median are not equal in number. The three measures of central tendency for each of these distribution are of different sizes.

A point of difference between the asymmetry of distribution *C* and that of *D* should be carefully noted. In distribution *C* where the mean (63.5) is greater than the Median (63) and the Mode (61.9) the curve is pulled more to the right. In distribution *D* where Mean (63.5) is lesser than the Median (64) and mode (65.1) the curve is pulled more to the left.

In other words, we may say that if the extreme variations in a given distribution are towards higher values they give the curve a longer

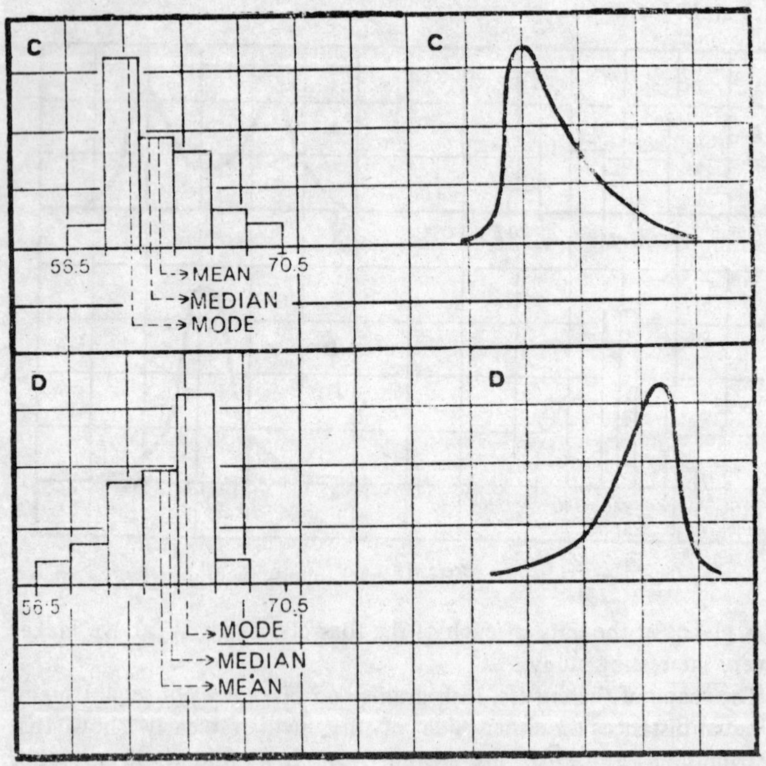

FIG. 11.4

tail to the right and this pulls the median and mean in that direction from the mode. If, however, extreme variations are towards lower values, the longer tail is to the left and the median and mean are pulled to the left of the mode.

It could also be shown that in a symmetrical distribution the lower and upper quartiles are equidistant from the median, so also are corresponding pairs of deciles and percentiles. This means that in a asymmetrical distribution the distance of the upper and lower quartiles from median is unequal.

From the above discussion, we can summarise the tests for the presence of skewness in the following words:

1. When the graph of the distribution does not show a symmetrical curve;
2. When the three measures of central tendency differ from one another;
3. When the sum of the positive deviations from the median are not equal to the negative deviations from the same value.
4. When the distances from the median to the quartiles are unequal;
5. When corresponding pairs of deciles or percentiles are not equi distant from the median.

Measures of Skewness

On the basis of the above tests, the following measures of skewness have been developed.

1. Relationship between 3 M's of central tendency—commonly known as the Karl Pearson's measure of skewness.
2. Quartile measure of skewness—known as Bowley's measures of skewness.
3. Percentile measure of skewness also called the Kelly's measure of skewness.
4. Measures of skewness based on moments.

All these measures tell us both the direction and the extent of the skewness.

1. **Karl Pearson's measure of skewness**. It has been shown earlier that in a perfectly symmetrical distribution, the three measures of central tendency, viz, mean, median and mode will coincide, As the distribution departs from symmetry these three values are pulled apart, the difference between the mean and mode being the greatest. Karl Pearson has suggested the use of this difference in measuring skewness. Thus absolute Skewness = Mean − Mode. (+) or (−) signs

obtained by this formula would exhibit the direction of the skewness. If it is positive the extreme variation in the given distribution are towards higher values. If it is negative, it shows that extreme variations are towards lower values.

Pearsonian coefficient of skewness. The difference between mean and mode, as explained in the preceding paragraph, is an absolute measure of skewness. An absolute measure cannot be used for making valid comparison between the skewness in two or more distributions for the following reasons: (*i*) The same size of skewness has different significance in distributions with small variation and in distributions with large variation, in the two series, and (*ii*) The unit of measurement in the two series may be different.

To make this measure as a suitable device for comparing skewness, it is necessary to eliminate from it the disturbing influence of 'variation' and 'units of measurements'. Such elimination is accomplished by dividing the difference between mean and mode by the standard deviation. The resultant coefficient is called *Pearsonian coefficient of skewness*. Thus the formula of Pearsonian coefficient of skewness is

$$\text{Coefficient of Skewness} = \frac{\text{Mean} - \text{Mode}}{\text{Standard deviation}} \qquad (11.7)$$

Since, as we have already seen, in moderately skewed distributions

$$\text{Mode} = \text{Mean} - 3\,(\text{Mean} - \text{Median})$$

We may remove the mode from the formula by substituting the above in the formula for skewness, as follows:

$$\text{Coefficient of Skewness} = \frac{\text{Mean} - [\text{Mean} - 3(\text{Mean} - \text{Median})]}{\text{Standard deviation}}$$

$$= \frac{\text{Mean} - \text{Mean} + 3\,(\text{Mean} - \text{Median})}{\text{Standard deviation}}$$

$$= \frac{3\,(\text{Mean} - \text{Median})}{\sigma} \qquad (11.8)$$

The removal of the mode and substituting median in its place becomes necessary because mode cannot always be easily located and is so much affected by grouping errors that it becomes unreliable.

Illustration 11.2. Find the skewness from the following data:

TABLE 11.2

Height (in inches)	*Number of persons*
58	10
59	18
60	30

61	42
62	35
63	28
64	16
65	8

Solution. Height is a continuous variable, and hence 58″ must be treated as 57.5″—58.5″, 59″ as 58.5″—59.5″, and so on.

Height (in inches)	*f*	*x′ from 61*	*fx′*	*fx′²*	*Cumulative, frequency*
58	10	—3	—30	90	10
59	18	—2	—36	72	28
59.5″—60—60.5″	30	—1	—30	30	58
60.5″—61—61.5″	42	0	—96		100
62	35	1	35	35	135
62.5″—63—63.5″	28	2	56	112	163
63.5″—64—64.5″	16	3	48	144	179
65	8	4	32	128	187
	187		171	611	
			+75		

$$\text{Mean}=61+\frac{75}{187}=61.4$$

$$\text{Mode}=60.5+\frac{35}{65}=61.04$$

$$\sigma=\sqrt{\frac{611}{187}-\left(\frac{75}{187}\right)^2}$$

$$=\sqrt{3.27-.16}$$

$$=\sqrt{3.11}=1.76,$$

$$\text{Skewness}=61.4-61.04$$

$$=0.36 \text{ inches.}$$

$$\text{Coefficient of Skewness}=\frac{.36}{1.76}=.205.$$

Alternatively, we can determine the median

$$\text{Median}=\text{the size of } \frac{187}{2} \text{th item}$$

$$=93.5\text{th item}$$

$$=60.5+\frac{1\times 35.5}{42}=61.35$$

$$\text{Skewness}=3(61.4-61.35)$$
$$=3(.05)$$
$$=.15.$$

$$\text{Coefficient of Skewness}=\frac{.15}{1.76}=.09.$$

The two coefficients are differents because of the difficulties associated with determination of mòde.

2. Bowley's (Quartile) measure of skewness. In the above two methods of measuring skewness, the whole of the series is taken into consideration. But absolute as well as relative skewness may be secured even for a part of the series. The usual device is to measure the distance between the lower and the upper quartiles. In a symmetrical series the quartiles would be equidistant from the value of the median, *i.e.*,

$$\text{Median}-Q_1=Q_3-\text{Median},$$

In other words, the value of the median is the mean of Q_1 and Q_3. In a skewed distribution, quartiles would not be equidistant from median unless the entire asymmetry is located at the extremes of the series. Bowley has suggested the following formula for measuring skewness, based on above facts.

$$\text{Absolute SK}=(Q_3-\text{Me})-(\text{Me}-Q_1)$$
$$=Q_3+Q_1-2\text{ Me} \qquad (11.9)$$

If the quartiles are equidistant from the median, *i.e.*, $(Q_3-\text{Med})=(\text{Med}-Q_1)$, then SK=0. If the distance from the median to Q_1 exceeds that from Q_3 to the median, this will give a negative skewness. If the reverse is the case; it will give a positive skewness.

If the series expressed in different units are to be compared, it is essential to convert the absolute amount into the relative. Using the interquartile range as a denominator we have for the coefficient of skewness the following:

$$\text{Relative SK}=\frac{Q_3+Q_1-2\text{ Med}}{Q_3-Q_1} \qquad (11.10)$$

or $$\frac{(Q_3-\text{Med})-(\text{Med}-Q_1)}{(Q_3-\text{Med})+(\text{Med}-Q_1)}$$

If in the series the median and lower quartiles coincide, then the SK becomes (+1). If the median and upper quartiles coincide, then the SK becomes (−1).

This measure of skewness is rigidly defined and easily computable. Further, such a measure of skewness has the advantage that it has value limits between (+1) and (−1), with the result that it is sufficiently sensitive for many requirements. The only criticism levelled against such a measure is that it does not take into consideration all the item of these series, *i.e.*, extreme items are neglected.

Illustration 11.3. Calculate the coefficient of skewness of the data of Table 11.2 based on quartiles.

Solution. With reference to Table 11.2

$$Q_1 = \text{the size of } \frac{N}{4} \text{th} \left(= \frac{187}{4} = 46.75 \text{ th} \right) \text{item}$$

$$= 59.5 + \frac{18.75}{30}$$

$$= 59.5 + .63 = 60.13.$$

$$Q_3 = \text{the size of } \frac{3N}{4} \text{th item} \left(= \frac{3 \times 187}{4} = 140.25\text{th} \right) \text{item}$$

$$= 62.5 + \frac{5.25}{28}$$

$$= 62.5 + .19 = 62.69.$$

$$\text{Skewness} = 62.69 + 60.13 - 2(61.35) = .12. \quad \text{(Formula 11.9)}$$

$$\text{Coefficient of Skewness} = \frac{.12}{62.69 - 60.13} \quad \text{(Formula 11.10)}$$

$$= \frac{.12}{2.56} = .047.$$

3. **Kelly's (Percentile) measure of skewness.** To remove the defect of Bowley's measure that it does not take into account all the values, it can be enlarged by taking two deciles (or percentiles), equidistant from the median value. Kelly has suggested the following measure of skewness:

$$\text{SK} = P_{50} - \frac{P_{90} + P_{10}}{2}$$

or

$$= D_5 - \frac{D_9 + D_1}{2}$$

Though such a measure has got little practical use, yet theoretically this measure seems very sound.

4. Measure of Skewness Based on Moments

It will be recalled that in symmetrical distribution, all the odd moments about mean, *i.e.* μ_3, μ_5, μ_7...etc., are equal to zero. If the odd moments (other than μ_1) are not equal to zero then it means that the distribution is skewed. But the computation of odd moments alone is not a satisfactory method of measuring skewness. To exhibit the degree of asymmetry we must relate these moments to the standard deviation. Thus, the various moments divided by the proper power of the standard deviation give us another family of useful coefficient which we denote by the Greek letter α (alpha).

Symbolically,

$$\alpha_1 = \frac{\mu_1}{\sigma} = 0$$

$$\alpha_2 = \frac{\mu_2}{\sigma^2} = 1$$

$$\alpha_3 = \frac{\mu_3}{\sigma^3} \tag{11.11}$$

$$\alpha_4 = \frac{\mu_4}{\sigma^4}$$

$$\alpha_5 = \frac{\mu_5}{\sigma^5}$$

Since μ_1 is always zero for every distribution $(\mu_1 = \frac{\Sigma x}{N} = 0)$, it is useless as a test of skewness. μ_3 is preferable to any higher moment as it is easier to calculate and also because the higher the moment the more will it vary from sample to sample. The positive and negative sign of α_3 will have the same significance as the sign of (mean$-$mode) has.

Another measure of skewness is obtained by the following formula:

$$\frac{\sqrt{\beta_1}(\beta_2+3)}{2\,(5\beta_2-6\beta_1-9)} \tag{11.12}$$

where $\beta_1 = \alpha_3^2$ and $\beta_2 = \alpha_4$.

This measure will be positive if the mean exceeds, and negative if the mean falls short of the mode.

Illustration 11.4. Calculate the Karl Pearson's coefficient of skewness from the following data:

Marks	*No. of students*	*Marks*	*No. of students*
above 0	150	above 50	70
,, 10	140	,, 60	30
,, 20	100	,, 70	14
,, 30	80	,, 80	0
,, 40	80		

Solution.

Marks	*f*	*Mid-point*	$x'=(X-A)/10$	fx'	fx'^2	*c.f.*
0—10	10	5	−3	−30	90	10
10—20	40	15	−2	−80	160	50
20—30	20	25	−1	−20	20	70
30—40	0	35	0	−130	—	70
40—50	10	45	1	10	10	80
50—60	40	55	2	80	160	120
60—70	16	65	3	48	144	136
70—80	14	75	4	56	224	150
	150			194	808	
				+64		

Since it is a bi-modal distribution Karl Pearson coefficient (Formula 11.8) is appropriate and we need to calculate $\bar{x}$, Me and σ.

$$\bar{x}=35+\frac{64}{150}\times 10=35+4.27=39.27$$

$$\text{Median}=\text{size of } \frac{150}{2} \text{ th item}$$

$$=40+\frac{10\times 5}{10}=45$$

$$\text{Standard deviation}=i\times\sqrt{\frac{\Sigma fx'^2}{N}-\left(\frac{\Sigma fx'}{N}\right)^2}$$

$$=10\times\sqrt{\frac{808}{150}-\left(\frac{64}{150}\right)^2}$$

$$=10\times\sqrt{5.387-.182}$$

$$=10\times 2.28=22.8$$

$$\text{Skewness}=\frac{3\,(\bar{x}-\text{Median})}{\sigma}$$

$$=\frac{3\,(39.27-45)}{22.8}$$

$$=\frac{3\,(-5.73)}{22.8}=\frac{-17.19}{22.8}=-.75$$

Illustration 11.5. Find the standard deviation and coefficient of skewness for the following distribution.

Variable	0—	5—	10—	15—	20—	25—	30—	30—40
Frequency	2	5	7	13	21	16	8	3

Solution. As nothing is specified skewness should be computed by Karl Pearson's method.

Variable	*x*	*Frequency*	$x'=\frac{x-22.5}{5}$	fx'	fx'^2
0—5	2.5	2	—4	— 8	32
5—10	7.5	5	—3	—15	45
10—15	12.5	7	—2	—14	28
15—20	17.5	13	—1	—13	13
20—25	22.5	21	0	—50	
25—30	27.5	16	1	16	16
30—35	32.5	8	2	16	32
35—40	37.5	3	3	9	27
				41	
		75		$\Sigma fx'=-9$	193

$$\bar{x}=22.5+\frac{-9}{75}\times 5=21.9$$

$$\sigma=5\times\sqrt{\frac{193}{75}-\left(\frac{-9}{75}\right)^2}$$

$$=5\sqrt{2.5733-.144}$$

$$=5\times 1.6=8.0$$

Mode lies in 20—25 class (evidently)

$$\text{Mode}=20+\frac{21-13}{2\times 21-13-16}\times 5$$

$$=20+\frac{8\times 5}{13}=23.1$$

$$\text{Skewness}=\frac{\text{Mean}-\text{Mode}}{\sigma}=\frac{21.9-23.1}{8}$$

$$=\frac{-1.2}{8}=-0.15.$$

Illustration 11.6. Calculate coefficient of skewness from the data as follows.

Life time (Haurs)	*No. of Tubes*
300—400	14
400—500	46
500—600	58
600—700	76
700—800	68
800—900	62
900—1000	48
1000—1100	22
1100—1200	6

Solution.

Class	*x*	*f*	$x'=\frac{x-750}{100}$	fx'	fx'^2
300—400	350	14	−4	− 56	224
400—500	450	46	−3	−138	414
500—600	550	58	−2	−116	232
600—700	650	76	−1	− 76	76
700—800	750	68	0	−386	
800—900	850	62	1	62	62
900—1,000	950	48	2	96	192
1,000—1,100	1050	22	3	66	198
1,100—1,200	1150	6	4	24	96
				+248	
		400		−138	1,494

$$\bar{x}=750+\frac{-138}{400}\times 100=750-34.5=715.5$$

$$\sigma=100\times\sqrt{\frac{\Sigma fx'^2}{N}-\left(\frac{\Sigma fx'}{N}\right)^2}$$

$$=100\times\sqrt{\frac{1,494}{400}-\left(\frac{-138}{400}\right)^2}$$

$$=100\sqrt{3.735-.1190}$$

$$=100\times 1.9015=190.15$$

Mode lies in 600—700 class interval

$$\text{Mode}=600+\frac{76-58}{2\times 76-58-68}\times 100$$

$$=600+\frac{18}{26}\times 100=669.23$$

$$\text{Skewness} = \frac{\text{Mean} - \text{Mode}}{\sigma}$$

$$= \frac{715.5 - 669.23}{190.15} = \frac{46.27}{190.15} = 0.243$$

Illustration 11.7. From the following data compute quartile deviation and the coefficient of skewness:

Size	5—7	8 10	11—13	14—16	17—19
Frequency	14	24	38	20	4

Solution.

Size	*Frequency*	*Cumulative Frequency*
4.5— 7.5	14	14
7.5—10.5	24	38
10.5—13.5	38	76
13.5—16.5	20	96
16.5—19.5	4	100

$$Q_1 = 7.5 + \frac{3 \times 11}{24} = 8.87$$

$$Q_3 = 10.5 + \frac{3 \times 37}{38} = 10.5 + \frac{111}{38} = 10.5 + 2.92 = 13.42$$

$$\text{Median} = 10.5 + \frac{3 \times 12}{38} = 10.5 + \frac{36}{38} = 10.5 + .947 = 11.447$$

$$\text{Quartile deviation} = \frac{Q_3 - Q_1}{2} = \frac{13.42 - 8.87}{2} = \frac{4.55}{2} = 2.275$$

$$\text{Skewness} = \frac{Q_3 + Q_1 - 2Me}{Q_3 - Q_1}$$

$$= \frac{13.42 + 8.87 - 22.89}{13.42 - 8.87}$$

$$= \frac{-.6}{4.55} = -0.13$$

Illustration 11.8. In a certain distribution the following **results** were obtained:

$\bar{x} = 45.00$; Median $= 48.00$

Coefficient of Skewness $= -.4$

You are required to estimate the value of **standard deviation**;

Solution.

$$\text{Skewness} = \frac{3\ (\text{Mean} - \text{Median})}{\sigma}$$

$$-.4 = \frac{3\ (45 - 48)}{\sigma}$$

$$-.4\sigma=-9$$

$$\sigma=\frac{9}{.4}=22.5.$$

Illustration 11.9. Karl Pearson's coefficient of skewness of a distribution is +0·32. Its standard deviation is 6.5 and mean is 29.6. Find the mode and median of the distribution.

Solution.

$$\text{Coefficient of Skewness}=\frac{\text{Mean}-\text{Mode}}{\sigma}$$

$$0.32=\frac{29.6-\text{Mode}}{6.5}$$

or $$6.5\times.32=29.6-\text{Mode}$$

$$\text{Mode}=29.6-2.08=27.52.$$

$$\text{Coefficient of Skewness}=\frac{3\ (\text{Mean}-\text{Median})}{\sigma}$$

$$0.32=\frac{3\ (29.6-\text{Median})}{6.5}$$

$$6.5\times0.32=88.8-3\ \text{Median}$$

$$\text{Median}=\frac{88.8-2.08}{3}$$

$$=28.91.$$

Illustration 11.10. For a distribution Bowley's coefficient of skewness is −0.36. Q_1=8.6 and Median=12.3. What is the quartile coefficient of dispersion?

Solution.

$$\text{Bowley's coefficient of Skewness}=\frac{Q_3+Q_1-2\ Me}{Q_3-Q_1}$$

$$-0.36=\frac{Q_3+8.6-24.6}{Q_3-8.6}$$

or $$-.36\ Q_3+8.6\times.36=Q_3+8.6-24.6$$

or $$3.096+16=Q_3+.36\ Q_3$$

or $$19.096=1.36\ Q_3$$

or $$Q_3=14.04$$

$$\text{Quartile coefficient of dispersion}=\frac{Q_3-Q_1}{Q_3+Q_1}$$

$$=\frac{14.04-8.6}{14.04+8.6}$$

$$=\frac{5.44}{22.64}=0.24.$$

Illustration 11.11. Compute quartile deviation and coefficient of skewness from the following values.

Median = 18.8 centimetres
Q_1 = 14.6 centimetres
Q_3 = 25.2 centimetres

Solution.

$$\text{Quartile deviation } \frac{Q_3-Q_1}{2}=\frac{25.2-14.6}{2}=\frac{10.6}{2}=5.3 \text{ centimetres}$$

$$\text{Coefficient of Skewness}=\frac{Q_3+Q_1-2\text{ Median}}{Q_3-Q_1}$$

$$=\frac{25.2+14.6-37.6}{25.2-14.6}$$

$$=\frac{2.2}{10.6}=0.207.$$

Illustration 11.12. You are given the position in a factory before and after the settlement of an industrial dispute. Comment on the gains or losses from the point of the workers and that of the management.

	Before	*After*
No. of workers	2,440	2,359
Mean wages	45.5	47.5
Median wages	49.0	45.0
Standard deviation	12.0	10.0

Solution.

Employment. Since the number of workers employed after the settlment is less than the number of employed before, it has gone against the interest of the workers.

Wages. The total wages paid after the settlement were 2,350×47.5 = Rs 1,11,625; before the settlement the amount disbursed was 2,400×45.5 = Rs 1,09,200.

This means that the workers as a group are better off now than before the settlement. and unless the productivity of workers have gone up this may be against the interest of management.

Uniformity in the wage structure. The extent of relative uniformity in the wage structure before and after the settlement can be determined by a comparison of the coefficient of variation.

$$\text{Coefficient of variation before}=\frac{12}{45.5}\times 100=26.4$$

$$\text{,, ,, after}=\frac{10}{47.5}\times 100=21.05.$$

This clearly means that there is comparatively lesser disparity in the wages received by the workers. Such a position is good for both the workers and the management.

Pattern of the wage structure. A comparison of the mean with the median leads to the obvious conclusion that before the settlement more than 50 per cent of the workers were getting a wage higher than this mean *i.e.* (Rs 45.5). After the settlement the number of workers whose wages were more than Rs 45.5 became less than 50 per cent. This means that the settlement has not been beneficial to all the workers. It is only 50 per cent workers who have been benefited as a result of an increase in the total wages bill.

11.6. KURTOSIS

So far we have characterised a frequency distribution by its central tendency, variability, and the extent of asymmetry. There remains but one more common type of attribute of freqency distribution *viz.*, its peakedness. Look at the following, Fig. 11.5 in which are drawn three symmetrical curves *A*, *B* and *C*.

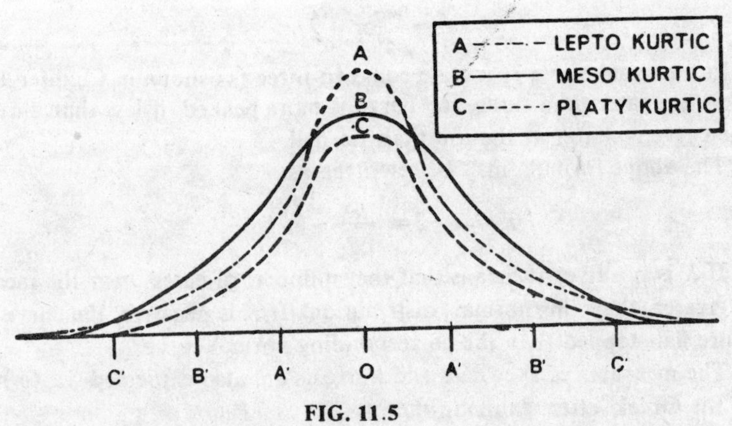

FIG. 11.5

The three curves differ widely in regard to convexity, an attribute to which Karl Pearson referred as 'Kurtosis.' The measure of kur tosis exhibits the extent to which the curve is more peaked or more flat-topped than the normal curve (See Chapter 17). *A* curve to be called a normal curve must have its convexity as shown in curve *B* [in addition to two other requisites, *viz.*, (*i*) Unimodal, (*ii*) Symmetrical]. When the curve of a distribution is relatively flatter than the normal curve, it is said to have kurtosis. When the curve or polygon is relatively more peaked, it is said to lack kurtosis,

Karl Pearson gave the name 'Mesokurtic' to a normal curve or a skewed curve that has the same degree of convexity as the normal curve. In Fig. 11.3 curve *B* is a mesokurtic curve. If some of the cases about one standard deviation from the mean move in towards the centre and other move out towards the tails, thus making the curve unusually peaked, we say that the result would be a 'Leptokurtic' curve, as curve *A*. If on the other hand, some of the cases around the mode move out a little towards each half of the curve, thus making the curve unusually flat-topped we say that the result would be a 'Platykurtic' curve, as curve *C*.

To quote Walker, "The terms platykurtic, leptokurtic, mesokurtic, are not particularly important but they are rather interesting and roll pleasantly under one's tongue." An amusing sentence written by 'Student' (a British Statistician) was quoted by him which reads "the platykurtic curves, like the platypus are squat with short tails, while leptokurtic curves are high with long tails like the Kangaroo which 'leps'."

The kurtosis is measured by

$$\beta_2 = \alpha_4 = \frac{\mu_4}{\mu_2^2} \text{ or } \frac{\mu_4}{\sigma^4}$$

In a normal curve β_2 will be equal to three (as shoun in Chapter 17) If β_2 is greater than three, the curve is more peaked, if less than three, the curve is flatter at the top than normal.

The above formula may be rewritten as:

$$K = \alpha_4 - 3 = \frac{\mu_4}{\sigma^4} - 3.$$

If *K* is positive, it means that the number of cases near the mean is greater than in normal distribution. If *K* is negative, the curve is more flat--topped than the corresponding normal curve.

The measures of skewness and kurtosis are also expressed in terms of the Greek letter, gamma (γ).

$$\gamma_1 = \sqrt{\beta_1} = \frac{\mu_3}{\sigma^3} = \alpha_3$$

$$\gamma_2 = \beta_2 - 3 = \frac{\mu_4}{\sigma^4} - 3 = \frac{\mu_4}{\mu_2^2} - 3 = \frac{\beta^2 - 3\mu_2^2}{\mu_2^2}$$

γ_1 and γ_2 are the measures of skewness and kurtosis respectively. If γ_1 is more than zero, then the conclusion will be the presence of positive skewness; if γ_1 is less than zero, it will mean negative skewness, and in case γ_1 is zero, then there will be absence of skewness.

Similarly, if a curve is Leptokurtic, γ_2 will be positive; if Platykurtic, γ_2 will be negative; and in case Mesokurtic, γ_2 will be exactly zero.

Sheppard's correction for grouping errors. Computation of mean, standard deviation, etc., from the grouped data of a frequency distribution are based upon the assumption that all the values in a class interval are concentrated at the *centre of that interval.* In unimodal (or single peaked) symmetrical distributions this assumption results in a systematic error in the calculation of the even moments. For the odd moments, the sum total of the errors on account of the above presumption is zero in such symmetrical distributions, because of the neutralizing effect of the errors as these appear with positive and negative signs.

Such errors may with advantage be corrected by the application of the formula propounded by W.F. Sheppard. The correction formulae for 2nd and 4th moments are:

$$\mu_2 \text{ (Corrected)} = \text{Uncorrected } \mu_2 - \frac{i^2}{12}$$

$$\mu_4 \text{ (Corrected)} = \text{Uncorrected } \mu_4 - \frac{1}{2}\, i^2 \mu_2 \text{ (Uncorrected)} + \frac{7}{240}\, i^4$$

(where i is the class interval).

The use of this correction be restricted to:

(1) grouped data, *i.e.*, series must not be discrete but continuous:

(2) symmetrical and moderately skewed distribution, *i.e.*, the distribution tapers off to zero in both directions;

(3) total frequency is sufficiently large, say 1,000; and

(4) class interval is more than $\frac{1}{20}$ th of the range.

Illustration 11.13. From the following frequency distribution calculate the first four moments, β_1 and β_2.

Class	*f*
10—14	1
15—19	4
20—24	8
25—29	19
30—34	35
35—39	20
40—44	7
45—49	1
50—54	5

Solution.

Let the assumed mean $A=32$

Class	Mid-point	f	x'	fx'	fx'^2	fx'^3	fx'^4	Cum. f
10–14	12	1	−4	−4	16	−64	256	1
15–19	17	4	−3	−12	36	−108	324	5
20–24	22	8	−2	−16	32	−64	128	13
25–29	27	19	−1	−19	19	−19	19	32
				−51		−255		
30–34	32	35	0		0		0	67
35–39	37	20	1	20	20	20	20	87
40–44	42	7	2	14	28	56	112	94
45–49	47	5	3	15	45	135	405	99
50–54	52	1	4	4	16	64	256	100
		100		+53	212	+275	1,520	
				$\Sigma fx'$ = +2	$\Sigma fx'^2$ = +212	$\Sigma fx'^3$ = +20	$\Sigma fx'^4$ = +1,520	

$$v_1=\frac{\Sigma fx'\times(c.i)}{\Sigma f}=\frac{2\times 5}{100}=\frac{10}{100}=.1.$$

$$v_2=\frac{\Sigma fx'^2\times(c.i.)^2}{\Sigma f}=\frac{212\times 5^2}{100}=\frac{212\times 25}{100}=53$$

$$v_3=\frac{\Sigma fx'^3\times(c.i)^3}{\Sigma f}=\frac{20\times 5^3}{100}=\frac{20\times 125}{100}=25$$

$$v_4=\frac{\Sigma fx'^4\times(c.i)^4}{\Sigma f}=\frac{1,520\times 5^4}{100}=\frac{1,520\times 625}{100}=9,500.$$

$\mu_1=v_1-v_1=0.$

$\mu_2=v_2-v_1^2=53-(.1)^2$

$=53-.01=52.99.$

$\mu_3=v_3-3v_2v_1+2v_1^3$

$=25-3\times 53\times(.1)+2(.1)^3=25-15.9+.002=9.102.$

$\mu_4=v_4-4v_3v_1+6v_2v_1^2-3v_1^4$

$=9,500-4\times 25\times(.1)+6\times 53\times(.1)^2-3(.1)^4$

$=9,500-10+3.18-.0003$

$=9,493.1797$ or $9,493.18.$

Corrected $\mu_2=\mu_2-\dfrac{h^2}{12}$ (where h is the class interval)

$$=52.99-\frac{5^2}{12}=52.99-2.083=50.91.$$

Corrected $\mu_3=\mu_3=9.1.$

Corrected $\mu_4 = \mu_4 - \frac{h^2\mu_2}{2} + \frac{7h^4}{240}$

$$= 9{,}493.18 - \frac{25}{2} \times 52.99 + \frac{7 \times (5)^4}{240}$$

$$= 9{,}493.18 - 662.375 + 18.23$$

$$= 8{,}849.03.$$

$$\beta_1 = \frac{\mu_3^2}{\mu_2^3}$$

$$= \frac{(9.1)^2}{(50.91)^3} = .000627.$$

$$\beta_2 = \frac{\mu_4}{\mu_2^2}$$

$$= \frac{8{,}849.03}{(50.91)^2} = 3.414.$$

$$\gamma_1 = \sqrt{\beta_1}$$

$$= \sqrt{.000627} = .025.$$

$$\gamma_2 = \beta_2 - 3$$

$$= 3.414 - 3 = .414.$$

Illustration 11.14. The first four central moments of a distribution are 0, 2.5, 0.7 and 18.75. Test the skewness and Kurtosis of the distribution.

Solution. Skewness is tested by μ_3 which should be equal to zero in a symmetrical distribution. Since in the given problem it is 0.7 we can conclude that the distribution is not symmetrical. But to measure the extent and direction of the skewness we make use of constant

$$\alpha_3 = \frac{\mu_3}{\sigma^3}$$

$$\alpha_3 = \sqrt{\frac{\mu_3^2}{\mu_2^3}}$$

$$= \sqrt{\frac{(0.7)^2}{(2.5)^3}}$$

$$= \sqrt{0.031} = 0.18$$

Since $\alpha_3 = 0.18$ we conclude that the distribution is not symmetrical but has a positive skewness $= 0.18$.

Kurtosis is tested by β_2. In normal case β_2 should be equal to three. If it is greater than three the curve is more peaked, if less than three the curve is more flat-topped.

$$\beta_2 = \frac{\mu_4}{\sigma^4} = \frac{\mu_4}{\mu_2^2}$$

$$= \frac{18.75}{(2.5)^2} = \frac{18.75}{6.25} = 3.$$

Since $\beta_1 = 3$ we conclude that the curve is mesokurtic.

Illustration 11.15. (*a*) The standard deviation of a symmetrical distribution is 3. What must be the value of the 4th moment about the mean in order that the distribution be mesokurtic?

(*b*) If the 4 moments of a distribution about the value 5 are equal to —4, 22, —117 and 560 determine the corresponding moments about the mean and about zero.

Solution. A normal curve is mesokurtic. In a normal curve (mesokurtic curve) the value $\beta_3 = 3$.

This means that 3 should be equal to $\frac{\mu_4}{\mu_2^2}$ in order that the curve be mesokurtic

i.e.
$$3 = \frac{\mu_4}{\mu_2^2}$$

$$3 = \frac{\mu_4}{81}$$

$$243 = \mu_4.$$

Thus the 4th moment should be equal to 243.

(*b*) Moment about the mean.

$$\mu_1 = (v_1 - v_1) = 0$$

$$\mu_2 = v_2 - (v_1)^2 = 22 - 16 = 6$$

$$\mu_3 = v_3 - 3v_2 \times v_1 + 2(v_1)^3$$

$$= -117 - 66 \times (-4) + 2 \times (-64)$$

$$= -117 + 264 - 128$$

$$= 19.$$

$$\mu_4 = v_4 - 4v_3 \times v_1 + 6v_2 \times v_1^2 - 3(v_1)^4$$

$$= 560 - (-468) \times (-4) + 132 \times 16 - 768$$

$$= 560 - 1872 + 2112 - 768$$

$$= 32.$$

The moments about zero are

$$v_1';\ v_2';\ v_3';\ v_4'$$

$$v_1' = \left[\begin{array}{c}\text{Assumed}\\ \text{Mean}\end{array}\right] + v_1$$

$$= 5 + (-4)$$

$$= 1$$

$$\nu_2' = \mu_2 + (\nu_1')^2$$
$$=6+(1)^2$$
$$=7$$
$$\nu_3' = \mu_3 + 3\nu_1'\nu_2' - 2\nu_1'^3$$
$$=19+3\times1\times7-2(1)^3$$
$$=19+21-2$$
$$=38$$
$$\nu_4' = \mu_4 + 4\nu_1'\nu_3' - 6\nu_1'^2\nu_2' + 3\nu_1'^4$$
$$=32+4\times1\times38-6(1)^2\times7+3(1)^4$$
$$=32+152-42+3$$
$$=145.$$

11.7. MISCELLANEOUS ILLUSTRATIONS

Illustration 11.16. Calculate quartile coefficient of skewness from the following:

Weight	*No. of persons*	*c.f.*
Under 100	1	1
100—109	14	15
110—119	66	81
120—129	122	203
130—139	145	348
140—149	121	469
150—159	65	534
160—169	31	565
170—179	12	577
180—189	5	582
190—199	2	584
200 and above	2	586

Solution.

$$\text{Coefficient of Skewness} = \frac{Q_3+Q_1-2Q_2}{Q_3-Q_1}$$

Weight	*No. of persons*	*c.f.*
Under 100	1	1
100—109	14	15
110—119	66	81
119.5—120—129—129.5	122	203
129.5—130—139—139.5	145	348
139.5—140—149—149.5	121	469
150—159	65	534
160—169	31	565

170—179	12	577
180—189	5	582
190—199	2	584
200 and above	2	586

Calculation of Median.

$$\frac{N}{2}=293.$$ This item lies in 129.5—139.5 group.

$$Q_2=L+\frac{\frac{N}{2}-c}{f}\times i=129.5+\frac{293-203}{145}\times 10$$

$$=129.5+\frac{900}{145}=129.5+6.2=135.7$$

Calculation of Q_1

$$\frac{N}{4}=146.5.$$ This item lies in 119.5—129.5 group.

$$Q_1=L+\frac{\frac{N}{4}-c}{f}\times i=119.5+\frac{146.5-81}{122}\times 10$$

$$=119.5+\frac{655}{122}=119.5+5.37=124.87.$$

Calculation of Q_2

$$\frac{3N}{4}=439.5.$$ This item lies in 139.5—149.5 group.

$$Q_3=L+\frac{\frac{3N}{4}-c}{f}\times i=139.5+\frac{439.5-348}{121}\times 10$$

$$=147.1$$

Coefficient of Skewness

$$\frac{147.1+124.87-2(135.7)}{147.1-124.87}=\frac{271.97-271.4}{22.23}=\frac{.57}{22.33}=.03$$

Illustration 11.17. Compute the Pearsonian measure of skewness for the following distribution:

Size in inches	*No. of observations*
30—33	2
33—36	4
36—39	26
39—42	47
42—45	15
45—48	6

Solution.

Size x	*f*	*Mid-point m*	$d' = \frac{m-37.5}{3}$	*fd'*	*fd'²*
30—33	2	31.5	−2	−4	8
33–36	4	34.5	−1	−4	4
36—39	26	37.5	0	0	0
39—42	47	40.5	1	47	47
42—45	15	43.5	2	30	60
45—48	6	46.5	3	18	54
	100			87	173

$$\text{Skewness}=\frac{\bar{X}-\text{Mode}}{\sigma}$$

Calculation of Mean

$$\text{Mean}=A+\frac{\Sigma fd'}{N}\times i=37.5+\frac{87}{100}\times 3=37.5+2.61=40.11$$

Calculation of S.D.

$$\sigma=\sqrt{\frac{\Sigma fd'^2}{N}-\left(\frac{\Sigma fd'}{N}\right)^2}\times i=\sqrt{\frac{173}{100}-\left(\frac{87}{100}\right)^2}$$

$$=\sqrt{.973}\times 3=2.96.$$

Calculation of Mode

By inspection modal class is 39—42.

$$\text{Mode}=L+\frac{f_1-f_0}{2f_1-f_0-f_2}\times i=39+\frac{47-26}{94-26-15}\times 3$$

$$=39+\frac{21\times 3}{53}=39+1.19=40.19$$

$$\text{Coefficient of Skewness}=\frac{40.11-40.19}{2.96}=\frac{-.08}{2.96}=-.027.$$

Illustration 11.18. It is known that $\bar{X}$ and the median of a distribution are 3.0 and 4.0. Is the distribution skewed?

Solution. In the symmetrical distribution the values of mean and median are alike. Skewness can be measured in absolute terms by taking the difference between mean and median. As the mean and median of this distribution are of different sizes, therefore. the distribution is skewed. Moreover, distribution is negatively skewed as Mean$<$Median.

Illustration 11.19. Consider the following distributions:

	A	*B*
$\bar{X}$	100	90
Median	90	80
S.D.	10	10

(*i*) Distribution A has the same degree of variation as distribution *B*. Do you agree? Give reasons.

(*ii*) Both the distributions have same degree of skewness; True/False. Give reasons.

Solution. (*i*) Coefficient of variation $\frac{\sigma}{\text{Mean}} \times 100.$

A | *B*

$$\text{C.V.}=\frac{10}{100}\times 100=10 \qquad \text{C.V.}=\frac{10}{90}\times 100=11.11$$

Coefficient of variation of distribution *B* is more than that of *A* and hence *A* does not have the same degree of variation as distribution *B*.

(*ii*) Coefficient of Skewness$=3\ \frac{(\text{Mean—Median})}{\sigma}$

A | *B*

$$\frac{3(100-90)}{10}=\frac{30}{10}=3 \qquad \frac{3(90-80)}{10}=\frac{30}{10}=3$$

As the coefficient of skewness is same, both distribution have same degree of skewness.

Illustration 11.20. Karl Pearson's coefficient of skewness of a distribution is +.40. Its standard deviation is 8 and mean is 30. Find mode and median of the distribution.

Solution. Coefficient of Skewness$=\frac{\bar{X}-\text{Mode}}{\sigma}$ or $\frac{3(\bar{X}-\text{Median})}{\sigma}$

$$.40=\frac{30-\text{Mode}}{8} \text{ and } .40=\frac{3(30-\text{Median})}{8}$$

$$\text{Mode}=30-3.2=26.8 \qquad 3\text{ Median}=90-3.2$$

$$\text{Median}=28.9$$

Illustration 11.21. In a moderately skewed frequency distribution mean is 50 and median is 53. If the coefficient of variation is 20%, find the Pearsonean coefficient of skewness.

Solution. $\text{C.V.}=\frac{\sigma}{X}\times 100$

Putting the values

$$20=\frac{\sigma}{50}\times 100$$

$$\therefore \quad \sigma=10$$

$$\text{Coefficient of Skewness} = \frac{3(\text{Mean} - \text{Median})}{\sigma}$$

$$= \frac{3(50-53)}{10} = -\frac{9}{10} = -.9.$$

Illustration 11.22. The sum of a set of 100 numbers is 4000 and the sum of its square is 162500 and Median is 41. Find the coefficient of skewness.

Solution. We are given $\Sigma X = 4000$ and $\Sigma X^2 = 1,62,500$
$N = 100.$

$$\text{Mean} = \frac{\Sigma X}{N} = \frac{4000}{100} = 40.$$

$$\text{S.D.} = \sqrt{\frac{\Sigma X^2}{N} - (\text{A.M.})^2} = \sqrt{\frac{162500}{100} - 1600}$$

$$= \sqrt{\frac{2500}{100}} = \sqrt{25} = 5.$$

$$\text{Coefficient of Skewness} = \frac{3(\bar{X} - \text{Median})}{\sigma}$$

$$= \frac{3(40-41)}{5} = \frac{-3}{5} = -.6.$$

Alternatively (Solving with the help of moments). Let the moments about origin *i.e.* zero are represented by v'

$$\text{Mean or } v_1' = \frac{\Sigma X}{N} = \frac{4000}{100} = 40.$$

$$v'_2 = \frac{\Sigma X^2}{N} = \frac{162500}{100} = 1625$$

Variance or $\mu_2 = v'_2 - (v'_1)^2 = 1625 - (40)^2 = 25.$

Therefore S.D. $= \sqrt{25} = 5.$

$$\text{Coefficient of Skewness} = \frac{3(\bar{X} - \text{Median})}{\sigma} = \frac{3(40-41)}{5} = -.6$$

Illustration 11.23. The first four moments of a distribution about the value 5 are 2, 20, 40 and 50. Obtain as far as possible, the various characteristics of the distribution on the basis of the information given.

(*M. Com., Delhi, 1974*)

Solution. v represents moments about arbitrary origin

v' " " " origin *i.e.* zero

μ " " " Mean

$\mu_1 = v_1 + v_1 = 2 - 2 = 0$

$\mu_2 = v_2 - (v_1)^2 = 20 - (2)^2 = 16$

$$\mu_3 = \nu_3 - 3\nu_1\nu_2 + 2(\nu_1)^3 = 40 - 3.2.20 + 2(2)^3 = 40 - 120 + 16$$
$$= -64$$

$$\mu_4 = \nu_4 - 4\nu_3\nu_1 + 6\nu_2\nu_1^2 - 3(\nu_1)^4$$
$$= 50 - 4.40.2 + 6.20.4 - 3(2)^4$$
$$= 50 - 320 + 480 - 48 = 162.$$

Mean = Assumed mean $+\nu_1 = 5 + 2 = 7$

S.D. $= \sqrt{\mu_2} = \sqrt{16} = 4.$

$$\text{Skewness} = \frac{-64}{\sqrt{16\times16\times16}} = \frac{-64}{64} = -1.$$

$$\beta_2 = \frac{\mu_4}{\mu_2^2} = \frac{162}{256} = .6.$$

So Kurtosis $= \beta_2 - 3 = 0.6 - 3 = -2.4$

Illustration 11.24. For a distribution mean is 10 and variance is 16. γ_1 (gama) is 1 and β_2 is 4. Obtain the first four moments about origin zero.

Solution. $\mu_1 = 0, \ \mu_2 = 16$

$$\gamma_1 = \frac{\mu_3}{\sqrt{\mu_2^3}} \text{ or } 1 = \frac{\mu_3}{\sqrt{16^3}} \text{ or } 64 = \mu_3,$$

$$\beta_2 = \frac{\mu_4}{\mu_2^2} \text{ or } 4 = \frac{\mu_4}{16^2} \text{ or } \mu_4 = 256\times4 = 1024$$

$$\nu_1' = \text{Mean} = 10$$
$$\nu_2' = \mu_2 + (\nu_1')^2 = 16 + (10)^2 = 116$$
$$\nu_3' = \mu_3 + 3\nu_1'\nu_2' - 2(\nu_1')^3$$
$$= 64 + 3.10.116 - 2(10)^3$$
$$= 64 + 3480 - 2000 = 1544$$
$$\nu_4' = \mu_4 + 4\nu_1'\nu_3' - 6\nu_2'\nu_1^2 + 3(\nu_1)^4$$
$$= 1024 + 4.10.1544 - 6.100.116 + 3.100.100$$
$$= 1024 + 61760 - 69600 + 30{,}000$$
$$= 23184$$

Illustration 11.25. Examine whether the following results of a piece of computation for obtaining the second (central) moment are consistent or not?

$N = 120$, $\Sigma x = -125$ and $\Sigma x^2 = 128$

Solution. Second (central) moment means second moment about arithmetic mean *i.e.* square of S.D. If deviations are taken from assumed mean, formula for its calculation is

$$\mu_2 = \sigma^2 = \frac{\Sigma x^2}{N} - \left(\frac{\Sigma x}{N}\right)^2$$

$$=\frac{128}{120}-\left(\frac{-125}{120}\right)^2$$

$=1.067-1.085=-0.018.$

But μ_2 *i.e.* square of standard deviation can never be **negative.** Hence there is inconsistency in the figures.

Illustration 11.26. The first four moments of a distribution about $x=4$ are 1, 4, 10, 45. Show that the Mean is 5 and calculate the moments about $x=0$

Solution. A=Arbitrary origin=4

$v_1=1=(\bar{X}-A)$

$\therefore$ Mean$=4+1=5$

Moments about the Mean

$\mu_2=v_2-v_1^2=4-1=3$

$\mu_3=10-3(1)(4)+2(1^3)=0$

$\mu_4=45-4(1)(10)+6(1^2)(4)-3(1^4)=26$

Moment about zero

$v_2=\mu_2+v_1^2=3+25=28$

$v_3=\mu_3+3v_1v_2-2v_1^3$

$=0+3(5)(28)-2(5)^3$

$=420-250=170$

$v_4=\mu_4+4v_1v_3-6v_1^2v_2+3v_1^4$

$=26+4\times(5)\times(170)-6\times25\times28+3\times625$

$=1091.$

Illustration 11.27. In a certain distribution the first four moments about the arbitrary origin (4) are -1.5, 17, -30 and 108. Calculate β_2 and state whether the distribution is Leptokurtic or Platykurtic.

Solution. Moments about mean are:

$\mu_1=v_1-v_1=-1.5-(-1.5)=0$

$\mu_2=v_2-v_1^2=17-(1.5)^2=17-2.25$

$=14.75$

$\mu_3=v_3-3v_2v_1+2v_1^3$

$=30-3(17)(-1.5)+2(-1.5)^3$

$=-30+76.5-6.75$

$=39.75$

$\mu_1=v_4-4v_3v_1+6v_2v_1^2-3v_1^4$

$=108-4(-30)(-1.5)+6(17)(1.5)^2-3(-1.5)^4$

$=108-180+229.5-15.1875$

$=142.313$

$$\beta_2 = \frac{\mu_4}{\mu_2^2}$$
$$= \frac{142.313}{217.56}$$
$$= .65$$

Since β_2 is less than 3, the distribution is Platykurtic.

Illustration 11.28. In a continuous frequency distribution grouped in intervals of 2, the first four moments about the Mean (=10) expressed in class-interval units are, 0, 2.8, −2 and 24.5 respectively. Obtain the first four moments about Mean and calculate β_1 and β_2. Also calculate the coefficient of skewness based on moments and comment on kurtosis.

Solution. Moments are given about Mean in terms of class interval units. These may be denoted as μ_1', μ_2', μ_3' and μ_4'. If the class-interval is represented by i the required moments are:

$$\mu_1 = \mu_1' \times i = 0 \times 2 = 0$$
$$\mu_2 = \mu_2' \times i^2 = 2.8 \times 4 = 11.2$$
$$\mu_3 = \mu_3' \times i^3 = -2 \times 8 = -16$$
$$\mu_4 = \mu_4' \times i^4 = 24.5 \times 16 = 392.0$$
$$\beta_1 = \frac{\mu_3^2}{\mu_2^3} = \frac{(-16)^2}{(11.2)^3} = 0.1823$$
$$\beta_2 = \frac{\mu_4}{\mu_2^2} = \frac{392}{(11.2)^2} = 3.12$$

Coefficient of skewness = $\sqrt{\beta_1} = \sqrt{0.1823} = +.427$

As β_2 is greater than 3, the distribution is Leptokurtic.

Illustration 11.29. The first three moments about the value 2 are 1, 16 and −40. Compute the moments about the mean and the origin.

Solution.

$$\mu_1 = \nu_1 - \nu_1 = 1 - 1 = 0$$
$$\mu_2 = \nu_2 - (\nu_1)^2 = 16 - (1)^2$$
$$= 16 - 1 = 15$$
$$\mu_3 = \nu_3 - 3\nu_2\nu_1 + 2(\nu_1)^3$$
$$= -40 - 3 \times 16 \times 1 + 2(1)^3$$
$$= -40 - 48 + 2 = -86.$$

Now moments about the origin are

$$\nu_1' = \frac{\Sigma fX}{N} = \bar{X} = A + \nu_1 = 2 + 1 = 3$$
$$\nu_2' = \frac{\Sigma fX^2}{N} = \mu_2 + \nu_1'^2 = 15 + 9 = 24$$

$$\nu_3' = \frac{\Sigma fX^3}{N} = \mu_3 + 3\nu_1'\nu_2' - 2\nu_1'^3$$

$$= -86 + 3\times3\times24 - 2\times27$$

$$= 76$$

Thus the first three moments about the mean stand at the figures of 0, 15 and −86 respectively; and about the origin are 3, 24 and 76 respectively.

It may be noted that as the first moment about the origin is always equal to the value of the mean, so the mean value of the above distribution is 3.

PROBLEMS

1. What is skewness? How would you find it in a non-symmetrical distribution?
2. Distinguish between
   (*a*) Dispersion and skewness.
   (*b*) Positive and negative skewness.
   (*c*) Quartile deviation method of measuring skewness and Pearson's measure of skewness.
3. Explain the three terms: dispersion, skewness and kurtosis.
4. Give a suitable formula for measuring the peakedness (kurtosis) of frequency distribution.
5. Find Bowley coefficient of skewness;
   According to the Census of 1941; the following are the population figures in 000' of the Ist 36 cities in India. 2488, 591, 437, 20, 213, 143, 1490, 407. 284, 176, 169, 181, 777, 387, 302, 213, 204, 153, 133, 391, 263, 176, 178, 142, 522, 360, 260, 193, 131, 92, 672, 258, 239, 160, 147, 151.
   (Ans. Bowley Coeff. of skewness = 0 47)
6. Calculate Bowley's measure of skewness from the following data:

Payment of Commission in Rs	*No. of Salesmen*	*Payment of Commission in Rs*	*No. of Salesmen*
100—120	4	200—220	80
120—140	10	220—240	32
140—160	16	240—260	23
160—180	29	260—280	17
180—200	52	280—300	7

   (Ans. Coeff. of Skewness = —.114)
7. Compute Coefficient of dispersion and skewness of the following data:

Central size	1	2	3	4	5	6	7	8	9	10
Frequency	2	9	11	14	20	24	20	16	5	2

   (Ans. Quartile Coeff. of dispersion = 0.27 and Bowley Coeff. of Skewness = 0.7)

8. Find mean, median, standard deviation and a coefficient of skewness from the following data of ages of students of a school.

Age	5—7	8—10	11—13	14—16	17—19
No. of students	7	12	19	10	2

(Ans. A.M. = 11.28 yrs, Median = 11.45 yrs, 26, Dev. = 3.15 yrs, and Coeff. of Skewness = 0.16).

9. Find the coefficient of skewness for the following distribution:

Variable	0—5—10—15—20 - 25—30—35—40
Frrquency	2 5 7 13 21 16 8 3

(**Ans.** Coefficient of Skewness = —0.15)

10. Find Karl Pearson's coefficient of skewness from the following data:

Marks above	0	10	20	30	40	50	60	70	80
No. of students	150	140	100	80	80	70	30	14	0

(**Ans.** Coefficient of skewness = —0.75)

11. The following facts were gathered before and after an industrial dispute:

	Before dispute	*After dispute*
No. of workers employed	516	508
Mean wages (in Rs)	49.50	51.75
Median wages (in Rs)	52.70	50.00
Variance of wages	100.00	121 00

Compare the position before and after the dispute in respect of *(a)* total wages *(b)* model wages *(c)* standard deviation *(d)* coefficient of variation *(e)* skewness

	Before dispute	*After dispute*
(**Ans.** *(a)* Total wages	= Rs 25,542	Rs 26,289
(b) Model wages	= Rs 59.1	Rs 46.5
(c) St. Dev	= Rs 10	Rs 11
(d) Coeff. of V.	= 20.02%	21,26%
(e) Skewness	= —0.96	0.477

12. From the following information regarding the marks obtained at college and the competitive examinations, find which group is more homogeneous in intelligence and which is more skew.

College Examination		*Competitive Examination*	
Marks	*No. of Students*	*Marks*	*No. of Students*
100—150	20	1200—1250	50
150—200	45	1250—1300	85
200—250	50	1300—1350	72
250—300	25	1350 - 1400	60
300—350	19	1400—1450	16

(**Ans.** College exam , coeff. of skewness = 0.182).

13. Compute the first four moments about an arbitrary origin (say, 67) from the following data of heights in inches adult Irishmen.

Height	59	60	61	62	63	64	65	66	67	68	69	70	71	72	73
Adults	9	0	2	2	7	15	33	58	73	62	40	25	15	10	3

(**Ans.** First moment about 67=0.341, Second moment about 67=4.821, Third moment about 67=4.468 and Fourth moment about 67=81.608)

14. For the distribution of heights of 347, adult Irish containing 15 classes of 1 unit ranging from 59″ to 73″, the following values are obtained for moments with 67″ as working origin $\Sigma fx=118$, $\Sigma fx^2=1168$, $\Sigma fx^3=1546$. Determine the second and third moments about the mean.

(**Ans.** $\mu_2=4.70$ approx., $\mu_3=-.377$ approx.)

15. The first four moments of a distribution about the value 4 are —1.7, 17, –30, 108. Calculate the moments about the mean.

(**Ans.** $\mu_1=0$, $\mu_2=14.75$, $\mu_3=38.75$ and $\mu_4=142.3125$ or 142.3 approx).

16. Compute the first four moments about arithmetic mean from the following data:

Variate value	5	10	15	20	25	30	35
Frequency	8	15	20	32	23	17	5

(**Ans.** $\mu_1=0$, $\mu_2=29.993$, $\mu_3=49.584$ and $\mu_4=8355.96$.)

17. Calculate the first four moments about the mean from the following data:

X (Mid point of the class interval)	1	2	3	4	5	6	7	8	9
f (Frequency)	1	6	13	25	30	22	9	5	2

(**Ans.** $\mu_1=0$, $\mu_2=2.488$, $\mu_3=.6753$ or 0.675 approx. and $\mu_4=18.344$ approx.)

18. Calculate first four moments for the following data:

Marks	10—30	30—40	40—50	50 - 60	60—70	70—80	80—90	90—100
Frequency	6	28	96	75	56	30	8	1

(**Ans.** $\mu_1=0$, $\mu_2=180.9$, $\mu_3=521.3$ approx. and $\mu_4=97146.6$ approx.)

19. In a continuous frequency distribution grouped in intervals of 2, the first four moments about the mean i.e. 10 expressed in class interval units are, 0, 2, 8, —2 and 24.5 respectively. Obtain the first four moments about mean and calculate β_1 and β_2. Also calculate the coefficient of skewness based on moments and comment on kurtosis.

(**Ans.** Moments about mean, $\mu_1=0$, $\mu_2=11.2$, $\mu_3=-76$ and $\mu_4=392$; $\beta_1=0.1823$. $\beta_2=3.12$ and coeff. of skewness= –.427.

20. For a distribution standard deviation is 2. What should be the value of μ_4 so that it is (*i*) mesokurtic (*ii*) leptokurtic (*iii*) platykurtic.

(**Ans.** For mesokurtic $\mu_4=48$, for leptokuric μ_4 must be greater than 48 and platykurtic $\mu_4<48$.)

21. Second, third and fourth central moments of a variable characteristics are 19.67, 29.26 and 866.0 respectively. Calculate the beta coefficients correct to three decimal places.

22. Calculate first four moments and γ_1 and γ_2 after applying Sheppard's corrections.

Weekly wages in Rs	0—10	10—20	20—30	30—40	40—50	50—60	60—70	70—80
No. of persons	15	23	35	49	32	28	12	6

(**Ans.** $\gamma_1=0.137$, $\gamma_2=-0.593$)

23. The first four central moments of a distribution are 0, 2.5, 0.7 and 18.75. Test the skewness and kurtosis of the distribution.

(**Ans.** As $\gamma_1=0.177$ the distribution is positively skewed and intensity of skewness is not very high. As $\beta_2=3$, the distribution is mesokurtic, i.e., as far as kurtosis is concerned it resembles the normal distribution.)

24. The first four moments of a distribution are 1, 4, 10 and 46, respectively. Compute the first four Central Moments and the Beta Constants. Comment upon the nature of the distribution.
(Ans. The distribution has β_1 as 0, so it has no skewness. The $\beta_2=3$, so the distribution is mesokurtic. The data is of symmetrical mesokurtic distribution *i.e.* normal distribution).

25. Find out the mean, mean deviation and Kurtosis of the following data:

Class interval	0—10	10—20	20—30	30—40
Frequencies	1	3	4	2

26. Calculate coefficient of skewness by all methods known for the following data on ages of 250 persons in a sample study.

Age below	10	20	30	40	50	60	70	80
No of persons	15	35	60	84	96	127	188	250

(Ans. Skewness=—20.2)

27. Following figures relate to marks obtained by 20 students of a class in a test. Find the first four moments about assumed mean 10, first four moments about arithmetic mean and the values of β_1 and β_2. 1, 9, 12, 2, 10, 15, 3, 16, 19, 5, 18, 17 6, 20, 0, 8, 5, 7, 11, 12.
(Ans. $\upsilon_1=-0.2$, $\upsilon_2=36.9$, $\upsilon_3=0.4$, $\upsilon_4=2483.7$. $\mu_1=0$, $\mu_2=36.38$, $\mu_4=22:52$ $\mu_4=2492.87$. $\beta_1=0.1014$, $\beta_2=1,835$.)

28. μ_2, μ_3 and μ_4 of a variable are 19.67, 29.26 and 866.0. Calculate β_1 and β_2.
(Ans. $\beta_1=0.113$, $\beta_2=2.29$.)

29. For a distribution of items $\bar{X}=54$, $\sigma=3$, $\beta_1=0$ and $\beta_2=3$. During calculations two items 64 and 50 were taken as 62 and 52. What are the correct constants.
(Ans. $\bar{X}=54$, $\sigma=2.97$, $\beta_1=.004$, $\beta_2=2.8$)

30. The first three moments of a distribution about the value 3 of a variable are 2, 10 and 30 respectively. Obtain the first three moments about zero. Show also that the variance of the distribution is 6.
(Ans. $\upsilon_1'=5$, $\upsilon_2'=31$, $\upsilon_3'=231$)

31. (*a*) Karl Pearson coefficient of skewness of a distribution is +0.32. Its standard deviation is 6.5 and mean is 29.6. Find the mode and median of the distribution.
(*b*) If the mode of the above distribution is 24.8, what will be the standard deviation?
(*c*) For a distribution Bowley's coefficient of skewness is −0.36, $Q_1=8.6$ and median=12.3. What is quartile coefficient of dispersion?
(*d*) Compute quartile deviation and coefficient of skewness given the following values: Me=18.8, $Q_1=14.6$, $Q_3=25.2$.
(Ans. (*a*) $M_0=27.52$, Me=28.91, (*b*) $\sigma=15$, (*c*) 0.24, (*d*) 5.3, 0.207

32. In a frequency distribution the coefficient of skewness based upon the quartiles is 0. If the sum of the upper and the lower quartiles is 100 and the median is 38, find the value of upper quartile.
(Ans. $Q_3=70$).

Chapter 12

ANALYSIS OF TIME SERIES

12.1. TIME SERIES PROBLEMS

A time series is a set of data pertaining to the values of a variable at different times. Typical time series are the population of India at each successive decennial census; daily business handled by a bank, monthly production statistics of a steel mill; annual rainfall in the gangetic plains; velocity of a satellite launched by milli- second from firing; monthly traffic-accident fatalities; and daily closing price of shares on stock-market.

Time-series are sometimes studied simply because of historical interest, but mostly because of the interest in future, in predicting the value of the variable at a future date. A stock-broker studies the daily movement of share-prices so as to be able to predict its future behaviour; demographers study census statistics to be able to forcast future population; insurance people study death-rate tables to decide on insurance rate on people buying coverage now; and meteorologists study rainfall patterns to be able to predict what kind of monsoon to expect this year and offer advice to farmers.

A time series has an important property which makes it quite distinct from any other kind of statistical data. If we were studying the height of boys in a class, the height of one boy is completely independent of that of the boy preceding him or following him, for we could take the boys in any order. But since the *time* is not an arbitrary variable, the adjacent values on a time series are not independent in the sense outlined above. Thus, the population in 1981 is related to that in 1971 and that in 1991. Similarly share price today is related to that tomorrow and yesterday.

In fact it is easy to see that any movement in a time series (*i.e.*, change in the value of a variable with time) is made up of four main contributions. These are referred to as the components of a time series, and are (*a*) trend, (*b*) seasonal changes, (*c*) cyclic changes and (*d*) irregular or random fluctuations. These four components are described and explained in the next section.

The task of the statistical analysis of time series lies in isolating each of the four components,, *viz.*, the trend, the seasonal, the cyclical and the irregular movements from the original composite series, so that each of these components may be measured, examined, analysed and described independently of the others, *i.e.*, by using the scientific procedure of 'holding other things constant.'

This composite series is symbolised by the following general terms:

$$O = T \times S \times C \times I$$

where O symbolizes the original data and T for trend, S for seasonal, C for cycle and I for irregular components have been used. This multiplicative model is to be used when S, C and I, are given in percentages. If, however, their true (absolute) values are known the model takes the additive form, *i.e.*, $O = T + C + S + I$.

It may be stated at this stage that though the analysis of time series involves the disentanglement of the four components, yet one should not conclude that the processes and techniques described above are absolutely accurate and definite. Even the seasonal variation can rarely be precisely repeated every year.

12.2. FOUR COMPONENTS OF A TIME SERIES

The *trend* is a broad long-term movement, such as a 'general' increase in the level. Thus even through the blue-chip shares show an apparantely random fluctuation in prices daily and hourly, a general upword motion may be apparent if we take a long-term view.

One important problem in the analysis of the time-series concerns isolating the trend so that reasonable prediction about the future can be made. The fluctuations of the series are then from this *trend line*.

The *cyclic* changes are generally long-period fluctuations about the trend and are caused by a complex combination of forces affecting the equilibrium of the phenomena under study. Thus the 3 to 10 years business cycles are the long term upswings followed by downswings at relatively regular intervals shaped by the complex equilibrium of supply and demand.

A business cycle may also be referred to as 'four phase' cycle, composed of prosperity, recession, depression and recovery. This swing from prosperity to recovey and back again to prosperity varies both in time and intensity. Statistical techniques are employed to isolate these disruptive oscillations and to analyse the conditions surrounding them, so that the Government may take such measures as are feasible to prevent the deterioration of mild recessions or early crises into deep depressions or collapsing economy and keep the swings of prosperity within reasonable limits without developing into stormy speculations. Such analysis is of great interest to the business enterprise also, as it can modify its programmes in accordance with the analysed predictions of the cyclical swing.

It is a common knowledge that consumption and production of many commodities, interest rates, bank clearings, etc. are marked by *seasonal* swings. Climate and custom together play an important role in giving rise to seasonal movements to almost all the industries—primary, secondary and tertiary. The yearly cycle of the weather directly affects agricultural production and agricultural marketing. There is almost a definite and limited period of growing and so also of the harvesting of crops. This more or less regularly recurring period of harvesting and marketing every year considerably affects the manufacturing, transport and other industries. Banks are called upon to provide a seasonal increase in their credits and the rail-roads are expected to be ready to have their peak loads because of harvesting. Even the retail and wholesale dealers feel the effect of the seasonality of farmer's income. Statisticians analysing time-series are called upon to provide an estimate of how much a given 'fluctuation' is due to seasonal variations, and how much it is because of other causes which may be the subject of investigation.

The *residual*, or the *random* fluctuations are what are beyond the three other components, and are obviously unpredictable. These occur because of random causes, too numerous to keep track of. Infact these random fluctuations mask the other three components and the purpose of the time-series analysis is to bring out the trend and cyclic variations from the 'noise' of the random fluctuations.

12.3. USES OF THE ANALYSIS OF TIME SERIES

1. It helps in understanding the past behaviour of a variable and in determining the rate of growth and the extent and direction of periodic fluctuations. Such information is needed by the sales manager who wishes to know how the volume of sales is fairing when it fluctuates and why, how it compares with the volume of production. To the economist also, who desires to trace the trend of prices and to examine the upward and downward movement of the price-level, such an analysis is of very great help.

2. The study of the past behaviour of a variable enables us to predict future tendencies. To business executives who are to plan their production programme such an analysis is, therefore, of great assistance, for it is with the help of analysis of this nature that approximately correct estimates of the future demand can be made. In fact the entire budgetary process is largely dependent upon forecasts which in their turn are made possible by analysis of past performance.

3. The knowledge of the behaviour of the variables may enable us to iron out intra-year variations. Thus seasonal ups and downs in sales may be reduced by making effective advertisements.

4. The determination of the impact of the various forces influencing different variables facilitates their comparison.

Having stated the nature and the perpose of the analysis of time series, we now proceed to isolate and measure the four components of the composite force which shapes a series in its movement through time.

12.4. EDITING TIME-SERIES DATA

The first step in this direction is to ensure comparability among the data. To the attainment of the requisite level of comparability adjustments may be needed for (1) calender variations, (2) price changes, (3) population change, and (4) other miscellaneous changes.

1. **Calendar variations**. If we have monthly data, it shall not be comparable because the number of days in each month is not the same. This difficulty can be overcome by expressing the monthly data per calendar day by dividing the amount for the month by the number of days of that month. The same purpose is served by expressing the data on a per week basis.

2. **Price changes**. Data pertaining to sales of a business unit is usually expressed in monetary terms, which is a product of quantity sold and the price at which sales have been effected. If the price has changed from one time to another no valid comparison of sales can be made unless we eliminate the disturbing influence of changing prices. This is done by dividing the sales fingures for given time periods by the prices of the respective time periods and in that process converting the sales data into the number of units sold. If the sales figure refer to more than one item, a price index (See Chapter 13) can be used for making the adjustment.

3. **Population changes**. A comparison of the total number of passengers carried by Indian Railways in 1948 and those in 1981

may not be useful since the number of users of railways facilities increased by many crores during these thirty-theee years. If however, we divide the total passengers in different years by the respective population for the year we would get the number of times an individual has made use of the passenger transport facilities provided by our railways.

4. **Other miscellaneous changes.** Besides the three factors named above there may be other changes which affect the comparability of the data and as such necessitate adjustments. Thus, units in terms of which data are reported may have changed during the span of the time series: or there may be change in the type of the products, or in the definitions of the terms that we use. It is necessary that proper adjustments are made in the time series for all these disturbing factors to ensure its comparability.

12.5 SECULAR TREND

A study of the series of economic and business statistics would reveal that most of them have a natural tendency to increase or decrease over a period of several years. For instance, a scrutiny of agricultural production in India during the last fifty years would show that the production has been in general on an increase and that the increase has been fairly regular. The same is true about the size of population, production of steel. bank deposits, currency in circulation etc. The underlying factor causing an upward trend in a time series may be application of natural sciences in the fields of agriculture and industry, the changes in the forms of business organisation facilitating accumulation of huge capital for specialisation and mass production, the introduction of automation, scientific management, quality control, improved marketing etc., to raise the standard of living, productivity, etc.

Not all time series show an upward trend. A declining rate is noticed in the data of epidemics, deaths and births, etc. owing to better and widely available medical facilities and higher standard of living. In economic series also a declining trend may be found due to keen competition—say, of Roadways against Railways, invention of cheaper or better substitute, *e.g.*, synthetic rubber for natural rubber, etc. Authorities like Raymond B. Prescott are of the view that Law of Growth [embracing (*i*) period of experimentation when growth is small, (*ii*) period of growth into the social fabric, (*iii*) period during which growth is retarded as a saturation point is approached and (*iv*) period of stability] applies to all industries and consequently not only does relative growth tend to decline, but eventually further expansion will be physically impossible.

This tendency of growth (positive or negative) is called secular trend, or secular movement. Thus, *secular movement is that irreversible movement which continues in general in the same direction for a considerable period of time.*

Let us discuss the parts of the definition. The words "irreversible movement" indicate that it does not change its direction as frequently as a so-called 'four phase' cycle composed of prosperity, recession, depression and recovery.

The words "continues in general in the same direction" tell that it has a gradual and presistent tendency to change in the same direction. But this does not mean that the rise or fall must continue each and every year throughout the period. If we are given the data for about 30 years during which production of a particular commodity tends generally to rise, we should say that there was a secular rise in production during the period, even though there might be a single year or two in which there was some fall in production.

The words "for a considerable period of time" convey the idea that the given movement must last for a period that one would call a long time—long for such data to continue to change uniformly. 'If we are counting the bacterial population of a culture every five minutes; and the population continues to increase fairly regularly for many days, we should say that this was a secular change. Again, if we are given production figures of a particular commodity only for 12 or 24 months, the mere fact of increase in production for two successive years, would not suggest a secular change." Thus a period of five days may be secular under one condition, whereas a period of two years may not be secular under other conditions.

The statistical problem in the trend analysis of a time series lies in (1) deciding type of trend which will fit the data satisfactorily, and (2) fitting the trend of the type decided. The problem of trend analysis arises because one may be interested either in the trend itself or one may wish to eliminate the trend statistically in order to get one more of other movements in the series.

There are a variety of methods of isolating the trend. We will discuss them one by one.

12.6. FREE HAND METHOD

The simplest, quickest and easiest method of estimating the secular trend is to plot the original data on a graph and then to draw a smooth curve through the points so that it may accurately describe the general long-run tendency of the data. While drawing such a curve the minor short-run fluctuations or abrupt variations are not taken into account. The use of flexible rulers may be made while drawing the smoothed line.

The obvious disadvantage of such a method is that it is highly subjective, since the trend line, obtained by this method, shows what an individual considers as the trend. This method should therefore be used only by experienced persons.

12.6. THE SEMI-AVERAGE METHOD

Another method of describing the secular trend is to divide the original data into two equal parts. The values of each part are then summed up and averaged. The average of each part is centred in the period of time of the part from which it has been calculated and then plotted on the graph. Thus, a line may be drawn to pass through the plotted points.

When the data consists of an even number of values its division into parts does not present any difficulty. Thus if there are ten values, each part would have five of them. But if there is an odd number of values, the easiest procedure would be to eliminate the middle value. Thus if we have eleven values, first five of them may be included in the first half, and the last five in the second half, thus omitting the middle value of the series. This is illustrated by the following example where the middle value corresponding to the year 1966 has been omitted for the purpose of computing averages of the two halves of the given data.

Illustration 12.1. Compute trend by the semi-average method of the following data:

Year	*Sales (lakhs of Rs)*	*Semi-total*	*Semi-average*
1961	**38**		
1962	**40**		
1963	**46**	224	44.8
1964	**49**		
1965	**51**		
1969	**55**		

1967	61		
1968	63		
1969	69	345	69.0
1970	72		
1971	80		

These two semi-averages **are plotted in the middle of the respective time spans**. Thus 44.8 is plotted against 1963; and 69.0 against 1969. These two points are then connected by a straight line as shown in Fig. 12.1.

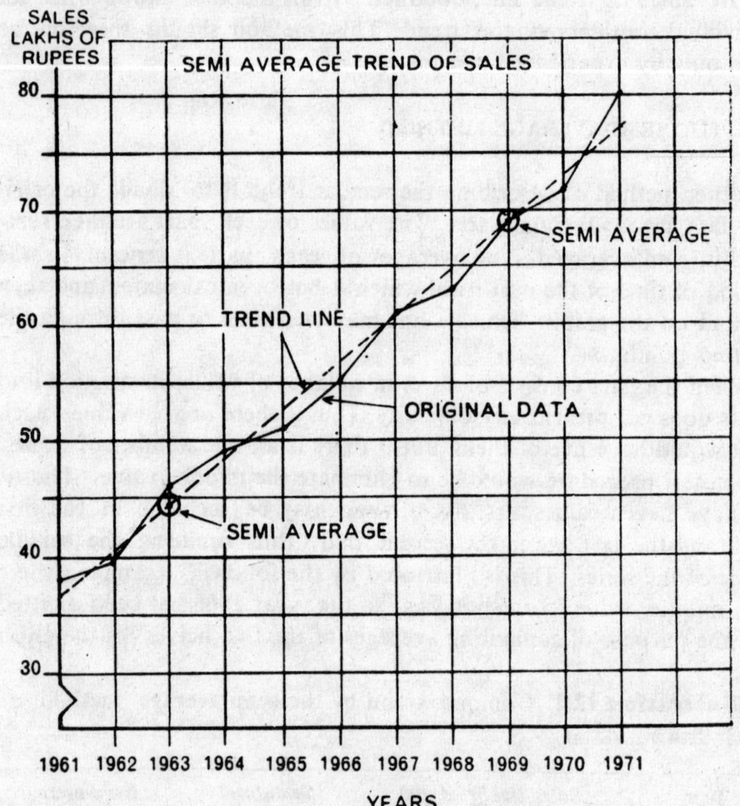

FIG. 12.1

In this illustration sales data related to 11 years, so that each half included a time period of 5 years and the semi-averages were plotted against the middle years of the respective halves. If, however, each half contains an even number of months (or years) a question arises as to against which month (or year) the average may be plotted. The answer to this question can be easily obtained if we remember that when we plot the sales for any particular year the point is put against the middle of that year, and not at the beginning (1st January) or the end (31 December) of that year. This means that the thick lines against which years 1961, 1962, 1963 etc. have been written in Fig. 12.1 represent the middle (night between 30th June to 1st July) of each year. From this it follows that the semi-average of four years would be plotted against that point on the X-axis where second year (in each of the two halves) comes to an end and the third year begins.

This can be seen from the following example:

Illustration 12.2. The sale of a commodity in tonnes varied from January 1961 to December 1961 in the following manner:

280	300	280	280	270	240
230	230	220	200	210	200

Fit a trend by the method of semi-average.

Solution.

	Sales	*Semi-average*		*Sales*	*Semi-average*
January	280		July	230	
Feb.	300		August	230	
March	280		Sept.	220	
		275			215
April	280		Oct	200	
May	270		Nov.	210	
June	240		Dec.	200	

These two semi-averages would be plotted against the middle of the time span to which they relate. The middle of the time span covered by the first half is the night between 31st March and 1st April. Likewise of the second half it is the night between 30th September and 1st October. This can be seen by Figure 12.2.

12.7. THE MOVING AVERAGE METHOD

This method is based on the principle that the random fluctuations can be removed from a data by averaging over a suitable period of time, and thus smoothening the data. The trend then stands out clearly. It is infact a logical extension of the semi-average method.

The moving average is a series of successive averages secured from a series of values by averaging groups of n successive values of the series. These groups are composed as follows. The first group consists of first n items of the series, the second group consists of the items from the second to the $(n+1)$th, the third consists of items from third to the $(n+2)$th, and so on. These averages give us the

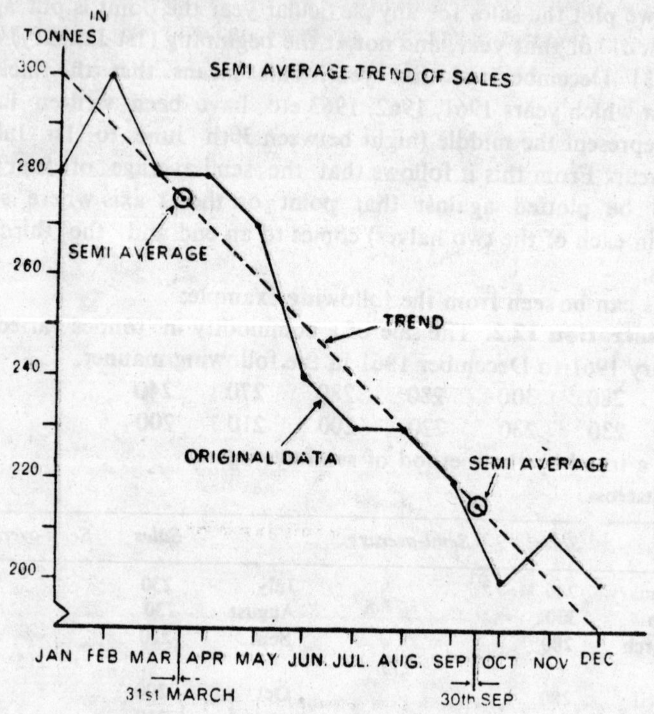

FIG. 12.2.

trend values for the middle period of each group from which they have been computed. Table 12.1 shows the results when a three-year moving average is computed. The three-year moving average means that the number of items included in the groups are three. It will be called a five-year, seven-year, nine-year moving average if the items included in the group are five, seven and nine respectively.

The moving average is calculated as follows:

The values for the first three years 1955, 57, 58 are added together and written in Column *B* against the mid-year 1957. This total is divided by three and the quotient written in Column *C*. Then the value of the first-year (1956) is dropped and the next three values are summed. This sum is written in Column *B* against 1958. Then the value of the second year (1957) is dropped and the next three

TABLE 12.1. THREE-YEAR MOVING AVERAGE APPLIED TO THE DATA OF CEMENT IN INDIA (IN '000 TONS)

Year	*Output* (*A*)	*Three-year moving total* (*B*)	*Three-year moving average* (*C*)
1956	1,542	—	—
1957	1,447	4,541	1,513.7
1958	1,552	5,101	1,700.3
1959	2,102	6,266	2,088,7
1960	2,612	7,909	2,636.3
1961	3,195	9,344	3,114.7
1962	3,537	10,299	3,433.0
1963 (Estimated)	3,567		

values are totalled and averaged. Thus, this process of totalling and averaging will continue till the series ends.

The calculated moving averages recorded in Column *C*, when plotted on a graph (Fig. 12.3) will give us the trend of cement production in Indian Union.

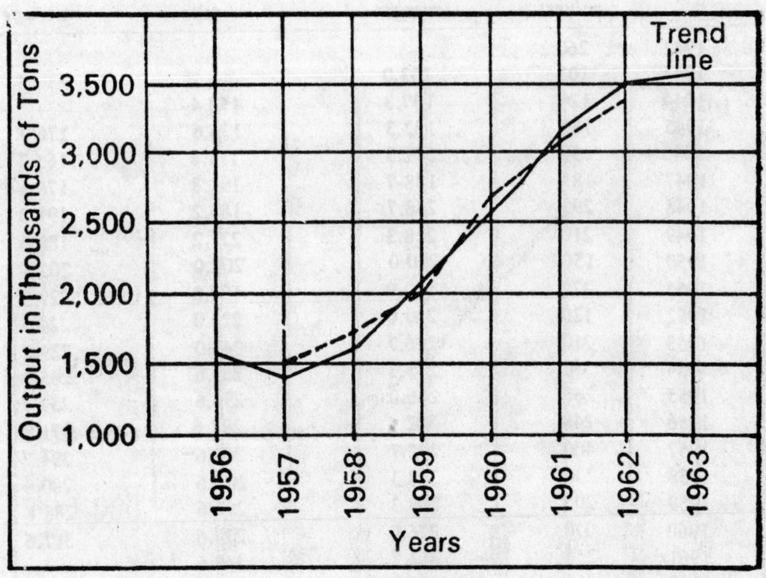

FIG. 12.3

The device of moving average can be advantageously employed for removing variations of a periodic type. So this device is best suited for the data which is characterised by periodic movements. Under certain circumstances, this method fully wipes off the periodic fluctuations and leaves the general trend of the data. But it is essential that the period selected for the moving average must coincide with the length of the cycle. If the former period differs from the latter, this method will not completely eliminate the cyclical movements. Very often the cycles would be of a uniform time-duration.* In such a situation the statistician is faced with the problem of selecting the proper period for calculating moving averages. It is suggested that in such cases we should take moving average period equal to or somewhat greater than the average period of the cycle in the data.

Centering a moving average. When the period selected for the

TABLE 12.2. TREND ESTIMATION BY METHOD OF MOVING AVERAGES

Year	*Annual values*	*Three-year moving average*	*Five-year moving average*	*Seven-year moving average*
1942	260			
1943	105	163.0		
1944	124	137.3	152.4	
1945	183	132.3	136.6	176.9
1946	90	151.3	174.6	169.7
1947	181	188-7	191.8	176,1
1948	295	288.7	185.2	197.0
1949	210	218.3	221.2	188.0
1950	150	210.0	208.0	205 1
1951	270	180.0	192.0	229.3
1952	120	200.0	220.0	224.3
1953	210	226.7	242.0	229.7
1954	350	273.3	237.6	265.4
1955	260	286.0	293.6	259.7
1956	248	302.7	297.6	271.8
1957	400	292.7	268.6	294.7
1958	230	278.3	290.6	280.8
1959	205	268.3	291.6	288.8
1960	370	276.0	274.6	317.6
1961	253	312.7	318 6	
1962	315	339.3		
1963	450			

*The time duration of a cycle is also called the periodicity of the cycle. Hence the period of moving average must be equal to the periodicity of the cycle in data.

computation of moving average consists of an odd number (3, 5, 7, 9, 11 etc.) of years or months there is no problem of centering it. The average obtained is written against the middle of the period as shown before. But when the period consists of an even number (2, 4, 6, 8, 10, etc.) of years, months or weeks there arises the problem of centring it. Thus, in Table 12.3 the average of four years 1956, 57, 58 and 59 cannot be written against the second or the third year since it falls between 31 December 1957 and 1 January 1958 and does not represent either 1957 or 1958.

In order to get over this difficulty we adjust or shift these averages so that they may coincide with years. This is done in the following manner:

1. Compute the four-year average in the usual manner and place them in between two years—thus, the first average is written between years 1957 and 1958, the second between 1958 and 1959 and so on.

2. Compute the two-year moving average of the averages obtained in step one, and write against the middle of the four-year average.

TABLE 12.3. PRODUCTION OF CLOTH IN INDIA (1956-63)
(Four-year moving average)

Year	*Output* (in '000 yds,)	*Four-year moving average*	*Moving average centred*
1956	3,908		
1957	3,762	3,973.0	
1958	4,318	3,912.0	3,942.5
1959	3,904	3,990.5	3,951.3
1960	3,664	4,060.5	4,025.5
1961	4,076	4,323.0	4,191.7
1962	4,598		
1963 (Estimated)	4,954		

This will coincide them with year. Thus, the first average will be written against 1958, second against 1959, and so on. What is done is, that the first and second four year moving averages are added and divided by two and the quotient it recorded in the next column against the year 1958. This process of adding the pairs of moving average is termed as "centering the moving average."

The moving average method will give a correct picture of the general long run tendency of the data only under certain conditions. These conditions are:

(1) The trend must be linear or approximately so.

(2) Cyclical variations affecting the data must be regular in their duration and their amplitude.

If the data contains the cyclical influences which are irregular in their periodicity and amplitude, the moving average method will not completely remove the cyclical influences and hence it cannot display a good picture of the general long-run movement.

If the basic trend in the data is not linear, this method will produce a bias in the trend. As Waugh has pointed out, "If the trend line is concave downward (like the side of a bowl), the value of the moving average will always be too high; the trend line in concave downward (like the side of a derby pot), the value of the moving average will always be too low."

The moving average method contains another disadvantage also, *viz.*, it cannot be extended to the extremes of the period in the data. While computing three-year moving average, we should have had to neglect one year on both extremes. If, on the other hand, we had computed five-year moving average, we should have had to neglect two years at both extremes. This type of defect can be avoided, if the moving average curve is extended both ways by free hand.

12.8 THE METHOD OF LEAST SQUARE

Perhaps the most commonly employed and a very satisfactory method to describe a trend is to use a mathematical equation. The type of equation chosen depends upon the nature of the phenomenon under study. If the values in the time series are expected to increase (or decrease) at a constant rate, a straight line 'fit' to the data is attempted. But if a constant percentage rate of change is expected, an exponential curve may be attempted. Various other curves may also be attempted depending upon the circumstances, but a straight line or on an exponential curve (which itself is a straight line on log-log plot) are the most common ones attempted.

If the variable X represents the independent variable of the time series, which is time, and Y represents the value of the dependent variable, then the linear trend equation can be written as

$$Y_c = a + bX$$

where the subscript c denotes that Y_c is the calculated trend value, and a and b are constants. The value of a represents the intercept on the Y-axis, and b represents the slope of the line, *i.e.*, the amount of change in Y for a unit change in the value of X. Different values of a and b generate different straight lines, for example line M and P in Fig. 12.4. For the straight line M, the value of $a = 5$ is the intercept on the Y axis, and $b = 4$ is the change in the value of Y for a unit change in the value of X. Thus, equation for line M is

$$Y_c = 5+4X,$$

while that for the line P is

$$Y_c = 10+3X$$

With this equation, it is possible for us to estimate the trend value, Yc for a given value of X.

It should be clear that all points of a time series will not lie along a straight line. Different straight lines can be drawn through the data points, but not all such lines would fit the data equally. Some will have better fit than others. The selection of the best fitting trend will then have to be made on the basis of some objective criterion of "goodness of fit". One such objective measure is the sum of the the **squares of deviation** of the data points from the trend values. Thus. $\Sigma\ (Y - Yc)^2$ measures the goodness of fit, the best fit being obtained when this sum of squared deviations is the least. This criterion of minimizing the sum of the squares of deviation is known as the 'principle of least squares' and the procedure of obtaining the best fitting straight line is known as the method of leas*zswt squares.

Fitting a straight line trend

We give below the procedure of fitting a straight line trend $Yc = a + bX$ to the given time series data Y for a number of time points X. The methods consists of choosing the values of a and b such that the sum of the squared deviations $(Y\text{-}Yc)^2 = (Y\text{-}a\text{-}bX)^2$ is the least.

For this we make use of the following normal equations:

$$\Sigma Y = Na + b\ \Sigma X \qquad \text{.......................(12.1)}$$

and $$\Sigma XY = a\Sigma X + b\ \Sigma X^2 \qquad \text{.......................(12.2)}$$

In these equations N stands for the total number of data pairs (Y,X). To calculate a and b, then, we need to obtain ΣX, ΣY, ΣXY and X^2. Calculation of the same may easily be organized as shown in the following examples.

An easy way to work out from memory these normal equations, is by writing down the general equation $Y = a + bX$, and summing up to obtain Eq. (12.1). We next multiply the equation by X and obtain

$XY = aX + bX^2$, and sum it up for all values of X to obtain $\Sigma XY = a\ \Sigma X + \text{b}\Sigma X^2$

By solving these two normal equations for a and b, the trend line is easily obtained, as demonstrated below.

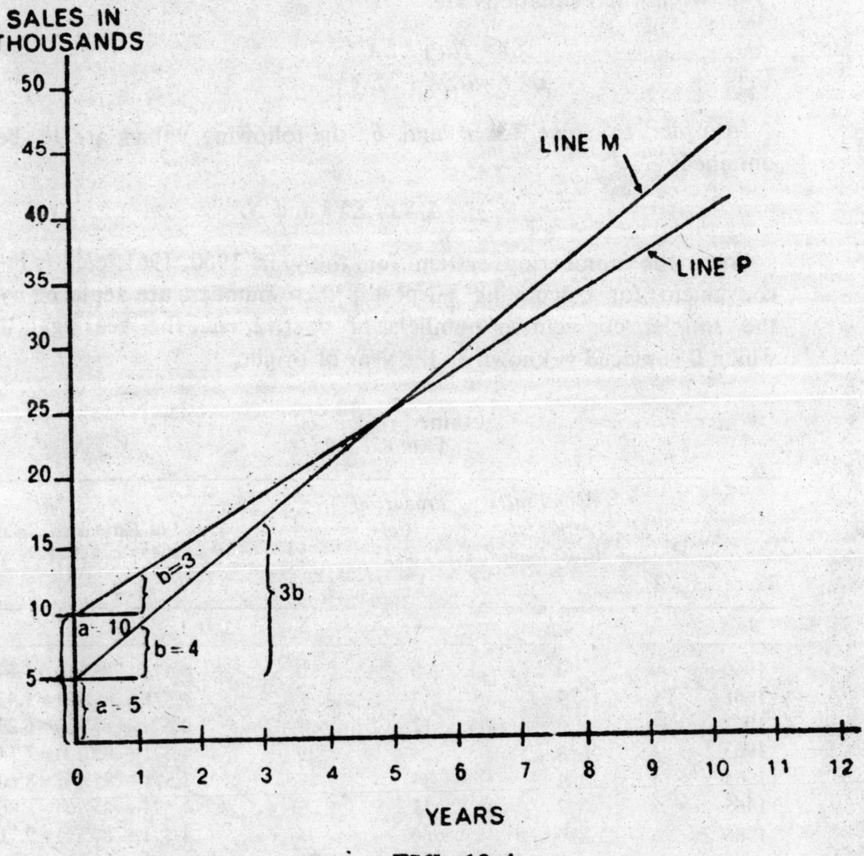

SALES IN THOUSANDS
50
45
40
35
30
25
20
15
10
5
LINE M
LINE P
b=3
a 10
b=4
3b
a 5
0
1
2
3
4
5
6
7
8
9
10
11
12
YEARS

FIG. 12.4

Illustration 12.3. Compute the straight line trend in cereal production from the data of Table 12.4.

Solution. The equation applied is of the type $Y_c = a + bX$.
The two normal equations are:

(*i*) $$\Sigma Y = Na + b\Sigma X$$
(*ii*) $$\Sigma XY = a\Sigma X + b\Sigma X^2$$

In order to solve for *a* and *b*, the following values are to be obtained:

$$\Sigma X, \Sigma Y, \Sigma XY, \Sigma X^2 \text{ and } N.$$

Since the numbering system for the years 1960, 1961, etc., is inconvenient for calculating purposes, these numbers are replaced by the simpler consecutive numbers of 0, 1, 2, etc., the year against which 0 is placed is known as the year of origin.

TABLE 12.4

Year	X	*Actual sales of cereals ('000 tonnes)* Y	*Product of Cols. (2) and (3)* XY	X^2	*Trend in Estimated Sales* $Y_c = 4.571 + .857\ X$
1	2	3	4	5	6
1960	0	3	0	0	4.571+.857(0)=4.57
1961	1	7	7	1	4.571+.857(1)=5.43
1962	2	6	12	4	4.571+.857(2)=6.29
1963	3	8	24	9	4.571+.857(3)=7.14
1964	4	9	36	16	4.571+.857(4)=8.00
1965	5	7	35	25	4.571+.857(5)=8.86
1966	6	10	60	36	4.571+.875(6)=9.71
$N=7$	$\Sigma X=21$	$\Sigma Y=50$	$\Sigma XY=174$	$\Sigma X^2=91$	Total of trend values=50

Note. The above data are for 7 years, therefore $N=7$,

Substituting the values in the normal equations,

$$\Sigma Y = Na + b\Sigma X \quad \ldots(i)$$
$$\Sigma XY = a\Sigma X + b\Sigma X^2 \quad \ldots(ii)$$

we get

$$50=7a+21b \quad \ldots(iii)$$
$$174=21a+91b \quad \ldots(iv)$$

Multiplying the equation (*iii*) by 3, we get

$$150=21a+63b \quad \ldots(v)$$

Deducting (*v*) from (*iv*), we obtain

$$24=28b; \text{ or } b=\frac{24}{28}=.857$$

Substituting the value of *b* in equation (*iii*), we get

$$50=7a+18; \text{ or } 7a=32, \text{ or } a=\frac{32}{7}=4.571$$

Thus, the equation of our best line for the above series is

$$Y_c=4.571+0.857X$$

In interpreting the equation it is necessary to mention the year of origin, time unit and the unit of the dependent variable, so that full information may be obtained. Thus, for the equation, just found,

$$Y_c=4.571+0.857X$$

Time unit: one year,

Year of origin or zero point: 1960,

Unit of dependent variable: thousand tonnes.

With the help of the above solved equation, the trend values for *Y* series can be found just by substituting the appropriate values of *X* in the trend equation as shown in the 6th column of the Table 12.4. It will be noted that the aggregate value of the actual sales tallies with the total of the trend estimates, although the individual values of the actuals and the trend differ. The reason is obvious. In the actuals, there may be ups and downs, but the trend deals with the overall growth or decline and that is why, we find that the trend values are constantly rising here though in the actuals we find even decline in the year 1962 and 1965.

These trend values, if plotted on a graph, will give a straight line and that is why the equation with which these trend values have been computed, is known as straight line equation. In plotting these trend values, at least two or perferably three of these trend values must be computed and joined by a straight line.

Short method of arithmetic straight line. In the Table 12.4 the year 1960, the first year with which the *X* series starts, has been taken as the origin year, *i.e.*, the zero point, and to obtain the unknown constants, the solution of the two simultaneous normal equations has been somewhat longish. The arithmetic involved in these two simultaneous equations can be greatly simplified if the origin is shifted from the first year to the middle year of the time series. Because the time unit of X series is uniformly spaced, *i.e.*, it increases by a single unit from one year to the next throughout the whole data, the sum of X's will be zero ($\Sigma X=0$) if zero point is placed exactly at the middle instead of first year of the time series and all the years preceding this origin year are consecutively given the numbers -1, -2, -3, etc., and so also all the years succeeding this origin year are consecutively designated as 1, 2, 3, etc. Here the values of ΣXY will also be smaller* as the XY values for the years preceding the middle year will be negative and the sum of which will be deducted from the positive sum of the XY values of the succeeding years. So also the values of ΣX^2 will considerably be reduced.

Thus, placing the zero value at the middle year reduces the sum of the X values to zero and consequently the solution of the normal equations is made very easy. For, if zero is put against the ΣX whenever occurring in the two normal equations, these equations will become

$$\Sigma Y=Na+b(0) \quad \text{or} \quad \Sigma Y=Na$$

$$\Sigma XY=a(0)+b\Sigma X^2 \quad \text{or} \quad \Sigma XY=b\Sigma X^2$$

Now, in each of these two simplified equations, there is only one unknown constant and so the values of a and b may directly be found. Thus, this simple or short-cut device considerably reduces the labour and time involved in calculations.

The expression, $\Sigma Y=Na$, may be written as

$$a=\frac{\Sigma Y}{N} \qquad \ldots(3)$$

This should be recognized as the formula for arithmetic mean of the Y series, and so a is the arthmetic mean of the Y series in this short-cut method.

Similarly, the expression $\Sigma XY=b\Sigma X^2$ may be written as

$$b=\frac{\Sigma XY}{\Sigma X^2} \qquad \ldots(4)$$

This value of b represents the average amount of change in the secular movement for one unit of time.

Illustration 12.4. (Odd number of years.) Taking the original values the actual sales for the years 1960-66 given in Table 12.4 the application of the short-cut method may be illustrated in Table 12.5.

*ΣXY will not be zero as ΣXY is the sum of the product of individual values of X and Y and is not the product of ΣX and ΣY.

Solution. Substituting the values in the already arrived expression:

$$a=\frac{\Sigma Y}{N}, \text{ and } b=\frac{\Sigma XY}{\Sigma X^2}$$

TABLE 12.5

Years	X	*Actual sales of cereals ,000 tonnes)* Y	*Product of Cols. (2) and (3)* XY	X^2	*Trend in Estimated Sales* $Y_c=7.143+.857X$
1	2	3	4	5	6
1960	−3	3	−9	9	7.143+.857(−3)=4.57
1961	−2	7	−14	4	7.143+.857(−2)=5.43
1962	−1	6	− 6	1	7.143+.857(−1)=6.29
1963	0	8	0	0	7.143+ 857(0) =7.14
1964	1	9	9	1	7.143+.857(1) =8.00
1965	2	7	14	4	7.143+.857(2) =8.86
1966	3	10	30	9	7.143+.857(3) =9.71
N =7	ΣX =0	ΣY =50	ΣXY =53—29=24	ΣX^2 =28	Total of the trend values=50

We get $a=\frac{50}{7}$, or 7.143.

$b=\frac{24}{28}$, or $\frac{6}{7}$, or .857.

Thus, the straight line equation for the above series becomes

$$Y_c=7.143+.857X$$

Time unit: one year,

Year of origin: 1963,

Unit of dependent variable: thousands of tonnes.

Looking at the trend values of both the above tables, we find them exactly identical and that is why the application of the short method is strongly favoured in the time series.

Illustration 12.5. Fit a straight line trend to the data given here by the method of least squares:

Year	1959	1960	1961	1962	1963	1964	1965
Gross exfactory Value of output (Rs crore)	672	824	968	1205	1464	1758	2058

Solution. Let the trend line be $y=a+bX$.

FITTING A TREND BY SHORT CUT METHOD

Year	*Output*	x	xy	x^2	*Trend Values*
1959	672	−3	−2016	9	579.64
1960	824	−2	−1648	4	812.57
1961	968	−1	−968	1	1045.50
1962	1205	0	0	0	1278.43
1963	1464	1	1464	1	1511.36
1964	1758	2	3516	4	1744.29
1965	2058	3	6174	9	1977.22
Total	8949	0	.6522	28	

$$a = \text{Trend for time denoted by '0'} = \frac{\Sigma y}{n}$$

$$= \frac{8949}{7} = 1278.43$$

$$b = \text{yearly increase} = \frac{\Sigma xy}{\Sigma x^2} = \frac{6522}{28} = 232.93$$

Trend equation is $y = 1278.43 + 232.93x$

The trend values for the year 1962 is 1278.43. For the year 1963 it will be 232.93 (yearly increase) more than the trend value for 1962. So it will be 1511.36. Similarly trend for 1961 will be 232.93 less than the trend value for 1962 *i.e.*, it will be 1045.50. Calculating in this way other trend values will be as shown above.

Explanation of constants a and b. It may be noted here that the value of b, also known as the coefficient of X in the trend equation, as obtained by short method is identical to that computed by the longer method, as it should be. This coefficient is of particular interest since it shows by how much the trend rises in case of growth or falls if it is a declining trend from one unit of time (one year in the above illustrations) to the next. It is evident from column 6 of the above two tables that if the value of X is increased by one unit, the value of Y increases by an amount equal to one time the value of b. Thus, b measures the steepness or *slope of the straight line.* Its sign may be positive or negative. If it is positive, the trend will have a rising tendency and if negative, a declining tendency will be observed in the trend.

The value of a, however, is different (7.143) in the shorter method than that (4.571) of the longer method. This is because the origin, in the shorter method, has been shifted from 1960 to 1963. But this change in origin has no effect upon the trend values, as it is amply evident from both the above tables, illustrating the application of short and long methods separately, that the trend values for all the years, including, of course, the origin years of 1960 and 1963, are exactly indentical. Thus, by the use of short method the change is effected only in the year of origin and not in the level or slope of the line. It may also be noted that for the year of origin, *a* is the trend value as when $X=O$, the second term of the general equation, *i.e.*, *bX* becomes zero. Thus, *a* locates the general height of the trend and is referred to as the *Y* intercept, *i.e.*, the computed value of *Y* when *X* equals zero.

Short method (even number of years). In a time series of even number of years, although there is no middle year opposite which can be assigned the value, $X=O$, yet the advantage of the short-cut device can still be had by taking the origin between the two middle years (as the mid-point of the even series lies between its two centre years). The first year on either side of the origin will be assigned 0.5, with the negative sign to the one preceding, and positive sign to the one succeeding, the origin year. The second year on either side of the origin will similarly be numbered 1.5, and so on. The resulting sum of *X* will be zero as in the odd number series. However, the decimal points may be cumbersome in their computations and consequently to eleminate these fractions, the time unit of *X* is changed from one year to half a year.

Illustration 12.6. Find the trend line for the first six years of data of Table 12.4 or Table 12.5.

Solution. Table 12.6 shows the trend calculations. Two middle years in Table 12.6 are 1962 and 1963 and the zero point will lie at the centre of these years, *viz.*, at the mid-night falling between 31 December 1962, and 1st January 1963. The data for 1962 is located at 1st July of that year and lies half a year before the zero point. But since half a year is being used as the time unit, so the *X* value for 1962 is assigned—1. For the year 1961, 1st July is one and a half year before the origin, and the *X*-value becomes—3. Similarly, the remaining *X* values, both preceding and succeeding the origin year, have accordingly been determined in the Table 12.6. The values of *a* and *b* have been found and trends have been computed exactly in the same fashion as explained above.

Substituting the values in the expression:

$$a=\frac{\Sigma Y}{N}, \quad \text{and} \qquad b=\frac{\Sigma XY}{\Sigma X^2},$$

we get

$$a=\frac{40}{6}=\frac{20}{3}=6.66, \qquad b=\frac{28}{70}=\frac{2}{5}=0.4$$

TABLE 12.6

Year, (Year)	X'	(Half Year) X	Actual sales of cereals ('000 tonnes) Y	Product of Columns (2) and (3) XY	X^2	Trend or Estimated Sales, i.e., $Y_c = 6.66 + .4X$
1		2	3	4	5	6
1960	−2.5	−5	3	−15	25	6.66+.4(−5)=4.66
1961	−1.5	−3	7	−21	9	6.66+.4(−3)=5.46
1962	−.5	−1	6	−6	1	6.66+.4(−1)=6.26
1963	.5	+1	8	+8	1	6.66+.4(+1)=7.06
1964	1.5	+3	9	+27	9	6.66+.4(+3)=7.86
1965	2.5	+5	7	+35	25	6.66+.4(+5)=8.66
N =6		ΣX =0	ΣY =40	ΣXY =70−42=28	ΣX^2 =70	Total of the trend value=40

Thus, the straight line equation for the above even series becomes

$$Y' = 6.66 + .4X$$

Time unit: half a year,

Year of origin: mid-1962-63,

Unit of dependent variable: thousands of tonnes.

Illustration 12.7. Below are given figures of production in 000's quintals of a Sugar factory.

Year	1941	1942	1943	1944	1945	1946	1947
Production	80	90	92	83	94	99	92

(*a*) Find the slope of a straight line trend to these figures.

(*b*) Plot these figures on a graph and show the trend line.

(*c*) Do these figures show a rising trend or a falling trend? How do you arrive at your conclusions?

Solution.

Year	X	Y	XY	X^2	Trend Values
1941	−3	80	−240	9	84
1942	−2	90	−180	4	86
1943	−1	92	−92	1	88
			−512		
1944	0	83	0	0	90
1945	1	94	94	1	92
1946	2	99	198	4	94
1947	3	92	276	9	96
			568		
N=7	ΣX=0	ΣY=630	ΣXY=56	ΣX^2=28	

Let the trend equation be $Y_c = a + bx$.

$$a = \frac{\Sigma Y}{N} = \frac{630}{7} = 90$$

$$b = \frac{\Sigma XY}{\Sigma X^2} = \frac{56}{28} = 2 = \text{slope of the line}$$

Trend eqnation is $Y_c = 90 + 2X$
When time unit = one year
Year of origin = 1944

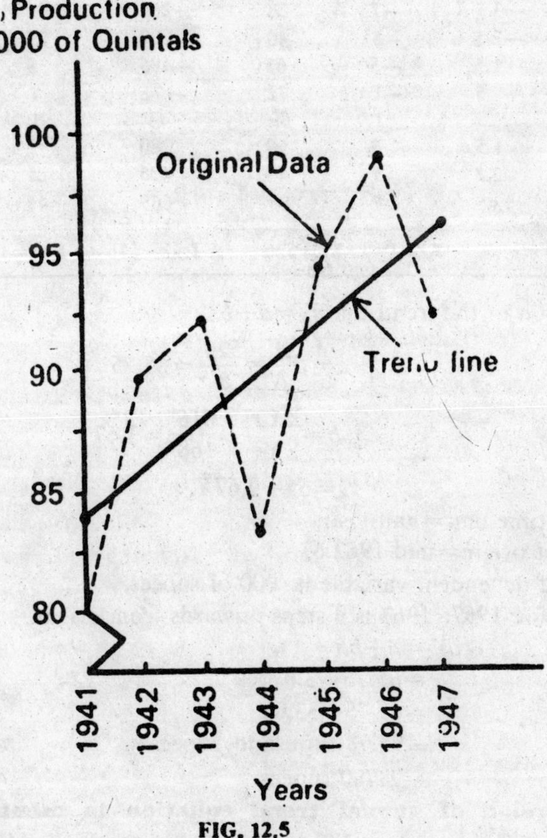

FIG. 12.5

Unit of dependent variable = 000's of quintals.

(*a*) Hence slope of the straight line trend is 2.

(*b*) The figures show a rising trend. Because the slope of the trend is positive trend values go on increasing as we move onward.

Illustration 12.8. Fit a straight line trend to the following data:

Year	1959	1960	1961	1962	1963	1964	1965	1966
Sales (in '000 rupees)	38	40	65	72	69	60	87	95

Estimate the sales for the year 1967.

Solution

Year	*X'* (*in years*)	*X* (*in half years*)	*Y*	*XY*	X^2	Y_c (*Trend Values*)
1959	—3.5	—7	38	—266	49	40.06
1960	—2.5	—5	40	—200	25	47.40
1961	—1.5	—3	65	—195	9	54.74
1962	— .5	—1	72	—72	1	62.08
1963	.5	1	69	69	1	69.42
1964	1.5	3	60	180	9	76.76
1965	2.5	5	87	435	25	84.10
1966	3.5	7	95	665	49	91.44
$N=8$		$\Sigma X=0$	$\Sigma Y=526$	$\Sigma XY=616$	$\Sigma X^2=168$	

Equation of the trend line $Y_c=a+bX$

$$a=\frac{\Sigma Y}{N}=\frac{526}{8}=65.75$$

$$b=\frac{\Sigma XY}{\Sigma X^2}=\frac{616}{168}=3.67$$

i.e., $Y_c=65+3.67X.$

When time unit = half year

Year of origin = mid-1962.63

Unit of dependent variable is '000 of rupees.

Sales for 1967: 1967 is 9 steps onwards from origin.

$$\begin{aligned} Y_c &= a+bX \\ &= 65.75+3.67\times 9 \\ &= 65.75+33.03 \\ &= 98.78 \text{ thousand rupees} \\ &= 98780 \text{ rupees.} \end{aligned}$$

Conversion of annual trend equation to monthly trend equation. There are two different possible situations which must be distinguished for deciding the method of conversion:

(*i*) When the values of the dependent variable (*Y*) are in annual totals, and

(*ii*) The values of the dependent variable (*Y*) are monthly averages.

When data are annual totals. Having fitted a line to annual data it is frequently necessary to make a transition to monthly units.

The constant a in the annual trend equation defines the trend value in the year taken as the origin. This constant must be divided by 12 to obtain the trend value for the month centring at the middle of the origin.

The constant b should be divided by 144 to obtain the increase or decrease that takes place from one month to the next.*

If the annual trend equation is

$$Y_c = 600 + 72X$$

Origin : 1960
X unit : one year
Y_c : Total annual quantity

The conversion to monthly equation would be

$$Y_c = \frac{600}{12} + \frac{72}{144}X$$

or

$$Y_c = 50 + .5X$$

Origin : 1 July 1960
X units : one month
Y_c : monthly quantity

When the data are given as monthly averages. In a situation like this the Y values are on a monthly level, since they are obtained by dividing annual totals by 12. As such the value of a constant in the trend equation measures the trend value for a month centring at the middle of the year of the origin. Therefore, the value of a constant remains unchanged in the conversion process.

The value of constant b in a case like this, shows the change on a monthly basis, but from one month in a year to the corresponding month in the following year. It is, therefore, necessary to convert the b value to make it represent the change from one month to the next following. To do this, we divide b by 12.

Thus, to convert the annual trend equation to a monthly trend

*When data are annual totals b represents annual increase for an entire year. If we divide b by 12, we obtain the monthly trend increment in the yearly totals. Since we still have yearly totals, we must divide again by 12 to reduce the increment to monthly totals.**

equation, when the annual data are given as monthly averages, a is left unchanged and b is divided by 12.

If the annual trend equation is

$Y_c = 40 + 6X$
Origin : 1960
X units : one year
Y units : average monthly amount,

The equation converted to monthly level would be

$$Y_c = 40 + \frac{6}{12}X = 40 + .5X$$

Origin : 1 July 1960
X units : one month
Y units : average monthly amount

When X units (time units) are one-half year, the conversion is made from a semi-annual level to monthly level.

Therefore the conversion would be as follows:

(*i*) When Y units are annual totals, divide a by 12; b by 72.

(*ii*) When Y units are monthly averages.
a remains unchanged, and
b is divided by 6.

Shifting the origin. From the above discussion it follows, that when an annual trend equation is converted to a monthly trend equation the origin always falls at the 1st of a month. A monthly trend equation also must have its origin at the centre of a month *i.e.*, at the 15th of month, for the same reason as the annual equation must have its origin at the centre of the year. Thus, a shifting of the origin becomes necessary whenever annual trend equations are converted to monthly trend equations. To do this, we have to find out the trend value for the date to which the origin is to be shifted. This trend value then becomes the value of a; the value of b would remain the same. Thus if the following monthly equation,

$Y_c = 40 + .5X$
Origin : 1 July 1960
X units : one month
Y units : monthly sales

is to be shifted to 15 July as the origin, it would be

$Y_c = 40.25 + .5X$
Origin : 15th July 1960
X units : one month;
Y units : monthly amount.

Semi-logarithmic or geometric straight line trend. The

straight line $Y_c=a+bX$ which has been illustrated so far, describes a constant amount of growth or decline per unit time. Many economic and business series, like the compound interest data, tend to show a constant rate of change, instead of a constant absolute amount of change from year to year. Such a trend is observed when the values of *X*-series are arranged in arithmetic progression, such as 1, 2, 3, 4, etc., and the data of the *Y* series are in geometric progression, such as 1, 2, 4, 8, 16, etc., showing that values change at a constant ratio, *viz.*, the tendency of the absolute amount to change more rapidly in succeeding years then in preceding years. If the trend values of such *Y*-series are fitted on the natural number scale a linear type of trend will not appear as these values are in geometric progression rather than in arithmetic progression. But when the same is fitted on a semi-logarithmic chart (the *X*-axis of which is on a natural number scale and the *Y*-axis is with a logarithmic ruling), a linear type of trend makes it appearance. This means that there is a straight line relationship between the natural values of *X*-series and logarithmic values of *Y*-series. This line is called geometric straight line or semi-logarithmic curve.

When the trend is geometric, the resulting curve is often referred to as the exponential curve and the formula of this geometric straight line trend is:

$$Y_c=ab^x$$

(It is referred to as the exponential curve of function because the *X* factor is used as a power.)

When expressed logarithmically, it takes the form:

$$\log Y_c=\log a+X \log b$$

To compute the specific values of *a* and *b* the method applied is similar to that of the arithmetic straight line, $Y_c=a+bX$, considered above. The only difference to be noted is that the logarithms of the original *Y* series (obtained by reducing all the *Y* values to their appropriate logarithmic values). instead of their natural numbers are used throughout.

The normal equations, therefore, become

$$\Sigma \log Y=N \log a+\log b\Sigma X$$

$$\Sigma X \log Y= \log a\, \Sigma X+ \log b\, \Sigma X^2$$

or when the middle year is the origin, the expression becomes

$$\log a=\frac{\Sigma \log Y}{N}.$$

It gives the logarithmic values of trend at the origin, which is also the logarithm of the geometric mean of the Y series.

$$\log b = \frac{\Sigma X \log Y}{\Sigma X^2},$$

It is the logarithm of the rate of change per unit of time, *i.e.*, the slope of the line in logarithmic terms.

To arrive at the natural numbers, these logs should be converted into antilogs. If, in the curve, $Y_c = ab^x$, b is a positive number greater then one the trend will be upward and the amount of change will indicate a constant percentage of increase and if b is less than one, though a positive number, the trend will be downward and the amount of change will show constant percentage of decline.

It may be observed that this geometric straight line is a least squares fit to the logarithms of the Y values, and not directly to the Y values. Here the sum of the squares of the deviations between the logarithms of the original values and the logarithmic trend values, $\Sigma(\log Y - \log Y_c)^2$ will be minimum.

Logarithmic straight line. There may be instances where a geometric progression is formed by the Y values when the X values are geometrically arranged and the semi-log chart may not give a linear trend. In terms of X and Y, the exponential function for these will be $Y_c = aX^b$

When this function is transformed in terms of logs, it becomes $\log Y_c = \log a + b \log X$, which is a straight line in terms of $\log X$ and $\log Y$. It may be noted here that the origin of X (original values of X when changed to logarithmic values) cannot be taken at the middle of the period, because, then, some values of X will become negative and logarithms of negative numbers are defined. Thus, the short-cut method may not be used in this case.

Non-linear trends—second degree curve. A linear trend may provide a reasonably good description of the trend of short period series of about a decade showing an increase or decrease throughout, but a non-linear trend may more appropriately be fitted to the series for longer periods of about a quarter or half a century, showing a turn from early years' rise to a fall in later years or exhibiting the turning points into the stages of early expansion, decline and a final attempt at recovery.

In order to represent a non-linear trend, polynomials of degrees higher than the first (the linear equation, $Y_c=a+bX$, is referred to as a polynomial of first degree as it involves only the first power of X) are used. The higher degree polynomials are obtained by adding further terms to the linear equation involving power of X, higher than the first. If, for example, second power of X *i.e.*, X^2 is added in the above equation, we will obtain the formula of the second degree polynomial or parabola.

$$Y_c=a+bX+cX^2$$

The method of fitting a curve of this type will also resemble the one applied in fitting the linear trend. As in the linear trend, so also in evolving the equation of second (or higher) degree parabola, the principle of least squares is applied, *viz.*, the condition that sum of the squared deviations from the curve selected shall be smaller than the sum of the same from any other second power curve fitted to the same data, is laid down. But since there are three unknowns a, b and c in the equation $Y_c=a+bX+cX^2$, three normal equations are required to obtain the values of these unknowns. The process of getting these three normal equations, described below, is very similar to that of the straight line and may also be easily memorized.

First, write the 'type' equation, with the original value of Y instead of computed value Y_c, by writing down

$$Y=a+bX+cX^2$$

Next, multiply each term of the equation by the co-efficient of each unknown constant (a, b, c) and sum up:

(*i*) For first normal equation, multiply the type equation by the co-efficient of a (which is 1) and get the sum, and the resulting equation will be of the form $\Sigma Y=Na+b\Sigma X+c\Sigma X^2$

(*ii*) For the second normal equation, multiply the type equation by the co-efficient of b (which is X) and sum up, the form thereof will be:

$$\Sigma XY=a\Sigma X+b\Sigma X^2+c\Sigma X^3$$

(*iii*) For the third normal equation, multiply the type equation by the coefficient of c (which is X^2) and sum up, thereby getting

$$\Sigma X^2Y=a\Sigma X^2+b\Sigma X^3+c\Sigma X^4$$

In this way, we get three normal equations:

$$\Sigma Y=Na+b\Sigma X+c\Sigma X^2$$
$$\Sigma XY=a\Sigma X+b\Sigma X^2+c\Sigma X^3$$
$$\Sigma X^2Y=a\Sigma X^2+b\Sigma X^3+c\Sigma X^4$$

By substituting the appropriate values in these normal equations, the desired values of the unknown constants can be obtained by the simultaneous solution of the equations.

This method of computing the value of a, b and c can be further simplified as the solution of the three simultaneous normal equations would generally involve considerable amount of labour. As in the linear trend, here also the short-cut device may be used by taking the zero value of X in the middle of the period. By taking the origin at the middle of the data, not only the sum of $X(\Sigma X)$ but also the sum of any odd power of X, such as ΣX^3, ΣX^5, etc., becomes zero, and therefore, by putting 0 for ΣX and ΣX^3 (here the highest odd power is ΣX^3) in the above three normal equations we get

$$\Sigma Y = Na + b(0) + c\Sigma X^2 \quad \text{or} \quad \Sigma Y = Na + c\Sigma X^2$$

$$\Sigma XY = a(0) + b\Sigma X^2 + c(0) \quad \text{or} \quad \Sigma XY = b\Sigma X^2 \text{ or } b = \frac{\Sigma XY}{\Sigma X^2}$$

$$\Sigma X^2 Y = a\Sigma X^2 + b(0) + c\Sigma X^4 \quad \text{or} \quad \Sigma X^2 Y = a\Sigma X^2 + c\Sigma X^4$$

Here the value of b is obtained directly, and the values of a and c will be obtained by the two simultaneous equations, and consequently the labour is considerably saved in solving the three simultaneous equations.

Computation of trend by second degree parabola. In the Table 12.7, we have taken a simple data of a few years just to illustrate the procedure of calculating the trend values for fitting a second degree curve by the short cut method.

Substituting the values in the normal equations

$$\Sigma Y = Na + c\Sigma X^2 \qquad \ldots(i)$$

$$\Sigma XY = b\Sigma X^2 \qquad \ldots(ii)$$

$$\Sigma X^2 Y = a\Sigma X^2 + c\Sigma X^4 \qquad \ldots(iii)$$

we get

$$100 = 7a + 28c \qquad \ldots(iv)$$

$$46 = 28b \text{ or } b = \frac{46}{28} = 1.64 \qquad \ldots(v)$$

$$334 = 28a + 196c \qquad \ldots(vi)$$

Multiplying the equation (*iv*) by 4, we get

$$400 = 28a + 112c \qquad \ldots(vii)$$

Deducting (*vi*) from (*vii*), we obtain

$$66 = -84c, \text{ or } c = -\frac{66}{84} = -\frac{11}{14} = -.79.$$

Substituting the value of c in equation (*iv*) we get

$$100=7a+28\left(\frac{-11}{14}\right) \text{ or } 100=7a-22$$

or $$7a=122 \text{ or } a=17.43.$$

Thus, the respective values of the constants a, b and c are 17.43, 1.64 and $-.79$.

The type equation of second degree parabola becomes

$$Y_c=17.43+1.64\ X-0.79X^2$$

The second degree parabola curve can be fitted by plotting the computed trend values of column 9 of Table 12.7. Here, all the values are to be plotted and not merely two or three values as in the case of straight line trend, for the simple reason that the second degree equation will not give a straight, but a curved line and consequently the desired curve can be obtained only when all, or at least most of the trend values are plotted.

From column 9 of Table 12.7, it is evident that the value of the constant a in this type of equation is the value of Y at the origin, the value of the constant b indicates the slope at the point where $X=0$ (at the origin); and the value of the constant c determines the extent of departure from linearity and also it indicates whether the curve is concave or convex to the base. When the value of c is positive tne curve will be convex and when negative, as we find in the above illustration, the curve will be concave to the base.

Third degree curve. The third degree parabola $Y=a+bX+cX^2+dX^3$ is obtained by adding one more unknown constant d (or raising X by one more power) to the type equation of the second degree parabola, and permits the fitted curve to change direction twice *viz*., allows two bends in the trend curve. A still more flexible curve may be obtained by using higher degree parabolas than the third degree, and the general formula for the type equation will be

$$Y^1=a+bX+cX^2+dX^3+eX^4+\ldots\text{etc.}$$

TABLE 12.7

Year	X	Actual Sales ('000 tons) Y	Product of Columns (2) and (3) XY	X^2	Product of Columns (3) and (5) X^2Y	X^3	X^4	Computed trend i.e., $Y_c = 17.43+1.64X-.79X^2$
1	2	3	4	5	6	7	8	9
1960	−3	7	−21	9	63	−27	81	17.43+1.64 (−3)−.79 (9)=5.40
1961	−2	9	−18	4	36	−8	16	17.43+1.64 (−2)−.79 (4)=10.99
1962	−1	13	−13	1	13	−1	1	17.43+1.64 (−1)−.79 (1)=15.00
1963	0	20	0	0	0	0	0	17.43+1.64 (0)−.79 (0)=17.43
1964	1	19	19	1	19	1	1	17.43+1.64 (1)−.79 (1)=18.28
1965	2	17	34	4	68	8	16	17.43+1.64 (2)−.79 (4)=17.55
1966	3	15	45	9	135	27	81	17.43+1.64 (3)−.79 (9)=15.24
$N=7$	$\Sigma X=0$	$\Sigma Y=100$	$\Sigma XY=+98$ -52 46	$\Sigma X^2=28$	$\Sigma X^2Y=334$	$\Sigma X^3=0$	$\Sigma X^4=196$	

The solutions for parabolas of third or higher degree can be arrived at by the extension of the method explained for the parabolas of first and second degrees.

It may be indicated at this stage that the parabolas of higher degree should be used more cautiously as the parabolas of fourth, fifth degree, etc. will respectively have 3, 4, etc., bends or changes of directions (*viz.*, several ups and downs) in the fitted curve, which may hardly coincide with the concept of secular trend.

Asymptotic growth curves. The straight line $Y_c = a + bX$ describes, as has been shown earlier, a constant amount of increase or decrease. The exponential curve, $Y = ab^x$ shows a constant ratio of change. Over long periods, time series are not likely to show either constant amount of change or constant ratio of change. It is, on the other hand, very likely that an increasing series will show an increasing amount of change but a decreasing ratio of change and approach an upper limit called the asymptote. It is also possible that an increasing series may show a decline in the amount of increase. Decreasing absolute growth, however, is not often encountered.

Such series can be best described up using growth curves. The three commonly referred types of these curves are: (*i*) modified exponential curve; (*ii*) gompertz curve and (*iii*) logistic curve.

1. Modified exponential curve. The curve not only describes a pattern of growth in which the amount of growth declines by a constant percentage but the curve reaches an upper limit.

The equation of the modified exponential is

$$Y_c = K + ab^x,$$

where K is the asymptote (the upper limit).

As this type of series do not frequently occur, the modified exponential is not of much practical importance.

2. The gompertz curve. It shows a pattern of growth in which trend would show a declining ratio of increase, but the ratio does not decrease by either a constant amount or a constant percentage. But the logarithms of the amount of growth tend to decrease by a constant percentage. The equation of the gompertz curve is

$$Y_c = Ka^{bx}$$

It may be put in logarithmic form as follows:

$$\log Y_c = \log K + (\log a)\, bx$$

The logistic curve. Also known Pearl Read Curve, the equation of the logistic curve is as given below:

$$\frac{1}{Y_c} = K + ab^x$$

This curve best describes a series in which first differences of reciprocals tend to decrease by a constant amount.

The Gompertz and the Logistic curves are similar in several respects, *viz*.

(*i*) Both take the form of an elongated *S*;

(*ii*) Both approximate zero at one limit and are asymptotic to a certain value at the other limit:

(*iii*) They do not assume negative values; and

(*iv*) Both show small absolute growth in early years, then rapid growth which tapers off gradually as retardation sets in and finally as the upper asymptote is approached very small increases take place.

12.10. ELIMINATION OF TREND

As stated earlier, the forces affecting time series are broadly divided

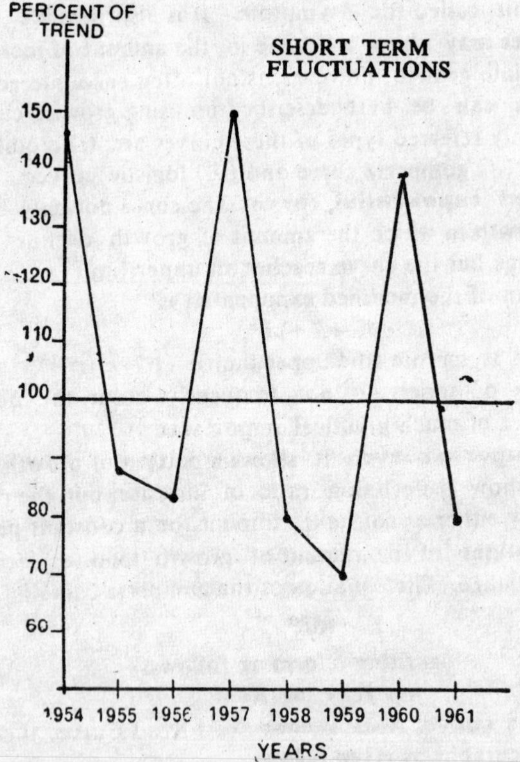

FIG. 12.6

into two categories, *viz.*, long-term forces and short-term forces. If we are interested in the long-term forces we will study only the trend. But when it is desired to study short-term movements we will have to get rid of the trend from the original data, *i.e.*, we will have to eliminate trend from the actual data. The task of eliminating is easy if trend figures are known to us.

To do so we must divide the original data, *TSCI*, by *T* (trend) for each time unit in the series. It would be remembered that in annual data there is no effect of seasons. As such annual data may be represented by *TCI*, and if it is divided by *T* we get an estimate of *CI* for each year, which may be called short-term fluctuations.

Reverting to Table 12.2 and assuming five-yearly moving average as representing trend we compute short-term fluctuations (eliminate trend) for years 1954 to 1961 in the following Table . These short term fluctuations as percent of trend are plotted in Fig.12.6

Year	Original Data TCI	Trend T	$CI = \frac{TCI}{T} \times 100$
1954	350	237.6	147
1955	260	293.6	88
1956	248	297.7	83
1957	400	268.6	150
1958	230	290.6	80
1959	205	291.6	70
1960	370	274.6	140
1961	253	318.6	80

12.11. SEASONAL VARIATIONS

It has been stated earlier that one of the component forces that determine the size of a variable at any point in time is the seasonal factor. Climate and custom together play an important role in giving rise to seasonal movements to almost all industries. The yearly cycle of weather directly affects agriculture production and agricultural marketing. Banks, wholesale and retail trade, all feel the effect of the seasonality of farmers' incomes as their purchases tend to concentrate mostly during the period of their marketings. Though the direct and main impact of nature is mainly on agriculture yet climatic conditions, including variations in rainfall, humidity, sunshine, heat, wind and snow do produce variations in almost all other types of industries.

Another factor responsible for most of the seasonal variations in time series is custom. Most of the festivals, holidays and marriages are attributable to customary parctices and many lines of retail trade in particular reach their peaks in these limited periods, regularly recurring year after year. As customs seldom die and seasons have little change in their periodicity, the variations produced by customs and seasons are more or less periodically regular. These more or less regular intra-year (within the year) movements recurring year after year are called seasonal variations.

Though seasonal variation generally deals with intra-year movements, yet periodic movements may also be characterised as intra-month, intra-week, intra-day, etc. The peak activity of a commercial bank around the first of each month may be cited as an example of intra-month movement. If we have the figures of the sales of retail stores in an industrial town day-by-day and if the intra-week sales are observed, it will be found that the sales are mostly concentrated on the local weekly pay day or the day after which the sales of the remainder of the week are very low. This type of periodic movement is repeated almost with the same regularity week after week, as the seasonal movement is repeated year after year. The business of a restaurant and the figures of automobile accidents in metropolitan centres are some of the examples of the intra-day movements, as these data will show the greatest concentration in the evening hours.

Yet another type of variation though fairly regular and respective, is caused by the use of Gregorian Calender. There is a 10 per cent range in the number of days falling between the shortest month (28 days) and the longest month (31 days). This variation is much more marked when the number of working days of different months is taken into account. Naturally, such a difference in the working days must lead to variations in the monthly figures of production, sales, etc.

Thus, the problem is essentially one of discovering what the regularities are. One may be interested in seasonal variations either because one wishes to separate the variations which are purely seasonal in order to study business cycles, etc., or because one may be interested in seasonal variation itself. The objectives behind the interest in seasonal variation itself may be:

(*i*) to take its advantage by purchasing such seasonal commodities which can be preserved, at the peak of the season when not only price is low but quality may also be high;

(*ii*) to iron out the seasonal so that the intra-year variation may be less marked *e.g.*, on the side of sales, by making effective advertisements, etc., and on the production side by stimulating increased production in the off season; and (*iii*) to curtail the activities of seasonal nature, say, of a manufacturing concern, by producing commodities with complimentary seasonals.

A knowledge of the seasonal impact is important to a business executive who is planning his production programme. For example an analysis of the seasonal factor helps him to plan for the hiring of personnel for peak periods, to accumulate an inventory of raw materials etc. Further it enables him to take action directed at leveling out these seasonal ups and downs so far as they affect his enterprise.

Again, an analysis of seasonal element helps us in making forecast. A prediction of the sales for a certain month is based not only on the trend factor but also on the seasonal position of that month.

Isolation of the seasonality and its elimination from the data becomes necessary to study the effect of cycles.

The Specific Seasonal and the Typical Seasonal

By specific seasonal we mean the effect of a season in a particular year. Thus if we are making a study of the seasonal element on the sales of 'rain coats' of a departmental store during the year 1970, we get at what is called a specific seasonal. We would make a statement somewhat like this 'During 1970 the sales of raincoats registered an increase of 30 per cent during the month of August over the trend of sales.'

If, however, we study seasonality with a view to determine the pattern of peaks and valleys, (that is to say the increases or decreases that are caused in general by different seasons) we study a number of years. The generalised expression of these variations is called the typical seasonal. Typical seasonal variation is, therefore, the average of specific seasonals for a number of years. If the effect of the month of August over the sales of raincoats for number of years is averaged the resultant figure would give us the typical seasonal.

Computation of Seasonal Variation. Before we proceed to discuss various methods of isolating seasonal variation it must be clearly understood that the effect of the seasons on a variable is present only when we have six-monthly, quarterly, monthly, weekly, daily data. Since seasonal fluctuations recur during the course of a year, we cannot see the effects of seasons if the data is lumped together by years or by longer time periods. This is so because if the values of a variable for any twelve consecutive months, or four consecutive quarters, or fifty-two consecutive weeks, is totalled up the effects of the seasons automatically disappear.

From this it follows that the problem of isolating the effects of the seasons arises only when the data is given for 'parts' of years.

The following are the different methods of measuring seasonal variations: (*a*) Simple average method; (*b*) Simple averages corrected to trend; (*c*) Link relative method; (*d*) Moving average method and (*e*) Ratio to trend method.

12.12 SIMPLE AVERAGE METHOD

Under this method (if we have monthly data for a series of years) a typical figure for each month is obtained. The steps in its computation are:

1. Averages the values for each month for all the years.

The averages for different months are given just below the totals for every month in Table 12.8.

2. Compute monthly indices in the following manner

$$\frac{\text{Average of the month} \times 1{,}200}{\text{Total of the monthly averages}}$$

Thuse, the seasonal index for the month of January is

$$\frac{82.33 \times 1{,}200}{1{,}082.22} = 91.3\,,$$

and for February is

$$\frac{82.11 \times 1{,}200}{1{,}082.22} = 91.$$

Thus, the last row gives indices of monthly fluctuation of the variable under consideration. This is quite a simple method. But the indices so obtained do not truly represent the seasonal variations in as much as they include the trend influences also. This method can, therefore, be usefully employed only if the original data has no long-term tendencies.

12.13. SIMPLE AVERAGES CORRECTED FOR TREND

In this method the trend is eliminated from the monthly averages and, thus, the main shortcoming of the previous method is removed. The main steps involved are:

1. Calculate the trend. In Table 12.9, we have computed the trend by the method of least squares. It gives us an yearly increase of 3.01. This figure divided by 12 gives us a monthly increase of .25.

TAELE 12.8. SEASONAL INDICES BY THE SIMPLE AVERAGE METHODS

Year	Jan.	Feb.	March	April	May	June	July	Aug.	Sep.	Oct.	Nov.	Dec.	Total	Yearly Average
1958	72	68	69	71	75	80	85	89	96	97	80	75	957	79.75
1959	82	77	84	73	86	86	90	96	97	100	88	72	1,031	85.92
1960	70	68	69	70	75	76	75	81	84	93	76	68	905	75.42
1961	70	76	82	72	78	84	82	87	93	99	94	83	1,000	83.33
1962	84	84	91	94	97	101	98	104	103	108	98	82	1,144	95.33
1963	85	90	91	87	89	90	89	97	104	109	97	84	1,112	92.67
1964	92	90	92	94	96	98	98	108	107	111	102	88	1,176	98.00
1965	92	91	96	95	103	102	104	110	115	120	107	90	1,225	102.08
1966	94	95	100	97	102	99	97	106	110	111	96	83	1,190	99.17
Total	741	739	774	753	801	816	818	878	909	948	838	725	9,740	811.67
Average	82.33	82.1	86	83.7	89	90.7	90.9	97.6	101	105.3	93.1	80.6	1,082.22	90.190
Sesonal Index	91.3	91.0	95.4	92.8	98.7	100.6	100.8	108.2	112.0	116.7	103.1	89.5	1,200.00	100.00

2. Calculate the mean of monthly averages. In our illustration it is 90.19.

3. This average (computed according to rule (2) 90.19 is the trend value of the middle point of this series. In our illustration this is the value on 1st July.

4. In order to compute the trend value for June (15th), $\frac{.25}{2}$ is deducted from 90.19. Thus, we get 90.065. To compute for July (15th), we add $-\frac{.25}{2}=.125$ to 90.19 and obtain 90.315.

The method of calculating trend values for other months has already been explained.

5. The seasonal indices have been obtained by finding the percentage of monthly average to the secular trend. Thus, for January the seasonal index is:

$$\frac{82.33}{88.815}\times 100=92.7.$$

12.14. LINK RELATIVE METHOD

This method, also called the "Pearson's Method," expresses each monthly figure as a relative of the immediately preceding month. The seasonal pattern is found by averaging all the link relatives for the same month and taking residual trend out of the chain relatives computed from these average link relatives. The steps involved under this method are given in Table 12.10.

1. Translate the original 'data into link relatives, expressing each monthly figure as percentage of the figure for the preceding month. Link relative for the month of April 1966 will be found as:

$$\frac{\text{Figure of April 1966}}{\text{Figure of March 1966}}\times 100$$

2. Sort out the link relatives by months and obtain the mean or median (median will be essier to locate) link relative for each month.

3. Convert the series of median link relatives into "series of chain relatives". The first chain relative (for January) must be fixed at 100. The chain relative of any month is obtained by multiplying the link relative of that month by the chain relative of the preceding month, and dividing by 100. The process is continued until we obtain chain relatives for all the 12 months and for January second time also. The chain relative for January was assumed to be 100.

TABLE 12.9. SEASONAL VARIATIONS (SIMPLE AVERAGE METHOD ADJUSTED FOR TREND) MONTHLY DATA FROM 1958 TO 1966

Year	Jan.	Feb.	March	April	May	June	July	Aug.	Sept.	Oct.	Nov.	Dec.	Total	Average	x	xy	x^2
(1)	(2)	(3)	(4)	(5)	(6)	(7)	(8)	(9)	(10)	(11)	(12)	(13)	(14)	(15)	(16)	(17)	(18)
1958	72	68	69	71	75	80	85	89	96	97	80	75	957	79.65	1	79.75	1
1959	82	77	84	73	86	86	90	96	97	100	88	72	1,031	85.92	2	171.84	4
1960	70	68	69	70	75	76	75	81	84	93	76	68	905	75.42	3	226.26	9
1961	70	76	82	72	78	84	82	87	93	99	94	83	1,000	83 33	4	333.32	16
1962	84	84	91	94	97	101	98	104	103	108	98	82	1,144	95.33	5	476.65	25
1963	85	90	91	87	89	90	89	97	104	109	97	84	1,112	92.67	6	556.02	36
1964	92	90	92	94	96	98	98	108	107	111	102	88	1,176	98.00	7	686.00	49
1965	92	91	96	95	103	102	104	110	115	120	107	90	1,225	102.08	8	316.64	64
1966	94	95	100	97	102	99	97	126	110	111	96	83	1,190	99.17	9	892.53	81
Total	741	739	774	753	801	816	818	878	909	948	838	725	9,740	811.67	45	4,239.01	285
Average	82.33	82.11	86.00	83.67	89.00	90.67	90.89	97.56	101.00	105.33	93.11	80.55	1,082.22	90.19			
Trend	88.815	89.065	89.315	89.565	89.815	90.065	90.315	90.565	90.815	91.0659	1.315	91.565					
Seasonal Index	92.7	92.2	96.3	93.4	99.1	100.7	100.6	107.7	111.2	115·67	102.0	88.0					

Let the trend be $Y=a+bX$.

Normal equations are $\Sigma y=Na+b\Sigma x$, and $\Sigma xy=a\Sigma x+b\Sigma x^2$

i.e., $811.67=9a+45b$, and $4239.01=45a+285b$,

Solving these, $b=3.01$

Yearly change in average $=3.01$

$\therefore$ Monthly Change $=\frac{3.01}{12}=.25$.

TABLE 12.10. SEASONAL VARIATION LINKED RELATIVE METHOD
(Data used is of Table 12.8)

Link Relatives

Year	*Jan.*	*Feb.*	*March*	*April*	*May*	*June*	*July*	*Aug.*	*Sept.*	*Oct.*	*Nov.*	*Dec.*	*Jan.*
1959	109.3	93.9	109.1	86.9	117.8	100	104.6	106.7	101.1	103.1	88.0	89.8	—
1960	97.2	97.1	101.5	101.5	107.2	100.1	98.7	108.0	103.7	110.7	81.7	89.5	—
1961	103.0	108.5	107.9	87.8	108.3	107.6	97.6	106.1	106.9	106.4	94.9	88.3	—
1962	101.2	100	108.3	103.3	103.2	104.1	97.0	106.1	99.0	104.9	90.7	83.7	—
1963	103.7	105.9	101.1	95.6	102.3	100.1	98.9	109.0	107.2	104.8	89.0	86.6	—
1964	109.5	97.8	102.2	102.2	102.1	102.1	100	110.2	99.9	103.8	92.0	86.3	—
1965	104.6	98.9	105.5	99.0	108.4	99.1	102.0	105.8	104.6	104.3	89.2	84.1	—
1966	104.5	101	105.2	97	105.2	97.1	98	109.2	103.8	100.1	86.5	86.4	—
Median...	104.1	99.5	105.3	98	106.2	100.6	98.8	107.4	103.8	104.6	89.0	86·4	—
Chain Relative...	100	99.5	104.8	102.7	109.0	109.7	108.4	116.4	120.8	126.4	112.6	97.31	101.3
Adjusted Chain Relative...	100	99.4	104.6	102.4	108.6	109.2	107.8	115.6	119.9	125.4	111.5	96.1	100
Seasonal Index ...	92.3	91.7	96.6	94.5	100.2	100.7	99.4	100.6	100.6	115.7	102.9	89.6	—

Sum of Adjusted Chain Relative = 1,300.5 ∴ Average = 108.4

The second chain relative for January will be:

$$\frac{\text{Chain relative for December} \times \text{Link relative for January}}{100}$$

4. The last chain relative (for January second time) must be 100. But this would not be usually so due to the presence of the element of trend. The difference between the two chain relatives for January will represent the trend increment or decrement. It is necessary to adjust the chain relative for the effect of the trend. If the last chain relative is greater than 100, the correction factor is to be deducted: if it is less than 100, it is to be added. The first month is kept at 100. But next month the correction factors should be added or subtracted, as the case may be. If we have monthly data the correction factor for the second month would be 1/12 of the amount by which the final chain relative differs from 100. The correction factor for the third month would be 2/12 of this amount, then, 3/12, 4/12, and so on. In our illustration correction factor for March is:

$$\frac{1.3}{12} \times 2.$$

5. The last step is to express the corrected relatives as percentages of their arithmetic mean. This is done by dividing each one of the chain relative by 1/12 of the total of twelve items (adjusted chain relatives) and multiplying by 100. These are the adjusted monthly indices of seasonal variation. The percentages thus obtained can be plotted on a chart showing diagrammatically the seasonal influence.

12.15. MOVING AVERAGE METHOD

This is the most commonly used method for finding an index of seasonal variation. The following steps are to be followed.

1. Compute a twelve-month moving total of the original data. The total for the months July 1959 to June 1960 is recorded in between December 1959 and January 1960, that of August 1959 to July 1960 in between January and February 1960. These two totals are then added and divided by 24. The result is written against January. The process is continued till we get the moving average for December 1966. The averages so obtained are shown in Table 12.11.

2. Divide each original figure by the corresponding moving average. State the result as a percentages (*see* Table 12.12).

3. Find out the mean percentage for each month.

4. Express these mean percentages as a percentage of their own average. This is the index of seasonal variation (*see* column 9, Table 12.12).

TABLE 12.11. SEASONAL VARIATION—MOVING AVERAGE METHOD—12-MONTH MOVING AVERAGES
[Data used is of Table 12 8]

	1959	*1960*	*1961*	*1962*	*1963*	*1964*	*1965*
Jan.	84.4	80.4	78.5	91.9	93.4	95.4	99.5
Feb.	84.9	79.1	79.1	93.0	92.6	96 3	99.8
March	85.2	77 9	79.7	94.1	92.3	96.8	100.3
April	85.4	77.1	80.4	94.9	92.4	97.0	101.0
May	85.9	76.3	81.4	95.4	92.4	97.3	101.6
June	86.1	75.6	82.7	95.6	92.4	97.7	101.9
July	85.5	75.4	83.9	95.6	92.8	97.8	102.0
August	84.6	75.8	85.0	95.8	93.1	97.9	102.3
Sept.	83.8	76.7	85.9	96.0	93.2	98.1	102.6
Oct.	82.8	77.3	87.1	95.7	93.5	98.3	102.9
Nov.	82.3	77.5	88.7	95.0	94.1	98.6	102.9
Dec.	81.4	77.9	90.[illegible]	94.3	94.7	98.1	102.8

TABLE 12.12. RATIOS TO MOVING AVERAGES (PERCENTAGE)

Months	*1959*	*1960*	*1961*	*1962*	*1963*	*1964*	*1965*	*Average of Ratio to moving average*	*Seasonal Index (adjusted)*
	1	2	3	4	5	6	7	8	9
Jan.	97.2	87.1	89.2	91.7	91.1	96.4	92.5	92.2	92.1
Feb.	90.7	85.9	96.1	90.3	97.2	93.5	91.2	92.1	92.1
March	98.6	87.3	102.9	99.7	98.6	95.0	95.7	96.4	96.3
April	85.5	97.9	89.6	96.1	94.2	96.9	94.1	93.9	93.8
May	100.1	98.3	95.8	101.7	96.3	98.7	101.4	98.9	98.8
June	99.9	100.5	101.6	105.7	97.4	100.3	100.1	100.8	100.7
July	105.3	99.5	97.7	102.5	95.9	100.2	101.9	100.4	100.3
Aug.	113.5	106.9	102.4	108.6	104.2	110.3	107.5	107.6	107.5
Sept.	115.8	109.6	108.3	107.3	111.6	109.1	112.1	110.5	110.4
Oct.	120.8	120.3	113.7	112.9	116.6	112.9	116.6	116.3	116.2
Nov.	106.9	93.9	105.9	103.6	103.8	103.5	103.9	103.5	103.4
Dec.	88.5	87.3	92.0	87.0	88.9	89.7	87.6	88.7	88.5

This method has been recommended as the most satisfactory method of estimating a typical seasonal index. **This method irons out the influences of trend and cyclical fluctuations from the index of seasonal variation. We should, however, remember that the moving average technique can be successfully employed if the data contains the variations of uniform periodicity and amplitudes, and seasonal variation is a regular one.**

This method is also termed as "ratio to moving average method" because the second step in its estimation is to divide the original figures by the corresponding moving average. By employing the second step, in fact, we have divided the raw data by trend and cycle and thereby have left in the seasonal and erratic variations.

The object of step (3) is to eliminate from these percentages the influence of erratic causes.

12.16. RATIO TO TREND METHOD

The following are the steps to compute seasonal index by the method of 'ratio to trend':

1. Compute monthly (or quarterly) trend figures by the method of least squares.
2. Express each original figure as the percentage of the corresponding trend figure.
3. Find out the mean (or median) percentage for each month.
4. Figures obtained in (3) above give seasonal variations. Seasonal index can be calculated from these mean percentages by expressing them as percentage of their own average.

Illustration 12.8. Calculate seasonal variation for the following data of sales in rupees of a firm by the ratio to trend method.

	1st Quarter	*2nd Quarter*	*3rd Quarter*	*4th Quarter*
1950	30	40	36	34
1951	34	52	50	44
1952	40	58	54	48
1953	54	76	68	62
1954	80	92	86	82

Solution: For determining the seasonal variation by ratio to trend method, we first calculate the quarterly trend by determining the average quarterly sales for each year. This eliminates the seasonal effect on the trend value.

Year	Yearly totals	Quarterly Avg. (y)	x	xy	x^2	Trend values
1950	140	35	—2	—70	4	32
1951	180	45	—1	—45	1	44
1952	200	50	0	0	0	56
1953	260	65	1	65	1	68
1954	340	85	2	170	4	80
Total		280		120	10	

Yearly increment in quarterly values $= \frac{\Sigma xy}{\Sigma x^2} = \frac{120}{10} = 12$, *Av.* of $y = \frac{280}{[illegible]} =$ [illegible]

∴ Trend value for the year represented by '0' *i.e.* 1952=56.

Yearly increment in quarterly sales = 12..

The yearly trend of the quarterly sales are given in the last column of the above table.

Calculation of quarterly trend values. Consider 1950. Trend values for the middle quarter *i.e.* half of 2nd and half of 3rd is 32. Quarterly increment is 3. So the trend value of 2nd quarter is $32 - \frac{3}{2}$ *i.e.*, 30.5 and for 3rd quarter is $32+\frac{3}{2}$ *i.e.*, 33.5. Trend value of 1st quarter is therefore 30.5—3 *i.e.* 27.5 and of 4th quarter is 33·5 +3 *i.e.* 36.5. Calculating in this manner we have the quarterly trend values and given values expressed as the percentages of corresponding trend value as follows:

TREND VALUES

Year	I Qrt	II Qrt	III Qrt.	IV Qrt.
1950	27.5	30.5	33.5	36.5
1951	39.5	42.5	45.5	48.5
1952	51.5	54.5	57.5	60.5
1953	63.5	66.5	69.5	72.5
1954	75.5	78.5	81.5	84.5

GIVEN VALUES AS % OF TREND VALUES

Year	I Qrt.	II Qrt	III Qrt.	IV Qrt.
1950	109.1	131.1	107.5	93.1
1951	86.1	122.4	109.9	90.7
1952	77.7	106.4	93.9	79.3
1953	85.0	114.3	97.8	85.5
1954	106.0	117.1	105.5	97.0
Total	463.9	591.3	514.6	445.6
Average	92.78	118.26	102.92	89.12
Seasonal Index	92.0	117.4	102.1	88.4

Total of Averages of percentage=(92.78+118.26+102.92+89.12) =403.08

Average of average percentages=403.08/4=100.77.

Now seasonal indices for various quarters are obtained by expressing the average percentages of the quarters as the percentages of their own average *i.e.*, 100.82.

Ratio to trend and ratio to moving average—A comparison: If trend component is important, use ratio to trend method because trend component is given better treatment in this method as compared to ratio to moving average method. But if cyclical component is important and trend component is relatively insignificant, use ratio to moving average because cyclical component is given better treatment in this method as compared to ratio to trend method. Other things being equal, ratio to trend method has got edge over ratio to moving average method, as there is no loss of data in this case.

Illustration 12.9. Assuming that trend is absent, determine if there is any seasonality in the data given below.

Year	*1st Quarter*	*2nd Quarter*	*3rd Quarter*	*4rth Quarter*
1937	3.7	4.1	3.3	3.5
1938	3.7	3.9	3.6	3.6
1939	4.0	4.1	3.3	3.1
1940	3.3	4.4	4.0	4.0

What are the seasonal indices for various quarters?

Solution. If the trend is absent, the difference in the averages of various quarters (if there is any) will be due to seasonal changes.

CALCULATION OF SEASONAL INDEX

Year	*1st Quarter*	*2nd Quarter*	*3rd Quarter*	*4th Quarter*
1937	3.7	4.1	3.3	3.5
1938	3.7	3.9	3.6	3.6
1939	4.0	4.1	3.3	3.1
1940	3.3	4.4	4.0	4.0
Total	14.7	16.5	14.2	14.2
Average	3.675	4.125	3.55	3.55
Seasonal Index	98.7	110.8	95.3	95.3

Explanation for calculating Seasonal Index:

$$\text{The average of averages} = \frac{3.675+4.125+3.55+3.55}{4}$$

$$= \frac{14.9}{4} = 3.725$$

$$\text{Seasonal Index} = \frac{\text{Quarter value}}{\text{General av.}} \times 100$$

∴ S. Index of 1st Quarter

$$= (3.675/3.725) \times 100 = 98.7$$

The above Table shows that there is some seasonality present in the data.

Illustration 12.10. (*a*) Calculate the seasonal index for the data given below by the Link relative method, and (*b*) ratio to moving averages.

Output of Coal in million tonnes

Year	*1st Quarter*	*2nd Quarter*	*3rd Qaarter*	*4th Quarter*
1	68	62	61	63
2	65	58	66	61
3	68	63	63	67
4	70	59	56	62
5	60	55	51	58

Solution

(*a*) CALCULATION OF SEASONAL INDEX (LINK RELATIVE METHOD)

	Link Relatives			
Year	*1st Quarter*	*2nd Quarter*	*3rd Quarter*	*4th Quarter*
1	—	91.2	98.4	103.3
2	103.2	89.2	96.6	108.9
3	111.5	92.6	100.0	106.3
4	104.5	84.3	94.9	110.7
5	96.8	91.7	92.7	113.7
Total	416.0	449.0	482.6	542.9
Arithmetic mean	104.0	89.8	96.5	108.6
				1st Qrt.
				98.0
Chain Relatives	100	89.8	86.5	94.2
Adjusted Chain Relatives	100	90.3	87.5	95.7
Seasonal Index	107.0	96.7	93.7	102.5

Explanation. Link relatives for a figure

$$= \frac{\text{The figure}}{\text{Previous quarter figure}} \times 100$$

∴ Link relative for 1st quarter of second year

$$= \left(\frac{65}{63}\right) \times 100 \text{ i.e., } 103.2.$$

Similarly other link relatives can be calculated. Here arithmetic means are calculated as the number of years are quite small, Median in such a case would be erratic.

Chain relative for 1st Quarter=100 (By assumption)

∴ The chain relative for 2nd Quarter

$$=\frac{\text{Link for II}\times 100}{100}=89.8$$

Chain relative for 3rd Quarter

$$=\frac{89.8\times 96.5}{100}=86.7$$

Chain relative for Quarter

$$=\frac{86.7\times 108.6}{100}=94.2$$

The second chain relative for 1st quarter

$$=\frac{94.2\times 104.0}{100}=98.0.$$

∴ Difference due to presence of trend

$$=98.0-100.0=-2.00$$

∴ Difference in one quarter$=-2.00/4=-0.5$

Adjusted chain relatives are found accordingly.

Average chain relative

$$=\frac{373.5}{4}=93.38$$

$$\text{Seasonal Index}=\frac{\text{chain relative}}{93.38}\times 100$$

(*b*) CALCULATION OF RATIO TO MOVING AVERAGES

Year	*Quarter*	*Given figure*	*4 figure moving totals*	*2 figure moving totals*	*4 figure moving average*	*Given figures as percentage of moving average*
I	I	68				
	II	62				
			254			
	III	61		505	63.1	96.7
			251			
	IV	63		498	62.2	101.3
			247			
2	I	65		489	61.1	106.4
			242			
	II	58		482	60.2	96.3
			240			
	III	56		483	60.4	92.7
			243			

	IV	61		491	61.4	99.3
			248			
3	I	68		503	62.9	108.2
			255			
	II	63		516	64.5	97.7
			261			
	III	63		524	65.5	96.2
			263			
	IV	67		522	65 2	102.8
			259			
4	I	70		511	63.9	109.5
			252			
	II	59		499	62.4	94.6
			247			
	III	56		484	60.5	92.6
			237			
	IV	62		470	58.8	105.4
			233			
5	I	60		461	57.6	104.2
			228			
	II	55		452	56.5	97.3
			224			
	III	51				
	IV	58				

CALCULATION OF SEASONAL INDEX

	Percentage to Moving average				
Year	*I Quarter*	*II Quarter*	*III Quarter*	*IV Quarter*	
1	—	—	96.7	101 3	
2	106.4	96.3	92.7	99.3	
3	108.1	97.7	96.2	102.8	
4	109.5	94.6	92.6	105.4	
5	104.2	97.3	—	—	Total
Total	428.2	385.9	378.2	408.8	1601.1
Average	107.1	96.5	94.5	102.2	400.3
Seasonal Index	107.0	96.4	94.4	102.1	400.0

Arithmetic average of averages

$$=\frac{400.3}{4}=100.1$$

Seasonal index is obtained by expressing each quarterly average as percentage of 100.1.

12.17. CYCLICAL FLUCTUATIONS

In a large number of time series of economic data it has been observed that there is somewhat periodic up and down movement. These movements are known as cyclical variations as they pursue an oscillating movement which, in general, takes the form of a wave, though the distances from peak to trough of the waves are uneven. Such cycles are generally repeated at intervals ranging from about 3 to 10 years and are caused by a complex combination of forces affecting the equilibrium of demand and supply. Cycles generally exhibit semi-regular periodicity as these are neither as regular as are the seasonal variations, nor as accidental as are the random fluctuations. Further, these movements, called *Business Cycles*, are of longer duration than the 12 months seasonal variations. These cycles also differ from the trend. Even when an industry may be showing an upward trend, it is nevertheless possible that general business conditions may at times take a prolonged adverse turn so as to depress the values in the time series well below those tending in earlier years. Similarly, in times of high level of production caused by war, etc., the values of the series may be well above those toward which the data seemed to be moving. Thus, business cycles, reflect the alternations in general prosperity and depression. "It is the cyclical variation which causes profits to become deficits overnight, which produces a public psychology which liquidates assets and casts doubt on the value of even the best investments as the wheels of industry suddenly slow down."*

We have so far discussed various methods of measuring trend (*T*) and (*S*) seasonal fluctuations. Now, we are concerned with the problem of identifying and measuring cyclical fluctuations in a time series. Major interest in many economic studies is attached to their cyclical changes, and their measurement is regarded by many as the central task in the analysis of time series. The residual method of isolating cyclical movements is the most widely used procedure. The essence of this method is the elimination of the trend and seasonal components of given time series. If we divide the actual values by their respective trend value and seasonal indices the result will represent the percentage cyclical fluctuations month by month. Thus, the steps involved in its computation are:

1. Calculate monthly trend values by the moving average method.
2. Calculate monthly seasonal indices preferably by moving average method.
3. Find the percentage ratios of the actual items to the trend values (dividing the actual by trend value and multiplying by hundred).
4. Divide the value obtained under step 3 by the seasonal index.

Their result will be the percentage cyclical fluctuation when the

*Neiwanger, *Elementary Statistical Method*.

data does not contain any random fluctuations.

Thus $$CI = \frac{T \times C \times S \times I}{T \times S}$$

The residue will, of course, also contain elements reflecting the play of irregular fluctuations. Irregular (Random) fluctuation can also be eliminated by applying the moving average method to $C \times I$ values obtained in manner stated above.

12.18. MISCELLANEOUS ILLUSTRATIONS

Illustration 12.11. Given below are the figures of production in thousand quintals of a sugar factory.

Year	1964	1965	1966	1967	1968	1969	1970
Production	77	88	94	85	91	98	90

(*i*) Fit a straight line trend by the 'least square' method. Tabulate the trend values.

(*ii*) What is the monthly increase in production?

Solution

Year	x	y	x^2	xy	*Trend Values*
1964	−3	77	9	−231	83
1965	−2	88	4	−176	85
1966	−1	94	1	94	87
				−501	
1967	0	85	0		89
1968	1	91	1	91	91
1969	2	98	4	196	93
1970	3	90	9	270	95
				+557	
	0	623	28	+ 56	

Substituting the values in the expression

$$a = \frac{\Sigma Y}{N} \text{ and } b = \frac{\Sigma XY}{\Sigma X^2}$$

$$a = \frac{623}{7} = 89 \text{ and } b = \frac{56}{28} = 2.$$

Thus, the straight line equation becomes:

$$Y = 89 + 2X$$

Time Unit: One year

Year of Origin: 1967

Unit of dependent variable: thousands of quintals

(ii) Monthly increase in production $=\frac{2000}{144}=14$ **quintals (approx).**

Illustration 12.12. Fit a trend line by the method of four yearly moving averages to the following time series data:

Production of sugar in India

Year	*Production(in lakhs of tonnes)*	*Year*	*Production*
1944	5	1951	10
1945	6	1952	9
1946	7	1953	10
1947	7	1954	11
1948	6	1955	11
1949	8		
1950	9		

Solution

Year	*Production*	*4 years moving totals*	*4 years moving average*	*4 years moving average centred*
1944	5			
1945	6	25	6.25	
1946	7	26	6.5	6.375
1947	7	28	7.0	6.75
1948	6	30	7.5	7.25
1949	8	33	8.25	7.875
1950	9	36	9.0	8.625
1951	10	38	9.5	9.25
1952	9	40	10.0	9.75
1953	10	41	10.25	10.125
1954	11			
1955	11			

Illustration 12.13. (*i*) Given the trend equation $Y_c=35+5X+3X^2$ where 1968=0 and X unit=1 year.

Change the origin of the equation to 1974.

(*ii*) Given the equation $Y_c=10\ (1.5)^x$ where 1968=0 and X unit=1 year. Shift the origin forward by two years.

Solution. *(i)* We are required to shift origin forward by 6 years. So new time variable X' = old time variable *(X-6)*

Therefore, we have to substitute $X'+6$ for X in the given equation:

$$Y_c=35+5\ (X'+6)+3\ (X'+6)^2$$
$$Y_c=35+5X'+30+3X'^2+36X'+108$$
$$Y_c=3X'^2+41X'+173. \text{ Origin 1970.}$$
$$X' \text{unit}=1 \text{ year}$$

(*ii*) We are required to shift origin forward by two years. The equation will be

$$Y_c=10(1.5)^{x+2}=22.5\ (1.5)^x$$

Illustration 12.14. Trend equation for yearly sales (in 1000 Rs) for a commodity with year 1971 as origin is $Y=81.6+28.8X$. Determine the trend equation to give the monthly trend values with Jan. 1972 as origin and calculate the trend for March 1972.

Solution. Annual trend equation is $Y=81.6+28.8\,X$.

Origin 1971, X unit=1 year.

Y unit=total yearly sales

The equation converted to monthly level, would be

$$Y=\frac{81.6}{12}+\frac{28.8}{144}X$$

$$X=6.8+.2X$$

Origin: June-July, 1971 or 1 July 1971

X unit: One month

Y unit=Average monthy amounts.

Shifting the origin to Jan. 1972.

$$Y=6.8+.2(X+6.5)$$

$$Y=6.8+.2X+1.3$$

$$Y=8.1+.2X$$

Origin: Jan 1972.

X: One month

Y unit: Average monthly amounts

Trend value of March 1972.

$$Y=8.1+.2\ (2)$$

$$Y=8.1+.4=8.5 \textbf{ Ans.}$$

Illustration 12.15. The sales of a company in thousands of Rs for the year 1965 through 1971 are given below:

Year	1965	1966	1967	1968	1969	1970	1971
Sales	32	47	65	92	132	190	275

Estimate the sales figure for the year 1972 using an equation of the form. $Y=ab^x$ where x=year and Y=sales.

Solution. Taking logs: $\log Y=\log a+x\log b$

Year	*x*	*y*	*log y*	x^2	*x.log y*
1965	−3	32	1.5051	9	−4.5153
1966	−2	47	1.6721	4	−3.3442
1967	−1	65	1.8129	1	−1.8129
					−9.6724
1968	0	92	1.8318	0	
1969	1	132	2.1206	1	+2.1206
1970	2	190	2.2788	4	+4.5576
1971	3	275	2.4393	9	+7.3179
					13.9961
			13.6606	28	4.3237

Substiuting the values in the expression

$$\log a = \frac{\Sigma \log Y}{N} \text{ and } \log b = \frac{\Sigma x. \log Y}{\Sigma x^2}$$

We get

$$\log a = \frac{13.6606}{7} = 1.9515 \text{ and } \log b = \frac{4.3237}{28} = 0.154$$

Thus the equation becomes

$$\log Y = 1.9515 + .154\, x.$$

Origin 1969 Time unit=1 year.

For 1972 x would be +4. Putting this value

$\log Y = 1.9515 + .154\,(4)$

$\log Y = 1.9515 + .6160 = 2.5675$

$\log Y = \text{Anti log. } 2.5675 = 369.4$

Hence the estimated sales for the year 1972 is 385.9 thousand rupees.

Illustration 12.16. The seasonal indices of the sale of readymade garments of a particular type in a certain store are given below:

Quarter	*Seasonal Index*
Jan.— March	98
April—June	89
July—Sep	82
Oct.—Dec.	130

If the total sales in the first quarter of the year be worth Rs 10,000, determine how much worth of garments of this type should be kept in stock by the store to meet the demand in each of the remaining quarters.

Solution

In this question we are required to calculate the estimated sales of the remaining quarters.

1st quarter—10,000

IInd " $-10{,}000 \times \frac{89}{98} = 9082$

IIIrd " $-10{,}000 \times \frac{82}{98} = 8367$

IVth " $-10{,}000 \times \frac{130}{98} = 13{,}265$

This assumes that there is no increasing or decreasing trend in the sales.

Illustration 12.17. An index of production of a particular article follows the trend $Y=100+6X$ with 1958 as $X=0$. Find the monthly trend equation with the origin at March 1960.

Solution. The given trend is an annual trend. We have to change it on monthly basis and for this the annual increment has to be changed to a monthly increment as follows. Since it is an equation for index and not totals, a will not change and b will reduce by 1/12.

Hence the equation for monthly trend would be $Y=100+0.5X$. It is with origin of X on July 1,1958.

ove the origin to 15th July, that is to the middle of the

To move the origin to 15th March 1960, *i.e.* to the middle of the 21st month, we have to replace X by $(X+20.5)$. Thus,

$$Y = 100+0.5(X+20.5)$$

$$= 100+10.25+0.5X$$

$$\text{or } Y = 110.25+0.5X$$

This is the monthly trend equation for the index of production with March 1960 as the origin for X.

Illustration 12.18. Assuming population grows with a geometric law, fit a trend to the following figures of population in millions for a certain place.

Year	1898	1908	1918	1928	1938	1948	1958	1968
Pop. in millions	3.9	5.3	7.3	9.6	12.9	17.1	23.2	30.

Solution. Increase with eometric law means that the curve $y=ab^x$ can explain the trend. Taking logarithms $\log y=\log a+x \log b$ or $Y = A+Bx$, where Y represents logarithms of y. To cut short calculaions we can take the origin of time at 1938 and write the values of time in steps of 10 *i.e.* we take $X = (x-1938)/10$. So the new curve will be $Y = A+BX$. This curve can now be fitted in same way as linear trend *i.e.* A and B will be given by normal equations $\Sigma Y = nA+B\ \Sigma X$, $\Sigma XY = A\Sigma X+B\ \Sigma X^2$.

Year X	Population y	Y $=\log Y$	X	XY	X^2
1898	3.9	0.5911	−4	−2.3644	16
1908	5.3	0.7243	−3	−2.1729	9
1918	7.3	0.8633	−2	−1.7266	4
1928	9.6	0.9823	−1	−0.9823	1
1938	12.9	1.1106	0	0	0
1948	17.1	1.2330	1	1.2330	1
1958	23.2	1.3655	2	2.7310	4
1968	30.5	1.4843	3	4.4529	9
Total		8.3544	−4	1.1707	44

Substituting the values in normal equation

$$8.3544=8A-4B \quad \ldots(i)$$

$$1.1707=-4A+44B \quad \ldots(ii)$$

Muitiplying (*ii*) by 2 and adding to (*i*)

$$10.6958=84B \quad \text{or} \quad B=0.12733X. \because A = 1.10796$$

$\therefore$ The trend equation is $Y=1.10796+0.12733X$

or $Y=1.10796+0.12733\left(\frac{x-1938}{10}\right)$ where x represents given years. To find out trend for any year we will first get Y for that year and then the trend value of population *i.e.* y will be antilog Y.

Illustration 12.19. Show that moving average is a device by which one can remove the periodic variations, provided the period of the moving average is same as that of the periodic variations. If the two periods are different the periodic variation will be persent in the data, though with reduced intensity.

Solution. Let us consider the following series of production figures given in table below). This has a clear cycle of three years and there is no trend in them. Calculating three years and 5 years moving averages we have

Year	*Production*	*Moving Totals 3 years*	*Moving Av. 3 years*	*Moving Totals 5 years*	*Maving Av. 5 years*
1960	1				
1961	2	6	2		
1962	3	6	2	9	1.8
1963	1	6	2	11	2.2
1964	2	6	2	10	2.0
1965	3	6	2	9	1.8
1966	1	6	2	11	2.2
1967	2	6	2	10	2.0
1968	3	6	2		
1969	1				

It is clear from the calculations that the periodic variations are completely removed when we took three years moving average *i.e.* with the same period as the period of variation. But when 5 year moving average was calculated periodic variations were still present though the intensity of variations is considerably reduced.

Illustration 12.20. A straight line trend fitted to pig iron production for the year 1930 was given by $y=2345+3.1\,x$, where unit of x is one month and the origin 1st of July.

(*i*) Tabulate the monthly trend figures.

(*ii*) Write down the corresponding trend equation for annual production (*i.e.* unit of x is one year) with reference to the same origin as above.

Solution (*i*) Origin 1st July means the month indicated by $x=0$, has the middle day as 1st July *i.e.*, it is 15th June to 15th July.

This means that the x value for June is -0.5 and of July 0.5, x value for May will then be -1.5 and August 1.5. Proceding in this manner the monthly production for various months are

Jan.	$=2345+3.1\times(-5.5)=2327.95$	July	$=2345+3.1\times(0.5)=2346.55$
Feb.	$=2345+3.1\times(-4.5)=2331.06$	Aug.	$=2345+3.1\times(1.5)=2349.65$
March	$=2345+3.1\times(-3.5)=2334.15$	Sept.	$=2345+3.1\times(2.5)=2352.75$
April	$=2345+3.1\times(-2.5)=2337.25$	Oct.	$=2345+3.1\times(3.5)=2355.85$
May	$=2345+3.1\times(-1.5)=2340.35$	Nov.	$=2345+3.1\times(4.5)=2358.95$
June	$=2345+3.1\times(-0.5)=2343.45$	Dec.	$=2345+3.1\times(5.5)=2362.05$

(*ii*) In a yearly trend equation the coefficient of x will give yearly increment. The yearly increment in the present case will be $3.1\times 12=37.2$. 1st July is the central day of the year also. So it means taking year 1930 as zero the trend equation for yearly averages is $y=2345+37.2x$.

If trend equation is required for the yearly totals a and b both values will be 12 times of their values in yearly average equation. Hence the trend equation for yearly total is

$$y=2345\times 12+37.2\times 12x, \quad i.e. \quad y=28140\times 446.4x.$$

Illustration 12.21. Following figures give the yearly trend averages of sales of a certain shop. Calculate the monthly figures for all the five years.

Years	1	2	3	4	5
Yearly Average in Rs	1200	1344	1488	1632	1776

What would be your figures had the above figures been the yearly totals.

Solution. In the present case trend values are given and not the trend equation. We notice that the difference between any two consecutive years is Rs 144 *i.e.*, in one year the yearly average increases by Rs 144. So the monthly increase will be 144/12 *i.e.*, Rs 12 per month. For any particular year the yearly average correspond to the middle one month *i.e.*, 15th June to 15th July (centred at 1st July). So the month June is half month earlier than the middle month and the July month comes half month after the middle month. In the case of 1st year therefore the June figure is $1200-12/2=1194$ and July figure is $1200+12/2=1206$. On either side of these months, monthly averages can now be found by adding and subtracting Rs 12 for each month moved. Hence the monthly averages for the various months are as below:

MONTHLY AVERAGE SALE (RS) OF A CERTAIN SHOP

Month	*1st year*	*2nd year*	*3rd year*	*4th year*	*5th year*
January	1134	1278	1422	1566	1710
February	1146	1290	1434	1578	1722
March	1158	1302	1446	1590	1734
April	1170	1314	1458	1602	1746
May	1182	1326	1470	1614	1758
June	1194	1338	1482	1626	1770
July	1206	1350	1494	1638	1782
August	1218	1362	1506	1650	1794
September	1230	1374	1518	1662	1806
October	1242	1386	1530	1674	1818
November	1254	1398	1542	1686	1830
December	1266	1410	1554	1698	1842

If the given figures were yearly totals, first we will have to find the yearly averages by dividing each of the yearly totals by 12. In the present case yearly averages would be Rs 100, 112, 124, 136, 148. This gives yearly increment at Rs 12 *i.e.* monthly increment at Re 1. Proceeding as above the monthly averages in this case would be

Month	*Jan*	*Feb.*	*March*	*April*	*May*	*June*	*July*	*Aug*	*Sept.*	*Oct.*	*Nov.*	*Dec.*
1st yr.	94.5	95.5	96.5	97.5	98.5	99.5	100.5	101.5	102.5	103.5	104.5	105.5
2nd yr.	106.5	107.5	108.5	109.5	110.5	111.5	112.5	113.5	114.5	115.5	116.5	117.5
3rd yr.	118.5	119.5	120.5	121.5	122.5	123.5	124.5	125.5	126.5	127.5	128.5	129.5
4th yr.	130.5	131.5	132.5	133.5	134.5	135.5	136.5	137.5	138.5	139.5	140.5	141.5
5th yr.	142.5	143.5	144.5	145.5	146.5	147.5	148.5	149.5	150.5	151.5	152.5	153.5

Illustration 12.22. From the following series of annual data, for which the trend values have already been calculated, isolate cyclical fluctuations.

Year	1931	1932	1933	1934	1935	1936	
Actual value	170	231	261	267	278	302	
Trend value	228	239	251	263	275	287	
Year	1937	1938	1939	1940	1941	1942	1943
Actual value	299	298	340	273	210	158	173
Trend value	299	311	323	334	346	358	370

Solution. In yearly figures seasonal variations will not be present. So the given actual figures contain trend, cyclical variation and irregular variation. Irregular variations cannot be separated out from the data. Trend values are already given. So the effect of cyclical variations along with irregular (if there are any) can be found by eliminating the trend effect from the given figures. For this, express every given actual value as the percentage of the corresponding trend value. Then subtracting 100 from each of these percentages the cyclical variations are given as follows:

Year	1931	1932	1933	1934	1935	1936	
% of trend	74.6	96.6	104.0	101.5	101.1	105.2	
Cyclical vari.	—25.4	—3.4	4.0	1.5	1.1	5.2	
Year	1937	1938	1939	1940	1941	1942	1943
% of trend	100.0	95.8	105.3	81.7	60.7	44.1	46.8
Cyclical vari.	0.0	—4.2	5.3	—18.3	—39.3	—55.9	—53.2

Illustration 12.23. The number (in hundreds) of letter posted in a certain city on each day in a typical period of five weeks was as follows:

	Sun.	*Mon.*	*Tue.*	*Wed.*	*Thu.*	*Fri.*	*Sat.*
1st week	18	161	170	165	153	181	76
2nd week	18	164	179	157	168	195	85
3rd week	21	162	168	153	137	185	84
4th week	21	171	182	170	162	179	95
5th week	27	162	186	170	170	180	120

Calculate the 'average fluctuations' indices within a week.

Solution. Fluctuations within a week means seasonal variations within a week. Totals for different weeks suggest that there is a trend present in the total posting for week to week. So variation within a week can be calculated by either averages adjusted to trend, or ratio to moving average method or by ratio to trend method. Link relative method may not be used as large amount of cyclical variations are not expected. As in every week, figures for Sunday are very low, it appears that variations within a week repeat with almost perfect regularity. So ratio to moving average method may be best. But as the question requires 'average fluctuations' so average adjusted to trend method may be used here.

CALCULATION OF WEEKLY VARIATIONS

Week	*Sun.*	*Mon.*	*Tue.*	*Wed.*	*Thu.*	*Fri.*	*Sat.*	*Total*	*Average y*	*x*	*xy*	x^2
1	18	161	170	165	153	181	76	924	132	−2	−264	4
2	18	164	179	157	168	195	85	966	138	−1	−138	1
3	21	162	168	153	137	185	84	910	130	0	0	0
4	21	171	182	170	162	179	95	980	140	1	140	1
5	27	162	186	170	170	180	120	1015	145	2	290	4
Total	105	820	885	815	790	920	460	4795	685		28	10
Average	21	164	177	163	158	184	92	959	137			
Av. adj. to Trend	22.2	164 8	177.4	163.0	157.6	183.2	90.8	959	137			
Indices	16	120	129	119	115	134	67	700	100			

$$\text{Weekly increase}=\frac{\Sigma xy}{\Sigma x^2}=\frac{28}{10}=2.8.$$

Daily increase$=2.8/7=0.4$.

General Average is 137. This represents the trend figure for the middle of the week *i.e.*, for Wednesday. Trend figure for other days can be calculated by adding or substracting daily increase to this. The easier method of adjustment, if the daily increase is not large, will be to bring every average figure to the level of Wednesday. This means we add the effect of daily rise to Sunday, Monday and Tuesday and substract it from Thursday, Friday and Saturday. Using this trend adjusted figures one as above which represents the average fluctuations. They can be put into indices from by expressing them in index numbers with general average 137 as base. The result is as shown above.

Illustration 12.24. The sales of a company for the last eight years are given as:

Year:	1973	1974	1975	1976	1977	1978	1979	1980
Sales '000:	52	45	98	92	110	185	175	220

Estimate the sales figures for 1981 using the trend $y=ab^x$ where $x=$ year and $y=$sales.

Solution. The equation is $y=ab^x$

Taking logs:

$$\log y=\log a+x \log b$$

So the normal equations are

$$\Sigma \log y = N \log a + \log b \Sigma x$$
$$\Sigma x \log y = \log a \Sigma x + \log b \Sigma x^2$$

Year	*x*	*Sales y*	*log y*	x^2	*x log y*
1973	−3	52	1.7160	9	−5.1480
1974	−2	45	1.6532	4	−3.3064
1975	−1	98	1.9912	1	−1.9912
1976	0	92	1.9638	0	0
1977	1	110	2.0414	1	2.0414
1978	2	185	2.2672	4	4.5344
1979	3	175	2.2430	9	6.7290
1980	4	220	2.3424	16	9.3696
Totals	4	977	16.2182	44	12.2288

Thus,

$$16.2182 = 8 \log a + 4 \log b \quad (i)$$
$$12.2288 = 4 \log a + 44 \log b \quad (ii)$$

Multiplying (*i*) by 11

$$178.4002 = 88 \log a + 44 \log b \quad (iii)$$

Subtracting (*ii*) from (*iii*)

$$166.1712 = 84 \log a$$

or $\log a = 1.9782$ and $\log b = 0.0981$

or $a = 95.111$ and $b = 1.2534$

Therefore, the trend equation is

$$y = 95.111 \times (1.2534)^x$$

Year 1981 corresponds to $x = 5$

$$y = 95.111 \times (1.2534)^5 = 294.26$$

This is the projected sale for 1981 in '000 rupees.

Illustration 12.25 Trend equation for yearly sales (in '000 Rs) for a commodity with year 1979 as origin is $y = 81.6 + 28.8x$. Determine the trend equation to give monthly sales with Jan. 1980 as origin, and calculate trend value for March 1980.

Sclution. Yearly trend equation is reduced to monthly trend equation by dividing *a* by 12 and *b* by 144. So monthly sales are

$$y = \frac{81.6}{12} + \frac{28.8}{144} x$$

or $y = 6,8 + 0.2\,x$

Its origin is the middle of 1979, *i.e.*. 1st July 1979. To shift origin to 15 Jan. 1980, we advance *x* by 6.5 months.

or $$y=6.8+0.2\,(x+6.5)$$
or $$y=8.1+0.2\,x$$

March 1980 is $x=+2$ with Jan. 1980 as origin. So March sales are $8.1+0.2\times2=8.5$ thousand rupees.

Illustration 12.26. The trend equation fitted to annual sales is given by $y=230+20\,x$, unit of x is one year, origin 30 June 1979. Adjust the trend equation for finding the monthly trend values for the month of January to March 1980.

Solution. Monthly trend equation with x measured in months from origin 30 June 1979 is

$$y=\frac{230}{12}+\frac{20}{144}\,x$$

or $$y=19.17+0.139\;x$$

To find monthly values, we shift origin to 15 July 1979 by replacing x by $(x+0.5)$ $\quad Y = 19.240+0.139X$

Now Jan 80 is $x=6$, Feb 80 is $x=7$
and March 80 is $x=8$

So monthly trend values for

Jan 80 = $19.240+0.139\times6=$ 20.074
Feb 80 = $19.240+0.139\times7=$ 20.213
and Mar 80 = $19.240+0.139\times8=$ 20.352

These are monthly averages.

Illustration 12.27. Population figures for a city are given below. Fit a curve of the type $y=ab^x$ and estimate the population for 1977.

Year	1971	1972	1973	1974	1975
Population ('000)	132	142	157	170	191

Solution. $y=ab^x$ becomes $\log y=\log a+x\log b$. The normal equations are

$$\Sigma\log y=N\log a+\log b\,\Sigma x$$
$$\Sigma x\log y=\log a\,\Sigma x+\log b\,\Sigma x^2$$

Year	x	*population* y	*log y*	x^2	*x log y*
1971	−2	132	2.1206	4	−4.2412
1972	−1	142	2.1523	1	−2.1523
1973	0	157	2.1959	0	0
1974	1	170	2.2304	1	2.2304
1975	2	191	2.2810	4	4.5620
Totals	0		10.9802	10	0.3989

Thus

$$10.9802 = 5 \log a$$
$$0.3989 = 10 \log b$$

or $a = \text{antilog } \frac{10.9802}{5}$

$= 157.05$

and $b = \text{antilog } \frac{0.3989}{10}$

$= 1.096$

So trend equation is

$$y = 157.05\ (1.096)^x$$

For 1977, $x = 4$, so that

$$y = 157.05\ (1.096)^4$$
$$= 226.77$$

This is the estimated population in '000.

Illustration 12.28. Given below are the figures of production in '000 quintals of a sugar factory:

Year:	1971	1972	1973	1974	1975	1976	1977
Production:	40	45	46	42	47	50	46

Fit a straight line trend and tabulate the trend values.

Solution.

Year	x	y	x^2	xy	*Trend values*
1971	−3	40	9	−120	42.032
1972	−2	45	4	−90	43.068
1973	−1	46	1	−46	44.104
1974	0	42	0	0	45.140
1975	1	47	1	47	46.176
1976	2	50	4	100	47.212
1977	3	46	9	138	48.248
Totals	0	316	28	29	316

Normal Equations

$$\Sigma y = Na + b\Sigma x$$
$$\Sigma xy = a\Sigma x + b\Sigma x^2$$

or $316 = 7a$

and $29 = 28b$

or $a = 45.14$

and $b = 1.036$

So the trend equation is $y = 45.14 + 1.036\,x$ with 1974 as origin, unit of x as 1 year and unit of y as 1000 quintal.

Illustration 12.29. Using 3-year moving average, determine the trend and short-term fluctuations of the given data:

year:	1968	1969	1970	1971	1972	1973	1974	1975
production:	21	22	23	25	24	22	25	26
year:	1976	1977						
production:	27	26						

Solution.

Year	*Production*	*3-year moving total*	*3-year moving average* (4)=(3)/3	*Short-term fluctuation* (5)=(2)—(4)
(1)	(2)	(3)	(4)	(5)
1968	21			
1969	22	66	22	0
1970	23	70	23.3	−0 3
1971	25	72	24	+1.0
1972	24	71	23.7	+0.3
1973	22	71	23.7	− 1.7
1974	25	73	24.3	+ 0.7
1975	26	78	26.0	0
1976	27	79	26.3	+0.7
1977	26			

PROBLEMS

1. Explain fully what is meant by 'Secular Trend'. Name the important types of forces influencing a time series.
2. Differentiate between the series—
   (*a*) which exists at a point of time.
   (*b*) which is spread over a period of time.
3. Write short notes on—
   (1) Periodicity,
   (2) Moving average,
   (3) Irregular Fluctuations, and
   (4) Line of best-fit.
4. What are the various methods of estimating seasonal variations in historical series?
5. Calculate 5 yearly moving averages of students studying in a commerce college as shown by the following figures:

Year	*No. of students*	*Year*	*No. of students*
1951	332	1956	405
1952	317	1957	410
1953	357	1958	427
1954	392	1959	405
1955	402	1960	431

6. Calculate the five yearly moving average for the following time series and plot it with the original figures on the same graph. Next calculate seven yearly moving average and plot it on the same graph. Comment on the reversal effect.

Year	1	2	3	4	5	6	7	8	9	10
Annual Fig.	110	104	78	105	109	120	115	110	115	122
Year	11	12	13	14	15	16	17	18	19	20
Annual Fig.	130	127	122	118	139	140	135	130	127	135
Year	21	22	23	24	25	26	27	28	29	30
Annual Fig.	146	142	138	135	145	155	150	148	143	155

7. For the following data of production of tea in India construct 4-year centred moving averages.

Year	1952	1953	1954	1955	1956	1957
(Production in million pounds)	614	615	652	678	681	655
Year	1958	1959	1960	1961	1962	
(Production in million pounds)	717	719	708	779	757	

8. What is weighted moving average. Find the trend for the series by three-year weighted moving average with weights 1, 2, 1.

Year	1	2	3	4	5	6	7
Production series	2	4	5	7	8	10	13

9. Find out an appropriate moving average for the following series.

Year	*Figs.*	*Year*	*Figs.*
1951	506	1962	818
1952	620	1963	745
1953	1036	1964	845
1954	673	1965	1276
1955	588	1966	898
1956	696	1967	814
1957	1116	1968	929
1958	738	1969	1360
1959	663	1970	61
1960	777	1971	925
1961	1189		

10. Find out the short term fluctuations of the following time series assuming a five yearly cycle:

Year	1910	1911	1912	1913	1914	1915	1916
Annual values	239	242	238	252	257	250	273
Year	1917	1918	1919	1920	1921	1922	1923
Annual values	270	268	288	284	282	300	303
Year	1924	1925	1926	1927	1928	1929	1930
Annual valurs	298	313	317	309	329	333	327
Year	1931	1932	1933	1934	1935		
Annual values	345	344	344	362	360		

11. Fit a trend line by the method of least squares to the following data:

Year	1941	1942	1943	1944	1945	1946	1947
Sales in Rs (thousand)	80	90	92	83	94	99	92

(*a*) Find the slope of a straight line trend to these figures.

(*b*) Plot these figures on a graph and show the trend line

(*c*) Do these figures show a rising trend or a falling trend? How do you arrive at your conclusions?

12. Fit a straight line trend to the data given below by the method of least squares:

Year	1959	1960	1961	1962	1963	1964	1965
Gross ex-factory value of output (Rs crore)	672	824	968	1205	1464	1758	2058

13 Calculate the trend by method of semi-averages for the data of Prob. 12. Compare this result with the one obtained in Prob. 12.

14. The number of units of a product exported during 1960-67 is given below. Fit a. straight line trend to the data. Findan estimate for the year 1968.

Year	1960	1961	1962	1963	1964	1965	1966	1967
No. of Units in 0^0's	12	13	13	16	19	23	21	23

15. Plot the data in Prob. 14, on graph paper and estimate the trend values by free hand curve method. Compare this result with the one obtained in Prob. 14 above.

17. Assuming that trend is absent, determine if there is any seasonality in the data given below:

	1st Quarter	*2nd Quarter*	*3rd Quarter*	*4th Quarter*
1967	3.0	4.1	3.3	3.5
1968	3.7	3.9	3.6	3.9
1969	4.0	4.1	3.3	3.1
1970	3.3	4.4	4.0	4.0

What are the seasonal indices for various quarters?

18. The following table gives the value of exports of merchandise from India during the years 1966-67 to 1970-71. Calculate the seasonal variation for each month during this period by simple average adjusted to trend method.

Months	*1966-67*	*1967-68*	*1968-69*	*1969-70*	*1970-71*
April	20	27	17	23	29
May	20	26	18	26	28
June	19	21	15	18	29
July	26	19	17	23	25
August	25	19	18	24	22
Sept.	30	21	19	20	23

October	28	19	17	21	25
Nov.	29	17	19	27	26
Dec.	26	18	21	26	30
Jan.	29	18	22	28	36
Feb.	26	17	21	30	35
March	30	18	26	31	40

19. Calculate the seasonal index for the data given below by the link relative method.

Year	*1st Quarter*	*2nd Quarter*	*3rd Quarter*	*4th Quarter*
	Output of Coal in million tonnes			
1	68	62	61	63
2	65	58	56	61
3	68	63	63	67
4	70	59	56	62
5	60	55	51	56

20. Calculate the seasonal indices for the data of Problem 19 by the method of ratio to moving averages.

21. Calculate seasonal variation for the following data of sales in Rs of a firm by the ratio to trend method.

	1st Quarter	*2nd Quarter*	*3rd Quarter*	*4th Quarter*
1950	30	40	36	34
1951	34	52	50	44
1952	40	58	54	48
1953	54	76	68	62
1954	80	92	86	82

22. Isolate cyclical variations for the data of Prob. 21 above.

23. Using the data given below, explain clearly how you would determine seasonal fluctuations of a time series.

Year	*Summer*	*Monsoon*	*Autumn*	*Winter*
1	30	81	62	119
2	33	104	86	171
3	42	153	99	221
4	56	172	129	235
5	67	201	126	302

24. Below are given the figures of production (in thousand quintals) of Sugar Factory.

Year	1953	1955	1956	1957	1958	1959	1962
Production	77	88	94	85	91	98	90

(*i*) Fit a straight line by the 'Least Squares' method and tabulate the trend values.

(*ii*) Eliminate the trend. What components of the time series are thus left over?

(*iii*) What is the monthly increase in the production of sugar?

25. Assuming population grows with a geometric law, fit a trend to the following figures of population in millions for a certain place.

Year	1898	1908	1918	1928	1938	1948	1958	1968
Population in million	3.9	5.2	7.3	9.6	12.9	17.1	23.2	30.2

26. Show that moving average is device by which one can remove the periodic variations, provided the period of the moving average is same as that of the periodic variation. If the two periods are different the periodic variation will be present in the data, though with reduced intensity.
27. Find the trend eliminated figures for the data of Prob. 5.
28. Remove the effect of season from the data of Prob. 19.
29. A straight line trend fitted to pig iron production for the year 1930 was given by $y=2345+3.1\ x$, where x is one month and the origin 1st of July.
    (*i*) Tabulate the monthly trend figures.
    (*ii*) Write down the corresponding trend equation for annual production *i.e.* x is one year) with reference to the same origin as above.
30. Compute the trend by four week moving averages for the following data:

Week	1	2	3	4	5	6	7	8	9	10
Production	82	73	74	75	73	72	76	76	74	75
Week	11	12	13	14	15	16	17	18	19	20
Production	75	73	75	76	75	75	78	76	78	79

31. Find the yearly moving averages from the following data and represent the original series and trend on a graph paper.

Year	*Index*	*Year*	*Index*
1940	105	1946	85
1941	115	1947	75
1942	180	1948	60
1943	90	1949	65
1944	80	1950	70
1945	95		

32. The following are the index of annual production of a certain commodity. Assume 5 yearly cycle, find out the trend values.

Year	*Index*	*Year*	*Index*
1941	225	1947	235
1942	210	1948	225
1943	201	1949	233
1944	215	1950	249
1945	223	1951	265
1946	245		

33. Calculate the seasonal indices by "the ratio to moving average method" from the following data on synthetic rubber production in appropriate units.

Year	*1st Quarter*	*2nd Quarter*	*3rd Quarter*	*4th Quarter*
1961	68	62	61	63
1962	65	58	66	61
1963	68	63	63	67

34. The number of units of a product exported during 1960-67 is given below. Fit a straight line trend to the data. Plot the given data showing also the trend line. Find an estimate for the year 1968.

Year	1960	1961	1962	1963	1964	1965	1966	1967
No. of Units in '000	12	13	13	16	19	23	21	23

35. Fit a linear trend to the following data:

x	10	12	14	16	18
y	1050	1100	1250	1300	1475

Chapter 13

INDEX NUMBERS

13.1. INTRODUCTION

Index number is a commonly used statistical device for measuring the combined fluctuations in a group of related variables. If we wish to compare the price level of consumer items today with that prevalent 10 years ago, we are not interestcd in comparing the prices of only one item, but in camparing some sort of average price levels as they obtain at two different times. We may wish to compare the agricultural production or industrial production today with that at the time of independence. Here again, we have to consider all items of production, and each item may have undergone a different fractional increase (or even a decrease). How then do we obtain a composite measure ?

This composite measure is provided by index number which may be defined as a *device for combining the variations that have come in a group of related variables over a period of time, with a view to obtain a figure that faithfully represents the 'net' result of the change in the constituent variables.*

Index numbers may be classified in terms of the variables that they are intended to measure. In business the different group of variables in the measurement of which index number techniques are commonly used are (i) price, (*ii*) quantity, (*iii*) value and (*iv*) business activity. Thus we have index of wholesale prices, index of consumer prices, index of industrial output, index of value of exports and index of business activity, etc.

Since the basic approach in the construction of all these types of index numbers is the same, we would give a detailed discussion of the method of constructing price index, and refer only in broad general terms about quantity index numbers.

13.2. METHOD OF COMBINING THE DATA

The principle problem in the construction of index numbers lies in deciding on the basis of combining the variables. Let us take, for example, the problem of obtaining a price index. Let us be given the prices of each variable in the base year, p_0, and in the current year, p_1. The prices p_0 and p_1 are given to us for each of the *n* commodities that we are interetested in.

A straight forward method of combining will consist of a *simple aggregate of actual prices*. Thus, for the base year Σp_0 represents simple aggregate of prices, and for the current year Σp_1 is the simple aggregate, and we may define a price index as

$$\text{Index} = \frac{\Sigma p_1}{\Sigma p_0} \times 100 \tag{13.1}$$

Table 13.1 shows the calculation of price index based on this formula.

TABLE 13.1. INDEX NUMBER BASED ON AGGREGATE OF ACTUAL PRICES

Commodity	*Unit*	*1961* *Rs* p_0	*1962* *Rs* p_1	*1963* *Rs* p_2
A	Per metre	100	90	110
B	,, ,,	10	11	9
C	,, ,,	5	6	4
D	,, ,,	4	5	2
E	,, gram	1	3	1
Totals		120	115	126
Index number with 1961 as base		100	95.83	105

The year 1961 has been selected as a point of reference (or base year), and the prices of five commodities *viz. A, B, C, D,* and *E* during 1962 and 1963 are compared with those of 1961 with a view to arrive at the net change that has taken place.

The following are the steps in this method:

(*i*) Total the prices for each year (base year as also the given years). In the table, the total of base year price (denoted by Σp_0) is 120 and those of given years 1962 and 1963 (denoted by Σp_1 and Σp_2) are 115 and 126 respectively.

(*ii*) Since the year 1961 is regarded as the base year (reference point) the price total of this year is taken as 100.

(*iii*) Index numbers for years 1962 and 1963 are computed by expressing their respective price totals as percentages of the price total in the base year.

Thus, the index of 95.83 for 1962 is computed by expressing Rs 115 (price total of 1962) as a per cent of Rs 120 (price total in the base year 1961), and the index of 105 in 1963 is computed by expressing Rs 126 as a per cent of Rs 120.

The simple aggregative index is easy to compute and understand, but it has some serious defects.

Defects of the simple aggregative method. The chief weakness of this type of index number is that those commodities which have large figu' quotations dominate the index. For instance, a decrease of 10 per cent in the price of commodity *A* for 1962 is enough to bring down the index in spite of the fact that there is an increase of 10 per cent, 20 per cent, 25 per cent and 200 per cent respectively in the prices of commodities *B*, *C*, *D* and *E*.

For 1963 an increase of only 10 per cent in A (above that in 1961) is more than enough to counter-balance a decrease of 10 per cent, 20 per cent and 50 per cent, respectively in the price of B, C and D. In other words, it can be said that the importance (weights) assigned to various commodities under this system is directly related to their price quotations. Large figure quotations imply large weights and small figure quotations small weights.

Thus, in Table 13.1 the price quotation of commodity *A* for 1961 is five times as large as those of the remaining four commodities put together. This means that the total weight assigned to *A* is five times that assigned to the other four commodities combined. Now, this system of weighting is illogical for the simple reason that it is unscientific and hence the index numbers thus obtained do not faithfully represent the variations of the group under consideration. This defect of unscientific weighting cannot be remedied by adopting uniform units, say quintal, for purposes of price quotations. Such a device will introduce new inequalities in place of the old ones and will not in any way be helpful. It can thus be concluded that this system is quite unreliable as a measure of price change.

The simple average of price relatives. As the name implies, this method, consists of finding price relatives and then averaging them.

A price relative is obtained by expressing the price of each commodity in a given year as a per cent of the price of the commodity in the base year. Thus, if the price of a commodity in 1960 (base year) is Rs 10 and in the given year (1961) Rs 11, the price relative would be $\frac{11}{10} \times 100 = 110$

Symbolically, price relative $= \frac{p_1}{p_0} \times 100.$

The next step is to average these price relatives of each given time period. The averages so obtained would be the indices for respective given periods.

For the purpose of averaging any one of the average—Mean, median, geometric mean, or harmonic mean may be used.

In practical statistical work, however, it is the arithmetic mean that is most commonly used for reasons of simplicity.

Expressed in the form of a formula, Mean of relatives

$$\frac{\Sigma\left(\frac{p_1}{p_0}\times 100\right)}{N}$$

where N represents the number of commodities included in the index.

TABLE 13.2. INDEX NUMBER BASED ON AGGREGATE OF ACTUAL PRICES

Commodity	*Unit*	*1961* *Rs*	*1962* *Rs*	*1963* *Rs*	*1964* *Rs*	*1965* *Rs*
A	per quintal	10	9	8	9	11
B	,, kilogram	4	5	3	6	4
C	,, gram	6	8	5	4	6
D	,, quintal	11	13	15	14	12
E	., kilogram	6	8	10	14	12
Total		37	43	41	47	45
Index Number with 1961 as base		100	116.2	110.8	127.0	121.62

TABLE 13.3. INDEX NUMBER BASED ON MEAN OF RELATIVES
(Data used is of Table 12.2

Commodity	*1961*	*1962*	*1963*	*1964*	*1965*
A	100	90	80	90	110
B	100	125	75	150	100
C	100	133	83	67	100
D	100	118	136	127	109
E	100	133	167	233	200
Total	500	599	541	667	619
Mean of Relatives	100	119.8	108.2	133.4	123.8

The calculation of index based on the price relatives is shown in Table 13.3.

The resulting composite index number shows that on an average these five prices rose to 119.8 in 1962, 108.2 in 1963, 133.4 in 1964 and 123.8 in 1965. Note that movement of this index is not dominated by movement of the price of commodities *A* and *D* as in the case with index based on actual prices (Table 13.2).

Simple average of relatives method is not influenced by units (quintal, kilogram) in which prices are quoted or by the largeness or smallness of a price quotation. The price relative of commodity *A* for 1965 with 1961 as base will remain the same whether we quote its price in quintal or kilogram.

Like the previous method, this method also involves a kind of illogical weighting. The quantity used as weight is the amount of the commodity which could be bought in the base year for Rs 100. Illustrating from the commodity *E*, if its price in base year is Rs 6 per kg, then we could purchase about 16.67 kilograms for Rs 100 in 1961 (base year). The same quantity will cost Rs 200 in 1965 when the rate is Rs 12 per kg. In other words, to purchase that quantity of commodity *E* which could be purchased in 1961 for Rs 100 it is necessary to pay Rs 200 in 1965. The weight assigned to *E* in this case is 16.7 kilogram (approx.); to *A*, 10 quantals to *B*, 25 kg; to C, 16.7 kg.; and to D, 9.1 quintals. This system of weighting is unscientific for the simple reason that weights have not been fixed is accordance with the relative importance of the commodities.

It is, therefore, necessary to evolve some other basis of weighting which assigns weights according to the importance of that commodity for the purpose of enquiry. Thus, if a person spends Rs 100 per month on cooking oil and Rs 2 per month on salt, a hundred per cent increase in the price of salt might pinch him less than even a five per cent increase in the price of cooking oil. This consideration must be reflected in the system of assigning weights.

13.3. SYSTEM OF WEIGHTING

It has been discussed in the preceding section that the two index numbers, 'the simple aggregative' and 'the simple mean of relatives', which apparently seem to be unweighted are not really so. The simple aggregative method is weighted by the magnitude of the prices (the higher price commodity has a greater influence on the result than a lower priced commodity). It is also influenced by units in which prices are quoted. Commodities whose prices are quoted in quintals exert greater influence than those whose prices are quoted in smaller units. *Simple mean of relatives* method gives equal weights to all commodities. Thus, a less important commodity has as much influence on the result as a very important commodity has. In order to eliminate this unscientific weighting it is necessary that a scientific system of weighting be introduced which will accord to each commodity price the importance it should have in the light of the object in view.

This difficulty is easily met and more representative index numbers will be obtained if we bring out the relative importance of the commodities by applying explicitly appropriate emphasis. The application of appropriate emphasis is known as *weighting*.

The decision about the appropriate weights depends upon answers to the following questions:

(*i*) **By what do we weight**? Is it to be production figures, consumption figures or distribution figures? The choice would depend upon whatever seems appropriate to bring out the economic importance of the commodities involved from the point of our object.

(*ii*) **What type of weight is to be used**? There are two types of weights—(*i*) quantity and (*ii*) value. A quantity weight (represented by symbol q) means the amount of commodity produced, consumed or distributed. A value weight is the product of price with quantity produced, consumed or distributed. Value is symbolized by $p \times q$.

Quantity weights are to be used if we adopt the method of aggregates because the product of price and quantity will always be in rupees. On the other hand, if the method used is of "price relatives' quantity cannot be used as weights for the product of price relatives and quantities expressed in different units would result in different units. Thus kilograms, multiplied by price relatives would give kilograms, tonnes multiplied by price relatives would result in tonnes and so on. This being so the products of price-relatives and quantity of different commodities cannot be added and averaged. To overcome this difficulty the price relatives are weighted by value figures (which are always expressed in rupees) and the product is always in rupees only.

In short it can be said that in the method of aggregates of actual price, weights used are quantity (q); and value weights ($p \times q$) must be used in the method of average of price relatives.

(*iii*) **The time from which weights should be taken**. The quantity weights may be quantities of the base period (q_0); or of the given period (q_1), or it may be the sum or average of these two. Sometimes the quantities of a typical year or the average of the quantities of several years which are thought as typical may also be used as weights.

Value weights are usually the product of the price and quantity of the base year symbolized by p_0q_0. These weights may also be obtained by the product of p in any time period and q in any other time period.

Based on these considerations, a number of index numbers, using different weighting systems have been evolved. All have their own strengths and weaknesses. We discuss below some of the common weighting schemes.

13.4. WEIGHTED AGGREGATES OF ACTUAL PRICES

It has been stated earlier that under this method the price of each commodity included in the group for the purposes of index number construction is weighted by the quantity (q). Within this framework many weighting schemes have been used.

1. When the base year quantities are used as weights, the method is called **Laspeyre's method** and can be defined symbolically as,

$$\text{Index Number of Prices} = \frac{\Sigma p_1 q_0}{\Sigma p_0 q_0} \times 100$$

2. When the quantities used as weights pertain to the given period, which is to be compared with the base period the method is **known** as **Paasche's method** and has the following formula:

$$\text{Index Number of Prices} = \frac{\Sigma p_1 q_1}{\Sigma p_0 q_1} \times 100$$

If the average (or total) quantities of base and given years are used as weights, the method is called **Marshall and Edgeworth**, and the formula takes the following form:

$$\text{Index Number of Prices} = \frac{\Sigma p_1 (q_0 + q_1)}{\Sigma p_0 (q_0 + q_1)} \times 100$$

4. When two index numbers for the same data, each with a different set of weights are prepared and averaged geometrically, the figure so obtained is **Fisher's ideal index** and can be defined symbolically as:

$$\text{Index Number of Prices} = \sqrt{\left(\frac{\Sigma p_1 q_0}{\Sigma p_0 q_0} \times \frac{\Sigma p_1 q_1}{\Sigma p_0 q_1}\right)} \times 100$$

5. When two index numbers for the same data each with a different set of weights are perpared and averaged arithmetically the figure so obtained is called **Bowley and Dorbish index** and can be defined symbolically as:

$$\text{Index Number of Prices} = \frac{\frac{\Sigma p_1q_0}{\Sigma p_0q_0} + \frac{\Sigma p_1q_1}{\Sigma p_0q_1}}{2} \times 100.$$

6. The above formula involves changing weights. In contrast to these we may have fixed weights. When fixed quantities are used as weights, the method is called aggregate index with fixed weights or *Kelly's method*, named after Truman L. Kelly. It can be defined symbolically as;

$$\text{Index Number of Prices} = \frac{\Sigma p_1q_0}{\Sigma p_0q_0} \times 100 \qquad \frac{\Sigma p_1q_0}{\Sigma p_0q_0} \times 100$$

Illustration 13.1. Construct index numbers of prices from the following data using (*a*) Laspeyre's (*b*) Paasche's (*c*) Marshall and Edgeworth, and (*d*) Fisher's methods.

Commodity	*1960*		*1970*	
	Price	*Quantity*	*Price*	*Quantity*
A	4	50	10	40
B	3	10	9	2
C	2	5	4	2

Solution

Commodity	*1960*		*1970*					
	Price (p_0)	*Qty.* (q_0)	*Price* (p_1)	*Qty.* (q_1)	p_0q_0	p_0q_1	p_1q_0	p_1q_1
A	4	50	10	40	200	160	500	400
B	3	10	9	2	30	6	90	18
C	2	5	4	2	10	4	20	8
Total					240	170	610	426

Index Number for 1970, by

(*i*) Laspeyre's Method $= \frac{\Sigma p_1q_0}{\Sigma p_0q_0} \times 100 = \frac{610}{240} \times 100 = 254.2$

(*ii*) Paasche's Method $= \frac{\Sigma p_1q_1}{\Sigma p_0q_1} \times 100 = \frac{426}{170} \times 100 = 250.6$

(*iii*) Marshall and Edgeworth's Method

$$= \frac{\Sigma p_1(q_0+q_1)}{\Sigma p_0(q_0+q_1)} \times 100$$

$$= \frac{\Sigma p_1q_0 + \Sigma p_1q_1}{\Sigma p_0q_0 + \Sigma p_0q_1} \times 100$$

$$= \frac{610+426}{240+170} \times 100 = \frac{1036}{410} \times 100 = 252.7$$

(*iv*) Fisher's Method$= \sqrt{\left(\frac{\Sigma p_1 q_0}{\Sigma p_0 q_0} \times \frac{\Sigma p_1 q_1}{\Sigma p_0 q_1}\right)} \times 100$

$$= \sqrt{\frac{610}{240} \times \frac{426}{170}} \times 100$$

$$= \sqrt{\frac{12993}{2040}} \times 100 = \sqrt{(6.369)} \times 100 = 252.4.$$

Illustration 13.2. Construct index number of prices from the following data using (*a*) Laspeyre's (*b*) Paasche's (*c*) Marshall and Edgeworth and (*d*) Fisher's methods.

Commodity	1960 Price p_0	1960 Quantity q_0	1970 Price p_1	1970 Quantity q_1
A	6	50	10	56
B	2	100	2	120
C	4	60	6	60
D	10	30	12	24
E	8	40	12	36

Solution

Commodity	Base year 1960 p_0	q_0	Current year 1970 p_1	q_1	p_0q_0	p_0q_1	p_1q_0	p_1q_1
A	6	50	10	56	300	336	500	560
B	2	100	2	120	200	240	200	240
C	4	60	6	60	240	240	360	360
D	10	30	12	24	300	240	360	288
E	8	40	12	36	320	288	480	432
					1360	1344	1900	1880

Index Number of Prices by

(*a*) Laspeyre's method$= \frac{\Sigma p_1 q_0}{\Sigma p_0 q_0} \times 100$

$$= \frac{1,900}{1,360} \times 100 = 139.7$$

(*b*) Paasch's method$= \frac{\Sigma p_1 q_1}{\Sigma p_0 q_1} \times 100$

$$= \frac{1880}{1344} \times 100 = 139.7$$

(*c*) Marshall and Edgeworth method

$$= \frac{\Sigma p_1(q_0+q_1)}{\Sigma p_0(q_0+q_1)} \times 100$$

$$=\frac{\Sigma p_1q_0+\Sigma p_1q_1}{\Sigma p_0q_0+\Sigma p_0q_1}\times 100$$

$$=\frac{1,900+1880}{1,360+1344}\times 100=139.8$$

(*d*) Fisher's Formula $=\sqrt{\frac{\Sigma p_1q_0}{\Sigma p_0q_0}\times\frac{\Sigma p_1q_1}{\Sigma p_0q_1}}\times 100$

$$=\sqrt{\frac{1900}{1360}\times\frac{1880}{1344}}\times 100$$

$$=100\times\sqrt{1.9542}$$

$$=100\times 1.398=139.8.$$

Comparison of Different Methods of Weighting

If the quantity of each commodity (included in the group) marketed or consumed changed from year to year in the same proportion the results would be identical irrespective of the period to which the weights referred. But when this is not so and the relative importance of different commodities changes due in part to changes in their relative prices the result would get influenced by the selection of the period from which the weights are drawn.

Under the Laspeyre's method, when base year quantities are used as weights, the index has in a sense an upward bias. This is so because there is a tacit assumption in this method that the same quantities are purchased in the given year as were done in the base year irrespective of the fact that the prices in the given year have changed. Ordinarily there is a likelihood that if the price of a commodity has risen in the given period the quantity purchased would be less and vice versa. Since this aspect is not considered under this method greater weight is assigned to those commodities where prices have gone up and as such the resultant index is likely to have an upward bias.

When given year quantities are used as weights (as under Paasche's method) the resultant index has a downward bias. This is because undue weight is given to the commodities that have declined in price. Besides this, certain new problems arise in using given year weights.

1. Since the weights change with each given year the base year figures have to be recomputed each time. Thus, if we have to compute weighted index for 1970 and 1971 with 1960 as a base, using given year weights, it would be necessary that 1960 prices be multiplied by 1970 quantities as base for 1970 index, and by 1971 quan-

tities as the base for the 1971 index. This means that far more work is required under this method than with base-year quantities as weights.

2. Since the quantities used as weights change with each year, index numbers for different years a e not comparable. Thus an index number for 1970 with 1960 as the base is comparable to 1960 and not to the index number of 1971, or any other year:

Marshall-Edgeworth formula. (Total quantities of base and given years). This is a compromise formula and has no bias in any known direction. But here again because of the shifting weights there shall be no comparability among the index numbers for different years.

4. **Fisher's ideal formula**. Index numbers computed according to this method also suffer from the lack of comparability.

13.5. WEIGHTED AVERAGE OF PRICE RELATIVES

It has been explained earlier in this chapter that an index number derived by the application of 'simple average of price relatives' does not reflect the true change in the prices of the group of commo- es because of the existence of an unscientific system of weighting inherent in this system. To overcome this defect these price relatives of each commodity are to be assigned appropriate weights. Since the price relatives are in percentages, the weights used must be *value weights* as has been explained earlier.

The following procedure should be adopted for the construction of weighted average of price relative index:

(*a*) Determine the base year.

(*b*) Express the price of each commodity for given year (p_1) as a percentage of its price in the base year $\left(\frac{p_1}{p_0} \times 100\right)$.

(*c*) Determine the value weights (p_0q_0) of each commodity in the group by multiplying its, price in base year (p_0) with its, quantity in the base (p_0). If, however, given values are to be used as, weights it would be represented by (p_1q_1).

(*d*) Multiply the price relative of each commodity with its value weight as obtained under (*c*) above.

(*e*) Sum up the products obtained under (*d*) above.

(*f*) Divide the sum under (*e*) by the sum of the value weights.

Symbolically. Index number obtained by the method of weighted average of price relative is

$$\text{Index Number of Prices} = \frac{\Sigma\left(\frac{p_1}{p_0} \times p_0q_0\right) \times 100}{\Sigma p_0q_0}$$

This formula holds good when the arithmetic mean is used as the average.

This formula can be transformed by cancellation

$$\frac{\Sigma\, p_1q_0}{\Sigma\, p_0q_0} \times 100$$

into

It will be seen that this transformed formula is the same as a weighted aggregate of actual prices. But this would be so only if the following two conditions are satisfied.

(*a*) Arithmetic mean is used for averaging the weighted relatives, and

(*b*) Base year values are used as weights.

Illustration 13.3. The price quotations for five different commodities for 1951 and 1965 are given below. Calculate index number for 1965 with 1951 as base by using the weighted average of price relatives.

Commodities	*1951*		*1955*
	Price per kgm	*Quantity kgm*	*Price*
A	6	50	10
B	2	100	2
C	4	60	6
D	10	30	12
E	8	40	12

Solution

Commodities	*1951*			*1965*		
	Price p_0	*Quantity* q_0	p_0q_0	p_1	$\frac{p_1}{p_0} \times 100$	$\frac{p_1}{p_0}(p_0q_0) \times 100$
A	6	50	300	10	166.667	50000
B	2	100	200	2	100	20000
C	4	60	240	6	150	36000
D	10	30	300	12	120	36000
E	8	40	320	12	150	48000
			Σp_0q_0 =1,360			$\Sigma\left(\frac{p_1}{p_0} \times p_0q_0\right)$ =1,90,000

$$\text{Index Number of Prices} = \frac{\sum\left(\frac{p_1}{p_0} \times p_0q_0\right)}{\Sigma p_0q_0} \times 100$$

$$= \frac{1{,}90{,}000}{1{,}360}$$

$$= 139.7$$

Note. The calculation work can be considerably curtailed if we take out from the value weights the 'highest common factor.' Thus if each item in column (p_0q_0) is divided by 20 the value weights for different commodities would be as under, and the computation would take the following form:

	p_0q_0	$\frac{p_1}{p_0} \times 100$	$\frac{p_1}{p_0} \times p_0q_0 \times 100$
A	15	166.67	2500
B	10	100	1000
C	12	150	1800
D	15	120	1800
E	16	150	2400
	68		9500

$$\text{Index Number} = \frac{9500}{68} = 139.7 \text{ approximately.}$$

Weighted geometric mean of relatives. The same system of value weights may be used in computing weighted geometric mean of relatives. In this case the logarithms of the relatives are multiplied by the weights. Then the products of the value weights and logarithms of relatives are added. Next the total for each year is divided by the total of the value weights. The quotients will give the logarithms of the index. If we find out the antilogarithms, we will get the indices of price based on weighted geometric mean of relatives.

Bias in weighted index numbers. We have seen earlier that in the so-called 'unweighted' index numbers the technique of construction will introduce a bais in the results. When weighting system is employed in the computation of index, another kind of bias, called 'weight bias', appears. In order to distinguish the two types of bias, the one present in the apparently 'unweighted' index is known as 'type bias.' Thus, a type bias is the outcome of implicit weighting, whereas the 'weight bias' is the result of explicit weighting.

13.6. QUANTITY INDEX NUMBERS

Just a price index numbers enable us to compare prices of certain goods, quantity index numbers permit comparison of the physical quantity of goods produced, consumed or distributed. The most common type of the quantity index is that of quantity produced.

The construction of a quantity index involves the same kinds of problems as are involved in price indexes and our approach in this respect is to be almost identical to the one already explained.

The essential point of difference, however, is that here we measure changes in quantity. Since quantities of different products are expressed in different units—tonnes, litres, metres; etc. they are not addable and as such the method of simple 'aggregates' cannot be used.

But if the given year quantities are expressed as relatives of the base year quantities this difficulty is overcome. The average of such relatives would give us the quantity index

$$\text{Symbolically } Q \text{ index} = \frac{\sum\left(\frac{q_1}{q_0}\right)}{N} \times 100$$

To overcome the defects of unscientific weighting, appropriate weights should be used in the construction of quantity Index Numbers. Weights may be **'prices'** or **'values.'**

In the weighted aggregate of quantities method 'prices' are to be used as the weights. If it is the base year price, the formula would be

$$\text{Quantity Index} = \frac{\Sigma q_1 p_0}{\Sigma q_0 p_0} \times 100$$

In the weighted mean of quantity relatives when base year values are used as the weights the formula would be

$$\text{Quantity index} = \frac{\sum\left(\frac{q_1}{q_0} p_0 q_0\right)}{\Sigma p_0 q_0} \times 100$$

Illustration 13.4. Compute a quantity Index for the following data, using weighted aggregate of quantities method.

Commodity	1960 Price	1960 Quantity	1970 Quantity
A	40	20	25
B	30	15	20
C	10	10	10
D	20	5	10

Solution.

Commodity	1960 Price p_0	1960 Quantity q_0	p_0q_0	1970 Quantity q_1	q_1p_0
A	40	20	800	25	1,000
B	30	15	450	20	600
C	10	10	100	10	100
D	20	5	100	10	200
			1,450 Σq_0p_0		1,900 Σq_1p_0

$$\text{Quantity index}=\frac{\Sigma q_1p_0}{\Sigma q_0p_0}\times 100$$

$$=\frac{1,900}{1,450}\times 100=131 \text{ approx.}$$

USING WEIGHTED MEAN OF QUANTITY RELATIVES

Commodity	1960 Price p_0	1960 q_0 quantity	p_0q_0	1970 quantity q_1	$\frac{q_1}{q_0}\times 100$	$\frac{q_1}{q_0}\times p_0q_0$
A	40	20	800	25	125	1,00,000
b	30	15	450	20	133.33	59,998
C	10	10	100	10	100	10,000
D	20	5	100	10	200	20,000
			1.450			1,89,998

$$\text{Quantity Index}=\frac{\sum\left(\frac{q_1}{q_0}\times q_0p_0\right)}{\Sigma q_0p_0}\times 100$$

$$=\frac{1,89,998}{1450}$$

$$=131.$$

13.7. MATHEMATICAL TESTS OF CONSISTENCY

Some people believe that if an index number formula meets certain mathematical tests it may be considered as an 'ideal' method for the construction of index numbers. Even though there is no finality about the validity of this statement it is necessary to understand the following three tests, which are commonly included in this category: (*i*) Time reversal test; (*ii*) Factor reversal test; and (*iii*) Circular test.

1. Time reversal test. This test is a device to determine if a method will work both ways in time 'backward and forward.' This means that the index number for (say) 1965 with (say) 1961 as the base should be the reciprocal of the index number for 1961 with 1965 as the base, *i.e.*, their product should be unity. Thus, if the price of a commodity has increased to two rupees per kg in 1964, as compared to one rupee per kgm in 1963, we would say that the 1964 price is 200 per cent of the 1963 price and the 1963 price is 50 per cent of the 1964 price. Now, these two figures are reciprocals of one another and their product (2.00×.50) is equal to unity. If the method does not work both ways, *i.e.*, if the index numbers for two years secured by the same method but with basis reversed are not reciprocals of each other there is an inherent bias in the method.

Algebraically, the test may be expressed as

$$p_{0\cdot1} \times p_{1\cdot0} = 1$$

where $p_{0\cdot1}$ stands for index for the current year on the base year omitting the factor 100, (*i.e.*, for price change in current year as compared with base year) and $p_{1\cdot0}$ stands for index for the base year on the current year without the factor 100 (*i.e.* for price changes in base year compared with current year).

According to Fisher, "The test is that the formula for calculating an index number should be such that it will give the same ratio between one point of comparison and the other, no matter which of the two is taken as base," or putting it another way, "the index number reckoned forward should be the reciprocal of that reckoned backward."

That Fisher's ideal index satisfies the 'Time Reversal Test' can also be seen from the following illustration:

Price index for the current year is

$$p_{0\cdot1} = \sqrt{\frac{\Sigma p_1 q_0}{\Sigma p_0 q_0} \times \frac{\Sigma p_1 q_1}{\Sigma p_0 q_1}}$$

Changing current year to base

$$p_{1\cdot0} = \sqrt{\frac{\Sigma p_0 q_0}{\Sigma p_1 q_0} \times \frac{\Sigma p_0 q_1}{\Sigma p_1 q_1}}$$

Time Reversal Test is: $p_{0\cdot1} \times p_{1\cdot0} = 1$.

$$\text{Now } p_{0\cdot1} \times p_{1\cdot0} = \sqrt{\frac{\Sigma p_1 q_0}{\Sigma p_0 q_0} \times \frac{\Sigma p_1 q_1}{\Sigma p_0 q_1} \times \frac{\Sigma p_0 q_0}{\Sigma p_1 q_0} \times \frac{\Sigma p_0 q_1}{\Sigma p_1 q_1}}$$

$$= 1.$$

Thus, we see that the indices prepared according to Fisher's ideal formula satisfy the Time Reversal Test.

2. Factor reversal test. We have already discussed one test, namelythe 'time reversal test,' in which 'the index number reckoned forward should be the reciprocal of that reckoned backward.' This test is not met by any of the index discussed above except the Fisher's ideal index. Irving Fisher has, however, suggested one more test, *viz.*, the 'factor reversal test' to be applied to weighted index numbers. Concerning this test he wrote:

"Just as our formula should permit the interchange of the two times without giving inconsistent results so it ought to permit interchanging the *prices and quantities* without giving inconsistent result, *i.e.*, the two results multiplied together should give the true ratio."

In simple words the test is satisfied if the product of the price index and the quantity index is equal to the ratio of the aggregate value (quantity × price) in the current year to the aggregate value in the base year.

Algebraically: $p_{0\cdot1} \times q_{0\cdot1} = \dfrac{\Sigma p_1 q_1}{\Sigma p_0 q_0}$

where $p_{0\cdot1}$ stands for the price change for the current year over the base year, $q_{0\cdot1}$ stands for the quantity change for the current year over the base year, $\Sigma p_1 q_1$ stands for the total value in the current year, and $\Sigma p_0 q_0$ stands for the total value in the base year.

That Fisher's ideal index satisfies this test can be seen from the following illustration:

TABLE 13.4

Commodity	*1961 Base year*		*1965 Current year*					
	p_0	q_0	p_1	q_1	p_0q_0	p_1q_0	p_0q_1	p_1q_1
A	10	3	11	3	30	33	30	33
B	4	15	4	12	60	60	48	48
C	6	3	6	4	18	18	24	24
D	11	8	12	7	88	96	77	84
E	6	17	12	12	102	204	72	144
					298	411	251	333

Factor Reversal Test is satisfied, if

$$p_{0\cdot1} \times q_{0\cdot1} = \frac{\Sigma p_1 q_1}{\Sigma p_0 q_0}$$

where $p_{0\cdot1}$ stands for the price change for the current year over the base year, and $q_{1\cdot 0}$ stands for the change for the current year over the base year.

Now according to Fisher's ideal index number formula:

$$p_{0\cdot 1}=\sqrt{\frac{\Sigma p_1q_0}{\Sigma p_0q_0}\times\frac{\Sigma p_1q_1}{\Sigma p_0q_1}}$$

and $$q_{0\cdot 1}=\sqrt{\frac{\Sigma p_0q_1}{\Sigma p_0q_0}\times\frac{\Sigma p_1q_1}{\Sigma p_1q_0}}$$

Hence,

$$p_{0\cdot 1}\times q_{0\cdot 1}=\sqrt{\frac{\Sigma p_1q_0}{\Sigma p_0q_0}\times\frac{\Sigma p_1q_1}{\Sigma p_0q_1}\times\frac{\Sigma p_0q_1}{\Sigma p_0q_0}\times\frac{\Sigma p_1q_1}{p_1q_0}}$$

Substituting the values from Table 14.4, we get

$$p_{0\cdot 1}\times q_{0\cdot 1}=\sqrt{\frac{411}{298}\times\frac{333}{251}\times\frac{251}{298}\times\frac{333}{411}}$$

$$=\sqrt{\frac{333\times 333}{298\times 298}}=\frac{333}{298}$$

Now $\frac{\Sigma p_1q_1}{\Sigma p_0q_0}$ is also equal to $\frac{333}{298}$.

Thus, it is proved that Fisher's ideal formula for index number satisfies the 'Factor Reversal Test.'

3. Circular test. In case of three given years a method is said to satisfy the circular test if

$$p_{12}\times p_{23}\times p_{31}=1$$

where p_{12} is the index number of 2nd year with 1st year as base,
p_{23} is the index number of 3rd year with 2nd year as base, and
p_{31} is the index number of first year with 3rd year as base.

It can be shown that Fisher's formula does not satisfy this test:

Commodity	*1961* Price	*Qty.*	*1962* Price	*Qty.*	*1963* Price	*Qty.*
A	2	5	4	1	5	2
B	3	4	5	2	6	1
C	4	3	6	5	8	4

Fisher's index No. ideal formula $=\sqrt{\frac{\Sigma p_1q_0}{\Sigma p_0q_0}\times\frac{\Sigma p_1q_1}{\Sigma p_0q_1}}$.

CALCULATION OF P_{12} *i.e.* I. NO. OF 1962 WITH 1961 BASE

Commodity	*1961* p_0	q_0	*1962* p_1	q_1	p_0q_0	p_0q_1	p_1q_0	p_1q_1
A	2	5	4	1	10	2	20	4
B	3	4	5	2	12	6	20	10
C	4	3	6	5	12	20	18	30
Total					34	28	58	44

$$\therefore \quad p_{12}=\sqrt{\frac{58}{34}\times\frac{44}{28}}=\sqrt{\frac{319}{119}}=\sqrt{2.681}=1.637.$$

CALCULATION OF p_{23} i.e., I. NO. OF 1963 WITH 1962 AS BASE

Commodity	1962 p_0	1962 q_0	1963 p_1	1963 q_1	p_0q_0	p_0q_1	p_1q_0	p_1q_1
A	4	1	5	2	4	8	5	10
B	5	2	6	1	10	5	12	6
C	6	5	8	4	30	24	40	32
Total					44	37	57	48

$$\therefore \quad P_{23}=\sqrt{\frac{57}{44}\times\frac{48}{37}}=\sqrt{\frac{684}{407}}=\sqrt{1.681}=1.296.$$

CALCULATION OF p_{31} i.e., I. NO. OF 1961 WITH 1963 AS BASE

Commodity	1963 p_0	1963 q_0	1961 p_1	1961 q_1	p_0q_0	p_0q_1	p_1q_0	p_1q_1
A	5	2	2	5	10	25	4	10
B	6	1	3	4	6	24	3	12
C	8	4	4	3	32	24	16	12
Total					48	73	23	34

$$\therefore \quad p_{31}=\sqrt{\frac{23}{48}\times\frac{34}{73}}=\sqrt{\frac{391}{1752}}=\sqrt{.2232}=0.472$$

$$\therefore \quad p_{12}\times p_{23}\times p_{31}=1{\cdot}637\times 1.296\times 0.472$$

or $\quad \log(p_{12}\times p_{23}\times p_{31})=\log 1.637+\log 1{-}296+\log 0.472$

$$=0.2141+.1126+\bar{1}.6739$$

or $\quad p_{12}\times p_{23}\times p_{31}=\text{antilog } 0.0006=1.001.$

Hence Fisher's method does not satisfy circular test.

13.8. FIXED AND CHAIN BASE INDICES

The base year used in the calculation of index number may be fixed or changing. It is said to be fixed when the indices for different periods are computed on the basis of the prices of a common base year. Thus, if the indices for 1962, 1963, 1964, 1965 are calculated with 1961 as the base year such indices will be called fixed-base indices. In the illustration given on preceding pages indices have been calculated on the fixed-base basis.

If, however, the whole series of index numbers is not related to any one base period, but the indices for different years are derived by relating each year's value to that of the immediately preceding year the indices so obtained are called link relative index numbers. Frequently, these link relatives are chained together to a common base. Such indices are known as chain indices. By applying this method to the series of quotation given in Table 13.2 the index numbers will be as given in Table 13.5.

The method of computation. According to the chain base method the price relatives for any year are computed on the basis of the prices for the year just preceding it. Thus, the relatives for 1962 are computed with 1961 prices as the base, relatives for 1963 with 1962 prices as the base, and so on. When the price relatives are computed according to this method they are called 'link relativas.' These link relatives are given from line 1 to 5 in Table 13.5. In line 6 are given their totals and in line 7 are given the averages of these link relatives for each year which are obtained by dividing the otal given in line 6 by 5 (the number of commodities). If now it is desired to relate them all to a common base (say, 1961 in our illustration) these averages may be placed in a chain. The chain relatives, so obtained, will be the indices on chain base method in respect to the year 1961. The method of chaining together the link relatives is as follows:

The average link relative for 1962 with 1961 as base is 120. This figure will remain the same for the simple reason that this is already related to 1961.

The average link relative for 1963 with 1962 as base is 90. This means that if the 1962 prices are represented by 100 the 1963 prices are represented by 90. If the 1962 prices are represented by 120 the figure for 1963 prices will be:

$$\frac{120}{100} \times 90 = 108$$

Similarily the chain relative for $1964 = \frac{108}{100} \times 125 = 135$, and

the chain relative for $1965 = \frac{135}{100} \times 102 = 137.7$.

Merits and demerits of the chain base method. The main advantages of the chain base method are two:

(1) Under this method the index for the current year is related to the year immediately preceding it. This enables us to know the extent of the change that has come in the current year as compared to the previous year. This is certainly more useful to business than a fixed index which is related to a year of the distant past.

(2) Under this method it is possible to introduce new items or

Table 13.5. Index Number on Chain Base Principle of Five Commodities for 1961-65

Commodities	Relatives Based on Preceding Years				
	1961	1962	1963	1964	1965
A	100	$\frac{9}{10}\times 100=90$	$\frac{8}{9}\times 100=89$	$\frac{9}{8}\times 100=112.5$	$\frac{11}{9}\times 100=122$
B	100	$\frac{5}{4}\times 100=125$	$\frac{3}{5}\times 100=60$	$\frac{6}{3}\times 100=200$	$\frac{4}{6}$ [illegible] $=67$
C	100	$\frac{8}{6}\times 100=133$	$\frac{5}{8}\times 100=62.5$	$\frac{4}{5}\times 100=80$	$\frac{6}{4}\times 100=150$
D	100	$\frac{13}{11}\times 100=118$	$\frac{15}{13}\times 100=115$	$\frac{14}{15}\times 100=93$	$\frac{12}{14}\times 100=86$
E	100	$\frac{8}{6}\times 100=133$	$\frac{10}{8}\times 100=125$	$\frac{14}{10}\times 100=140$	$\frac{12}{14}\times 100=86$
Total of Link Relative	500	599	451.5	625.5	511
Mean of link Relative	100	120	90	125	102
Chain Reiatives	100	120	108	135	137.7

drop out old ones without having to re-calculate the whole series. This is because of the fact that the index of any one year is related only to the year just preceding it and the changes occurring in neighbouring periods are never so great as to impair comparability. Thus, if the list of commodities needs frequent change the chain base method is preferable to the method of fixed base.

This method, however, involves lengthy calculations and if any error is committed it tends to be perpetuated in chaining process.

13.9. BASE SHIFTING, SPLICING AND DEFLATING

Base shifting. Many times it becomes necessary to shift the base of a series of index numbers from one period to another. For instance, let a series of indices, say, of cost of living, have 1949 as its base and its value in 1952 and 1960 be 150 and 300 respectively. Let another series of indices, say, of production, have a base 1952 and its value in 1960 be 200. From these figures one may conclude that as the change in production from 1952 to 1960 is of 100 points (200—100) and the change in cost of living is of 150 points (300—150), the change in the latter series is greater. But this conclusion is not correct as the two series have different base periods. To have valid comparisons it will be necessary to correct the cost of living series into a new series with 1952 as the base year, *i.e.*, the base of this series should be shifted to 1952.

The best method of base shifting which will give correct results is to reconstruct the series with the new base. This means that for each year relatives corresponding to each commodity included in that index number are recomputed on the new base and then averaged out. This new average will give the appropriate index number. But this process is very lengthy and may not be possible to apply in all cases. Another method may be followed, which gives nearly the correct results when arithmetic mean is used for averaging and gives exactly the same results as the first method when geometric mean is used for averaging. The method is as under:

Divide each index number of the series by the index number of the time period selected as new base and multiply the result so obtained by 100. The figure thus obtained will give the required series with the new base.

Let us explain it further with the help of an example. Let the index numbers for various years with 1939 as base for a certain commodity be as follows:

Year	1939	1940	1945	1950	1955	1960
Index No.	100	111	126	150	162	180

It is desired to shift the base to the year 1950.

Let us make the calculations for 1955. If 1950 is to be the new base, its index number must be 100. But in the old series it is 150 and index number for 1955 is 162. The problem stated in simple terms is to determine the index number for 1955 onwards if index number for 1950 is changed from 150 to 100, *i.e.*,

If the figure for 1950 is 150, the figure 1955=162.

If the figure for 1950 is 100, the figure for $1955=\frac{162\times 100}{150}=108.$

Therefore, 108 is the index number for 1955 with 1950 as base.

Similarly, the index number for 1960 with 1950 as base will be 180×100/150, *i.e.*, 120; index number for 1945 will 126×100/150, *i.e.*, 84 and so on. The new series with its base shifted to 1950 is thus:

Year	1939	1940	1945	1950	1955	1961
Index No.	67.7	74.0	84.0	100.0	108 0	120.0

Splicing two index number series. It is usually found that in course of time some articles included in an index number series may go out of the market. New ones may come in. Their relative imoortance may also change. When these changes become sufficiently important their inclusion in the index number becomes necessary. As a consequence the old series of index number is discontinued and a new series is construqted with the year of discontinuation of the first as base. This means that we now have two series of index numbers for the same phenomenon—one of them coming up to the year from which the other begins. Thus, the index numbers contained in two series are not directly comparable for the simple reason that they are prepared on different bases. In order to facilitate the comparison these two series are put together in one continuous series *i.e.*, the two series are *spliced* together. The method for doing this is:

Multiply the various indices of the new series by the index number of the last year in the old series and divide the result so obtained by 100.

Let us explain it with the help of an example

Year	1939	1940	1945	1950	1955
Series A	100	120	150	—	—
Series B	—	—	100	112	136

Here series *A* was discontinued in 1945 and in that year a new series was started. It is desired to splice the two series.

Let us make calculations for 1950:

When I.N. for 1945 is 100, I.N. for 1950=112 (given by series *B*)

$\because$ When I.N. for 1945 is 150, I.N. for 1950= $\frac{112\times150}{100}$ =168.

$\therefore$ 168 becomes the index number for 1950 in the spliced series. The two series spliced in this way give the result as follow:

Year	1939	1940	1945	1950	1955
Spliced Series	100.0	120.0	150.0	168.0	204.0

More frequently instead of carrying series *A* forward, series *B* may be brought backwards. In this case every figure of series *A* is divided by the index number of the year in which change takes place and the result so obtained is multiplied by 100. In the present example the two series spliced in this way give the result as follows:

Year	1939	1940	1945	1950	1960
Spliced Series	66.7	80.0	100.0	112.0	136.0

Deflating. Deflating means making allowance for the effect of changing price levels. Over a period of time wages may be rising. But side by side the cost of living may also be increasing. The real wages in this case would be less than the money wages. To get the real wage figure one may reduce the money wage figure to the extent the prices have risen. The rise in price in this case may be best represent by cost-of-living index number. If the cost-of-living index number in a certain year is double the base year figure, then real wages for that year (woges in terms of the price level as in base year) would be half the money wages. The process of decreasing a figure with the help of index numbers as to allow for change in the price level is called *deflating*. In deflating only that index number should be used which is appropriate to the given case. In the above example, if one decrease the actual wages in the same proportion as the rise in gold prices, the correct real wages will not be obtained.

The method for deflating a series of figures to the base year level of a suitable index number series is to divide the figures corresponding to various time periods of the given series by the corresponding figure of the index number series and multiply the result to be obtained by 100. The example given below will illustrate it further:

Year	1949	1950	1955	1957	1961
Wages per month (Rs)	120	125	150	178	215
Cost of Living Index No.	100	105	130	142	208

It is desired to deflate the monthly wages by cost of living index number.

Let us calculate the figure for 1955. In this year the wages are Rs 150 p.m. and ccst of living index number is 130. To get the deflated income one has to proceed as below:

When index of cost of living is 130, wages=Rs 150.

If index of cost of living was 100, wages= $\frac{150}{130} \times 100$

=Rs 115.4 approx.

Similarly, the deflated income for 1961= $\frac{215 \times 100}{208}$

=Rs 103.4.

Thus, deflated Incomes for various years are

Year	1949	1950	1955	1957	1961
Deflaied income (Rs)	120	119.0	115.4	125.3	103.4

13.10. CONSUMER PRICE INDEX NUMBERS*

Concept and scope. Consumer price index numbers are designed to measure by means of appropriate weighting the average change over time in the prices paid by the ultimate consumers for a specified quantity of goods and services. It should, however, be clearly understood that the customer price indices measure changes in the cost of living of workers due to changes in the retail prices only. The measurement of changes in the cost of living due to change in the living standards is not included in the usual concept of the consumer price index.

In defining the scope of a consumer price index it is necessary to specify:

(1) The population groups covered, *e.g.*, working class, middle class, etc., and

(2) The geographical areas covered *e.g.*, urban areas, rular areas, a city, town etc.

Functions of consumer price indices. The main function of a consumer price index is to serve as a measure of change in retail prices of a specified quantity of goods and services. But such indices

*For detailed study consult 'Consumer Price Index Numbers' monograph issued by Labour Bureau, Ministry of Labour, Government of India (1959).

Consumer Price Index was formerly called 'Cost of Living Index'. The change in the name was made in accordance with internal recommendations and the growing parctices in other countries.

are useful in many other ways. They help in wage negotiations and dearness allowance adjustments, etc. Governments can make use of such indices in framing wage policy, price policy, rent control, taxation and general economic policies. Changes in the purchasing power of money and real income can be measured, and markets for particular kinds of goods and services can be analysed with the help of these indices.

Precautions in the use of a consumer price index. Before making use of a consumer price index it is necessary to inquire into the following:

(1) **Scope of the index.** The class of the people and the area to which the consumer price index is related must be carefully determined.

(2) **Reliability of the index number should be carefully ascertained.** The reliability of an index number depends mainly upon the reliability of the price data used and sampling technique adopted. The sample of the households covered in the course of a family budget inquiry should be representative of the population group. Similarly the items selected for pricing should be representative of all the items in the average budget. The localities for which price data are collected should be representative of all localities from which the population group makes its purchases, and the retail outlet from which prices are collected should be representative of all retail outlets used by the population group.

Problem in the construction of consumer price index number. The main problems in the construction of Consumer Price Index consist of: (*i*) The Determination of Weights and (*ii*) The Collection of retail prices.

(i) **Determination of weights.** In general, weights are determined on the basis of the consumption pattern of the class of population to which the index relates. This means that the weights that are assigned to different commodities are related to the actual consumption expenditure upon them. Statistical data relating to consumption or expenditure is derived from family budget inquiries. It is, therefore, necessary that such inquiries be properly planned. Since a complete count of all the families included in the area is not parcticable, it is of vital importance that a sound sampling method be adopted. On the basis of the results of the family budget inquiries, an average budget of the expenditure on different items consumed by families of different size and composition included in the study and quantities of the different items consumed by them is derived. This average budget is representative of the population group to which the consumer price index is to finally relate. It covers all groups of consumption expenditure—food, housing, clothing and miscelleneous. It has been recommended by the seventh International Conference of Labour Statisticians that for the purpose of international comparison the classification of consumption expenditure should be made in such a way as to make possible a grouping or regrouping of items in the following groups and subgroups:

(*a*) a group of food, including, as separate items, food consumed away from home and alcoholic beverages;

(*b*) a group of housing, including, as separate items, rent, fuel and light, and household furnishings and appliances;

(*c*) a group of clothing; and

(*d*) a group of miscellaneous, including, as separate itemes, the following ten sub-groups—medical care, personal care, insurance and other contributions, education and reading, postage, recreation, tobacco, gifts and charities.

The conference also recommended that items of non-consumption outgo (income and similar taxes and interest on personal debts) should be separated from the items of consumption expenditure: and, thus items like taxes, interest on debts, purchase of savings certificates, insurance premiums etc., should be excluded from the items of consumption expenditure.

The various items that are included in each consumption group of the average budget are then assigned weights in proportion to their importance within that group, on the basis of the figures of either expenditure or quantity of consumption in the average budget.

Though all important items of expenditure are included in the index, the sample of items selected for pricing has to be limited, because the larger the list of items, the greater the time and labour involved in the collection of prices and the computation of the index. Moreover, no loss in accuracy will result if out of a few items having similar price trends only one is selected for pricing, and the weights of all such items is assigned to the priced item. Thus, if from a study of price behaviour and other factors, it is established that a particular unpriced item has a price trend similar to that of a priced item, the weight of the unpriced item is added to that of the priced item.

On this principle, the weights of individual items and consumption groups, based on expenditure data, can be derived as follows:

To the expenditure on each priced item is added the expenditure on unpriced items known to have similar price movements. The resultant expenditure on each item is expressed as a percentage of the total expenditure accounted for by all the items included in a group to yield the weight of the item within the group. The weight of a consumption group is obtained by expressing the total expenditure on the group as a percentage of the total expenditure on all groups as recorded in the average budget.

(ii) Collection of retail prices. The second main problem in the construction of the cost of living index number is the collection of retail prices of the items in the index. The prices are to be collected both for the base period and the current period.

Some of the principles that should be observed in the collection of retail prices are mentioned below.

1. The work on the collection of retail prices should commence simultaneously with the conduct of family, budget inquiry, because the retail prices are required both for the base period and the current period, and the base period has to synchronise more or less with the period of the family budget inquiry. Since the selection of the items to be priced can be finished only after the completion of the average budget, retail prices, to begin with, are to be collected for a list of items on the basis of general knowledge of the consumption habits of the population group covered. This initial list should be sufficiently large. Unimportant item can be dropped at a later stage.

2. When the average budget is ready it would be possible to fix a list of items that are to be priced. For each item that is priced, a standard of quality should be fixed by means of suitable specifications.

It may be pointed out here that since in practice neither market conditions nor consumers' preferences remain unchanged over a period of time, it becomes difficult to conform to a fixed list and constant qualities and quantities of goods and services. Now the quantities consumed (which means the same thing as weights assigned to different commodities) cannot be changed without conducting fresh family budget inquiries. The Sixth International Conference of Labour Statisticians recommended that the pattern of consumption should be examined and the weights adjusted, if necessary, at intervals of not more than ten years to correspond with the changes in the consumption pattern. The Conference also recommened the use of small sample studies of consumer purchases in the intervals between the more comprehensive surveys for discovering significant changes in consumption pattern to indicate the need for revision in the weights.

Changes in the quality of priced goods and services are more frequent and when a marked change in the quality of an item occurs, an appropriate adjustment has to be made to ensure that index takes into account only real changes in prices. Such adjustment can be made in the following ways:

If prices are not available for old qualities over a period, the method of linking may be adopted *i.e.*, the prices of the old quality may be estimated on the basis of the trends of the prices of the new quality. Quality differences can be evaluated in terms of prices in consultation with the traders and only that part of the difference between the quotations for old and new qualities which represents a real price movement may be taken in account.

It is, however recognised that the detection of certain changes in in the quality which are not sudden but take place gradually is difficult. No allowance can therefore, be made for such quality in computing consumer price indices.

3. Another importont requisite in the collection of retail prices is that they should be those actually charged to consumers for cash sales. Account should be taken of discounts (if any) given automatically to all customers, and sales tax, etc., payable by them. It is necessary to see that the retail outlets chosen for the collection of retail prices should be such as can yield an average price, representative of the price that is being paid by the population group to which the index relates.

4. If during a period of rationing or price control exorbitant prices are charged openly to the groups to which the index applies, such prices should be taken into consideration along with the controlled prices.

5. Attention should also be paid to the methods of price collection and the price collection personnel. Prices are collected usually by special agents or mailed questionnaires. Where special agents are employed, it essential to give them intensive training. They may be supplied with a manual of instructions and a manual of specifications of items to be priced. The collected price should be checked making actual purchase of the goods priced.

Method of Compilation of Consumer Price Index

After determining the weights and collecting the prices of the selected commodities and services, the Consumer Price Index Number is compiled with the help of Laspeyre's formula which is a weighted average of price relatives. In this formula the weights are based on the values of expenditure during the base year.

The formula is: $I_n = \frac{\Sigma p_n q_0}{\Sigma p_0 q_0} \times 100$

were p_n's are the prices of the current period, and p_0's the prices of the base period. The formula can be re-written as:

$$I_n = \frac{\sum \left(\frac{p_n}{p_0} \times p_0 q_0 \right)}{\Sigma p_0 q_0} \times 100.$$

In practice, average price of each item for the base period is calculated, known as the 'base price' of a particular item. Again, for the current period an average price in the form of a simple average of weekly quotations is taken. Taking the base price as 100, the ratio of price change for each item is expressed as a percentage, and that is called the 'price relative' of a particular item. Where there are different varieties of a particular commodity price relative is calculated for each variety, and then a simple average of such price relatives is taken. The price relative of each item, thus arrived at, is multiplied by its corresponding weight and the sum of these products for all item is divided by the sum of the weights of the items, thus, giving us the group index number.

Each one of the group index number is then multiplied by its corresponding group weights, and sum of the products of different groups is divided by the sum of the group weights to give us the consumer price index number for that period.

13.11. PROBLEMS IN THE CONSTRUCTOIN OF INDEX NUMBERS

The construction of index numbers involves the consideration of the following important problems: (*i*) Definition of the purpose for which index nuber is being constructed, (*ii*) Selection of commodities for inclusion in the group, (*iii*) Selection of the sources of data, (*iv*) Method of collecting the data, (*v*) Selection of base year, (*vi*) Method of combining the data, and (*vii*) System of weighing.

1. Definition of the purpose. As in all statistical enquiries a clear understanding of the purpose for which index numbers are to be computed is very necessary. Thus, if it is desired to construct an index of consumer's prices we must know the class of consumers whose cost of living we intend to measure—whether it is the cost of living of the middle class people, agriculturists, artisans or industrial workers. Such definiteness is necessary for the importance of various items consumed by the different categories of people may be very different. The price of luxury articles shall have no relevance in measuring the cost of living of the poor people, but will certainly have an effect on the cost for the richer section of the community. A definite awareness of the purpose for which index numbers are to be used will have a determining influence on the selection of commodities, selection of sources of data, selection of the base year, and the system of weighting.

2. Selection of commodities for inclusion in the group. In the matter of selection of commodities attention should be directed to the following points:

(*a*) Selected Commodities should be fairly representative of the phenomenon under investigation. Thus, if it is desired to construct an index for the measurement of changes in the purchasing power of money, the commodities selected should be such as represent a large majority of the transactions that take place for money in the given country. The number of commodities should be enough to permit the inertia of large numbers. (The larger the size of the sample the greater is the possibility of its being representative of the whole population). But the number included must not be so large as to make the work of computation uneconomical and even difficult.

(*b*) The commodities should be such as remain uniform in quality from year to year. The inclusion of a commodity whose quality is likely to charge rapidly would take away from the index number its essential quality as a tool of comparison. It is, therefore, essential that the commodities included are such as remain uniform in quality for reasonably long periods.

(*c*) Problem of varieties. There may be several varieties of a selected commodity. It is the most popular one which should be included in the group. If, however, different varieties are more or less equally important, all such varieties may be included without bringing any 'bias' in the result.

3. Selection of sources of data. The problem of collecting suitable price quotations for the commodities selected is somewhat more difficult. Since it is neither possible nor necessary to collect the price of commodity from all the markets in the country where it is bought and sold we should take a sample of the markets. In selecting a sample care should be taken to see that the markets included are such as are well-known for trading in that particular commodity.

4. Method of collecting the data. Once we have decided about the sources from where the data (quotations) are to be obtained, the next thing to decide is about the method of collection. It must be remembered that the labour involved in the collection of this kind of data is of quite large. Moreover, collection is not a one-time task, since index numbers of prices are ordinarily computed monthly, or weekly and even daily.

The usual practice for obtaining this data is to select suitable price reproducing agency. As there may be a number of agencies that may be reproducing price quotations *viz.* chambers of commerce, news correspondents, etc. our endeavour should be to select an agency which may be most reliable. To check the accuarcy of price quotations supplied by the agency it is advisable to obtain such quotations from more than one reproducing agency. It must be carefully seen that the price quotations are always of the same quality of the commodity, for a charge in quality may mean considerable difference in price.

In order to facilitate the construction of index number, prices should always be quoted as so much money per unit of commodity, *e.g.* Rs 20 per quintal, or 50 paise per kilogram, and not as so many units of a commodity per unit of money (*e.g.*, 2 kg per rupee). Another important point to be decided is whether the prices to be used in the construction of index numbers should be wholesale or retail. Wholesale prices should be preferred to the retail ones unless their is a special reasons to suggest to contrary because they fluctuate less and are more sensitive to conditions of demand and supply as compared to retail prices.

5. **The selection of base**. Since the base period serves as a reference period and the prices for a given year are expressed as percentages of those for the base year, it is necessary that (*i*) the base period should be normal and (*ii*) it should not be too far in the past.

The base year should be normal because if it is not so (*i.e.* if it is influenced by some unusual factors), all the other indices that are related to this year will be distorted as a result of the abnormal condition then prevailing. It is not easy to select a year which may strictly be called as normal. If a year is normal in one respect it may very possibly be abnormal in some other respect. In order to overcome the difficulty of this type an average of a number of years, is generally taken as the base. This average is more representative and is less affected by chance variations.

It must, however, be remembered that a particular base year may be satisfactory for a number of years, but it becomes less meaningful with the lapse of time. This is because of many reasons such as (*i*) the quality of many commodities changes with the lapse of time, (*ii*) there may be a change in the consumption pattern to such an extent that the aggregates of commodities in the 'given' period may not be comparable to the aggregate of the base year, (*iii*) the dispersion of prices may become so great that no average is reliable, or (*iv*) due to certain reasons such as growth of population, currency depreciation or technological improvements, new levels may have been attained by production, income, price and consumption.

6. **Method of combining the data**. The prices of the various commodities that constitute our raw material for the construction of index numbers may be combined either by (*i*) totalling, or by (*ii*) averaging. These two methods of combining the data lead respectively to different methods of constructing index numbers, *viz.*,

(*a*) Simple aggregate of actual prices.

(*b*) Simple average of price relatives.

7. **System of weighting**. It has been discussed in the preceding pages that the two index numbers, 'the simple aggregative' and 'the simple mean of relatives', which apparently seem to be unweighted are not really so. The simple aggregative method is weighted by the magnitude of the prices (the higher price commodity has a greater influence on the result than a lower priced commodity). It is also influenced by units in which prices are quoted. Commodities whose prices are quoted in quintals exert greater influence than those whose prices are quoted in smaller units. Simple mean of relatives method gives equal weights to all commodities. Thus a less important commodity has as much influence on the result as a very important commodity has. In order to eliminate this unscientific weighting it is necessary that scientific system of weighting be introduced which will accord to each commodity price the importance it should have in the light of the object in view.

That a system of weighting is necessary to appreciate fully the impact of the change can be illustrated by the behaviour of prices of two commodities *X* and *Y*. If the price of *X* rose by 10 per cent and that of *Y* fell by 10 per cent, the mean of price relative would show no change—and yet the consumers might complain of a rise in the cost of living if they were spending more on *X* than on *Y*. For them the increased price of *X* (which is a necessity) is not offseted by the lowered price of *Y* (which is a luxury).

This difficulty can be easily met and more representative index numbers obtained if we bring out the relative importance of the commodities by applying explicitly appropriate emphasis. The application of appropriate emphasis openly is known as weighting.

The decision about the appropriate weights depends upon answers to the following questions:

(*i*) **By what do we weight**? Is it to be production figures, consumption figures or distribution figures. The choice would depend upon whatever seems appropriate to bring out the economic importance of the commodities involved from the point of our object.

(*ii*) **What type of weight is to be used**? There are two types of weights—(*i*) quantity and (*ii*) value. A quality weight (represented by symbol q) means the amount of commodity produced, consumed or distributed. A value weight is the product of price with quantity produced, consumed or distributed value is symbolized by $p \times q$.

Quantity weights are to be used if we adopt the method of aggregates because the product of price and quantity will always be in rupees. On the other hand, if the method used is 'price relatives' quantity cannot be used as weights for the product of price relatives and quantities expressed in different units would result in different units. Thus kilogram, multiplied by price relatives would given kilograms, tonnes multiplied by price relatives would result in tonnes and so on. This being so the products of price-relatives and quantity of different commodities cannot be added and averaged. To overcome this difficulty the price relatives are weighted by value figures (which are always expressed in rupees) and the product is always in rupees only.

In short it can be said that in the method of aggregates of actual price weights used are quantity (q): and value weights ($p \times q$) must be used in the method of average of price relatives.

(*iii*) **The time from weich weights should be taken**. The quantity weights may be quantities of the base period (q_0); or of the given period (q_1), or it may be the sum or average of these two. Sometimes the quantities of a typical year of the average of the quantities of several years which are thought as typical may also be used as weights.

Value weights are usually the product of the price quantity of the base year symbolized by p_0q_0. These weights may also be obtained by the product p in any time period and q in any other time period.

13.12. MISCELLANEOUS ILLUSTRATION

Illustration 13.5. Construct the cost of living index number for the year 1965 based on 1960 from the following data by assigning the given weights.

Group	*Group Index No. per 1965 with 1960 base*	*Weight*
Food	152	48
Fuel and Lighting	110	5

Clothing	130	15
House Rent	100	12
Miscellaneous	80	20

Solution. The index number of cost of living *i.e.*, weighted arithmetic mean of given figures is:

$$\text{Index Number} = \frac{152\times48+110\times5+130\times15+100\times12+80\times20}{48+5+15+12+20}$$

$$= \frac{12596}{100} = 125.96.$$

Illustration 13.6. The price quotations of four different commodities for 1951 and 1965 are given below. Calculate the index number of 1965 with 1951 as base by using (*i*) simple average of price relatives: (*ii*) weighted average of price relatives.

Commodity	*Weight*	*(Price in Rs)* 1965	1951
A	5	4.5	2.0
B	7	3.2	2.5
C	6	4.5	3.0
D	2	1.8	1.0

Solutian.

CALCULAIION OF INDEX NUMBER

Commodity	*Prices* *1961*	*1965*	*Price relative 1965 (R)*	*Weight (w)*	*Rw*
A	2.0	4.5	225	5	1125
B	2.5	3.2	128	7	896
C	3 0	4 5	150	6	900
D	1.0	1.8	180	2	360
Total			683	20	3281

Index Numbers of 1965 using

(*i*) Simple average of relatives $= \frac{\Sigma R}{n} = \frac{683}{4} = 170.75$

(*ii*) Using weighted average of relatives $= \frac{\Sigma Rw}{\Sigma w} = \frac{3281}{20} = 164{\cdot}05$

Note. When weights are given weighted arithmetic mean is used. As against weighted average, simple average oridinarily indicates simple arithmetic mean and not geometric mean. So here under (*i*) simple arithmetic mean is calculated even though it is not a good method of averaging index number using in weights.

Illustration 13.7. Construct the cost of living index number for the year 1966 from the following data:

1953 (BASE)

Commodity	*Price* (p_0)	*Quantity Consumed* (q_0)	*1966 Price* p_n
A	25	16	35
B	36	7	48
C	12	3.5	16
D	6	2.5	10
E	28	4	28

Solution. As the base year quantities are given, and we have to construct price index number, weighted aggregate method with base year weights can be used. Relative method with base years weights can also be used. But the two methods will give identical results. From the table below $I_n = \frac{1089}{821} \times 100 = 132.6$

CALCULATION OF INDEX NUMBERS

Commodity	*1953 (base)*		*1966*		
	p_0	q_0	p_n	p_0q_0	p_nq_0
A	25	16	35	400	560
B	36	7	48	252	336
C	12	3.5	16	42	56
D	6	2.5	10	15	25
E	28	4	28	112	112
Total				821	1089

Illustration 13.8. Compute index numbers from the following data using (*i*) Laspeyre's (*ii*) Paasche's and (*iii*) Ideal formula.

Commodity	*Base Year*		*Current Year*	
	Quantity	*Price*	*Quantity*	*Price*
A	12	10	15	12
B	15	7	20	5
C	24	5	20	9
D	5	16	5	14

Solution. It is not mentioned whether index number of price is required or quantity. Let us calculate only one, say price index number. The other can be calculated in the same manner.

CALCULATION OF INDEX NUMBERS

Commodity	Basic Year p_0	q_0	Current Year p_1	q_1	p_0q_0	p_1q_0	p_1q_1	p_0q_1
A	10	12	12	15	120	144	180	150
B	7	15	5	20	105	75	100	140
C	5	24	9	20	120	216	180	100
D	16	5	14	5	80	70	70	80
Total					425	505	530	470

Laspeyre's Index of Price $(L)=\frac{\Sigma p_1q_0}{\Sigma p_0q_0}\times 100=\frac{505}{425}\times 100=118.8.$

Paasche's Index of price $(P)=\frac{\Sigma p_1q_1}{\Sigma p_0q_1}\times 100=\frac{530}{470}\times 100=112.8.$

Fishers Ideal Index of price $=\sqrt{L\times P}=\sqrt{118{\cdot}1\times 112.8}=115.7.$

It may be noted that Fisher's Ideal Index is geometric mean of Laspeyre's and Paasche's Index Numbers.

Illustration 13.9. From the following data on clothing prices, show that the arithmetic mean of relatives (unweighted) does not meet the time reversal test.

Items	*Price in Rs* 1950	1960
Hats	5.00	6.00
Ties	1.00	1.50
Shoes	8.00	8.00

Solution. Time reversal test is satisfied when $P_{01}\times P_{10}=1$

CALCULATION OF P_{01} AND P_{10}

Items	*1950* (p_0)	*1960* (p_1)	p_1/p_0	p_0/p_1
Hats	5.00	6.00	1.2	.833
Ties	1.00	1.50	1.5	.677
Shoes	8.00	8.00	1.0	1.00
$N=3$			3.7	2.5

$p_{01}=\frac{\Sigma p_1/p_0}{N}=\frac{3.7}{3}=1.233$ $\qquad P_{10}=\frac{\Sigma p_0/p_1}{N}=\frac{2.5}{3}=.833$

$p_{01}\times p_{10}=1.233\times .833=1.028.$ $\qquad$ *i.e* $\neq 1.$

Hence time reversal test in not sdatisfied by arithmetic mean of the relatives.

Illustration 13.10. The following table gives the per capita income and the cost of a living index of a particular community Calculate the real income takihg into account the rise in the cost of living.

Year	1949	1950	1951	1952	1953	1954	1955	1956
Cost of living Index (1949=100)	100	104	115	160	210	260	300	320
Per Capita income (Rs)	360	400	480	520	550	590	610	650

Solution.

$$\text{Real Income} = \frac{\text{Per Capita Income}}{\text{Cost of Living Index}} \times 100$$

So for $1949 = \frac{360}{100} \times 100 = 360.0$ For $1953 = \frac{550}{210} \times 100 = 261.9$

" " $1950 = \frac{400}{104} \times 100 = 384.6$ " $1954 = \frac{590}{260} \times 100 = 226.9$

" " $1951 = \frac{480}{115} \times 100 = 417.3$ " $1955 = \frac{610}{300} \times 100 = 203.3$

" " $1952 = \frac{520}{160} \times 100 = 325.0$ " $1965 = \frac{650}{320} \times 100 = 203.1.$

Illustration 13.11. From the following data relating to working class consumer price index of a city calculate Index Number for 1972 and 1973.

Group	*Weight*	*Group* *1972*	*Indices* *1973*
Food	48	110	130
Clothing	8	120	125
Fuel and Light	7	110	120
Room Rent	13	100	100
Miscellaneous	14	115	135

Solution. Consumer Price Index=weighted arithmetic mean of group index No.

$$\text{Index for } 1972 = \frac{110\times48+120\times8+110\times7+100\times13+115\times14}{48+8+7+13+14}$$

$$= \frac{5280+960+770+1300+1610}{90} = \frac{9920}{90} = 110.22$$

$$\text{Index No. for } 1973 = \frac{130\times48+125\times8+120\times7+100\times13+135\times14}{48+8+7+13+14}$$

$$= \frac{6240+1000+840+1300+1890}{90} = \frac{11270}{90}$$

$$= 125.22.$$

Illustration 13.12. From some given data, the consumer price index based on 5 groups was calculated as 205. Ths percentage increase in prices over base period is given below :

Rent and taxes 60, Clothing—210, Fuel and light—120, Miscellaneous—130.

Calculate the percentage increase in the food group given that the weights of the different groups were as follows:

Food—60, Rent and taxes—16, Fuel and light—12, Clothing—8, Miscellaneous—4.

Solution. Assuming indexforthefoodgroupis x, group indices and corresponding weights are as follows:

Food	x	60
Rent and Taxes	160	16
Clothing	310	8
Fuel and Light	220	12
Miscellaneous	230	4

$$205=\frac{60x+160\times 16+310\times 8+220\times 12+230\times 4}{60+16+8+12+4}$$

$$20500=60x+2560+2480+2640+920$$

$$=60x+8600$$

$$60x=11900$$

$$x=198 \text{ approximately.}$$

Increase in food group over base period = 198 − 100 = 98 %.

Illustration 13.13. Following information relating to workers in an industrial town is given.

Items of Consumption	*Consumer Group Index in 1970 (1960 = 100)*	*Proportion of expenditure*
Food, drinks and tobacco	225	52%
Clothing	175	8%
Feul and Light	155	10%
Housing	250	14%
Miscellaneous	150	16%

Average wage in 1960 was Rs 200. What should be the average wage per worker per month in 1970 in that town so that the standard of living of the worker does not fall below the 1960 level.

Solution.

$$\text{Index for 1970}=\frac{225\times 52+175\times 8+155\times 10+250\times 14+150\times 16}{52+8+10+14+16}$$

$$=\frac{11700+1400+1550+3500+2400}{100}$$

$$=\frac{20550}{100}=205.5.$$

Average wage required per worker

$$=\frac{\text{Index for 1970}}{\text{Index for 1960}}\times\text{Wage in 1960}$$

$$=\frac{205.5}{100}\times 200=411$$

Illustration 13.14. Owing to change in prices the consumer price index of the working class in a certain area rose in a month by one quarter of what it was before to 225. The index of food became 252 from 198, that of clothing from 185 to 205, that of fuel and lighting from 175 to 195 and that of miscellaneous from 138 to 212. The index of rent, however, remained unchanged at 150. It was known that the weights of clothing, rent and fuel and lighting were same. Find out the exact weight of all groups.

Solution. New Index=225

$$225=\frac{1}{4}\text{ old Index}+\text{old Index}=\frac{5}{4}\text{ Old Index}$$

$$\therefore\ \text{Old Index}=225\times\frac{4}{5}=180.$$

Group	*Index*		*Weight*
	New	*Old*	
Food	252	198	x
Clothing	205	185	y
Fuel	195	175	y
Miscellaneous	212	138	z
Rent	150	150	y

Generally weights are given as percentage, therefore

$$x+3y+z=100 \quad \ldots(1)$$

$$225=\frac{252x+205y+195y+212z+150y}{100}$$

$$252x+550y+212z=22,500 \quad \ldots(2)$$

$$180=\frac{198x+185y+175y+138z+150y}{100}$$

$$198x+510y+138z=18,000 \quad \ldots(3)$$

Solving equation (1), (2) and (3)

Multiplying (1) by 252 and subtracting from it *(2)*

$$206\,y+40z=2700 \quad \ldots(4)$$

$$\text{or } 103\,y+20z=1350..$$

Multiplying (1) by 198 and subtracting from it *(3)*

$$84y+60z=1800 \quad \ldots(5)$$

Multiplying (4) by 3 and subtracting equation (5) from it

We get $225y=2250$ or $y=10$

Putting value of y in equation (5)

$$60z=1800-840=960$$
$$z=16$$

Putting the values of y and z in equation (1)

$$x+30+16=100$$
$$x=54.$$

Thus, the exact weights are 54, 10, 10, 16 and 10 respectively.

Illustration 13.15. Construct the index no. of business activity from the following data using (*a*) A.M. and (*b*) G.M.

Item	*Weight*	*Index*
Industrial Production	36	250
Mineral Production	7	135
Internal Trade	24	200
Financial Activity	20	135
Exports and Imports	7	325
Shipping Activity	6	300

Solution.

(*a*) Index no. of business activity $= \dfrac{\Sigma \text{ Group Index} \times \text{Weight}}{\Sigma \text{ Weights}}$

$$= \frac{250\times36+135\times7+200\times24+135\times20+325\times7+300\times6}{36+7+24+20+7+6}$$

$$=\frac{21520}{100}=215.2.$$

(*b*) Using G.M.

$$\text{Index no.} = \text{Antilog}\left[\frac{\Sigma \log \text{Index} \times \text{Weight}}{\Sigma \text{ Weight}}\right]$$

$$=\text{Antilog}\left[\frac{\log 250\times36+\log 135\times7+\log 200\times24+\log 135\times20+\log 325\times7+\log 300\times6}{36+7+24+20+7+6}\right]$$

$$=\text{Antilog}\left[\frac{86.3244+14.9121+55.224+42.606+17.5833+14.8626}{100}\right]$$

$$=\text{Antilog}\,\frac{231.5124}{100} \quad \text{or Antilog } 2.315124$$

$$=206.6.$$

Illustration 13.16. In a working class consumer price index no.

of a particular town the weights corresponding to different groups of items were as follows:

Food—55. Fuel—15, clothing—10, Rent—8 and Miscellaneous—12.

In Oct. 1972, the, D.A. was fixed by a mill of that town at 182 per cent of the workers wages which fully compensated for the rise in prices of food and rent but did not compensate for anything else. Another mill of the same town paid D.A. of 46.5 per cent which compensated for the rise in fuel and miscellaneous groups. It is known that rise in food is double that of fuel and the rise in miscellaneous group is double that of rent.

Find the rise of food, fuel, rent and miscellaneous groups.

Solution. Assuming rise in fuel is x and rent is y, the rise in food and miscellaneous group are $2x$ and $2y$ respectively.

From data for mill A

where an index number of (100+182) accounts for increases in prices of food from 100 to (100+2x) and in rent from 100 to (100+y), but does not account for any other pice change, we get

$$282=\frac{55(100+2x)+15\times 100+10\times 100+8(100+y)+12\times 100}{100}$$

$$110x+8y=18200 \quad \ldots(1)$$

From data for mill B

where an imdex of (100+46.5) accounts for increase in prise of fuel to (100+x) and of miscellaneous group to (100+2y). we get

$$146.5=\frac{55\times 100+15(100+x)+10\times 100+8\times 100+12(100+2y)}{100}$$

$$15x+24y=4650 \quad \ldots(2)$$

Multiplying (1) by 3 and subtracting equation (2) from it

$$315x=49950 \text{ or } x=158.6$$

Putting the value of x in equation (2)

$$y=94.5.$$

Hence the rise are as follows.

Food $=2x=158.6\times 2=317.2$, Rent $=y=94.5$

Fuel x = 158.6. Miscellaneous = $2y$ = 189.0. all in percentages.

Illustration 23.17. Given the data:

Commodities	p_0	q_0	p_1	q_1
A	1	10	2	5
B	1	5	x	2

Where p and q respectively stand for price and quantity and subscripts for time periods. Find x if the ratio between Laspeyre's index and Paasche's index no. is 28:27.

Solution. Laspeyre's Price Index for year (1)

$$=\frac{\Sigma P_1Q_0}{\Sigma P_0Q_0}\times 100=\frac{10\times 2+5x}{10\times 1+5\times 1}\times 100$$

Paasche's price Index for year (1)

$$=\frac{\Sigma P_1Q_1}{\Sigma P_0Q_1}\times 100=\frac{2\times 5+2x}{5\times 1+2\times 1}\times 100$$

$$\therefore \quad \left(\frac{20+5x}{10+5}\times 100\right) \Big/ \left(\frac{10+2x}{5+2}\times 100\right) = 28/27$$

$$\frac{20+5x}{15}\times\frac{7}{10+2x}=\frac{28}{27}$$

$$4200+840\,x=3780+945\,x$$

or $$105\,x=420 \text{ or } x=4.$$

Illustration 13.18. Show that the Fisher's Ideal Index satisfies both the time reversal and factor reversal test. Use the data given below:

Commodity	*1970* Price	*1970* Quantity	*1972* Price	*1972* Quantity
A	6	50	10	56
B	2	100	2	120
C	4	60	6	60
D	10	30	12	24
E	8	40	12	36

Solution. According to time reversal test

$$p_{0\cdot 1}\times p_{1\cdot 0}=1$$

Commodity	*Year 1970* p_0	q_0	*Year 1972* p_1	q_1	p_0q_0	p_0q_1	p_1q_0	p_1q_1
A	6	50	10	56	300	336	500	560
B	2	100	2	120	200	240	200	240
C	4	60	6	60	240	240	360	360
D	10	30	12	24	300	240	360	288
E	8	40	12	36	320	288	480	432
					1360	1344	1900	1880

$$p_{0.1}=\sqrt{\frac{\Sigma p_1q_0}{\Sigma p_0q_0}\times\frac{\Sigma p_1q_1}{\Sigma p_0q_1}}=\sqrt{\frac{1900}{1360}\times\frac{1880}{1344}}$$

$$p_{1.0}=\sqrt{\frac{\Sigma p_0q_1}{\Sigma p_1q_1}\times\frac{\Sigma p_0q_0}{\Sigma p_1q_0}}=\sqrt{\frac{1344}{1880}\times\frac{1360}{1900}}$$

$$p_{0.1}\times p_{0.1}=\sqrt{\frac{1900}{1360}\times\frac{1880}{1344}\times\frac{1344}{1880}\times\frac{1360}{1900}}=1$$

Thus time reversal test is satisfied.

According to Factor Reversal Test

$$p_{0\cdot 1}\times q_{0\cdot 1}=\frac{\Sigma p_1q_1}{\Sigma p_0q_0}$$

$$p_{0.1}=\sqrt{\frac{\Sigma p_1q_0}{\Sigma p_0q_0}\times\frac{\Sigma p_1q_1}{\Sigma p_0q_1}}=\sqrt{\frac{1900}{1360}\times\frac{1880}{1344}}$$

$$q_{0.1}=\sqrt{[illegible]\times\frac{q_1p_1}{q_0p_1}}=\sqrt{\frac{1344}{1360}\times\frac{1880}{1900}}$$

$$p_{0\cdot1}\times q_{0\cdot1}=\sqrt{\frac{1900}{1360}\times\frac{1880}{1344}\times\frac{1344}{1360}\times\frac{1880}{1900}}=\sqrt{\frac{1880}{1360}\times\frac{1880}{1360}}$$

$$=\frac{1880}{1360}=\frac{\Sigma p_1q_1}{\Sigma p_0q_0}$$

Thus, factor reversal test is satisfied.

Hence Fisher's Ideal Index satisfies the time reversal and factor reversal test.

Illustratio 13.19. Expenditure of a family on 3 items are in the ratio 2:5:3. The prices of these commodities rise by 30 per cent, 20 per cent, 40 per cent respectively. By what per cent has total expenditure increased ?

Solution. The price of a commodity rise by 30 per cent means that with suitable time as base, index no. of a commodity is 100+30 *i.e.*, 130.

Item	*Index No. (I)*	*Weight (W)*	*IW*
A	130	2	260
B	120	5	600
C	140	3	420
		10	1280

$$\text{Price Index No.}=\frac{1280}{10}=128.0$$

∴ The total expenditure has increased by 28 per cent (128-100).

Illustration 13.20. In a working class budget enquiry in town *A* and *B* it was found that an avarage working class family expenditure on food and other items was as follows:

	A	*B*
Food	64%	50%
Other items	36%	50%

In 1971 Consumer Price Index stood at 279 for town *A* and 265 for town *B* (1961=100). It was known that the rise in the prices of all the articles consumed by the working class was same for *A* and *B*. Obtain the rise in price of the two groups separately

Solution. Let rise in prices of food articles be x

and ,, ,, ,, other items be y.

Items	*Weights*		*1971 Index*	
	Twon A	*Town B*	*Town A*	*Town B*
Food	64	50	$100+x$	$100+x$
Other items	36	50	$100+y$	$100+y$
	100	100		

$$279 = \frac{64\,(100+x)+36\,(100+y)}{100} \text{ and } 265 = \frac{50\,(100+x)+50\,(100+y)}{100}.$$

$$27900 = 6400+64x+3600+36y \text{ and } 26500 = 5000+50x+5000+50y$$

$$17900 = 64x+36y \text{ and } 330 = x+y$$

Solving these two simultaneously, we get

$$x=215;\ y=115$$

Index for food=315; Index of other items=215.

Illustration 13.21. From the chain base index no. given below, prepare fixed base index no.

Year	1971	1972	1973	1974	1975
Index	110	160	140	200	150

Solution. Current year's Fixed Base Index

$$= \frac{\text{Current year Chain Base Index} \times \text{Previous year Fixed Base Index}}{100}$$

Year	*Chain Index*	*Fixed Index*
1971	110	110
1972	160	$\frac{110\times160}{100}=176$
1973	140	$\frac{140\times176}{100}=246.4$
1974	200	$\frac{200\times246.4}{100}=492.8$
1975	150	$\frac{492.\ \times150}{100}=739.2$

Illustration 13.22. A price index series was started with 1961 as base. By 1965 it rose by 20 per cent, the link relative for 1966 was 90. In this year a new series was started. This new series rose by 12 points by next year. But during the next three years the rise was not rapid. During 1970 the price level was only 10 per cent higher than that of 1967. Splice the two series and calculate the index no for various years by shifting the base to 1967.

Solution.

Year	*Index No. (1961=100)*	*Index No, (1966=100)*	*Old series spliced to new (base 1966)*
1961	100		$100\times\frac{100}{108}=92.6$
1965	120		$120\times\frac{100}{108}=111.1$
1966 $\left(120\times\frac{90}{100}\right)$	108	100	100
1967		112	112
1970		123.2	123.2

Base shifting to 1967

1961	$92.6\times\frac{100}{112}=82.7$
1965	$111.1\times\frac{100}{112}=99.2$
1966	$100\times\frac{100}{112}=89.3$
1967	$112\times\frac{100}{112}=100$
1970	$123.2\times\frac{100}{112}=110$

Illustration 13.23. The following table gives the monthly wages and cost of living index no. based on 1968:

	68	69	70	71	72	73	74
Wages	65	70	75	80	90	100	120
Index No.	100	110	120	130	150	200	250

Calculate the real wages.

Solution. Real wages$=\frac{\text{Wages}}{\text{Cost of Living Index}}\times 100$

Year	*Wages*	*Index No.*	*Real Wages*
68	65	100	$\frac{65}{100}\times100=65$
69	70	110	$\frac{70}{110}\times100=63.6$
70	75	120	$\frac{75}{120}\times100=62.5$
71	80	130	$\frac{80}{130}\times100=61.5$
73	90	150	$\frac{90}{150}\times100=60$
74	100	200	$\frac{100}{200}\times100=50$
74	120	250	$\frac{120}{250}\times100=48$

Illustration 13.24. The price quotations of four different commodities for 1976 and 1977 are given below calculate the index number for 1977 with 1976 as base by using

(*i*) Simple average of price relatives,

(*ii*) Weighted average of price relatives.

Commodity	*Unit*	*Weight*	*Price*	
			1976	*1977*
A	**kg**	5	2.00	4.50
B	quintal	7	2.50	3.20
C	dozen	6	3.00	4.50
D	**kg**	2	1.00	1.80

Solution. Calculate of Index number

Commodity	*Weight*	p_0	p_1	$I=\frac{p_1}{p_0}\times 100$	*IW*
A	5	2.00	4.50	225	1125
B	7	2.50	3.20	128	896
C	6	3.00	4.50	150	900
D	2	1.00	1.80	180	360
Total	20			683	3281

$$\text{Simple average of price relatives}=\frac{683}{4}$$

$$=170.75$$

$$\text{Weighted average of price relative}=\frac{3281}{20}$$

$$=164.05$$

Illustration 13.25. Construct the index number of business activity from the following data:

Group	*Index*	*Weight*
1. Industrial Production	250	36
2. Mineral Production	135	7
3. Internal Trade	200	24
4. Financiai Activity	135	20
5. Exports and Imports	325	7
6. Shipping	300	6

Solution.

COMPUTATION OF BUSINESS ACTIVITY INDEX

Group	*Index, I*	*Weight, W*	*I W*
1	250	36	9000
2	135	7	950
3	200	24	4800
4	135	20	2700
5	325	7	2275
6	300	6	1800
Total		100	21520

Illustration 13.25 Compute by suitable method the index number of quantity from the given data:

Commodity	1901 Price	1901 Value	1982 Price	1982 Value
A	8	80	10	110
B	10	90	12	108
C	16	256	20	340

Solution.

CALCULATION OF QUANTITY INDEX NO.

Commodity	p_0	p_0q_0	q_0	p_1	p_1q_1	q_1	p_0q_1	p_1q_0
A	8	80	10	10	110	11	88	100
B	10	90	9	12	108	9	90	108
C	16	256	16	20	340	17	272	320
Total		426			558		450	528

$$\text{Fisher's } Q_{01} = \sqrt{\frac{\Sigma q_1 p_0}{\Sigma q_0 p_0} \times \frac{\Sigma q_1 p_1}{\Sigma q_0 p_1}} \times 100$$

$$= 100\sqrt{\frac{450}{426} \times \frac{558}{528}} = 105.66$$

Illustration 13.26 Calculate the cost of living index.

Item	Quantity used in given year	p_0	p_1
1. Rice	30 quantals	12	25
2. Pulses	36 kg.	0.40	0.60
3. Oil	24 kg	1.5	2.2
4. Clothing	72 m	0.75	1.0
5. Housing	unit	20 p.m.	30 p.m.
6. Misc		10 p.m	15 p.m.

Solution.

CALCULATION OF COST OF LIVING INDEX

Item	q_1	p_0	p_1	p_1q_1	p_0q_1
1	30	12	25	750	360
2	36	0.4	0.6	21.6	14.4
3	24	1.5	2.2	52.8	36.0
4	72	0.75	1.0	72.0	54.0
2	12	20	30	360	240
6	12	10	15	180	120
Total				1436.4	824.4

Cost of Living Index

$$= \frac{1436.4}{824.4} \times 100$$

$$= 174.24$$

PROBLEMS

1. Define index number and show the importance and the use of general index numbers.
2. What are economic barometers? Show their importance in forecasting economic events.
3. Describe briefly the problems that are involved in the construction of an index number of prices.
4. Distinguish between fixed base and chain base methods of constructing index numbers and describe their relative merits and demerits.
5. Describe briefly the various methods employed for constructing an index number of prices.
6. What do you mean by reversibility of an index number? Which index numbers are reversible?
7. What is meant by weighting in Statistics? What are the various ways of assigning weights in the construction of index numbers?
8. What is meant by value weights? Describe with an example the weighted index number of wholesale prices.
9. What do you understand by Time Reversal Test and Factor Reversal Tests?
10. Explain Fisher's 'Ideal' method of weighting index numbers and describe the difficulties that are to be faced in using it.
11. Write short notes on:
    (*a*) Link relatives,
    (*b*) Chain relatives,
    (*c*) Base shifting, and
    (*d*) Implicit and Explicit weighting.
12. Explain the use of index numbers with the help of the following table which gives the average price of a commodity:

Year	*Price in dollars*	*Year*	*Price in dollars*
1960	7.8	1965	7.6
1961	6.7	1967	9.9
1962	7.2	1968	7.5
1963	9.8	1969	10.2
1964	8.8	1971	11.2

3. Compare the index numbers of sales of two commodities *A* and *B* by taking (*i*) average of first three yeaes as base, and (*ii*) 1970 as base given in appropriate units.

Year	*Sales* *A*	*Sales* *B*	*Year*	*Sales* *A*	*Sales* *B*
1960	70	60	1966	79	75
1961	66	66	1967	78	73
1962	63	69	1968	78	74
1963	67	68	1969	82	75
1964	70	70	1970	80	75
1965	75	71			

14. Following gives the figures of the index of Indian industrial activity. Calculate the Industrial Activity index number by using (*i*) simple average (*ii*) weighted mean, (*iii*) simple geometric mean, (*iv*) weighted geometric mean.

Commodity	*Weight*	*Index nos.*	*Commodity*	*Weight*	*Index nos.*
Cotton	9	133.1	Wagons loaded	24	142.6
Jute	6	105.5	Cheque clearance	20	94.3
Steel	5	192.8	Notes in circulation	6	163.5
Pig Iron	8	221,0	Consumption of		
Cement	5	223.2	Electricity	7	135.8
Paper	3	175.9			
Coal	7	178.6			

(**Ans.** (*i*) By simple average = 151.5, (*ii*) By weighted mean = 138.96 and (*iii*) By simple geometric mean = 155.1)

15. Calculate from data given below the Index No. for 1965 and 1971 with year 1959 as base by using (*i*) arithmetic mean, (*ii*) geometric mean, (*iii*) median respectively as the method of averaging. Also show that index number calculated on the basis of arithmetic mean is not reversible while index no. calculated on the basis of geometric mean is

Commodity	*A*	*B*	*C*	*D*	*E*	*F*	*G*
1959 Prices	3.2	4.4	2.4	6.0	1.0	8.4	1.0
1965 Prices	1.6	5.5	3.6	6.0	0.9	6.3	1.1
1971 Prices	6.4	4.4	1.2	2.0	3.0	2.1	4.0

(**Ans.** (*i*) A.M. Index No. for 1965 = 100 and for 1971 = 158.3, (*ii*) G.M. Index No. for 1965 = 94.95 and 1971 = 100.00 and (*iii*) Median for 1965 = 100.00 and 1971 = 100.0).

16. The following table gives the prices of 8 commodities iu the base year. Find the unweighted index number of prices of the current year as also the weighted number, weights being proportional to the value.

Quality in Units		*Price per Unit*	
Commodity	*Base Year*	*Base Year*	*Current Year*
1	2692	64.2	72.3
2	831	119.8	111.5
3	1247	39.8	45.0
4	185	57.5	67.8

5	345	141.4	96.5
6	8989	10.9	19.6
7	87	141.0	113.5
8	1298	18.2	21.2

Also find the index number of current year using quantities as welghts. (Ans. Unweighted Index No. = 110.2 approx. Weighted Index No. = 170.0 approx. Weighted Index No. using base year quantities as weights = 117.0 approx).

17. Calculate price index no. for 1945 (*i*) Laspeyre's methhd, (*ii*) Paasche's method, (*iii*) Bowley's method, (*iv*) Marshall and Edgworth's method and (*v*) Fisher's method from the following data:

Commodity	*Price 1935*	*Quantity 1935*	*Price 1945*	*Quantity 1945*
A	4	50	10	40
B	3	10	9	2
C	2	5	4	2

(Ans. (*i*) Laspeyre's method = 254.2, (*ii*) Paasehe's method = 250.6, (*iii*) Bowley's metiiod = 252.4, (*iv*) Marshall and Edgeworth's method = 252.7 and (*v*) Fisher's method = 252.4.)

18. From the data given below, compute the index number of price by an appropriate method.

Base year			*Current year*	
Commodity	*Price*	*Total value*	*Prjce*	*Total value*
A	40 units	240 units	30 units	210 units
B	45 ,,	180 ,,	50 ,,	250 ,,
C	90 ,,	45 ,,	40 ,,	60 ,,

(Ans. By Fisher's Method. Index No = 83.6.)

19. Taking 1953 as base period calculate an index number of prices for the year 1971 following the data given in appropriate units.

	1953		*197!*	
Commodity	*Quantity*	*Price*	*Quantity*	*Price*
Wheat	562	170	632	72
Rice	535	192	756	70
Sugar	639	195	926	95
Gbee	128	187	255	92
Fuel	542	185	632	92
Gold	217	150	314	180

(Ans. Fisher's ideal index No. = 50.1)

20. Calculate the index number of p:ice using weighted relative method taking (*i*) base year weights, (*ii*) current year weights for the data of Problem 17. (Ans. (*i*) Index number by relativc metbod using case year weights = 254.2 and (*ii*) Index number by relative method using current year weights = 251.2.)

21. Prove using the following data that the Factor Reversal Test is satisfied by Fisher's Ideal formula for index number.

Commodity	*Base Year*		*Current Year*	
	Price	*Qty.*	*Price*	*Qty.*
A	6	50	10	56
B	2	100	2	120
C	4	60	6	60
D	10	30	12	24
E	8	40	12	36

22. Using the data of Prob. 21 above show that Fisher's method of calculating index number satisfies both the time reversal and factor reversal test and no other method mentioned in Prob. 17 satisfies both the tests.

23. Use the following data, compare the annual fluctuation in Indian industrial activity by chain base method.

Year	*I. No.*	*Year*	*I. No.*
1961	120	1966	137
1962	122	1967	136
1963	116	1968	149
1964	120	1969	15[illegible]
1965	120	1970	137

Assuming 1960 to be 100 show that for the above series chain indices are same as fixed base index numbers given above.

24. You are given the following series of index numbers of price of four commodities and an index number of the four taken together based on average. Calculate new indices for seven years based on Chain Method.

Year	*Sugar*	*Milk*	*Coffee*	*Tea*	*Total*	*Average*
1961	81	77	119	55	332	83.0
1962	62	54	128	83	326	81.5
1963	104	87	111	100	402	100.5
1964	98	75	154	96	418	104 5
1965	60	43	165	88	356	89.0
1966	60	44	159	89	352	88 0
1967	62	47	139	84	332	83.0

(**Ans.** Chain Index for 1961=100, 1962=100.82, 1963=135.48, 1964=139.00 1965=111.41, 1966=111.35 and 1967=109 10.)

25. Calculate the chain index number and fixed base index number from the following data of price in rupees of four commodities and show that two series give different results.

Commodity	*Price in rupees*				
	1967	*1968*	*1969*	*1970*	*1971*
A	2	3	4	3	7
B	3	6	9	1	3
C	4	12	20	8	19
D	5	7	18	11	22

(**Ans.** Average *i.e.*, Fixed base)

Index No.	100	197.5	340.0	163.3	322.5
Chain Index No.	100	197.5	349.2	170.4	315.3

26. From the data given below calculate the cost of Living Index number for the current year by the aggregate Expenditure and Family Budget methods separately.

Article	*Quantity Base Year*	*Unit*	*Price in Rs per Unit Base Year*	*Current Year*
Rice	5	40 kg	6	8
Millets	5	,,	4	5
Wheat	1	,,	5	10
Gram	1	,,	3	6
Arhar	½	,,	4	6
Other Pulses	2	,,	3	4
Ghee	4	kg	1.25	2
Gur	2	40 kg	2.50	5
Salt	12½	,,	4	5
Oil	24	kg	20	25
Clothing	20	m	0.25	0.50
Firewood	10	40 kg	0.50	0.80
Kerosene oil	1	tin	4	6
House rent	—	—	12	15

(**Ans** Cost of living index number by family budget method = 146, Cost of living index number by aggregate expenditure method = 146.)

27. Construct the cost of Living Index Number from the table given below:

Group	*Index for 1963*	*Expenditure*
Food	550	46%
Clothing	215	10%
Fuel and lighting	220	7%
House rent	150	12%
Miscellaneous	275	23%

(**Ans.** Cost of living index number = 376.95)

28. On a certain date the Ministry of Labour retail price index was 104.6. The price indexes in July 1914 were:
Rent and Rates 65' Clothing 220, Fuel and Light 110, Miscellaneous 125. What was the price index of the food group? Given that the weights of different items in the group were as follows:
Food 60, Rent and rates 16, Clothing. 12, Fuel and Light 8 and Miscellaneous 4.
(**Ans.** Index No. of food group 90)

29. An average family of industrial workers in a certain town consumed during Aug. 1939, 60 kg of foodgrains, 10 metre of cloth, 80 kg of fuel and 1 tin of kerosene oil and paid Rs 15 as house rent. Foodgrains then sold at an average price of Rs 6 per 40 kg, cloth at 8 as per metre, and fuel at Rs 2/4 per 40 kg while a tin of kerosene oil at Rs 5. By Aug. 1943, the average of foodgrains and cloth had risen to three times and five times the prewar average,

respectively, fuel rose to Rs 4 per 40 kg and house rent to Rs 20. The solitary exception was kerosene whose price fell by 50 p per tin.

Express in quantitative terms the rise that took place in the cost of living of industrial workers in the given town in Aug. 1943, as compared with of Aug. 1939, making clear your method approach.

(**Ans.** The cost of iiving of Aug. 1943, is 224.7 based on the level of Aug. 1939.)

30. Rewrite the following index numbers compiled by Labour Bureau for certain cities by changing the base to year 1966.

Year	*Ajmer*	*Jharia*	*Cuttack*	*Jabalpur*
1961	100	100	100	100
1962	110	97	102	95
1963	118	122	106	101
1964	152	139	117	123
1965	161	153	137	146
1966	161	159	147	151
1967	168	182	163	153
1968	178	184	181	168
1969	174	175	160	150
1970	168	166	157	151

31. Expenditure of a family on three items are in the ratio 2:5:3. The prices of these commodities rise by 30 per cent, 24 per cent and 40 per cent respectively. By what percentage the total expenditure has increased.

(**Ans.** Rise in the percentage of expenditure=28 per cent).

32. In 1920 a Statistical Bureau started an index of production based on 1914 with the following results, Index Nos. 1914—100, 1920—120 and 1929—200. In 1930 the Bureau reconstructed the Index on a new plan with the 1929 as base *i.e.*, 1929=100, and found index of 1935 as 150. In 1939 the Bureau again reconstructed the Index on yet another plan with base 1935, *i.e.*, Index No. 1935=100 and calculated Index No. of 1939 as 120. Splice the three series together so as to give a continuous series with base 1935=100. Draw up a working table in parallel columns.

33. The following table gives per capita income and cost of living index for India from 1962-63 to 1970-71. Deflate the per capita income with reference to the cost of Living Index.

Year	*Per Capita Index*	*Cost of Living Index (Base 1962-63)*
1962—63	67	100
1963—64	70	105
1964—65	78	117
1965—66	112	160
1966—67	139	217
1967—68	139	216
1968—69	137	219
1969—70	143	243
1970—71	160	258

(**Ans.** Deflated income 1962—63=67.00; 1963—64=66.67; 1964—65=66.67;

1965—66 = 70.00; 1966—67 = 64.06; 1967—68 = 64.35; 1968—69 = 62.57: 1969—70 = 59.09; 1970—71 = 62.02).

34. Prepare index number of price for three year with the average price as base.

	Rate per Rupee		
	Wheat	*Cotton*	*Oil*
1st year	10 kg	4 kg	3 kg
2nd year	9 ,,	$3\frac{1}{3}$,,	3 ,,
3rd year	9 ,,	3 ,,	$2\frac{1}{2}$,,

(**Ans.** Required Index Nos. are: 1st year—91.0, 2nd year—98.6 and 3rd year —110.3)

35. The average of wholesale price was higher in 1967 than in 1966 by 15.1%. The index number for the twoyears being 108.7 and 94.4 respectively (1960 = 100). The increase followed rises to 6.1, 1.0 and 2.28 per cent each year compared with the preceding. In 1963 prices were the same as in 1962 but 2.5 per cent below 1961. Prices in 1961 were 13.2 per cent below 1960. From these data compute the index number for each year from 1960 to 1967.

(**Ans.** Year	1960	1961	1962	1963	1964	1965	1966	1967
Index No,	100	86.8	85.7	85.7	90.93	91.83	94.4	108.7)

36. The table below gives the per capita income and the cost of living indices. Determine the real income taking into account the rise in the cost of living.

Year	1949	1950	1951	1952	1953	1954	1955	1956
Cost of living index (1949-100)	100	104	115	160	210	260	300	320
Per Capita income (Rs)	360	400	480	520	550	590	610	650

(**Ans.** 1949 = 360.0, 1950 = 384.6, 1951 = 417 3, 1952 = 325.0, 1953 = 216.9, 1954 = 226.9, 1955 = 203.3 and 1956 = 203.1.)

37. Using the data of Prob. 17 show that aggregate method and price relative method give (*i*) same results if base year weights are used (*ii*) different results if current year weights are used, but give same results if harmonic mean is used for the average price relatives.

38. Given the data:

Commodities	P_0	Q_0	P_1	Q_1
A	1	10	2	5
B	1	5	x	2

where P and Q respectively stand for price and quantity and subscripts stand for time period: Find x if the ratio between Laspeyre's (L) and Paasche's (P) index number is $L:P::28:27$.

(Ans. $x = 4$.)

39. Following data relate to construction of Industrial Production Index. From 1954 chemicals are to be included in the index and from 1956 onwards, Pig iron is to be replaced by non-ferrous metals. Construct a suitable series which can be used to compare changes in production for the various years

Commodity	*Weights*	*1952*	*1953*	*1954*	*1955*	*1956*	*1957*	*1558*
		Production						
Cotton	9	4	5	5	4.5	4	6	6.7
Jute	6	3	4	3.5	5	5.5	4.5	3.6
Steel	5	2	2.3	2.8	2.9	3.0	3.2	4
Pig iron	8	4	4.8	5	6	7	—	—
Cement	5	7	7.3	7.7	8.1	8.5	8.6	9
Chemicals	3	—	—	4	4.5	5	5.8	5.9
Non-ferrous metals	4	—	—	—	—	6	8	7.5

40. Using the data of Prob. 15, show that the Index No. calculated by geometric mean method of averaging satisfies circular test, but that by arithmetic mean method of averaging does not satisfy it.

41. Show that ideal index fails to satisfy the circular test but satisfies factor reversal and time reversal tests.

42. Show that Fisher's Method of calculating index number satisfies both the time reversal and factor reversal test and Laspeyre's, Paasche's, Bowley's Marshall's method, simple aggregate method and simple relative method do not satisfy either both or one of them.

43. Price relative for the year 1966 with 1968 as base is 80 while the price relative for the year 1967 with 1966 as base is 150. Find the price relative 1968 with (*a*) 1967 as base, (*b*) 1966 as base.

44. A Textile worker in the city of Bombay earns Rs 350 per month. The cost of living index for a particular month is given as 136. Using the following data, find out the amounts he spends on house rent and clothings.

Group	*Expenditure*	*Group Index*
Food	140	180
Clothing	?	150
House rent	?	100
Fuel and Lighting	56	110
Miscellaneous	63	80

(Ans. Expenditure in clothing is Rs 42 and on house rent is Rs 49.)

45. In 1951 for working class people food was selling at an average price of Rs 16 per quintal., cloth at Rs 2 per metre, house rent Rs 30 per house and other items (Misc.) at Rs 10 per unit. By 1957, cost of foodstuffs rose by Rs 4 per quintal house rent by Rs 15 and other items (Misc.) doubled in price. The working class cost of living index number (found from above four categories) with 1951 as base, for 1957 was 155. Find by how much the cloth rose in price during the period 1951 — 1957. (Ans. Increase in cost of cloth during 1951 to 1957, 90 paise per metre.)

46. An enquiry into the budgets of middle class families in a city gave the following information.

Expenses on	*Food*	*Rent*	*Clothing*	*Fuel*	*Offices*
	30%	15%	20%	10%	25%
Prices in 1967 Rs	100	20	70	20	40
Prices in 1968 Rs	90	20	60	15	55

What is the change in the cost of living figure in 1968, as compared with 1967.
(Ans. Index No. 100.12)

47. From the following data calculate a price index for the year 1970 by using simple geometric mean.

Commodity	A	B	C	D	E	F
Average price 1960	16.1	9.2	15.1	5.6	11.7	100.0
Axerage price 1970	14.2	8.7	12.5	4.8	13.4	117.0

Now reverse the process, taking 1970 as base year and 1960 as current year and show the two results are constant.
(Ans. Index No. for 1970=96.20, for 1960=104.0)

48. Prepare Index No. for 1967 with 1965 as base;

Commodity	*1965 Price Rs*	*1967 Price Rs*	*1967 Quantity in appropriate Units*
1	8	12	100
2	6	7.5	25
3	5	5.25	10
4	48	52	20
5	15	16.5	65
6	19	27	30

(Ans. Index No. 124.5)

49. Construct Index No. of prices from the following data by applying (*i*) Laspeyre's Method, (*ii*) Paasche's Method, and (*iii*) Bowley's Method.

	Base Year		*Current Year*	
Commodity	*Price*	*Quantity*	*Price*	*Quantity*
A	2	8	4	6
B	5	10	6	5
C	4	14	5	10
D	2	19	2	13

Chapter 14

INTERPOLATION

14.1. INTRODUCTION

Interpolation may be defined as the estimation of the most likely figure of a dependent variable from the given relavant facts. If we are given two variables x and y simultaneously and if one of them is known to be the function of the other, the one which is the function is called the dependent variable and the other one is denoted as independent. A variable is said to be the function of the other if for any values of the independent variable (say x) we can always find a definite value of the dependent variable(say y). Thus if y is the function of x as $y=x^2$, the value of y would be 25 when $x=5$. We generally use y_x to denote the general value of y: the suffix x denoting the value of y at x: (y_5 denotes the value of y when $x=5$).

If (as many times happen) exact function is not known instead we know some values of y for certain values of x, the value of y for any other x cannot be exactly determined. In a case like this we can, at best make an estimate on the basis of the data; that is available. Thus, if the different values of y for x_1, x_2, x_3 and x_5 are 16, 25, 36 and 64, we can assume that $y_x=(x+3)^2$ and from this the value of y for any given x can be estimated, *e.g.*, $y_4=49$ or $y_7=100$.

This process of estimation is known as 'Interpolation.' It is of great value in statistical work.

Whenever the method of interpolation is applied it is based on the assumption that the variable whose value is to be estimated is the function of the other variable, *i.e.*, there is some regular law connecting the two variables. one frequent assumption is that y can be expressed as a polynomial in x.

The process of extrapolation is the same as that of interpolation and the underlying assumption also is similar. The only difference between the two is that whereas intepolation refers to the estimation of a figure within the given limits of the variable x, the extrapolation denotes the estimation beyond these limits. If from the population figures for 1941, 1951, 1961, 1971 and 1981, we are to make an estimate for 1956 the technique used is called 'interpolation.' But if it is desired to estimate it for any year after 1981, the process will be termed as 'extrapolation'

14.2 GRAPHICAL METHOD OF INTERPOLATION

The graphical method is applicable in all types of data, and the rules of drawing the curve for this purpose are the same as discussed earlier in this book. The independent variable is represented on the abscissa or the axis of x, and the dependent variable is plotted on the ordinate or the axis of y. When the points have been plotted they are joined by a straight line. The line so obtained is then smoothed. The smooth curve will enable us to determine the value of y for any x within the given limits. Thus, if figures of population are available for 1921, 1931, 1941, 1951, 1961 and 1971, and it is desired to find the population for 1966, the method would be as follows:

(1) Mark years along the x axis.

(2) Represent population along the y axis; plot points and connect them.

(3) Smoothen the curve.

(4) Draw a perpendicular from point 1966 on the x axis and extend it till it cuts the smoothed curve.

(5) From the point of intersection obtained in (4) draw a line parallel to x axis and extend it to the left till it cuts the y axis.

(6) This point on the y axis (given in 5) will give us the estimate of population for 1966.

If it is desired to extrapolate for the year 75 the procedure would be as under:

(1) Extend the smoothed curve to the required point, and

(2) Adopt the same procedure as is described in the case of interpolation.

The graphical method is simple and it gives a broad idea of the relationship. But it requires graphic skill and the results given by this may vary from individual to individual.

14.3. PARABOLIC CURVE METHOD

This is an algebraic method in which a parabolic curve is algebraically fitted to the data. The equation of this parabolic curve is then used as the interpolating function. The general form of the parabolic curve is

$$y=a+bx+cx^2+dx^3+...+nx^n \quad (14.1)$$

where x is the independent variable, y the dependent variable and a, b, c, d, n are n constants to be determined. The n chosen must be equal to the number of data points available. These data points then provide n simulaneous equations in a, b, c, etc. Which can be solved to obtain the interpolating function, The following illustration explains the method.

Illustration 14.1. The following are the sales of a retail store in Delhi. It is required to interpolate the sales for 1969·

Year	*Sales (in thousands)*
1967	200
1968	240
1969	?
1970	350
1971	400

Solution. Since four values of the dependent variable are known, we would take the curve of four constants or (N—1) th order (4—1), *i.e.*, of 3rd order,

$$Y=a+bx+cx^2+dx^3 \quad ...(2)$$

Now the four known values of the dependent variable y would be sufficient to find out values of four constant a, b, c and d and consequently the sales of 1969. For convenience, we define the independent variable x as (year—1969):

x	-2	-1	0	1	2
y	200	240	y_0	350	400

where y_0 is the number to be estimated. Since all the points would be on the curve with equation (2), we substitute the above values of x in the equation and get

$$200=a-2b+4c-8d \quad ...(i)$$

$$240=a-b+c-d \quad ...(ii)$$

$$y_0=a \quad ...(iii)$$

$$350=a+b+c+d \quad ...(iv)$$

$$400=a+2b+4c+8d \quad ...(v)$$

Equation (*iii*) tells us that the value of y_0 $=a$. We have now to find out from the remaining four simultaneous equations the value of a.

Adding (*i*) and (*v*), we have

$$2a+8c=200+400=600 \quad ...(vi)$$

Adding (*ii*) and (*iv*), we get

$$2a+2c=240+350=590 \quad \ldots(vii)$$

Multiplying (*vii*) by 4, we get

$$8a+8c=2{,}360 \quad \ldots(viii)$$

Subtracting (*vi*) from (*viii*)

$$6a=1{,}760$$

$$a=y_0=293.3$$

The sales of 1969 as interpolated, therefore, are 293.3.

The above method involves the formation and solution of simultaneous equations and due to this reason it is sometimes called the method of simultaneous equations. This involves a lot of algebraic work.

14.4. NEWTON'S METHOD FOR EQUAL INTERVALS

This method is applicable when the independent variable advances by equal intervals and gives the best estimate for interpolation near the beginning of the table. The method is fairly easy in calculation and the students would be able to follow from the given examples.

Illustration 14.2. The following table shows expectation of life at different ages. You are required to find out the expectation of life at the age of 16.

Age (in years)	*Expectation of life*
10	35
15	33
20	29
25	27
30	22
35	20

Solution. Here the value of the independent variable x changes by equal intervals (for the known data points). We first construct the difference table as shown below. Snice there are six known values of the variable x, we calculate five diffirences (five *deltas*) as shown below. These five deltas then give a fifth order equation for interpolation.

Age (in years)	Expectation of life	First Δ^1	Second Δ^2	Third Δ^3	Fourth Δ^4	Fifth Δ^5
10	35 y^0					
		$-2\Delta^1_0$				
15	33 y^1		$-2\Delta^2_0$			
		$-4\Delta^1_1$		$+4\Delta^3_0$		
20	29 y^2		$+2\Delta^2_1$		$-9\Delta^4_0$	
		$-2\Delta^1_2$		$-5\Delta^3_1$		$+20\Delta^5_0$
25	27 y^3		$-3\Delta^2_2$		$+11\Delta^4_1$	
		$-5\Delta^1_3$		$+6\Delta^3_2$		
30	22 y^4		$+3\Delta^2_3$			
		$-2\Delta^1_4$				
35	20 y^5					

Each entry in the difference columns is obtained by taking the algebraic difference of the entries on the left. Thus,

$$\Delta^1_0 = y_1 - y_0 = 33 - 35 = -2,$$
$$\Delta^1_1 = y_2 - y_1 = 29 - 33 = -4,$$
$$\Delta^2_0 = \Delta^1_1 - \Delta^1_0 = -4 - (-2) = -2,$$
$$\Delta^2_1 = \Delta^1_2 - \Delta^1_1 = -2 - (-4) = +2.$$

In this manner, all the entries in this table have been calculated. The number of differences that will be required for this purpose can be found out from the last index of x, *i.e.*, x_5.

The following is the Newton's formula:

$$Y_x = y_0 + x\Delta^1_0 + \frac{x(x-1)}{1\times 2}\Delta^2_0 + \frac{x(x-1)(x-2)}{1\times 2\times 3}\Delta^3_0 + \frac{x(x-1)(x-2)(x-3)}{1\times 2\times 3\times 4}\Delta^4_0 + \frac{x(x-1)(x-2)(x-3)(x-4)}{1\times 2\times 3\times 4\times 5}\Delta^5_0 \quad (14.3)$$

where Y_x is the figure to be interpolated. Δ's are the differences and x is calculated as follows:

$$\frac{\text{Year of interpolation—Year of origin}}{\text{Time distance between adjoining years}}$$

For the table, we find

$$y_0 = 35;\; x = \frac{16-10}{5} = \frac{6}{5}$$

$$\Delta^1_0 = -2;\; \Delta^2_0 = -2;\; \Delta^3_0 = 4;\; \Delta^4_0 = -9;\; \Delta^5_0 = 20$$

$$y_x = 35 + (-2)\times\frac{6}{5} + \frac{\left(\frac{6}{5}\right)\left(\frac{6}{5} - \frac{5}{5}\right)\times -2}{1\times 2} + \frac{\frac{6}{5}\left(\frac{6}{5} - \frac{5}{5}\right)\left(\frac{6}{5} - \frac{10}{5}\right)}{1\times 2\times 3}\times\frac{4}{1}$$

$$+\frac{\frac{6}{5}\left(\frac{6}{5}-\frac{5}{5}\right)\left(\frac{6}{5}-\frac{10}{5}\right)\left(\frac{6}{5}-\frac{15}{5}\right)}{1\times2\times3\times4}\times-9$$

$$+\frac{\left(\frac{6}{5}\right)\left(\frac{6}{5}-\frac{5}{5}\right)\left(\frac{6}{5}-\frac{10}{5}\right)\left(\frac{6}{5}-\frac{15}{5}\right)\left(\frac{6}{5}-\frac{20}{5}\right)}{1\times2\times3\times4\times5}\times20$$

$$=35-\frac{12}{5}-\frac{6}{25}-\frac{16}{125}-\frac{81}{625}-\frac{504}{3,125}$$

$$=35-2.4-.24-.13-.13-.16 = 35-3.06$$

$$=31.94 \text{ years.}$$

[**Note**. If there are five differences in the table, then the Newton's equation should be extended only up to fifth differences order; if more or less, then up to that difference,

This method can also be used for estimating values beyond the limits].

Illustration 14.3. From the following figures find the premium payable at the age of 40.

Age (in years)	20	25	30	35
Annual premium (in Rs	28	31.25	35	41

Solution

X	x	y	Δ^1	Δ^2	Δ^3
20	0	28			
			3.25		
25	1	31.25		.50	
			3.75		1.75
30	2	35		2.25	
			6		
35	3	41			

For age 40,

$x=4$

$$y_x=y_0+x\Delta^1{}_0+\frac{x(x-1)}{1\times2}\Delta^2{}_0+\frac{x(x-1)(x-2)}{1\times2\times3}\Delta^3{}_0$$

$$=28+4\times3.25+\frac{4\times3}{1\times2}\times.50+\frac{4\times3\times2}{1\times2\times3}\times1.75$$

$$=28+13+3+7=51.$$

Premium payable at age 40 = Rs 51.

There is yet another method which can be used when the value to be estimated is for the *next step beyond the given limits.* This method consists in obtaining differences in the usual manner and assuming the last differences to be constant and building up differences by addition backwards:

Δ^3_1 is taken as Rs 1.75; Δ^2_2 then should be Rs (2.25+1.75)=Rs 4.00; Δ^1_3 then would be Rs (6.00+4.00)=Rs 10.00; and Y_4=Rs (41+10)=Rs 51, as. obtained above.

Illustration 14.4. Estimate the number of persons whose income is between Rs 400 and 500 from the following figures:

Income (in Rs)	*Number of persons (in '000)*
0—200	120
200—400	145
400—600	200
600—800	250
800—1000	150

Solution. In such a problem we work with cumulative frequencies and calculate for incomes below 400 and 500. The difference of the two gives the required result.

Income up to Rs	*x*	*Number of persons*	Δ^1	Δ^2	Δ^3	Δ^4	Δ^5
0	0	0					
			120				
200	1	120		25			
			145		30		
400	2	265		55		—35	
			200		—5		—110
600	3	465		50		—145	
			250		—150		
800	4	715		—100			
			150				
1000	5	865					

For income up to Rs 500,

$$x=\frac{500-0}{200}=\frac{5}{2}$$

$$y_x=y_0+x\Delta^1_0+\frac{x(x-1)}{1\times2}\Delta^2_0+\frac{x(x-1)(x-2)}{1\times2\times3}\Delta^3_0$$

$$+\frac{x(x-1)(x-2)(x-3)}{1\times2\times3\times4}\Delta^4_0$$

$$+\frac{x(x-1)(x-2)(x-3)(x-4)}{1\times2\times3\times4\times5}\Delta^5_0$$

Substituting the values,

$$y_x = 0 + \frac{5}{2}\times\frac{120}{1} + \frac{\frac{5}{2}\left(\frac{5}{2}-\frac{2}{2}\right)}{1\times2}\times 25$$

$$+\frac{\frac{5}{2}\left(\frac{5}{2}-\frac{2}{2}\right)\left(\frac{5}{2}-\frac{4}{2}\right)}{1\times2\times3}\times 30$$

$$+\frac{\frac{5}{2}\left(\frac{5}{2}-\frac{2}{2}\right)\left(\frac{5}{2}-\frac{4}{2}\right)\left(\frac{5}{2}-\frac{6}{2}\right)}{1\times2\times3\times4}\times(-35)$$

$$+\frac{\frac{5}{2}\left(\frac{5}{2}-\frac{2}{2}\right)\left(\frac{5}{2}-\frac{4}{2}\right)\left(\frac{5}{2}-\frac{6}{2}\right)\left(\frac{5}{2}-\frac{8}{2}\right)}{1\times2\times3\times4\times5}\times(-110)$$

Simplifying, we get

$$y_x = 0+300+46.9+9.4+1.4-1.3$$

$$= 356.4 \quad \text{or } 356$$

Up to income of Rs 400 there are 265 thousand persons.

Therefore, persons with income between Rs 400 and 500 are 356 − 265 = 91 thousand.

14.5. BINOMIAL EXPANSION METHOD

This method is applicable when the independent variable moves by equal steps and value to be interpolated is for one of these steps. Before coming to the actual formula let us introduce a new symbol *E* which is used in this connection.

E of any entry represents the entry at the next higher step. Thus. $Ey_0=y_1$, $Ey_5=y_6$, etc.

Just as powers of Δ represent the order of the difference, in a similar manner *E* can also be used in powers. E^n of any entry represents the entry at *n* steps after the given entry.

Thus, $E^n\, y_0=y_n$, $E^n\, y_5=y_{n+5}$, $E^6y_0=y_6$, $E^4y_3=y_7$, etc.

We have $Ey_0=y^1$ and $\Delta y_0=y_1-y_0$

$\because \quad \Delta y_0=Ey_0-y_0$ or $\Delta y_0=(E-1)y_0$

$\because$ Symbolically, $\Delta=E-1$ (14.4)

As stated earlier, if 5 entries are given, differences up to 4th order can be calculated. Rest of the terms in the Newton's formula are not used, *i.e.*, they are assumed to be zeros. This assumption the same as if we assumed that the differences higher than the one calculated (5th and above in the above case) are all zeros. The binomial expansion method is based on this fact. If we are given n entries we will assume that the nth order difference is zero,

i.e., $\Delta^n y_0=0 \because (E-1)^n y_0=0.$

Expanding the binomial $(E-1)^n$ by binomial series, we get,

$$\left(E^n-nE^{n-1}+\frac{n(n-1)}{2!}E^{n-2}+\ldots\right)y_0=0$$

or

$$\left(E^n y_0-nE^{n-1}y_0+\frac{n(n-1)}{2!}E^{n-2}y_0+\ldots 0\right.$$

or

$$y_n-ny_{n-1}+\frac{n(n-1)}{2!}y_{n-2}+\ldots=0 \quad (14.5)$$

The illustrations given below will explain the procedure clearly.

Illustration 14.5. Compute the population of 1941 from the following table:

(in millions)

Year	*Population*
1921	253 y_0
1231	287 y_1
1941	? y_2
1951	315 y_3
1961	319 y_4

Solution. We have to find out the value of y_2.

Since four values are given we will assume $\Delta^4 y_0=0$.

$\because$ $$(E-1)^4 y_0=0$$

or $$(E^4-4E^3+6E^2-4E+1)y_0=0$$

or $$y_4-4y_3+6y_2-4y_1+y_0=0$$

Substituting the values, we have

$$319-4\times315+6y_2-4\times287+253=0$$

or $$319-1260+6y_2-1148+253=0$$

$\therefore$ $$6y_2=1836$$

$$y_2=306$$

Hence the population of 1941 = 306 millions.

Illustration 14.6. Extrapolate the cost of living index for 1972 from the following data:

Year	*Cost of Living Index*
1966	328
1967	378
1968	471
1969	478
1970	434
1971	451

Solution. Since six values are known, we have

$$\Delta^6 y_0 = 0 \text{ or } (E-1)^6 y_0 = 0$$

$\therefore$ $(E^6-6E^5+15E^4-20E^3+15E^2-6E+1)y_0=0$

or $(E^6y_0-6E^5y_0+15E^4y_0-20E^4y_0+15E^2y_0-6Ey_0+y_0=0$

or $y_6-6y_5+15y_4-20y_3+15y_2-6y_1+y_0=0$

or $y_6-2706+6510-9560+7065-2268+328=0$

or $y_6=2706-6510+9560-7065+2268-328$

or $y_6=631$

$\therefore$ 631 is the cost living index for 1972.

Illustration 14.7. Interpolate the missing figures in the following table of rice cultivation.

Year	*Acres (in millions)*
1961	76.6
1962	78.7
1963	—
1964	77.7
1965	78.7
1966	—
1967	80.6
1968	77.6
1969	78,6

Solution. Since seven values of the dependent variable are known, all the seventh order differences and higher order differences are assumed to be zero. As two values are missing we will require two equations.

$\therefore$ Let $\Delta^7 y_0=0$ and $\Delta^7 y_1=0$

or $(E-1)^7 y_0=0$ and $(E-1)^7 y_1=0$

or $(E^7-7E^6+21E^5-35E^4+35E^3-21E^2+7E-1)y_0=0$

and $(E^7-7E^6+21E^5-35E^4+35E^3-21E^2+7E-1)y_1=0$

$\therefore$ The two equations are:

$$y_7-7y_6+21y_5-35y_4+35y_3-21y_2+7y_1-y_0=0$$

and $y_8-7y_7+21y_6-35y_5+35y_4-21y_3+7y_2-y_1=0$

Substituting the values,

$$77.6-7\times 80.6+21y_5-35\times 78.7+35\times 77.7 -21y_2+7\times 78.7-76.6=0 \quad \ldots(i)$$

$$78.6-7\times 77.6+21\times 80.6-35y_5+35\times 78.7 -21\times 77.7+7y_2-78.7=0 \quad \ldots(ii)$$

Simplifying (*i*) and (*ii*) equations, we get

$$21y_5-21y_2=47.3 \quad \ldots(iii)$$

$$35y_5-7y_2=2,272.1 \quad \ldots(iv)$$

Mnltiplying (*iv*) by 3

$$105y_3-21y_2=6,816.3$$

Subtracting (*iii*) from (*v*), we have

$$84y_3=6,769$$

$$\therefore \quad y_3=80.6$$

Substituting the value of y_3 in equation (*iii*),

$$1,692.6-21y_2=47.3$$

$$\therefore \quad -21y_2=47.3-1,692.6$$

$$\therefore \quad 21y_2=1,645.3$$

$$y_2=78.3.$$

Hence the rice cultivation in 1963 and 1966 was 78.3 and 80.6 million acres respectively.

14.6. LAGRANGE INTERPOLATION FORMULA

Lagrange method can be used even when the independent variable advances by unequal amount.

The formula is as under:

$$y_x=y_0\frac{(x-x_1)(x-x_2)...(x-x_n)}{(x_0-x_1)(x_0-x_2)...(x_0-x_n)}$$

$$+y_1\frac{(x-x_0)(x-x_2)...(x-x_n)}{(x_1-x_0)(x_1-x_2)...(x_1-x_n)}$$

$$+\ldots\ldots\ldots\ldots\ldots\ldots\ldots\ldots$$

$$+y_n\frac{(x-x_0)(x-x_1)...(x-x_{n-1})}{(x_n-x_0)(x_n-x_1)...(x_n-x_{n-1})} \quad (14.6)$$

Here x is the value of independent variable corresponding to which y_x is to be interpolated. $x_0, x_1, x_2 \ldots x_n$ are the given values of the variable x and y_0, y_1, y_n are the corresponding values of variable y.

Illustration 14.8. The observed values of a function are respectively 168, 120, 72 and 63 at the four positions 3, 7, 9 and 10 of the independent variable. What is the best estimate you can give of the value of the function at the position 6 of the independent variable?

Solution. Applying Lagrange's formula, we get

Independent Variable	*Function*
3x_0	168y_0
6x	?y_x
7x_1	120y_1
9x_2	72y_2
10x_3	63y_3

$$y_x = 168\frac{(6-7)(6-9)(6-10)}{(3-7)(3-9)(3-10)} + 120\frac{(6-3)(6-9)(6-10)}{(7-3)(7-9)(7-10)} + 72\frac{(6-3)(6-7)(6-10)}{(9-3)(9-7)(9-10)} + 63\frac{(6-3)(6-7)(6-9)}{(10-3)(10-7)(10-9)}$$

or

$$y_x = 168\frac{(-1)(-3)(-4)}{(+4)(-6)(-7)} + 120\frac{(+3)(-3)(-4)}{(4)(-2)(-3)} + 72\frac{(3)(-1)(-4)}{(6)(2)(-1)} + 63\frac{(3)(-1)(-3)}{(7)(3)(1)}$$

or

$$y_x = 12 + 180 - 72 + 27 = 147$$

Thus, the value of the function for the value 6 of the independent variable is 147.

14.7. MISCELLANEOUS ILLUSTRATIONS

Illustration 14.9. Estimate the expectation of life at the age of 16 year from the following data:

Age in yrs:	10	15	20	25	30	35
Expection:	35.4	32.3	29.2	26.0	23.2	20.4

Solution. As the values are given at equal intervals, Newton's difference method is appropriate

Age	*Expectation*	Δ^1	Δ^2	Δ^3	Δ^4	Δ^5
10	35.4					
		−3.1				
15	32.3		0			
		−3.1		−0.1		
20	29.2		−0.1		+0.6	
		−3.2		+0.5		−1.5
25	26.0		+0.4		−0.9	
		−2.8		−0.4		
30	23.2		0			
		−2.8				
35	20.4					

Age 16 corresponds to $x = \frac{16-10}{5} = 1.2$

Then

$$y_x = y_0 + x\Delta_0^1 + \frac{x(x-1)}{2}\Delta_0^2 + \ldots$$

$$= 35.4 + (1.2)(-3.1) + \frac{(1.2)(0.2)}{2}(0) + \frac{(1.2)(0.2)(-0.8)}{2.3}(-0.1)$$

$$+ \frac{(1.2)(0.2)(-0.8)(-1.8)}{2.3.4}(+0.6)$$

$$+ \frac{(1.2)(0.2)(-0.8)(-1.8)(-2.8)}{2.3.4.5}(-1.5)$$

$$= 35.4 - 3.72 + 0 + 0.0032 + 0.00864 + 0.012096$$

$$= 31.70$$

This is the expection of life at age 16.

Illustration 14.10. Using binomial expansion method, estimate the average number of children born to mothers aged 30--34 years:

Age of mother	15—19	20—24	25—29	30—34	35—39	40—44
Average no. of children	0.7	2.1	3.1	?	5.7	5.8

Solution. Since we have 5 known values, we use

$$\Delta^5 y_0 = (E-1)^5 y_0 = 0$$

or $$E^5 - 5E^4 + 10E^3 - 10E^2 + 5E - 1)y_0 = 0$$

or $$E^5 y_0 - 5E^4 y_0 + 10E^3 y_0 - 10E^2 y_0 + 5Ey_0 - y_0 = 0$$

or $$y_5 - 5y_4 + 10y_3 - 10y_2 + 5y_1 - y_0 = 0$$

Puting the values

$$5.8 - 5(5.7) + 10y_3 - 10(3.1) + 5(2.1) - 0.7 = 0$$

or $$y_3 = 4.39$$

This is the estimated average no. of children of mothers aged 30-34.

Illustration 1411. Following are the values of x (independent variable) and y (dependent variable). Find the value of y when $x=2$.

x :	0	1	3	4
y :	2	5	35	97

Solution. As the values of x are not equally spaced, Lagrange formula is appropriate

$$y_x = y_0 \frac{(x-x_1)(x-x_2)\ldots(x-x_n)}{(x_0-x_1)(x_0-x_2)\ldots(x_0-x_n)}$$

$$+ y_1 \frac{(x-x_0)(x-x_2)\ldots(x-x_n)}{(x_1-x_0)(x_1-x_2)\ldots(x_1-x_n)} + \ldots$$

So

$$y_2 = 2\frac{(2-1)(2-3)(2-4)}{(0-1)(0-3)(0-4)} + 5\frac{(2-0)(2-3)(2-4)}{(1-0)(1-3)(1-4)}$$

$$+ 35\frac{(2-0)(2-1)(2-4)}{(3-0)(3-1)(3-4)} + 97\frac{(2-0)(2-1)(2-3)}{(4-0)(4-1)(4-3)}$$

$$=2\left(-\frac{1}{6}\right)+5\left(\frac{2}{3}\right)+35\left(\frac{2}{3}\right)+97\left(-\frac{1}{6}\right)$$

$$=-0.33+3.33+23.33-16.17$$

$$=10.16$$

PROBLEMS

1. Interpolate by the curve fitting method the quantity supplied for the price Rs 5 and extrapolate for Rs 7

price in Rs	2	4	6
Qty. Supplied in ton	8	15	35

(Ans. Interpolated = 23.4 ton. approx, and extrapolated value = 49.9 ton approx.

2. The following table gives the death rate per one lakh population on account of a certain ailment.

Year	1956	1958	1960
Death rate	160	175	180

Estimate by algebraic method the death rate for 1949.

(Ans. Estimate for 1949 = 28.75)

3. The following values are given in a table.

x	1	2	3	4	5
y	21600	226981	–	250047	262144

Using any suitable algebraic method find the value y for $x=3$. Also draw a graph of above points on a piece of squared paper, and form this graph find the value of y for $x=4.4$.

(Ans. Missing item corresponding to $x=3$ is 238328.)

4. A life assurance company advertises the following immediate life annuities per Rs 100 paid:

Age in years	50	60	65	70
Annuity Rs	6.25	8.30	9.90	12.10

By graphical means or otherwise estimate the corresponding values for ages 62 and 67 years.

(Ans. Interpolated annuity at 62 yrr. = 8.50 and at 67 years = 10.00)

5. Find by curve fitting method the number of persons having their income below Rs 175 from the data given below:

Income in Rs below	50	100	200	375
No. of persons 000's	2	8	12	15

(Ans. Required result is 11.7 thousands.)

6. By constructing the difference table, find the 7th term as well as the general term of the sequence 0, 0, 2, 6, 12, .., 20, ...

(Ans. $Y^n=n^2-3n+2$.)

7. For a certain dependent variable y, values corresponding to independent variable x are given in the table below. Find the values corresponding to $x=2.5$ and 5.

x	0	1	2	3	4
y	4	12	32	76	156

(Ans. $Y_5=284$, $Y_{2.5}=50.25$)

8. The expectation of life at different ages for males (All India) is shown below:

Age (in years)	20	25	30	35.	40
Expectation of Life (in years)	33.0	29.8	26.6	23.5	20.5

Use Newton's formula to estimate the expectation of life at age 32.
(Ans. 25.34 years).

9. If L_x represents the number of living persons at the age X in a life table. Find the value of L_x for the value $X=60$, given $L_{20}=512$, $L_{30}=439$, $L_{40}=346$, and $L_{50}=243$.
(Ans. Required Value = 140)

10. Exterpolate population in 1971

Year	1901	1911	1921	1931	1941	1951	1961
Population	165	167	143	126	150	175	291

11. Estimate the number of candidates who get more than 48 but not more than 50 marks from the following:

Marks up to	45	50	55	60	65
No. of Candidates	447	484	505	511	514

(Ans. No. of candidates getting marks between 48 and 50 = 13)

12. Estimate from the following data the number of workers earning Rs 34 or more but less than Rs 35.

Earning less than Rs	25	24	30	35	40
No. of workers	296	599	804	91\	966

(Ans. The required nu mber is 19 workers)

13. From the following data estimate the number of persons' earning wages between Rs 60 and 7o.

Wages in Rs	40	40-60	60-80	80-100	100-120
No. of persons in thousands	250	120	100	70	50

(Ans. Wages between Rs 60 and 70 = 53.6 thousand persons)

14. From the information provided by the following table prerare a table giving estimate of the total consumption in each vear 1941-1948.

Two-year period	1941-42	1943-44	1945-46	1947-48
Monthly consumption ,000 tonnes	2.29	2.98	3.09	3.06

15. The following figures give the number of student admitted in a Coaching School for I.A.S. in the first few years of its starting. Estimate the number of students in 1951 and 1953 and comment on the results.

Year	1948	1950	1952	1954
No. of students	50	79	102	113

16. The following table gives the average monthly production of pig iron in India over a number of years. Find by suitable interpolation the production figures for 1947.)

Year	1942	1944	1946	1948	1950	1952
Production of pig iron in, 000 tonnes	150.1	118.8	112.2	117.1	130.4	140.3

Discuss the usefulness of your figures as an estimate of the actual production 1947.

(**Ans.** The estimated figures production of pig iron in 1947 is 113.42 thousand tonnes)

17. Following table gives the values of X and Y calculated from mathematical equation $Y=2^x$. Find by interpolation the values of Y when $X=4$. Why the calculated value is different from 2^4, *i.e.*, 16?

X	1	3	5	7	9
Y	2	8	32	128	512

(**Ans.** 18.17 approx.)

18. Interpolate the value of y for $x = 1.5$ from the following data:

x	0	1	2	3
y	2	5	13	35

What will be your result if it is known that given data follows an exponential curve.

(**Ans.** Value of y for $x=1.5$ is 8.0 approx.)

19. The following table gives the normal weight of a baby during the first six months of life.

Age in months	0	2	3	5	6
Weight in lbs.	5	7	8	10	12

Estimate the weight of a baby at the age of 4 months.

(**Ans.** 8.9 lbs. approx).

20. Determine by Lagrange's formula the percentage number of criminals under 35 years.

Age in years under	25	30	40	50
%age of criminals	52.0	67.3	84.1	94.4

(**Ans.** 77.4)

21. Calculate Median from the data given below using an explicit interpolation scheme, and comment on the results.

Marks below	10	20	30	40
No. of candidates	10	40	60	100

(**Ans.** 25.8 approx.)

22. From the following data, estimate the number of persons having income between 1000 and 1500 in groups A and B:

Income below	500	500-1000	1000-2000	2000-3000	3000-4000
Group A	6000	4250	3600	1500	650
Group B	5000	4500	4800	2200	1500

(**Ans.** For Group A is 2321 approx.
For Group B is 2920 ,, .)

23. Given $\log_{10} 654 = 2.8156$, $\log_{10} 658 = 2.8182$, $\log_{10} 660 = 2.8195$, $\log_{10} 662 = 2.8209$. find $\log_{10} 656$ using two different interpolation formulae available for observations at equal intervals, say Lagrange's formula and the formula for divided difference.

(**Ans**. Interpolated value of $\log_{10} 656$ by Newton's divided difference formulas = 2.8168 approx. Interpolated value of $\log_{10} 656$ Lagrange's method = 2.8168.)

24. Following table gives the value of x (independent variable) and y (dependent variable). Calculate by Newton's divided difference method the value of y

when $x=4$.

x	0	1	3	6	7
y	4	5	13	40	53

(Ans. The value of y when $x=4$ is 20.)

25. The function 3^x gives, as it should, the values, 1, 3, 9 and 81 when x equals 0, 1, 2 and 4 respectively. Applying any method of finite difference obtain the value corresponding so $x=3$. Explain why the resulting value differ from 3^3 or 27.

26 Find the missing value of the population for the year 1961 from the following census data of a certain town using Lagrange's method.

Year	1931	1941	1951	1961	1971
Population in 000's	252	241	275	?	497

(Ans. The missing value of the population of 1961 is 358.75 thousand persons.)

27 The working class cost of living index numbers for a certain place are given below for certain years. Interpolate the missing number.

Year	1. No.	Year	1. No.
1949	320	1953	280
1950	300	1958	278
1951	?	1954	250

(Ans. Missing Index No. is 284)

28 The table below gives the profits of a concern for few years. Estimate the profits for the years 1953 and 1956.

Year	1951	1952	1953	1954	1956
Profits lakhs of Rs	8.5	12.0	?	10.0	?

29. Complete the following table with the help of any suitable method of interpolation.

X	2.0	2.1	2.2	2.3	2.4	2.5	2.6
Y	0.135	?	0.111	0.100	?	0.082	0.074

(Ans. Required value of Y when $x=2.1$ is .123 gnd when $x=2.4$ is .0904.)

30. Estimate the missing term in the following table

X	1	2	3	4	5
U	2	4	8	?	32

(Ans. Estimated $U_4=16·5$)

31. From a difference table from the following steam data where p is pressure in lbs per square inch (dependent variable) and θ is temperature in centigrade (independent variable).

θ	93.0	96.2	100.0	164.2	108.7
p	11.38	12.80	14.70	17.07	19.91

Calculate p when$=99.1$ and θ when $p=15$.
(Ans. When temperature is 99.1°C, the presure is 14.23 lbs/sq. inch : when pressure$=15$ is 100.4°C)

32. Estimate the value of Y when $X=2.5$ from the data given below:

X	0	1	2	3	4
Y	4	12	32	76	156

(Ans. The interpolated value of $Y=50.25$).

33. From the data given below estimate the number of persons earing wages between rupees 60 and 70.

Wages below	40	60	80	100	120
No. of persons in thousands	250	370	470	540	590

(Ans. No. of persons earning wages between 60 to 70=73.59 thousands.)

34. The following table shows the value of an immediate life annuity for every Rs. 100 paid

Age in years	40	50	60	70
Annuity Rs	6.2	5.7	9.2	12.0

Interpolate for age 42.
(Ans. Rs. 6.31)

35. Find expectation of age at 16 from the following figures:

Age in years	10	15	20	25	30	35
Expectation of life	35.4	32.3	29.2	26.0	23.2	20.4

(Ans. 31.7 years)

36. Find the net annual premium at age 25 from the table given below

Age	20	24	28	32
Premium	.01427	.01581	.01772	.01996

(Ans. .016255)

37. From the following temperature readings taken on a particular day, estimate the temperature at 9 A.M. and 5 P.M.

Time	4 A.M.	8 A.M.	12 Noon	4 P.M
Temp°C	42.2	44.4	53.0	46.6

38. The observed values of a function are respectively 168, 120, 72, 63 at four values of x : 3, 7, 9, and 10. Estimate the value for x = 6.
(Ans. 147)

39. The table below gives the profits of a concern for a few years. Estimate the profits for the years 1953 and 1956.

Year	1951	1952	1953	1954	1956
Profits (in lakhs of Rs)	8.5	12.0	?	10.0	?

(Ans. Profits of 1953 is 12.5 lakhs of rupees. Estimated profits for year 1956 is—4.0 lakhs of rupees.

40. Calculate by two methods possible, say Lagrange's method and binomial expansion method the missing figure when $X=4$.

X	0	2	4	6
Y	11	15	?	29

(Bns. Binomal expansion method: Missing figure of Y for $X=4$ is 21 and Lagrange's method: Missing figure of $X=4$ is 21.)

Chapter 15

RELATIONSHIP BETWEEN VARIABLES — REGRESSION AND CORRELATION

15.1. RELATIONSHIP BETWEEN VARIABLES

The statistical methods discussed so far have been concerned with only one single variable. There are, however, many situations where we are interested in the relationship between two or more variables occurring together. For example, we may be interested in studying the effect of fertiliser consumption on the yield of wheat, or the effect of the strength of the El Nino in South Atlantic on the intensity of the monsoons in the north Indian plains. The variables are said to be *correlated* if a relationship exists between the two. We will introduce in this chapter some statistical concepts and techniques which are useful in analyzing the relationship between such multiple variables.

In such cases we start with some data in the form of a set of paired values, each of the pair representing the corresponding values of the two variables. Consider, as a simple example, the data in Table 15.1, which gives the distribution of marks obtained by 7 students in some tests in Physics and Chemistry.

TABLE 15.1

Student	*Marks in*	
	Physics	*Chemistry*
A	82	71
B	68	63
C	76	61
D	39	45
E	54	48
F	75	72
G	47	42

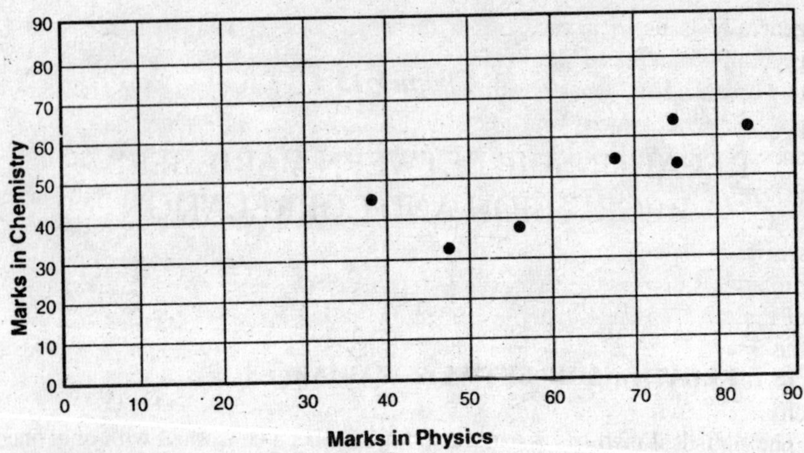

Fig. 15.1 Data of Table 15.1

Even a cursory look at this data reveals that generally a student who scores high in one test, does so in the other test as well. But this in not an iron-clad relationship: for example, student C scores higher than student B in Physics but lower than her in Chemistry. If we plot the seven pairs of marks on a graph with one axis representing the marks in one subject and the other axis the marks in the second subject, we obtain a plot as shown in Fig. 15.1. Each of the seven points in this graph represents the marks of a specific student in the two tests. If all the plotted points lay along a straight line we can immediately conclude that the two variables are firmly related. But most often we get a plot as in Fig. 15.1, where the points are scattered about in what can be imagined as a narrow or a wide band. It is for this reason that such a graph is termed as a *scatter* diagram.

It is obvious that the closer the various points are to a straight line, i.e., the narrower is the band within which the various points lie, the more definitive is the relationship between the two variables, and the more they are scattered about, the weaker is this relationship. The measure of the magnitude of scatter can thus be used as a measure of the strength of correlation between the two variables. We shall develop some appropriate measure for the same.

A correlation is termed as a positive correlation when an increase in the value of one variable leads generally to an increase in the value of the other variable as well. Such a situation is evidenced on a scatter diagram with the band of values running generally from the bottom left-hand corner to the top right-hand corner, i.e., having a positive slope. On the other hand, the variables are said to be negatively correlated if the increase in one generally leads to a decrease in the other, or the band of values runs generally from the top left-hand corner to the bottom right-hand corner. In this case the band has a negative slope. Thus, the variables of Fig. 15.1 are positively correlated, and the variables in Fig. 15.2a are negatively correlated, while that in Fig. 15.2b are not correlated at all since the points in it are well scattered with no trend discernible.

Besides measuring the strength of the relationship between the variables wherever it exists, we are interested in also developing some tools or procedures to *predict* the most likely value of one variable given the value of the other variable. This is termed as the *regression* of one variable on the other. The straight line about which the various points may be considered as scattered in called the regression line, and later in this chapter we will explore the procedures for obtaining such lines which best approximate the relationship between the variables.

It should be clear that one can predict exactly only if the two variables are perfectly related. In that case, there is no scatter in the data, and the various points lie exactly on the regression line. But when the correlation is less than perfect, i.e., there is a scatter of points on the scatter diagram, then the regression line in only a representation of the general trend. It should be clear that one can draw any number of different straight lines on Fig. 15.1 to represent the general trend. We, therefore, have to develop a method for defining the 'best' line, and then a procedure for constructing it.

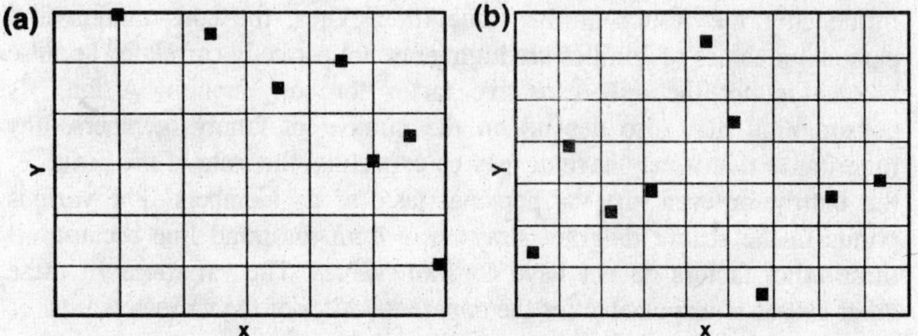

Fig. 15.2 (a) Negative correlated variables (b) No apparent correlation

15.2. CORRELATION AND CAUSATION

The presence of correlation between two variables does not necesarily mean that there is a cause and effect relationship between the two. In the data of Table 15.1, the marks in the two subjects are correlated, but it can in no way be said that a student gets high marks in Chemistry *because* he gets high marks in Physics, or *vice versa.* In fact, a correlation implies only that the two variables move together in the same (or the opposite) direction(s). Two variables can be correlated because one is the cause of the other, or because both are the results of some other factor. Thus, one may find a very high level of correlation between the vocabulary and the shoe-size of children in an elementary school. Clearly, neither of the two variables can be a causative factor for the other. It is simply that both are highly dependent on a third variable, the age of the child. Even the correlation of the data in Table 15.1 can be explained through the most likely common cause, the intelligence or the diligence of the students involved. Although a causal relationship implies a correlation, even a very high correlation cannot in itself imply a causation.

When one of the variables is the direct cause of the other, it is important to recognize which is the direct cause of which. The yield of wheat in north American prairie is seen to be directly and heavily correlated with the number of sunshine hours in the growing season. The sunshine-hours is clearly the cause, i.e., the independent variable, and the yield is the effect or the dependent variable. No increase in yield is ever going to increase the sunshine hours !

One may rightly wonder why the correlation is not perfect (i.e., the 'scatter' in a scatter diagram is not completely absent) when one variable is the cause of the other. A lack of perfect correlation between variables is most often the result of many explanatory or causative variables influencing the results at the same time. Thus, the consumption and personal incomes of families are highly but not perfectly correlated because income is not the only causative factor for consumption. A family's consumption may also depend on the number of family members, any inheritance that it may have or may be expecting, the state of the health of the family, or even just the personal taste of its members. The various points in the scatter diagram stray away from the trend line because all these other factors do not have constant values. The variations in these other causes is responsible for the apparent scatter of the various points.

15.3. LINEAR REGRESSION ANALYSIS

As stated above, regression is the process by which we obtain a functional relationship between the two variables under consideration. The most commonly used regression line[1] is a straight line whose equation is

---

[1]The term regression line was first used by Sir Francis Galton in describing his findings of a study of hereditary characteristics. He found that the height of descendency has a tendency to 'regress' (i.e., to return) to the average for the race. He, therefore, termed the line of average relationship as the regression line. The usage has stuck and the term is used even where it has no literal relevance.

$$Y = a + b\,x \tag{15.1}$$

where Y and x are the dependent and independent variables, a and b are constants. This equation with appropriate values of a and b may be used to *predict* the values of the dependent variable given a value of the independent variable x. We have advisedly used here the capital letter Y to distinguish it as a calculated or the predicted values of the variable whose actual values are denoted by the lower letter y. The regression problem then is to find the values of a and b such that the line given by Eq. 15.1 describes the 'average relationship' between the two variables. Once we know the values of a and b, we have the equation which we can use to predict the value Y of the dependent variable for any given value of x. The actual value of y that obtains may or may not be equal to the value predicted by the regression equation. It should be clear that the actual values of y and the expected values of Y are more likely to be closer together when the correlation of the two variables is high (positive or negative).

When we talk of an average relationship we are not being exact enough. It is clear that we can draw many lines through the points in a scatter diagram like Fig. 15.1. Which one of those should we designate as the regression line? One way to resolve this issue is to consider the deviations of the actual data y from that predicted by the regression line. Consider the data of Table 15.2 which has been plotted in the scatter diagram of Fig. 15.3. The nature of this scatter diagram suggests that the two variables are correlated, but the correlation is far from perfect. It is because of this reason that one is not sure which of the infinity of possible lines passing through the scatter may be used to represent the *average* relationship envisaged by the term *regression line*.

To answer this question in a rational manner, let us use the concept of *deviations* defined as the differences between the actual values y of the dependent variable from the values Y predicted by a given equation for identical values of x. One obvious criterion of a good regression is that the sum of the deviations for all given points, i.e., $\Sigma\ (y-Y) = 0$. A little arithmetic will show that this translates to the condition that the line should pass through the centroid of the data, i.e., through the point (x, y). But this condition does not give a unique line. In fact, it can be verified that this condition is met for each one of the eight equations proposed for the data of Table 15.2.

We, therefore, have to look for improvement. One obvious improvement is to consider absolute values of the deviations i.e., to ignore the positive and negative signs and to just add up the magnitudes. It stands to reason that a cancellation of one large positive deviation with another large but negative deviation does not make for a good regression. Instead, we should attempt to minimize the sum of the absolute values of deviations Σ (y–Y).

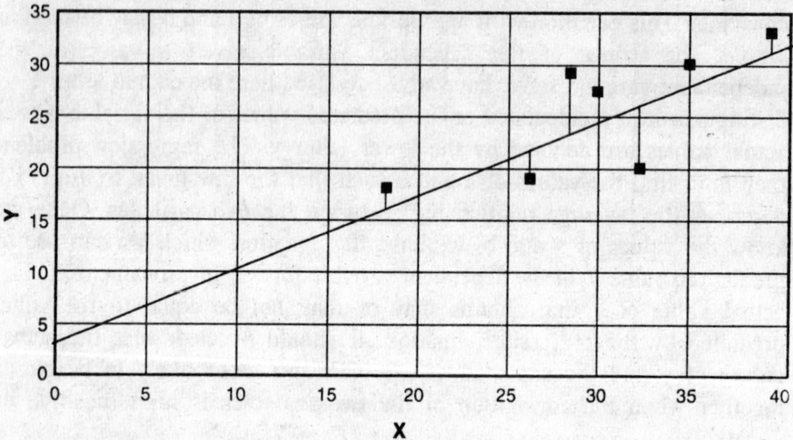

Fig. 15.3 Data of Table 15.2. The length of vertical lines is the deviations of the actual values from the values predicted by the trend line

In Table 15.2, we have compared the sum of the absolute deviations with respect to a few lines. The following lines have been used:

$$Y = 10 + 0.5x \quad (15.2)$$

$$Y = 7 + 6x \quad (15.3)$$

$$Y = 4 + 0.7x \quad (15.4)$$

$$Y = 2.5 + 0.75x \quad (15.5)$$

$$Y = 1 + 0.8x \quad (15.6)$$

$$Y = 5 + x \quad (15.7)$$

$$Y = -20 + 1.5x \quad (15.8)$$

$$Y = -35 + 2.0x \quad (15.9)$$

Notice that each of the equations above is of the form of Eq. 15.1, where b, the coefficient of the independent variable x, represents the slope of the equation in the x-y plane. A positive value of b denotes a positive slope or the fact that as x increases so does y. The opposite is the case with a negative value of b. The more is the value of b, faster does y change with x. Also note that a positive value of b corresponds to a positive correlation and a negative value of b to a negative correlation. But a higher value of b does *not* signify a higher correlation, positive or negative. The value of a denotes the intercept the line makes at the y axis (where the value of x is zero). Since each of the above lines passes through a fixed point (x, y), an increasing value of slope b results in a decreasing value of the intercept a.

Notice also that in each of above eight lines the algebraic sum of the deviations of the actual values y from the corresponding predicted values Y

TABLE 15.2

		line 15.2		*line 15.3*		*line 15.4*		*line 15.5*		*line 15.6*		*lime 15.7*		*line 15.8*		*line 15.9*	
x	*y*	*Y*	*d*	*Y*	*d*	*Y*	*d*	*Y*	*d*	*Y*	*d*	*Y*	*d*	*Y*	*d*	*Y*	*d*
30	27	25	2	25	2	25	2	25	2	25	2	25	2	25	2	25	2
35	30	27.5	2.5	28	2	28.5	1.5	28.75	1.25	29	1	30	0	32.5	–2.5	35	–5
19	18	19.5	–1.5	18.4	–0.4	17.3	0.7	16.75	1.25	16.2	1.8	14	4	8.5	9.5	3	15
39	32	29.5	2.5	30.4	1.6	31.3	0.7	31.75	0.25	32.2	–0.2	34	–2	38.5	–6.5	43	–11
26	19	23	–4	22.6	–3.6	22.2	–3.2	22	–3	21.8	–2.8	21	–2	19	0	17	2
29	29	24.5	4.5	24.4	+4.6	24.3	+4.7	24.25	4.75	24.2	4.8	24	5	23.5	5.5	23	6
32	20	26	–6	26.2	–6.2	26.4	–6.4	26.5	–6.5	26.6	–6.6	27	–7	28	–8	29	–9
Σd			0		0		0		0		0		0		0		0
Σ|d|			23		20.4		19.2		19		19.2		22		34		60
Σd^2			91		83.28		80.5		81.0		82.72		102		237		496

is exactly zero, but the sum of the absolute values of deviation changes. While it is as high as 60 in the case of Eq. 15.9, it is as low as 19 for Eq. 15.5. Clearly, Eq. 15.5 is a better regression equation for this data than Eq. 15.9.

Further compare the predictions of Eq. 15.7 with those of Eq. 15.2. Though the sum of absolute deviations is less in the case of Eq. 15.7 (22 against 23 for Eq. 15.2), the latter equation is definitely a better fit because the magnitudes of individual deviations are lower in that case compared to those of the other. We, therefore, devise a criterion which weights against large values of absolute deviations. Such a criterion is provided by the *method of least squares,* wherein we select that line as the line of regression the *sum of the squared deviations* from which has the minimum value. Thus we minimize

$$\Sigma\,(y - Y)^2$$

Using very simple calculus one can show that the coefficient a and b of the least square regression line are given by the following equations:

$$\Sigma\, y = Na + b\Sigma x \tag{15.10}$$

$$\Sigma xy = a\Sigma x + b\Sigma x^2 \tag{15.11}$$

where the summations extends over all of the data considered and N denotes the total number of cases included. These two equations are known as the *normal equations.* Simplification of the above yields,

$$b = \frac{\Sigma\, xy - \dfrac{(\Sigma\, x)\,(\Sigma\, y)}{N}}{\Sigma\, x^2 - \dfrac{(\Sigma\, x)^2}{N}} \tag{15.12}$$

and,

$$a = \frac{\Sigma\, y}{N} - b\frac{\Sigma\, x}{N} \tag{15.13}$$

If we use the fact that the regression line passes through (x, y), we can show easily that

$$b = \frac{\dfrac{\Sigma xy}{N} - \overline{x}\,\overline{y}}{\dfrac{\Sigma x^2}{N} - \overline{x}^2} \tag{15.14}$$

$$a = \overline{y} - b\overline{x} \tag{15.15}$$

The denominator in Eq. 15.14 is recognised as the variance of the variable x. The numerator is defined as the *covariance* of the variables x and y, so that Eq. 15.14 can be written as

$$b = \frac{\text{cov}\,(x, y)}{\text{var}\,x} \tag{15.16}$$

In the above example Eq. 15.4 is the line of least squared deviations, and is therefore the fit candidate to be designated as the *regression equation for y on x*.

Since a causal relationship is not necessary for regression analysis we can attempt to predict the value of x for a given value of y. This is termed as the *regression of x on y,* and the corresponding equation is

$$x = a + b\,y \tag{15.17}$$

Here we have used the subscript xy with b to indicate that it represents the slope of the regression line for x on y. Similarly, a subscript yx can be used with b in the equation for regression for y and x. The determining equation for b_{xy} can easily be written as

$$b_{xy} = \frac{\text{cov}\,(x, y)}{\text{var}\,y} \tag{15.18}$$

Illustration 15.1. Given the bivariate data:

X :	1	5	3	2	1	1	7	3
Y :	6	1	0	0	1	2	1	5

(*a*) Fit a regression line of y on x and hence predict Y if $x = 10$.
(*b*) Fit a regression line of x on y and hence predict X if $y = 2.5$.

Solution.

x	y	x^2	y^2	xy
1	6	1	36	6
5	1	25	1	5
3	0	9	0	0
2	0	4	0	0
1	1	1	1	1
1	2	1	4	2
7	1	49	1	7
3	5	9	25	15
23	16	99	68	36

(*a*) Regression equation of y on x is

$$Y = a + bx$$

The values of a and b are given by normal equations

$$\Sigma y = \text{N}a + b\Sigma x$$

$$\Sigma xy = a\Sigma x + b\Sigma x^2$$

Substituting the values from the table, we have

$$16 = 8a + 23b \quad \text{...}(i)$$

$$36 = 23a + 99b \quad \text{...}(ii)$$

Solving (*i*) and (*ii*), we get

$$a = 2.874$$
$$b = -0.304$$

$\therefore$ $Y = 2.874 - 0.304x$ is the regression equation of *y* on *x*.

when $x = 10$, Y is given by

$Y = 2.874 - 0.304 \times 10 = -0.166$

Hence the predicted value of Y is -0.17 for $x = 10$.

(*b*) Regression of *x* on *y* is

$$X = C + b_{xy}\, y$$

The values of *c* and b_{xy} are given by the normal equation

$$\Sigma x = Nc + b_{xy}\, \Sigma y$$
$$\Sigma xy = c\Sigma y + b_{xy}\, \Sigma y^2.$$

Substituting the values from table, we get

$$23 = 8c + 16b_{xy} \qquad \ldots (iii)$$
$$36 = 16c + 68b_{xy} \qquad \ldots (iv)$$

Solving (*iii*) and (*iv*), we get

$$c = 3.431 \text{ and } b_{xy} = -0.278$$

$X = 3.431 - 0.278y$ is regressing equation of *x* on *y*

If $y = 2.5$, X is given by

$X = 3.431 - 0.278 \text{ X } 2.5 = 2.736$

Hence the value of X if 2.74 for Y = 2.5

Illustration 15.2. From a sample of 200 pairs of observations the following quantities were calculated.

$\Sigma x = 11.34$ $\qquad \Sigma y = 20.72$ $\qquad \Sigma y^2 = 12.16$

$\Sigma y^2 = 84.96$ $\qquad \Sigma xy = 22.13$

From the above data show how to compute the coefficients of the equation. $\quad y = a + bx$.

Solution. We can compute the coefficients of the equation

$Y = a + bx$ by solving normal equations:

$\Sigma y = na + b\Sigma x$ and $\Sigma xy = a\Sigma x + b\Sigma x^2$

Putting the values

$20.72 = 200a + 11.34b$ and $22.13 = 11.34a + 12.16b$.

or $\quad a = \dfrac{20.72 - 11.34b}{200}$

or $\quad a = .1036 - .0567b$ and $22.13 = 11.34\,(.1036 - .0567b) + 12.16b$

or $\quad 22.13 = 1.175 - .643b + 12.16b$.

or $\quad 20.955 = 11.517b$

or $\quad b = \dfrac{20.955}{11.517} = 1.82$

Then $a = .1036 - .0566\,(1.82) = 0.00059$

Thus, the regression equation of *y* on *x* is

$Y = 0.00059 + 1.82x$

15.4. SIMPLIFIED DETERMINATION OF REGRESSION ANALYSIS

As has been stated before, all regression lines pass through the centroid of the data $(\bar{x}, \bar{y})$, and therefore the equation of the regression line can be written as

$$Y - \bar{y} = b_{yx}(x - \bar{x}) \tag{15.19}$$

for regression of y on x, and

$$X - \bar{x} = b_{xy}(y - \bar{y}) \tag{15.20}$$

for regression of x on y.

These form lead to the following

Simplified equations

$$b_{yx} = \frac{\dfrac{\Sigma(x-\bar{x})(y-\bar{y})}{N}}{\dfrac{\Sigma(x-\bar{x})^2}{N}} \tag{15.21}$$

Note that the numerator and denominator in Eq. 15.21 are nothing but the basic definition of covariance (x, y) and variance x, respectively.

Following the same approach as was used in the short-cut method, we define deviations d_x and d_y as

$$d_x = (x - A_x), \text{ and } d_y = (y - A_y)$$

where A_x and A_y are appropriate 'middle values' which simplify calculations.

Then

$$\text{var}(x) = \frac{\Sigma d_x^2}{N} - \left(\frac{\Sigma d_x}{N}\right)^2 \tag{15.22}$$

and

$$\text{cov}(x, y) = \frac{\Sigma d_x d_y}{N} - \left(\frac{\Sigma d_x}{N}\right)\left(\frac{\Sigma d_y}{N}\right) \tag{15.23}$$

So that

$$b_{yx} = \frac{\dfrac{\Sigma d_x d_y}{N} - \left(\dfrac{\Sigma d_x}{N}\right)\left(\dfrac{\Sigma d_y}{N}\right)}{\dfrac{\Sigma d_x^2}{N} - \left(\dfrac{\Sigma d_x}{N}\right)^2} \tag{15.24}$$

When A_x and A_y are the actual means $\bar{x}$ and $\bar{y}$, $\Sigma d_x = \Sigma d_y = 0$, and Eq. 15.24 reduces to Eq (15.21). Further, when $A_x = A_y = 0$, $d_x = x$ and $d_y = y$, and Eq. 15.24 reduces to the original expression for b_{yx} given by Eq. 15.14.

A equation similar to Eq. 15.24 can be written for b_{xy}..

Illustration 15.3. In the following table S is weight of Potassium bromide which will dissolve 100 gm of water at T °C. Fit an equation of the form $S = mT + b$ by the method of least squares. Use this relation to estimate S when $T = 50°$

T	0	20	40	60	80
S	54	65	75	85	95

Solution. Here the values of T and S are pretty large. To reduce the calculation effort required, we define new variables x and y as

$$x = (T–40)/10 \text{ and } y = (S–75)/10$$

and reconvert into T and S at the end.

T	$x = \dfrac{T-40}{10}$	x^2	S	$y - \dfrac{S-75}{10}$	y^2	xy
0	–4	16	54	–2.1	4.41	8.4
20	–2	4	65	–1.0	1.00	2.0
40	0	0	75	0	0.00	0.0
60	+2	4	85	+1.0	1.00	2.0
80	+4	16	96	+2.1	4.41	8.4
200	0	40	375	0	10.82	20.8

The two normal equations are

$$\Sigma y = Na + b\Sigma x; \text{ and } \Sigma xy = a\Sigma x + b\Sigma x^2$$

So $\quad 0 = 5a + b.0$

and $\quad 20.8 = a.0 + 40b$

Which give $a = 0$ and $b = \dfrac{20.8}{40} = 0.52$

So the regression equation is

$$Y = 0.52x$$

or

$$\frac{S-75}{10} = 0.52\left(\frac{T-40}{10}\right)$$

which simplify to

$$S = 0.52T + 54.2$$

For $T = 50°$, $S = 0.52 \text{ X } 50 + 54.2 = 80.2$

Illustration 15.4. A panel of two judges P and Q graded seven dramatic performances by independently awarding marks as follows :

Performance	1	2	3	4	5	6	7
Marks by P	46	42	44	40	43	41	45
Marks by Q	40	38	36	35	39	37	41

The eight performance, which judge Q would not attend, was awarded 37 marks by judge P. If judge Q had also been present, how many marks would be expected to have been awarded by him to the eighth performance.

Solution. We first calculate the means.

Mean marks by *P* (*x* variable) = $\bar{x}$ = 301/7 = 43

and mean marks by *Q* (*y* variable) = $\bar{y}$ = 266/7 = 38.

To calculate b_{yx} we use formula (15.21)

No.	x	$(x-\bar{x})$	$(x-\bar{x})^2$	y	$(y-\bar{y})$	$(x-\bar{x})(y-\bar{y})$
1	46	+3	9	40	+2	+6
2	42	–1	1	38	0	0
3	44	+1	1	36	–2	–2
4	40	–3	9	35	–3	+9
5	43	0	0	39	+1	0
6	41	–2	4	37	–1	+2
7	45	+2	4	41	+3	+6
	301		28	266		21

So $$b_{yx} = \frac{\Sigma(x-\bar{x})(y-\bar{y})}{\Sigma(x-\bar{x})^2} = \frac{21}{28} = 0.75$$

So $Y-38 = 0.75\ (x-43).$

Simplifying, $Y = 0.75x + 5.75$

If $x = 37$, them

$Y = 0.75\ (37) + 5.75 = 33.5$

So the judge *Q* was *likely* to award 33.5 marks.

Illustration 15.5. For 10 observations on price (*x*) and supply (*y*) the following data was obtained (in appropriate units) :

$$\Sigma x = 130,\ \Sigma y = 220,\ \Sigma x^2 = 2288,\ \Sigma y^2 = 5506,\ \Sigma xy = 3467.$$

Obtain the line of regression of *y* on *x* and estimate the supply when the price is 16 units.

Solution. The regression of *y* on *x* is

$$Y - \bar{y} = b_{yx}(x - \bar{x})$$

Now

$$\bar{x} = \frac{\Sigma x}{N} = \frac{130}{10} = 13$$

$$\bar{y} = \frac{\Sigma y}{N} = \frac{220}{10} = 22$$

and by Formula (15.14)

$$b_{yx} = \frac{\Sigma xy - \frac{(\Sigma x)(\Sigma y)}{N}}{\Sigma x^2 - \frac{(\Sigma x)^2}{N}} = \frac{3467 - \frac{130 \times 220}{10}}{2288 - \frac{(130)^2}{10}}$$

$$= 1.02$$

$\therefore$ $Y - 22 = 1.20\,(x - 13)$

or $Y = 1.20x + 8.74$

When $x = 16$, Y is given by

$Y = 1.20 X 16 + 8.74 = 25.06$

Hence the supply is expected to be 25 units corresponding to the price 16 units.

Illustration 15.6. On the basis of figures recorded below for "supply" and "price" for nine years, build a regression of "price" on "supply". Calculate, from the equation established, the most likely price when supply = 90.

Years :	1981	1982	1983	1984	1985	1986	1987	1988	1989
Supply :	80	82	86	91	83	85	89	96	93
Price :	145	140	130	124	133	127	120	110	116

Solution. Let X represent supply and Y represent price.

Year	*Supply* x	$x' = x - 90$	x'^2	*Price* y	$y' = y - 127$	y'^2	$x'y'$
1981	80	−10	100	145	+18	324	180
1982	82	−8	64	140	+13	169	−104
1983	86	−4	16	130	+3	9	−12
1984	91	+1	1	124	−3	9	−3
1985	83	−7	49	133	+6	36	−42
1986	85	−5	25	127	0	0	0
1987	89	−1	1	120	−7	49	+7
1988	96	+6	36	110	−17	289	−102
1989	93	+3	9	116	−11	121	−33
$N = 9$	785	−25	301	1145	+2	1006	−469

Regression equation of y on x is

$$Y - \bar{y} = b_{yx}\,(x - \bar{x})$$

$$\text{where } \bar{x} = 90 + \frac{\Sigma x'}{N} = 90 - \frac{25}{9} = 90 - 2.8 = 87.2, \text{ and}$$

$$\bar{y} = 127 + \frac{\Sigma y'}{N} = 127 + \frac{2}{9} = 127 + 0.2 = 127.2$$

$$\text{and } b_{yx} = \frac{\Sigma x'y' - \dfrac{(\Sigma x')(\Sigma y')}{N}}{\Sigma x'^2 - \dfrac{(\Sigma x')^2}{N}} = \frac{-469 - \dfrac{(-25)(+2)}{9}}{301 - \dfrac{(-25)^2}{9}}$$

$$= \frac{-469 + 5.56}{301 - 69.44} = \frac{-463.44}{231.56} = -2.00$$

$\therefore$ $\qquad Y - 127.2 = -2.00\,(x - 87.2)$

or $\qquad Y = 2x + 301.6$

when $x = 90$, Y is given by

$$Y = -2 \times 90 + 301.6 = 121.6$$

Hence the most likely price is 121.6 corresponding to the supply of 90.

5.5. REGRESSION ANALYSIS OF GROUPED DATA

The regression analysis of grouped data follows the same procedure as above,. except the fact that all 'items' falling within a specified group are approximated as having a value equal to the mid-point value of the group. Since the grouping may be on both variables x and y, the data is usually organized in a 2 way matrix. The following example illustrates the method. The following formulae may be used for the value of b_{yx}

$$b_{yx} = \frac{\dfrac{\Sigma fxy}{N} - \dfrac{\Sigma fx}{N} \cdot \dfrac{\Sigma fy}{N}}{\dfrac{\Sigma fx^2}{N} - \left(\dfrac{\Sigma fx}{N}\right)^2} \tag{15.25}$$

$$= \frac{\dfrac{\Sigma fd_x d_y}{N} - \dfrac{\Sigma fd_x}{N} \dfrac{\Sigma fd_y}{N}}{\dfrac{\Sigma fd_x^2}{N} - \left(\dfrac{\Sigma fd_x}{N}\right)^2} \tag{15.26}$$

where the f associated with each term has the appropriate meaning. Thus, in Eq. 15.25, the f with the first term in the numerator is the frequency or the count of those items which have their values in the specified group for the term variables x and y, while in the second term, the f in Σfx is the count of items with values in the appropriate group for x and *all* values of y. In formal terms, this f is the marginal frequency of f. Similarly, f in Σfy is the frequency for the specified group for y and *all* values of x.

Illustration 15.7 Following is the distribution of students according to their height and weight.

Weight/Height	*90-100*	*100-110*	*110-120*	*120-130*
50—55	4	7	5	2
55—60	6	10	7	4
60—65	6	12	10	7
65—70	3	8	6	3

Calculate *(i)* the two coefficients of regression and *(ii)* obtain the two regression equations.

Solution. Product fdxdy is on top right of each cell.

Weight in lbs/ Height in inches	*m*	*dy/ dx*	*90-100* 95 —2	*100-110* 105 —1	*110-120* 15 0	*120-130* 125 1	*f*	*fdx*	*fdx²*	*fdxdy*
			[8]	[7]	[0]	[-2]				
50-55	52.5	–1	4	7	5	2	18	–18	18	13
			[0]	[0]	[0]	[0]				
55-60	57.5	0	6	10	7	4	27	0	0	0
			[–12]	[–12]	[0]	[7]				
60-65	62.5	1	6	12	10	7	35	35	35	–17
			[–12]	[–16]	[0]	[6]				
65-70	67.5	2	3	8	6	3	20	40	80	–22
		f	19	37	28	16	100	57	133	–26
		fdy	–38	–37	0	16	–59			
		fdy²	76	37	0	16	129			
		fdxdy	–16	–21	0	11	–26			

Regression equation of *y* on *x*

$$\bar{y} = A_y + \frac{\Sigma fdy}{N} \times i_y = 115 - \frac{59}{100} \times 10 = 109.1$$

$$\bar{x} = A_z + \frac{\Sigma fdx}{N} \times i_x = 57.5 + \frac{57}{100} \times 5 = 60.35$$

$$b_{yx} = \frac{\Sigma fdxdy - \dfrac{\Sigma fdx \times \Sigma fdy}{N}}{\Sigma fdx^2 - \dfrac{(\Sigma fdx)^2}{N}} \times \frac{i_y}{i_x}$$

$$= \frac{-26 - \dfrac{(-59)(57)}{100}}{133 - \dfrac{(57)^2}{100}} \times \frac{10}{5} = 0.145$$

Therefore $Y - 109.1 = 0.145\,(x - 60.35)$

$X = 0.145x + 10.35.$

Regression equation of *X* on *Y*

$$b_{xy} = \frac{\Sigma fdxdy - \dfrac{\Sigma fdx.\Sigma fdy}{N}}{\Sigma fdy^2 - \dfrac{(\Sigma fdy)^2}{N}} \times \frac{i_x}{i_y}$$

$$= \frac{-26 - \dfrac{(-59)(57)}{100}}{129 - \dfrac{(-59)^2}{100}} \times \frac{5}{10} = .044$$

$$X - 60.35 = 0.44\,(Y - 109.1) = .044Y - 4.8$$

$$X = 0.44Y + 55.55$$

Note that in equation marked with (*), a factor i_y/i_x or i_x/i_y have been used because of the fact that *dx* and *dy* need to be multiplied by i_x and i_y, respectively, to obtain the variables *x* and *y*.

15.6. CORRELATION ANALYSIS

Correlation analysis is the mathematical tool that is used to describe the *degree* to which one variable is *linearly* related to the other. It, therefore, is directed towards measuring the *degree* of *association* of the two variables. All measures of correlation are defined in such a fashion that a measures of zero signifies no correlation at all, and the perfect correlation (a mathematically linear relation) is indicated by a magnitude of one, positive for a direct linear relationship, and negative for an inverse linear relationship.

Most measures of correlation are based on finding the sum of squares of the deviations from the regression line and relating it to the inherent variability of the data itself. Thus, one measure, termed as the *coefficient of determination* is defined as

$$r^2 = 1 - \frac{\text{variations of } y \text{ values from the regression line}}{\text{variations of } y \text{ values from their own mean}}$$

or,

$$r^2 = 1 - \frac{\Sigma(y-Y)^2}{\Sigma(y-\bar{y})^2} \qquad (15.27)$$

This is based on treating the variations of the sample values from the regression-predicted values $\Sigma(y-Y)^2$ (refer to Fig. 15) as random or the *unexplaind variations*, so that part $\Sigma(y-\bar{y})^2 - \Sigma(y-Y)^2$ can be taken to be the variation which can be *explained* by the observed relationship. Its ratio with the total variation $\Sigma(y-\bar{y})^2$ is then a natural measure of the strength of correlation of the two variables as revealed by the data.

It can be shown quite easily that on substituting the value of Y by using the regression equation this reduces to

$$r^2 = \frac{\text{cov}^2(x, y)}{\text{var}(x) \,.\, \text{var}(y)} \qquad (15.28)$$

The square root of the coefficient of determination, or the value of r itself is termed as the *Karl Pearson's coefficient of correlation.*

The following set of formulae may be used of the determinaiton of the coefficient of correlation.

The basic formula:

$$r = \frac{\Sigma(y-\bar{y})(x-\bar{x})}{\sqrt{\Sigma(y-\bar{y})^2\,\Sigma(x-\bar{x})^2}} \qquad (15.29)$$

The long formula:

$$r = \frac{\Sigma xy - (\Sigma x)(\Sigma y)/N}{\sqrt{\Sigma x^2 - (\Sigma x)^2/N}\sqrt{\Sigma y^2 - (\Sigma y)^2/N}} \quad (15.30)$$

The short-cut formula :

$$r = \frac{\Sigma d_x d_y - (\Sigma d_x)(\Sigma d_y)/N}{\sqrt{\Sigma d_x^2 - (\Sigma d_x)^2/N}\sqrt{\Sigma d_y^2 - (\Sigma d_y)^2/N}} \quad (15.31)$$

The formula for grouped data :

$$r = \frac{\Sigma f(x-\bar{x})(y-\bar{y})}{\sqrt{\Sigma f(x-\bar{x})^2}\sqrt{\Sigma f(y-\bar{y})^2}} \quad (15.32)$$

The short-cut formula for grouped data :

$$r = \frac{\Sigma fd_x d_y - (\Sigma fd_x)(\Sigma fd_y)/\Sigma f}{\sqrt{\Sigma f(d_x)^2 - (\Sigma fd_x)^2/\Sigma f}\sqrt{\Sigma f(d_y)^2 - (\Sigma fd_y)^2/\Sigma f}} \quad (15.33)$$

In Eqs. 15.31 and 15.33, *dx* and *dy* stand for deviation from some appropriate values (also termed as *assumed means*) A_x and A_y, and *f* has a meaning appropriate for the term antaining it.

It can further be shown that the two regression coefficients b_{yx} and b_{xy} are related to the coefficient of correlation *r* by a simple formula :

$$r^2 = b_{yx} \text{ X } b_{xy} \quad (15.34)$$

Illustration 15.8. Calculate the coefficient of correlation of the marks in statistics (*y*) and in Maths (*x*) for eight students given in column (2) and (5) below.

CALCULATION OF KARL PEARSON'S COEFFICIENT OF CORRELATION BETWEEN MARKS IN STATISTICS AND IN MATHEMATICS.

	Marks in Statistics y	*Deviation from Mean(68)* $(y-\bar{y})$	*Deviation squared* $(y-\bar{y})$	*Marks in Math.* x	*Deviations from Mean(69)* $(x-\bar{x})$	*Deviations squared* $(x-\bar{x})^2$	*Product of Deviations. i.e..Cols (3) X(6)* $(x-\bar{x})(y-\bar{y})$
(1)	(2)	(3)	(4)	(5)	(6)	(7)	(8)
1	65	–3	9	67	–2	4	6
2	66	–2	4	68	–1	1	2
3	67	–1	1	65	–4	16	4
4	67	–1	1	68	–1	1	1

5	68	0	0	72	+3	9	0
6	69	+1	1	72	+3	9	3
7	70	+2	4	69	0	0	0
8	72	+4	16	71	+2	4	8
8	544	0	36	552	0	44	24

Solution. We calculate the means and take deviations from the means to use Eq.(15.29)

$$\bar{y} = \frac{544}{8} = 68.$$

$$\bar{x} = \frac{552}{8} = 69.$$

Applying Karl Pearson's formula (15.29) to our data we get

$$r = \frac{24}{\sqrt{36 \times 44}} = +0.60$$

Illustration 15.9. Calculate Karl Pearson's correlation coefficient for the following paired data.

X	28	41	40	38	35	33	40	32	36	33
Y	23	34	33	34	30	26	28	31	36	38

What inference would you draw from the estimate.

Solution.

COMPUTATION OF CORRELATION COEFFICIENT

	Deviation from assumed mean	*Square of deviation*		*Deviation from assumed mean*	*Square of deviations*	*Product of deviation*
X	$x' = X–35$	x'^2	Y	$y' = Y–31$	y'^2	$x'y'$
28	–7	49	23	–8	64	56
41	+6	36	34	+3	9	18
40	+5	25	33	+2	4	10
38	+3	9	34	+3	9	9
35	0	0	30	–1	1	0
33	–2	4	26	–5	25	10
40	+5	25	28	–3	9	–15
32	–3	9	31	0	0	0
36	+1	1	36	+5	25	5
33	–2	4	38	+7	49	–14
	$N = 10$	$\Sigma x' = 6$	$\Sigma x'^2 = 162$	$\Sigma y' = 3$	$\Sigma y'^2 = 195$	$\Sigma x'y' = 79$

Applying to the above data, the formula, 15.31

$$r = \frac{\Sigma x'y' - \dfrac{\Sigma x' \Sigma y'}{N}}{\sqrt{\left[\Sigma x'^2 - \dfrac{(\Sigma x')^2}{N}\right] \times \left[\Sigma y'^2 - \dfrac{(\Sigma y')^2}{N}\right]}}$$

$$= \frac{79 - \dfrac{(6) \times (3)}{10}}{\sqrt{162 - \dfrac{(6)^2}{10}}\sqrt{195 - \dfrac{(3)^2}{10}}} = 0.45$$

Illustration 15.10. Find the coefficient of correlation between X and Y,

X	1	2	3	4	5	6	7	8	9
Y	12	11	13	15	14	17	16	19	18

TABLE 15.3

X	Y	X^2	Y^2	XY
1	12	1	144	12
2	11	4	121	22
3	13	9	169	39
4	15	16	225	60
5	14	25	196	70
6	17	36	289	102
7	16	49	256	112
8	19	64	361	152
9	18	81	324	162
45	135	285	2085	731

$$\bar{X} = \frac{45}{9} = 5 \qquad \bar{Y} = \frac{135}{9} = 15$$

Applying the formula: 15.30

$$r = \frac{\Sigma(XY) - N(\bar{X})(\bar{Y})}{\sqrt{[\Sigma X^2 - N(\bar{X})^2][\Sigma^2 Y - N(\bar{Y})^2}}$$

$$= \frac{731 - 9 \times (5 \times 15)}{\sqrt{[285 - (9 \times 25)][2085 - (9 \times 225)]}}$$

$$= \frac{56}{\sqrt{60 \times 60}} = \frac{56}{60} = \frac{14}{15} = .933$$

Illustration 15.11. Twenty-five pairs of values of variate x and y led to the following results.

$$N = 25, \Sigma x = 127, \Sigma y = 100, \Sigma x^2 = 760, \Sigma y^2 = 449 \text{ and } \Sigma xy = 500.$$

A subsequent scrutiny showed that two pairs of values were copied down as

x	y
8	14
8	6

whereas the correct values where

x	y
8	12
6	8

obtain the correct value of coefficient.

Solution. We are given

$$N = 25, \Sigma x = 127, \Sigma y = 100, \Sigma x^2 = 760, \Sigma y^2 = 449 \text{ and } \Sigma xy = 500.$$

Let us first make corrections for the errors.

Incorrect values					*Correct values*				
x	y	x^2	y^2	xy	x	y	x^2	y^2	xy
8	14	64	196	112	8	12	64	144	96
8	6	64	36	48	6	8	36	64	48
16	20	128	232	160	14	20	100	208	144

$\therefore$ Corrected $\Sigma x = 127 - 16 + 14 = 125$

Corrected $\Sigma y = 100 - 20 + 20 = 100$ (unchanged)

Corrected $\Sigma x^2 = 760 - 128 + 100 = 732$

Corrected $\Sigma y^2 = 449 - 232 + 208 = 425$

and Corrected $\Sigma xy = 500 - 160 + 144 = 484$

$$\therefore r = \frac{\Sigma xy - \frac{(\Sigma x)(\Sigma y)}{N}}{\sqrt{\Sigma x^2 - \frac{(\Sigma x)^2}{N}}\sqrt{\Sigma y^2 - \frac{(\Sigma y)^2}{N}}} = \frac{484 - \frac{100 \times 125}{25}}{\sqrt{732 - \frac{(125)^2}{25}}\sqrt{425 - \frac{(100)^2}{25}}}$$

$$= \frac{484 - 500}{\sqrt{107 \times 25}} = \frac{-16}{51.7} = -0.3$$

Illustration 15.12. Find the correlation between the ages of husbands and wives from the following data.

	Age of wives						*Total*
Age of husband	*15—25*	*25—35*	*35—45*	*45—55*	*55—65*	*65—75*	
15—25	1	1	...	...	...	...	2
25—35	2	12	1	...	...	...	15
35—45	...	4	10	1	...	...	15
45—55	...	...	3	6	1	...	10

55—65	...	...	...	2	4	2	8
65—75	...	...	...	...	1	2	3
Total	3	17	14	9	6	4	53

Solution. Each figure in the table indicates the number of married couples where the husband's age was within the range given at the left of the row and the wife's age within the range given at the top of the column. Thus, there are 4 couples with husband's age between 35 and 45 and wife's age between 25 and 35. Eq. 15.35 is the appropriate formula for this case. The following table shows one way of organizing the calculation of various sums involved.

Age Groups Husband \ *Wife*	*Middle value*	x' / y'	*5—25*	*25—35*	*35-45*	*45-55*	*55-65*	*65—75*				
			20	*30*	*40*	*50*	*60*	*70*				
			—2	*—1*	*0*	*+1*	*+2*	*+3*	*Total* $fx'y'$	f	fy'	fy'^2
15—25	20	—2	+4 / 1 / 4	+2 / 1 / 2	...	...	...		6	2	−4	8
25—35	30	—1	+2 / 2 / 4	+1 / 12 / 12	0 / 1 / 0	...	...	—	16	15	−15	15
35—45	40	0	...	0 / 4 / 0	0 / 10 / 0	0 / 1 / 0	...	...	0	15	0	0
45—55	50	+1	...	...	0 / 3 / 0	+1 / 6 / 6	+2 / 1 / 2	...	8	10	10	10
55-65	60	+2	...	...	...	+2 / 2 / 4	+4 / 4 / 16	+6 / 2 / 12	32	8	16	32
65-75	70	+3	...	...	...	...	+6 / 1 / 6	+9 / 2 / 18	24	3	9	27
Total		$fx'y'$	8	14	0	10	24	30	$\Sigma fx'y'$ 86	53 N	$\Sigma fy'$ =−16	$\Sigma fy'^2$ =92
		f	3	17	14	9	6	4	53N	In each cell : Top left $= x'y'$ Middle $= f$ Bottom right $= fx'y'$		
		fx'	—6	—7	0	9	12	12	$\Sigma fx'$=10			
		fx'^2	12	17	0	9	24	36	$\Sigma fx'^2$=98			

Each group interval here is represented by its middle value, and the deviation have been obtained from $A_x = 40$ and $A_y = 40$. The variable x' and y' have been defined as

$$x' = \frac{d_x}{10} = \frac{x - A_x}{10}, \text{ and}$$

$$y' = \frac{d_y}{10} = \frac{y - A_y}{10}$$

Plugging in x' and y' in Eq. 15.35 instead of d_x and d_y result in the cancellation of the factor of 10 all around, giving

$$r = \frac{\Sigma fx'y' - (\Sigma fx')(\Sigma fy')/N}{\sqrt{(\Sigma fx'^2 - (\Sigma fx')^2/N}\sqrt{\Sigma fy'^2 - (\Sigma fy')^2/n}}$$

Applying this formula, we get for the present case,

$$r = \frac{86 - 10 \times 16/53}{\sqrt{98 - (10)^2/53}\sqrt{92 - (16)^2/53}}$$

$$= 0.906$$

A very strong correlation.

Illustration 15.13. Calculate the coefficient of correlation between the age and the sum assured.

	Sum assured in rupees(y)				
Age(x)	*50*	*100*	*200*	*500*	*1,000*
15—24	18	20	6	2	—
25—34	21	26	6	5	1
35—44	10	9	3	6	1
45—54	7	8	5	4	—
55—64	8	3	1	—	—

Find also the coefficient of correlation for the ages 15-44 and discuss the result.

Solution. We use similar definition for x' and y' as in the previous illustration.

$$x' = (x-30)/10, \text{ and } y' = (y-100)/50.$$

The calculation of the sums can now be organized as in the table shown.

The frequency of each cell is written on the right within the cell, and $fx'y'$ is written on the left.

Calculation of $\Sigma fy'$ and $\Sigma fy'^2$ is shown in the table given below.

Age group	Sum assured	50	100	200	500	1,000			
	x' \ y'	−1	0	2	8	18	f	fx'	fx'^2
15–24	−1	18 18	0 20	−12 6	−16 2	–	46	−46	46
25–34	0	0 21	0 26	0 6	0 5	0 1	59	0	0
35–44	1	−10 10	0 9	6 3	48 6	18 1	29	29	29
							134	$\Sigma fx'$ = −17	$\Sigma fx'^2 = 75$
45–54	2	−14 7	0 8	20 5	64 4	–	24	48	96
55–64	3	−24 8	0 3	6 1	–	–	12	36	108
Total	—	−30 64	0 66	20 21	96 17	18 2	$\Sigma fx'y'=104$	67	279
Total up to 44		8 49	0 55	−6 15	32 13	18 2	$\Sigma fx'y'=52$		

Solution See table on the next page for calculation of fx', fx'^2 and — $fx'y'$. Take $i_y = 50$ and $i_x = 10$.

	Full Data			*Up to 44*		
y'	f	fy'	fy'^2	f	fy'	fy'^2
–1	64	–64	64	49	–49	49
0	66	0	0	55	0	0
2	21	42	84	15	30	60
8	17	136	1,088	13	104	832
18	2	36	648	2	36	648
	170	150	1,884	134	121	1,589

We can Substitute these values directly in Eq. 15.35 since both the numirator and the denominator need to be corrected by a factor of 10 × 50, being the width of the two intervals.

Therefore,

$$r = \frac{104 - \dfrac{67 \times 150}{170}}{\sqrt{\left(279 - \dfrac{67^2}{170}\right)\left(1{,}884 - \dfrac{150^2}{170}\right)}}$$

$$= \frac{104 - 59.1}{\sqrt{252.6 \times 1{,}751.6}}$$

$$\log r = 1.65225 - \frac{1}{2}(2.4024 + 3.24435)$$

$$= \bar{2}.8293$$

or, $$r = 0.0675$$

For ages up to 44 years :

$$r = \frac{52 - \dfrac{(-17)(121)}{134}}{\sqrt{\left(75 - \dfrac{17^2}{134}\right)\left(1{,}589 = \dfrac{121^2}{134}\right)}}$$

$$= \frac{67.4}{\sqrt{72.9 \times 1{,}479.7}}$$

$$\log r = 1.8287 - \frac{1}{2}(1.8621 + 3.1703)$$

$$= 1.3125$$

or, $$r = .205.$$

Correlation for the whole data is very small and for the data up to 44 years is quite significant and positive. This shows that the correlation between 45 to 64 is significant and negative. In other words, sum assured increase with age up to 44 and then decreases.

15.7. CORRELATION ANALYSIS IN HISTORICAL SERIES

The coefficients of correlation in the preceding examples have been computed from the data which exist at a point of time. Coefficient of correlation may also be computed for such data which are spread over time. As explained earlier the variation in the long term series are often the result of the long-term forces. Now, our object may be to study correlation either between long-term changes of *two* historical variables or between short-term change only.

The method as explained so far, if applied to original historical data as they stand, without any preliminary analysis, would show the correlation between the combined long and short-time changes of the subject and the relative. If we are required to study correlation between long-term changes only, it is necessary that our data should be got rid of short-time fluctuations.

The best way to do this is to compute the correlation *not* between the trend values which suppress the short-term fluctuations.

One of the most convenient way to accomplish this is to obtain moving averages over an appropriate duration to obtain the trend values and to calculate the correlation between these trend values. See the following illustration for an application of this procedure.

The short-term fluctuations are the deviations of the actual values from the trend values, and a correlation of the short-term fluctuations is obtained by analysing the correlations between these deviations.

Illustration 15.14. Calculate the correlation between the long-term trends and the short-term fluctuations in supply and price index numbers u and v given in the Table below for the years 1981 to 1989.

Solution. The calculation of the correlations in the longterm trends and the short-term fluctuations can be organized as shown in the Table. Here u and v are the actual index number series. The trend values are obtained as the 3-year moving averages and are denoted as variables x and y. The correlation for long-term variables is obtained as the correlation in

Table showing the computation of the coefficient of correlation for the long-term and short-term variations between supply and price of an item.

Year	Supply price		Moving averages(3 yr)		Long-term changes					Short-term Fluctuation				
	u	v	Supply x	Price y	$x'=(x-87)$	$y'=(y-138)$	$x'y'$	x'^2	y'^2	$X=u-x$	$Y=v-y$	XY	X^2	Y^2
1981	80	146	–	–	–	–	–	–	–	–	–	–	–	–
1982	82	140	83	143	–4	5	–20	16	25	–1	–3	4	1	9
1983	87	143	87	139	0	1	0	0	1	0	+4	0	0	16
1984	92	134	88	136	1	–2	–2	1	4	+4	–2	–8	16	4
1985	85	131	87	135	0	–3	0	0	9	–2	–4	+8	4	16
1986	84	140	85	139	–2	1	–2	4	1	–1	+1	–1	1	1
1987	86	146	88	137	1	–1	–1	1	1	–2	+9	–18	4	81
1988	94	125	93	141	6	3	18	36	9	+1	–16	–16	1	256
1989	99	142	–	–	–	–	–	–	–					
Σ					2	4	–7	58	50	–1	–11	–31	27	383

variables x and y. Note that the total number of items is reduced to 7 in this case. The correlation has been calculated by the short-cut method with the new-variables defined as $x' = (x - 87)$ and $y' = (y - 138)$.

Using Eq. 15.31, we obtain for long-term trend

$$r = \frac{\Sigma x'y' - (\Sigma x')(\Sigma y')/N}{\sqrt{[\Sigma x'^2 - (\Sigma x')^2/N][\Sigma y'^2 - (\Sigma y')^2/N}}$$

$$= \frac{-7 - 2 \times 4/7}{\sqrt{(58 - 2 \times 2/7)(50 - 4 \times 4/7)}}$$

$$= -0.11$$

A slight negative correlation, slowing that in the long run, there is only a marginal inverse effect of prices on supply.

For the short-term fluctuations, we use the variables X and Y as the deivations of the actual time-series variables from the relevant trend-values. The coefficient of correlation has been calculated using the long method, since the numbers involved are small anyway. Thus,

$$r = \frac{\Sigma XY - (\Sigma X)(\Sigma Y)/N}{\sqrt{[\Sigma X^2 - (\Sigma X)^2/N][\Sigma Y^2 - (\Sigma Y)^2/N}}$$

$$= \frac{-31 - (-1)(-11)/7}{\sqrt{[27 - (-1)^2/7][383 - (-11)^2/7}}$$

$$= -0.33$$

Thus, the correlation between short-term fluctuations is better than that between long-term variations.

15.8. METHOD OF CONCURRENT DEVIATION

In such historical variables where no trends are apparent or where the trend is considered unimportant, the method of coefficient of concurrent deviations may be employed satisfactorily. In a majority cases this method gives accurate results similar to Pearson's coefficient with much less calculation. This method of concurrent deviation is unsuitable for studying correlation between long-term variations, as it does not consider the general long-run tendency of the data.

The important characteristics of this method are:

(*a*) The deviations of items are measured not from the true or assumed mean or moving mean, but from the preceding items.

(*b*) Only the directions of the deviation (*i.e.*, positive or negative) and not the extent of deviations are considered.

The steps involved under this method are:

(*i*) Examine the fluctuations of each series and find whether each item increases or diminishes in comparison with the item just preceding. All increases are noted as plus and all decreases as minus.

(*ii*) Count the number of items, where increase concurrently occur in the subject and the relative. Such case are considered as concurrences. These are devoted by the letter c.

The correlation coefficient r is then given by:

$$r = \pm\sqrt{\pm\frac{2c-n}{n}} \tag{15.35}$$

where n is the number of comparisons

The use of the signs requires a word of explanation. If the quantity $(2c-n)$ is negative, the negative signs are introduced at the two places indicated, otherwise both signs are taken as positive.

We illustrated below the application of the method of concurrent deviation.

Illustration 15.15. The table below gives the data regarding number of students appearing the pass percentage. Is there any correlation between the two?

Year	*No. of students appearing x (lakhs)*	*Pass Percentage y*	*Deviation in x*	*Deviation in y*	*Concurrence*
1985	2.0	47	0	0	0
1986	2.2	46	+	—	—
1987	2.4	44	+	—	—
1988	2.3	43	—	—	+
1989	2.6	46	+	+	+
1990	2.8	45	+	—	—
1991	2.7	46	—	+	—
1992	2.9	42	+	—	—
1993	2.8	40	—	—	+
1994	2.9	43	+	+	+

Solution. The deviations in x and y, and the concurrences are shown in the table above. Here $c = 4$ and $n = 9$, so that

$$r = -\sqrt{-\frac{8-9}{9}}$$

We have retained negative signs because $(2c - n)$ is negative. This gives $r = -0.33$, denoting a moderate negative correlation between the pass percentage and the number of candidates appearing.

15.9. LAG

When there is a casual relationship existing between two time series it will be frequently noted that a change in the independent variable takes some time to have its effect upon the dependent variable. For instance, the quantity of money in circulation and the cost of living index have a high degree of positive correlation. But an increase in the money supply will make itself felt on the cost of living index only after the expiry of several months. This tendency on the part of effect to occur some time after the occurrence of the cause is known as the 'lag'.

When 'lag' is known to exist it is necessary that a reasonable allowance be made for it if a correct coefficient of correlation is to be computed. This will involve the determination of the 'time lag' *i.e.,* estimating the time which a change in one variable ordinarily takes to have its effect upon the magnitude of the other variable. This can be done easily by plotting the two series on a graph paper and reading the time distance between the peaks or troughs of the two curves. If the peak in variable A comes six months after the peak in variable B we can conclude that there is a six month time lag between the two variables.

When once the time lag is determined the next step would be to push backwards the dependent series in such a manner that the time lag is totally eliminated. Thus, if an increase in the money supply in January results in a rise in the living cost in July, the cost of index should be pushed back in such a manner that the figure for July is paired with the figure of money supply for January. It is only after the series have been adjusted like this that will be possible to have a correct estimate of the correlation coefficient.

15.10. COEFFICENT OF RANK CORRELATION

The Karl Pearson's coefficient of correlation, as discussed before, cannot be used in cases where the direct quantitative measurement of the phenomenon under study is not possible, for example, efficiency, honesty, intelligence etc. In such cases one may rank the different items and apply the Spearman method of rank differences for finding out the degree of correlation. The formula for computing rank correlation by the method is:

$$R = 1 - \frac{6 \sum D^2}{N(N^2 - 1)} \tag{15.36}$$

where R denotes coefficient of rank correlation, D denotes the differences between paired ranks, and N stands for the number of pairs.

The greatest use of this method lies in the fact that one could use it to find correlation of qualitative variables, but since the method reduces the amount of labour of calculation drastically, it is at times used also where quantitative data is available, and we are willing to suffer a slight loss of accuracy. The following are some examples of the use of rank correlation.

Illustration 15.16. A group of ten workers of a factory is ranked according to their efficiency by two different judges as follows:

Name of Worker	*Judgement of Judge A*	*Judgement of Judge B*
A	4	3
B	8	9
C	6	6
D	7	5
E	1	1
F	3	2
G	2	4
H	5	7
I	10	8
J	9	10

Compute the coefficient of rank correlation.

Solution.

COMPUTATION OF THE RANK CORRELATION COEFFICIENT

Name of Worker	R_1	R_2	$R_1 - R_2$ D	D^2
A	4	3	1	1
B	8	9	−1	1
C	6	6	0	0
D	7	5	2	4
E	1	1	0	0
F	3	2	1	1
G	2	4	−2	4
H	5	7	−2	4
I	10	8	2	4
J	9	10	−1	1
				$\Sigma D^2 = 20$

$$R = 1 - \frac{6\sum D^2}{N(N^2-1)}$$

$$= 1 - \frac{6 \times 20}{10(10^2-1)}$$

$$= 1 - \frac{120}{990} = .88$$

This indicates that the opinion of the two judges with regard to the efficiency of the workers shows great similarity.

This method can also be used where the actual values (and not the ranks) are given. In such case it will be necessary to first rank the different items and then proceed in the same manner as above.

Illustration 15.17. A group of eight students get the following percentage of marks in test in Statics and Accountancy.

Roll numbers of Students	*% of marks in Statistics*	*% marks in Accountancy*
11	50	80
12	60	71
13	65	60
14	70	75
15	75	90
16	40	82
17	70	70
18	80	50

Compute the rank correlation coefficient.

Solution. Since we are given the actual marks and not the rank it will be necessary to rank the different values. Ranking may be done either from largest to the smallest or *vice versa.* Assigning rank from the highest to the lowest, we get:

Roll No. of Students	*Rank in Students*	*Rank in Accountancy*	*Difference in Ranks D*	D^2
11	7	3	4	16.00
12	6	5	1	1.00
13	5	7	–2	4.00
14	3.5	4	–0.5	0.25
15	2	1	1	1.00
16	8	2	6	36.00
17	3.5	6	–2.5	6.25
18	1	8	–7	49.00
				$\Sigma D^2 = 113.5$

$$R = 1 - \frac{6\Sigma D^2}{N(N^2-1)}$$

$$= 1 - \frac{6 \times 113.5}{8(8^2-1)}$$

$$= 1 - \frac{681}{504} = -.35 \text{ approx.}$$

It may be noted that where two or more items of a group have the same value, the rank in such a case is determined taking the average of the ranks which these items would have occupied had they differed slightly from each other. Thus, in the above example, Roll Nos. 14 and 17 get the same marks in Statistics, *i.e.*, 70 each. The value 70 stands third in the rank but since it is repeated twice we will take the average of 3 and 4, *i.e.*, 3.5.

15.11. MULTIPLE AND PARTIAL CORRELATION

So far we have been concerned with methods of determining correlation between a dependent variable and a single independent variable. But it is obvious that the fluctuations in a variable may be due to a number of factors. Thus, the yield of crop is influenced by the quality of seed, rainfall, irrigation facilities, temperature, quantity of fertilizers, etc. As such if our analysis is to be complete we have to use methods that whould enable more than two independent variables to be handled at a time. We need instruments that will assist us in measuring the relation of a single variable (*i*) to a combination of two or more other variables and (*ii*) to the individual elements of such a combination. The former relationship can be studied with the help of multiple correlation analysis and the later with the help of partial correlation analysis.

In brief, 'multiple correlation' study the relation of a dependent variable to a whole group of independent variables. Multiple correlation co-efficient is denoted by $R_{1.234....n}$. The subscript of R shows that the relation studied is between the dependent variable X_1 (the subscript before the point) and the independent variables X_2, X_3, X_4... X_n (the subscripts after the point). In partial correlation analysis we aim at measuring the relation between a dependent variable and a particular independent variable by holding all other variables constant. Partial correlation coefficient is denoted $r_{12.345...n}$. The subscripts of r show that the relation studied is between variable X_1 and X_2 (the subscripts before the point) by holding all other variables (X_3, X_4, X_5, X_n) constant (the subscripts after the point).

15.12. MULTIPLE CORRELATION ANALYSIS

In multiple correlation it is assumed that the dependent variable is related to a number of independent variables and the degree of association between the dependent variable and a number of independent variables, taken together, is measured. For example, we may wish to measure the relationship between the yield of wheat on the one hand and the amount of rainfall and temperature, taken together, and the other. In this case, yield of corn is dependent variable and rainfall and temperature are independent variables. Thus, multiple correlation analysis help us in measuring correlation between the dependent and the whole group of independent variables.

Multiple correlation coefficients, for trivariate case, may be defined in terms of simple correlation coefficients as given below.

$R_{1\cdot23}$ = Mulitple correlation between X_1 on the one hand, and X_2 and X_3

on the other = $\sqrt{\dfrac{r_{12}^2 + r_{13}^2 - 2r_{12}r_{13}r_{23}}{1 - r_{23}^2}}$ (15.37)

$R_{2\cdot13}$ = Multiple correlation between X_2 on the one hand, and X_1 and X_3

on the other = $\sqrt{\dfrac{r_{21}^2 + r_{23}^2 - 2r_{12}r_{13}r_{23}}{1 - r^2{}_{12}}}$ (15.38)

$R_{3\cdot12}$ = Multiple correlation between X_3 on the one hand, and X_1 and X_2

on the other = $\sqrt{\dfrac{r_{31}^2 + r_{32}^2 - 2r_{12}r_{13}r_{23}}{1 - r_{12}^2}}$ (15.39)

15.13. PARTIAL CORRELATION ANALYSIS

In partial correlation analysis, the degree of association between the dependent variable and a particular independent variable is studied by holding other variables constant. Given trivariate data, in determining a partial correlation coefficient between X_1 and X_2, we attempt to remove the influence of X_3 from each of the two variables and then measure the relationship between X_1 and X_2. The partial correlation co-efficient thus obtained will help us in knowing whether the correlation between X_1 and X_2 is due merely to the fact that both are affected by X_3, or is there a net covariation between X_1 and X_2 over and above the association due to common influence of X_3.

The usefulness of this concept may be explained by an example. Suppose we have the data giving the weight, the height and the age of some persons. If it is desired to study the degree of relationship of weights and heights, calculation of the simple correlation coefficient between the weight and height figures will not serve our purpose. Because the persons vary in age and the younger persons, have, on the average, lower weights and heights than the older persons, this would introduce into the data a tendency for low weights to be associated with low heights, and high with high. But what is the degree of relationship of weights and heights quite aside from the effect if the variation in age? To achieve this end, we adjust the weight and height data for the variation in age, and then a correlation of weights and heights with the effect of age removed is calculated. It is a net 'or' partial correlation.

Partial correlation coefficient can be defined in terms of simple correlation coefficient. Partial correlation coefficients for a trivariate case are as follows.

$r_{12 \cdot 3}$ = Partial correlation between X_1 and X_2, when X_3 is held constant = $\dfrac{r_{12} - r_{13}r_{23}}{\sqrt{(1-r_{13}^2)(1-r_{23}^2)}}$ (15.40)

$r_{13 \cdot 2}$ = Partial correlation between X_1 and X_3, when X_2 is held constant = $\dfrac{r_{23} - r_{12}r_{23}}{\sqrt{(1-r_{12}^2)(1-r_{23}^2)}}$ (15.41)

$r_{23 \cdot 1}$ = Partial correlation between X_2 and X_3, when X_1 is held constant = $\dfrac{r_{23} - r_{12}r_{13}}{\sqrt{(1-r_{12}^2)(1-r_{13}^2)}}$ (15.42)

The simple correlation coefficients are called zero order coefficients, since no variable is held constant. The partial correlation coefficients are called first order coefficients, when one variable is held constant, second order, when two variables are held constant. Stated generally, the order designation indicates the number of variables that have been held constant statistically.

Illustration 15.18. The following correlation coefficients are given for a trivariate data:

$$r_{12} = 0.98,\ r_{13} = 0.44 \text{ and } r_{23} = 0.54.$$

Calculate the multiple correlation coefficient treating the first variable as dependent and the second and third variables as independent.

Solution. $R_{1 \cdot 23}$ = Multiple correlation between X_1 on the one hand, and X_2 and X_3 on the other is given by Eq. 15.37.

$$R_{1\cdot23} = \sqrt{\frac{(0.98)^2 + (0.44)^2 - 2(0.98)(0.44)(0.54)}{1-(0.54)^2}}$$

$$= +0.986.$$

Illustration 15.19. The following bi-variate correlation coefficients are given for a trivariate data

$$r_{12} = 0.98,\ r_{13} = 0.44 \text{ and } r_{23} = 0.54$$

Calculate the partial correlation coefficient between the first and the third variables keeping the effect of the second variable constant.

Solution. $r_{13\cdot2}$ = Partial correlation between X_2 and X_3 when X_2 is held constant is given by Eq. 15.41 as

$$\frac{r_{13} - r_{12}r_{23}}{\sqrt{\left(1-r_{12}^2\right)\left(1-r_{23}^2\right)}}$$

Putting the values:

$$\frac{0.44 - (0.98 \times 0.54)}{\sqrt{\left[1-(0.98)^2\right]\left[1-(0.54)^2\right]}} = -0.533.$$

Illustration 15.20. On the basis of observations made on 30 cotton plants, the total correlation of yield of cotton (X_1), number of balls *i.e.*, seed vessels (X_2), and height (X_3) are found to be

$$r_{12} = 0.8,\ r_{13} = 0.65 \text{ and } r_{23} = 0.7$$

Calculate the partial correlation between yield of cotton and the no. of balls, eliminating the effect of height.

Solution.

$$r_{12\cdot3} = \frac{r_{12} - r_{13}r_{23}}{\sqrt{\left(1-r_{13}^2\right)\left(1-r_{23}^2\right)}}$$

$$r_{12\cdot3} = \frac{0.8 - (0.65 \times 0.7)}{\sqrt{\left[1-(0.65)^2\right]\left[1-(0.7)^2\right]}} = 0.635$$

Illustration 15.21. (*i*) Given that $r_{12} = 0.6$, $r_{13} = 0.7$ and $r_{23} = .65$ determine $R_{1\cdot23}$. Calculate also the partial correlation coefficient $r_{23\cdot1}$

(*ii*) If $R_{1\cdot23} = 0$ does it follow that $R_{2\cdot13} = 0$.

(*iii*) If $R_{1\cdot23} = 1$ does it follow that $R_{2\cdot13} = 1$.

Solution.

(*i*) $R_{1\cdot 23} = \sqrt{\dfrac{r_{12}^2 + r_{11}^2 - 2r_{12}r_{13}r_{23}}{1-r_{23}^2}}$

$= \sqrt{\dfrac{(0.6)^2 + (0.7)^2 - 2(0.6)(0.7)(0.65)}{1-(0.65)^2}} = 0.726$

$r_{23\cdot 1} = \dfrac{r_{23} - r_{12}r_{13}}{\sqrt{(1-r_{12}^2)(1-r_{13}^2)}} = \dfrac{0.65-(0.6)\times(0.7)}{\sqrt{(1-0.6^2)(1-0.7^2)}} = .403$

(*ii*) If $R_{1.23} = 0$ then $r_{12}^2 + r_{13}^2 - 2r_{12}r_{13}r_{23} = 0$

i.e., $r_{12}^2 + r_{13}^2 = 2r_{12}r_{13}r_{23}$

The value of $R_{2\cdot 13}^2 = \dfrac{r_{12}^2 + r_{23}^2 - 2r_{12}r_{13}r_{23}}{1-r_{13}^2}$

Putting form (*i*)

$$R_{2\cdot 13}^2 = \frac{r_{12}^2 + r_{21}^2 - r_{12}^2 - r_{13}^2}{1-r_{13}^2}$$

$$= \frac{r_{23}^2 - r_{13}^2}{1-r_{13}^2}$$

So it is not necessary that this value is zero. In fact it will have some positive value.

(*iii*) If $R_{1.23}^2 = 1$ then $r_{12}^2 + r_{13}^2 - 2r_{12}r_{23}r_{13} = 1 - r_{23}^2$

or $r_{12}^2 + r_{23}^2 = 1 - r_{13}^2 + 2r_{12}r_{13}r_{23}$ ...(*i*)

The value of $R_{2\cdot 13}^2 = \dfrac{1-r_{13}^2 + 2r_{12}r_{13}r_{23} - 2r_{12}r_{13}r_{23}}{1-r_{13}^2}$

Hence if $R_{1\cdot 23} = 1$, then $r_{2\cdot 13}$ will definitely be 1.

Illustration 15.21. If all the correlations of order zero are equal to *r*, what are the values of the partial correlation of first order.

Solution. Correlation of zero order means correlation of the type r_{12}, r_{13} etc. All of these are equal of *r*. The correlations of order one, *i.e.,* the correlation of the type $r_{12\cdot 3}$ $r_{13\cdot 2}$ etc. are

$$r_{12\cdot 3} = \frac{r_{12} - r_{13}r_{23}}{\sqrt{(1-r_{13}^2)(1-r_{23}^2)}} = \frac{r - r.r}{\sqrt{(1-r^2)(1-r^2)}}$$

$$= \frac{r(1-r)}{1-r^2} = \frac{r(1-r)}{(1-r)(1+r)} = \frac{r}{1+r}$$

15.14. STANDARD ERRORS OF ESTIMATES OF REGRESSION COEFFICIENTS

As is true in all statistical calculations, the calculated values of b and r in any data are at best *estimates* of the population values obtained from limited sample data. If we take a number of samples of size N each from a given population, the sample statistics b and r (as well as any other sample statistics) will have a variability of their own.

Therefore, the value of b or r obtained from the data pertaining to the sample are termed as sample statistics are at best *estimates* of the corresponding statistics pertaining to the whole population, and may or may not be equal to each other. Thus, the estimate of the population statistics obtained by calculations on the sample data may be in error.

This error is likely to be small when the variability of the sample statistics is small and *vice-versa*. As will be shown later, that the variability of the sample statistics measured as its standard deviation is termed as the standard error of estimate of the statistics concerned. In general, the larger the sample size, lesser is the variability of the sample statistics, the less is the standard error of estimate and more accurate is the estimate.

Calculating standard error of estimate for Y from x using the regression equation is relatively easily. It is defined as

$$S_{Y.x} = \sqrt{\frac{\Sigma(y-Y)^2}{N-2}} \tag{15.43}$$

Note that this is of the same time as the formula for standard deviation, except that the deviations in this case are taken from the predicted value Y (for each x), rather than from the mean $\bar{y}$, and that we have used (N–2) in the denominator, rather than N. Since Y is given by equation

$$(Y-\bar{y}) = b(x-\bar{x}) \text{ with}$$

$$b = \Sigma(y-\bar{y})(x-\bar{x}) / \Sigma(x-\bar{x})^2$$

we can easily show that,

$$S_{Y.x} = \sqrt{\frac{\Sigma(y-\bar{y})^2 - b\,\Sigma(x-\bar{x})(y-\bar{y})}{n-2}} \tag{15.44}$$

or

$$S_{Y.x} = \sqrt{\text{var}(y)\left[1-r^2(x,y)\right]} \tag{15.45}$$

It can be postulated that the deviation of individual values of the variable y from the regression equation are distributed *normally* with mean zero and a standard deviation equal to S_{Yx}. Thus, the points are as likely to lie below the regression line, as they are above the regression line. Also, the probability of the point lying farther and farther from the regression line decrease like the normal distribution, or some 66 per cent points are expected to lie within a band of ± 0.96 S_Y about the regression line, some 95% points within a band ± 1.96 S_Y about it, and some 99% points within ± 2.57 S_Y about the regression line.

Note that all this is in line with the concept discussed in Sec 15.6 wherein it was argued that $\Sigma (y-Y)^2$ is the unexplained variation and that $\Sigma(y-\bar{y})^2 - \Sigma(y-Y)^2$ can be taken as the explainable variation due to the correlation that exists. It is a simple matter to show that the proportion of the total variation $\Sigma(y-\bar{y})^2$ that is explained is r^2, the coefficient of determination (or the square of the coefficient of correlation) and then $(1-r^2)$ is the fractional unexplained variations. It is this unexplained variation which gives the error in estimate, and thus,

$$S_{Y.x} = \sqrt{\text{var}\,(y).(1-r^2)}, \text{ and}$$

$$S_{X.y} = \sqrt{\text{var}\,(x).(1-r^2)}$$

To illustrate this concept, consider the data of Table 15.2 reproduced below. The regression line for this data is given by Eq. 15.4, *i.e.*,

$$Y = 0.7x + 4$$

x	y	$(x-\bar{x})$	$(y-\bar{y})$	$(x-\bar{x})(y-\bar{y})$	$(y-\bar{y})^2$	Y	$(y-Y)$	$(y-Y)^2$
30	27	0	2	0	4	25	2	4
35	30	+5	5	25	25	28.5	1.5	2.25
19	18	–11	–7	+77	49	17.3	0.7	0.49
39	32	+9	+7	+63	49	31.3	0.7	0.49
26	19	–4	–6	+24	36	22.2	–3.2	10.24
29	29	–1	+4	–4	16	24.3	4.7	22.09
32	20	+2	–5	–10	25	26.4	–6.4	40.96
210	175	0	0	175	204	175		80.52

We calculate, $S_{Y.x}$ based on the two formula.

By Eq. 15.43,

$$S_{Y.x} = \sqrt{\frac{80.52}{5}} = 4.01$$

and by Eq. 15.44

$$S_{Y.x} = \sqrt{\frac{204 - 0.7 \times 175}{5}} = 4.02$$

Thus, both the equations give the same result, the difference is due to rounding off in the value of *b*.

The data is plotted in Fig. 15.4, along with the regression equation, and the band lines for 65 per cent, 95 per cent and 99 per cent confidence.

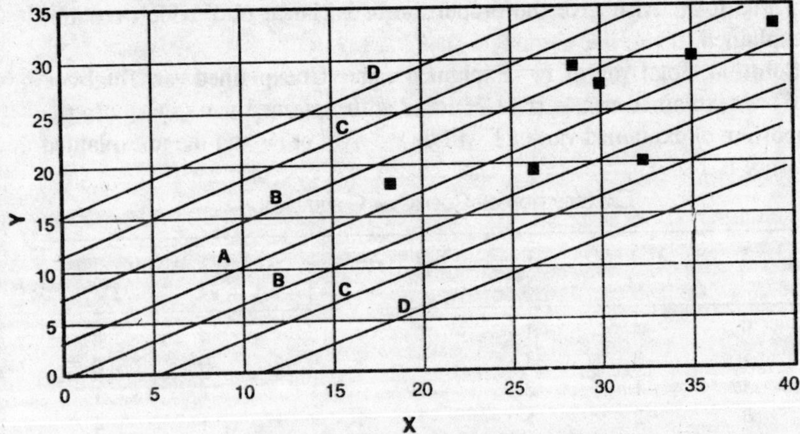

Fig. 15.4 Line A is the trend line. Pairs of lines *B*, *C* and *D* define the 1–S_Y, 2–S_Y and 3–S_Y bands about the trend lines.

Illustration 15.22. Given $\bar{x} = 79$, $\bar{y} = 56$, $\sigma_x = 4$, $\sigma_y = 3$ and $r(x, y) = +0.2$, find the range of values of x when $y = 50$.

Solution. In this, we will have to work with the regression equation for x on y.

$$(x-\bar{x}) = \frac{r\sigma_x}{\sigma_y}(y-\bar{y}),$$

$$\text{or } (X-79) = \frac{0.2\times 4}{3}(y-56),$$

$$\text{or } X = 0.267y + 64.0$$

When $y = 50$, this gives a regression value of $X = 77.4$. Thus, the actual value of X will lie within $77.4 \pm 2.57\ S_X$ with 99% probability. Here S_X is the standard error of estimate of X gives by any of the formulae 15.43, 15.44 or 15.45. Using Eq. 15.45 here, we get

$$Sx = \sqrt{(\text{var}\,x)[1-r^2(x,y)]}$$

$$= \sqrt{16(1-0.2)^2} = 3.92$$

The 99 per cent band, therefore, is $77.4 \pm 2.57 \times 3.92$ or 67.3 to 87.5.

Illustration 15.23. You are given the following observations on X and Y

X	0	1	3	6	8
Y	1	3	2	5	4

Using linear regression of Y on X, estimate the proporation of variance of Y due to X. Also give the proportion of variance of Y which remains unexplained.

Solution. Total var. of Y = Explained var. + Unexplained var. Total var. $= \sigma^2_y$. Unexplained var. $= \sigma^2_y(1-r^2)$, and so. Explained var. $= r^2\sigma_y$. So the proportion of explained var. of Y will be $r^2\sigma_y^2/\sigma_y^2$ or r^2, and the unexplained variance is $1-r^2$.

CALCULATION OF COEFF. OF CORRELATION

X	$X-4$ x'	Y	$Y-3$ y'	x'^2	y'^2	$x'y'$
0	–4	1	–2	16	4	8
1	–3	3	0	9	0	0
3	–1	2	–1	1	1	1
6	2	5	2	4	4	4
8	4	4	1	16	1	4
Total	–2		0	46	10	17

$$\therefore r = \frac{n\Sigma x'y' - \Sigma x'\Sigma y'}{\sqrt{n\Sigma x'^2 - (\Sigma x')^2}\sqrt{n\Sigma y'^2 - (\Sigma y')^2}}$$

$$= \frac{5\times 17 - (-2)\times 0}{\sqrt{5\times 46 - (-2)^2}\sqrt{5\times 10 - (0)^2}}$$

$$= \frac{85}{\sqrt{226\times 50}} = 0.8.$$

$\therefore$ Proportion of explained variance $= r^2 = .8^2 = .64$ or 64%

" " unexplained " $= 1 - r^2 = 1 - .64 = .36$ or 36%

Illustration 15.24. A correlation coefficient of 0.5 does not mean that 50% of the variauce is explained. Comment.

Solution. As is clear from the discussion in the illustration above, it is r^2 and not r, which is the coefficient of determination and gives an idea as to what proportion of variation of y is explained by the variations in x. Looking it from this angle we can say that $(.5)^2$ or 25% of variations of 'y' are explained by the linear regression fitted to the data.

15.15. MISCELLANEOUS ILLUSTRATIONS

Illustration 15.25. Find the most likely price in Bombay corresponding to the price of Rs. 70 at Calcutta from the following data:

Average price at Calcutta ($\bar{x}$)	65
Average price at Bombay ($\bar{y}$)	67
Standard deviation of Calcutta price (σ_x)	2.5
Standard deviation of Bombay price (σ_y)	3.5

$r = +.8$ between the price in the two towns.

Solution. The price at Bombay can be found by the help of the following regression equation:

$$Y - \bar{y} = \frac{\sigma_y}{\sigma_x} r(x - \bar{x})$$

or $$(Y - 67) = .8\frac{3.5}{2.5}(70 - 65)$$

or $$Y - 67 = 5.6$$

or $$Y = 67 + 5.6 = 72.6$$

Thus, 72.6 is the most likely price in Bombay.

Illustration 15.26. In a partially destroyed laboratory record of an analysis of correlation data the following results only are legible.

Variance of $x = 9$

Regression Equations:

$$8x - 10x + 66 = 0$$
$$40x - 18y = 214$$

What are:

(*a*) the mean value of x and y,

(*b*) the coefficient of correlation between x and y,

(*c*) the standard deviation of y?

Solution. (*a*) If the two given regression equations are solved as simultaneous equations the value of x and y obtained will be respectively the values of $\overline{X}$ and $\overline{Y}$.

$$8x - 10y = -66 \quad ...(i)$$
$$40x - 18y = 214 \quad ...(ii)$$

Multiplying (*i*) by 5, we get

$$40x - 50y = -330 \quad ...(iii)$$
$$40x - 18y = 214 \quad ...(iv)$$
$$32y = 544$$
$$y = 17, \text{ i.e., } \bar{y} = 17$$

Substituting the value of y in equation (i) we get

$$8x-170 = -66$$

or $$8x = 104$$

or $$x = 13, \text{ i.e., } \bar{x} = 13.$$

(b) $r = \sqrt{b_{yx} \times b_{xy}}$

We shall have to determine the values of these two regression coefficients for determining the value of r. From the question itself however, it is not clear as to which equation gives the regression of y on x and, therefore, we assume the first equation as giving the regression of y on x and the second as giving the regression of x on y. On this assumption the first of the regression equation given becomes

$$10Y = 8x + 66$$

or $$Y = \frac{8x + 66}{10}$$

or $$Y = .8x + 6.6$$

$\therefore$ the regression of y on x, i.e., $b_{yx} = .8$

The second of the regression equations will take the form of

$$40X = 18y + 214$$

or $$X = \frac{8x + 214}{40}$$

or $$X = .45y + 5.35$$

$\therefore$ the regression of x on y, i.e., $b_{xy} = .45$

$$r = \sqrt{.8 \times .45} = \sqrt{.360} = 0.6.$$

(It may be mentioned that if we had assumed the first equation as the regression of X on Y and the second of Y on X then the calculated value of r would have been more than unity, which would have indicated that we have made a wrong assumption.)

(c) Calculation of the standard deviation of y:

$$b_{yx} = r\frac{\sigma_y}{\sigma_x} = .8.$$

Substituting the values of r and σ_x ($\sqrt{9}$ given), we get

$$0.6\frac{\sigma_y}{3} = .8.$$

or $$\sigma_y = \frac{2.4}{0.6} = 4$$

Illustration 15.27. Are the following consistent?

(*a*) the value of b_{yx} and b_{xy} are 0.7 and 3.2, respectively.

(*b*) the values of b_{yx} and b_{xy} are 4 and –0.2, respectively.

Solution. (*a*) If the data were correct,

$$b_{yx} \times b_{xy} = r^2 = 0.7 \times 3.2 = 2.24$$

This would suggest a value of $r > 1.0$ which is not possible. Hence the data is inconsistent.

(*b*) If the data were correct;

$r = 4 \times (-0.2) = -0.8$, which again is not possible since r should be positive (and less than 1 in magnitude). Hence the data is inconsistent.

Illustration 15.28. For 50 students of a class the regression equation of marks in Statistics (X) on the marks in Accountancy (Y) is $3Y–5X + 108 = 0$. The mean marks of Accountancy is 44 and the variance of marks in Statistics is $\frac{9}{16}$th of the variance of marks in Accountancy. Find the mean of Statistics and the coefficient of correlation between marks in two subjects.

Solution. If in the regression equation, mean values for X and Y are substituted for X and Y, the equation is satisfied. In other words, if in the regression equation, mean value of Y is put down for Y, the value of X obtained will be the mean of X.

∴ Putting the value 44 for Y in the given equation

$$3 \times 44 - 5x + 108 = 0 \quad \text{or} \quad 5x = 132 + 108 = 240$$

or $X = 48$. So Mean of Statistics marks is 48.

The given regression equation of X on Y can be put as:

$$5X = 3Y + 108 \quad \text{or} \quad X = \frac{3}{5}Y + \frac{105}{5}$$

∴ Regression coefficient of X on Y *i.e.*, $b_{xy} = \frac{3}{5}$

Now variance of $X = \frac{9}{16}$ variance of Y

i.e. $$\sigma_{x^2} = \frac{9}{16}\sigma_{y^2} \text{ or } \frac{\sigma_{x^2}}{\sigma_{y^2}} = \frac{9}{16} \text{ or } \frac{\sigma_x}{\sigma_y} = \frac{3}{4}, \quad b_{xy} = \frac{r\sigma_x}{\sigma_y}$$

Putting the known values in it. $\frac{3}{5} = r.\ \frac{3}{4}$ or $r = \frac{4}{5} = 0.8$.

Illustration 15.29. Regression of savings S_1 of a family over income Y_1 may be expressed as $S_1 = A + Y_{1/m}$, where A and m are constants. In a random sample of 100 families the variances of the savings is one quarter of the variance of the income and the coefficient of correlation between S_1 and Y_1 is found to be 0.40. Obtain the estimate of m.

Solution. If the regression equation is of the type $y = a + b_x$, b is called the regression coefficient of y on x and its value is rb_y/b_x.

Comparing $y = a + bx$ and the given equation $S_1 = A + Y_{1/m}$ we find S_1 comes instead of y, Y_1 come instead of x, A comes instead of a and $1/m$ comes instead of b.

$\therefore$ The value of $1/m = r\sigma_{s1}/\sigma_{y1}$.

Or the estimated value of $m = \dfrac{\sigma_{y1}}{r\sigma_{s1}}$ ...(*i*)

Now variance of savings = ¼ variance of income

i.e., $\sigma_{s1^2} = \dfrac{1}{4}\sigma_{y1^2}$ or $\sigma_{y1}/\sigma_{s1} = 2$

Putting the known values in (*i*) above,

$$m = \frac{1}{0.4} \times 2 = 5$$

$\therefore$ The estimated value of $m = 5$.

Illustration 15.30. Given that $x = 4y + 7$, $y = kx + 12$ are the regression lines, show that $0 < 4k < 1$. For $k = \dfrac{1}{8}$ find the means of the two variables and r. What is the value of $\sigma_x : \sigma_y$?

Solution. The form of writing the equations suggests that $x = 4y + 7$ can be taken as regression of x on y and $y = kx + 12$ as regression of y on x.

$\therefore$ $b_{xy} = 4$ and $b_{yx} = k$.

$r^2 = b_{xy} \times b_{yx} = 4 \times k$.

Since $0<r^2<1$, so we have $0<4k<1$.

If $k = \dfrac{1}{8}$, then $r^2 = 4k = 4 \times \dfrac{1}{8} = \dfrac{1}{2}$.

$r = \sqrt{\dfrac{1}{2}} = .7$ approx.

We know $b_{xy} = \dfrac{r\sigma_x}{\sigma_y}$ So $4 = .7\dfrac{\sigma_x}{\sigma_y}$

$\therefore$ $\dfrac{40}{7} = \dfrac{\sigma_s}{\sigma_y}$ or $\sigma_x : \sigma_y$ is 40:7.

The two regression equations are

$$x = 4y + 7 \quad ...(i). \quad y = \frac{1}{8}x + 12 \text{ or } \quad 8y = x + 96 \quad ...(ii)$$

The mean of two variables can be found by solving the two regression equations

$$x - 4y - 7 = 0$$
$$x - 8y + 96 = 0$$

Subtracting $\quad 4y - 103 = 0$

$$y = \frac{103}{4} \text{ or } 25.75$$

Substituting this in (*i*)

$$x = 4 \times \frac{103}{4} + 7 \text{ or } x = 10$$

∴ Mean of x = 110 and mean of y = 25.75.

Illustration 15.31. For the variables X, Y and Z following data has been observed:

$\bar{X} = 79$	$\sigma_x = 4$	$r_{xy} = +.2$
$\bar{Y} = 56$	$\sigma_y = 3$	$r_{yz} = +.7$
$\bar{Z} = 120$	$\sigma_x = 6$	$r_{zx} = +.5$

Find the regression equations of x on y, y on z and z on x.

Solution. Regression equation of X on Y is

$$(X - \bar{X}) = r_{xy}\frac{\sigma_x}{\sigma_y}(Y - \bar{Y})$$

Putting the given values, regression of X on Y is

$$\bar{X} - 79 = \frac{.2 \times 4}{3}(Y - 56) \text{ or } X - 79 = 0.267(Y - 56)$$

or $\quad X = 0.267Y + 64.0.$

Regression equation of Y on Z is

$$(Y - \bar{Y}) = r_{yz}\frac{\sigma_y}{\sigma_z}(Z - \bar{Z})$$

By putting the given values, we get

$$Y - 56 = \frac{.7 \times 3}{6}(Z - 120) \text{ or } Y - 56 = .35(Z - 120)$$

or $\quad Y = 0.35\,Z + 14.0$

Regression equation of Z on X is

$$Z - \bar{Z} = r_{zx}\frac{\sigma_z}{\sigma_x}(X - \bar{X})$$

By putting the given values, we get

$$Z-120=\frac{0.5\times 6}{4}(X-79) \text{ or } Z-120=0.75\,(X-79)$$

or $\quad Z = 0.75x + 60.8.$

Illustration 15.32. Calculate the rank correlation from the data given below:

Sales (in lacs):	45	56	39	54	45	40	56	60	30	36
Advertising cost (Rs. 000):	40	36	30	44	36	32	45	42	20	36

Solution.

X	Rank X	Y	Rank Y	d	d^2
45	5.5	40	4	+1.5	2.25
56	2.5	36	6	–3.5	12.25
39	8	30	9	–1	1.00
54	4	44	2	+2	4.00
45	5.5	36	6	–0.5	0.25
40	7	32	8	–1	1.00
56	2.5	45	1	+1.5	2.25
60	1	42	3	–2	4.00
30	10	20	10	0	0.00
36	9	36	6	+3	9.00
$n = 10$				Total	36.00

$$r=1-\frac{6\sum d^2}{n(n^2-1)}=1-\frac{6\times 36}{10(100-1)}=0.7$$

Illustration 15.33. The following table gives the 'age' of cars of a certain make and their annual maintenance costs. Obtain the regression equation for costs related to age:

Age of cars, year	2	4	6	8
Maintenance costs, 00 Rs:	10	20	25	30

Solution. Let x denote 'age' and y 'cost'.

	x	y	x^2	xy
	2	10	4	20
	4	20	16	80
	6	25	36	150
	8	30	64	240
Totals	20	85	120	490

$$\overline{X} = \frac{20}{4} = 5$$

$$\overline{Y} = \frac{85}{4} = 21.25$$

$$(Y - \overline{Y}) = b_{yx}(X - \overline{X})$$

with $$b_{yx} = \frac{N\sum xy - (\sum x)(\sum y)}{N\sum x^2 - (\sum x)^2}$$

$$= \frac{4 \times 490 - 20 \times 85}{4 \times 120 - 20 \times 20} = 3.25.$$

So the regression equation is

$$(Y - 21.25) = 3.25\,(X - 5)$$

or $$Y = 3.25\,X + 5.$$

Illustration 15.34. Find out the regression coefficients b_{xy} and b_{yx} if

$$\Sigma x = 50,\ \overline{x} = 5,\ \Sigma y = 60,\ \overline{y} = 6,$$
$$\Sigma xy = 350,\ \sigma_x = 4,\ \sigma_y = 9$$

Solution

$$\overline{x} = \frac{\Sigma x}{N} \text{ or } 5 = \frac{50}{N} \text{ So } N = 10$$

$$r = \frac{\frac{\Sigma xy}{N} - \overline{xy}}{\sigma_x \sigma_y} = \frac{\frac{350}{10} - 5 \times 6}{4 \times 9} = \frac{5}{36} = 0.139$$

So $$b_{xy} = r\frac{\sigma_x}{\sigma_y} = \frac{0.139 \times 4}{9} = 0.0617$$

and $$b_{yx} = r\frac{\sigma_y}{\sigma_x} = \frac{0.139 \times 9}{4} = 0.312.$$

Illustration 15.35. In trying to evaluate the effectiveness of its advertising campaign, a firm compiled the following data:

Year:	1973	1974	1975	1976	1977	1978	1979	1980
Expenditure ('000):	12	15	15	23	24	38	42	48
Sale (lacs):	5.0	5.6	5.8	7.0	7.2	8.8	9.2	9.5

Calculate the regression equation of sales on expenditure. Find also the possible sale when the expenditure is Rs. 60 thousand.

Solution. We work with assumed means $A_x = 25$ and $A_y = 7$.

$$\bar{x} = A_x + \frac{\sum x'}{N} \times i = 25 + \frac{17}{8} \times 1 = 27.125$$

$$\bar{y} = A_y + \frac{\sum y'}{N} \times i = 7 + \frac{21}{8} \times 0.1 = 7.26$$

$$b_{yx} = \frac{N\sum x'y' - \sum x' \sum y'}{N\sum x'^2 - (\sum x')^2} \times \frac{i_y}{i_x} = \frac{8 \times 1701 - 17 \times 21}{8 \times 1361 - 17 \times 17} \times$$

$$= 0.125$$

	Expenditure		*Sales*				
Year	x	$x' = x-25$	y	$y' = \frac{y-7}{0.1}$	x'^2	y'^2	$x'y'$
1973	12	–13	5.0	–20	169	400	260
1974	15	–10	5.6	–14	100	196	140
1975	15	–10	5.8	–12	100	144	120
1976	23	–2	7.0	0	4	0	0
1977	24	–1	7.2	2	1	4	–2
1978	38	13	8.8	18	169	324	234
1979	42	17	9.2	22	289	484	374
1980	48	23	9.5	25	529	625	575
Totals		17		21	1361	2177	1701

Regression equation:

$$(y - 7.26) = 0.125\,(x - 27.125)$$

or $\quad y = 0.125x + 3.87$

when $\quad x = 60,\ y = 0.125 \times 60 + 3.87 = 11.37$

Illustration 15.36. From the given data find (*i*) the two regression equations, (*ii*) coefficient of correlation.

Marks in										
Economics:	25	28	35	32	31	36	29	38	34	32
Statistics:	43	46	49	41	36	32	31	30	33	39

Solution. Let x denote marks in Economics and y in Statistics.

x	$(x-\bar{x})$	$(x-\bar{x})^2$	y	$(y-y)$	$(y-y)^2$	$(x-\bar{x})\,(y-y)$
25	–7	49	43	+5	25	–35
28	–4	16	46	+8	64	–32
35	+3	9	49	+11	121	+33
32	0	0	41	+3	9	0
31	–1	1	36	–2	4	+2
36	+4	16	32	–6	36	–24
29	–3	9	31	–7	49	+21
38	+6	36	30	–8	64	–48
34	+2	4	33	–5	25	–10
32	0	0	39	+1	1	0
320	0	140	380	0	398	–93

$$\bar{x} = \frac{\Sigma x}{N} = \frac{320}{10} = 32; \quad \bar{y} = \frac{\Sigma y}{N} = \frac{380}{10} = 38$$

$$b_{xy} = \frac{\Sigma(x-\bar{x})(y-\bar{y})}{\Sigma(y-\bar{y})^2} = \frac{-93}{398} = -0.234$$

$$b_{yx} = \frac{\Sigma(x-\bar{x})(y-\bar{y})}{\Sigma(x-\bar{x})^2} = \frac{-93}{140} = -0.664$$

Regression equation of x on y:

$$(x-\bar{x}) = b_{xy}\,(y-\bar{y})$$

or $$(x-32) = -0.234\,(y-38)$$

or $$x = 0.234y + 40.89$$

Regression equation of y on x:

$$(y-38) = -0.664\,(x-32)$$

or $$y = -0.664x + 59.25$$

Correlation coefficient,

$$r = \frac{\Sigma(x-\bar{x})(y-\bar{y})}{\sqrt{\Sigma(x-\bar{x})^2\,\Sigma(y-\bar{y})^2}}$$

$$= \frac{-93}{\sqrt{398 \times 140}} = -0.394.$$

Illustration 15.37. You are given the following data:

	x	y
Arithmatic mean	36	85
S.D.	11	8

Coefficient of correlation = 0.66.

(*i*) Find the two regression equations.

(*ii*) Estimate the value of x when $y = 75$.

Solution.

$$(i)\quad b_{xy} = r\frac{\sigma_x}{\sigma_y} = \frac{0.66\times 11}{8} = 0.9075$$

$$b_{yx} = r\frac{\sigma_y}{\sigma_x} = \frac{0.66\times 11}{8} = 0.48$$

Regression equation of x on y:

$$(x-36) = 0.9075\ (y-85)$$

or

$$x = 0.9075y - 41.1375$$

Regression equation of y on x:

$$(y-85) = 0.48\ (x-36)$$

or

$$y = 0.48x + 67.72$$

When $y = 75$,

$$x = 0.9075 \times 75 - 41.1375 = 26.925.$$

PROBLEMS

1. Discuss fully what is meant by correlation and distinguish between positive and negative correlation.
2. What is correlation? Explain how you will use the following methods in determining correlation: *(i)* Graph, *(ii)* Correlation table, and *(iii)* Karl Pearson's Coefficient of Correlation.
3. Write short notes on:
   (i) Positive correlation,
   (ii) Coefficient of current deviation, and
   (iii) Lag.
4. Two variates X and Y when expressed as deviation from their respective means are given as follows. Find the coefficient of correlation between them.

x	–3	–2	–1	0	1	2	3
y	–3	–1	0	–2	–3	1	2

(Ans. r = 0.786)

5. Following are the heights and weights of 10 students of a B.Com, class:

Height (inches)	62	72	68	58	65	70	66	63	60	72
Weight (kgs)	50	65	63	50	54	60	61	55	54	65

Draw the scatter diagram and indicate whether the correlation is positive or negative.

6. Draw the scatter diagram for the data given below and comment on the correlation.

x'	1	2	3	4	5	6
y'	25	16	9	4	1	0

[5.2]

7 Two series of X and Y with 50 items each have standard deviations 4.5 and 3.5 respectivly. If the sum of the products of deviation of X and Y series from respective arithmetic mean be 420.0. Find the coefficient of correlation between X and Y.

(**Ans.** r = 0.53)

8. Calculate the coefficient of correlation from the data given below:

x	12	9	8	10	11	13	7
y	14	8	6	9	11	12	3

(**Ans.** r = 0.95)

9. Two variates x and y when expressed as deviation from their respective means are given as follows. Find their standard deviations and the coefficient of correlation.

x	–4	–3	–2	–1	0	1	2	3	4
y	–3	–3	–4	0	4	1	2	–2	–1

10. From the following frequency table calculate the coefficient of correlation.

x \ y	–3	–2	1	0	1	2	3
–3							5
–2					8	3	4
–1				9	7	4	
0			2	5	9	3	
1		2	3	6			
2	3	3					
3	4						

(**Ans.** r = – 0.857)

11. Calculate the coefficient of correlation between the values of X and Y given below:

X	78	89	96	69	59	79	68	61
Y	125	137	156	112	107	136	123	108

(**Ans.** r = 0.954)

12. The following table shows the marks obtained by 10 students in Accountancy and Statistics. Find the coefficient of correlation.

Student No.	1	2	3	4	5	6	7	8	9	10
Accountancy	45	70	65	30	90	40	50	75	85	60
Statistics	35	90	70	40	95	40	60	80	80	50

(**Ans.** r = 0.903)

13. Calculate the value of coefficient of correlation from the following table giving demand and price of a certain luxury goods.

Price in Rs	2	4	6	10	16
Demand in mds.	66	48	30	12	6

(**Ans.** r = 0.923)

14. Compute the coefficient of correlation of the short–time oscillations from the following ignoring decimals.

Year	1961	1962	1963	1964	1965	1966	1967	1968	1969
Index of supply	80	82	86	91	83	85	89	96	93
Index of Price	146	140	130	117	133	127	115	95	100

(**Ans.** = –.99)

15. From the following data compute the coefficient of correlation between X and Y.

	x series	*y* series
No. of items	15	15
Arithmetic Mean	25	18
Squares of deviation from mean	136	138

Summation of product of deviation of x and y from their respective arithmatic means = 122.

(**Ans.** r = 0.891)

16. A sample of five items is taken from the production of a firm. Length and weight of the five items are given below:

Length (cm)	3	4	6	7	10
Weight (gm)	9	11	14	15	16

By comparing the coefficients of variations of two characters conclude which of them is more variable.

Calculate Karl's Pearson's correlation coefficient between length and weight in the above sample and interpret the value of this coefficient.

(**Ans.** r = 0.939)

17. Calculate the coefficient of correlation between the marks obtained by 80 students in the terminal (X) and Annual (Y) Examination in 'Trade and Statistics.'

Y/X	*21–30*	*31–40*	*41–50*	*51–60*	*61–70*	*71–80*	*81–90*	*Total*
21–30	4							4
31–40	5	3	11					19
41–50		2	10	8				20
51–60		6	10	5				21
61–70				4	3			7
71–80					2	2	1	5
81–90						3	1	4
Total	9	11	31	17	5	5	2	80

Also calculate the arithmetic mean and standard deviations of X and Y.

(**Ans.** r = 0.801, For X: AM = 48.1 marks σ =13.9 marks, For Y: AM = 50.4 marks, σ = 14.8 marks.)

18. Calculate Karl Pearson's coefficient: of correlation between the following pairs of values:

X	100	110	115	116	120	125	130	135
Y	18	18	17	16	16	15	13	10

(**Ans.** r = – 0.915)

19. Following are the ranks obtained by 10 students in two subjects: Statistic and Mathematics. To what extent is the knowledge of students in the two subjects related?

Statistics	1	2	3	4	5	6	7	8	9	10
Mathematics	2	4	1	5	3	9	7	10	6	8

(**Ans.** $r = 0.76$)

20. Calculate Spearman's rank correlation coefficient for the following data:

X	35	37	38	42	44	46	51	54	55	56
Y	30	32	39	42	41	31	50	52	46	55

(**Ans.** $r = 0.03$)

21. Calculate the coefficient of concurrent deviation of
(a) $n=40$, C=12, (b) $n=10$, $c=8$,
(c) $n=12$, C=0 and (d) $n=18$, $c=18$.

(**Ans.** $r = 1.0$)

22. Calculate the coefficient of concurrent deviation for the data of

X	58	60	62	64	66	68	70
Y	90	81	99	108	126	117	135

(**Ans.** $r = 0.577$)

23. From the data given below find the number of item *i.e.*, $n : r = .5$, $\Sigma xy = 120$, $\sigma y = 8$, $\Sigma x^2 = 90$.

(**Ans.** $n = 10$)

24. Calculate Spearman's coefficient of rank correlation and Pearson's coefficient of correlation from the following data on height and weight of seven students. Why is it that the two values are same.

Height	58	60	62	64	66	68	70
Weight	90	81	99	108	126	117	135

25. The following table gives the distribution of the total population and those who are wholly or partially blind among them. Find out if there is any correlation between age and blindness.

Age	0–10	10–20	20–30	30–40	40–50	50–60	60–70	70–80
No. of Pearsons (in 000)	100	60	40	36	24	11	6	13
Blind	55	40	40	40	36	22	18	15

26. Ten competitors in a voice test are ranked by three judges in the orders given:

First Judge:	1	6	5	10	3	2	4	9	7	8
Second Judge:	3	5	8	4	7	10	2	1	6	9
Third Judge:	6	4	9	8	1	2	3	10	5	7

Use the method of rank correlation to gauge which pair of judge have the nearest approach to common likings in voice.

(**Ans.** Between First and Second Judge = – 0.212
" " " Third " = 0.636
" Second and Third " = – 0.297

27. The coefficient of the rank correlation of the marks obtained by 30 students in English and Economics was found to be 0.5. It was later discovered that the difference in ranks in the two subjects obtained by one of the students was wrongly taken as 3 instead of 7. Find the correct coefficient of rank correlation.

(**Ans.** $r = 0.258$)

28. The coefficient of rank correlation between marks in Statistics and marks in Accountancy obtained by a certain group of students is 0.8. If the sum of squares of the difference in rank is given to be 33 find the number of students in the group.

(**Ans.** 10).

29. In a problem on correlation it was found that standard deviation of x and y are 2 and 3 respectively and the coefficient of correlation between them is 0.5. What is the value of standard deviation of series formed by taking the differences of x and y values.

(**Ans.** St. Dev. of $(x—y) = 2.65$)

30. Calculate coefficient of correlation for the following data:

Husband's age	23	27	28	29	30	31	33	35	36	39
Wife's age	18	22	23	24	25	26	28	29	30	32

(**Ans.** $r = .995$)

31. In an aptitude test two judges rank the 10 competitions in the following order.

Individual	1	2	3	4	5	6	7	8	9	10
Ranking by I	6	4	3	1	2	7	9	8	10	5
Ranking by II	1	1	6	7	5	8	10	9	3	2

Is there any concordance between the two judges.

(**Ans.** Yes, $r = .17$)

32. Find the coefficient of correlation from the following data:

X	78	36	98	25	75	82	90	62	65	39
Y	84	51	91	60	68	62	86	58	53	47

(**Ans.** $r = 0.78$)

33. Given No. of pairs 8.

	X series	Y series
Arithmetic mean	74.50	125.50
Assumed mean	69.00	112.00
Standard deviation	13.07	15.85

Summation of products of corresponding deviations of x and y series $(\Sigma x'y') =$ 2176. Calculate the correlation between x and y.

34. Find the regression equation of Y on X from the following data using normal equations.

X	10	20	30	40	50
Y	3	2	0	5	4

(**Ans.** $Y = 1.3 + 0.05\ X$)

35. Calculate Karl Peason's Coefficient of correlation and the regression equations from the following data:

Age of husband	18	19	20	21	22	23	24	25	26	27
Age of wife	17	17	18	18	18	19	19	20	21	22

36. On the basis of the figures recorded below for 'Supply' and 'Price' for nine years, build a regression of 'Price' on 'Supply'. Calculate from the equation established, the most likely 'Price,' when 'Supply' = 90.

Years	1951	1952	1953	1954	1955	1956	1957	1958	1959
Supply	80	82	86	91	83	85	89	96	93
Price	145	140	130	124	133	127	120	110	116

(**Ans.** $Y = 291.646 - 2.01\ X$, and Price = 110.75 approx.)

37. The following table gives the various values of two variables:

X	42	44	58	55	89	98	66
Y	56	49	53	58	64	76	58

Determine the regression equations which may be associated with these values and calculat Karl Person's coefficient of correlation.

(**Ans.** $Y = .372\,X + 35.27$, $X = 2.2\,Y - 65.9$, $r = .905$)

38. The following calculations have been made for closing prices of twice stocks (X) on the Bombay stock exchanges on a certain day, along with the volume of sales in thousands of shares (Y). From these calculations find the regression equation.

$\Sigma x = 580$, $\Sigma y = 370$, $\Sigma xy = 11494$, $\Sigma x^2 = 41658$, $\Sigma y^2 = 17206$

39. Draw two regression lines on a graph paper:

Mean height = 50.07 : Mean age = 9.98

St. deviation of height = 5.26, Standard deviation of age = 2.59 and r = 0.898.

(**Ans.** $Y = 0.422\,X - 12.15$, $X = 1.825\,Y + 31.86$).

40. In the following table are recorded data showing the test scores made by salesmen on an intelligence test and their weekly sales:

Salesman	1	2	3	4	5	6	7	8	9	10
Test scores	40	70	50	60	80	50	90	40	60	60
Sales (000)	2.5	6.0	4.5	5.0	4.5	2.0	5.5	3.0	4.5	3.0

Calculate the regression line of sales on test score and estimate the most probable weekly sales if a salesman makes a score of 70.

(**Ans.** $Y = 0.058\,X + 0.57$)

41. Calculte the coefficient of correlation and obtain the lines of regression of the following data:

X	1	2	3	4	5	6	7
Y	9	8	10	12	11	13	14

Obtain an estimate of Y which should correspond on the average to $X = 6.2$

(**Ans.** $Y - .929\,X + 7.284$, $X = .929\,Y - 6.219$. $r = .929$)

42. The following table shows the mean and standard deviation of the series of two shares on Delhi Stock Exchange.

Share	*Mean*	*St. Deviation*
A. Co. Ltd.	Rs. 39.5	Rs. 10.8
B. Co. Ltd.	Rs. 47.5	Rs. 16.8

If the coefficient of correlation between the prices of two shares is 40.42 find most likely prices of share A corresponding to a price of Rs. 55 of share B.

(**Ans.** Most likely prices of share A is Rs. 41.5 approx.)

43. Given the following data calculate the expected values of Y where $X = 12$.

Average	$X = 7.6$	$Y = 14.8$	
St. deviation	$X = 3.6$	$Y = 2.5$	and $r = 0.99$

(**Ans.** Expected value of Y = 17.8 approx.)

44. A study of wheat prices at Hapur and Kanpur yield the following data:

	Hapur	*Kanpur*	
Average price	Rs 2.463	Rs 2.797	
St. Deviation	Rs 0.326	Rs 0.207	$r = 0.74$

Estimate from the above data the most likely price of wheat (*a*) at Hapur corresponding to the price of Rs 2.334 per kg. at Kanpur, (*b*) at Kanpur corresponding to the price of wheat at Hapur = 1.899

,, ,, ,, ,, Kanpur = 3.086

45. For two variables X and Y the regression of X on $Y = 5Y - 7$ and regression equation of Y on X is $Y = 0.1x + 1.7$. Find the means of X and Y and the coefficient of correlation between X and Y.

(**Ans.** Mean of $X = 3$, mean of $Y = 2$ and r between them $= 0.707$)

46. Find the means of the variable X and Y and the correlation coefficient given the following regression equations:

$2Y - X - 50 = 0, 3Y - 2X - 10 = 0.$

(**Ans.** Mean of $X = 130$, Mean of $Y = 90$, $r = .866$)

47. A computer obtained the regression coefficient of Y on X as 3.5 and the regression coefficient of X on Y as 0.5. Comment on the values.

(**Ans.** As r is never more than one, this is impossible. Hence there is some mistake in calculation).

48. You are given the following observation on X and Y.

X	0	1	3	6	8
Y	1	3	2	5	4

Using linear regression of Y on X, estimate the proportion of variance of Y due to X. Also give proportion of variance of Y which remains unexplained.

(**Ans.** Proportion of explained variance = 64 per cent

,, unexplained ,,= 36 per cent

49. A correlation coefficient of 0.5 does not mean that 50 per cent of the data are explained. Comment.

50. If the regression lines of x and y are given by $2x–3y = 0$ and $4y - 5x - 8 = 0$, find two regression coefficients and also the coefficients of correlation. If the standard deviation of x is 3, find the standard deviation of y.

(**Ans.** $b_{xy} = .67, b_{xy} = .8, r = y = 2.7$)

51. Mean soil temperature and germination interval for winter wheat for 12 places as are below:

Mean soil temp.	57	42	38	42	45	42	44	40	46	44	40	40
No. of days	10	26	41	29	27	27	19	18	19	31	29	33

Obtain the regression equation of germination interval on mean soil temperature and comment on your result.

(**Ans.** $y = - 1.203x + 80, 797$)

52. Find our the equations of two regression lines from the following data of prices of a certain commodity at two places.

Average price	Calcutta (X) – 65,	Bombay (Y) – 67
St. Deviation	,, – 2.5,	,, – 3.5

r between two places = 0.8.

(**Ans.** Reg. of Y on X is $28X– 25Y = 145$

,, X on Y is $7X - 4Y = 147$)

53. From 120 items, computations are summarized as follows:

$\Sigma x = 510, \Sigma y = 7140, \Sigma x^2 = 4150, \Sigma xy = 549000, \Sigma y^2 = 740200.$

Estimate the regression equations from the above data.

Chapter 16

PROBABILITY

16.1 INTRODUCTION

In ordinary language probability, chance, likelihood, odds etc. are interchangeable. We all have some notion of what is meant by these terms though we may not be able to define them mathematically. There is a wide variety of usages to which these terms are put. We may come across people referring to the chance of winning the cricket match by the home team against the Aussies or to the likelihood of it raining today. Here the reference is to a *single* event that would happen just once. We may also talk of the probability of a past event about which we don't have sufficient information. Or, we may refer to the probability of a boy born this year completing university education. Here we are referring not to any one boy but to a kind of generalised boy.

In statistics we are not concerned with what happens in a single case. The probability of it raining today is a subjective statement showing how much confidence one has in the truth of the given proposition. The probability of the second type where the reference is to a kind of generalised situation, is what interests us in statistics.

Before explaining the concept of probability as is used here, let us define a few terms commonly used.

Random experiment. It is a statistical process which can be repeated and in any single trial of which the outcome is unpredictable. Tossing of a coin and rolling of an unloaded dice are examples of random experiments.

Event. It is a statement about one or more outcomes of an experiment. for example, 'a number greater than 4 appears' is an event for the experiment of throwing a dice.

Consider an experiment of drawing a card from a pack containing just four cards: Ace of Spades, Ace of Hearts, King of Spades and King of Hearts. We draw any one of these four cards. So there are four possible outcomes or events—drawing of *SA*, *HA*, *SK*, or *HK*. These events are called simple events because they cannot be decomposed further into two or more events.

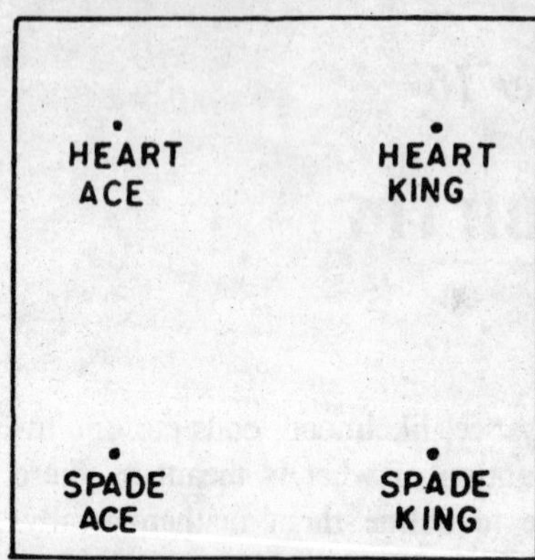

FIG. 16.1

Any set of *simple events* can be represented on diagram like Fig. 16.1. The collection of all possible simple events in an experiment is called a *sample space* or a *possibility space*. Thus the sample space of drawing a card from the pack described earlier consists of four points.

An event is termed *compound* if it represents two or more simple events. Thus, the event 'Spade' is a compound event as it represents two simple events *SA* and *SK* (Fig. 16.2).

Similarly, 'not *HA*' is also a compound event made up of all event except *HA*, that is made up of *SA*, *SK* and *HK* (Fig. 16.3).

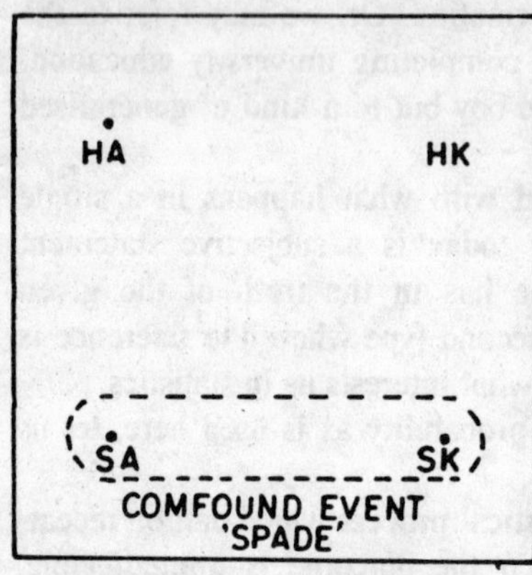

FIG. 16.2 FIG. 16.3

The sample may be *discrete* or *continuous*. If we are dealing with discrete variable, the sample space is discrete and if we are dealing with continuous variables it is continuous. The sample space for rolling of two dices is a discrete one consisting of 36 points (Fig. 16.4) and that for the weights of individuals selected at random would be continuous.

Mutually exclusive events. Two events are said to be mutually exclusive if the occurrence of one precludes the occurrence of another.

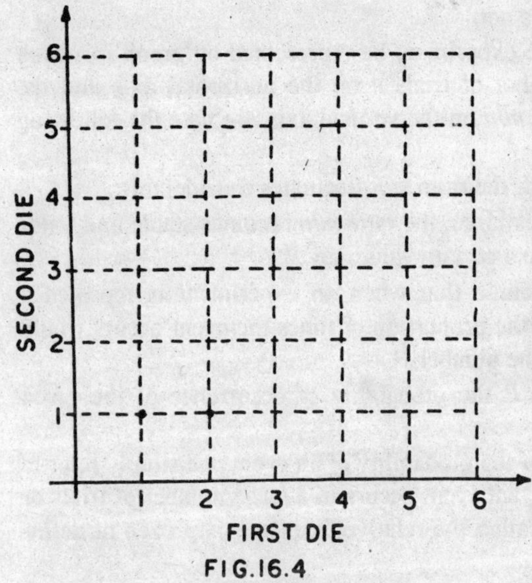

FIG.16.4

All simple events are mutually exclusive. Compound events are mutually exclusive only when they contain no simple event in common. Thus, in drawing of a card from a normal pack, the events 'a red card' and 'a spade' are mutually exclusive as a card cannot be red and a spade at the same time. But 'a heart' and 'a king' are not mutually exclusive as both the compound events include the elementary event 'King of Heart'.

Complementary event. Two mutually exclusive events are said to be complementary if they between themselves exhaust all possible outcomes. Thus, 'not Ace of Heart' and 'Ace of Heart' are complementary events. So also are 'no head' and 'at least one head' in repeated flipping of a coin. If one of them is denoted by A, the other (complementary) event is denoted by $\overline{A}$.

Sum of events. Sum of two events A_1 and A_2 is the compound event 'either A_1 or A_2 or both' *i.e.*, at least one of A_1 and A_2 occurs. This is denoted by $A_1 + A_2$. In general, $A_1 + A_2 + ... + A_n$ is the event which means the occurrence of at least one of A_i's.

Product of events. Product of two events $A_1 A_2$ is the compound event 'A_1 and A_2 both occur'. This is denoted by $A_1 A_2$. Obviously, if A_1 and A_2 are two mutually exclusive events than $A_1 A_2$ is impossible event.

16.2 PROBABILITY

Suppose a person is required to calculate the possibility of the occurrence of one outcome (simple or compound) of an experiment. One method to do this is to try the experiment a large number of times under exactly similar circumstances. If an outcome occurs m times in n trials, $\frac{m}{n}$ is called its relative frequency. It is conventional to use the term *success* whenever the event under consideration takes place and *failure* whenever it does not.

If the outcome of the experiment be represented by graph in which we have the total number of trials n on the horizontal axis and the proportion of successes m/n on the vertical axis, we note the following points:

1. When n is small, the ratio m/n fluctuates considerably.
2. When n becomes large, the ratio m/n becomes stable and tends to settle down to a certain value, say P.

From these, we conclude that when an experiment is repeated a large number of times, the proportion of times the event occurs would be practically equal to the number P.

We call the number P the probability of occurrence of the given event.

Thus, when we talk of the probability of an event, we simply refer to the proportion of times that event occurs in a *large* number of trials or *in a long run*. This is called the relative frequency approach of defining probability.

So the probability of getting a six in a single rolling of a die is the proportion of times a six would show up in a large number of rollings of a single die under exactly similar circumstances.

Note carefully that P and the proportion of successes m/n are not the same things. The ratio m/n changes with n while P does not. It is a fixed number. However, when n is large and P is not known, m/n is taken as an estimate of P.

16.3 A PRIORI PROBABILITY

According to the definition given above, the probability of an even can be found out only *after* conducting a large number of similar experiments. If we are interested in finding the probability of an event prior to conducting experiments, we use the *principle of indifference* or the *principle of insufficient reason*. The principle proposes that whenever there is no basis for preferring one simple outcome over others, all simple outcomes should be treated as *equally likely* and for such cases, the ratio of the number of outcomes favourable to an event to the total number of outcomes is the probability of occurrence of that event. This approach of defining probability is called the 'classical approach' or 'a priori definition of probability'.

Definition. If an experiment results in n equally likely, mutually exclusive and exhaustive outcomes and r of them favour a particular event A, then probability of occurrence of A is $\frac{r}{n}$.

In rolling of 2 dices, each point in the 36 point sample space is equally likely. If we are to find the probability of rolling a total or 7, the points enclosed in the dotted curve (Fig. 16.5) correspond to favourable cases. Therefore, probability of rolling a total of 7 is,

$$P(7) = \frac{\text{No. of favourable outcomes}}{\text{Total No. of equally likely outcomes}}$$

$$= \frac{6}{36} = \frac{1}{6}$$

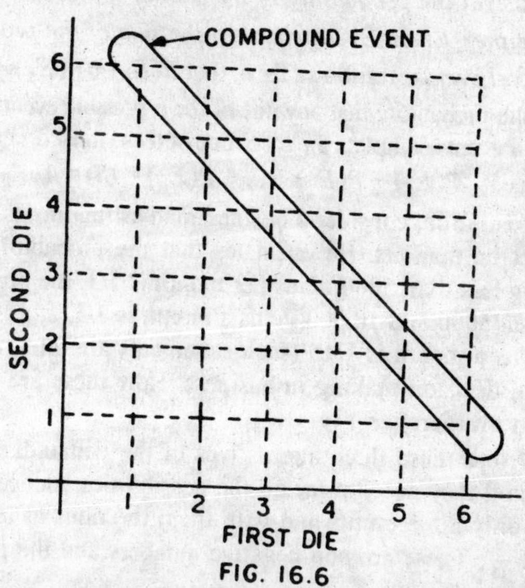

FIG. 16.6

In using the principle of insufficient reason, caution must be exercised in selecting equally likely cases. In a simultaneous toss of 2 coins, a person might say that there are 3 events; both heads; one head, other tail and both tails; and that there is insufficient reason for preferring one over the other. Hence he might conclude that probability of each one of them is $\frac{1}{3}$. But this argument is fallacious in so far as it does not recognise that 'one head, other tail' is a compound event consisting of 2 simple events: 'first head, second tail' and 'first tail, second head'. The 4 events thus recognised equally likely and each has a probability of 1/4, so that probability of 'one head, other tail' is 1/2.

16.4 MATHEMATICS OF PROBABILITY

Let there be a sample space S defined over a random experiment with n possible mutually exclusive outcomes, E_1, E_2, ..., E_n. We can associate with each event a number $P(E_1)$, $P(E_2)$, etc. denoting the probability of its occurrence. These numbers may vary between 0 and 1 such that a probability of 1 signifies that the event in question is sure to occur. The higher the value of P more is the likelihood of the occurrence of the corresponding event. It should be noted that P cannot be negative.

For mutually exclusive events (simple events are necessarily mutually exclusive) the probability of the sum of two events E_1 and E_2, *that is, of either* E_1 or E_2, *is equal to the sum of the probabilities of the individual events.* It follows then, that $P(E_1) + P(E_2) + ... + P(P_n)$ represents the probability that any one of the n possible events occurs, and since these n events complete the set S, one of these must occur. Thus

$$P(E_1) + P(E_2) + ... + P(E_n) = (S) = 1 \tag{16.1}$$

As an illustration, consider a businessman estimating the demand of the product he markets. He estimates that the probability of the demand being less than 5000 items per month is 1/3, the probability that it is between 5000 and 10,000 items a month is 3/5, and that it is over 10,000 items a month is 1/15 (such statements are often made for the purposes of decision making in business. How these are arrived at is not our concern here).

We find that these three alternatives of the demands are mutually exclusive and they account for all the possibilities. Hence this sample space has only three events and with them the number associated are $\frac{1}{3}$, $\frac{3}{5}$ and $\frac{1}{15}$. These are non negative numbers and the probability of total sample space [*i.e.*, $P(S)$] of all the three events is $\frac{1}{3}+\frac{3}{5}+\frac{1}{15}$ $= \frac{5+9+1}{15} = 1$. Further if we have to find out the probability of any combination of event, say demand below 10,000, this is the combination of first two events and logically the probability with this will be $\frac{1}{3}+\frac{3}{5}=\frac{14}{15}$. Thus, these three numbers associated with the demands can be the probabilities for the three events.

If for the last event the businessman associates a chance of one out of ten then $P(S) = \frac{1}{3}+\frac{3}{5}+\frac{1}{10} \neq 1$. This means that the probability of all the events in the sample space does not add to one, and hence we cannot say that these numbers give the probabilities of the three events.

16.5 ADDITION RULE OF PROBABILITY

The simplest and most important rule used in calculation of probabilities is the addition rule. It states:

For any two events A_1 and A_2 (not necessarily simple) defined on a given sample space,

$$P(A_1 + A_2) = P(A_1) + P(A_2) - P(A_1 A_2) \quad (16.2)$$

Proof. Let us first try to write down $A_1 + A_2$ as sum of mutually exclusive events with the help of the Fig. 16.6.

We can see that the event $A_1 + A_2$ is made up of A_1 and $A_2 \overline{A}_1$ where $A_2 \overline{A}_1$ means that A_2 occurs but A_1 does not occur. In the diagram 16.6 it is represented by the dotted portion. It is clear that A_1 and $\overline{A}_1 A_2$ are mutually exclusive and therefore we have

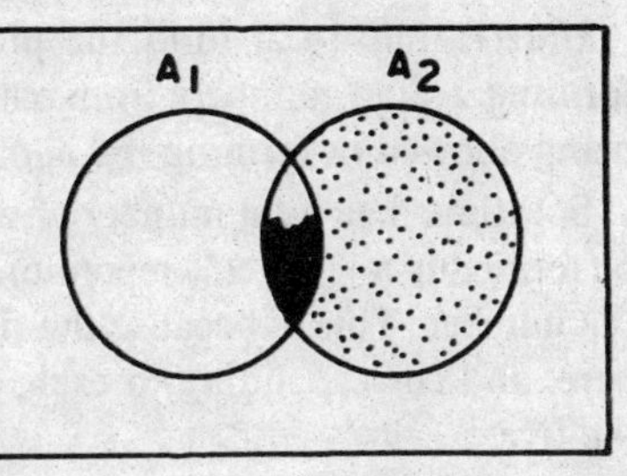

$$p(A_1 + A_2) = P(A_1) + P(A_2 \overline{A}_1)$$

Event A_2 can be written as a sum of two mutually exclusive events in the following way:

$$A_2 = A_2 A_1 + A_2 \overline{A}_1,$$

where $A_2 A_1$ is the shaded portion representing the occurrence of both A_1 and A_2.

Therefore $\quad P(A_2) = P(A_2 A_1) + P(A_2 \overline{A}_1)$

Substituting the value of $P(A_2 \overline{A}_1)$ so obtained, we get the result.

Cor. If A_1 and A_2 are mutually exclusive then we know that $A_1 A_2$ is impossible event. Therefore, $P(A_1 A_2) = 0$ and the addition rule reduces to

$$P(A_1 + A_2) = P(A_1) + P(A_2) \quad (16.3)$$

Remark. By constructing a similar diagram for three events A_1, A_2 and A_3 we can see that $P(A_1 + A_2 + A_3) = P(A_1) + P(A_2) + P(A_3)$

$$-P(A_1 A_2) - P(A_2 A_3) - P(A_3 A_1) + P(A_1 A_2 A_3) \quad (16.4)$$

Illustration 16.1. (*a*) If a die (*a* cube of six marked 1 to 6) is rolled without bias, find the chance that it may not show up 1.

Solution. The total number of ways in which the cube can show up = 6.

But any one face can show up only in 1 way,

The probability of showing up 1 = $\frac{1}{6}$

Hence the probability of not showing up 1 = $1 - \frac{1}{6} = \frac{5}{6}$

(*b*) What is the chance of throwing a number greater than 3 in a single throw with a die.

Solution. There are 6 ways in which the die can fall. It may show up 1 to 6. Of these 3 are favourable to the event, '4, 5 and 6'.

$\therefore$ The required chance $= \frac{3}{6} = \frac{1}{2}$.

Illustration 16.2. Find the probability of getting 37 in randomly forming 2-digit numbers from out of 1, 3, 5, 7, 9, repetitions of a digit being allowed in forming the number.

Solution. The total number of ways in which a two digit number can be formed is $5 \times 5 = 25$, repetitions of a digit being allowed in forming the number. This is because the first digit could be any one of the five here, and corresponding to each, the second could also be any one of the five.

The number 37 can be formed in one way.

Hence the required probability $= \frac{1}{25}$.

Illustration 16.3. Find the chance of picking an even number from the series of Natural numbers 1 to 100.

Solution. The number of even numbers in series 1 to 100 = number of odd numbers = 50.

The chance of picking an even number $= \frac{50}{100} = \frac{1}{2}$.

Illustration 16.4. What is the chance of throwing not less than 16 in a single throw with three dice?

Solution. Maximum points on three dice is 18. The number of points may be 16, 17 or 18.

(*i*) 16 can occur in 6 ways *viz*., (6, 6, 4), (6, 4, 6), (4, 6, 6), (6, 5,5), (5, 6, 5), (5, 5, 6).

(*ii*) 17 can occur in 3 ways, *viz*., (6, 6, 5), (6, 5, 6), (5, 6, 6).

(*iii*) 18 can occur in only one way *viz*. (6, 6, 5).

The total number of favourable cases = 6 + 3 + 1 = 10

Total number of all possible events is $6 \times 6 \times 6 = 216$

Hence the required probability $= \frac{10}{216}$.

Illustration 16.5. A bag contain 20 balls marked 1 to 20. One ball is drawn at random. What is the probability that it is marked with a number multiple 5 or 7?

Solution. The sample space in this illustration consists of 20 points.

The number of events favourable to drawing of a number multiple of 5 is 4 (5, 10, 15 and 20). So probability of a 'multiple of 5' is 4/20 = 1/5. Similarly, number of events favourable to drawing of a number multiple of 7 is 2 (7 and 14). and the probability of it is 2/20 = 1/10.

Hence probability of union of the two, *i.e.*, probability of drawing a number 'multiple of 5 or 7' is

$$\left(\frac{1}{5}+\frac{1}{10}\right)=\frac{3}{10}$$

because there is no number which is a multiple of both 5 and 7.

Illustration 16.6. In the above illustration, find the probability of drawing a number multiple of 3 or 5.

Solution. Probability of drawing a number multiple of 5 = 4/20 or 1/5 (as above). the number of events favourable to drawing a number multiple of 3 is (3, 6, 9, 12, 15 and 18).

Hence probability of drawing a multiple of 3 = 6/20 or 3/10.

But to find the probability of union of the two we cannot simply add the two probabilities as drawing of a 'multiple of 3' and that of 5 are *not mutually exclusive*. They contain one event (15) in common. Hence probability of *intersection of the two events* is 1/20, so that the probability of the *union of two events* is

$$P(\text{multiple of 3 or 5}) = \frac{1}{5}+\frac{3}{10}-\frac{1}{20}$$
$$= \frac{4+6-1}{20}=\frac{9}{20}.$$

An important note. The addition theorem can be validly applied only when the mutually exclusive events belong to the same set. To show this point Von Mises has given following example:

Suppose the probability of a man dying in his 41st year is 0.011 and the probability of a man marrying in his 42nd year is 0.009. These events are mutually exclusive but it cannot be said that the probability of a man dying in his 41st year and marrying in 42nd year is 0.011 + 0.009 = 0.02. The two events do not belong to the same set.

16.6 CONDITIONAL PROBABILITIES

Suppose we have a group of 20 persons of which 14 are males and 6 females. Of the 14 males, 8 are above 40 years of age and 6 below 40 years of age. Of 6 females, 3 are above 40 and 3 are below 40. In all 11 are above 40 and 9 below 40.

If a person is selected at random, what is the probability that he is above 40 years of age?

Clearly of 20 possible outcomes (choice of a particular person constitutes a distinct outcome) 11 are favourable, and probability of a person being above 40 is 0.55. Now, if we have an additional information that the person selected is a female, how does the probability of her being above 40 modified? Obviously, it would be different. The additional condition imposed restricts the relevant sample space to only 6 outcomes of which 3 are favourable to the given proposition. So probability that a 'female selected is above 40 years of age' is 0.50.

Thus, the probability of an event changes as conditions regarding its sample space change. It is meaningful to talk about the probability of an even *only* if we specify a space in which the event is represented by a simple or compound event. To ask for the probability that a student is 18 years old is meaningless unless we specify whether we are referring to entire student population, or of only colleges or only schools, whether we are referring to students in one city or in a state.

Since there are many problems in which we are interested in probabilities relative to different sample spaces, let us use the symbol $P(A/B)$ and read its conditional probability of A relative B.

In fact, every probability is a conditional probability. Whenever we use the simplified notation $P(A)$, it is understood that the sample space is specified in advance and that there is no scope of confusion if it is not specifically stated.

In the above illustration if 'above 40 years of age' is denoted by A and below 40 years of age is denoted by B, and also if F represents a female and M a male, than $P(A/F)$ reads the *conditional probability of over 40 years of age given a female*.

Let us consider another example. In an urn containing 10 balls, 4 are red and the remaining green. What is the probability of drawing a red ball after drawing a green ball without replacement?

Now, as one green ball has been drawn, the sample space consists of only 9 points—4 representing the outcome of drawing a red ball and 5 the green ball. Number or cases favourable to drawing a red ball are 4. So probability of drawing a red ball *after* drawing a green ball is 4/9.

Let us come back to first example. We know the probability of getting female who is above 40 years of age is

$$P(AF) = \frac{3}{20}.$$

Also, the probability of getting a female, of any age, is

$$P(F) = \frac{6}{20}.$$

To get the probability of a *person* above 40, given a female, we can apply the following rule reducing it to a problem in proportion

$$P(A/F) = \frac{P(AF)}{P(F)} = \frac{3/20}{6/20} = 0.5.$$

In general, we adopt the following definition:

$$P(A/B) = \frac{P(AB)}{P(B)}. \text{ Provided } P(B) \neq 0 \qquad (16.5)$$

Thus, probability of an event A given B is equal to the probability of event A *and* event B (occurring simultaneously) divided by the probability of event B.

16.7 THE MULTIPLICATION RULE

By the definition of conditional probability

$$P(A/B) = \frac{P(AB)}{P(B)}$$

So, $P(AB) = P(B).P(A/B)$ (16.6)

Similarly, $P(BA) = P(A).P(B/A)$ (16.7)

In words, probability of joint occurrence of events A and B is equal to the probability of occurrence of A and the conditional probability of occurrence of B given A.

Before illustrating the use of the multiplication rule, let us introduce the concept of statistical independence. Two events A and B are said to be *independent* if the occurrence of one does not affect the probability of the occurrence of the other. In other words, probability of occurrence of one remains same whether the other has occurred or not.

The above condition can be represented by

$$P(A) = P(A/B) \text{ and } P(A/\overline{B})$$

The condition of independency can be illustrated by taking the example of a deck of cards. The probability of any one card being an ace is independent of whether it is black or red, because while there are 4 aces in all cards, two are red and two black. So if we are given a red card, the probability of it being an ace is the same 1/13, as when we didn't know if it was red or black.

By virtue of this fact, in case of *independent events*, the law of multiplication of probability acquires a simple from

$$P(AB) = P(A).P(B) \qquad (16.9)$$

For more than two independent events occurring simultaneously,

$$P(ABC.......N) = P(A).P(B).P(C)......P(N) \qquad (16.10)$$

Illustration. 16.7. Moroney, in facts from figures has explained the law of multiplication of probabilities by the following illustration:

"Consider the case of a man who demands the simultaneous occurrence of many virtues of an unrelated nature in his young lady. Let us suppose that the insists on a Grecian nose, plantinum-blonde hair, eyes of odd colours, one blue, one brown and finally, a first class knowledge of statistics. What is the probability that the first lady he meets in the street will put ideas of marriage in his head ?"

Solution. To solve the problem we must know the probabilities for the several different requisites. Suppose that the

Probability of finding a lady with Grecian nose = 0.01

Probability of finding a lady with platinum-blonde hair = 0.01

Probability of finding a lady with odd eyes = 0.001 and

Probability of finding a lady with first class knowledge of Statistics = 0.00001.

In order to calculate the probability that all these desirable independent attributes to be found in one lady, we use the multiplication rule (16.10) Multiplying together the individual probabilities, we find for our result that the probability of the first young lady he meets coming up to his expectations is equal to 0.000000000001, or precisely one in a billion.

Illustration 16.8 (*a*). What is the chance of getting two sixes in two rollings of a single die?

Solution. The probability of a six in first rolling = $\frac{1}{6}$.

The probability of a six in second rolling = $\frac{1}{6}$.

As the two rollings are independent, the probability of getting two sixes in two rolls is $\frac{1}{6} \times \frac{1}{6} = \frac{1}{36}$

(*b*) If in a population of adults 1/5 are classified as short and 1/20 as superior in intelligence what is the probability of selecting a person at random who is both short and superior in intelligence?

Solution. Assuming the two characteristics, and thus the two events, being short and being superior in intelligence, as independent,

$$= P \text{ (short and superior intelligence)}$$
$$= \frac{1}{6} \times \frac{1}{20} = \frac{1}{100}.$$

Illustration 16.9. (*a*) A bag contains 6 white and black balls. Three drawings of 1 ball each are made such that (*i*) the balls are replaced before the next drawing, and (*ii*) the balls are not replaced before the next drawing. Find the probability that all the black balls are drawn in each case.

(*b*) Under certain conditions, the probability of hitting a target is only 0.5. Find the probability of destroying the target if 5 trials are made.

Solution. (*i*) Probability of drawing a black ball = $\frac{9}{15}$.

If the balls are replaced before the next drawing is made, the drawing of consecutive balls are independent and probability of each of them being black is $\frac{9}{15}$. So the probability of all the three being black is

$$P(\text{3 blacks}) = \frac{9}{15} \times \frac{9}{15} \times \frac{9}{15} = \frac{729}{3375}$$

(*ii*) In this case, as the balls are not replaced before the next drawing is made, the sample space for second and third drawing changes.

$$P(\text{I black}) = \tfrac{9}{15}$$

$$P(\text{II black/I black}) = \tfrac{8}{14}$$

as there are only 14 events of which 8 are favourable.

Similarly, $P(\text{III black/I and II both black}) = \frac{7}{13}$.

So, $P(\text{all black}) = P(\text{I}\,B) \times P(\text{II}\,B\,/\,\text{I}\,B) \times P(\text{III}\,B\,/\,\text{I}\,B \text{ and II}\,B)$

$$= \frac{9}{15} \times \frac{8}{14} \times \frac{7}{13} = \frac{504}{2730}.$$

Thus, we see that the probability of all 'black' is quite different in the two cases.

(*b*) Assuming only one bullet is sufficient to destroy the target if it hits the mark, we are to find the probability that *at least* one bullet hits the mark.

This can be found by nothing that the events 'at least one hitting the mark.' and 'none hitting the mark' from a complete sample space for given experiment, the sum of their probabilities should be 1

So, probability of at least one hitting the mark is 1 *minus* probability of none hitting the mark.

Probability of a bullet not hitting the mark is 1 *minus* probability of it hitting the mark $= 1 - 0.5 = 0.5$.

As firing of 5 bullets is independent, probability of none of the trials being successful is $(0.5)^5 = .03125$.

So the probability of *at least* one hitting the mark is $1 - 0.03125 = 0.96875$.

Thus, although the probability of destroying the target is only $\frac{1}{2}$ if one short is fired, with the firing of 5 shots, the destroying of the target is nearly definite.

Illustration 16.10. Find the chance of throwing head or tail alternately in 3 successive tossing of a coin.

Solution. The two mutually exclusive possibilities of throwing head and tail are (*i*) *HTH* (*ii*) *THT*.

Now $P\,(i) = P(HTH) = P(H)P(T)P(H) = \frac{1}{2} \times \frac{1}{2} \times \frac{1}{2} = \frac{1}{8}$.

Similarly $P\,(ii) = P(THT) = \frac{1}{2} \times \frac{1}{2} \times \frac{1}{2} = \frac{1}{8}$.

Hence the required probability $= \frac{1}{8} + \frac{1}{8} = \frac{1}{4}$.

Illustration 16.11. A problem in statistic is given to three students A, B and C whose chances of solving it are $\frac{1}{2}$, $\frac{1}{3}$, $\frac{1}{4}$ respectively. What is the probability that the problem will be solved?

Solution. Probability that A will *not* solve the problem $= 1-\frac{1}{2}=\frac{1}{2}$

" B " $= 1-\frac{1}{3}=\frac{2}{3}$

" C " $= 1-\frac{1}{4}=\frac{3}{4}$

Probability that all the three will not solve the problem $= \frac{1}{2}\times\frac{2}{3}\times\frac{3}{4}$

" " $= \frac{1}{4}$

The probability that all the three will be solved $= 1-\frac{1}{4}=\frac{3}{4}$.

Illustration 16.12. A and B play 12 games of chess. A wins 6, B wins 4 and two are drawn. They agree to play three games more. Find the probability that:

(*a*) A wins all the three games,

(*b*) Two games end in a tie,

(*c*) A and B win alternately,

(*d*) B wins at least one game,

(*e*) A wins at least one game.

Assuming that the probabilities of win, loss and draw are the same as the proportion in first 12 games.

Solution. The probability that A wins anyone game $= \frac{6}{12}=\frac{1}{2}$

The probability that B wins any one game $= \frac{4}{12}=\frac{1}{3}$

" " game ends in a draw $= \frac{2}{12}=\frac{1}{6}$

" " does not end in draw $= 1-\frac{1}{6}=\frac{5}{6}$

(*a*) P (A wins all the three games) $= \frac{1}{2}\times\frac{1}{2}\times\frac{1}{2}=\frac{1}{8}$

(*b*) The two games which end in a tie may be 1st and 2nd or the 2nd and 3rd or the 1st and 3rd.

Hence the required probability,

$$P=\left(\frac{1}{6}\times\frac{1}{6}\times\frac{5}{6}\right)+\left(\frac{5}{6}\times\frac{1}{6}\times\frac{1}{6}\right)+\left(\frac{1}{6}\times\frac{5}{6}\times\frac{1}{6}\right)=\frac{5}{72}$$

(*c*) A and B can alternately win in two mutually exclusive ways.

(*i*) A wins, B wins and A wins.

(*ii*) B wins, A wins and B wins.

Hence $$P=\left(\frac{1}{2}\times\frac{1}{3}\times\frac{1}{2}\right)+\left(\frac{1}{3}\times\frac{1}{2}\times\frac{1}{3}\right)=\frac{5}{36}.$$

(*d*) The probability that B wins no game $=\left(1-\frac{1}{3}\right)^3=\left(\frac{2}{3}\right)^3$

Hence $$P=1-\left(\frac{2}{3}\times\frac{2}{3}\times\frac{2}{3}\right)=\frac{19}{27}$$

(*e*) The probability that A wins no game $=\left(1-\frac{1}{2}\right)^3$

Hence $$P=1-\left(\frac{1}{2}\times\frac{1}{2}\times\frac{1}{2}\right)=\frac{7}{8}.$$

Illustration 16.13 (*a*) The odds against A solving a problem are 7 : 5 and the odds in favour of B solving it are 12 : 9. What is the probability that if both of them try it, it will be solved?

(*b*) The odds that A speaks the truth is 3 : 2 and the odds that B speaks the truth is 5 : 3. In what percentage of cases are they likely to contradict each other on an identical point?

Solution. (*a*) The probability of A not solving the problem $=\frac{7}{12}$

" B " $=\frac{9}{21}$

Hence the prob. of both not solving the problem $=\frac{7}{12}\times\frac{9}{21}=\frac{1}{4}$, as the two are independent.

Hence the prob. of both solving the problem, either singly or together, is $\left(1-\frac{1}{4}\right)=\frac{3}{4}$.

(*b*) The probability that A speaks the truth and B, the lie $=\frac{3}{5}\times\frac{3}{8}=\frac{9}{40}$.

The prob. that B speaks the truth and A, the lie $=\frac{2}{5}\times\frac{5}{8}=\frac{1}{4}$.

Hence required prob. $=\frac{9}{40}+\frac{1}{4}=\frac{19}{40}$.

Hence the percentage of cases in which they contradict each other $\frac{19}{40}\times 100=47.5\%$.

Illustration 16.14. From a bag containing 4 white and 6 black balls, two balls are drawn at random.

(*a*) If the balls are drawn one after the other without replacement, find the probability that (*i*) both the balls are white, (*ii*) both the balls are black, (*iii*) the first ball is white and the second is black and (*iv*) one ball is white and the other is black.

(*b*) Find the probability in each of the above four cases, if the balls are drawn one after the other with replacement.

Solution. (*a*) (*i*) Probability that the first ball is white $=\frac{4}{10}$

Probability that the second ball is white $=\frac{3}{9}$

$\therefore$ required probability $=\frac{4}{10}\times\frac{3}{9}=\frac{2}{15}$.

(*ii*) Probability that the first ball is black $= \frac{6}{10}$

Probability that the second ball is black $= \frac{5}{9}$.

$\therefore$ required probability $= \frac{6}{10} \times \frac{5}{9} = \frac{1}{3}$.

(*iii*) Probability that the first ball is white $= \frac{4}{10}$

Probability that the second ball is black$= \frac{6}{9}$

$\therefore$ required probability $= \frac{4}{10} \times \frac{6}{9} = \frac{4}{15}$

(*iv*) The following are the mutually exclusive situations satisfying the problem:

(1) The first ball is white and the second is black;

(2) the first ball is black and the second is white.

Probability of (1) $= \frac{4}{10} \times \frac{6}{9} = \frac{24}{90}$

Probability of (2) $= \frac{6}{10} \times \frac{4}{9} = \frac{24}{90}$

$\therefore$ required probability $= \frac{24}{90} + \frac{24}{90} = \frac{8}{15}$

(b) (*i*) Probability that the first ball is white $= \frac{4}{10}$

Probability that the second ball is white $= \frac{4}{10}$

$\therefore$ required probability $= \frac{4}{10} \times \frac{4}{10} = \frac{4}{25}$.

(*ii*) Probability that the first ball is black $= \frac{6}{10}$

Probability the second ball is black $= \frac{6}{10}$

$\therefore$ required probability $= \frac{6}{10} \times \frac{6}{10} = \frac{9}{25}$.

(*iii*) Probability the first ball is white $= \frac{4}{10}$

Probability that the second ball is black $= \frac{6}{10}$

$\therefore$ required probability $= \frac{4}{10} \times \frac{6}{10} = \frac{6}{25}$.

(*iv*) The following are the mutually exclusive situations satisfying the problem:

(1) the first ball is white and the second is black;

(2) the first ball is black and the second is white.

Probability of (1) $=\frac{4}{10}\times\frac{6}{10}=\frac{24}{100}$

Probability of (2) $=\frac{6}{10}\times\frac{4}{10}=\frac{24}{100}$

$\therefore$ Required probability $=\frac{24}{100}+\frac{24}{100}=\frac{12}{25}$.

16.8. BAYE'S THEOREM ON INVERSE PROBABILITY

Let $A_1, A_2, \ldots, A_n$ be n mutually exclusive and exhaustive events defined on a sample space S and let H be an arbitrary event defined on S such that $P(H) \neq 0$, then

$$P(A_i/H)=\frac{P(H/A_i)P(A_i)}{P(H/A_1)P(A_1)+P(H/A_2)P(A_2)+\ldots P(H/A_n)P(A_n)} \quad (16.11)$$

Proof. Let the events A_i's and H be depicted as in Fig. 16.7.

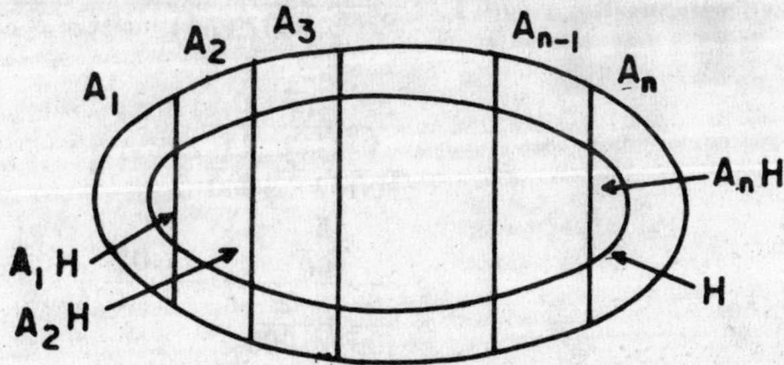

FIG. 16.7.

By definition of conditional probability we have

$$P(A_i/H)=\frac{P(A_iH)}{P(H)} \quad (16.12)$$

Let us try to get the value of $P(A_iH)$ and $P(H)$.

We know that $$P(H/A_i)=\frac{P(A_iH)}{P(A_i)} \quad (16.13)$$

or $$P(A_iH)=P(H/A_i)P(A_i).$$

Also H can be written as sum of mutually exclusive events in the following way;

$$H = A_1H + A_2H + \ldots\ldots + A_nH$$

$$\begin{aligned}\therefore P(H) &= P(A_1H)+P(A_2H)+\ldots+P(A_nH) \quad \text{by axiom of sum}\\ &= P(H/A_1)P(A_1)+P(H/A_2)P(A_2)+\\ &\quad \ldots+P(H/A_n)P(A_n) \quad \ldots (16.14)\end{aligned}$$

using(16.13).

The result follows on substituting the values of $P(A_iH)$ and $P(H)$ from (16.12) and (16.13) respectively in (16.14).

Illustration 16.15. Suppose that 5 men out of 100 and 25 women out of 1000 are colour blind. A colour blind person is chosen at random. What is the probability of his being male (assuming that males and females are in equal proportion).

Solution. Let us first define a few events.

M : A person is male

F : A person is female

C : A person is colour blind.

Obviously, we want to calculate $P(M/C)$. Also it is given that $p(M) = p(F) = \frac{1}{2}$ each, as males and females are in equal proportion.

$$P(C/M) = \frac{5}{100} \text{ and } P(C/F) = \frac{25}{1000}.$$

By Baye's theorem

$$P(M/C) = \frac{P(C/M)P(M)}{P(C/M)P(M) + P(C/F)P(F)}$$

$$= \frac{\frac{5}{100}\cdot\frac{1}{2}}{\frac{5}{100}\cdot\frac{1}{2} + \frac{25}{1000}\cdot\frac{1}{2}}$$

$$= \frac{\frac{5}{100}}{\frac{5}{100} + \frac{25}{1000}} = \frac{.05}{.05+.025}$$

$$= \frac{.05}{.075} = \frac{2}{3}.$$

Illustration 16.16. Suppose that a day's production schedule calls for 9000 items. Three machine each with a daily production capacity of 4000 are available and the probability that an item is defective is 1, 2 and 4 percent for machines A, B and C respectively. On a given day 4000 items were produced on machine A, 4000 on B and 1000 on C. One item is selected at random and found defective. What is the probability that it was produced on machine A.

Solution. Let us define the following events :

A : item is produced on machine A

B : item is produced on machine B

C : item is produced on machine C

D : item is defective.

Obviously, we want $P(A/D)$.

It is given that $P(A) = \frac{4000}{9000} = \frac{4}{9}$

$$P(B) = \frac{4000}{9000} = \frac{4}{9}$$

$$P(C) = \frac{1000}{9000} = \frac{1}{9}$$

$$P(D/A) = \frac{1}{100} = .01$$

$$P(D/B) = \frac{2}{100} = .02$$

and $$P(D/C) = \frac{4}{100} = .04$$

By Baye's theorem :

$$P(A/D) = \frac{P(D/A)P(A)}{P(D/A)P(A) + P(D/B)P(B) + P(D/C)P(C)}$$

$$= \frac{(.01)\frac{4}{9}}{(.01)\frac{4}{9} + (.02)\frac{4}{9} + (.04)\frac{1}{9}}$$

$$= \frac{.04}{.04 + .08 + .04} = \frac{.04}{.16} = \frac{1}{4}.$$

Illustration 16.17. The contents of 3 vessels 1, 2 and 3 are as follows :

1 white, 2 red, 3 black balls
2 white, 3 red, 1 black balls, and
3 white, 1 red, 2 black balls.

A vessel is chosen at random and from it 2 balls are drawn at random. The two balls are one red and one white. What is the probability that they come from the second vessel.

Solution. Let us define the following events :

V_1 : vessel selected is 1
V_2 : vessel selected is 2
V_3 : vessel selected is 3

and A: Balls drawn are one red and one white.

Obviously, we want the value of $P(V_2/A)$. As the vessel is selected at random, therefore,

$$P(V_1) = P(V_2) = P(V_3) = \frac{1}{3} \text{ each}$$

Also $$P(A/V_1) = \frac{{}^1C_1 \times {}^2C_1}{{}^6C_2} = \frac{2}{15}$$

$$P(A/V_2) = \frac{{}^2C_1 \times {}^3C_1}{{}^6C_2} = \frac{2}{5}$$

and $$P(A/V_3) = \frac{{}^3C_1 \times {}^1C_1}{{}^6C_2} = \frac{1}{5}$$

Applying Baye's theorem :

$$P(V_2/A) = \frac{P(A/V_2)P(V_2)}{P(A/V_1)P(V_1)+P(A/V_2)P(V_2)+P(A/V_3)P(V_3)}$$

$$= \frac{\frac{1}{3}\times\frac{2}{5}}{\frac{1}{3}\times\frac{2}{15}+\frac{1}{3}\times\frac{2}{5}+\frac{1}{3}\times\frac{1}{5}} = \frac{6}{11}.$$

Illustration 16.18. There is one urn which contains either 40% black balls or 60% black balls. Both these possibilities are considered equally likely. A ball is drawn at random and found to be black. (*a*) What is the probability that the urn has 60% black balls. (*b*) Next, another ball is drawn (after replacement) and this one is also found to be black. What is the new probability that the urn contains 60% black balls.

Solution. Once again, let us define the following events :

B_1 : Urn contains 40% black balls

B_2 : Urn contains 60% black balls, and A: The balls drawn is black.

(*a*) We want $P(B_2/A)$. By Baye's theorem

$$P(B_2/A) = \frac{P(A/B_2)P(B_2)}{P(A/B_1)P(B_1)+P(A/B_2)P(B_2)}$$

It is given that $P(B_1) = P(B_2) = \frac{1}{2}$

as the two cases are equally likely.

Also $\quad P(A/B_1) = 40\% = .4$

and $\quad P(A/B_2) = 60\% = .6$

Therefore, $P(B_2/A) = \frac{(.6)(.5)}{(.6)(.5)+(.4)(.5)} = \frac{.3}{.3+.2} = .6$

(*b*) In this case, $P(B_2)$ has changed from .5 to .6 on the basis of the information that first ball drawn is black. Thus, Now $P(B_1) = .4$ and $P(B_2) = .6$

And $P(A/B_1)$ and $P(A/B_2)$ remain as they are.

So $\quad P(B_2/A) = \frac{(.6)\times(.6)}{(.6)(.6)+(.4)(.4)} = \frac{.36}{.36+.16} = \frac{36}{52} = .69.$

Illustration 16.19. A bag contains 4 white and 5 black balls and another bag contains 6 white and 10 black balls. A bag is selected at random and a ball is drawn out of it. If the ball so drawn happens to be white, find the probability that it came from the second bag.

Solution. Let B_1 and B_2 denote the events of selecting the first bag and the second bag, respectively. Let A denote the event of drawing a white ball. Now, we have the following probabilities :

$$P(B_1)=\frac{1}{2}, \qquad P(B_2)=\frac{1}{2}$$

$$P(A/B_1)=\frac{4}{9}, \qquad P(A/B_2)=\frac{6}{16}.$$

We are required to find the value of $P(B_2/A)$ which is given by

$$P(B_2/A) = \frac{P(B_2).P(A/B_2)}{P(B_1).P(A/B_1)+P(B_2).P(A/B_2)}$$

$$= \frac{\frac{1}{2}.\frac{6}{16}}{\frac{1}{2}.\frac{4}{9}+\frac{1}{2}.\frac{6}{16}} = \frac{27}{59}.$$

Illustration 16.20. If a machine is correctly set up, it will produce 90% acceptable items. If it is incorrectly set up, it will produce only 40% acceptable items. Experience shows that the 70% of the setups are correctly done. Find the probability that a setup has been correctly done if refer the setup (*i*) the first item produced is accepted; (*ii*) the first two items produced are both acceptable; and (*iii*) the first three items produced are acceptable, acceptable and unacceptable, respectively.

Solution. Let B_1 and B_2 denote the events correct setup and incorrect setup, respectively. We are given that $P(B_1)=0.7$ and $P(B_2) = 1-0.7=0.3$.

(*i*) Let A denote the event that the first item produced is acceptable. Now, we are given that $P(A/B_1)=0.9$ and $P(A/B_2)=0.4$. We are to find the value of $P(B_1/A)$, which is given by

$$P(B_1/A) = \frac{P(B_1).P(A/B_1)}{P(B_1).P(A/B_1)+P(B_2).P(A/B_2)}$$

$$= \frac{0.7\times 0.9}{0.7\times 0.9+0.3\times 0.4} = \frac{0.63}{0.75} = \frac{21}{25}.$$

(*ii*) Let A denote the event that the first two items produced are both acceptable. Now, we have

$$P(A/B_1) = 0.9\times 0.9 = 0.81$$

$$P(A/B_2) = 0.4\times 0.4 = 0.16.$$

The required probability is given by

$$P(B_1/A) = \frac{P(B_1).P(A/B_1)}{P(B_1).P(A/B_1)+P(B_2).P(A/B_2)}$$

$$= \frac{0.7\times 0.81}{0.7\times 0.81+0.3\times 0.16} = \frac{0.567}{0.615} = \frac{189}{205}.$$

(*iii*) Let A denote the event that the first three items produced are acceptable, acceptable and unacceptable, respectively. Now, we have

$$P(A/B_1) = 0.9 \times 0.9 \times 0.1 = 0.081$$
$$P(A/B_2) = 0.4 \times 0.4 \times 0.6 = 0.096.$$

The required probability is given by

$$P(B_1 / A) = \frac{P(B_1).P(A/B_1)}{P(B_1).P(A/B_1) + P(B_2).P(A/B_2)}$$
$$= \frac{0.7 \times 0.081}{0.7 \times 0.081 + 0.3 \times 0.096}$$
$$= \frac{0.0567}{0.0855} = \frac{63}{95}.$$

Illustration 16.21. There are three plants in a factory producing the same product independently. Experience shows that 2%, 4% and 5% of the items produced by the respective plants are defective. The daily outputs of the respective plants are 2000, 3000 and 5000 units. If an item selected at random from a day's output of the three plants is found to be defective, find the probability that it was produced by the second plant.

Solution. Let B_1, B_2 and B_3 denote the events of an item being produced by the first, second and third plant, respectively. Let A denote the event of an item produced being defective. Now, we have the following probabilities:

$$P(B_1) = \frac{2000}{10000} = 0.2,$$
$$P(B_2) = \frac{3000}{10000} = 0.3$$
$$P(B_3) = \frac{5000}{10000} = 0.5$$
$$P(A/B_1) = \frac{2}{100} = 0.02$$
$$P(A/B_2) = \frac{4}{100} = 0.04$$
$$P(A/B_3) = \frac{5}{100} = 0.05.$$

We are to find the value of $P(B_2 / A)$ which is given by

$$P(B_2 / A) = \frac{P(B_2)P(A/B_2)}{P(B_1)P(A/B_1) + P(B_2)P(A/B_2) + P(B_3)P(A/B_3)}$$
$$= \frac{0.3 \times 0.04}{0.2 \times 0.02 + 0.3 \times 0.04 + 0.5 \times 0.05}$$
$$= \frac{0.012}{0.041} = \frac{12}{41}.$$

16.9 PERMUTATIONS AND COMBINATIONS

Many problems in probability require the enumeration of how many points there were in sample space and how many of them are favourable to the given event. For instance, in flipping of 2 coins, we can construct a tree diagram (Fig. 16.8) and count the number of cases to be 4. This procedure is all right when we are dealing with small number of events increases, it becomes impractical to construct such tree diagrams. There exist some general methods for determining the number of possible cases without actually counting them.

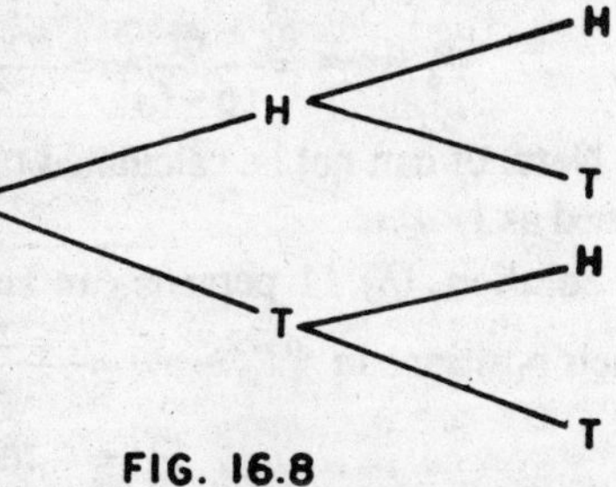

FIG. 16.8

Permutations of dissimilar things. suppose we are to count in how may ways can 4 cards be drawn from a pack of 52 cards without replacement. Obviously, first card can be any of the 52 cards. The second card can be any one of the remaining 51 cards. As can be easily visualised, there are 52 × 51 *ordered* two card sets possible, because for every one card drawn in 50 ways. So 52 × 51 × 50 ordered sets of 3 cards can be drawn from a pack of 52 cards. similarly, 52 × 51 × 50 × 49 sets of 4 cards are possible if cards are drawn without replacement.

The argument of the above illustration can be generalised to the following theorem :

The number of ordered sets of r elements taken from a set of n different elements without replacement is

$$n(n-1)(n-2).....(n-r+1)$$

and is represented by ^nP_r, *read* permutations of n things taken r at a time. Using factorial notation

$$n! \text{ for } n(n-1)(n-2).....3.2.1$$

We can write ^nP_r as

$$^nP_r = n(n-1).....(n-r+1)$$

$$= \frac{n(n-1).....(n-r+1)n(n-r).....(n-r-1).....2\times 1}{(n-r).....(n-r-1).....2\times 1}$$

$$= \frac{n!}{(n-r)!}. \qquad \text{... (16.15)}$$

Illustration 16.22 (*a*). There are six seats available in a compartment. In how many ways can six persons be seated?

(*b*) In how many different ways can a 11-men football team be chosen from a squad of 20 men if the positions are not ignored.

Solution. (*a*) Six persons can be seated in

$$^{6}P_6 \text{ ways} = \frac{6!}{(6-6)!} = \frac{6\times5\times4\times3\times2\times1}{1} = 720 \text{ ways.}$$

Note. 0! can not be calculated the way, we have defined *n*!. It is defined as 1.

Solution. (*b*) 11 persons can be chosen from 20 men not ignoring their positions in $^{20}P_{11}$ ways $= \frac{20!}{(20-11)!} = \frac{20!}{9!}$ ways

$$= 6.704 \times 10^{12} \text{ ways approx.}$$

Permutation of *n* things not all of which are dissimilar. If of *n* things r_1 are of one type, r_2 of another kind, and so on, the number of permutation would be different. For example, the number of ways in which letters of the word 'Exercise' can be arranged is not given by 8!. Distinguishing at the moment between three E's by subscripts 1, 2 and 3, there would be 8! ways of arranging the letters. However, if we drop the subscripts, a few of the permutations would look alike. In above arrangements, interchanging the three E's, we get 3! permutations with subscripts while only 1 without the subscript. So we must divide 8! by 3! to obtain the true permutations possible. More generally, if on *n* things r_1 are of one kind, r_2 of other, and so on, the total number of permutations possible is $\frac{n!}{r_1!r_2!...}$.

Illustration 16.23. In how many ways can the letters of the word 'statistics' be arranged?

Solution. Here *t* and *s* occur thrice each, while *i* occurs twice. The total numbers of letters is 10. So number of permutations possible are

$$\frac{10!}{3!\ 3!\ 2!...} = \frac{10\times9\times8\times7\times6\times5\times4\times3\times2\times1}{3\times2\times1\times3\times2\times1\times2\times1}$$

$$= 50{,}400 \text{ ways.}$$

Combinations. In permutations, we are concerned with order of the items. If order is immaterial, the arrangements, thus, resulting are called combinations. Suppose of *n* different things *r* were to be selected. Then there are *n*!(*n* – *r*)! ways of selecting ordered sets. In this number, each combination has been counted *r*! permutations possible of a group or *r* items. So the number of combinations of *n* things taken *r* at a time is

$$^{n}C_r = \frac{^{n}P_r}{r!} = \frac{n!}{(n-r)!\ r!}.$$

Illustration 16.24. In how many ways can a 11-men football team be selected from a squad of 20, if positions be disregarded?

Solution. The combination of 20 items taken 11 at a time are

$$\frac{20!}{11!\ 9!} = 167{,}960 \text{ ways.}$$

Illustration 16.25. What is the probability of getting 3 white balls in a draw of 3 balls from a box containing 5 white and 4 red balls?

Solution. The number of favourable cases to the selection of 3 white balls is the same as the number of ways of getting 3 balls out of the 5, *i.e.*, ${}^5C_3$.

Similarly, the number of all possible mutually exclusive and equally likely cases is the same as the number of ways of selecting any three balls (white or red) out of the total of 5 + 4 or 9 balls in all ; *i.e.*, ${}^9C_3$.

The required probability is :

$${}^5C_3 / {}^9C_3 = \frac{5\times4\times3}{3\times2\times1} \div \frac{9\times8\times7}{3\times2\times1} = \frac{5}{42}.$$

Illustration 16.26. In a game of bridge, what is the chance that a specified player gets all the four Kings?

Solution. A player gets 13 cards in the game. The total number of ways in which a particular person can get any of the 13 cards = ${}^{52}C_{13}$. The favourable cases to the given event are those where the player gets out of 4 Kings all the four, and any 9 out of the remaining cards.

4 Kings can be obtained in ${}^4C_4$ ways and any 9 out of the remaining 48 cards can be obtained in ${}^{48}C_9$ ways.

∴ The number of favourable cases are ${}^4C_4 \times {}^{48}C_9$

∴ The required probability $= {}^4C_4 \times {}^{48}C_9 / {}^{52}C_{13}$.

Illustration 16.27. In an urn, there are 5 white and 4 black balls. What is the probability of drawing the first ball white, the second black, the third white, the fourth black, and so on, if they are drawn one at a time?

Solution. The problem is similar to the one relating to the arrangement of 9 balls (5 white + 4 black) in 9 places, which can be arranged in 9 ways.

5 white balls are to occupy 5 odd places, *viz.*, 1, 3, 5, 7 and 9, which can be arranged in 5! ways, and the 4 black balls are to occupy 4 even places, *viz.*, 2, 4, 6, and 8, which can be arranged in 41 ways.

So the required probability $= \frac{6\times5\times4\times3\times2\times1}{1}.$

Illustration 16.28. A bag contains 3 green and 8 white balls. If one ball is drawn from it, find the chance that the ball drawn is green.

Solution. Total number of ways, in which one ball can be drawn out of 11 balls is ${}^{11}C_1$.

Favourable number of ways in which one green ball can be drawn out of 3 green balls is ${}^{3}C_1$.

Hence, the required probability $= \frac{{}^{3}C_1}{{}^{11}C_1} = \frac{3}{11}$.

Illustration 16.29. From a bag containing 5 white and 5 black balls, 4 balls are drawn at random.

(*i*) What are the odds against these being all white?

(*ii*) What is the chance that 2 are white and 2 black?

Solution. The total number of ways in which 4 balls can be drawn out of 5 + 5 of 10 balls is ${}^{10}C_4$.

The number of ways of drawing 4 white balls $= {}^{5}C_4$.

Hence the probability of drawing 4 white balls is $\frac{{}^{5}C_4}{{}^{10}C_4} = \frac{5.4.3.2}{10.9.8.7} = \frac{1}{42}$

∴ The odds against the particular event are 41 to 1.

(*i*) 2 white balls can be drawn in ${}^{5}C_2$ ways and 2 black balls can be drawn in ${}^{5}C_2$ ways.

∴ 2 white and 2 black balls can be drawn in ${}^{5}C_2 \times {}^{5}C_2$ ways.

Hence the required chance $= \frac{{}^{5}C_2 \times {}^{5}C_2}{{}^{10}C_4} = \frac{10}{21}$.

Illustration 16.30. From a pack of 52 cards, 1 card is drawn at random. Find the chance of drawing or not drawing a diamond, drawing or not drawing an ace.

Solution. Since a full pack consists of 13 hearts, 13 diamonds, 13 spades and 13 clubs.

(*i*) the chance of drawing a diamond $= \frac{13}{52}$

(*ii*) the chance of not drawing a diamond $= 1 - \frac{13}{52} = \frac{39}{52}$

(*iii*) the chance of drawing an ace $= \frac{4}{52}$

(*iv*) the chance of not drawing an ace $= 1 - \frac{4}{52} = \frac{48}{52}$.

Illustration 16.31. From a pack of 52 cards, 4 cards are drawn at random. What is the probability that (*i*) all of them are spades.

(*ii*) there is one of each suit ?

Solution.

(*i*) The total number of ways in which 4 cards are drawn in $^{52}C_4$ and the number of ways of drawing 4 spades is $^{13}C_4$.

Hence the required probability $= \dfrac{^{13}C_4}{^{52}C_4} = \dfrac{13.12.11.10}{52.51.50.49} = \dfrac{11}{4165}$.

(*ii*) The total number of ways of drawing 4 cards $= {}^{52}C_4$.

But one card, of each suit can be drawn in $^{13}C_1$ or 13 ways.

∴ 4 cards, one of each suit, can be drawn in $13 \times 13 \times 13 \times 13$ or 13^4 ways.

Hence the required probability $= \dfrac{13^4}{^{52}C_4} = \dfrac{2197}{20825}$.

Illustration 16.32. Five cards are drawn from a pack. What is the probability of three being 4 kings.

Solution. The number of ways of selecting 4 kings $= {}^4C_4$. The number of ways of selecting 1 card from remaining 48 cards $= {}^{48}C_1$.

Hence the number of ways of selecting four kings and 1 other card of pack is $^4C_4 \times {}^{48}C_1$.

Hence the required probability $P = {}^4C_4 \times {}^{48}C_1 / {}^{52}C_5 = \dfrac{1}{54145}$.

Illustration 16.33. What is the chance of drawing, without bias four aces successively from a pack of 52 cards?

Solution. The chance of drawing an ace at the second drawing $= \frac{4}{52}$.

When 1 ace is drawn, 3 aces and 51 cards are left.

The chance of drawing an ace at the second drawing$= \dfrac{3}{51}$.

Now 2 aces and 50 cards are left.

The chance of drawing a third ace at the third drawing $= \dfrac{2}{50}$.

Now 1 ace and 49 cards left.

The chance of drawing the fourth ace at the last drawing $= \dfrac{1}{49}$.

Hence, by multiplication law of probability, the chance of the compound event is $= \dfrac{4}{52} \times \dfrac{3}{51} \times \dfrac{2}{50} \times \dfrac{1}{49} = \dfrac{1}{270725}$.

Illustration 16.34. In a bag there are 4 white and 8 yellow balls. Two balls are drawn at random. What is the probability that of these two balls one is white and the other yellow.

Solution. The total number of ways of selecting 2 balls out of 12 balls $= {}^{12}C_2$.

One white ball can be drawn out of 4 white balls in ${}^4C_1$ ways. also one yellow can be drawn out of 8 yellow balls in ${}^8C_1$ ways.

$\therefore$ ${}^4C_1 \times {}^8C_1$ are the number of ways of drawing two balls of different colours.

Hence the required probability $= \dfrac{{}^4C_1 \times {}^8C_1}{{}^{12}C_2} = \dfrac{16}{33}$.

Illustration 16.35. Two drawings each of 4 balls are made from a bag containing 6 white and 7 green balls. What is the chance that the first drawing will give 4 white, and the second 4 green balls?

(*a*) The balls drawn are being replaced before the second draw.

(*b*) The balls drawn are not being replaced before the second draw.

Solution. (*a*) The balls are being replaced

4 white balls can be drawn out of 6 white balls in ${}^6C_4$ ways in the first drawing.

$$\therefore \quad P\text{ (drawing 4 white balls)} = \frac{{}^6C_4}{{}^{13}C_4} \quad \ldots (i)$$

and 4 green balls can be drawn out of 7 green balls in ${}^7C_4$ ways in the second draw.

$$\therefore \quad P\text{ (drawing 4 green balls)} = \frac{{}^7C_4}{{}^{13}C_4} \quad \ldots (ii)$$

Since (*i*) and (*ii*) are independent, the required probability $= \dfrac{{}^6C_4}{{}^{13}C_4} \times \dfrac{{}^7C_4}{{}^{13}C_4}$

$= \dfrac{21}{20449} = 0.00102.$

(*b*) The balls are not replaced. If the balls are not replaced, then the required probability $= \dfrac{{}^6C_4}{{}^{13}C_4} \times \dfrac{{}^7C_4}{{}^9C_4} = \dfrac{105}{18018} = 0.00582$.

Illustration 16.36. An urn contain 9 balls, two of which are white, three blue and four black. Three balls are drawn at random from the urn *i.e.*, every ball has an equal chance of being included in the three. What is the chance that

(*i*) the three balls are of different colours?

(*ii*) two balls are of the same colour and third of different?

(*iii*) the balls are all of the same colour?

Solution.

(*i*) The total number of exhaustive mutually exclusive and equally likely cases = ${}^9C_3 = 84$.

The number of ways in which a white, a blue and a black ball can be drawn = $2 \times 3 \times 4 = 24$.

Therefore the required chance = $\frac{24}{84} = \frac{2}{7}$.

(*ii*) Two white balls can be drawn in ${}^2C_2$ ways and then a blue or black ball in 7 ways, *i.e.*, the number of ways in which two white balls and one another coloured ball can be drawn is $7 \times {}^2C_2 = 7$ ways. Two blue balls and another coloured ball can be drawn in $6 \times {}^3C_2 = 18$ ways. Similarly, 2 black balls and another coloured ball can be drawn in $5 \times {}^4C_2 = 30$.

∴ The total number of ways in which two balls of the same colour and a third of a different colour may be drawn is 7 +18 + 30 = 55.

Hence the required chance $= \dfrac{55}{84}$.

(*iii*) Three blue balls can be drawn in 1 way, and 3 black balls in ${}^4C_3 = 4$ ways, so that the total number of ways of drawing three balls of the same colour is 5.

The required chance $= \dfrac{5}{84}$.

Illustration 16.37. A bag contains, identical except for colour, of which 5 are red and 3 white. A man draws two balls at random. What is the probability that :

(*i*) one of the balls shown is white and the other red?

(*ii*) both are of the same colour?

What would be the values of these probabilities if a ball is drawn and replaced, and, then, another ball is drawn.

Solution. When the ball drawn is not replaced.

The total number of ways in which 2 balls can be drawn out of 8 balls is ${}^8C_2$.

(*i*) Now in ${}^5C_1$ ways a red ball and in ${}^3C_1$ ways a white can be drawn. Therefore in ${}^5C_1 \times {}^3C_1$ ways, two balls, both having different colours, can be drawn.

∴ The probability of drawing a white and a red ball $\frac{{}^5C_1 \times {}^3C_1}{{}^5C_1 \times {}^3C_1} = \frac{30}{56} =$ 0.536.

(*ii*) Now in ${}^5C_2$ ways two red balls and in ${}^3C_2$ ways two white balls can be drawn. Therefore, in $({}^5C_2 + {}^3C_2)$ ways two balls of the same colour can be drawn.

∴ The probability of drawing two balls of the same colour

$$\frac{({}^5C_2 + {}^3C_2)}{{}^8C_2} = \frac{26}{56} = 0.464$$

When the ball drawn is replaced.

Since the ball is replaced, the number of balls remains same before drawing the balls.

∴ The probability of drawing a red ball $= \frac{5}{8}$

and probability of drawing a white ball $= \frac{3}{8}$.

(*i*) Now two balls both of different colours can be drawn in 2 ways, *i.e.*, first red and then white, and first white then red.

The probability of drawing a white ball and a red ball $= 2 \times \frac{3}{8} \times \frac{5}{8}$

$= \frac{30}{64} = 0.469.$

(*ii*) Now two balls of the same colour can be drawn as both are red or both are white.

∴ The probability of drawing two balls of the same colour

$$= \frac{5}{8} \times \frac{5}{8} + \frac{3}{8} \times \frac{3}{8} = \frac{34}{64} = 0.531.$$

Illustration 16.38. From a bag containing 4 white and 6 red balls, two balls are drawn together. Find the probability when (*i*) both the balls are white, (*ii*) both the balls are black, (*iii*) one is white and other is black.

Solution. Total number of ways of drawing two balls together out of 10 balls $= {}^{10}C_2$.

(*i*) The number of favourable ways, *i.e.*, the number of ways of drawing 2 white balls out of 4 $= {}^4C_2$

∴ required probability $= \frac{{}^4C_2}{{}^{10}C_2} = \frac{2}{15}$.

(*ii*) The number of favourable ways, *i.e.*, the number of ways of drawing 2 black balls out of 6 $= {}^6C_2$.

∴ required probability $= \frac{{}^6C_2}{{}^{10}C_2} = \frac{1}{3}$.

(*iii*) The number of favourable ways, *i.e.*, the number of ways of drawing 1 white ball out of 4 and one black ball out of 6 $= {}^4C_1 \times {}^6C_1$.

$\therefore$ required probability $= \dfrac{{}^4C_1 \times {}^6C_1}{{}^{10}C_2} = \dfrac{8}{15}$.

Illustration 16.39. From a pack, containing 52 cards, two cards are drawn at random. Find the probability that (*i*) both are of different colours, (*ii*) both are from a particular suit and (*iii*) both are from the same suit.

Solution. Total number of ways of drawing 2 cards out of 52 $= {}^{52}C_2$

(*i*) The number of favourable ways, *i.e.*, the number of ways of drawing 1 black card out of 26 and 1 red card out of 26 $= {}^{26}C_1 \times {}^{26}C_1$.

$\therefore$ required probability $= \dfrac{{}^{26}C_1 \times {}^{26}C_1}{{}^{52}C_2} = \dfrac{26}{51}$.

(*ii*) The number of favourable ways, *i.e.*, the number of ways of drawing 2 cards from a particular suit $= {}^{13}C_2$.

$\therefore$ required probability $= \dfrac{{}^{13}C_2}{{}^{52}C_2} = \dfrac{1}{17}$.

(*iii*) The number of favourable ways, *i.e.*, the number of ways of drawing 2 cards from the same suit (*i.e.*, from any one of the four suits) $= {}^4C_1 \times {}^{13}C_2$.

$\therefore$ required probability $= \dfrac{{}^4C_1 \times {}^{13}C_2}{{}^{52}C_2} = \dfrac{4}{17}$.

Illustration 16.40. *A* and *B* keep on cutting a pack of cards one after the other till one cuts a diamond. If *A* makes the first cut, find their respective chances of cutting a diamond.

Solution. Calculation of *A*'s chance of cutting a diamond :

A will cut a diamond in the following infinite mutually exclusive situations :

(1) *A* cuts the diamond in the fast cut ;

(2) *A* and *B* fail in order and then *A* cuts a diamond ;

(3) *A*, *B*, *A*, *B* fail in order and then *A* cuts a diamond ; and so on.

$$\text{Probability of (1)} = \frac{13}{52} = \frac{1}{4};$$

$$\text{Probability of (2)} = \frac{3}{4} \cdot \frac{3}{4} \cdot \frac{1}{4} = \left(\frac{3}{4}\right)^2 \frac{1}{4};$$

$$\textbf{Probability of (3)} = \left(\frac{3}{4}\right)^4 \frac{1}{4}\text{; and so on.}$$

$$\therefore \quad A\text{'s chance of cutting a diamond} = \frac{1}{4}+\left(\frac{3}{4}\right)^2\frac{1}{4}+\left(\frac{3}{4}\right)^4\frac{1}{4}+\ldots..$$

$$= \frac{\frac{1}{4}}{1-\left(\frac{3}{4}\right)^2} = \frac{4}{7}$$

$$\therefore \quad B\text{'s chance of cutting a diamond} = 1-\frac{4}{7} = \frac{3}{7}.$$

Illustration 16.41. Six dice are thrown together and the appearing of 3 on a die is counted as a success. Find the probability that there will be 4 successes.

Solution. The probability of a success on single die is $\frac{1}{6}$ and that of failure is $\frac{5}{6}$. We want success on any 4 of the 6 dice and the failure on the remaining two.

$$\therefore \quad \text{Required probability} = {}^6C_4\left(\tfrac{1}{6}\right)^4\left(\tfrac{5}{6}\right)^2 = \frac{375}{(6)^6}.$$

Illustration 16.42. An urn contains 4 white and 6 black balls and another urn contains 8 white and 9 black balls. A ball is drawn from the first urn and put into the second and then a ball is drawn from the second. Find the probability that the ball drawn from the second urn is white.

Solution. The following are the mutually exclusive situations to get a white ball from the second urn:

(1) White ball is drawn from the first urn and then a white ball from the second.

(2) Black balls is drawn from the first urn and then a white ball is drawn from the second.

$$\text{Probability of (1)} = \frac{4}{10}\times\frac{9}{18} = \frac{36}{180}$$

$$\text{Probability of (2)} = \frac{6}{10}\times\frac{8}{18} = \frac{48}{180}$$

$$\therefore \quad \text{Required probability} = \frac{36}{180}+\frac{48}{180} = \frac{7}{15}.$$

Illustration 16.43. *A* can hit a target 4 times in 5 shots, *B* 3 times in 4 shorts and *C* 2 times in 3 shots. If they fire a volley, find the chance that the target is hit.

Solution. Probability of *A*'s failing to hit the target $= 1-\frac{4}{5} = \frac{1}{5}$

Probability of *B*'s failing to hit the target $= 1-\frac{3}{4} = \frac{1}{4}$

Probability of C's failing to hit the target $= 1-\frac{2}{3}=\frac{1}{3}$.

$\therefore$ Probability of all three failing to hit the target $=\frac{1}{5}\cdot\frac{1}{4}\cdot\frac{1}{3}=\frac{1}{60}$

$\therefore$ Probability that the target is hit (by at least one) $=1-\frac{1}{60}=\frac{59}{60}$.

Illustration 16.44. What is the probability that, in a group of three persons, atleast, two were born in the same month (disregard year)?

Solution. Probability that any two were born in the same month

$$={}^3C_2\ {}^{12}C_1\left(\tfrac{1}{12}\right)^2\left(\tfrac{11}{12}\right)=\frac{33}{144}$$

Probability that all three were born in the same month

$$={}^3C_3\ {}^{12}C_1\left(\tfrac{1}{12}\right)^3\left(\tfrac{11}{12}\right)^0=\frac{1}{144}$$

$\therefore$ Probability that, atleast, two were born in same month $=\frac{33}{144}+\frac{1}{144}=\frac{17}{72}$.

Illustration 16.45. From a sales force of 150 persons, one will be chosen to attend a special sales meeting. If 52 are single, and 72 are college graduates and the three-fourths of the 52 (that are single) are college graduates, what is the probability that a sales person selected at random will be neither single nor college graduate?

Solution.

	Single	*Married*	*Total*
Graduate	39	33	72
Non-graduate	13	65	78
Total	52	98	150

On the basis of the given figures (put in circles), this table has been prepared. The table shows that, out of 150 persons, there are 65 persons who are neither single nor graduate.

$\therefore$ Required probability $=\frac{{}^{65}C_1}{{}^{150}C_1}=\frac{65}{150}=\frac{13}{30}$.

Illustration 16.46. A and B stand in a ring with 10 other persons. If the arrangement of the 12 persons is at random, find the chance that there are three persons between A and B.

Solution. Total number of ways in which 12 persons can stand in a ring $=\underline{|12-1}=\underline{|11}$.

Let A and B occupy positions (with three persons in between). Now the arrangement of the remaining 10 persons can be made in $\underline{|10}$ ways. But an equal number of ways is possible if A and B interchange their positions. Hence, the number of favourable ways is $2\underline{|10}$.

$\therefore$ Required probability $=\frac{2\underline{|10}}{\underline{|11}}=\frac{2}{11}$.

16.10 RANDOM VARIABLES AND PROBABILITY DISTRIBUTIONS

In the introductory section of this chapter, we introduced the concept of random experiment as a statistical process the outcome of which is predictable. Tossing of two dice simultaneously, counting the number of vehicles that pass by on a road in a five-minute interval, ascertaining the proportion of boys in new born babies in a week, measuring the daily yield of a milk cow, etc., are all random experiments. The outcome of each such experiment is a number which is termed 'random variable'. The random variable is more or less unpredictable and varies from trial to trial over a specified range in each experiment. Thus, the sum obtained on two dice is a random variable which can take any one of the eleven integral values between 2 and 12, both inclusive. Proportion of boys in new-born babies is also a random variable which can take any fractional value between 0 an 1.

Though a random variable can take any value over its specified range, not all values are equally likely. In the rolling of two dice, the probability of obtaining a 7 is one in six (six favourable sample events in a sample space of 36) while that of obtaining a 12 is only one in thirty-six. Thus, with each value which a random variable may take, is associated a corresponding probability of its occurrence. *A probability distribution gives the probability associated with each value that a random variable may take.* Thus, the probability distribution for the random variable x denoting the number of heads in two tosses of a coin is

$$P(x) = \begin{cases} 1/4 \text{ for } x = 0 \\ 1/2 \text{ for } x = 1 \\ 1/4 \text{ for } x = 2 \end{cases}$$

and the probability distribution for the random variable Y denoting the sum obtained on simultaneous losses of two dice is

$$P(Y) = \begin{cases} 1/36 \text{ for } Y = 2 \text{ and } 12 \\ 2/36 \text{ for } Y = 3 \text{ and } 11 \\ 3/36 \text{ for } Y = 4 \text{ and } 10 \\ 4/36 \text{ for } Y = 5 \text{ and } 9 \\ 5/36 \text{ for } Y = 6 \text{ and } 8 \\ 6/36 \text{ for } Y = 7 \end{cases}$$

It should be noted that the sum of all probabilities in a probability distribution equals one.

Such distributions as are obtained by calculating probabilities for random variables by theoretical considerations (and not by actually conducting the experiment) are called theoretical probability distributions.

The probability distribution can be specified in the manner stated above only in case of discrete variables, *i.e.*, those variables which take

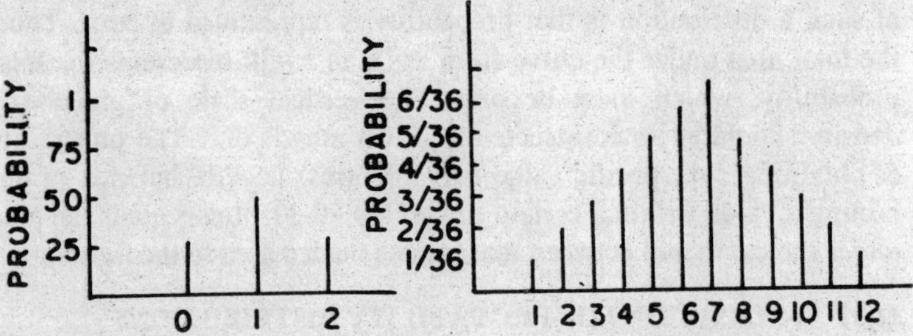

FIG. 16.9.

specific discrete values such as 1/4, 1/2, 1, 2, 3 etc. The random variable in a two-dice experiment is discrete because it can take only integral values between 2 and 12. any other number, say 2.4 or 6.5 cannot be obtained. However, the daily yield of a milk cow is a random variable since it can take any value within the specified range. It can be 12.4 litre or even 12.41 litres. If the specified range is between 0 and 30 litres, in principle, the random variable can take any of the infinite values between these two limits. Such a variable is termed as a continuous random variable. It should be obvious that since there are infinite points in such a sample space, the probability of each point cannot be anything but zero, and, therefore, it is meaningless to talk of a probability distribution in the sense explained above. The probability distribution of a continuous variable is instead specified by a *density function*.

The mathematical expression representing the probability density function can also be obtained like discrete distribution from theoretical considerations governing the behaviour of the random variable under study. This can also be expressed by numerical figures and by graph as in Fig. 16.10.

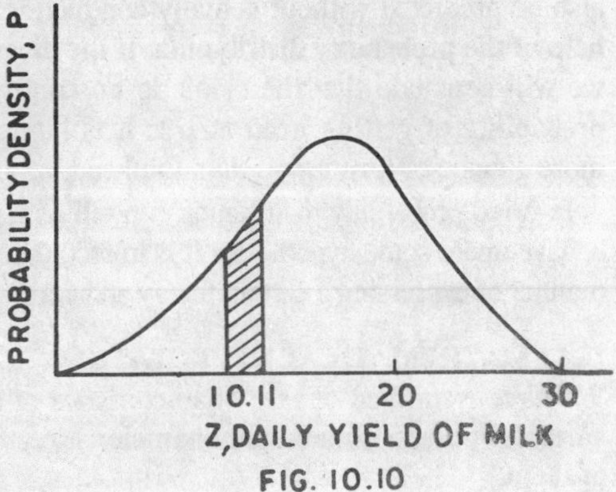

FIG. 10.10

Fig. 16.10 denotes a possible probability density distribution of the continuous random variable *z* representing the daily yields of milk cow. The basic principle of such a distribution is that probability is represented by area. Thus, the total area under the curve from $z = 0$ to $z = 30$ represents the total probability, which must be one. The vertical scale of probability density should be so constructed that this area is one. The probability of obtaining any specific value (say 10 litres) is zero, but that of obtaining a value within a certain range, say 10-11 litres, equals the area *under* the curve and between that range (shaded area in the figure).

16.11 APPLICATIONS OF PROBABILITY DISTRIBUTIONS

The probability distribution has a very wide application in statistical investigations and inferences. Let us explain it by a simple example. We have seen above that the probability distribution of number of heads in tossing of two coins (assumed fair) is

No. of Head	0	1	2	Total
Probability	1/4	1/2	1/4	1

If we toss two coins say 400 times, the expected frequencies of the number of heads if the coins can be assumed fair, will be ;

No. of Heads	0	1	2	Total
Expected No. of Heads	$1/4 \times 400 = 100$	$1/2 \times 400 = 200$	$1/4 \times 400 = 100$	400

If we actually toss two coins 400 times the number of heads may be as above or different from this. Let us assume that in one such experiment, we observe them to be as below :

No. of Head	0	1	2	Total
Actual No. of Heads	95	203	102	400

The divergence between these two may be significant or may be negligible. What constitutes 'negligible divergence', can be determined by statistical techniques which we will study later. If the divergence is negligible, we can say the coins conform to our belief that probability of getting head for each is 1/2 or that coins are unbiased. If they are unbiased, how will they turn up in any other experiment can also be predicted without actually conducting the experiment with the help of the probability distribution. If the divergence is significant then we will conclude that the coins do no confirm to the hypothesis that probability of getting head is 1/2. It is biased and either the tail has more chance of turning up than the head, or vice versa.

In brief, probability distribution can tell us how a random variable will behave under some hypothesisv It is in fact, the numerical or the statistical manner of expressing a certain theory about a random variable.

16.12 MATHEMATICAL EXPECTATION

We have examined certain characteristics of random variables. These variables possess some more characteristics, *expected value* being one of them.

The expected value of a random variable X is generally denoted be $E(X)$ and is read as "the expected value of X." The expected value is also interrupted as the 'mean' or the 'average' value. It can be defined for discrete as well as continuous variables, but we shall confine to the case of discrete variables only.

Definition. Let X be a discrete variable which can take the values $x_1, x_2 ..., x_n$ with probabilities $P_1, P_2 ..., P_n$ respectively. Then $E(X)$ is defined as

$$E(X) = \sum_{i=1}^{n} x_i P_i$$

Thus, expected value of a random variable is obtained by considering the various values that the variable can take multiplying these by their corresponding probabilities and then summing these products.

Remark. Let us consider the following discrete frequency distribution :

Variable	*Frequency*
x_1	f_1
x_2	f_2
:	:
x_n	f_n

$$\text{Then Arithmetic Mean} = \frac{x_1 . f_1 + x_2 . f_2 + ... + x_n . f_n}{f_1 + f_2 + ... + f_n (= N)}$$

$$= x_1\left(\frac{f_1}{N}\right) + x_2\left(\frac{f_2}{N}\right) + ... + x_n\left(\frac{f_n}{N}\right)$$

$$= x_1 P_1 + x_2 P_2 + ... + x_n P_n$$

$$= E(X)$$

Thus, the expected value is nothing but the mean.

Laws of Mathematical Expectation

We now state some useful laws of mathematical expectation without actually proving them. These laws enable us to calculate expected values in terms of other known expected values which are easily computable. All the laws listed below are valid for discrete as well continuous variables.

(1) The expected value of a constant is the constant itself

i.e., $E(C) = C$ for every constant C.

(2) The expected value of the product of a constant and a random variable is equal to the product of the constant with expected value of the random variable *i.e.*,

$$E(CX) = CE(X) \qquad (16.18)$$

(3) The expected value of the sum or difference of two random variables is equal to the sum or difference of the expected value of the individual random variables,

i.e., $$E(X \pm Y) = E(X) \pm E(Y) \quad (16.19)$$

(4) The expected value of the product of two *independent* random variables is equal to the product of their individual expected values,

i.e., $$E(XY) = E(X).E(Y) \quad (16.20)$$

provided X and Y are independent.

(5) $$E[X - E(X)] = 0 \quad (16.21)$$

(6) $$E[\phi(X)] = \sum_{i=1}^{n} \phi(x_i) P(X = x_i) \quad (16.22)$$

In particular $$E(X^2) = \sum_{i=1}^{n} x_i^2 \, P(X = x_i) \quad (16.23)$$

16.13 MISCELLANEOUS ILLUSTRATIONS

Illustration 16.47. A total of 100 people was surveyed. 60 of them were smokers — 40 being male and 20 being female. 40 of them were non-smokers — 10 being male and 30 being female. What is the probability of picking up a male given he is a smoker? Also calculate the probability of picking up a non-smoker given she is a female.

Solution. (a) Probability that person selected is male $P(M) = \frac{50}{100}$

Probability that person selected is male smoker $P(S) = \frac{60}{100}$

Probability of selecting a smoker male $P(SM) = \frac{40}{100}$

Hence the probability of selecting a male who is a smoker = $P(M/S)$

$$= \frac{P(SM)}{P(S)} = \frac{40}{100} \Big/ \frac{60}{100} = \frac{4}{6}.$$

(*b*) Here a female is selected first and we want to calculate her chance of being non-smoker.

This is also a case of conditional probability.

$$P(NS/F) = \frac{P(NS.F)}{P(F)} = \frac{30/100}{50/100} = 3/4$$

where $P(N.S.F)$ is the probability of selecting non-smoker who is a female.

Illustration 16.48. Proprietor of a food stall has invented a new item of food delicacy which he call WHIM. He has calculated that the cost of manufacture is Re 1 per piece and that because of its novelty and quality it would be sold for Rs 3 per piece. It is, however, perishable, and any goods unsold at the end of a day are a dead loss. He expects the demand to be variable and has drawn up the following probability distribution expressing his estimates :

No. of piece demand	10	11	12	13	14	15
Probability	.07	.10	.23	.38	.12	.10

Calculate his expected net profit or loss if he manufactures 12 pieces.

Unit produced = 12. Manufacturing cost Re 1 per unit selling price = Rs 3 per unit.

Solution.

Unit demanded	*Profits*
10	$(10 \times 3) - 12 = 18$
11	$(11 \times 3) - 12 = 21$
12	$(12 \times 3) - 12 = 24$
13	$(12 \times 3) - 12 = 24$*
14	$(12 \times 3) - 12 = 24$*
15	$(12 \times 3) - 12 = 24$*

* Unit sold can't exceed unit produced even if unit demanded are more.

$$\begin{aligned}\text{Expected Net Profit} &= \Sigma \text{ Profit} \times \text{Probability}\\ &= 0.07 \times 18 + 0.10 \times 21 + .23 \times 24\\ &\qquad + .38 \times 24 + .12 \times 24 + .10 \times 24\\ &= 1.26 + 2.10 + 0.83 \times 24\\ &= 1.26 + 2.10 + 19.92 = 23.28.\end{aligned}$$

Illustration 16.49. There are two series with 5,000 lottery tickets in each. There is only one prize of Rs 3000 and tickets are sold at Rs 2 each. A person buys two tickets what will be his expected gain if he buys (*i*) One ticket of each series (*ii*) two tickets of the same series.

Solution. (*i*) Probability of winning the prize $= \frac{1}{10,000}$

Probability of not winning the prize $= \frac{9999}{10,000}$

If he wins, he gains = 3000 – 2 = Rs 2998.
If he does not win, he gains = – Rs 2

$$\therefore \text{Expectations of one ticket } = \frac{1}{10000} \times 2998 + \frac{9999}{10,000}(-2)$$

$$= \frac{-17000}{10,000} = \text{Rs } 1.7 \text{ loss}$$

Expectations of both the tickets (using addition law of expected values) = 1.7 + 1.7 = loss of Rs 3.4.

(*ii*) **In this case, the result will not change as the probability of winning on each ticket will remain same.**

∴ The expectation will be a loss of Rs 3.4.

Illustration 16.50. *A* and *B* throw one dice for a prize of Rs 11 which is to be won by the player who first throws 6. If *A* has first throw what are their respective expectations?

Solution. Chance of throwing up six in any one throw = 1/6

Chance of not turning up six $= 1 - \frac{1}{6} = \frac{5}{6}$

If *A* is to win, 6 should turn up either in first throw or 3rd throw (*A* and *B* not getting six and then *A* getting six) or 5th throw ... etc. As all these will be mutually exclusive cases, the chance of winning of *A* will be total of probabilities for each of them.

$$\therefore \quad \text{Chance of winning of } A = \frac{1}{6} + \left(\frac{5}{6}\right)^2 \frac{1}{6} + \left(\frac{5}{6}\right)^4 \frac{1}{6} + \ldots \infty$$

$$= \frac{\frac{1}{6}}{1 - \left(\frac{5}{6}\right)^2} = \frac{\frac{1}{6}}{\frac{11}{36}} = \frac{6}{11}.$$

The game is to end only when any one of the two wins. So the total chance of winning either *A* or *B* is certainly *i.e.*, 1. So the chance of winning of *B*

So the chance of winning of *B* = 1 – chance of winning of *A*

$$= 1 - \frac{6}{11} = \frac{5}{11}.$$

If *A* wins he receives Rs 11 and if he does not, neither receives nor pays anything. In other words, *A* gets Rs 11 with probability $\frac{6}{11}$ and gets zero with probability $= 1 - \frac{6}{11}$ *i.e.*, $\frac{5}{11}$. so, the expectations of

$$A = 11 \times \frac{6}{11} + 0 \times \frac{5}{11} = Rs\ 6.$$

Similarly, *B* gets Rs 11 on winning *i.e.*, with probability $\frac{5}{11}$ and gets nothing on losing *i.e.*, with probability $\frac{6}{11}$.

$$\therefore \quad \text{Expectations of } B = 11 \times \frac{5}{11} + 0 \times \frac{6}{11} = Rs\ 5.$$

Illustration 16.51. A store manager states that the probability for sale of less than 70 cans of orange juice on a Wednesday is 0.40, for

70 to 100 cans is 0.75, and for more than 100 cans is .05. comment on his assignment of probabilities.

Solution. The probabilities add to 1.20. Since the events are mutually exclusive and collectively exhaustive they should add to 1.00.

Hence the statement is incorrect.

Illustration 16.52. Three ships A, B, and C sail from England to India. Odds in favour of their arriving safely are 2 : 5, 3 : 7 and 6 : 11 respectively. Find the chance that they all arrive safely.

Solution. Probability of ship A arriving safely $= \frac{2}{2+5} = \frac{2}{7}$

Probability of ship B arriving safely $= \frac{3}{3+7} = \frac{3}{10}$

Probability of ship C arriving safely $= \frac{6}{6+11} = \frac{6}{17}$

Chance that all the three arrive safely $= \frac{2}{7} \times \frac{3}{10} \times \frac{6}{17} = \frac{18}{595}.$

Illustration 16.53. It is 8 : 5 against a person who is now 40 years old living till he is 70 and 4 : 3 against a person now 50 living till he is 80. Find the probability that at least one of these persons will be alive 30 years hence.

Solution. Probability that first person will live 30 years is $\frac{5}{(8+5)}$ *i.e.*, $\frac{5}{13}$ and the probability that he will die within 30 years is $\frac{8}{(8+5)} = \frac{8}{13}$.

The probability that second person will live up to 30 years is $\frac{3}{(4+3)}$ *i.e.*, $\frac{3}{7}$ and the probability that he will die within 30 years $= \frac{4}{(4+3)} = \frac{4}{7}$.

Probability that at least one of them live up to 30 years

= Prob. 1st lives, 2nd does not

+ Prob. 2nd lives, 1stdoes not + Prob. both lives

$$= \frac{5}{13} \times \frac{4}{7} + \frac{8}{13} \times \frac{3}{7} + \frac{5}{13} \times \frac{3}{7}$$

$$= \frac{20+24+15}{91} = \frac{59}{91}.$$

Illustration 16.54. A piece of equipment will function only when all the three components A, B and C are working. The probability of A failing during one year is 0.15, that of B failing is 0.05, and of C failing is 0.10. What is the probability that the equipment will fail before the end of one year?

Solution. Probability that A will not fail = 1 – 0.15 = .85

Probability that B will not fail = 1 – 0.05 = .95

Probability that C will not fail = 1 − 0.10 = .90

Probability that the equipment will not fail = .85 × .95 × .90 = .73

Probability that the equipment will fail before the end of year = 1 − .73 = .27.

Illustration 16.55. A salesman has a 60% chance of making a sale to any one customer. The behaviour of successive customers is independent. If two customers A and B enter, what is the probability that the salesman will make a sale to A or B.

Solution. If $P(A)$ is the probability of making a sale to A and $P(B)$ is the probability of making a sale to B, then $P(A \text{ or } B)$ is the probability of making a sale to A or B. The two events are not mutually exclusive as the occurrence of A does not preclude the occurrence of B.

$$\therefore \quad P(A \text{ or } B) = P(A) + P(B) - P(AB)$$

$$P(A) = .6$$

and $$P(B) = .6$$

$$P(AB) = P(A) \,.\, P(B) = .6 \times .6 = .36$$

[because the behaviour of successive customers is independent]

Hence $$P(A \text{ or } B) = .60 + .60 - .36 = .84.$$

Illustration 16.55A. What is the chance that a leap year, selected at random, will contain 53 Sundays?

Solution. A leap year consists of 52 complete weeks and two days over ; so it will contain 53 Sundays if Sunday is one of the days out of the remaining two days. these two days can be any of these possibilities :

(*i*) Monday and Tuesday
(*ii*) Tuesday and Wednesday
(*iii*) Wednesday and Thursday
(*iv*) Thursday and Friday
(*v*) Friday and Saturday
(*vi*) Saturday and Sunday
(*vii*) Sunday and Monday

Hence the favourable cases for 53 Sundays in the above 7 are only 2.

Hence the required probability is $\frac{2}{7}$ **Ans.**

Illustration 16.56. Urn A contains two white and 4 black balls, another urn B contains 5 white and 7 black balls. A ball is transferred from urn A to the urn B. Then a ball is drawn from urn B. Find the probability that it will be white?

Solution. These are two mutually exclusive ways of transferring a ball from the first urn to second, a white or a black ball.

Probability of drawing a white ball from the Ist urn $=\frac{1}{3}$. On transferring this ball, the second urn has 6 white and 7 black balls. So the probability of drawing a white ball from this will be

$$\frac{6}{6+7}=\frac{6}{13}.$$

Thus the probability of the compound event, drawing ball (white) from the first urn and then drawing a white ball from the second urn is

$$\frac{1}{3}\times\frac{6}{13}=\frac{2}{13}.$$

Similarly, the probability of transferring a black ball from the Ist urn is $\frac{2}{3}$.

The probability of then drawing a white ball from the second urn is $\frac{5}{13}$.

So the probability of the compound event, drawing a ball (black) from the 1st urn and then drawing white ball from the IInd urn is $\frac{2}{3}\times\frac{5}{13}=\frac{10}{39}$.

Thus, the total probability of drawing a white ball from the second urn is $\frac{2}{13}+\frac{10}{39}=\frac{16}{39}$.

Illustration 16.57. Two men M_1 and M_2 and three women W_1, W_2 and W_3 are in a chess tournament. Those of the same sex have equal probabilities of winning, but each man is twice as likely to win as any women. (1) Find the probability that a woman wins the tournament. (2) If M_1 and W_1 are husband and wife, find the probability that one of them wins the tournament.

Solution. (1) To make all cases equally likely count each man twice and each woman once. This means there are in all 7 cases out of which 3 are favourable to women :

$\therefore$ The probability a woman wins $=\frac{3}{7}$

(2) Probability that either M_1 (which is equal to two cases) and W_1 (which is equal to one case) wins $=\frac{2}{7}+\frac{1}{7}=\frac{3}{7}$.

Illustration 16.58. *A*, *B* and *C* in order toss a coin. The first one to throw a head wins. What are their respective chances of winning?

Solution. Probability of getting head $=\frac{1}{2}$

$$\text{tail} = \frac{1}{2}.$$

If A is to win, a head should turn up either in first throw or 4th throw (A, B and C all getting tail and then A getting head) or 7th throw or 10th throw of 13th throw etc.

$$\therefore \quad \text{Chances of } A \text{ winning} = \frac{1}{2} + \frac{1}{2}\left(\frac{1}{2}\right)^3 + \frac{1}{2}\left(\frac{1}{2}\right)^6 + \frac{1}{2}\left(\frac{1}{2}\right)^9 + \ldots$$

This forms a G.P. Therefore is equal to $\dfrac{\frac{1}{2}}{1-\left(\frac{1}{2}\right)^3} = \dfrac{1}{2} \times \dfrac{8}{7} = \dfrac{4}{7}.$

If B is to win, a head should turn up in either of the 2nd, 5th and 8th throw and no other.

$$\therefore \text{ Chance of winning } B = \left(\frac{1}{2}\right)^2 + \left(\frac{1}{2}\right)^2 \times \left(\frac{1}{2}\right)^3 + \left(\frac{1}{2}\right)^2 \times \left(\frac{1}{2}\right)^6 \ldots$$

$$= \frac{\frac{1}{2} \times \frac{1}{2}}{1-\left(\frac{1}{2}\right)^3} = \frac{1}{4} \times \frac{8}{7} = \frac{2}{7}.$$

Chances of winning C can be found as above or it can be found by subtracting from the total chances of winning the chances of winning by A or $B = \dfrac{\frac{1}{2} \times \frac{1}{2}}{1-\left(\frac{1}{2}\right)^3} = \dfrac{1}{4} \times \dfrac{8}{7} = \dfrac{2}{7}.$

Illustration 16.59. A man speaks truth 3 times out of 5. He states that in a toss of 6 coins there were 2 heads. What is the probability that actually this event happened?

Solution. $P \text{ (speaking truth)} = \dfrac{3}{5}$

$$P \text{ (getting head)} = \frac{1}{2}$$

$$\therefore \quad P \text{ getting 2 heads in a toss of 6 coins } = 6C_2\left(\frac{1}{2}\right)^4\left(\frac{1}{2}\right)^2 = \frac{15}{64}$$

$$\therefore P \text{ getting 2 heads and man speaking truth} = \frac{3}{5} \times \frac{15}{64} = \frac{9}{64}.$$

$$P\text{(speaking lie)} = 1 - \frac{3}{5} = \frac{2}{5}.$$

$$P\text{ (not getting two heads) } = 1 - \frac{15}{64} = \frac{49}{64}.$$

$$\therefore P \text{ not getting two heads and man speaking lie} = \frac{49}{64} \times \frac{2}{5} = \frac{49}{160}.$$

These are the only two mutually exclusive ways of informing that in a toss of 6 coins there are two heads. So its total probability is

$$\frac{49}{160} + \frac{9}{64} = \frac{143}{320}.$$

Out of the total probability of $\frac{143}{320}$, the probability in favour of speaking truth is $\frac{9}{64}$ and in favour of telling a lie $\frac{49}{160}$.

So given that person informed of obtaining two heads, the probability that he speaks truth is

$$\frac{\frac{9}{64}}{\frac{143}{320}} = \frac{45}{143}.$$

Illustration 16.60. A bank has a test designed to establish the credit rating of a loan applicant. If the persons, who default (D), 90% fail the test (F). Of the persons, who will repay the bank (ND), 5% fail the test. Furthermore, it is given that 4% of the population is not worthy of credit ; *i.e.*, $P(D) = .04$. Given that someone failed the test, what is the probability that he actually will default (when given a load)?

Solution. Since we are given that 4% of the applicants are defaulters, we have

$$P(D) = .04$$

$$\therefore \quad P(ND) = 1 - .04 = .96.$$

Next, of the persons who will repay the bank (ND), 5% fail the test (F). Of the persons who default (D), 90% fail the test (F). So the conditional probability given are

$$P(F/D) = .90$$

and $$P(F/ND) = .05.$$

Next, we need the probability of the event F.

$$P(F) = P(D \text{ and } F) + P(ND \text{ and } F)$$

$$P(D \text{ and } F) = P(D).\ P(F/D)$$
$$= .90 \times .04 = .036$$

$$P(ND \text{ and } F) = P(ND).\ P(F/ND)$$
$$= .05 \times .96 = .048$$

$$\therefore \quad P(F) = .036 + .048 = .084.$$

What we seek is to find out is the possible probability $P(D/F)$ *i.e.*, given that someone failed the test, he actually will default.

$$P(D/F) = \frac{P(D \text{ and } F)}{P(F)}$$

$$= \frac{.036}{.084} = \frac{3}{7}$$

Similarly, $P(ND/F) = \frac{.048}{.084} = \frac{4}{7}$.

Illustration 16.61. A manufacturing firm produces steel pipes in three plants with daily production volumes of 500, 1000 and 2000 units respectively. According to the past experience, it is known that the fraction of defective output produced by the three plants are respectively 0.005, 0.008 and 0.010. If a pipe is selected from a day's total production and found to be defective, find out :

(*a*) From which plant the pipe comes?

(*b*) What is the probability that it came from the first plant.

Solution. Let A denote defective and 'B' the plants

$$P(B_1) = \frac{500}{500+1000+2000} = \frac{1}{7}$$

$$P(B_2) = \frac{1000}{3500} = \frac{2}{7}$$

$$P(B_3) = \frac{2000}{3500} = \frac{4}{7}.$$

Also, $P(A/B_1) = 0.005$

$P(A/B_2) = 0.008$

and $P(A/B_3) = 0.010$

$$P(A) = P(B_1)P(A/B_1) + P(B_2)P(A/B_2) + P(B_3)P(A/B_3)$$

$$= \frac{1}{7} \times 0.005 + \frac{2}{7} \times 0.008 + \frac{4}{7} \times 0.01$$

$$= \frac{0.061}{7}$$

$$P(B_1/A) = \frac{P(B_1)P(A/B_1)}{P(A)}$$

$$= \frac{0.005/7}{0.061/7} = \frac{5}{61}.$$

This is the probability that the defective pipe come from the first plant. Also

$$P(B_2/A) = \frac{0.016/7}{0.061/7} = \frac{16}{61}$$

$$P(B_3/A) = \frac{0.040/7}{0.061/7} = \frac{40}{61}.$$

We obviously cannot tell from above as to which plant the defective pipe comes from, but the chances that it comes from the third plant are pretty high (40 in 61).

Illustration 16.62. At tennis A defeats B 3 times out of 4, A defeats C 2 times out of 3, A defeats D 9 times out of 10, C defeats D half the times. To win a tournament A must defeat B and the winner of a match between C and D. What is the probability that A will win the tournament?

Solution.

Probability for A winning against $B = \frac{3}{4}$

" A " $C = \frac{2}{3}$

" A " $D = \frac{9}{10}$

" C " $D = \frac{1}{2}$

" D " $C = \frac{1}{2}$

In the match between C and D :

Probability of C winning $= \frac{1}{2}$

" D " $= \frac{1}{2}$

Match between A and winner of C and D :

Probability that C wins and then A wins $C = \frac{1}{2} \times \frac{2}{3} = \frac{2}{6}$

Probability that D wins and then A wins $D = \frac{1}{2} \times \frac{9}{10} = \frac{9}{20}$.

There are mutually exclusive ways in which A can win the tournament. First if C wins in match of C and D — A wins over C and then wins over B also. Second if D win in the match of C and D — A wins over D and then wins over B also.

The probability of first alternative $= \frac{2}{6} \times \frac{3}{4} = \frac{1}{4}$

The probability of second alternative $= \frac{9}{20} \times \frac{3}{4} = \frac{27}{80}$

Total probability of A winning the tournament $= \frac{1}{4} + \frac{27}{80} = \frac{47}{80}$.

Illustration 16.63. The products of three factories producing radio valves are know to be 2%, 3% and 5% defective. One valve is selected at random from the product of each factory. Find the probability that at least two of them will not be defective.

Solution. Probability that valve of first factory is defective is 2/100 and not defective is 98/100. The probability that valve of second factory is defective is 3/100 and not defective is 97/100. The probability that valve of third factory is defective 5/100 and not defective is

95/100. At least two of them not defective means the total of four mutually exclusive cases *i.e.*, (*i*) All three good ; (*ii*) 1st, 2nd good ; 3rd defective (*iii*) 2nd, 3rd good ; 1st defective (*iv*) 3rd, 1st good ; 2nd defective.

$$\text{Probability of } (i) = \frac{98}{100} \times \frac{97}{100} \times \frac{95}{100}$$

$$\text{Probability of } (ii) = \frac{98}{100} \times \frac{97}{100} \times \frac{5}{100}$$

$$\text{Probability of } (iii) = \frac{97}{100} \times \frac{95}{100} \times \frac{2}{100}$$

$$\text{Probability of } (iv) = \frac{95}{100} \times \frac{98}{100} \times \frac{3}{100}$$

$$\therefore \text{ Required result} = \frac{98 \times 97 \times 95 + 98 \times 97 \times 5 + 97 \times 95 \times 2 + 95 \times 98 \times 3}{100 \times 100 \times 100}$$

Illustration 16.64. Three houses of the same type were advertised to be let in a locality. Three men made separate applications for a house. What is the probability (*i*) that all three made applications for the same house, (*ii*) that each of the three applied for a different house and (*iii*) that two of them applied for the same house and third for one of the other house.

Solution. First person can apply to any of the three houses, second can also apply to any of the three houses and third one can also apply to any of the three houses. So total number of ways of making applications = 3 × 3 × 3 = 27.

(*i*) As there are 3 houses, all can apply to the same houses in 3 different ways. so the probability $= \frac{3}{27} = \frac{1}{9}$.

(*ii*) Number of favourable cases for this case will be the same as number of ways of arranging 3 things in three places, keeping due consideration of order *i.e.*,

$${}^3P_3 = 3 \times 2 \times 1 = 6$$

$$\therefore \qquad \text{Required probability} = \frac{6}{27} = \frac{2}{9}.$$

(*iii*) The total probability of 3 persons applying to 3 houses is 1.

The various ways in which applications can be made are only three, the first two and the present case. All these are mutually exclusive and exhaustive cases. So the probability of this case will be

$$1 - \left(\frac{1}{9} + \frac{2}{9}\right) = 1 - \frac{1}{3} = \frac{2}{3}.$$

Illustration 16.65. During war one ship in ten was sunk in the average in making a certain voyage. What was the probability that at least three out of a convoy of six ships would arrive safely?

Solution. Chance that a ship was sunk $= \frac{1}{10}$

$$\text{arrived safely} = 1 - \frac{1}{10} = \frac{9}{10}.$$

It is easy to see that the number of ways in which r ships out of 6 can be selected is ${}^6C_r$. So probability that exactly r ships survived is

$$ {}^6C_r\left(\frac{9}{10}\right)^r\left(\frac{1}{10}\right)^{6-r} $$

Chance that at least 3 ships arrived safely = sum of chances for 3, 4, 5 and 6 ships arriving safely.

$$ \therefore \text{ Required probability } = {}^6C_3\left(\frac{1}{10}\right)^3\left(\frac{9}{10}\right)^3 + {}^6C_4\left(\frac{1}{10}\right)^2\left(\frac{9}{10}\right)^4 + {}^6C_5\left(\frac{1}{10}\right)\left(\frac{9}{10}\right)^5 + \left(\frac{9}{10}\right)^6 $$

$$ = \frac{20\times 729}{10^6} + \frac{15\times 9\times 729}{10^6} + \frac{6\times 81\times 729}{10^6} + \frac{729\times 729}{10^6} $$

$$ = \frac{998730}{1000000}. $$

This result could also be found as 1 – chance of 0, 1, 2 ships arriving safely $= 1 - \left[\left(\frac{1}{10}\right)^6 + 6\left(\frac{1}{10}\right)^5 \times \left(\frac{9}{10}\right) + 15\left(\frac{1}{10}\right)^4 \times \left(\frac{9}{10}\right)^2\right]$

$$ = 1 - \frac{1276}{1000000} = \frac{998730}{1000000}. $$

Illustration 16.66. A lady declares that by tasting a cup of tea made with milk, she can discriminate whether milk, or tea infusion was first made into the cup. It is proposed to test this claim by means of an experiment with 12 cups of tea, 6 made in one way and presenting them in random order to her.

Calculate the probability, on the null hypothesis, that the lady would judge correctly all the 12 cups, it being known to her that 6 are of each kind.

If however, the 12 cups were presented to the lady in 6 pairs, each pair to consist of either kind and the presentation be again in random order ; how will the probability of correctly judging with every cup on the null hypothesis be altered?

Solution.

(*a*) When 12 cups are presented in random order :

The number of ways of arranging 12 things 6 of each kind 12! / 6! 6! = 924

The cups are presented in any one manner out of these 924. The lady has to select which one is that manner. So number of favourable cases is one out of a total of 924. So the required probability $= \frac{1}{924}$.

(*b*) When cups are presented in 6 pairs.

Each pair consists of one cup prepared by taking milk first (*M*) and one cup prepared by taking insulin first (*I*). There are two ways of presenting a pair (*M*, *I* ; *I*, *M*). When a pair is presented, lady has to find, out of these two, which one is the method used. So probability of correct judging the six pairs $= \frac{1}{2} \times \frac{1}{2} \times \frac{1}{2} \times \frac{1}{2} \times \frac{1}{2} \times \frac{1}{2} = \frac{1}{64}$.

Illustration 16.67. In a lottery n^2 tickets, are sold and n prizes are awarded. If a person buys n tickets, what is the probability of his winning a prize.

Solution. As there are n prizes and person buys n tickets, he can win prize on 1 or 2 or all the n tickets. So it will be easier to find the probability of not winning and then subtract it from 1 to find the probability of winning.

If a ticket does not fetch a prize, it has to be any one out of $n^2 - n$ ticket on which there is no award.

$\therefore$ The probability that one ticket does not bring prize is $(n^2 - n)/n^2$ or $(n - 1)/n$.

$\therefore$ The probability of the compound event that all the n tickets bought do not fetch a prize $= \left(\frac{n-1}{n}\right)^n$.

$\therefore$ The required probability $= 1 - \left(\frac{n-1}{n}\right)^n$

Illustration 16.68. A number x is chosen at random from the integers 1, 2, 3, ..., n and A and B denotes the events that x is a multiple of 3 and 4, respectively. Show that A and B respectively. Show that A and B are independent events when $n = 96$, but not when $n = 100$.

Solution. *When* $n = 96$. The number are 1, 2, 3, ... 96.

Total count of numbers which are multiple of 3= 32

Total count of numbers which are multiple of 4= 24

Total count of numbers which are multiple of 12= 8.

The probability that x is multiple of 3 $= \frac{32}{96} = \frac{1}{3}$

The probability that x is multiple of 4 $= \frac{24}{96} = \frac{1}{4}$

The probability that x is multiple of 12 $= \frac{8}{96} = \frac{1}{12}$.

Two events are said to be independent of one another if the probability of joint occurrence is equal to product of probabilities of individual events.

$P(x \text{ mult. of } 3) \times P(x \text{ mult. of } 4) = \frac{1}{3} \times \frac{1}{4} = \frac{1}{12}$ (*i*)

$P(x \text{ is mult. of 3 as well as 4 } i.e.\ 12) = \frac{1}{12}$ (*ii*)

As (*i*) and (*ii*) are same two events are independent.

When $n = 100$. Preceding as above

$$P(x \text{ mult. of } 3) = \frac{33}{100},$$

$$P(x \text{ mult. of } 4) = \frac{25}{100}.$$

$$P(x \text{ mult. 3 as well as 4 } i.e., 12) = \frac{8}{100}$$

$$P(x \text{ mult. of } 3) \times P(x \text{ mult. of } 4) = \frac{33}{100} \times \frac{25}{100} \neq \frac{8}{100}$$

i.e., $P(x \text{ mult. of } 12)$

∴ Two events are not independent.

Illustration 16.69. *A*, *B*, *C* are three mutually exclusive events such that $P(A) = (2/3)\, P(B) = (1/5)\, P(C)$. Comment on the statement.

Solution. $P(A) = \frac{1}{6}$, $\frac{2}{3} P(B) = 1/6$ or $P(B) = \frac{1}{4}$

$\frac{1}{5} P(C) = \frac{1}{6}$ or $P(C) = \frac{5}{6}$.

For mutually exclusive events $P(A + B + C) = P(A) + P(B) + P(C)$

∴ Here $P(A + B + C) = \frac{1}{6} + \frac{1}{4} + \frac{5}{6} = 1\frac{1}{4}$.

But probabilities can not exceed one.

Hence there is some mistake in the calculation of probabilities.

Illustration 16.70. Ten pairs of shoes are in a closest. Four shoes are selected at random. Find the probability that there is at least one pair among the four selected.

Solution. Any four-shoes can be selected out of 20 shoes (10 × 2) in $^{20}C_4$ ways.

To find the probability of at least one pair, we can do it easily by first finding the probability of no pair and then subtract it from 1.

The first shoe selected could be any one of the 20 shoes. The next shoe could not be the matching to the first, or one out of 19 is ruled out, or 18 of 19 are allowed. Similarly for the third shoe, two matching shoes are disallowed, or 16 of 18 remaining are allowed. Similarly, for the fourth shoe, 14 of 17 remaining shoes are allowed, if none of the four shoes is to match. Thus

$$P(\text{none matching pair}) = \frac{20}{20} \times \frac{18}{19} \times \frac{16}{18} \times \frac{14}{17} \quad \text{and}$$

$$P(\text{at least one matching pair}) = 1 - \frac{20}{20} \times \frac{18}{19} \times \frac{16}{18} \times \frac{14}{19} = \frac{99}{323} = 0.3065.$$

Illustration 16.71. A bag contains 4 balls which are just as likely to be white as coloured. Two balls are drawn from the bag. (*a*) What is the probability that they are white? (*b*) Knowing that two balls drawn are white, what is the probability that the remaining are coloured?

Solution. The event that two balls drawn are white is possible only when bag contains either 4 or 3 or 2 white balls. Probability that a ball is white is $\frac{1}{2}$ and it is coloured is also $\frac{1}{2}$.

$\therefore$ Probability that all the 4 balls are white $= \left(\frac{1}{2}\right)^4 = \frac{1}{16}$

Probability that 3 balls are white $= {}^4C_3\left(\frac{1}{2}\right)^3\left(\frac{1}{2}\right) = \frac{4}{16} = \frac{1}{4}$

Probability that 2 balls are white $= {}^4C_2\left(\frac{1}{2}\right)^2\left(\frac{1}{2}\right)^2 = \frac{6}{16} = \frac{3}{8}$

Chance of drawing 2 white balls when 3 are white 1 coloured $= \dfrac{{}^4C_2}{{}^4C_2} = 1$

Chance of drawing 2 white balls when 3 are white 1 coloured $= \dfrac{{}^3C_2}{{}^4C_2} = \dfrac{1}{2}$

Chance of drawing 2 white balls when 2 are white 2 coloured $= \dfrac{{}^2C_2}{{}^4C_2} = \dfrac{1}{6}$

$\therefore$ The probability of the compound event that there were

4 white balls and 2 white balls are drawn $= 1 \times \frac{1}{16} = \frac{1}{16}$

3 white balls and 2 white balls are drawn $= \frac{1}{2} \times \frac{1}{4} = \frac{1}{8}$

2 white balls and 2 white balls are drawn $= \frac{3}{8} \times \frac{1}{6} = \frac{1}{16}$

(*a*) All the above events are mutually exclusive. so the total probability of getting two white balls $= \frac{1}{16} + \frac{1}{8} + \frac{1}{16} = \frac{1}{4}$.

(*b*) After drawing two white balls, the remaining two balls are coloured means that bag contained 2 white and 2 coloured balls. The probability of drawing 2 white balls is $\frac{1}{4}$. Out of this total $\frac{1}{16}$ is due to the event that bag contained 2 white and 2 coloured.

So knowing that two white balls are drawn chance that is due to bag containing two white and two coloured (By Bayes theorem).

$$\frac{\frac{1}{16}}{\frac{1}{4}} = \frac{4}{16} = \frac{1}{4}.$$

Illustration 16.72. A bag contains 4 rupee, 10 one paise, 6 fifty paise coins and 3 twenty-five paise coins. A man draws a coin

(*a*) What is the expectation of his draw?

(*b*) What is his expectation if he draws three coins?

Solution. Total number of coins in the bag

$$4 + 10 + 6 + 3 = 23.$$

Chance of getting a rupee is $\frac{4}{23}$, of getting a one-paisa is $\frac{10}{23}$, of getting fifty-paise coin is $\frac{6}{23}$ and of getting a twenty-five paise coin is $\frac{3}{23}$.

(*a*) In making one draw, he thus gets a rupee (100 paisa) with a probability $\frac{4}{23}$, one paisa with probability $\frac{10}{23}$, a fifty-paisa coin with probability $\frac{6}{23}$ and a twenty-five paise coin with probability $\frac{3}{23}$.

$\therefore$ Expectation of his draw of one coin

$$= 100 \times \frac{4}{23} + 1 \times \frac{10}{23} + 50 \times \frac{6}{23} + 25 \times \frac{3}{23}$$

$$= \frac{785}{23} \, P \text{ or } 34\frac{3}{23} \text{ P.}$$

(*b*) If 3 coins are drawn, he can expect to draw three times more than what he expects in one draw.

$\therefore$ Expectation in drawing 3 coins $= 3 \times \frac{785}{23} = 102\frac{9}{23}$ P

or $\quad = \text{Rs } 1.024$ approximately.

Note. Expectation of a draw of one coin can also be found in a simple manner. all the 23 coins total

$$= 100 \times 4 + 10 \times 1 + 50 \times 6 + 25 \times 3 = 785 \text{ P.}$$

So one coin is expected to have a value of $\frac{785}{23}$ P *i.e.*, several drawings of one coins each are made and their value noted, the average value per draw will be $\frac{785}{23}$ P.

Illustration 16.73. There are 12 coins in a bag. Four of them are a rupee each. Remaining coins consist of two types of coins, four of each type. The ratio of the value of the two types of coins is 1 : 5. If the expectation of a draw of one coin is $53\frac{1}{3}$ P. Find the values of the coins.

Solution. Let the value of one of the type of the coin be Rs x. So the value of the other type of coin will be Rs $5x$.

$\therefore$ The total value of all the 12 coins $= \text{Rs } (4 \times 1 + 4 \times x + 4 \times 5x)$

$$= 24x + 4 \text{ rupees.}$$

$\therefore$ The expectation of a draw of one coins $= \text{Rs } \dfrac{(24x+4)}{12}$

$$\therefore \quad \text{Rs } \frac{24x+4}{12} = 53\frac{1}{3} \text{ P}$$

or $$\text{Rs}\ \frac{24x+4}{12} = \text{Rs}\ \frac{8}{15}$$

or $$x = 0.1 \text{ or } 10 \text{ P}$$

$\therefore$ One type of coin is 10 P and the other is 50 P.

Illustration 16.74. How many throws with a single die must a man have in order that his chance of throwing a six may be $\frac{1}{2}$?

Solution. Let x be the required number of throws.

Chance of not throwing six = $\frac{5}{6}$

$\therefore$ The chance of not throwing a six in all the x throws is $\left(\frac{5}{6}\right)^x$. So the chance of throwing six, at least once is $1-\left(\frac{5}{6}\right)^x$. This is $\frac{1}{2}$ (given).

$\because$ $$1-\left(\frac{5}{6}\right)^x = \frac{1}{2}$$

or $$\left(\frac{5}{6}\right)^x = \frac{1}{2}$$

or $$x \log \frac{5}{6} = \log \frac{1}{2}$$

or $$x = \frac{\log 0.5}{\log 0.8333}$$

or $$x = \frac{\bar{1}.6990}{\bar{1}.9208} = \frac{-0.3010}{-0.0792} = 3.8 \text{ (approx.)}$$

No. of throws can only be in whole numbers. So the required result is 4.

Illustration. 16.75. If A_1, A_2 and A_3 be any three events prove that

$$P(A_1 + A_2 + A_3) = \sum_{i=1}^{3} P(A_i) - \sum_{i<j} P(A_i A_j) + P(A_1 A_2 A_3)$$

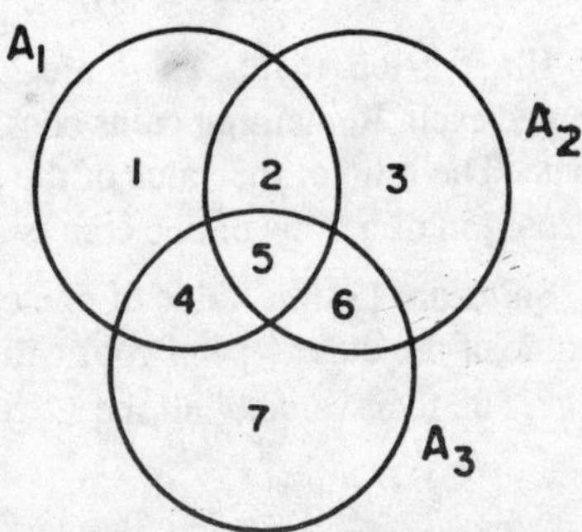

Solution. If A_1, A_2, A_3 are any events let us assume that they are not mutually exclusive events. Let there be in all n events. Out of this A_1, a_2 are favourable to A_2 and a_3 are favourable to A_3. If A_1 and A_2 are not mutually exclusive then they will have some common items. Let their number be a_{12}. Similarly let a_{13} be number of items common between A_1 and A_3 and a_{23} be number of items common between A_2 and A_3. Finally let a_{123} be number of items common to all A_1, A_2 and A_3.

$$\therefore P(A_1) = \frac{a_1}{n}, P(A_2) = \frac{a_2}{n}, P(A_3) = \frac{a_3}{n}, P(A_1 A_2 A_3) = \frac{a_{123}}{n}$$

$$P(A_1 A_2) = \frac{a_{12}}{n}, P(A_1 A_3) = \frac{a_{13}}{n}, P(A_2 A_3) = \frac{a_{23}}{n}.$$

To count the number of items which are either A_1 or A_2 or A_3 (counting common items only once) we have to subtract from the sum of a_1, a_2 and a_3, the common items in pairs of two events. By doing this, the items which are common to all the three events will be completely excluded. So they will have to be added. This is clear from the diagram and the explanation given below.

Let event A_1 be the area of a circle A_1, which is composed of the areas 1 + 2 + 4 + 5 = a_1 area. Event A_2 be circle A_2 which is 2 + 3 + 5 + 6 *i.e.* a_2 area, and the event A_3 which is equal to areas 4 + 5 + 6 + 7 or a_3 area.

$\therefore$ Either A_1 or A_2 or A_3 counting each area once the-total figure will be 1 + 2 + 3 + 4 + 5 + 6+ 7 which can be written as

(1 + 2 + 4 + 5) + (2 + 3 + 5 + 6) + (4 + 5 + 6 + 7) – (2 + 5) – (5 + 6) – (4 + 5) + (5)

Now area 2 + 5 is a_{12}, area 5 + 6 is a_{23}, area 4 + 5 is a_{13} and area 5 is a_{123}.

Hence the total case common to $(A_1 + A_2 + A_3) = a_1 + a_2 + a_3 - a_{12} - a_{23} - a_{13} + a_{123}$.

$$\begin{aligned} P(A_1 + A_2 + A_3) &= \frac{a_1 + a_2 + a_3 - a_{12} - a_{23} - a_{13} + a_{123}}{n} \\ &= P(A_1) + P(A_2) + P(A_3) - P(A_1 A_2) \\ &\quad - P(A_2 A_3) - P(A_1 A_3) + P(A_1 A_2 A_3) \end{aligned}$$

Illustration 16.76. A card is taken out from a well shuffled pack of cards. Events A_1, A_2, A_3 and defined as A_1 = the card is a club, A_2 = the card is a spade, and A_3 = the card is an ace. find $P(A_1 + A_2 + A_3)$.

Solution. Here the events A_1 *i.e.* card is a club and event A_2 that card is a spade are mutually exclusive. But event A_3 that card is an ace is not mutually exclusive to both A_1 and A_2.

$$\therefore \quad P(A_1 A_2) = 0 \text{ and } P(A_1 A_2 A_3) = 0$$

$$P(A_1) = 13/52, \; P(A_2) = 13/52, \; P(A_3) = 13/52$$

$P(A_1 A_3)$ = Prob. of an ace of club = 1/52 and

$P(A_2 A_3)$ = Prob. of an ace of spade = 1/52.

When events are not mutually exclusive

$$P(A_1+A_2+A_3) = P(A_1)+P(A_2)+P(A_3)-P(A_1A_2) -P(A_2A_3)-P(A_1A_3)+P(A_1A_2A_3)$$

$$=\frac{13}{52}+\frac{13}{52}+\frac{4}{52}-0-\frac{1}{52}-\frac{1}{52}+0=\frac{28}{52}.$$

Illustration 16.77. Out of $2n + 1$ tickets consecutively numbered, three are drawn at random. find the chance that the numbers on them are in arithmetical progression.

Solution. Suppose the first selected number is 1. Then the possible groups of three tickets drawn can be

1, 2, 3 ; 1, 3, 5 ; 1, 4, 7, ...1, $n + 1$, $2n + 1$.

These are n in all.

The first selected number can be any of 1, 2, 3, 4, ..., $2n - 1$. Numbers $2n$ and $2n + 1$ cannot occur as three tickets forming an A.P. are to be drawn. Writing out as above, the number of cases favourable to tickets in A.P. can be easily found.

When first cards drawn are 2, 3, 4, 5, ..., $2n - 2$, $2n - 1$, the numbers of groups in A.P. are respectively $n - 1$, $n - 2$, $n - 2$,..... 1, 1, 1.

∴Total number of favourable cases $= 2[(n - 1) + (n - 2) + ...1] + n$

$$= 2\times\frac{(n-1)(n)}{2}+n=n^2.$$

Total number of ways of drawing 3 tickets $= {}^{2n+1}C_3$

∴ Required probability, $P = n^2 / {}^{2n+1}C_3... = \dfrac{3n}{4n^2-1}$.

Illustration 16.78. If 4 integers taken at random are multiplied together, shows that the chance that the last digit in the product is 1, 3, 7 or 9 is 16/625.

Solution. The whole numbers can end with any of the 10 digits 0, 1, 2, 3, 4, 5, 6, 7, 8, 9. The whole numbers having last digits an even number *i.e.*, 2, 4, 6 and 8 when multiplied together will end with the even numbers. similarly, numbers ending with 0 or 5 will not give 1, 3, 7 or 9 when multiplied together.

So only the numbers which end with 1, 3, 7 or 9 when multiplied together will give 1, 3, 7 or 9.

Probability that a number end with 1, 3, 7 or 9, out of 10 possibilities 0, 1, 2, 3, 4, 5, 6, 7, 8 or 9 is $\frac{4}{10}$ or $\frac{2}{5}$.

∴ The probability that all the four numbers end with any of 1, 3, 7 or 9

$$=\left(\frac{2}{5}\right)^4=16/625$$

Illustration 16.79. A six faced die is so biased that it is twice as likely to show an even number as an odd number when thrown. It is thrown twice. What is the probability that the sum of the two numbers thrown is even.

Solution. The die has six faces 1, 2, 3, 4, 5 and 6. Out of these even numbers 2, 4 and 6 have double the chance of turning up as compared to odd numbers *i.e.*, 1, 3, 5. So we can say that each odd number has one favourable case and each even has 2 favourable cases out of the total of 9 cases. So the chance of getting 1 or 3 or 5 is $\frac{1}{9}$ each and chance of getting 2 or 4 or 6 is $\frac{2}{9}$ each.

When thrown twice, the sum will be even, when in both the throws an even number turns up or in both the throws an odd number turns up. The various mutually exclusive of both odd numbers are 9 in all *i.e.*, (1, 1), (1, 3), (1, 5), (3, 1), (3, 3), (3, 5), (5, 1), (5, 3), (5, 5). The chance for each of them is $\left(\frac{1}{9}\right)\left(\frac{1}{9}\right)$ *i.e.*, $\frac{1}{81}$ (By theorem on multiplication of probabilities).

So the total chance of getting both digits odd are $9 \times \frac{1}{81} = \frac{1}{9}$.

Similarly, there are 9 mutually exclusive cases for getting both even digits with chance $\frac{2}{9} \times \frac{2}{9}$ *i.e.*, $\frac{4}{81}$ for each case.

So the total chance of both even digits is $\frac{4}{81} \times 9 = \frac{4}{9}$.

So the total chance of getting the sum as even number $= \frac{1}{9} + \frac{4}{9} = \frac{5}{9}$.

Illustration 16.80. There are four letters and four addressed envelopes. If the letters are put at random in the envelopes, find the chance that no letter will occupy its correct place.

Solution. Four letters can be placed in four envelopes in 4! or 24 ways. The first letter can be placed in a wrong envelope in 3 ways. Next, taking the letter corresponding to the envelope occupied by first, this can also be placed in wrong envelopes in 3 ways.

$\therefore$ The total number of ways up to this stage $= 3 \times 3 = 9$.

Now, in these 9 ways, there are two envelopes and two letters remaining.

There are two types of cases :

(*i*) Two letters (say C and D) and corresponding envelopes (C and D) are remaining. so in this case, there is only one way of putting both of them in wrong place *i.e.*, letter C in envelope D and letter D in envelope C.

(*ii*) Two letters (say C and D) and two envelopes are corresponding to one of the letters (say D) and one any other (say A). Even in this case, there is only one way of putting both the letters wrongly *i.e.*, letter C in envelope D and letter D in enveloped A. Hence, there are only 9 ways of placing all the letters wrongly.

$\therefore$ The required probability $= \frac{9}{24}$.

Illustration 16.81. Two packs of cards are made up in such a way that the first pack consists of 39 red cards and 13 black cards, second pack consists of 39 black cards and 13 red cards. A sampling experiment is carried out in the following ways : A card is drawn from the first pack, if it is red, a second card is drawn from the same pack after replacing the first red card. The colour of the second card drawn from the first pack is noted. If the first card drawn from the first pack is black, then the second card is drawn from the second pack and the colour of the second card is noted. Both the cards are then replaced in their respective packs. What is the probability that the second card is red?

Solution. The chance of drawing a red from 1st pack $= \frac{39}{39+13} = \frac{3}{4}$.

The chance of drawing a red from 2nd pack $= \frac{13}{39+13} = \frac{1}{4}$.

The chance of drawing a red from 2nd pack $= \frac{13}{39+13} = \frac{1}{4}$.

The second card red can come in two ways (*i*) 1st red, 2nd red, (*ii*) 1st black, 2nd red. Case (*i*) can happen, when both the red come from 1st pack and case (*ii*) can happen when 1st black comes from the 1st pack and 2nd red from the 2nd pack.

The compound event (*i*) has the chance $= \frac{3}{4} \times \frac{3}{4} = \frac{9}{16}$.

The compound event (*ii*) has the chance $= \frac{1}{4} \times \frac{1}{4} = \frac{1}{16}$.

Both these are mutually exclusive cases. So, the total chance of getting 2nd card red is $\frac{9}{16} + \frac{1}{16} = \frac{10}{16}$, *i.e.*, $\frac{5}{8}$.

Illustration. 16.82. In a throw of three dice what is the possibility of getting the total 11.

Solution. A dice has six faces marked with 1, 2, 3, 4, 5, 6.

$\therefore$ Total number of cases $= 6^3 = 216$.

If we start counting the possibility of 11 total in 3 dice it will be quite difficult. This can be done by mathematical method easily. The number of times 11 can be thrown is the coefficient of x^{11} in

$$(x^1+x^2+x^3+x^4+x^5+x^6)^3$$

or $$x^{11} \text{ in } x^3(1+x+x^2+x^3+x^4+x^5)^3$$

or $$x^8 \text{ in } (1+x+x^2+x^3+x^4+x^5)^3$$

or $$x^8 \text{ in } \left(\frac{1-x^6}{1-x}\right)^3$$

or $$\text{coeff. of } x^8 \text{ in } (1-x^6)^3(1-x)^{-3}$$

or $$\text{coeff. of } x^8 \text{ in } (1-3x^6+...)(1+3x+6x^2+...+45x^8+...)$$

or $$45 - 18 = 27.$$

$\therefore$ Required probability $= \dfrac{27}{216}$.

Illustration 16.83. A circle is divided in three zones and marked with 1, 0, –1. It is revolved and on stopping the number of the zone which comes in front of a fixed pointer is noted. This is repeated 5 times. What is the chance that the total number is zero.

Solution. Total number of ways = $3^5 = 243$.

The number of favourable ways can be found by coefficient of x^0 in the expansion of $(x^{-1}+x^0+x^1)^5$.

or $$\text{coeff. of } x^0 \text{ in } (x^{-1})^5(1+x+x^2)^5$$

or $$\text{in } x^{-5}\left(\frac{1-x^3}{1-x}\right)^{+5}$$

or $$\text{coeff. of } x^5 \text{ in } (1-x^3)^5(1-x)^{-5}$$

or coeff. of x^5 in $(1-5x^3+10x^6+...)(1+5x+15x^2+...+126x^5+...)$

$$= 126 - 75 = 51.$$

$\therefore$ Required probability $= \dfrac{51}{243}$.

Illustration 16.84. An urn contains 3 one-rupee coins, 5 fifty paise coins and 12 ten-paise coins. If a coin is drawn at random, find the expected value of the draw, assuming that all coins are equally likely.

Solution.

Type of coin	*Value of coin (Rs)* x_i	*Probability* p_i	$x_i \cdot p_i$
One-rupee	1	3/20	3/20
50-paisa	1/2	5/20	5/40
10-paisa	1/10	12/20	12/200
	Total	1	$\Sigma\, x_i \cdot p_i = \frac{67}{200}$ $= 0.335$

$\therefore$ Expected value of the draw = Re 0.335.

Illustration 16.85. A bag contains 4 white and six black balls. Three balls are drawn from it at random.

(*i*) Find the expected number of the white balls drawn.

(*ii*) If each white ball drawn carries a reward of Rs 4 and each black ball Rs 6, find the expected reward of the draw.

Solution. (*i*)

Number of white balls drawn x_i	*Probability* p_i	$x_i \cdot p_i$
0	$\frac{{}^4C_0 \times {}^6C_3}{{}^{10}C_3} = \frac{20}{120}$	0
1	$\frac{{}^4C_1 \times {}^6C_2}{{}^{10}C_3} = \frac{60}{120}$	$\frac{60}{120}$
2	$\frac{{}^4C_2 \times {}^6C_1}{{}^{10}C_3} = \frac{36}{120}$	$\frac{72}{120}$
3	$\frac{{}^4C_3 \times {}^6C_0}{{}^{10}C_3} = \frac{4}{120}$	$\frac{12}{120}$
Total	1	$\Sigma x_i p_i = \frac{144}{120} = 1.2$

$\therefore$ expected number of white balls drawn = 1.2.

(*ii*)

Number of white balls drawn	*Amount of reward (Rs)* x_i	*Probability* p_i	$x_i \cdot p_i$
0	18	$\frac{{}^4C_0 \times {}^6C_3}{{}^{10}C_3} = \frac{20}{120}$	$\frac{360}{120}$
1	16	$\frac{{}^4C_1 \times {}^6C_2}{{}^{10}C_3} = \frac{60}{120}$	$\frac{960}{120}$
2	14	$\frac{{}^4C_2 \times {}^6C_1}{{}^{10}C_3} = \frac{36}{120}$	$\frac{504}{120}$
3	12	$\frac{{}^4C_3 \times {}^6C_0}{{}^{10}C_3} = \frac{4}{120}$	$\frac{48}{120}$
	Total	1	$\Sigma x_i \cdot p_i = \frac{1872}{120} = 15.60$

$\therefore$ expected amount of reward = Rs 15.60.

Illustration 16.86. An insurance company offers a 45 year old man a one-year policy of Rs 1000 for an annual premium of Rs 12. The number of deaths per 1000 is 5 in one year for the persons of this age group. What is the expected gain to the insurance company on such policy.

Solution.

Case	*Gain to Insurance company (Rs)* x_i	*Probability* p_i	$x_i \cdot p_i$
No death	12	$\frac{995}{1000}$	$\frac{11940}{1000}$
Death	– 988 (i.e., loss of Rs. 1000 – 12)	$\frac{5}{1000}$	$-\frac{4940}{1000}$
	Total	1	$\Sigma\, x_i \cdot p_i = \frac{7000}{1000} = 7$

∴ expected gain to the insurance company = Rs. 7.

Illustration 16.87. A person applies for 200 equity shares of Rs.10 each to be issued at a premium of Rs.6 per share, Rs.8 being payable along with the application and the balance at the time of the allotment. The company may issue 100 shares or 50 shares to those who apply for 200 shares, the probability of issuing 100 shares being 0.6 and that of issuing 50 shares 0.4. In either case, the probability of an applicant being selected for the allotment of any shares is 0.2. The allotment usually takes three months and the market price per share is expected to be Rs.25 at the time of allotment.

Find the expected rate of return of the investor.

Solution. Calculation of expected number of share to be allotted

Number of shares to be allotted (x_i)	*Probability* (p_i)	$x_i\, p_i$
0	0.8	0
50	$0.2 \times 0.4 = 0.08$	4
100	$0.2 \times 0.6 = 0.12$	12
Total	1	$\Sigma\, x_i\, p_i = 16$

Gain per share (after 3 months) = Rs (25 – 16) = Rs 9

∴ gain on 16 shares (after 3 months) = 9 × 16 = Rs 144

Money invested (for 3 months) = 200 × 8 = Rs 1600

$$\therefore \quad \text{Annual rate of return} = \frac{144}{1600} \times \frac{12}{3} \times 100 = 36\% .$$

PROBLEMS

1. **In a certain town the ratio of males to females is 1000 : 987. If this tendency is to continue, what is the chance that a newly born baby is male?** **(Ans.** 1000/1987)
2. What is the probability that a digit selected at random from the logarithmic table is (*i*) 1. (*ii*) 3 or 7.
[**Ans.** (*i*) Probability of getting 1 = 1/10. (*ii*) The number of favourable cases for getting 3 or 7 is 2. ∴ required probability is 2/10.]
3. The following mortality table shows the number of survivors to various ages of 100,000 newly born males.

Age X	*Survivors*	*Age X*	*Survivors*
0	100,000	60	67,787
10	93,601	70	46,739
20	92,293	80	19,860
30	90,092	90	2,812
40	86,880	100	6
50	80,521	—	—

(*i*) Find the probability of a newly born infant in this population living to be 60 years old. (*ii*) Find the probability of a 20 years old in this population living until he is 50 years old.
[**Ans.** (*i*) 67,787/100,000. (*ii*) 80,521/92,293]

4. The following table gives distribution of wages :

Weekly wages	30—35	35—40	40—45	45—50	50—55	55—60	60—65	65—70
No. of wage Earners	9	108	488	230	112	30	16	7

An individual is taken at random from the above group. Find the probability (*i*) his wages were under 40, (*ii*) his wages were 55 or over, (*iii*) his wages were either between 45 50 or 34 40.
[**Ans.** (*i*) 117/100,000, (*ii*) 53/1000, (*iii*) 338/1000].

5. Tickets numbered 1 to 100 are well shuffled and a ticket is drawn. What is the probability that the drawn ticket has (*i*) an odd number, (*ii*) a number 5 or multiple of 5, (*iii*) a number which is a multiple of 10.
[**Ans.** (*i*) 1/2, (*ii*) 1/5, (*iii*) 1/10]
6. Find out the probability of forming of 563 and 169 with the digits 1, 2, 3, 4, 5, 6, 7, 8, 9 when only numbers of three digits are formed and when (*i*) repetitions are not allowed (*ii*) when repetitions are allowed.
[**Ans.** (*i*) 1/552, (*ii*) 2/729]
7. A bag contains 7 red, 12 white and 4 green balls. What is the probability that (*i*) a ball drawn is white, (*ii*) 3 balls drawn are all white and (*iii*) 3 balls drawn are one of each colour.
[**Ans.** (*i*) 12/23, (*ii*) 220/1771, (*iii*) 336/1771]

8. Find the chance of throwing an ace with a perfect die. **[Ans.** 1/6]

9. Find the probability of obtaining a total of 10 or more points when two ordinary dice are thrown. **[Ans.** 6/36]

10. Compare the chance of throwing 5 with one dice, 10 with two dice and 15 with three dice. **[Ans.** (*i*) 1/6, (*ii*) 1/12, (*iii*) 5/108]

11. What is the probability of getting 9 cards of same suit in one hand at a game of bridge? **[Ans.** Required probability = ${}^{13}C_9 \times {}^{39}C_4 / 4\ {}^{52}C_{13}$]

12. The letters of the word "Statistics" are written on 10 identical cards. If two cards are drawn at random, what is the probability that one *s* and one *i* will occur. **[Ans.** Required probability = 6/45]

13. In a race where 12 horses are running the chance that horse *A* will win is 1/6, that *B* will is 1/10 and that *C* will win is 1/8. Assuming that a dead heat is impossible, find that chance that one of them will win.
[Ans. So the chance that one of them will win the race is = 47/120]

14. Three ships, *A*, *B* and *C* sail from England to India. Odds in favour of their arriving safely are 2 : 5, 3 : 7 and 6 : 11 respectively. Find the chance that they all arrive safely. **[Ans.** 18/595]

15. It is against 8 : 5 against a person who is now 40 years old, living till he is 70 and 4 : 3 against a person now 50 living till he is 80. Find the probability that at least one of these persons will be alive 30 years hence. **[Ans.** 58/91]

16. The odds in favour of '*A*' winning a game of chess against *B* are 5 : 2. If three games are to be played, what are the odds in favour of *A*'s winning at least one game. **[Ans.** 335/343]

17. A bag contains 6 white and 9 black balls. Two drawings of 4 balls are made such that (*a*) the balls are replaced before the second trail, (*b*) the balls are not replaced before the second trail. Find the probability that the first drawing will give 4 white and the second 4 black balls in each case. **[Ans.** 6/5915, 3/715]

18. A bag contains 5 and 4 black balls. another contains 3 red and 7 black balls. A ball is drawn from the first bag and is placed in the second. A ball is then drawn from the second bag. What is the probability that it is red? **[Ans.** 32/99]

19. In a college, there are 40 M.A. Previous students and 35 M.A. Final students. Three students selected at random from these are to speak on a topic. What is the chance that they are alternately of different classes? **[Ans.** The required probability is = 2044/8103]

20. Three urns identical in appearance contain respectively 2 white and 2 black balls, 1 white and 3 black balls, 2 white and 3 black balls. One ball is drawn at random from each urn. Calculate the probability of obtaining one black and two white balls.
[Ans. Required probability = 11/40]

21. A coin is tossed three times. find the chances of throwing (*i*) 3 heads, (*ii*) 2 heads one tail, (*iii*) head and tail alternately.

[Ans. (*i*) 1/8, (*ii*) 3/8, (*iii*) 1/4]

22. Find the chance of throwing 4 or 5 (*i*) three times and (*ii*) at least once in 5. throws of a single die. **[Ans.** (*i*) 40/243, (*ii*) 211/243]

23. *A* and *B* stand in a ring with ten other persons. If the arrangements of 12 persons is at random, find the chance that there are exactly three persons between *A* and *B*. **[Ans.** 2/11]

24. From an urn containing 9 balls numbered 1 to 9. Two balls are drawn successively, the first ball is returned to the urn if its number is 1. Otherwise it is not returned. Find the probability that the ball numbered 2 is drawn at the second trail. **[Ans.** 17/640]

25. In 120 families each of five children, number of males were distributed as follows :

No. of male children per family	0	1	2	3	4	5
No. of families	4	18	40	36	20	2

Calculate the probability of male birth.

[Ans. Probability of male birth = 296/600]

26. A letter is taken at random out of 'assistant' and a letter out of 'Statistics'. What is the chance that they are the same letter?

[Ans. 19/90]

27. What is the probability that each of the four players of cards hold 13 cards of the same suit? **[Ans.** 13! 13! 13! 13! 4! 52!]

28. Licence plates of cars in a city numbered serially beginning with 1. If 30,000 cars are registered. What is the probability that the number on the plate of car selected at random is 3 or begins with the digit 3.

[Ans. 1111/30,000]

29. *A* can hit a target 8 times in 5 shots, *B* 2 times in 5 shots, *C* 3 times in 4 shots. They fire a volley. What is the probability that shots hit 2.

[Ans. 2 shots hit = 9/20]

Chapter 17

SAMPLING DISTRIBUTIONS

17.1. NATURE OF SAMPLING DISTRIBUTIONS

Consider a large bag of balls of which a fraction π are white (*W*), and the rest, a fraction $(1-\pi)$ are red (*R*). When we pick any *n* balls at random, we are said to have drawn a *sample*. The composition of any one sample is unpredictable, and will vary from sample to sample. If x denotes the *random* number equal to the number of, say, white balls in the sample, x is a discrete random variable which can take any integral value from 0 to n. If we conduct the experiment of drawing a sample a large number of times and note down the value of x in each case, we get a probability distribution which in this case is termed a *sampling distribution* for the given problem. A sampling distribution is *the probability distribution associated with a random variable denoting any statistics of a sample drawn randomly from a given population.* We can, thus, have a sampling distribution for the number of red balls (*R*), of the fraction of red balls, or any other statistics associated with the sample.

As is obvious, these sampling distributions should depend on the nature of the sample statistics considered, on the sample size, and on the nature of the population itself. Thus the sampling distribution in our problem will depend on the exact meaning of x (*i.e.*, whether it is number or the fraction of red or white balls), on the sample size n, and on the population parameter π.

In many cases, it may be possible to predict the sampling distribution from purely theoretical considerations. In other cases, we can obtain these by actual repeated samplings from the population. When we do this, we find that many sampling distributions can be approximated quite well by certain very simple theoretically obtained distributions. This one fact is quite remarkable and on this foundation stands the grand edifice of the statistical techniques of estimation and inference. We will return to these techniques in the following chapters.

17.2 BINOMIAL DISTRIBUTION

Suppose an experiment can have two results, A and B. Let us term A as the 'success' and B, which is the non-occurrence of A as the 'failure'. Let the probability of a 'success' be π, so that the probability of a failure is $(1-\pi)$.

Now each experiment is like drawing one ball from a large bag containing a fraction π of red (R) balls and $(1-\pi)$ of white (W) balls. Success, then, is drawing R.

If we draw two balls, *i.e.*, have two trials, then probability of two successes is $\pi.\pi=\pi^2$; probability of one success and one failure (in either order) is $\pi.(1-\pi)+(1-\pi).\pi=2\pi(1-\pi)$; and the probability of both failures is $(1-\pi)^2$. Thus, in a sample of 2 balls, the probability of no R, $1R$ and $2R$ are $(1-\pi)^2$, $2(1-\pi)\pi$ and π^2, respectively.

Similarly, if a sample of 3 balls are drawn the probability of no R, $1R$, $2R$ and $3R$ are $(1-\pi)^3$, $3(1-\pi)^2\pi$, $3(1-\pi)\pi^2$, and π^3 respectively.

It can be seen by inspection, that the probability of 0, 1, 2, 3, ... successes are given by the successive terms of the binomial expansion $[(1-\pi)+\pi]^n$ where n is the number of trials, or items in the sample. Thus, the probabilities of 0, 1, 2, 3, ... successes are given by successive terms of

$[(1-\pi)+\pi]^1$; namely $(1-\pi)$, π; for $n=1$,

$[(1-\pi)+\pi]^2$; namely $(1-\pi)^2$, $2(1-\pi)\pi$, π^2; for $n=2$,

$[(1-\pi)+\pi]^3$; namely $(1-\pi)^3$, $3(1-\pi)^2\pi$, $3(1-\pi)\pi^2$, π^3; for $n=3$,

and so on.

In general, for n trials: $[(1-\pi)+\pi]^n$, so that

No. of success x	*Probability* p
0	$(1-\pi)^n$
1	$n.(1-\pi)^{n-1}\pi^1$
2	$\frac{n(n-1)}{1.2}.(1-\pi)^{n-2}.\pi^2$
3	$\frac{n(n-1)(n-2)}{1.2.3}.(1-\pi)^{n-3}.\pi^3$
⋮	⋮
r	$nCr.(1-\pi)^{n-r}.\pi^r$
⋮	⋮
	$\pi^r(1-\pi)^{n-r}$

In this table ^nC_r stands for $\frac{n!}{(n-r)!\,r!}$

The above probability distribution is known as the binomial probability distribution or more commonly as binomial distribution.

The general term of this distribution is. ${}^nC_r(1-\pi)^{n-r}\pi^r$

Exaplanation. Consider n independent trials. Let the probability of success in 1 trial be π and the probability of failure be $(1-\pi)$. Let us calculate the probability of r successes and $n-r$ failures out of these n trials. Let first r trials be all successes and next $n-r$ trials be all failures. The probability of joint occurrence of r successes is the product of the individual probabilities of the success, *i.e.*, $\pi\times\pi\times\ldots\ldots r$ times$=\pi^r$. Similarly, the probability of $n-r$ failures is $(1-\pi)^{n-r}$. Thus, the probability of getting exactly r successes and $n-r$ failures in a specific order is $\pi^r(1-\pi)^{n-r}$. But all these r successes need not occur in the first r successive trials, they may be spread over all the n trials. The number of ways in which we can get r successes and $n-r$ failures out of n trials is

$$\frac{n!}{r!(n-r)!}\ i.e., {}^nC_r$$

All these ${}^nC_r$ ways of occurrences are mutually exclusive, with probability π^r $(1-\pi)^{n-r}$ each time. So that total probability of r successes and $n-r$ failures in n trials is

$$\pi^r(1-\pi)^{n-r}+\pi^r(1-\pi)^{n-r}\ldots\ldots{}^nC_r \text{ times, } i.e., {}^nC_r(1-\pi)^{n-r}\pi^r.$$

This means that any term of the binomial expansion can be obtained by making use of the general term also. If in the general term the values of r are taken as 0, 1, 2 and 3......etc., we shall be able to obtain the first, second, third, fourth......etc. terms of the binomial expansion. Thus, if it is desired to find the chance of 0 success in n trials, r should be taken as 0, for chance 1 success r be taken as 1, and so on.

The coefficient for the successive values of exactly 1, 2, 3...... successes can also be obtained from what is known as *Pascal's Triangle* (Blaise Pascal—1623-62) by writing on successive lines in the form of a triangle as given on the next page.

It is clear from the triangle that in any line each entry is the sum of the two entries in the line above to its immediate left and right. Thus, in the seventh line (exponent=6), we have 6=1+5, 15=5+10, and so on.

Fig 17.1 *a* and *b* show the sampling destribution for x, the number of successes in a sample of 5 (*i.e.*, $n=5$) when the population proportion of successes π is (*a*) 0.25 and (*b*) 0.50. Note that the destribution changes as the value of π changes. For $\pi=0.25$, the distribution is skewed, with most values concentrated towards the lower end,

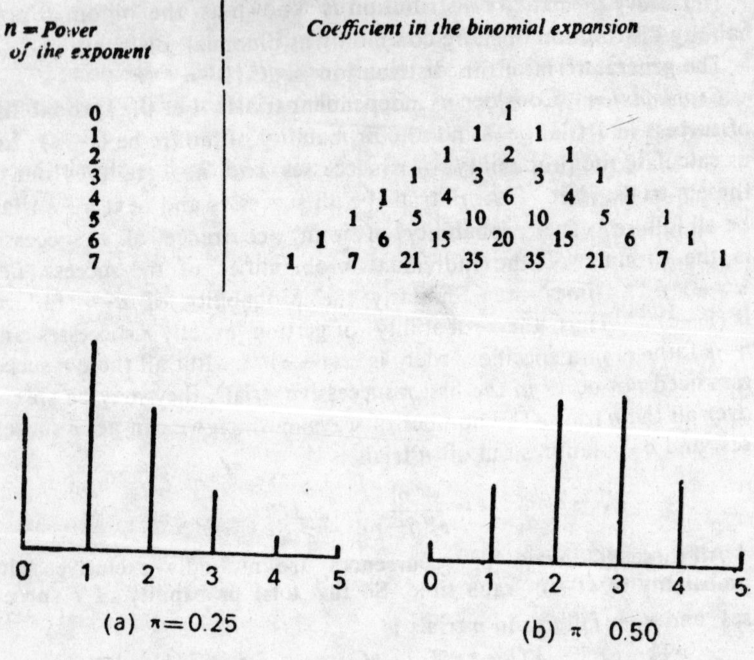

n = Power of the exponent	Coefficient in the binomial expansion
0	1
1	1 1
2	1 2 1
3	1 3 3 1
4	1 4 6 4 1
5	1 5 10 10 5 1
6	1 6 15 20 15 6 1
7	1 7 21 35 35 21 7 1

FIG 17.1

and a long tail to the right. For $\pi=0.50$, *i.e.*, with success as probable as failure, the *distribution* is symmetrical, as expected.

Even for π different from 0.5, the skewness of the distribution can be reduced by increasing n, the sample size.

Notice that a binomial distribution, is completely specified by two parameters, the population proportion, π, and the sample size n.

Illustration 17.1. Eight coins are thrown simultaneously. Find the chance of obtaining:

(*i*) at least 6 heads
(*ii*) no heads
(*iii*) all heads.

Solution.

The probability of head = The probability of a tail $=\frac{1}{2}$.

(*i*) Probability of getting 6 heads $={}^8C_6\left(\frac{1}{2}\right)^6\left(\frac{1}{2}\right)^2=\frac{28}{256}$

" " 7 " $={}^8C_7\left(\frac{1}{2}\right)^7\left(\frac{1}{2}\right)=\frac{8}{256}$

" " 8 " $={}^8C_8\left(\frac{1}{2}\right)^8=\frac{1}{256}$

$\therefore$ The probability of getting at least 6 heads $=\frac{28}{256}+\frac{8}{256}+\frac{1}{256}=\frac{37}{256}$

(*ii*) **The probability of getting no head** $={}^8C_0\,(\frac{1}{2})^8=\frac{1}{256}$

(*iii*) **The probability of getting all heads** $={}^8C_8\,(\frac{1}{2})^8=\frac{1}{256}$.

Illustration 17.2. In how many cases should we expect to get 6 heads and 4 tails if 10 coins are simultaneously tossed 1,000 times?

Solution. Here $\pi=\frac{1}{2}$, $N=1{,}000$, $n=10$

So the expansion of $N[(1-\pi)+\pi]^n$ will give the numbers of successes 0, 1, 2...10.

Number for r successes is given by $N\,{}^nC_r(1-\pi)^{n-r}\pi^r$.

Here r is 6. So the required number is $1{,}000\,{}^{10}C_6\,(\frac{1}{2})^4\,(\frac{1}{2})^6$

$$1{,}000\,\frac{10\times9\times8\times6}{4\times3\times2\times1}\times\frac{1}{1{,}024}=\frac{13{,}125}{64}=205 \text{ approx.}$$

Note. Instead of 6 heads, if at least 6 heads would have been asked in the question, then one should get the sum of terms with 6, 7, 8, 9 and 10 successes.

Illustration 17.3. The normal rate of infection of a certain disease in animals is known to be 40%. In an experiment with 6 animals injected with a new vaccine it was observed that none of the animals caught infection. Calculate the probability of the observed result if the vaccine was ineffective.

Soluiion. Here we have to find the probability of 0 infections.

$$\text{With } \pi=40\%=\frac{2}{5}$$

We know that the probability of 0 success in n trials is given by $(1-\pi)^n$.

$$\text{So } p(0 \text{ success})=\left(\frac{3}{5}\right)^6=\frac{729}{15{,}625}=0.0467.$$

Illustration 17.4. Assuming that 20% of the population of a city are literate, so that the chance of an individual being literate is $\frac{1}{5}$ and assuming that 100 investigators each take 10 individuals to see whether they are literate, how many investigators would you expect to report that three people or less were literate?

Solution. Let π=Probability of a person being literate$=\frac{1}{5}$

Here we have 100 sets of 10 trials each. The number of investigators reporting 0, 1, 2,..., 10 literates are given by the successive terms of the binomial

$$100\left(\frac{4}{5}+\frac{1}{5}\right)^{10}=100[(0.8)^{10}+{}^{10}C_1\,(0.8)^9\,(0.2)+{}^{10}C_2(0.8)^8\,(.2)^2+\ldots]$$

The number of investigators who are expected to report three or less literates shall be the sum of the first four terms. *i.e.*

$$=100[(0.8)^{10}+{}^{10}C_1(0.8)^9(0.2)+{}^{10}C_2(0.8)^8(0.2)^2+{}^{10}C_3(0.8)^7(0.2)^3]$$

$=10.74+26.85+30.21+20.14$

$=87.9=88$ nearly.

Illustration 17.5. Find the chance of getting 3 successes in 5 trials when the chance of getting a success in one trial is 2/3.

Solution. Here $n=5$, $\pi=2/3$, $r=3$, $1-\pi=1-2/3=1/3$. The probability of 3 successes is given by the 4th term of the binomial expansion with $n=5$. The coefficient from Pascals triangle (or by ${}^5C_3$) is 10. Therefore,

$$p\text{ (3 successes)}=10\,(1-\pi)^2\,(\pi)^3$$
$$=10\,(1/3)^2\,(2/3)^3=0.33.$$

Illustration 17.6. Assuming that half of the population is vegetarian so that the chance of an individual being a vegetarian is ½ and assuming that 100 investigators take a sample of 10 individuals to see whether they are vegetarians, how many investigators would you expect to report that 3 people or less were vegetarians?

Solution. We have $\pi=\frac{1}{2}$ and $n=10$ for each investigator so that the probability of getting 0, 1, 2, 3...10 vegetarians will be given by the terms of the expansion of $(\frac{1}{2}+\frac{1}{2})^{10}$.

∴ Probability of 3 or less vegetarians

$=(\frac{1}{2})^{10}+10C_1(\frac{1}{2})^9(\frac{1}{2})+10C_2(\frac{1}{2})^8(\frac{1}{2})^2+10C_3(\frac{1}{2})^7(\frac{1}{2})^3.$

$=(\frac{1}{2})^{10}[1+10+45+120]=176/1024$

For all the $100(N)$ investigators the probability will be same. So the expected no. of investigators finding 3 or less vegetarians will be $100\times176/1024=17.2$.

Or 17 Investigators approximately.

Illustration 17.7. In a precision bombing attack there is a 50% chance that any one bomb will strike the target. Two direct hits are required to destroy the target completely. How many bombs must be dropped to give a 99% chance or better of completely destroying the target?

Solution. Let success denote hitting of target

∴ $\pi=50\%=\frac{1}{2}$

∴ Out of n bombs dropped the probability of hitting the target 0, 1, 2, 3...n bombs is given Ist, IInd,...etc. terms in the expansion of $(\frac{1}{2}+\frac{1}{2})^n$.

The target is not destroyed if 0 or one bomb falls over it.

∴ The probability of not destroying the target

$=(\frac{1}{2})^n+n(\frac{1}{2})^{n-1}(\frac{1}{2})=(\frac{1}{2})^n(1+n)$

Probability of destroying should be 99% or better *i.e.*, n should be such that the probability of not destroying should be $100-99=1\%$

or less *i.e.* .01 or less.

$$\left(\tfrac{1}{2}\right)^n(n+1) < .01.$$

Up to $n = 10$, this expression is not true and when $n = 11$ for the first time $\left(\tfrac{1}{2}\right)^n(n+1)$ is less than 0.01 and will remain less for $n > 11$.

Least no. of bombs to be dropped = 11.

17.3. MEAN AND STANDARD DEVIATION OF BINOMIAL DISTRIBUTION

If we take a large number N of samples of size n, we expect that the number of samples which have exactly x successes are $f(x) = N.p(x)$ where $p(x)$ is the probability of x successes. Then the mean number of successes is

$$\bar{x} = \frac{\Sigma x\, f(x)}{\Sigma\, f(x)} = \frac{\Sigma x \cdot N\, p(x)}{\Sigma N\, p(x)} = \Sigma x\, p(x)$$
$$= E(x)$$

which is the expected value of x,

For a binomial distribution

$$\Sigma x\, p(x) = \Sigma x.{}^nC_x\,(1-\pi)^{n-x}\,\pi^x$$
$$= 0.(1-\pi)^n + 1.n(1-\pi)^{n-1}\,\pi + 2.\ \frac{n(n-1)}{1.2}(1-\pi)^{n-2}.\pi^2 + \ldots\ldots + n\pi^n$$
$$= n(1-\pi)^{n-1}\,\pi + n(n-1)(1-\pi)^{n-2}\,\pi^2 + \ldots\ldots + n\pi^n$$
$$= n\pi[(1-\pi)^{n-1} + (n-1)(1-\pi)^{n-2}\,\pi + \ldots\ldots\pi^{n-1}]$$
$$= n\pi[(1-\pi)+\pi]^{n-1} = n\pi \qquad (17.1)$$

So the mean of binomal distribution is $n\pi$.

Similarly we can obtain the standard deviation as $\sqrt{n(1-\pi)\pi}$.

Proof. $\sigma^2 = \mu_2 = \nu_2 + \nu_1^2$

Where ν_2 and ν_1 and moments about the origin. We known $\nu_1 = \text{mean} = n\pi$.

$$\nu_2 = \Sigma x^2\, p(x)$$
$$= 0^2(1-\pi)^n + 1^2 n(1-\pi)^{n-1}\,\pi + 2^2\,\frac{n(n-1)}{1.2}\,(1-\pi)^{n-2}\,\pi^2$$
$$+ 3^2\,\frac{n(n-1)(n-2)}{1.2.3}\,(1-\pi)^{n-3}\,\pi^3 + \ldots + n^2\pi^n$$
$$n\pi[(1-\pi)^{n-1} + 2(n-1)(1-\pi)^{n-2}\,\pi + \frac{3(n-1)(n-2)}{2\times 1}(1-\pi)^{n-3}\pi^2 + \ldots + n\pi^{n-1}]$$

Breaking the second and third terms into two parts, we get

$$\nu_2 = n\pi\{[(1-\pi)+\pi]^{n-1} + (n-1)\pi[(1-\pi)^{n-2} + (n-2)(1-\pi)^{n-3}\pi + \ldots\pi^{n-2}]\}$$
$$= n\pi\{1 + (n-1)\pi[(1-\pi)+\pi]^{n-2}\}$$
$$= n\pi[1+(n-1)\pi]$$

So

$$\sigma^2 = \nu_2 - \nu_1^2 = n\pi + n^2\pi^2 - n\pi^2 - n^2\pi^2$$
$$= n\pi(1-\pi)$$

or $$\sigma = \sqrt{n\pi(1-\pi)} \quad (17.2)$$

If we define the random variable x as the *proportion* of successes in a sample of size n, then clearly the resulting sampling distribution is also binomial with the general term as

$$^nC_r\,(1-\pi)^{n-r}\pi^r \quad (17.3)$$

It is easy to show that the mean or the expected value and standard deviation of this distribution are π and $\sqrt{\pi(1-\pi)/n}$, respectively.

We summarize this in the following form:

TABLE 17.1

Sample Statistics	*Expected Value*	*Standard Deviation*
Number of successes	$n\pi$	$\sqrt{n\pi(1-\pi)}$
Proportion of successes	π	$\sqrt{\pi(1-\pi)/n}$

When we recall that the standard deviation measures the spread of values of the random variable, and that we expect hardly any observation to lie outside the interval (mean$\pm 3\sigma$), we notice an interesting significance of the standard deviation of the sampling distribution of proportion of successes varying *inversely* as $\sqrt{n}$. As the sample size increases, the standard deviation decreases. Thus, the range within which the *sample* proportion may be expected to lie can be made as narrow as one likes, by taking a large enough sample. This is used to great advantage in estimation of population proportion π when the sample proportion x is measured, a problem common enough to be within everybody's experience.

Suppose we want to estimate the proportion of Indian men taller than 1.65 m *without* taking a census of all men. We can take a sample and determine the proportion x of men taller than 1.65 m in that sample. What information does it provide about the population proportion π? We know that the sample proportion x will differ from π according to the binomial distribution, but the spread of the values will decrease as $\sqrt{n}$ increases. In other words, as n increases, x cannot differ very much from π and the maximum expected deviation varies as $1/\sqrt{n}$. Therefore x can be taken as an

estimate **of π with increasing *level of confidence* as the sample size n increases.**

This means that the probable error in the estimate of π as x decreases is $1/\sqrt{n}$. It is for this reason that the standard deviation of a sampling distribution is termed as the *standard* error of *estimate.* We will go into this in more details in the next chapter.

This is also in line with the law of large numbers discussed in chapter 3. When the sample size n becomes increasingly large the proportion of successes settles down to the population value.

Illustration 17.8. For a binomial distribution the mean is 6 and standard deviation is $\sqrt{2}$. Write out all the terms of the distribution.

Solution. If n is the number of trials, and π, is probability of successes the mean of the binomial distribution is $n\pi$ and standard deviation is $\sqrt{n\pi(1-\pi)}$.

$$\therefore \qquad n\pi=6 \qquad \ldots(i)$$

$$\text{and} \qquad n\pi(1-\pi)=2 \qquad \ldots(ii)$$

Dividing (*ii*) by (*i*),

$$(1-\pi)=2/6=1/3$$

$$\text{or} \qquad \pi=2/3$$

Putting the value of π in (*i*),

$$n\times\tfrac{2}{3}=6 \quad \text{or} \quad n=9$$

The various terms of binomial distribution are given by expansion of $(\frac{1}{3}+\frac{2}{3})^9$ Hence the binomial distribution is:

No. of successes x	*Porbability* P	*No. of successes* x	*Probability* P
0	$\left(\frac{1}{3}\right)^9$	5	$126\left(\frac{1}{3}\right)^4\left(\frac{2}{3}\right)^5$
1	$9\left(\frac{1}{3}\right)^8\left(\frac{2}{3}\right)$	6	$84\left(\frac{1}{3}\right)^3\left(\frac{2}{3}\right)^6$
2	$36\left(\frac{1}{3}\right)^7\left(\frac{2}{3}\right)^2$	7	$36\left(\frac{1}{3}\right)^2\left(\frac{2}{3}\right)^7$
3	$84\left(\frac{1}{3}\right)^6\left(\frac{2}{3}\right)^3$	8	$9\left(\frac{1}{3}\right)\left(\frac{2}{3}\right)^8$
4	$126\left(\frac{1}{3}\right)^5\left(\frac{2}{3}\right)^4$	9	$\left(\frac{2}{3}\right)^9$

Illustration 17.9. Seven coins are tossed and the number of heads noted. This experiment is repeated 128 times and the following is the observed frequency distribution of the 128 throws according to the number of heads:

No. of heads:	0	1	2	3	4	5	6	7
Throws:	7	6	19	35	30	23	7	1
Total	128							

Fit a binomial, distribution under the hypothesis that the coins are unbiased. Also find its mean and standard deviation.

Solution. The frequencies of the successive values 0, 1, 2, 3,...7 will be obtained by the expansion of

$$128(\tfrac{1}{2}+\tfrac{1}{2})^7$$

i.e., $128\{(\frac{1}{2})^7+7(\frac{1}{2})^7+21(\frac{1}{2})^7+35(\frac{1}{2})^7+35(\frac{1}{2})^7+21(\frac{1}{2})^7+7(\frac{1}{2})^7+(\frac{1}{2})^7\}$

$$=128\left(\frac{1}{128}+\frac{7}{128}+\frac{21}{128}+\frac{35}{128}+\frac{35}{128}+\frac{21}{128}+\frac{7}{128}+\frac{1}{128}\right)$$

$$=1+7+21+35+35+21+7+1=128$$

Thus, the theoretical frequencies for

	0	1	2	3	4	5	6	7 heads
will be	1	7	21	35	35	21	7	1.

Mean of the distribution is given by $n\pi$ where n is the number of coins and π is the probability of success. $n\pi=7\times\frac{1}{2}=3.5$. Standard deviation or $\sigma=\sqrt{n\pi(1-\pi)}=\sqrt{7\times\frac{1}{2}\times\frac{1}{2}}=\sqrt{1.75}=1.323$. So the mean and standard deviation of the distribution are 3.5 and 1.323 respectively.

Illustration 17.10. Is it possible to have a binomial distribution with $\bar{x}=6$ and $\sigma=3$?

Solution. $\bar{x}=n\pi=6$

$$\sigma=\sqrt{n\pi(1-\pi)}=3$$

or $$n\pi(1-\pi)=3^2=9$$

Dividing by $\bar{x}$

$$1-\pi=\frac{9}{6}=1.5 \quad \text{or} \quad \pi=-0.5$$

Which is not possible.

17.4. NORMAL DISTRIBUTION

If we take a continuous random variable, say the height of individuals, the length of various pieces cut by an automatic machine, or even the yearly rainfall data over a long period of time, and plot the *probability density fnuction**, one curious fact emerges: most of these

*Recall from chapter 16 that the probability density of a continuous variable impiies that the *area under the density curve* between the variable value x_1 and x_2 represents the *probability* of the value of the variable lying in the interval x_1 to x_2.

density curves resemble one another *in shape*. This common shape is a bell-shaped symmetrical one as shown in Fig. 17.2. It is a unimodel curve with most of the density concentrated near the mode (which is also the mean and the median), and very little density (though not zero) lying towards the lower and the upper extremes.

Most of the frequency distributions, specially of natural phenomena, vary approximately in accordance with this curve. Many of the statistical methods of describing and interpreting data are directly attributable to the properties of normal curve. Without it, the techniques of sampling could not have developed to its present stage. In fact it is a tool indispensable to modern statistics.

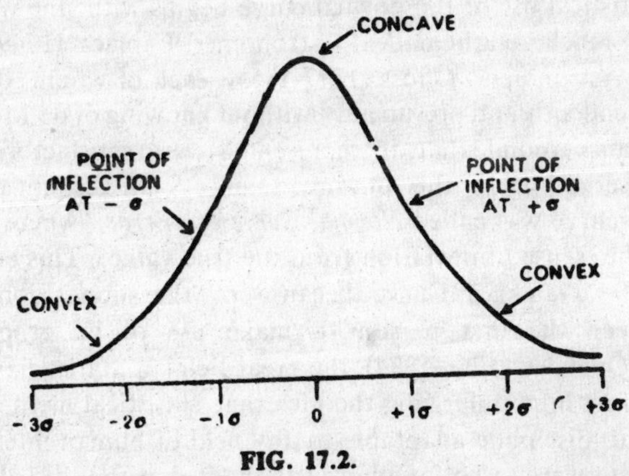

FIG. 17.2.

The physical conditions which lead to the emergence of a normal curve are:

1. The forces which affect individual events are independent of one another.

2. These forces are numerous and of equal weights.

3. Such forces affect the individual event in such a way that the maximum frequencies are clustered around the mean value and give rise to a symmetrical curve. The positive and negative deviation from this mean value are equally likely and large deviations are less likely to occur than the small ones.

Discovery of the normal curve. This curve was originally developed by Abraham de Moivre (1667-1754) a french mathematician who spent 66 years of his life in England as a refugee mathematician, and the equation of this curve was published in 1733. He discovered that the random variation in the number of heads appearing on throws of n coins corresponds to the terms of the binomial expansion of $(\frac{1}{2}+\frac{1}{2})^n$ and as n becomes larger this distribution approaches a definite from, while working on certain problems in games of chance. Because of this origin and because the data from certain coin throwing experiments closely approached it in form, it is often called the *Normal Probability Curve.* To de Moivre, it was a purely mathematical exercise, completely unconnected with the application to empirical data.

Jacques Bernoulli (1654-1705), a Swiss mathematician, suggested for the application of the theory of probability in economic and moral affairs, but could not personally investigate the applications because of the paucity of numerical data.

Actual statistical use of the normal curve begins with the work of the famous French Mathematical Astronomer, Laplace (1749-1827) and German Astronomer, Gauss (1777-1855), each of whom derived the law independently and presumably without knowing of de Moivre's treatment. Gauss found that it represented very satisfactorily the errors of observation in the physical sciences. On account of this reason, this curve was called *Normal Curve of Error*, where error is used in the sense of deviation from the true value. This curve is also called the Gaussian Curve because of the supposition that Gauss has been the first person to make use of its properties. Lambert A. Quetelet (1796-1874), the great Belgian Statistician, was probably the first in popularizing the idea that statistical method was a fundamental discipline adaptable to any field of human interest in which mass data were to be found, and thought that the distribution of the sufficient and trustworthy data of the measuremant of mental and moral traits would be found to be in accordance with the normal curve of error.

Since that time, experience has shown that it serves quite well to describe many of the distributions, such as stature and other anthropometric measurements, age, etc., which arise in the field of biology, education and sociology. Much of theory of statistics is built around it.

17.5. MATHEMATICAL FORM OF NORMAL DISTRIBUTION

It can be shown that if we take a binomial distribution with proportion of success π and the sample size n, and we increase the value of n, then the probability distribution that we obtain has a shape which is similar to the normal curve. (Note that binomial distribution is a discrete variable probability distribution while normal distribution is a continuous variable probability density distribution). When n becomes very large and approaches infinity, the sample proportion x can have so many values that x/n can be treated as a continuous variable, and then it is proper to define a probability density distribution. This density distribution is a normal distribution. If we carry out the mathematical steps indicated, we obtain the following density function

$$p(x) = \frac{1}{2.5066\,\sigma}\, e^{-\frac{(x-\mu)^2}{2\sigma^2}} \tag{17.4}$$

where $\mu=\pi$, the population proportion and σ is the sample standard deviation equal to $\sqrt{\pi(1-\pi)/n}$. It can be shown (and it should be obvious) that the mean (or the expected value) of the normal distribution is μ and its standard deviation is σ.

It should be noted that we need to specify only two parameters μ and σ to completely specify or fix the normal curve. The parameter μ which measures the expected value, fixes the location of the curve along the x-axis. A change in the value of μ(with σ held constant) just shifts the curve along the x-axis. The parameter σ on the

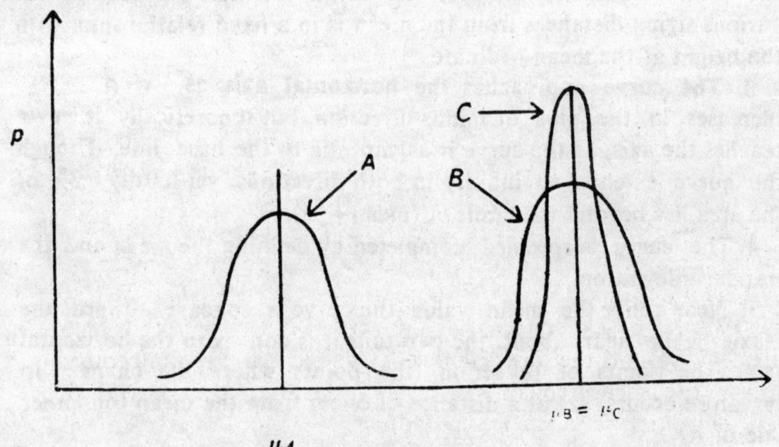

FIG. 17.3.

other hand fixes the spread, so that a change in the value of σ (with μ held constant) just changes the spread along the x-axis. Curve A and B of Fig. 17.3 shown two normal distributions with same σ's but different μ's, while curves B and C have same μ's but different σ's.

Properties of the normal curve. A study of the normal curve reveals the under-mentioned significant properties:

1. It is a symmetrical curve, *i.e.*, it is a single peaked or unimodal symmetrical curve. It has only one point of maximum ordinate and so if a curve is bi-modal, it cannot be called normal though it may be symmetrical. It means that the distribution of the frequencies on either side of the maximum ordinate of this curve is exactly the same. In other words, if we fold the paper in such a way that the folding line is on the ordinate drawn from the mean, then the curve on one side of this will exactly fall on the other side of the ordinate at mean.

2. There are few items on the extremes and most of the observations are clustered around the mean or the central value. The mean, the median and the mode coincide because the distribution is symmetrical and single peaked. The maximum ordinate frequency is at the centre of the distribution, *i.e.*, at the mean. In other words, we can say that the height of the ordinate at the mid-point of the base below the apex of the curve is the greatest, *i.e.*, at the point $x=\mu$, the ordinate has the maximum height. The height of the ordinate at a distance of one σ (sigma) from the mean is 60.653% of the height of the mean ordinate. Likewise, the height of the other ordinate at various sigma distances from the mean is in a fixed relationship with the height at the mean ordinate.

3. **The curve approaches the horizontal axis as $x-\mu$** increases in the plus or minus direction, but theoretically, it never reaches the axis, as the curve is asymptotic to the base line. Though the curve extends to infinity in both directions, yet hardly 0.3% of the area lies beyond the limits of (mean$+3\sigma$).

4. The curve is specified completed by defining the mean and the standard deviation.

5. Near about the mean value the curve is concave towards the x-axis while near about the two tails it is convex to the horizontal axis. The points of inflection (the points where the change in curvature occurs) are at a distance of one σ from the mean (on either side of it).

6. In normal curve the values of the odd moments are all zeroes. (as normal curve is symmetrical and for symmetrical distribution the value of odd moment will always be zero, for the sum of the positive deviations will always equal the sum of the negative deviations, and thus will cancel each other). If n is even, we have

n is even, we have

$$\mu_n = \frac{n!}{2^{n/2}\left(\frac{n}{2}\right)!}\sigma^n \quad \text{(where } n \text{ is even)}$$

It follows that $\beta_1=\frac{\mu_3^2}{\mu_2^3}=\alpha_3^2=\gamma_1^2=0$, *i.e.*, **skewness is zero and** $\beta_2=\frac{\mu^4}{\mu_2^2}=\alpha_2=3$, **and** $\gamma_2=0$, ***i.e.* normal curve has zero kurtosis. In other words, normal distribution is mesokurtic.**

7. The mean deviation of the normal curve is

$$\sigma\sqrt{\frac{2}{\pi}}=0.79798\sigma \text{ or approx. } \tfrac{4}{5}\sigma.$$

8. The first and third quartiles are equidistant from mean and at a distance of .6745σ. Thus, the semi-interquartile range is equal to 0.6745 of the standard deviation.

Of all the important properties of the normal curve the most important property is the area relationship. The area lying between the curve, the horizontal x-axis and any two ordinates is said to be the area under the curve and is equal to the probability for the variable to be in the interval marked on x-axis by two ordinates. The area lying between the mean ordinate and an ordinate at a certain sigma distance from the mean will always be the same proportion of the total area of the curve. Thus, the area enclosed between mean ordinate at sigma distance from the mean is always 34.134% of the total area of the curve. It means that the area lying between $\pm 1\sigma$ from the mean *i.e.*, between two ordinates at one sigma distance from the mean on either side will always be 68.268% or approx. 68.27% of the total area. Similarly, the area lying between two ordinates at $\pm 0.6745\sigma$ from the mean will always be 50% of the whole area. Ordinates at $\pm 1.96\sigma$, $\pm 2\sigma$ $\pm 2.5758\sigma$ and $\pm 3\sigma$ from the mean will respectively cover 95%, 95.45%, 99% and 99.73% of the total area. It shows that practically all observations of a normal frequency distribution lie within a range of $\pm 3\sigma$ or six standard deviations.

17.6. STANDARD NORMAL DISTRIBUTION

The properties of the normal curve permit us to define a standardized distribution in terms of the variable z defined as

$$z=\frac{x-\mu}{\sigma} \tag{17.5}$$

This is equivalent to measuring the distance x from the mean μ using the standard deviation σ as the unit of measuring distance. The variable z is termed as the standard normal variate and plays a very important role in statistics. The probability density function in terms of z is

$$p(z)=\frac{1}{\sqrt{2\pi}}\ e^{-z^2/2}$$

where $\pi=3.14156$.

If we know the mean μ and the standard deviation σ, then we can calculate z corresponding to any x and find the corresponding $p(z)$.

The values of $p(z)$, the ordinates of the standard normal curv as a function of z, the standard normal variate are tabulated in the appendix.

Note that $p(z)$, the probability density is in itself of little significance in practical calculations, since we are mostly interested in obtaining probabilities and not densities. It is, therefore, desirable to give tables of areas under the normal curve. There are various ways in which the areas can be specified. We give here two way in

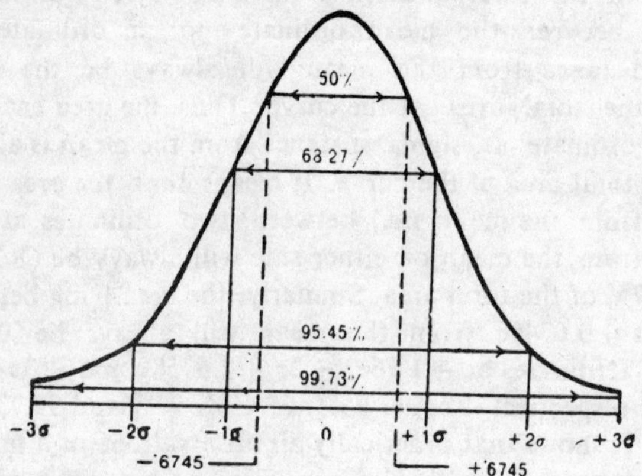

FIG. 17.4. Standard Normal Curve.

which it is done. Both tables are given as appendix.

In the first table, we specify the area between zero and z as a function of z. This represents the probability that the value of the variate lies between o and z (with z positive).

In the second table, we specify the area to the right of z and thus represents the probability that the value of the standard normal variate is larger than z.

Fitting a normal curve to an observed distribution. If the observed distribution appears to be reasonably symmetrical and bell shaped, it may be worth while to attempt to fit a normal curve for it. The normal curve idealizes the observational data and smoothes out the irregularities due to sampling fluctuations. In case the fit is good, the statisticians proceed to draw various inferences about the behaviour of sample from a normal parent population, and can confidently feel that their assumptions apply reasonably well to the actual population sampled.

In fitting the curve, the mean and standard deviation of the observed distribution are treated as the population mean and population standard deviation. The procedure for fitting the curve should become clear through the following illustrations:

Illustration 17.11. You are given below a frequency distribution. Fit a normal curve to the data.

Mid value	412	417	422	427	432	437	442	447	452	457
Frequency	6	34	132	179	218	183	146	69	30	3

Solution. We first calculate the mean and standard deviation of the data. Interval 5, Assumed $A=432$.

Mid Value	*Class interval*	*f*	*d*	*fd*	*fd*²
412	409.5—414.5	6	—4	—24	96
417	414.5—419.5	34	—3	—102	306
422	419.5—424.5	132	—2	—264	528
427	424.5—429.5	179	—1	—179	179
432	429.5—434.5	218	0	0	0
437	434.5—439.5	183	1	183	183
442	439.5—444.5	146	2	292	584
447	444.5—449.5	69	3	207	621
452	449.5—454 5	30	4	120	480
457	454 5—459.5	3	5	15	75
		1000		248	3052

Here $d=(x-A)/5$

$$\mu=\bar{X}=432+\frac{248}{1000}\times 5=433.24.$$

$$\sigma=5\times\sqrt{\frac{3052}{1000}-\left(\frac{248}{1000}\right)^2}=5\sqrt{3.052-0.061504}=8.647$$

Calculation of Expected Normal Frequencies

Class interval	St. normal variate $\frac{\text{upper limit} - \bar{X}}{\sigma} = Z$	Area to the left of Z	Area for class interval (P)	Expected frequencies P × 1000
below 409.5	−2.746	.0030	.0030	3.0 or 3
409.5—414.5	−2.167	.0151	.0121	12.1 or 12
—419.5	−1.589	.0560	.0409	40.9 or 41
—424.5	−1.011	.1560	.1000	100.0 or 100
—429.5	—0.433	.3325	.1765	176.5 or 177
—434.5	−0.146	.5580	.2255	225.5 or 226
—439.5	+0.724	.7654	.2074	207.4 or 207
—444.5	+1.302	.9035	.1381	138.1 or 138
—449.5	+1.880	.9700	.0665	66.5 or 67
—454.5	+2.459	.9931	.0231	23.1 or 23
—459.5	+3.036	.9988	.0057	5.7 or 6
459 and above	+α	1.0000	.0012	1.2 or 1
			1.0000	1000.3

Note. Values of *P* for a class interval is found by subtracting from the area to the left of upper limit the area to the left of the lower limit given in column 'Area to the left of *z*'.

Illustration 17.12. The Delhi Municipal Committee instals 5,000 electric lamps in the streets of the city. If these lamps have an average life of 800 burning hours with a standard deviation of 150 hours, what proportion of the lamps might be expected to fail in the first 600 hours?

Solution. In this case we want the area lying between the ordinate at 0 and at 600 hours, *i.e.*, between $Z = \frac{0-800}{150}$ and $\frac{600-800}{150}$. Consulting the table we find that the area between these two ordinates =.09176 (*i.e.* .50000 − .40824). Hence the expected number of failures will be .09176 of 5,000 = 459.

Illustration 17.13. In the above illustration what number of lamps may be expected to fail between 700 and 900 burning hours?

Solution. In this case we are interested in calculating the area lying between $Z = \frac{700-800}{150}$ and $\frac{900-800}{150}$, *i.e.*, between ±.666. Consulting our table we find that the area between these two ordinates is equal to .49074 (*i.e.*, .24537 + .24537). Therefore, the expected number of failures are .49074 × 5,000 = 2,454.

Another important point which is of very great interest (in the significance tests specially) is that form the area property of the curve, we can find out the probability of certain event deviating by a stated amount in either direction from the mean of a normal distribution. The probability of a certain event deviating from the mean by $\pm 1\ \sigma$ is 1 out of three, by $\pm 2\ \sigma$, 1 out of 20 and by $\pm 3\sigma$, 1 out of 370.

Illustration 17.14. How would you use normal distribution to find approximately the frequency of exactly 5 successes in 100 trials, the probability of success in each trial being $\pi=0.1$?

Solution. Let n be the number of trials.

So $n=100$, and $\pi=0.1$

The given problem is a case of binomial distribution and so

mean$=n\pi$ or $100\times 0.1=10$, and $\sigma=\sqrt{n\pi(1-\pi)}$
$=3$

We know that in the limit when n is large the binomial distribution approaches normal distribution which is a continuous distribution. So the frequency of exactly 5 successes in the normal dirtribution will correspond to the frequency of class interval 4.5 to 5.5, and the mean and σ of the binomial distribution will correspond to the mean and σ of the normal distribution.

The normal variate for X = 4.5

$$Z=\frac{X-\bar{X}}{\sigma}=\frac{4.5-10}{3}=-1.83$$

and the normal variate for X = (5.5)

$$Z=\frac{5.5-10}{3}=-1.5.$$

Now area to the left of ($Z=-1.83$) is equal to area to the right of ($Z=+1.83$) which is 0.03362. Similarly area to the left of ($Z=-1.5$) is equal to the area to the right of ($Z=+1.50$) which is 0.0668. So the area between -- 1.83 and --1.50 is (0.06681 -- 0.03362) = 0.03321

If N be the total number of frequencies, the frequency of the class interval 4.5 to $5.5=N\times 0.0332$. Approximate frequency of exactly 5 successes in 100 trials when the probability of success in each trial is 0.1 is each trial is 0.1 is 3.32 or 3 appromately.

Illustration 17.15. If two normal universes A and B have the same total frequencies, but the σ of the universe A is x time that of the universe B, show that the maximum frequency of universe A is $1/x$ that of universe B.

Solution. Let the distributions A and B have the means μ_1 and μ_2 respectively, σx and σ respectively as their standard deviations and total frequencies in both cases be N.

Maximum ordinate in each case is the ordinate corresponding to $Z=0$.

So $p_A(x=\mu_1)=\dfrac{1}{\sigma_x\sqrt{2\pi}}$,

and $p_B(x=\mu_2)=\dfrac{1}{\sigma\sqrt{2\pi}}$

Since the maximum frequency equals maximum ordinate times the class interval, so the maximum frequencies of A and B are in the ratio of p_A and p_B or $1/x$, as required.

Illustration 17.16. In an aptitude test administered to 900 college students, the mean is 50 and the standard deviation is 20. Find

(*i*) the number of students securing between 30 and 70;

(*ii*) the number of students exceeding the score of 65;

(*iii*) the value of the score exceeding by the top 90 students.

Solution. As some children will have test score more than average and others less than average, the distribution of test scores is expected to be normal distribution having mean (μ) 50 and standard deviation (σ) 20.

Standard normal variate corresponding to a score

$$Z=\frac{X-\mu}{\sigma}$$

(*a*) Standard normal variate corresponding to score 30

$$=\frac{30-50}{20}=-1.$$

Standard normal variate corresponding to score 70

$$=\frac{70-50}{20}=1$$

$\therefore$ Area to the left of ordinate -- 1.0 from normal tables is the same $=0.1587$

$\therefore$ Area to the left of ordinate 1.0 from normal tables is the same $=0.1587$

$\therefore$ Probability of having children between scores 30 and 70 $=1-(0.1587+0.1587)=.6826$

$\therefore$ Number of children having score between 30 and 70 is

$$=900\times 0.6826=614.34 \text{ or } 614$$

(*b*) Standard normal variate corresponging to score 65

$$=\frac{65-50}{20}=0.75$$

Area to the right of ordinate at $0.75=0.2266$

$\therefore$ Probability of getting a child above score $65=0.2266$

$\therefore$ Expected number of children above score 65

$$=900\times 0.2266=203.94 \text{ or } 204$$

(*c*) Probability corresponding top 90 students

$$=\frac{90}{900}=0.1$$

∴ Standard normal variate having 0.1 area to its right

$$=1.281$$

or $$1.281=\frac{X-50}{20}$$

or $X=50+20\times1.281=50+25.62=75.62$ or 76.

Illustration 17.17. Mean and standard deviation of chest measurement of 1200 soldiers are 85 and 5 cm respectively. How many of them are expected to have their chest measurements exceeding 95 cm, assuming the measurements follow the normal pattern?

Solution. Since the distribution is normal with mean 85 cm and variance 5 cm.

Standard normal variate corresponding to the chest measurement 95 cm is

$$Z=\frac{95-85}{5}=2$$

∴ Area to the right of ordinate at 2 from the normai tables

$$=0.02275.$$

∴ Probability of getting a soldier above chest measurement of 95 cm.

$$=0.02275.$$

∴ Expected number of soldiers of chest measurement of above 95 cm out of 1200 soldiers.

$$=1200\times0.02275=27.30000 \text{ or } 27 \text{ approx.}$$

Illustration 17.18. The results of a particular examination are given below in a summary form:

Result	% of candidates
(*i*) Passed with distinction	10
(*ii*) Passed without distinction	60
(*iii*) Failed.	30

It is known that candidate fails if he obtains less then 40 marks (out of 100), while he must obtain at least 75 marks in order to pass with distinction. Determine the mean and standard deviation of the distribution of marks, assuming this to be normal.

Solution Let the mean be μ and standard deviation be σ. 30% of candidates get marks less than 40.

∴ Area to the right of ordinate at 76 is 0.10.

∴ Standard normal variate corresponding to this from the tables

$$=1.28$$

30% of the candidates get marks below 40. As the percentage of candidateis is less than 50%; 40 marks are less than mean marks and the standardised normal variate corresponding to it will be negative.

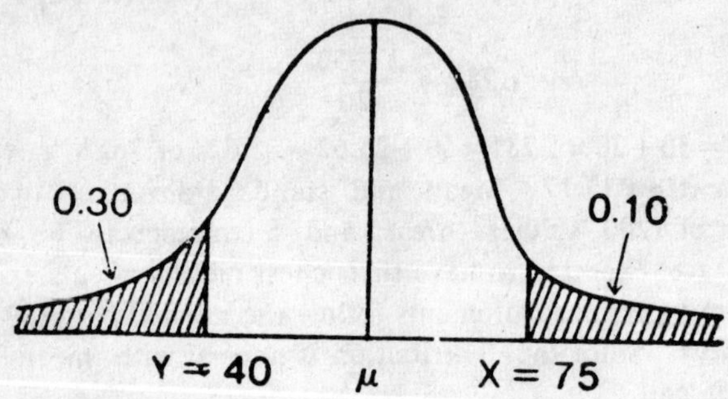

Area to the right of corresponding positive ordinate = 0.30.

∴ Positive standardised normal variate = 0.527.

∴ Standardised normal variate corresponding to the marks 40 = −0.527

$$\because \quad \frac{75-\mu}{\sigma}=1.28 \text{ and } \frac{40-\mu}{\sigma}=-0.527$$

or $$\mu+1.28\sigma=75 \qquad \ldots(i)$$

and $$\mu-0.527\sigma=40 \qquad \ldots(ii)$$

Subtracting (*ii*) from (*i*); we get

$$\sigma\times 1.807=35$$

or $$\sigma=\frac{35}{1.807}=19.37.$$

Substituting the value of σ in (*i*)

$$\mu+1.28\times\frac{35}{1.807}=75$$

or $$\mu=75-1.28\times\frac{35}{1.807}$$

$$=75-24.23=50.77.$$

∴ Mean of the distribution = 50.77.

Standard deviation of the distribution = 19.37.

Illustration 17.19. Draw on the same graph the normal curve

$$Y_c=\frac{1}{\sigma\sqrt{2\pi}}\,e^{\frac{-(x-\mu)^2}{2\sigma^2}}$$

when (*i*) $\mu=0,\ \sigma=1$, (*ii*) $\mu=0,\ \sigma=2$, (*iii*) $\mu=2,\ \sigma=1$, (*iv*) $\mu=2,\ \sigma=2$.

You are not expected to plot the points from the table.
Fitting normal curve for:

(*i*) $$p=\frac{1}{\sqrt{2\pi}}\, e^{\frac{-x^2}{2}}$$

(*ii*) $$p=\frac{1}{2\sqrt{2\pi}}\, e^{\frac{-x^2}{8}}$$

(*iii*) $$p=\frac{1}{\sqrt{2\pi}}\, e^{\frac{-(x-2)^2}{2}}$$

(*iv*) $$p=\frac{1}{2\sqrt{2\pi}}\, e^{\frac{-(x-2)^2}{8}}$$

Solution.
Taking logs in the first equation:

$$\log p = -\frac{1}{2}(\log 2 + \log \pi) - \frac{x^2}{2}\log e$$

$$=-\tfrac{1}{2}(0.3010+0.4972)-\frac{x^2}{2}(0.4343)$$

$$=-\tfrac{1}{2}(0.7982)-\frac{x^2}{2}(0.4343)$$

$$=-0.3991-x^2(0.2171)$$

Putting for x 0, ± 1, ± 2, ± 3, we get

$\log p$	1.6009	$\bar{1}.3837$	$\bar{2}.7323$	$\bar{3}.6470$
p	0.3989	0.2419	0.0539	0.0044

Thus, when

x	0	± 1	± 2	± 3
p	0.3989	0.2419	0.0539	0.0044

are plotted on a graph paper, we will get the normal curve corresponding to (*i*) of the problem.

To obtain the plot for part (*ii*) we can use the above obtained information as below:

At the maximum ordinate $x=0$, (as $\mu=0$), we get $Y=\frac{1}{2\sqrt{2\pi}}$ being just half of $\frac{1}{\sqrt{2\pi}}$ the maximum ordinate of case (*i*)

Similarly, the ordinates at $\pm 1\sigma$, $\pm 2\sigma$ and $\pm 3\sigma$ will also be respectively half of case (*i*).

As $\sigma=2$

x	0	± 2	± 4	± 6
p	0.1995	0.1210	0.0269	0.0022

Fitting these figures of x and y on the graph, the normal curve corresponding to (ii) will be obtained.

The only difference between case (i) and case (iii) is that whereas in the former the power of e is $\frac{-x^2}{2}$, here it is $\frac{-(x-2)^2}{2}$.

This means that value of Y at

x	−3	−2	−1	0	1	2	3

of case (i) are similar to the values of

x	−1	0	1	2	3	4	5

of case (iii)

So plotting for

x	−1	0	1	2	3	4	5
p	.0044	.0539	.2419	.3989	.2419	.0539	.0044

we get the normal curve for case (iii).

On a comparison of case (iv) with case (ii), we find the value of Y corresponding

x	−6	−4	−2		2	4	6

of case (ii) are similar to

x	−4	−2	0	2	4	6	8

of case (iv).

So p =.0022 .0269 .1210 .1995 .1210 .0269 .0022

So plotting the values of x and y, we would get the required graph.

Illustration 17.20. It is known that the weight of a group of 10,000 persons is distributed normally. Draw up a table showing the number of persons you expect within the range mean $\pm\sigma$ in various groups of class intervals of 0.5σ.

Solution. Let mean$=\bar{X}$.

Class interval	*Standard normal deviate for upper limit*	*Area to the left for (—) and to the right for(+)*	*Area of Proportion for the class interval*	*Required frequency i.e., Proportion×10,000*
Below $\bar{x}-3\sigma$	−3	.00135	.00135	13.5
$\bar{x}-3\sigma$ to $\bar{x}-2.5\sigma$	−2.5	.0062	.00485	48.5
$\bar{x}-2.5\sigma$ to $\bar{x}-2.0\sigma$	−2	.0228	.0166	166
$\bar{x}-2.0\sigma$ to $\bar{x}-1.5\sigma$	−1.5	.0668	.0440	440
$\bar{x}-1.5\sigma$ to $\bar{x}-1.0\sigma$	−1	.1587	.0919	919
$\bar{x}-1.0\sigma$ to $\bar{x}-0.5\sigma$	−.5	.3085	.1498	1,498
$\bar{x}-0.5\sigma$ to $\bar{x}$	0	.5000	.1915	1,915
$\bar{x}$ to $\bar{x}+0.5\sigma$	−.5	.3085	.1915	1,915
$\bar{x}+0.5\sigma$ to $\bar{x}+1.0\sigma$	1	.1587	.1498	1,498

$\bar{x}+1.0\sigma$ to $\bar{x}+1.5\sigma$	1.5	.0668	.0919	919
$\bar{x}+1.5\sigma$ to $\bar{x}+2.0\sigma$	2	.0228	.0440	440
$\bar{x}+2.0\sigma$ to $\bar{x}+2.5\sigma$	2.5	.0062	.0166	166
$\bar{x}+2.5\sigma$ to $\bar{x}+3.0\sigma$	3	.00135	.00485	48.5
				9986.5

We notice here that the total of fitted frequency is 9986.5 and not 10,000. The reason is obvious. We have not included the cases lying above 3 σ the probability of which is .00135 and if this is multipiied by 10,000, then we get 13.5. When this figure is added to 9986.5 we get a total figure of exactly 10,000.

It may be noted that instead of finding the area to the left for (−) and to the right for (+), the total area for each of the normal standard deviates may be found, and with it also area of class interval can be similarly calculated.

Illustration 17.21. There are 600 business students in the graduate department of a university, and the probability for any student to need a copy of a particular textbook from the university library on any day is 0.05. How many copies of the book should be kept in the university library so that the probability may be greater than .90 that none of the students needing a copy from the library has to come back disappointed. (Use normal approximation to the binomal probability law)

Solution. Let n be the number of students and π the probability for any student to need a copy of a particular textbook from the university library.

Mean $= n\pi = 600 \times .05 = 30 =$ Mean no. of copies of a textbook required on any day

$$\text{S.D.} = \sigma = \sqrt{n\pi(1-\pi)} = \sqrt{600 \times .05 \times .95} = \sqrt{28.5} = 5.3 \text{ approx.}$$

Let number of copies of a textbook required on any day be x. Then x is distributed normally with $\mu = 30$ and $\sigma = 5.3$.

The number of copies X to be stocked should be such that probability of x being greater than X be less than 0.1 so that the demand will be met on 90% of the days. This means that area to the right of X be 0.1. This requires that $z = 1.28$ (from Tables). Thus

$$\frac{X-30}{5.3} > 1.28$$

or

$$X > 36.784$$

i.e., 37 copies are required.

Illustration 17.22. Sacks of grain packed by an automatic machine loader have an average weight of 100 kg. Its standard deviation is 0.38 kg. (a) Find the chance of getting a bag over 101.5 kg and below 98.48 kg. (b) If a dealer rejects a bag below 99 kg how many bags do you expect would be rejected in a lot of 1000 bags.

Solution. Machine may fill a bag more than average weight or less than average weight. So let us assume that weight of the bags follows normal distribution having mean (μ) 100 kg and standard deviation (σ) 0.38.

Standard normal variate corresponding to value $X=(X-\mu)/\sigma$.

(a) Standard normal variate for 101.5 kg. $=\dfrac{101.5-100}{0.38}=3.95.$

" " " 98.48 $=\dfrac{98.48-100}{0.38}=-4.00$

Area to the right of ordinate at 3.95 from the standard normal tables

$=.00004.$

∴ Probability of getting a bag above 101.5 kg. $=.00004$

Normal tables do not give areas for negative values of X. As normal curve is a symmetrical curve, area to the left of the ordinate at -4.00 will be same as area to the right of ordinate at 4.00.

∴ Area to the right of 4.00 from standard normal tables

$=.00003$

∴ Probability of getting a bag below

98.48 kg$=.00003$

∴ Probability of getting a bag above 101.5 kg and below

98.48 kg$=.00004+.00003=.00007.$

(b) Standard normal variate corresponding to 99

$=\dfrac{99-100}{0.38}=-2.63.$

Bags are rejected below 99 kg of weight.

Area to the left of ordinate at -2.63=area to the right of ordinate at

2.63=,00427

∴ Probability for rejecting a bag

$=.00427.$

∴ Number of bags rejected out of 1000 bags

$=1000\times.00427$

$=4.27$ or 4 approx.

Illustration 17.23. The charge accounts at a certain departmental store have an average balance of Rs 120 and standard deviation of Rs 40. Assuming the accounts balances are normally distributed

(*i*) what proportion of accounts is between Rs 100 and Rs 150·
(*ii*) what proportion of accounts is over Rs. 150;
(*iii*) what proportion of accounts is between Rs 60 and Rs 90.

Solution.

(*ii*) Standard normal variate corresponding to 150

$$= \frac{150-120}{40} = .75$$

The probability that the balance is over Rs 150 is the area to the right of Z =0.75 under

(*i*) St. normal variate for $100 = \frac{100-120}{40} = -.5.$

The required probability is the area between the ordinates at −.5 and .75.

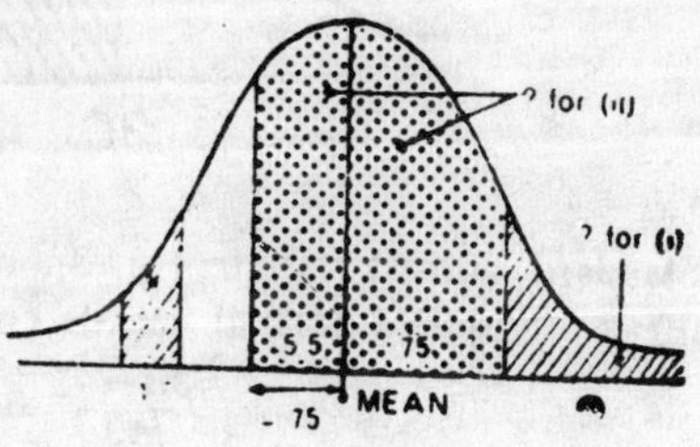

From the tables, area between mean and .75 is .2734 and area between mean and .5 is .1915. Area between mean and −.5 will also be .1915.

So required probability
= .2734+.1915 = .4649

(*iii*) Standard normal variates for $90 = \frac{90-120}{40} = -.75$

" " " " $60 = \frac{60-120}{40} = -1.5$

The required proportion is the area between the ordinates −.75 and −1.5. As normal distribution is symmetrical, area between +1.5 and +.75 will be same as area between −1.5 and −.75.

From tables area between mean and 1.5=.4332

" " " " .75=.2734

∴ Required area of proportion =.4332−.2734
=.1598.

Illutration 17.24. In an intelligence test administered on 1000 childeren, the average was 42 and standard deviation 24.

(*a*) Find the number of children exceeding a score 50.

(*b*) Find the number of children lying between 30 and 54.

(*c*) Find value of score exceeded by the top 100 childeren.

Solution. As some childeren will have test score more than average, the distribution of test scores is expected to be normal distribution having mean (μ) 42 and standard deviation (σ) 24.

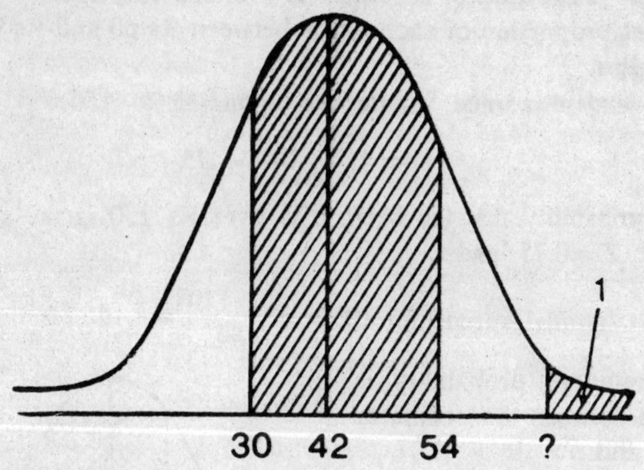

Standard normal variate corresponding to a score $X = (X - \mu)/\sigma$.

(*a*) $\therefore$ Standard normal variate corresponding to a score 50 is

$$\frac{50-42}{24}=0.33$$

Area to the right of ordinate at 0.33=0.3707.

$\therefore$ Probability of getting child above score 50 is 0.3707.

$\therefore$ Expected number of childeren above score 50 is $1000 \times .3707$ $=370.7$ or 371 approx.

(*b*) Standard normal variate corresponding to score 30 is

$$\frac{30-42}{24}=-0.5.$$

Standard normal variate corresponding to score 54 is

$$\frac{54-42}{24}=+0.5.$$

Area to the right of ordinate at 0.5 from normal tables is 0.30854.

$\therefore$ Area to the left of ordinate at $-0.5=0.30854$

$\therefore$ Probability of having children with score below 30 and above 54 is $0.30854+0.30854 = .61708$.

$\therefore$ Probability of having children between scores 30 and 54

$$=1-0.61708$$
$$=0.38292.$$

$\therefore$ Number of children expected to have score between 30 and 54

$$=1000 \times 0.38292$$
$$=382.92 \text{ or } 383.$$

(*c*) Probability corresponding to top 100 students=100/1000=0.1.

∴ **Standard normal variate having 0.1 area to its right=1.281.**

So $$1.281=\frac{X-42}{24} \text{ or } X=24\times1.281+42=72.7.$$

Illustration 17.25. The income distribution of a group of 10,000 persons was found to be normal with mean Rs 750 p.m. and standard deviation Rs 50. What percentage of this group had income (*i*) exceeding Rs 668 (*ii*) exceeding Rs 832 ?

Solution. St. normal variate for $832=\frac{832-750}{50}=\frac{82}{50}=1.64$

„ „ „ „ $668=\frac{668-750}{50}=\frac{-82}{50}=-1.64.$

(*i*) For this part we need area for 668 and higher *i.e.* area between ordinate at −1.64 and mean plus area above mean=.4495+.5000=.9495.

∴ Percentage of persons=94.95%. ...(*i*)

(*ii*) For this we need area above ordinate at 832 or +1.64 in standard normal curve.

Area above 1.64 is .0505.

∴ Percentage of persons = 5.05% ; ...(*ii*)

Illustration 17.26. In a normal distribution whose mean=2 and standard deviation 3, (*a*) find the value of the variate such that the probability of the interval from the mean to that value is 0.4115. (*b*) find another value such that the probability for the interval from 3.5 to that value is 0.2307.

Solution. Standard normal variate=$(X-\mu)/\sigma$, where μ is the mean and σ the standard deviation of the distribution.

(*a*) Let the required value be X.

∴ Probability from mean to X=0.4115.

∴ „ „ X and above=0.5−0.4115=0.0885

Standard normal variate having 0.0885 area to the right of it
=1.35

∴ $$1.35=\frac{X-2}{3} \qquad \therefore X=3\times1.35+2=6.05$$

As normal curve is symmetrical curve −1.35 will have 0.0885 area to the left of it *i.e.*, 0.4115 area between mean and −1.35.

∴ Second value of X is given by

$$-1.35=\frac{X-2}{3} \qquad \therefore X=3\times(-1.35)+2=-2.05 \quad (i)$$

(*b*) Standard normal variate corresponding to 3.5

$$=\frac{3.5-2}{3}=0.5$$

$\therefore$ Area to the right of 0.5,
$=0.3085$

Let the required value be X.

As probability between 3.5 and X is 0.2307, area between 3.5 and X is also 0.2307.

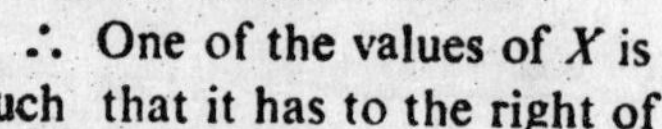

$\therefore$ One of the values of X is such that it has to the right of it $0.3085-0.2307$ *i.e.*, 0.0778 area and the other value of X is such that it has to the right of it $.3085+0.2307$ *i.e.*, 0.5392 area.

Standard normal variate having 0.0778 area on its right side from the tables$=1.42$.

$$1.42=\frac{X-2}{3} \quad \text{or } X=3\times 1.42+2=6.26.$$

In the second case as the area to the right of X is greater than 0.5 the point X lies on the left of mean. So the standard normal variate corresponding to this value of X will have negative value The area to the left of it is $1-0.5392=0.4608$. The positive standard normal variate from the tables having 0.4608 area to the right of it$=0.102$.

$\because$ The standard normal variate corresponding to $X=-0.102$.

$$\therefore \quad -0.102=\frac{X-2}{3} \quad \text{or } X=3\times(-0.102)+2=1.694.$$

Illustration 17.27. The mean yield of 1000 normally distributed plots of corn was 20.50 kg per plot. It was observed that 200 of the plots yielded 17.75 kg or less. Estimate the standard deviation of the yield.

Solution. Let the plot yield follow normal distribution. It has mean (μ) = 20.50 and standard deviation (σ) is to be determined.

Proportion of plots having yield less than 17.75 kg

$$=\frac{200}{1000}=0.2.$$

$\therefore$ Area to the left of the ordinate at $17.75=0.2$.

As 17.75 is less than mean, standardised normal variate corresponding to it will be negative.

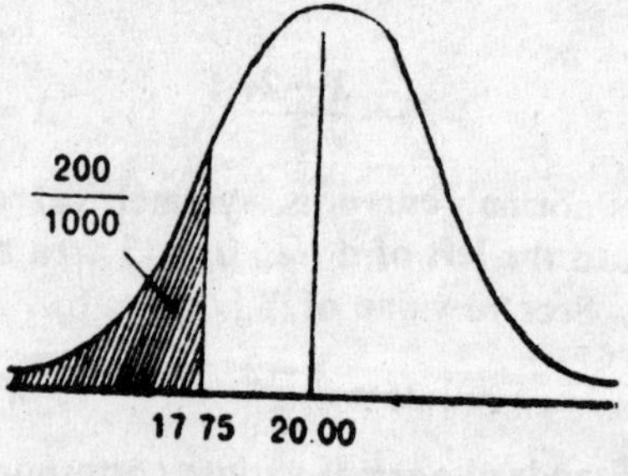

The positive ordinate from the tables having area 0.2 to its right $=0.842$.

Standardised normal variate corresponding to 17.75 kg yield $=-0.842$.

Standardised normal variate corresponding to $X=(X-\mu)/\sigma$.

$$\therefore \quad -0.842=\frac{(17.75-20.50)}{\sigma} \qquad \therefore \ \sigma=\frac{2.75}{0.842}=3.266.$$

Illustration 17.28. The distribution of marks in a certain examination was found to be normal with 23% of the candidates scoring above 60 and 21% below 40. Find the mean and the standard deviation of the distribution of marks.

Solution. Let the mean be μ and standard deviation be σ. 23% of candidates get marks above 60.

$\therefore$ Area to the right of ordinate at 60 is 0.23.

$\because$ Standardised normal variate corresponding to 60 from the tables$=0.739$.

21% of candidates get marks below 40. As the percentage of candidates is less than 50%; 40 marks are less than mean marks and the standardised normal variate corresponding to it will be negative.

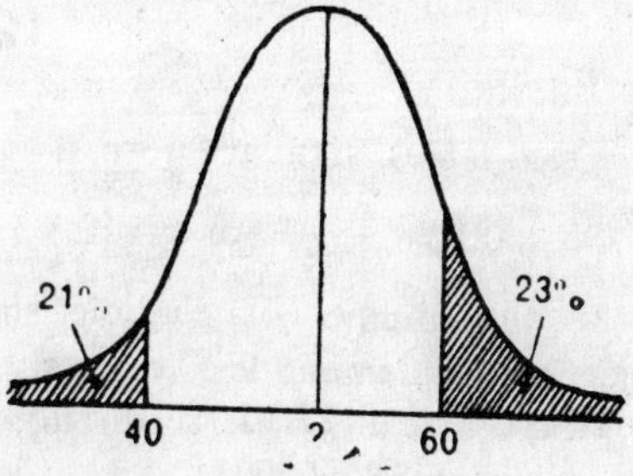

Area to the right of corresponding positive ordinate $=0.21$

$\therefore$ Positive standardised normal variate$=0.807$.

$\therefore$ Standardised normal variate corresponding to marks 40 $=-0.807$.

$$\therefore \quad 0.739=\frac{60-\mu}{\sigma} \quad \text{and} \quad -0.807=\frac{40-\mu}{\sigma}.$$

$$\therefore \quad \mu+\sigma\times(.739)=60 \qquad \dots(i)$$

$$\text{and} \quad \mu-\sigma\times.807=40 \qquad \dots(ii)$$

Subtracting (*ii*) from (*i*)

$$\sigma\times1.546=20$$

$$\sigma=20/1.546=12.94$$

Substituting the value of σ in (*i*)

$$\mu+12.94\times.739=60$$

$$\text{or} \quad \mu=60-9.56=50.44,$$

∴ Mean of the distribution $=50.44$.

Standard deviation $=12.94$.

Illustration 17.29. Daily demand for gladiole at a flower shop is approximately normally distributed with mean sales 12 per day and standard deviation of 4 gladiole. How many gladiole must be on hand in the morning to assure no more than one chance in 5 of running out of gladiole during the day.

Solution Probability of not meeting the demand $=1/5=0.2$.

If the number of gladiole kept is x, the demand will not be fully met if it is more than x. So we require a number x such that the area above this in the demand distribution curve is .2.

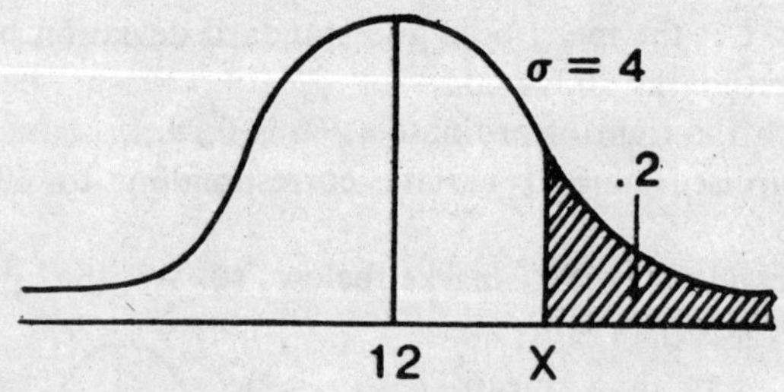

From tables, standard normal variates for area .2 is 0.84,

So $0.84=\dfrac{x-12}{4}$ or $x=12+3.36=15.36$.

So the number of gladiole should be 16 so that chance of not meeting the demand is .2 or less.

Illustration 17.30. A distribution is known to be normal. The quartiles are 8.64 and 14.32. Calculate the mean and the standard deviations.

Solutien. Let the mean be μ and standard deviation be σ

∴ Quartiles are $\mu+0.6745\ \sigma$ and $\mu-0.6745\ \sigma$

∴ $\mu+0.6745\ \sigma=14.32$ ...(*i*)

and $\mu-0.6745\ \sigma=8.64$ ...(*ii*)

Subtracting (*ii*) from (*i*)

$1.3490\ \sigma=5.68$

$\sigma=5.68/1.3490$ or 4.21

Adding (*i*) and (*ii*) $2\mu=22.96$ or $\mu=11.48$.

Mean $=11.48$ and standard deviation $=4.21$.

Illustration 17.31. The mean height of 18 year old boys is 70 inches. The standard deviation of heights is 2.5 inches. The distribution is known to be normal. Find the probability that two students selected at random should be taller than 74 inches.

Solution Standardised normal variate corresponding to height 74 inches=(74−70)/2.5=1.6.

Area to the right of 1.6=.0548

∴ Probability of getting a student with height more than 74 inches=.0548.

∴ Probability of getting two students both having height above 74 inches=.0548×.0548=.003003.

17.7. TESTING THE NORMALITY OF A DISTRIBUTION

There are two methods of testing the normality of a certain curve. The first method involves the calculation of the values of β_1, β_2, γ_1 and γ_2. β_1 denotes whəther the curve is symmetrical or skewed. It is zero when the curve is symmetrical. β_2 shows the peakedness of the curve. The peak is normal or the curve is mesokurtic when $\beta_2=3$. If β_2 is less than three, the curve is described as platykurtic and if it is more than three then it is called leptokurtic. γ_1 also is a measure of whether or not the distribution is symmetrical. γ_2 measures the departure of peakedness from normality. The two standard errors of γ_1 and γ_2 denote that if the values of γ_1 and γ_2 are less than twice their standard errors, then the distribution is not significantly different from the normal form. If they are greater than twice their standard errors the distribution is not normal.

$$\beta_1=\frac{\mu_3^2}{\mu_2^3};\ \beta_2=\frac{\mu_4}{\mu_2^2};\ \gamma_1=\pm\sqrt{\beta_1},\ \gamma_2=\beta_2-3;$$

$$\text{S.E. of } \gamma_1=\sqrt{\frac{6}{N}};\ \text{S.E. of } \gamma_2=\sqrt{\frac{24}{N}}.$$

The second method is the fitting of the normal curve and testing the goodness of fit by the χ^2 (chi-square,) method. The first method is illustrated in the example.

Illustration 17.33. The following is the frequency distribution of 1,000 students of a college:

Height (in inches)	*Frequency*
59.5—60.5	2
60.5	9
61.5	28
62.5	75
63.5	125
64.5	200
65.5	214

66.5	160
67.5	110
68.5	50
69.5	20
70.5	5
71.5—72.5	2
	1,000

Test the normality of the distribution.

Solution.

Mid points	*Deviation from 66* x'	f	fx'^1	fx'^2	fx'^3	fx'^4
60	—6	2	— 12	72	— 432	2,592
61	—5	9	— 45	225	—1,125	5,625
62	—4	28	—112	448	—1,792	7,168
63	—3	75	—225	675	—2,025	6,075
64	—2	125	—250	500	—1,000	2,000
65	—1	200	—200	200	— 200	200
66	—0	214	0	0	0	0
67	1	160	160	160	160	160
68	2	110	220	440	880	1,760
69	3	50	150	450	1,350	4,050
70	4	20	80	320	1,280	5,120
71	5	5	25	125	625	3,125
72	6	2	12	72	432	2,592
		1,000	—197	3,687	—1,847	40,467

$$\nu_1=\frac{\Sigma f(x')}{N}=\frac{-197}{1,000}=-.197.$$

$$\nu_2=\frac{\Sigma f(x')^2}{N}=\frac{3,687}{1,000}=3.687.$$

$$\nu_3=\frac{\Sigma f(x')^3}{N}=\frac{-1,847}{1,000}=-1.847.$$

$$\nu_4=\frac{\Sigma f(x')^4}{N}=\frac{40.467}{1000}=40.467.$$

$$\mu_2=\nu_2-(\nu_1)^2-\tfrac{1}{12} \text{ (corrected)}$$
$$=3.687-.039-.083$$
$$=3.565.$$

$$\mu_3=\nu_3-3\nu_2\nu_1+2\nu_1^3$$
$$=-1.847-3\times3.687\times(-.197)+2\times(-.197)^3$$
$$=.317.$$

$$\mu_4 = v_4 - 4v_3v_1 + 6v_1v_2^2 - 3v_1^4 - \tfrac{1}{2}\mu_2 + \tfrac{7}{240}$$
$$= 40.467 - 1.455 + .860 - .005 - 1.783 + .029$$
$$= 38.113$$

$$\beta_1 = \frac{\mu_3^2}{\mu_2^2} = \frac{(.317)^2}{(3.565)^2} = .002.$$

$$\beta_2 = \frac{\mu_4}{\mu_2^2} = \frac{38.113}{(3.565)^2} = 2.994.$$

$$\gamma_1 = \sqrt{\beta_1} = \sqrt{.002} = .045.$$

$$\gamma_2 = \beta_2 - 3 = 2.994 - 3 = -.006$$

$$\text{S.E. of } \gamma_1 = \sqrt{\frac{6}{N}} = \sqrt{\frac{6}{1,000}} = .077.$$

$$\text{S.E. of } \gamma_2 = \sqrt{\frac{24}{N}} = \sqrt{\frac{24}{1,000}} = .15.$$

It is clear that the curve is symmetrical as γ_1 is less then twice its standard error. As γ_2 is also considerably less than twice its standard error the curve does not depart from normality.

17.8. CENTRAL LIMIT THEOREM

The normal distribution discussed above is not confined to just making approximations to the binomial distributions. In general, the normal curve results whenever there are a large number of independent small factors influencing the final outcome. It is for this reason that many practical distributions, be it the distribution of annual rainfall, the weight at birth of babies, the daily production statistics, the height of individuals, the daily attendance in the out-patient department of a large hospital, are all more or less normal, if sufficiently large number of items are included in the population.

The significance of the normal curve is much more than this. It can be shown that *even when the original population is not normal*, if we draw samples of n items from it and obtain the distribution of sample means, we notice that the distribution of the *sample means becomes more and more normal* as the sample size n increases. This fact can be proved mathematically, and is stated as the *central limit theorem* which lays down that:

If we take samples of size n from any arbitrary population (with any orbitrary distribution) and calculate $\bar{x}$, thhe sampling distribution of $\bar{x}$ will approach the normal distribution as the sample size n increases.

If the original population were itself normal, the sampling distribution of the mean would be exactly normal, no matter how small the sample. The sampling distribution of the standard deviation will, however, not be exactly normal even if the original population were normal, but it would rapidly approach normality as n increases. Even when the original population is far from normal, the distributions derived from it ordinarily approach normality.

The fact of central limit theorem results in enormous simplification. It means that a wide class of important problems can be dealt with by taking samples of large enough size and using normal curve to predict the sampling distribution.

17.9. POISSON DISTRIBUTION

We have seen earlier that when number of trials or sample size n becomes large, the binomial distribution can be treated as a normal distribution. But this was true only under the condition that πn was large enough so that the possible values of x were very many, making it possible for us to treat the proportion of successes as a continuous variable.

In the other limit when π is so small that even when n is large πn is not very large so that the expected number of successes is quite small, the continuous variable approximation is not suitable. Under such conditions, a new limit is approached. Such a case will arise in connection with rare events, *e.g.*, number of persons killed per day in road accidents, number of mistakes per page in the final proof of a book etc. The occurrence of such events is not haphazard. Quetelet and Van Bortkiwiez pointed out that their behaviour can also be explained by mathematical law. Poisson, in 1837 obtained a limiting form of the bionomial distribution. Under the conditions that π becomes small, n becomes large, such that the produced $n\pi$ is a finite value. This mathematical limit was later found useful by Van Bortkiewiez in explaining the behaviour or such rare events and was given the name Poisson Distribution.

The Poisson distribution of the probabilities of occurrence of various rare events (successes) are given as follows:

No. of Success (x)	0	1	2	3	......	r	......
Probabilities (P)	e^{-m}	me^{-m}	$\frac{m^2e^{-m}}{2!}$	$\frac{m^3e^{-m}}{3!}$	......	$\frac{m^re^{-m}}{r!}$	...

where $e=2.718$ (base of the Napierian Logarithm) and m is the constant for a given distribution, but changes from one distribution to the next. It is termed as the parameter of the distribution and is equal to the arithmetic mean or the square of standard deviation of the distribution.

The binomial distribution is completely known if two things π (the probability of success in one trial) and n (the total number of trials) are known. Similarly, normal distribution is known when two things, *i.e.*, μ (the mean) and σ (the standard deviation) are known. But Poisson distribution is completely known when only one value m (the mean of the distribution) is known. As in bionomial distribution, the variate of the Poisson distribution is a discrete one. It is the number of success (x) and has only integral values.

Utility of Poisson distribution. It is clear from the above that Poisson distribution can explain the behaviour of the discrete variate arising in a case where probability of occurrence of the event is small and the total number of possible events is sufficienlly large. It has been applied in connection with death by rare diseases or events like road accidents or horse kicks. It has also been found useful in connection with several biological problems, problems connected with telephone traffic, quality control and sampling etc.

Properties of the Poisson distribution. The Poisson distribution is a distribution formed by discrete variate with a long tail towards right hand side. The various constants of this distribution (with parameter m) are as follows:

(1) Mean of the distribution is m.

(2) Standard deviation is $\sqrt{m}$.

(3) Skewness, given by β_1, is $1/m$, *i.e.*, as the value of m becomes larger and larger the distribution becomes less skewed.

(4) Kurtosis, given by ν_2 is also $1/m$. Thus, as the value of m becomes larger and larger ν_2 approaches zero and the Poisson distribution approaches normal distribution.

(5) The first four moments about mean are:

$$\mu_1=0,\ \mu_2=m,\ \mu_3=m,\ \mu_4=3m^2+m.$$

(6) β-coefficients of the distribution are:

$$\beta_1=\frac{1}{m} \text{ and } \beta_2=3+\frac{1}{m}.$$

Fitting a Poisson distribution. The two examples given below will explain the method of fitting the Poisson distribution (*i*) when value m is given, and (*ii*) when m is to be determined from the given data.

Illustraiion 17.34. Mean of a distribution is 2.0. What are the probabilities corresponding to various successes 0, 1, 2,..., assuming that the Poisson distribution explains the data. If the total frequency

is 1,000 what are various frequencies?

Solution. Probability of r successes $= P(r) = \dfrac{m^r e^{-m}}{r!}$

Probability of $r-1$ successes $= P(r-1) = \dfrac{m^{r-1} e^{-m}}{(r-1)!}$

$$\therefore\ P(r) = \frac{m^{r-1} e^{-m}}{(r-1)!} \cdot \frac{m}{r} = \frac{m}{r} P(r-1) \qquad \ldots(1)$$

Now,

$$P(0) = \frac{m^0 e^{-m}}{0!} = e^{-m} = e^{-2} \text{ (in the present case}$$

$$\therefore\ \log P(0) = -2 \log e$$
$$= -2(.4343)$$
$$= -.8686$$
$$= \bar{1}.1314$$
$$\therefore\ P(0) = 0.1353.$$

From the relation (1) above, putting $r=1, 2, 3$ etc.,

$$P(1) = \frac{m}{1} P(0) = 2P(0) = 2 \times 0.1353 \qquad = 0.2706$$

$$P(2) = \frac{m}{2} P(1) = \frac{2}{2} P(1) = P(1) \qquad = 0.2706$$

$$P(3) = \frac{m}{3} P(2) = \frac{2}{3} P(2) = \frac{2}{3}(0.2706) \qquad = 0.1804$$

$$P(4) = \frac{m}{4} P(3) = \frac{2}{4} P(3) = \frac{1}{2}(.1804) \qquad = 0.0902$$

$$P(5) = \frac{m}{5} P(4) = \frac{2}{5} P(4) = \frac{2}{5}(.0902) \qquad = 0.0361$$

$$P(6) = \frac{m}{6} P(5) = \frac{2}{6} P(5) = \frac{1}{3}(0.0361) \qquad = 0.0120$$

$$P(7) = \frac{m}{7} P(6) = \frac{2}{7} P(6) = \frac{2}{7}(0.0120) \qquad = 0.0034$$

$$P(8) = \frac{m}{8} P(7) = \frac{2}{8} P(7) = \frac{1}{4}(0.0034) \qquad = 0.0009$$

This process may be carried to infinite number of terms. But, practically speaking, after 10 or 15 terms, the probability of success will be so small that there will hardly be any interest in calculating terms beyond that.

In the present case, Poisson probability distribution is as follows:

No. of Successes (x)	0	1	2	3	4
		5	6	7	8 or more
Probabilities (P)	0.1353	0.2706	0.2706	0.1804	0.0902
		0.0361	0.0120	0.0034	0.0009

As the theoretical distribution has infinite number of terms, the last term is written as '8 or more'. Its probability has been found by subtracting the total of probabilities corresponding to 0 to 7 successes, *i.e.*, 0.9986 from the total probability, *i.e.*, 1.

Frequencies corresponding to various successes can be found by multiplying the respective probabilities by the total frequency. In the present case, thus, we have

No. of Successes (x)	0	1	2	3	4	5	6	7	8 and more
Frequency (f)	135.3	270.6	270.6	180.4	90.2	36.1	12.0	3.4	0.9

Illustration 17.35. The frequency distribution of the number of men killed by the kicks of horse in a certain Prussian army corps in twenty years is given as below (Von Bortkiewicz data):

Deaths (x)	0	1	2	3	4
Frequency (f)	109	65	22	3	1

Fit a Poisson distribution to the data

Solution. Here mean of the distribution is not given and has to be calculated from the given data.

From the data, we have $N=\Sigma f=200$ and $\Sigma xf=122$

$$\text{Mean}=\frac{122}{200}=0.61=m$$

Therefore, the frequency corresponding to any successess, say r. is given by $F(r)=N\times P(r)=Ne^{-m}\ m^r/r!$ or by $200\times e^{-0.61}\ (0.61)^r/r!$ in the present case. Here also the procedure followed in the last case can be adopted. But that will involve long multiplications. Here we adopt a different method:

$$\log F(r)=\log Ne^{-m}\frac{m^r}{r!}=\log N-m\log e+r\log m-\log r!$$

$$=\log 200-0.61\times 0.4343+r\log 0.61-\log r!$$

(in the present case)

$$=2.3010-0.2649+r(\bar{1}.7853)-\log r!$$

$$=2.0361+r(\bar{1}.7853)-\log r!$$

Putting $r=0, 1, 2, 3, 4$, etc., we have

No. of Successes	0	1	2	3	4 and more
$\log F(r)$	2.0361	1.8214	1.3057	0.5139	
$F(r)$	108.6	66.3	20.2	3.3	.63

Note. (1) log $F(r)$ cannot be determine for '*4 and more*'.

Frequency corresponding to this is determined by subtracting from total frequency, *i.e*, 200, the total of frequencies up to 3, *i.e.*, 199.3.

(2) Here we observe that the given frequencies and the frequencies obtained by Poisson law are quite close to one another. This confirms the fact stated earlier that Poisson distribution in the law explaining the behaviour of rare events such as men dying by horse kicks etc.

Problem involving the use of Poisson Distribution

Illustration 17.36. The mean of the Poisson distribution is 2.5. Find the other constants.

Solution. The mean of the Poisson distribution is m, *i.e.*, $m=2.5$. The other constants are:

(1) Standard deviation $=\sqrt{m}=\sqrt{2.5}=1.58$

(2) $\mu_1=0$

(3) $\mu_2=m=2.5$

(4) $\mu_3=m=2.5$

(5) $\mu_4=3m^2+m=21.25$

(6) $\beta_1=\dfrac{1}{m}=0.4$

(7) $\beta_2=3+\dfrac{1}{m}=3.4$

Illustration 17.37. In a certain Poisson frequency distribution, the frequency corresponding to 2 successes is half the frequency corresponding to 3 successes. Find its mean and standard deviation.

Solution. Let the parameter of the distribution be m and the total frequency N.

$\therefore$ Frequency corresponding to 2 successes $=\dfrac{e^{-m}m^2}{2!}N$

and ,, ,, ,, 3 successes $=N\dfrac{e^{-m}m^3}{3!}$

Therefore, we have

$$N\frac{e^{-m}m^2}{2!}=\tfrac{1}{2}N\frac{e^{-m}m^3}{3!}$$

or $$1=\frac{m}{6}$$

or $$m=6$$

or $$\text{Arithmetic Mean}=6$$

$\therefore$ Standard deviation $=\sqrt{m}=\sqrt{6}=2.45$ approx.

Note. When m is known, other remaining constants and also the probability corresponding to various number of successes can also be determined.

Illustration 17.38. A book contains 100 misprints distributed randomly throughout its 100 pages. What is the probability that a page observed at random contains atleast two misprints. Assume Poisson Distribution.

Solution. m = Mean number of misprints per pages

$$= \frac{\text{Total No. of misprints}}{\text{Total No. of pages}}$$

$$= \frac{100}{100} = 1.$$

Probability that a page contains at least two misprints $P(r \geqslant 2)$ $= 1 - [P(0) + P(1)]$

$$P(r) = \frac{m^r e^{-m}}{r!}$$

$$P(0) = \frac{1^0 . e^{-1}}{0\,!} = e^{-1} = \frac{1}{e} = \frac{1}{2.7183}$$

$$P(1) = \frac{1^1 e^{-1}}{1\,!} = e^{-1} = \frac{1}{e} = \frac{1}{2.7183}$$

$$P(0) + P(1) = \frac{1}{2.7183} + \frac{1}{2.7183} = 2 \times .368 = .736$$

$$P(r \geqslant 2) = 1 - [P(r=0) + P(r=1)]$$

$$= 1 - .736 = .264$$

Illustration 17.39. In a Poisson distribution, the probability $P(x)$ for $x=0$ is 10 per cent. Find the mean of the distribution.

Solution. If m is the parameter of the Poisson distributions, then

$$P(x) = \frac{e^{-m} m^x}{x!}$$

$\therefore$ for $x=0$, $P(0) = e^{-m} = 0.1$

or $\quad -m \log e = \log 0.1 = -1$

or $\quad m = \frac{1}{0.43429} = 2.30259$

Illustration 17.40. In a certain factory turning out fountain pens, there is a small chance 1/500 for any pen to be defective. The are supplied in packets of 10. Use Poisson's distribution to calculate the approximate number of packets containing no defective, one defective, two defective and three defective pens, respectively in a consignment of 20,000 packets.

($e^{-0.02} = 0.9802$ approx.)

Solution. Here $N=20,000$, $n=10$, $\pi=\frac{1}{500}$

$\therefore\ m = n\pi = 10 \times \frac{1}{500} = 0.02.$

$\therefore\ e^{-m}=e^{-0.02}=0.9802.$

The probability of 0, 1, 2, 3 defective is given by

$$\frac{e^{-0.02}\ m^r}{r!} \quad \text{where } r = 0, 1, 2, 3$$

$\therefore$ Number of packets containing no defective $=NP(0)$

$=N e^{-m}=20,000\times0.9802=19604$

Number of packets containing one defective $=NP(1)$

$=Nme^{-m}=20,000\times0.02\times0.9802$

$=392.08=392$

Number of packets containing two defective $=NP(2)$

$$=\frac{N.m^2e^{-m}}{2!}=20,000\times\frac{(0.02)^2}{2}\times0.9802$$

$=3.9208\simeq4$

Number of packets containing three defective $=NP(3)$

$$=N\frac{m^3e^{-m}}{3!}=20,000\times\frac{(0.02)^3}{6}\times0.9802$$

$=0.0261\simeq0$

Hence the approximate number of packets containing no defective, one defective, two defective and three defective are 19604, 392, 4 and 0 respectively.

Illustration 17.41 A firm produces articles of which 0.1 per cent are usually defective. It packs them in cases each containing 500 articles. If a wholsaler purchases 100 such cases, how many cases are expected to be free of defective items and how many are expected to have one defective each?

Solution. Here $\pi=0.1\%$, *i.e.* 0.001 and $n=500$. With these constants Binomial distribution can be applied. But this will involve long calculations. In this case, as π is small and n is sufficiently large, we can also apply Poisson distribution. The parameter of Poisson distribution $=m=n\pi=.001\times500=0.5$.

$\therefore$ The probability of 0, 1, 2, 3......defectives is given by

$$e^{-0.5}\ m^r/r! \quad \text{where } r = 0, 1, 2........$$

$\therefore$ The probability that a case contains no defective

$=e^{-0.5}=0.6065$

and the probability that a case contains one defective

$$=\frac{.5e^{-0.5}}{1}=0.30325.$$

Therefore, in a lot of 100 cases the number of cases expected to have no defective article $=100\times0.6065=61$ approx. and the number of cases having one defective $=100\times0.30325=30$ approx.

Illustratian 17.42. It is 1 in 1,000 that a birth is a case of twins. If there are 100 births in town in one day, what is the chance that two or more pairs of twins are born? Compare the results obtained by using (*a*) Binomial distribution, (*b*) Poisson distribution.

Solution. (*a*) Here probability of a twin birth $(\pi)=\frac{1}{1000}=.001$ and the maximum number of possible births $(n)=100$

$\therefore$ The probabilities of 0, 1, 2......twins are given by the 1st, 2nd, 3rd......terms in the expansion of $(.999+.001)^{100}$.

$\therefore$ Probability (no twin) = $(.999)^{100}$ = .9048

and Probability (one twin) $=100\,(.999)^{99}(.001)=.0906$.

Now, Probability of two or more pairs of twin births

$=1-\{$(Probability (no twin)+Probability (one twin)$\}$

$= 1 - \{ 0.9048+ 0.0906 \} = 1 - 0.9954 = .0046$

$\therefore$ The required probability using binomial distribution

$= .0046$

(*b*) Parameter of Poisson distribution $=np=100\times.001=0.1$

Therefore, the probabilities of 0, 1, 2, 3...twin births are given by various terms obtained by $r=0, 1, 2, 3$...........etc., in $e^{-m}m^r/r!$, *i.e.*, $e^{-0.1}\,(0.1)^r/r!$ (in the present case).

$\therefore$ Probability (no twins) $=e^{-0.1}=(2.718)^{-0.1}$

$= 0.9048$

and Probability (one twin) $= (0.1)\,e^{-0.1} = 0.09048$

$=0.0905$ approx.

$\therefore$ The required probability of two or more twin births

$= 1 - (0.9048 + .0905) = 1 - (0.9953) = 0.0047.$

Comparing the two results we find that the two methods give similar results.

Illustrotion 17.43. The number (x) of items of a certain kind demanded by customers follows the Poisson law with parameter 9. What stock k of this item should a retailer keep in order to have a probability of 0.99 of meeting all demands made on him. Use normal approximation to Poisson Law.

Solution. The mean and standard deviations of the corresponding normal distribution will be same as mean and standard deviation of the Poisson distribution.

Mean=9. Standard Dev. $=\sqrt{9}=3.$

Demand will not be met if it is more than the stock 'k'. So as in Illustration 17.29. We have to find the value k such that the chance of getting a value higher than it is 1—.99 or .01 or less.

From normal tables standard normal variate for tail area .01 is 2.33.

$$\therefore \quad \frac{k-9}{3}=2.33 \text{ or } k=16 \text{ approx.}$$

17.10. SOME MATHEMATICAL PROOFS RELATING TO POISSON DISTRIBUTION

We give below calculations of mean and standard deviation of Poisson distribution and give proof that the binomial distribution indeed does tend to Poisson distribution as π decreases and n increases such that the product $n\pi$ is constant. These proofs use the following mathematical identity:

$$e^m=1+m+\frac{m^2}{2!}+\frac{m^3}{3!}+\ldots \text{infinite number of terms.}$$

Mean:

The Poisson distribution is given as:

No. of Successes (x)	0	1	2	3	4	
Probability	e^{-m}	me^{-m}	$\frac{m^2e^{-m}}{2!}$	$\frac{m^3e^{-m}}{3!}$	$\frac{m^4e^{-m}}{4!}$	...

The arthmetic mean is given by $\Sigma xp/\Sigma p$

$$\text{Now, } \Sigma p=e^{-m}+me^{-m}+\frac{m^2e^{-m}}{2!}+\frac{m^3e^{-m}}{3!}+\ldots\ldots$$

$$=e^{-m}\left(1+m+\frac{m^2}{2!}+\frac{m^3}{3!}+\ldots\ldots\right)$$

$$=e^{-m}e^m=e^0=1.$$

$$\Sigma xp=0.e^{-m}+1.me^{-m}+2.\frac{m^2e^{-m}}{2!}+3.\frac{m^3e^{-m}}{3!}+4.\frac{m^3e^{-m}}{4!}+\ldots$$

$$=0+me^{-m}+m^2e^{-m}+\frac{m^3e^{-m}}{2!}+\frac{m^4e^{-m}}{3!}+\ldots\ldots$$

$$=me^{-m}\left(1+m+\frac{m^2}{2!}+\frac{m^3}{3!}+\ldots\ldots\right)$$

$$=me^{-m}.e^{+m}=me^0=m.1=m$$

$\therefore$ The mean of the Poisson distribution$=m$

Standard Deviation

Let the moments be taken from origin, *i.e*, 0.

Then, $\mu_2 = \nu_2 - \nu_1^2$

where $\nu_1 = \frac{\Sigma xp}{\Sigma p}$ = Arithmetic Mean and $\nu_2 = \frac{\Sigma x^2 P}{\Sigma p}$.

$$\therefore \Sigma x^2 p = 0.e^{-m} + 1^2.me^{-m} + 2^2.\frac{m^3 e^{-m}}{2!} + 3^2.\frac{m^3 e^{-m}}{3!} + 4^2.\frac{m^4 e^{-m}}{4!} + \ldots$$

$$= 0 + me^{-m} + 2.m^2 e^{-m} + 3.\frac{m^4 e^{-m}}{2!} + 4.\frac{m^4 e^{-m}}{3!} + \ldots$$

$$= me^{-m}\left[1 + 2m + 3\frac{m^2}{2!} + 4\frac{m^3}{3!} + \ldots\right]$$

Breaking each term within the brackets into two parts, we get

$$\Sigma x^2 p = me^{-m}\left[\left(1 + m + \frac{m^2}{2!} \times \frac{m^3}{3!} + \ldots\right) + \left(m + 2.\frac{m^2}{2!} + 3.\frac{m^2}{3!} + \ldots\right)\right]$$

$$= me^{-m}\left[(e^m) + \left(m + m^2 + \frac{m^3}{2!} + \ldots\right)\right]$$

$$= me^{-m}\left[\left(e^m\right) + m\left(1 + m + \frac{m^2}{2!} + \ldots\right)\right]$$

$$= me^{-m}(e^m + me^m) = me^{-m}.e^m(1+m)$$

$$= me^0(1+m) = m(1+m) = m + m^2$$

$\therefore \mu^2 = \Sigma xp^2 / \Sigma p = (m+m^2)/1 = m + m^2$

$\therefore \mu_2 = \nu_2 - \nu_1^2 = (m + m^2) - (m^2)$ (as ν_1 = Artithmetic *Mean* = m)

$\therefore \sigma^2 = \mu_2 = m.$

$\therefore$ Standard deviation $= \sqrt{m}.$

Proof for Poisson's distribution as a limiting case of binomial distribution. Before giving the proof it will not be out of place to point out two results of higher mathematics which are used in this connection.

(1) The exact value of e, the base of Napierian Logarithm, is given as $e = \underset{n \to \infty}{\text{Lt}} \left(1 + \frac{m}{n}\right)^{n/m}$. It can be shown by expansion that this is approximate equal to 2.718.

(2) We know $n! = n(n-1)(n-2)\ldots\ldots1$. Stirling has shown that the approximate value of $n!$ in terms of e and powers of n is $\sqrt{2\pi n}\, n^n e^{-n}$.

Proof. If p is the probability of success in one trial, n the number of trials and $q = 1 - p$, the various terms of the binomial distribution are given by putting $r = 0, 1, 2, 3 \ldots\ldots\ldots\ldots n$ in the general term

$$[P(r)] = {}^nC_r\, q^{n-r}\, p^r \quad or \quad \frac{n!}{r!(n-r)^r}(1-p)^{n-r}\, p^r \ldots$$

If we can show that under certain conditions this term approaches the general term of the Poisson distribution, *i.e.*, $e^{-m}m^r/r!$, we can say Binomial distribution tends to Poisson distribution, under those conditions. The conditions as stated earlier are that n approaches a very large number, *i.e.*, ∞ and $x \to 0$, such that $np \to m$, a finite value.

Putting $np=m$, *i.e.*, $p=\frac{m}{n}$ in $P(r)$, we get

$$P(r) = \frac{n!}{r!\,(n-r)!}\left(1-\frac{m}{n}\right)^{n-r}\left(\frac{m}{n}\right)^r$$

$$=\frac{n!}{(n-r)!n^r}\left[1+\left(-\frac{m}{n}\right)\right]^n\left[1-\frac{m}{n}\right]^{-r}\times\frac{m^r}{r!}$$

Applying Stirling's approximation and taking limits:

$$\underset{n\to\infty}{\text{Lt}}\ P(r)=\underset{n\to\infty}{\text{Lt}}\ \frac{\sqrt{2\pi n}\ n^n e^{-n}}{\sqrt{2\pi(n-r)}\ (n-r)^{n-r}e^{-(n-r)}n^r}$$

$$\times\left[1+\left(-\frac{m}{n}\right)\right]^{\frac{-n}{m}\times(-m)}\left(1-\frac{m}{n}\right)^{-r}\times\frac{m^r}{r!}$$

Now,

$$(n-r)^{n-r}=n^{n-r}\left(1-\frac{r}{n}\right)^{n-r}$$

$$=n^{n-r}\left[1+\left(-\frac{r}{n}\right)\right]^n\left[1-\frac{r}{n}\right]^{-r}$$

$$=n^{n-r}\left[1+\left(-\frac{r}{n}\right)\right]^{-\frac{n}{r}\times(-r)}\left(1-\frac{r}{n}\right)^{-r}$$

As $e^m=\underset{n\to\infty}{\text{Lt}}\left(1+\frac{m}{n}\right)^n$, in the limiting case as $n\to\infty$

$$\underset{n\to\infty}{\text{Lt}}\left[1+\left(-\frac{r}{n}\right)\right]^{-\frac{n}{r}\times-r}\to e^{-r}\ \text{and}$$

$$\underset{n\to\infty}{\text{Lt}}\left[1+\left(-\frac{m}{n}\right)\right]^{-\frac{n}{m}\times(-m)}\to e^{-m}$$

$$\underset{n\to\infty}{\text{Lt}}\ P(r)=\underset{n\to\infty}{\text{Lt}}\ \frac{\sqrt{2\pi}\ n^n e^{-n}e^{-m}}{\sqrt{2\pi}\ n^{n-r}\ e^{-r}\left(1-\frac{r}{n}\right)^{-r}e^{-n}e^r n^r}$$

$$\times\sqrt{\frac{n}{n-r}}\left[1-\frac{m}{n}\right]^{-r}\times\frac{m^r}{r!}$$

$$= \underset{n\to\infty}{\text{Lt}} \left[\frac{n^n}{n^{n-r+r}e^{-r+r}}\right] \times \left[\frac{\left(1-\frac{m}{n}\right)^{-r}}{\left(1-\frac{r}{n}\right)^{-r}} \sqrt{\frac{n}{n\left(1-\frac{r}{n}\right)}}\right] \times \left[\frac{e^{-m}m^r}{r!}\right]$$

Now, the terms in the first bracket simplify to 1 and the terms in the second bracket will approach 1 as $n\to\infty$. So we get,

$$\underset{n\to\infty}{\text{Lt}} \quad P(_r)=\frac{c^{-m}m^r}{r!}.$$

Hence, each and every term of the Binomial distribution tends to the corresponding term of the Poisson distribution, provided $n\to\infty$, $p\to 0$, such that $np\to m$, a finite value.

PROBLEMS

1. The chance of success of an event is p and of its failure is $q=1-p$. Assuming that this chance remains constant in each trial obtain the expected frequencies of 0, 1, 2, . . . success in N sets of n trials.
   Derive the binomial distribution with parameter n and p.
   (Ans. So the frequencies of getting 0, 1, 2. . .success in N sets of n trial each is given by various terms obtained by putting $r=0, 1, 2, \ldots n$ in N. ${}^nC_r\, p^r q^{n-r}$. *i.e.*, Nq^n, $N{}^nC_1\, pq^{n-1}$, $N{}^nC_2\, p^2q^{n-2}$, etc.
2. Obtain the mean and variance of binomial distribution.
3. For a binomial distribution mean is 6 and standard deviation is $\sqrt{2}$. Write out all the terms of the distribution.
4. What is probability of picking three defective items from a lot of 100 items when it is known that the probability of an item being defective is only 5%.
   (Ans. ${}^{100}C_3\,(.95)^{97}\,(.05)^3$).
5. The incidence of occupational disease in an industry is such that the workmen have a 20% chance of suffering from it. What is the probability that out of six workmen 4 or more will contract disease.
   (Ans. 53/3125).

6. Twelve dice are thrown and the number of dice showing a 5 or 6 is noted. The experiment is repeated 26306 times and the following is frequency distribution of the 26306 throws according to the number of dice showing either 5 or 6. Fit a binomial distribution under the hypothesis that the dice a perfect cubes.

No. of dice	*No. of throws*	*No. of dice*	*No. of throws*
0	185	7	1331
1	1149	8	403
2	3265	9	105
3	5475	10	14
4	6174	11	4
5	5194	12	0
6	3067		

7. The following is the distribution of 53680 German families of eight children each, according to the number of male children.

No. of boys	0	1	2	3	4	5	6	7	8
No. of families	215	1485	5331	10649	14955	11929	6678	2092	342

Fit a binomial distribution calculating the sex ratio from the given data.

8. (*a*) Derive Poisson distribution as the limit of the Binomial distribution. Show that the mean and variance of the Poisson distribution are equal.
(*b*) Define Poisson distribution. Calculate its first four moments.

9. Mean of a Poisson distribution is 4. Find out its first four moments about the mean and other constants of the distribution derived from them.
10. γ is a Poisson variate. It is known that frequency of γ taking the values 3 and 4 are equal. Calculate the probabilities with which γ can take the values 0 and 1.
11. It is given that 3% of the electric bulbs manufactured by a company are defective. Using the Poisson approximation, find the probability that a sample of 100 bulbs will contain (*i*) no defective, (*ii*) exactly one defective. [Ans. (*i*) 05 (*ii*) 15]
12. In a Poisson distribution the probability $p(X)$ $n=0$ is 10%. Find the mean of the distribution. Also calculate $p(1)$, $p(2)$, $p(3)$ and $p(4)$.
13. The following table shows the distribution of the number of faulty units produced in a single shift in a factory. The data is for 100 shifts.

No. of faults	0	1	2	3	4	5	6 and more
No. of shifts	4	14	23	23	18	9	9

Fit a Poisson distribution to the data.

[Ans.

No. of faults (n)	0	1	2	3	4	5	6	7 and above
No. of shifts (d)	5.0	15.0	22.5	22.5	16.9	10.1	5.00	5

The total frequency up to 6 is 97. So the rest *i.e.* 3 is for 7 and above.

14. When first proof of 200 pages of a book of 5000 pages were read the distribution of printing mistakes were found to be as follows:

No. of mistakes in page	0	1	2	3	4	5
No. of pages	112	63	20	3	1	1

Fit a Poisson law to the above distribution and apply the suitable test of goodness of fit.

15. For a statistical study of the agricultural income of Bombay Province 105 villages were selected, supposed to be randomly drawn. But the selected villages tended to be grouped in clusters. To test this assertion, the whole area was divided into 50 rectangles of equal area and a count was made of the rectangles containing 0, 1, 2 villages.

No. of villages	0	1	2	3	4	5	6 and over	Total
No. of rectangles	6	15	11	9	5	3	1	50

Do you conclude that the villages were selected at random?

16. The mean and standard deviation of a normal distribution are 60 and 5 respectively. Find the inter quartile range and the mean deviation of the distribution.
[Ans. (*i*) 6.745 (*ii*) 4].

17. Balls are tested for dropping from a height and measuring the height of bounce. A ball is 'fast' if rises above 35 inches. The average height of bounce was 32 inches. What is the probability of picking a fast ball?
(Ans. Probability of getting a fast ball = .02275).

18. The mean and standard deviation of normal distribution of 2484 measurements are respectively 20.6 cm. and 1.2 cm. A measurement is picked at random. The probability of this measurement being between 21 cm. and ω cm. inclusive is 0.2268. Find the number ω if the measurements were made to the nearest units.

19. The probability that a man aged 40 years will die within next year is 0.001. What is the probability that out of 100 such persons at least 99 will survive till next year. You are not to use logarithm tables.
(Ans. .99055 Approx.)

20. A manager accepts the work submitted by his typist only when there is no mistake in the work. The typist has to type on an average 20 letters per day of about 200 words each. Find the chance of his making a mistake, (*a*) if less than 1% of the letters submitted by him are rejected, (*b*) if on 90% of days all the work submitted by him is accepted.
(Ans. 0.0000026 Approx.)

21. It is conjectured that in recent mid-term elections the percentage of voting in individual booths is distributed normally with mean 50 and standard deviation 25. 1000 booths are now selected at random. Find in how many of these booths the percentage of voting is expected to be between (*i*) 20 and 30 (*ii*) 60 and 70 (*iii*). In how many booths do you expect the percentage to exceed 90.
(Ans. 55 Approx.)

Chapter 18

ESTIMATION AND TESTING OF HYPOTHESIS

18.1. ESTIMATION AND CONFIDENCE LEVEL

Suppose we are to find the average life of some electric switches in a large shipment of switches. Testing every switch to failure will give the exact average life, but the knowledge will most certainly be useless since we won't have any switches left. The correct procedure is to take out a sample of n switches, test them to failure and take their average life $\bar{x}$. Then, we can take $\bar{x}$ as an estimate of the population mean μ. But how much confidence do we have or our estimate? If $\bar{x}=2500$ hrs, surely μ can be greater than or less than 2,500. It is for this reason that it is meaning ful to take of an *interval, i.e.*, a range of values about $\bar{x}$ in which μ may lie. This procedure is known as *interval estimation.* But how wide is this interval to be is the basic problem. To be absolutely sure, we have to have an almost infinitely wide interval for there is always a chance, howsover remote, that the actual μ is wider from $\bar{x}$ than any prespecified value.

There is a way out of this dilemma by prescribing what is termed as the *level of confidence.* This refers to the degree of confidence we have, *i.e.*, the probability that the actual population parameter, lies within a specified range about the sample statistics. The population parameter is that what we are trying to estimate (which in our case in μ), and the sample statistics is what we know by measurment ($\bar{x}$ here).

Thus, a wide range or interval has a higher level of confidence as the estimator than a narrower interval.

18·2. STANDARD ERROR OF ESTIMATE

It was seen in chapter 17 that if a sample of n items is taken from a population whose proportion of success is π, than the expected value of the sample proportion of success x is also π. But as seen before, this is the expected value in an averaged sense only and, any single actual value will most probably differ from it. It too was shown that the probability distribution of the sample proportion x (termed as a sampling distribution of x) is a binomial distribution with mean π and standard deviation $\sqrt{\pi(1-\pi)/n}$. Thus, if the sample proportion x is uses as an estimate of π, we are going to be off from the actual π with a probability distribution that can be obtained from the binomial distribution.

It was also seen that as n becomes very large, the binomial distribution tends to become a normal distribution, so that for *'large enough' samples*, the sampling distribution can be treated as a normal distribution with mean $\mu=\pi$ and the standard deviation $=\sqrt{\pi(1-\pi)/n}$. Since the sampling standard deviation measures the spread about the mean, if $\bar{x}$ is to be used as an estimate of π, the sampling σ measures the order of error that we are likely to make. It is for this reason that sampling standard deviation is termed as *standard error of estimate*, or simply *standard error*.

This standard error has an important function in constructing the confidence interval for estimation. Since we know that in a normal distribution (which represents our sampling distribution) about 68.2% of the items lie within mean $\pm$ one σ, it is obvious that the probability that the actual π lies within ($\bar{x} \pm 1\times$standard error) is 68.2%. So ($\bar{x} \pm 1\times$standard error) is a 68.2 per cent *confidence interval* in the terminology of the last section. Similarly, ($\bar{x} \pm 1.96$ $\times$standard error) is a 95% confidence interval, ($\bar{x} \pm 2.0\times$Std. error) is a 95.4 % confidence interval), ($\bar{x}\pm 2.58\times$Std. error) is a 99 % confidence interval, and ($\bar{x} \pm 3.0\times$Std. error) is a 99.73% confidence interval.

For our problem, the standard error is estimated by $\sqrt{x(1-x)/n}$, and as is to be expected, this decreases with n, the sample size. Thus, a given confidence interval shrinks in width as n increases, or conversly we get more confidence in a given inteval.

Illustration 18.1. The measured proportion of success in a sample of 25 is 0.5. (*a*) what is a 99% confidence interval for π? (*b*) what should be the sample size for width of this interval to be 0.02?

Solution. Here $\bar{x}=0.5$. An estimate of std. error S is provided by $\sqrt{x(1-x)/n}$

(*a*) for $n=25$, $S=\sqrt{0.5(1-0.5)/25}=0.1$

Therefore, a 99% confidence interval which is a $\pm$ 2.58 S interval is $0.5 \pm 2.58 \times 0.1$, or 0.242 to 0.758 Thus, we can have 99% confidence only over an interval as large as $\pi = 0.242$ to 0.758.

(*b*) To reduce the width of this interval to 0.02, S must be given by

$$\frac{0.02}{2} = 2.58\,S \text{ or } S = .00388$$

The required n is then obtained from

$$0.00388 = \sqrt{0.5(1-0.5)/n}$$

or

$$n = 16641$$

Thus, only with a sample of size 16641 we can have a 99% confidence interval of width 0.02.

18.3. STATISTICAL INFERENCES OR TESTING OF HYPOTHESIS

Consider the claim of a 'doctor' who advertises that he has discovered a better cure of a certain type of cancer where the recovery rate with the already existing cures is only 20 per cent. To verify his claim, an experiment is set up where his medication is given to 10 patients with the result that 4, that is, 40% patients recover. Is his cure really better than the old cure, or is the evidence 'insufficient' to justify the claim. There is no straight forward or simple answer because it is known that the sample proportion of success does vary (from the population proportion), and even if the actual cure rate (population value) is less than even 20%, it is possible to draw a sample of 10 with sample proportion of 40%, or even more.

We are frequenty called upon to make inferences of this nature, based on such envidence. We outline here the basis on which such inferences are made. The procedure for such 'test' starts out by laying down that the claim is wrong and the new treatment does no better than the old. The given sample, then, can be taken as coming from the 'population' where the cure rate is the old 0.20. This is termed as the *null hypothesis*.

We next determine the probability of obtaining the given sample from a population satisfying the null hypothesis (referred to as 'null' population, for brevity). Let this probability be P, and the probability that it does not come from such a population is $1-P$. The basic question then is: at what value of P should we reject the null hypothesis and say that it is beyond 'reasonable' doubt that the sample cannot come from the 'null' population.

It is clear that in making either decision we take the risk of being wrong. There are two distinct types of error we can make, as is made clear by reference to Table 18.1.

TABLE 18.1. TYPE OF ERRORS IN DECISIONS

True State	*Decision Taken*	
	Null accepted	*Null rejected*
Null hypothesis correct	Correct Decision	Type I Error
Null hypothesis incorrect	Type II Error	Correct Decision

When the null hypothesis is correct and we reject it, the error is termed as the type I error. Type II error occurs when we accept the null hypothesis even when it is wrong. In the present example, type I error is when we accept the claim of better treatment when in fact it is no better, and type II error when we reject his claim even though it is correct. In rejecting the claim, we state that the observed difference between the sample statistics and the (expected) population parameter is *insignificant* in the sense that it can reasonably arise purely from the chance sampling variations.

The permissible value of probability above which we do not reject the null hypothesis is termed the level of significance, and we denote it by α. A low value of α, *i.e.*, of the permissible type I error means the test is conservative, or that the null hypothesis is rejected only when the evidence piles up against it. We usually test at 5% or 1% level of signifieance.

To sum up, the procedure of obtaining statistical inferences about population parameters from sample statistics consists of the following steps:

1. A null hypothesis H_0 is set up. This is a conservative statement about the population parameter, and is so termed because it almost invariably states that the given sample comes from a population which is better or no worse than some standard population, or that no change has occured.

2. On the assumption that the null hypothesis is correct, the probability of obtaining the given value of the sample statistics is calculated. For this purpose the knowledge of the sampling distri-

bution for the relevent statistics is required. The fact of the central limit theorem helps to a large extent because it permits the use of normal distribution which is very simple to use.

3. If the probability P obtained in (2) above is more than the chosen level of tolerable risk α for type I error, the null hypothesis is accepted stating that *there is insufficient evidence to reject the null hypothesis at* α *level of significance*, which means that it is not unreasonable to obtain such a sample from the null population at α level of significance.

4. If the probability P obtained in (2) above is less than the chosen level of significance α, the null hypothesis is rejected saying that it has been shown beyond 'reasonable doubt' (*i.e.*, beyond α level of significance) that such a sample is not expected of the 'null' population and, therefore, a change in the population parameters has occured.

A knowledge of sampling distributions for the statistics considerd is 'then' the first requirement of setting up the tests. For large samples, the normal distribution is a statisfactory approximation for most statistics. What is required than is the information about the standard error to completely specify the sampling distribution.

In chapter 19, we give some tests based on large samples where normal curve can be used, and in chapter 20 we give tests on small samples where other distributions have to be used.

Chapter 19

TESTS ON LARGE SAMPLES

19.1 INTRODUCTION

It has been noted earlier that as the sample size increases, the sampling distributions of many of the sample statistics approach the normal distribution, even when the original population is far from normal. Since the normal distribution is a very well understood and tabulated distribution, this enables one to design very quick and efficient tests based on large sample statistics.

A normal distribution requires specification of two parameters, namely, the mean and standard deviation. The mean of a sampling distribution is that about which the null hypothesis is set up. The standard denation of the sampling distribution is termed the standard error, and it depends on the sample statistics being considerd and the sample size. We have already noted that the standard error of sample proportion is $\sqrt{\pi(1-\pi)/n}$ when n is the sample size. Similarly, we can determine the standard errors of varions other statistics. In this chapter we develop some tests based on these considerations

19.2. TESTING ASSUMPTIONS ABOUT POPULATION MEAN

Suppose we wish to test the claim of a baby-food manufacturer that babies taking his formulation have a higher growth rate than babies fed on the other leading brand. It is known that the mean growth rate (μ) on the other leading brand is 0.3 kg/week. We take a sample of 25 babies fed on the new baby food and measure the sample mean growth rate ($\bar{x}$) as 0.35 kg/week. Is the variation entirely due to chance or does the evidence prove the manufacturers, claim (at some level of significance)? To answer this question we need to find the probability that a sample with $\bar{x}=0.35$ kg/week comes from a population with $\mu=0.30$ kg/week. If this probability is less than α, the *pre*-specified level of significance, we claim that there is sufficient evidence to accept the claim, otherwise the variations are not large enough not to arise purely because of chance.

To calculate this probability we need the standard error of the distribution of $\bar{x}$. A theorem from mathematical statistics states that if random sample of size n are taken from a population with standard deviation σ, the standard error of the mean $\bar{x}$ is

$$\text{S.E.} = \frac{\sigma}{\sqrt{n}} \qquad (19.1)$$

Thus, if we know σ of the population, we can find the S.E. of the mean. But we usually do not know σ. Under the circumstance we estimate σ from the standard deviation of the sample data. It can be shown that the best estimate of σ from a sample of n items is obtained by

$$S = \sqrt{\frac{\Sigma(x-\bar{x})^2}{n-1}} \qquad (19.2)$$

The test procedure then consists of the following steps

1. Set up the null hypothesis H_0 stating that the sample comes from a population with no change, *i.e.*, with $\mu=0.30$ kg/week.
2. Determine the standard error of the sample mean. This requires calculation of S by Eq. (19.2) and then use of Eq. (19.1) to determine S.E. If $S=0.15$kg/week, then $\text{S.E.} = \frac{0.15}{\sqrt{25}} = 0.03$.
3. Specify the level of significance required: 'Say' 5% in this case.
4. Calculate standard normal vairate correspoding to the observed sample mean $\bar{x}$:

$$Z = \frac{\bar{x}-\mu}{\text{S.E.}} = \frac{0.35-0.30}{0.03} = 1.67$$

5. 5% level of significance requires that area beyond Z should be less than 5%. The value of Z for tail area to be 5% is 1.645.
6. Since the observed value of Z is larger than that needed for 5% level of significance, we have enough evidence at this level to say that such a variation cannot arise because of chance above and hence the null hypothesis can be rejected.
7. If the value of Z turns out to be less than that required at the given level of significance, we do not have sufficient evidence to reject the null hypothesis.

19.3. ONE-TAIL AND TWO-TAIL TESTS

Consider the case of a soap manufacturer who makes cakes of 200 g by an automatic machine. He wants to test the average weight of the cakes by taking out a sample. The null hypothesis H_0 is that $\mu=200$ g. Now there are two distinct view points: one of the government inspector in which he wants to be sure that the true mean is at least 200 g (he does not care if it is more), and the other of the manufacturer: he wants μ to be neither too high nor too low. A 5% level of significance in the two cases will mean different things. From the manufacturer's view point we reject the null hypothesis when $\bar{x}$ is either too low or too high, and, therefore, the cut-off value of Z should be chosen such that the combined probability of the too 'tails' is 0.05, which means that of one 'tails' is 0.025. From the table of normal areas this corresponds to $Z=-1.64$ for the first case and $Z=\pm 1.9$ in the two- tail case as shown in Fig 19.1.

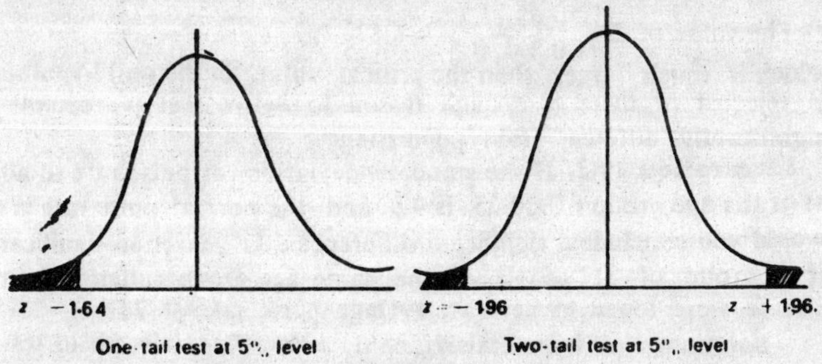

FIG. 19.1

If we were to ensure only that μ is not greater than μ_0 we take the upper tail area into occount and the critical value of Z will be $+1.64$.

We summarize in the folloying table the critical values of Z for the various tests:

TABLE 19.1. CRITICAL VALUES OF Z

Level of Significance	*Two-tail Test*	*One-tail Test*	
		upper	*lower*
5 per cent	$Z=\pm 1.96$	$Z=+1.64$	$Z=-1.64$
1 per cent	$Z=\pm 2.57$	$Z=+2.33$	$Z=-2.33$

The following examples illustrate the tests on the mean of largo samples.

Illustration 19.1. Suppose the standard deviation of the heights of university male students is 3". One hundred male students of a university are measured and their mean height is found to be 68". Ascertain if this (sample) mean height represents a significant divergence from the population mean of 70".

Solution. Set the null hypothesis H_0 as μ = 70". Since we are interested in just determing if the sample is different, we conduct a 2-tail test.

At 5% level of significance critical $Z = \pm 1.96$

Standard error of mean $\frac{\sigma}{\sqrt{n}} = \frac{3}{\sqrt{100}} = 0.3$

The value of $Z = \frac{70\text{-}68}{0.3} = \frac{2}{0.3} = 6.67$

which is much larger than the critical value, so the null hypothesis is rejected or there is enough reason to belive that the sample is significantly different from population

Illustration 19.2. If the standard deviation of pulse rate in adults of the age group of 20-25 is 9.9, and the normal pulse rate is 69, would you conclude a significant difference at 2% level of significance if a group of 81 people of the same age group suffering from a disease were found to have an average pulse rate of 74?

Solution. Null hypothesis H_0 is μ_0 = 69. Since we are to test if the rate is 'different', it is a two-tail test. At 2% level, critical value of Z from the tables (corresponding to one-tail area of 0.01) is ± 2.32.

Standard error $= \frac{\sigma}{\sqrt{n}} = \frac{9.9}{\sqrt{81}} = 1.1$

The sample value of $Z = \frac{\bar{x} - \mu_0}{\text{S.E.}} = \frac{74 - 69}{1.1}$

$$= \frac{5}{1.1} = 4.54$$

Hence the difference is significant at 2% level.

Illustration 19.3. The data concerning height measurement for a random sample of individual from a given population are as follows: Mean is equal to 172, S.D. is equal to 12, N is ebual to 65.

If a large number of samples of the same size were selected at random from the given population, what would be the limits within which the true mean almost certainly lies?

Solution. Standard error of mean $=\frac{\text{S.D.}}{\sqrt{n}}=\frac{12}{\sqrt{65}}=1.49$

We know that 99% confidence interval is $\bar{x}\pm 2.56$ SE. Therefore 99% confidence interval is $172\pm 2.56\times 1.49=172\pm 3.8=168.2$ to 175.8.

Illustration 19.4. If it costs a rupee to draw one member of a sample, how much would it cost, in sampling from a universe with mean 100 and standard deviation 10, to take sufficient members to ensure that the mean of the sample in all probability would be within .01 per cent of the true value? Also find the additional cost to double the precision.

Standard error of mean $=\frac{10}{\sqrt{n}}$.

We want a 0.01 per cent accuracy or the interval should be

$$100\pm 100\times 0.0001=99.99 \text{ to } 100.01$$

Let us take the interval mean ± 3 S.E. as the interval of all proba-
The width 0.02 is then equal to 6 S.E.

or $$6\times \frac{10}{\sqrt{n}}=0.02$$

or $$n=\left(\frac{60}{.02}\right)^2=9{,}000{,}000$$

So we require a sample of size 9 million to ensure 0.01 per cent accuracy. To double the accuracy, we need 4 times this size, *i.e.*, $n=36$ million. So the *additional* cost is Rs 27 million.

Illustration 19.5. The standard deviation of a large number, N of observations on chest measurement is 0.5″. If random samples of 400 individuals are drawn from the "population of N individuals", what is the probability that the sample mean will differ from the the parameter mean by 0.05″ or more?

Solution. Standard Error of the Mean $=\frac{\sigma_p}{\sqrt{N}}$

$$=\frac{0.5}{\sqrt{400}}=\frac{0.5}{20}=0.025$$

The difference (between sample mean and parameter mean) is 2 times that of the standard error of mean, viz., $\frac{.05}{.025}=2$. In other words, the difference of .05 is equal to 2 S.E. which covers 95.45% of the total area (in both directions from the mean) of the normally distributed population, and leaves only 4.55% area out of ± 2 S.E. from the mean.

Consequently the probability that the sample mean will differ from the parameter mean by 0.05 or more in the above case is .0455 or $\frac{1}{22}$ approx.

19.4. TESTING SIGNIFICANCE OF DIFFERENCE IN MEANS

For large samples the difference in means of two samples of size n_1 and n_2 are also distributed normally with mean $\mu=0$ and standard error

$$\text{S.E.}=\sigma\sqrt{\frac{1}{n_1}+\frac{1}{n_2}}$$

where σ is the standard deviation of the population. If standard deviation of the population is not given, we estimate it from standard deviation S_1 and S_2 of the sample, so that,

$$\text{S.E.}=\sqrt{\frac{S_1^2}{n_1}+\frac{S_2^2}{n_2}}$$

The null hypothesis in each of such problem is that μ of the the two samples come from the same population.

Illustratian 19.6. The mean produce of wheat of a sample of 100 fields comes to 200 kg. per acre with a standard deviation of 10 kg. Another somple of 150 fields gives the mean at 220 kg with a standard deviation of 12 kg. Assuming the standard deviation of the yield at 10 kg. for the universe, find out if there is significant difference between the mean yields of the two samples.

Solution. As the standard deviation of the universe is given, so we will use this in spite of the fact that the standard deviation of each of the samples is given separately. Standard error of difference between two sample means is

$$\text{S.E.}=\sigma\sqrt{\frac{1}{n_1}+\frac{1}{n_2}}$$

where σ is the standard deviation of the universe;

n_1 is the number of the first sample; and

n_2 is the number of the second sample.

Substituting the values, we get

$$\text{S.E.} = 11\sqrt{\frac{1}{100}+\frac{1}{150}}$$

$$=11\sqrt{\frac{1}{60}}=1.42.$$

The actual difference between the two sample means is 220—200 $=20$, which is $\frac{20}{1.42}$ or about 14 times the S.E. of the difference of means, leading us to conclude that this difference is very unlikely to arise on account of chance and so the difference is highly significant.

Illustration. 19.7. In a survey of buying habits 400 women shoppers are chosen at random in super-market '*A*' located in a certain section of the city. Their average weekly food expenditure is Rs 250 with a standard deviation of Rs 40. For 400 women shoppers chosen at random in super-market '*B*' in another section of the city, the average weekly food expenditure is Rs 220 with a standard deviation of Rs 55. Test at 1% level of significance whether the average weekly food expenditures of the populations of shoppers are equal.

Solution. Let us assume that the average weekly food expenditures of the two populations of shoppers are equal. Then S.E. of the difference in means

$$\text{S.E.}=\sqrt{\frac{\sigma_1^2}{n_1}+\frac{\sigma_2^2}{n_2}}$$

when, in the usual notations, we are given

$$n_1=400,\ \bar{x}_1=250,\ \sigma_1=40$$

$$n_2=400,\ \bar{x}_2=220,\ \sigma_2=55$$

Substituting the values, we get

$$\text{S. E}\ (\bar{x}_1-\bar{x}_2)=\sqrt{\frac{40^2}{400}+\frac{55^2}{400}}=\sqrt{\frac{4625}{400}}$$

$$=\frac{68.01}{20}=3.37$$

The observed difference between the two means is 250—220=30 which is about 8.8 times the standard error.

Since it is much greater than 2.56, we conclude that the average weekly expenditures of two populations of shoppers differ significantly at 1% level.

Illuctration 19.8. Intelligence tests of two groups of boys and girls give the following result. Examine if the difference is significant.

Girls: Mean 84, S.D. 10 Number 121

Boys: Mean 81, S.D. 12, Number 81.

Solution. The standard error of the difference between their means would be

$$\text{S.E.}_{\bar{x}_1-\bar{x}_2}=\sqrt{\frac{\sigma_1^2}{n_1}+\frac{\sigma_2^2}{n_2}}$$

where σ_1 and n_1 represent respectively the S.D. number of girls, and σ_2 and n_2 represent respectively the S.D. and number of boys.

Substituting the values, we get

$$\text{S.E}\ (\bar{x}_1-\bar{x}_2)=\sqrt{\frac{10^2}{121}+\frac{12^2}{81}}=\sqrt{\frac{100}{121}+\frac{144}{81}}$$

$$=\sqrt{\frac{8100+17424}{9801}}=\sqrt{\frac{25524}{9801}}$$

$$=\sqrt{2.064}=1.61.$$

The observed difference between the two means is 84−81=3, which is about 1.9 times the $\text{S.E.}_{\bar{x}_1-\bar{x}_2}$. *i.e.*, $\frac{3}{1.61}=1.9$ approx. This difference is significant at 5% level but not so at 1% level.

Illustration 19.9. A random sample of 600 villages was taken from district Kanpur and average population per village was found to be 480 with a S.D. of 48. Another random sample of 400 villages from the same district gave an average population of 500 per village with a S.D. of 54. Find out if the mean of the first sample differs significantly from the second.

Solution. For null hypothesis, that the difference in means is zero, standard error is given by

$$\text{S.E.}=\sqrt{\frac{\sigma_1^2}{n_1}+\frac{\sigma_2^2}{n_2}}$$

$$=\sqrt{\frac{48^2}{600}+\frac{54^2}{400}}=3.34$$

So the observed difference corresponds to

$$z=\frac{480-500)-0}{3.34}=-5.98$$

Since this is much larger than critical for even 1% level of significance, we reject the null hypothesis and say that the difference is significant.

Illustration 19.10. A simple sample of height of 6400 Englishmen has a mean of 67.85″ and a standard deviation of 2.56″. While a sample of height of 1600 Australians has a mean 68.55″ and a standard deviation of 2.52″. Do the data indicate that Australians are on average taller than Englishmen?

Solution. We test the null hypothesis that the difference in the mean is zero.

Estimate of S.E. is provided by

$$\text{S.E.}=\sqrt{\frac{\sigma_1^2}{n_1}+\frac{\sigma_2^2}{n_2}}=\sqrt{\frac{(2.56)^2}{6400}+\frac{(2.52)^2}{1600}}$$

$$=0.071$$

So the z corresponding to given difference is

$$z=\frac{(67.85-68.55)-0}{0.071}=-9.86$$

which is very large for any reasonable level of significance. Thus, the average height of Englishmen is significantly different from Australians.

Illustration 19.11. It is desired to know whether urban families consume significantly more tea than rural families. The average purchase per urban family is found to be 3.0 kg per year with a variance of 1.4, whereas sample of 150 rural families reveals their average annual tea purchase to be 2.4 kg with a variance of 1.5. What is your conclusion?

Solution. Here we cannot test on the difference of sample means since the data for urban families appears to be for the whole urban population, since sample size is not given. So the relevant null hypothesis is that the rural sample also comes from a population with mean 3.0 and variance $\sigma^2=1.4$.

$$=\sqrt{\frac{1.4}{150}}=0.097 \text{ kg}$$

So the corresponding $z=\dfrac{3.0-2.4}{0.097}=-6.18$

Illustration 19.12. Electric bulbs manufactured by X and Y companies gave the following results:

	X Co.	*Y Co.*
No. of bulbs used	100	100
Mean life in hours	1300	1248
Standard Deviation	82	93

Using the standard error of the difference between the means, state whether there is any significant difference in the mean life of the two makes.

Solution. $S.E._{\bar{x}_1-\bar{x}_2} = \sqrt{\dfrac{\sigma_1^2}{n_1}+\dfrac{\sigma_2^2}{n_2}}$

$$=\sqrt{\frac{82^2}{100}+\frac{93^2}{100}}=\sqrt{67.24+86.49}$$

$$=\sqrt{153.73}=12.4.$$

The observed difference between the two means is 1300—1248=52 which is about 4.2 times the S.E. Since it is greater than 2.57, the difference in the mean life of bulbs of the two makes is significant at 1% level

Illustration 19.13. In a certain factory there are two indepent processes manufacturing the same item. The average weight in a sample of 250 items produced from one process is found to be 120 gm and with a standard deviation of 12 gm. While corresponding figures in a sample of 400 items from the other process are 124 and 14. Is the difference between the mean weight significant at 1% level of significance?

Solution. Let us assume that there is no real difference between the mean weights

$$S.E._{\bar{x}_1-\bar{x}_2} = \sqrt{\frac{\sigma_1^2}{n_1}+\frac{\sigma_2^2}{n_2}}$$

$$=\sqrt{\frac{12^2}{250}+\frac{14^2}{400}}=\sqrt{1.066}=1.03.$$

The observed difference between the two means is 4 which is about 3.8 times the S.E.

Since 3.8 is greather than 2.57, the difference is singificant (at 1% level of significance).

19.5. TESTS RELATING TO PROPORTION

It has been shown earlier that the sampling distribution of proportion x in large samples from a binomial population with proportion π is a normal distribution with $\mu=\pi$ and S.E.$=\sqrt{\pi(1-\pi)/n}$. This fact can be used to set up tests on proportions in a sample as illustrated below.

Similarly, we can test on difference in proportions in two large samples by using the fact that its sampling distribution tends to be normal with $\mu=0$ and

$$\text{S.E.}=\sqrt{\pi(1-\pi)\left(\frac{1}{n_1}+\frac{1}{n_2}\right)}$$

Illustration 19.14. A railway company installed two sets of 50 Burma ties each. The two sets were treated with creosate by two different processes. After a number of years of service it was found that 22 ties of first set and 18 ties of the second set were still in good condition. Are we justified in claiming that there is no real difference between the preserving properties of the two processes.

Solution. Let us assume that there is no real difference between the preserving properties of the two processes.

Proportion of ties of first set in good condition $p_1=\frac{22}{50}$.

Proportion of ties of the second set in good condition $p_2=\frac{18}{50}$

Proportion of ties in good condition (estimate of π for population):

$$\pi=\frac{22+18}{50+50}=0.4$$

Standard error of the difference of proportion

$$=\sqrt{\pi(1-\pi)}\left[\sqrt{\frac{1}{n_1}+\frac{1}{n_2}}\right]$$

$$\sqrt{0.4\times0.6}\sqrt{\frac{1}{50}+\frac{1}{50}}=0.098$$

The z corresponding to observed difference in proportion is

$$z=\frac{\left(\frac{22}{50}-\frac{18}{50}\right)-0}{0.098}=0.816$$

Since this is less than the critical z for even 5% level of significance, there does not exist sufficient evidence to reject the null hypythesis, or we are justified in making the said claim.

Illustration 19.15. A firm found with the help of a sample survey of a city, (size of sample 900) that 3/4 th of the population consumes things produced by them. The firm then advertised the goods in paper and on radio. After one year, sample of size 1000 reveals that proportions of consumers of the goods produced by the firm is now $\frac{4}{5}$th. Is this rise significant indicating that the advertisement was effective?

Solution. Let us assume that advertisement was not effective *i.e.*, two proportions differ only due to sampling and there is no significant difference between them.

Proportions of persons preferring before advertisement $(p_1)=\frac{3}{4}$

Proportions of persons preferring after advertisement $(p_2)=\frac{4}{5}$.

Here the population π is not known. To estimate that we pool the two samples:

$$\pi=\frac{n_1p_1+n_2p_2}{n_1+n_2}=\frac{\frac{3}{4}\times 900+\frac{4}{5}\times 1000}{900+1000}$$
$$=0.776$$

and the standard error of difference in proportion is

$$\text{S.E.}=\sqrt{\pi(1-\pi)\left(\frac{1}{n_1}+\frac{1}{n_2}\right)}=\sqrt{0.776\times 0.224\times\left(\frac{1}{900}+\frac{1}{1000}\right)}$$
$$=0.019$$

The corresponding z is then

$$z=\frac{(0.80-0.75)}{0.019}=2.61$$

This is slightly less than the critical z for 1% level of significance so the difference is significant at 5% level, but not at 1% level.

Illustration 19.16. Random samples of 200 bolts manufactured by machine A and 100 bolts manufactured by machine B showed 19 and 5 defctive bolts respectively. Is there a significant difference between the performance of the two machines?

Solution. Proportion of defective bolts in $A=\frac{19}{200}=.095\ (p_1)$

" " " " $B=\frac{5}{100}=.050\ (p_2)$

Estimate of population proportion

$$\pi=\frac{n_1p_1+n_2p_2}{n_1+n_2}.$$

$$=\frac{19+5}{200+100}=\frac{24}{300}=.08$$

$$\text{S.E.}=\sqrt{\pi(1-\pi)\left(\frac{1}{n_1}+\frac{1}{n_2}\right)}=\sqrt{.08\times.92\times\left(\frac{1}{100}+\frac{1}{200}\right)}$$

$$=0.033$$

So $z=\dfrac{(0.095-0.050)-0}{0.033}=1.36$

which is less than the critical z for 5% level of significance. Hence the null hypothesis cannot be rejected.

Illustrstion 19.17. The subject under investigation is the measure of dependence of Tamil on words of Sanskrit origin. One newspaper article reporting the proceedings of Constituent Assembly contained 2025 words of which 729 words were declared by a literary critic to be of Sanskrit origin. A second article by the same author describing atomic research contained 1600 words of which 640 words were declared by the same critic to be of Sanskrit origin. Assuming that simple sampling conditions held, estimate the limits for the proportion of Sanskrit terms in the writer's vocabulary, and examine whether there is any significant difference in the independence of this writer in words of Sanskrit origin in writing on these two subjects.

Solution. Here the same writer is writing two articles. So for proportion of Sanskrit words for calculating S.E. pooled estimate should be used.

Total number of words in the two articles

$$=2025+1600=3625$$

Total no. of words of Sanskrit origin

$$=729+640=1369$$

∴ Proportion of Sanskrit terms in writer's vocabulary

$$=\frac{1369}{3625}=.378$$

Proportion of Sanskrit terms in first article

$$=\frac{729}{2025}=0.36$$

Proportion of Sanskrit terms in the second article

$$=\frac{640}{1600}=0.40$$

∴ Difference between proportion in the two articles

$$=0.40-0.36=0.04.$$

Standard error of difference of propotions

$$=\sqrt{\pi(1-\pi)}\sqrt{\frac{1}{n_1}+\frac{1}{n_2}}$$

where π is the pooled proportion, here 0.378.

∴ S.E. of difference in the present case

$$=\sqrt{.378\times.622}\times\sqrt{\tfrac{1}{2025}+\tfrac{1}{1600}}$$

$$=\sqrt{0.2352}\times\sqrt{.00049+.00063}=0.016$$

∴ observed difference between the two proportions is 0.04/0.016 *i.e.*, 2.5 times the standard error. This is more than 1.96×S.E. Therefore, the difference is significant at 5% level of significance.

Hence there is significant difference in the vocabulary of the writer in words of Sanskrit origin in writing on the two subjects.

Since the two vocabularies differ, it is pointless to determine the limits for the wirters vocabulary.

Illustration 19.18.. A sample study of the population of two districts gives that in district *A* the percentage of male population is 52% while this is the figure for the female percentage in the *B* district. If the size of the sample selected was in district *A* and *B* as 400 and 625 respectively, can both of these samples be taken to come from a population where the percentage of males and females is equal?

Solution. Percentage of males in district $A=52$

Percentage of females in district $A=100-52=48$.

Let both the samples come from a population with male percentage $(\pi)=50$.

So S.E. of difference between percentages

$$=\sqrt{\pi(1-\pi)\left(\frac{1}{n_1}+\frac{1}{n_2}\right)}\times 100$$

$$=\sqrt{.5\times.5\left(\frac{1}{400}+\frac{.1}{625}\right)}\times 100=3.2$$

The corresponding $Z=\dfrac{(52-48)-0}{3.2}=1.25$

As this is less than the critical value, the difference is not significant.

Illustration 19.19. A person throws 10 dice 500 times and obtains 4, 5 or 6 a total of 2560 times. Can this be attributed to sampling?

Solution. Treating the throw of each die as an independent event there are a total of $500\times10=5000$ items in the sample. For a fair die, the probability of getting a 4, 5 or 6 is 0.5 so that the S.E. of proportion is

$$\sqrt{\pi(1-\pi)/n}=\sqrt{\frac{0.5\times0.5}{5000}}=0.007$$

The observed proportion p is $2560/5000=0.512$.

The corresponding $Z=\frac{0.512-0.500}{0.007}=1.71$

For a 2-tail test the critical Z at 5% level of significance is ± 1.96, therefore, the observed difference can be attributed to sampling.

Illustration 19.20. While throwing 5 dice 30 times, a person obtained 6 a total of 23 times. Can we consider the difference between observed and expected results as being significantly different?

Solution. Proceeding exactly as in Illustration 19.19

$n=5\times 30=150$. Expected $\pi=\frac{1}{6}$ so

S.E. of proportion $=\sqrt{\frac{1}{6}\times\frac{5}{6}/150}=0.03$

Observed proportion $=23/150=0.1533$

So corresponding $Z=\frac{0.1533-0.1666}{0.03}=0.444$

Which is too low to be considered significant.

Illustration 19.21. In a sample of 400 oranges from a large consignment 40 fruits were considered defective. Estimate the % of defective oranges in the whole consignment and assign limits within which the % will probably lie?

Solution. Percentage of defective oranges $=\frac{40}{400}\times 100=10\%$

Standard error of proportion of defectives $\sqrt{\pi(1-\pi)/n}$

$n=100$, $\pi=$ probability of defective item $=\frac{1}{10}$,

$\therefore$ S.E. $=\sqrt{\frac{1}{10}\times\frac{9}{10}\times\frac{1}{400}}=\sqrt{\frac{9}{40000}}=\frac{3}{200}=.015$

S.E. of the percentage of defective oranges

$=.015\times 100=1.5.$

The limits within which percentage of oranges lies are (99.7% confidence interval)

$$10\pm 3\times 1.5$$

i.e.,

$$10\pm 4.5$$

$$5.5 \text{ and } 14.5.$$

Illustration 19.22. In a certain district A 450 persons are considered as regular consumer of tea out of a sample of 1000 persons. In another district B, 400 were regular consumers of tea out of a sample of 800 persons.

Do these facts reveal a significant difference between the two districts (use 5% level)?

Solution. Let us assume that there is no real difference between the two districts.

$$\text{Proportion of regular consumers of tea in district } A = \frac{450}{1000} = .45$$

$$\text{Proportion of regular Consumers of tea in district } B = \frac{400}{800} = .5$$

$$\pi = \text{Pooled proportion} = \frac{450+400}{1000+800} = .47$$

Standard error of the difference of proportion

$$= \sqrt{.47 \times .53\left(\frac{1}{1000}+\frac{1}{800}\right)} = \sqrt{\frac{47 \times 53 \times 9}{100 \times 100 \times 4000}} = .0237$$

Difference between proportion

$$= .5 - .45 = .05$$

$$\therefore Z = \frac{.05}{.0237} = 2.1 \text{ approximately.}$$

Since $|Z| > 1.96$. the difference is significant. Therefore, the hypothesis stands rejected at 5% level of significance. Hence, the difference between the two districts is significant.

Illustration 19.23. In a random sample of 600 persons from a certaian large city. 450 are found to be smokers. In one of 900 from another large city. 450 are smokers. Do the data indicate that the cities are significantly dlfferent with respect to the prevalence of smoking.

Solution. Let us assume that there is no real difference between the two cities.

$$\text{Proportion of smokers in first city} = \frac{450}{600} = .75$$

$$\text{Proportion of smokers in second city} = \frac{450}{900} = .50$$

$$\pi = \text{Pooled proportion} = \frac{450+450}{900+600} = \frac{900}{1500} = \frac{3}{5} = .6$$

$$\text{S.E.} = \sqrt{.6 \times .4\left(\frac{1}{900}+\frac{1}{600}\right)} = .026$$

Difference between proportions

$$.75 - .50 = .25$$

$$Z = \frac{.25}{.026} = 9.6$$

Since $|Z| > 1.96$, the difference is signicant. Therefore, the hypothesis stands rejected at even 1% level of significance.

Illustration 19.24. In a random sample of the 500 persons from Maharashtra, 200 are found to be consumers of vegetable oil. In another sample of 400 persons from Gujrat, 200 are found to be consumers of vegetable oil. Discuss whether the data reveal a significant difference between Maharashtra and Gujarat so far as the proportion of vegetable oil consumers is concerned.

Solution. Let us assume that there is no real difference between Maharashtra and Gujarat so far as the proportion of vegetable oil consumers is concerned.

Proportion of consumers of vegetable oil in Maharashtra

$$\frac{200}{500} = .4$$

Proportion of consumers of vegetable oil in Gujarat

$$= \frac{200}{400} = .5$$

Pooled $\pi = \dfrac{200+200}{500+400} = \dfrac{400}{900} = \dfrac{4}{9}$

$$\text{S.E.} = \sqrt{\frac{4}{9} \times \frac{5}{9}\left(\frac{1}{500} + \frac{1}{400}\right)} = \sqrt{\frac{1}{900}} = \frac{1}{30} = .033$$

Difference between proportion $= .5 - .4 = .1$ which is about 3 times the S.E.

Since 3 is greater than 2.58, the difference is significant (at 1% level of significance).

9.6. SOME OTHER TESTS ON LARGE SAMPLES

Many other statistics related to large samples can be tested in the manner outlined above since the sampling distribution of these statistics can be approximated as 'normal'. Table 19.2 on p. 634 gives the relevant formulae for the standard errors of the various statistics of large samples.

Illustrion 19.25. 1,800 persons of a certain age group were observed to have a standared deviation of 9.2 beats per minute. Assign the limits for the standard deviation of the population, assuming that the above sample of 1,800 persons came from a normally distributed universe.

TABLE 19.2. LARGE SAMPLE TESTS

Statistics	*Standard Error*
Mean, $\bar{x}$	$\sigma/\sqrt{n}$ or $S/\sqrt{n}$ where $S=\sqrt{\frac{\Sigma(x-\bar{x})^2}{n-1}}$
Median, Me	$1.25331\sigma/\sqrt{n}$
Quartiles	$.1.3626\sigma/\sqrt{n}$
Proportion, p of success	$\sqrt{\pi(1-\pi)/n}$
Number of success	$\sqrt{n\pi(1-\pi)}$
Standard deviation	$\sigma/\sqrt{2n}$
Mean deviation	$0.6028\ \sigma/\sqrt{n}$
Variance	$\sigma^2\ \sqrt{\frac{2}{n}}$
Quartiale diviation	$7.8672\sigma/\sqrt{n}$
Coefficient of skewness	$\sqrt{\frac{3}{2n}}$
Coeffiienct of variation	$\frac{V}{\sqrt{2n}}\sqrt{1+\frac{2V^2}{10^4}}$ where V is coefficient of variation
Coefficient of Corre-lation	$\frac{1-r^2}{\sqrt{n}}$
Coefficient of regession	$\frac{\sigma_x\sqrt{1+r^2}}{\sigma_y\sqrt{n}}$
Difference of sample means	$\sqrt{\frac{S_1^2}{n_1}+\frac{S_2^2}{n_2}}$ or $\sigma\sqrt{\frac{1}{n_1}+\frac{1}{n_2}}$
Difference of sample standard deviations	$\sigma\sqrt{\frac{1}{2}\left(\frac{1}{n_1}+\frac{1}{n_2}\right)}$ or $\sqrt{\frac{s_1^2}{2n_1}+\frac{s_2^2}{2n_2}}$
Difference of sample proportions	$\sqrt{\pi(1-\pi)\left(\frac{1}{n_1}+\frac{1}{n_2}\right)}$ or $\sqrt{\frac{p_1(1-p_1)}{n_1}+\frac{p_2(1-p_2)}{n_2}}$

Solution. Standard Error of Standard Deviation

$$SE = \frac{\sigma}{\sqrt{2N}}$$

Substituting the values, we get

$$SE = \frac{9.2}{\sqrt{2 \times 1,800}} = \frac{9.2}{60} = 0.153$$

As thrice the standard error covers almost the total number of cases (to be exact, 99.73% cases), so the population standard deviation should not differ by more than ± 3 S.E. or $9.2 \pm 3(.153)$

$$9.2 - 3\ (.153) = 9.2 - .459 = 8.741$$

$$9.2 + 3\ (.153) = 9.2 + .459 = 9.659$$

Hence the limits of the population standard deviation are 8.741—9.659.

Illustration 19.26. Compute the standard errors of Median, Quartiles, Mean Deviation, Variance and Quartile Deviation if standard deviation is 10 and number of cases included in a sample is 2,500, presuming that the sample has been drawn from a normal population.

Solution. S.E. of Median $= 1.25331 \times \frac{\sigma}{\sqrt{n}}$

$$= 1.25331 \times \frac{10}{\sqrt{2,500}}$$

$$= 1.25331 \times \tfrac{1}{5} = .25066$$

S.E. of Quartiles (both 1st and 3rd):

$$= 1.36263 \frac{\sigma}{\sqrt{n}}$$

$$= 1.36263 \times \tfrac{1}{5} = 0.27253.$$

S.E. of Mean Diviation:

$$= .6028 \times \frac{\sigma}{\sqrt{n}}$$

$$= .6028 \times \frac{10}{\sqrt{2,500}}$$

$$= .12056.$$

S.E. of variance $= \sigma^2 \sqrt{\frac{2}{n}}$

$$= (10)^2 \sqrt{\frac{2}{2,500}}$$

$$= 2.8.$$

S.E. of Quartile Deviation:

$$=7.8672\frac{\sigma}{\sqrt{n}}$$

$$=.78672\times\frac{10}{\sqrt{2{,}500}}$$

$$=.15734.$$

Illustration 19.27. A sample study of 2,500 couples gives a correlation coefficient of 0.45. Estimate the limits to the correlation in the universe.

Solution. The S.E. of the Correlation Coefficient is $\frac{1-r^2}{\sqrt{n}}$

where r is the correlation cofficient.

Substituting the values, we get.

$$=\frac{1-(0.45)^2}{\sqrt{2{,}500}}$$

$$=\frac{1-0.2025}{50}=.016$$

In all probability, the population coefficient of correlation should not differ by more than thrice the S.E. from sample correlation coefficient, as $r\pm3$ S.E. would cover 99.73% of the total population. So the limits to the coefficient correlation are

$$0.45-3(.016)=0.45-.048=0.402$$
$$0.45+3(.016)=0.45+0.048=0.498$$

Then, we can confidently expect that the population correlation coefficient should be within the limits of 0.402—0.498.

Illustraiion 19.28. A correlation coefficient of 0.2 is obtained from a random sample of 1,600 pairs of observations. Do you think this value of correlation coefficient is signinificant?

Solution. To conclude whether the value of $r=.2$ is significant, *i.e.*, whether the observed pairs are really correlated, it is necessary to find out the value of r which may arise on account of chance when 1,600 pairs are observed, presuming that the observed pairs are uncorrelated.

On the hypothesis that the pairs are uncorrelated, viz., $r=0$.

$$\text{S.E.}=\frac{1-r^2}{\sqrt{n}}=\frac{1-0}{\sqrt{1{,}600}}=\frac{1}{40}$$

$$=0.025.$$

We know that 3(S.E.) covers 99.73% cases, thererfore, the upper limit of r will be 3 (.025)=.075 on account of sampling fluctuations.

But the value of the observed r is .2 which is many times this value, so we can safely conclude that the value of r viz., .2 is significant, *i.e.*, the observed pairs are really correlated.

Illustration 19.29. Standard deviations In two samples of size 100 each were found to be 13.9 and 17.6. If population standard deviation is 15, is there any significant difference between two standard deviations?

Solution. S.E. of difference of two standard deviations when population σ is given $=\sigma\sqrt{\frac{1}{n_1}+\frac{1}{n_2}}$ (σ is the population standard deviation)

$\therefore$ S.E. of difference $=15\sqrt{\frac{1}{100}+\frac{1}{100}}=15\sqrt{0.02}=2.12.$

Difference between sample σ's $=17.6-13.9=3.7.$

$\therefore$ The observed difference less than twice the S.E.

Therefore the difference is not significant at even 5% level.

Illustration 19.30. Hardness was measured on 140 pieces of wood of which 100 pieces were stored inside a warehouse and 40 outside. The means and the sums of squares of deviations about mean of the measurements on hardness are given below:

	Inside	Outside
Sample size	100	40
Mean	132	117
Sums of squares	27244	8655

Test whether the variability of hardness is effected by weathering.

Solution. The coefficient of variations is the best measure of varibility

C.V. for 'Inside' $(V_1)=100\times\sqrt{\frac{27244}{100}}\Big/132=\frac{1654}{132}=12.5\%$

C.V. for 'Outside' $(V_2)=100\times\sqrt{\frac{8655}{40}}\Big/117=\frac{1471}{117}=12.6\%$

St. Error of C.V. $=\frac{V}{\sqrt{2N}}\sqrt{+\frac{2V^2}{10000}}$

$\therefore$ St. Error of the difference of C.V. $=\sqrt{(\text{S.E. I})^2+(\text{S.E. II})^2}$

Square of St. Errror of I C.V. $=\frac{(12.5)^2}{2\times 100}\left(1+\frac{2(12.5)^2}{10000}\right)=0.8$ app.

" " " " II C.V. $=\frac{(12.6)^2}{2\times 40}\left(1+\frac{2(12.6)^2}{10000}\right)=2.0$ app.

$\therefore$ St. Error of difference of C.V. $=\sqrt{.8+2.0}=1.7$ approx.

Difference of C.Vs. $=12.6-12.5=0.1$.

As the observed difference is less then S.E., the difference in two coefficient of variations is insignificant.

Hence there is no evidence against the hypothesis of equality of variability.

Illustration 19.31. For the data given in Illustration 19.6 examine: Is the difference between standard deviations significant? (*a*) 1% level of significance (*b*) 2% level of significance.

Solution. Let us take the hypothesis that there is no difference between the S.D. of the two samples.

S.E. of the difference between two S.D.

$$=\sqrt{\frac{\sigma_1^2}{2n_1}+\frac{\sigma_2^2}{2n_2}} \text{ or } \sqrt{\frac{\sigma^2}{2}\left(\frac{1}{n_1}+\frac{1}{n_2}\right)}$$

$\sigma=11,\ n_1=100,\ n_2=150.$

$$\text{S.E.}=\sqrt{\frac{11^2}{2}\left(\frac{1}{100}+\frac{1}{150}\right)}=\sqrt{\frac{121\times 5}{2\times 300}}=1.004$$

Difference of standard deviations

$=12-10=2.$

$$|Z|=\frac{2}{1.004}=1.99.$$

5% level

Since $|Z|>1.96$, the difference is significant. Therefore, the hypothesis stands rejected at 5% level of significance. Hence the difference between to standard deviations is significant at this level.

1% level.

Since $|Z|<2.58$, the difference is insignificant. Therefore the hypothesis may be accepted at 1% level of significance Hence the difference between two standard deviations is not significant at this level.

19.7. MISCELLANEOUS ILLUSTRATIONS

Illustration 19.32.. Calculate standard error of mean from the following data showing amount paid by 100 firms in Calcutta on the occasion of Durga Puja:

Midvalue (Rs)	39	49	59	69	79	89	99
No. of Firms	2	3	11	20	32	25	7

Solution.

CALCULATION OF STANDARD DEVIATIONS

Mid-value x	f	$x' = (x-69)/10$	fx'	fx'^2
39	2	−3	−6	18
49	3	−2	−6	12
59	11	−1	−11	11
69	20	0	0	0
79	32	+1	+32	32
89	25	+2	+50	100
99	7	+3	+21	63
Totals	100		80	236

Standard deviation

$$\sigma = i \times \sqrt{\left(\frac{\Sigma fx'^2}{n}\right) - \left(\frac{\Sigma fx'}{n}\right)^2}$$

$$= 10 \times \sqrt{\frac{236}{100} - \left(\frac{80}{100}\right)^2}$$

$$= 13.11$$

∴ Standard errror

$$\text{S.E.} = \frac{\sigma}{\sqrt{n}} = \frac{13.11}{\sqrt{100}} = 1.311$$

Illustration 19.33. An examination was given to 50 students at College *A* and to 60 students at Collage *B*. At *A*, the mean grade was 75 with S.D.=9. At *B*, the mean was 79 with SD=7. Is there a significant difference in the performance at the two Colleges at 5% level of significance?

Solution. Let the null hypothesis be that there is no difference in means. Since the samples are large we can use normal curve.

Standard error of difference in means is

$$\text{S.E.} = \sqrt{\frac{\sigma_1^2}{n_1} + \frac{\sigma_2^2}{n_2}} = \sqrt{\frac{81}{50} + \frac{49}{60}}$$

$$= 1.56$$

The corresponding normal ordinate is

$$Z = \frac{(79-75)-0}{1.56} = 2.56$$

As this is larger than the critical value at 5% level of significance, the difference in performance is significant at that level.

Illustration 19.34. A person buys 100 electric tubes each of two well-known makes, taken at random from the stocks for testing purpose. He finds that make *A* has a mean life of 1300 hrs with SD=82 hrs, and make *B* has a mean life of 1248 hrs with SD=93 hrs. Discuss the significance of these results.

Solution. Let us take the null hypothesis that there is no difference in means.

Standard error of the difference in means is

$$\text{S.E.}=\sqrt{\frac{\sigma_1^2}{n_1}+\frac{\sigma_2^2}{n_2}}=\sqrt{\frac{82^2}{100}+\frac{93^2}{100}}$$

$$=12.4$$

The corresponding value of *Z* is

$$Z=\frac{(1300-1248)-0}{12.4}=\frac{52}{12.4}=4.18$$

As this is larger than 2.57, the critical value at 1% level, the life of make *B* is significantly less than that of make *A*.

Illustration 19.35. It was found that the correlation coefficient between two variables calculated from a sample of size 25 was 0.37. Does this show a significant correlation?

Solution. Take null hypothesis H_0 as $n=0$,

$n=25$ can be treated as large sample (though, barely so).

Standard error of correlation based on a null $r=0.0$ is

$$\text{SE}=\frac{1-r^2}{\sqrt{n}}=\frac{1}{\sqrt{25}}=0.20$$

The corresponding normal variate is

$$Z=\frac{0.37-0}{0.20}=1.85$$

For a two tail test, the critcal value of Z=1.96 at 5% level. Thus the correlation is insignificant at 5% level.

PROBLEMS

1. A sample of 144 bricks has a mean weight 7.1 pounds and a standard deviation of 0.30 pounds. Is it likely that this sample comes from a brick yard that produces bricks with a mean weight 7 pounds?
2. The credit manager for an oil company claims that the average balance on statements mailed to credit card holders is at least Rs 32. To check

his claims an auditor takes a sample of 64 statements and finds that the average amount owed is Rs 30 with a standard deviation of Rs 12. On the basis of sample evidence what can you say about the credit manager's claim.

3. The following table gives the distribution of familes according to the number of persons in each.

No. of persons	1	2	3	4	5	6	7	8	9
No. of families	107	245	228	174	106	65	35	21	19

Calculate the mean and standard deviation of the above distribution. Does the average differ significantly from 5.
(Ans. a.m. = 3.481 persons.
Standard deviation = 1.84 persons, and
Normal variate = 26.1)

4. A sample cf 50 pieces of a certain type of string was tested. The mean breaking strength turned out to be 14.5 lbs. Test whether the sample is from a batch of string having a mean breaking strength of 15.6 lbs and standard deviation 2.2 lbs.

5. Medial income of 150 labourers selected at random from a certain province is Rs 105.75 p.m. with standard deviation of Rs 25.47. Do you have reasons to believe that average income of the labour community in this district is not Rs 100?

6. A sample of size 400 is selccted from a population of very large size. Quartiles of the sample are found to be 17 and 29 and standard deviation 10. Can the sample come from a population with quartiles 15 and 30?

7 In a sample of 5000 births in a hospital, 2550 were male and the rest female. Test the hypothesis that the male proportion is 0.5 in the population.

8. In a large city 327 men out of 600 men were found to be smokers. Does this information support the view that the majority of men in this city are smokers?

9. In 315672 throws of a die 5 or 6 turned up 106602 times. Is this sufficient evidence that die is biased?

10. It is claimed that 60% of the unmarried women employed as lower division clerks in Bombay Secretariat get married and quit the job within two years after they are employed. Tcst this hypothesis at 5% level of significance if it is observed that out of 400 unmarried women employed as lower division clerks in Bombay Secretariat 256 got married and quit the job within two years after they were employed.

11. In a sample of size 250, the standard deviation is found to be 23.8. The coefficient of variation is 78%. Can the population standard deviation be 15 and C.V. 75%?

12. In a sample of size 100, coefficient of correlation was found to be 0.1. Is it a significant figure?

13. For a sample of size 400 sample correlation was found to be 0.62. Can this sample come from a population with correlation 0.4? If not, what is the range within which the population parameter will lie?

14. It was found that the correlation coefficient between two variables calculated from a sample of size 25 was 0.37. Does this show evidence of having come from a population with zero correlation?

15. The sample of size 186 gives the value of γ_1 as 0.56, of γ_2 as 0.29. Can the population values be 0.25 and 0.75 respectively?
16. A sample of size 2400 gives γ_1 and β_2 as 0.37 and 3.48. Can this sample come from normal population?
17. 900 samples are taken from a large consignment and 100 are found to be bad. Assign the the limits within which the percentage of bad apples in the consignment lies.

    The 99% confidence limits will be 11.1 ± 3 (1.03) or 14.2% to 8.0%).
18. Out of a consignment of 50,000 articles, 200 were selected at random and examined. Of the latter 20 were found to be defective. How many sound articles can you reasonably expect to have in the whole consignment?

    (No. of sound articles reasonably expected in the whole consignment =42880 to 47120).
19. Following statistical coefficients have been reduced in the course of an examination of the relationship between *X*, the yield of wheat per acre and *Y*, the rainfall in inches.

	X	*Y*
Mean	900 lbs.	12″
St. deviation	80 lbs.	2″

The coefficient of correlation between yield and rainfall =0.5.

From the above data calculate the range in which

(*a*) yield is expected to lie when rainfall is 9″,

(*b*) rainfall is expected to lie when yield is 100 lbs.

(Ans. (*a*) Expected range of *X* = 701.4 lbs. to 978.6 lbs.

(*b*) ,, ,, *Y*=9.79″ to 16.71″.)

20. In a sample study following results are obtained:

    Size = 330, A.M. = 50.6, σ = 16.8.

    Does A.M. difference from 48.0 significantly?

    (Ans. Yes at 5% level)
21. In a sample of pea plant the number of round peas is 336 and of angular 101. Is it in agreement of Mendelian theory which gives the ratio 3 : 1.

    (Ans. Yes)
22. A sample of 600 gives the percentage of male births as 53. Is it sufficient evidence that chance of male and female birth are not same?

    (Ans. No)
23. A random sample of 500 pineapples was taken from a large consignment and 65 found bad Estimate the proportion of bad fruits in the lot.

    (Ans 99% limits are 8.5% to 17.5%)
24. Sample I: $n=125$, $\bar{x}=18.65$, $\sigma=4.72$

    ,, II: $n=100$, $\bar{x}=14.72$, $\sigma=3.45$.

    Have you reason to believe that two $\bar{X}$'s are different.

    (Ans. Yes)
25. A random sample of 200 gives mean population 485 with standard deviation 50. Another sample of same size gives average 510 and standard deviation 40. Is the difference significant.

    (Ans. No)

26. In a consumers preference survey 58% gives preference in city *A* out of 600 and 52% in city *B* out of 500. What is your conclusion about the difference in preferences in two cities?
(**Ans.** No difference)
27. Two investigators investigate the possibility of thought reading. One threw a die unseen to other and thought of number turned up for 30 seconds. The other person noted down. On checking, figures were in agreement 118 times out of 600. Does this result reveal thought reading?
(**Ans.** Yes at 5% level but not at 1% level)

Chapter 20

TESTS ON SMALL SAMPLES AND GOODNESS OF FIT

20.1. SMALL SAMPLES

The methods of statistical analysis for testing the significance of sample statistics discussed so far were based on two assumptions viz.,

(*i*) Sample standard deviation is close to population standard deviation and as such can be used in its place for the computation of standard error. Thus, in the compution of the standard error of the mean, the standard deviation of the sample is used in the absence of the standard deviation of the population.

(*ii*) The distribution of sample statistics is normal. Because of this it is possible to assign limits within which the difference between sample statistics and population parameters is likely to lie.

These assumptions do not hold good when the size of the sample is small (say, less then 30). In fact, for small values of *n* (number of items included in the sample) the standard deviation of the sample is subject to a definite bias, tending to make it consistently lower than the standard deviation of the population. Thus, if the standard deviation of a small sample is used in the computation of the standard error of the mean, the result will also have a downward bias. It can, therefore, be said that the methods, discussed so far, when applied with small samples, the sampling errors to which our estimates are subject, are consistently under-estimated. This under-estimation of the sampling error takes away a part of its utility for purposes of statistical inference.

It is for this reason, that tests for small samples are not based on normal curve, but on other theoretically obtained sampling distributions. We give in this chapter a few of the more commonly used theoritical distributions and their use.

20.2. THE t-DISTRIBUTION

When we take samples of size n from a normal population, the variable t defined as

$$t=\frac{\bar{x}-\mu}{S/\sqrt{n}}$$

where S is the sample standard deviation $\sqrt{\Sigma(x-\bar{x})^2/(n-1)}$ has an interesting distribution. This distribution is known as *Student -t distribution* (named after W S. Gosset, its discoverer who wrote under the name Student). The t-distribution is not a single distribution, but a family of symmetrical distributions distinguished by various values of the parameter v. This parameter is recognised as the 'degrees of freedom' and is equal to the number of observations that can be freely chosen under some overall constraints. Suppose we have ten (10) values of x which average to $\bar{x}$. Under the constraint that $\bar{x}$ is the same, the individual values of x may vary, but only 9 of them may do so independently. If we chose the values of 9 arbitrarily, the tenth is automatically fixed because of the reqiurement that $\bar{x}$ is fixed. Thus, with a sample of size n, the variable 't' above will have a t-distribution with $v=(n-1)$ degrees of freedom.

The values of the variate t for different values of v and different 'tail' areas are tabulated in the appendix. Various tests concerning means and their differences based on small samples can then be comstructed.

For testing of mean under the null hypothesis that $\mu=\mu_0$, the variable t with $(n-1)$ degrees of freedom is given by

$$t=\frac{\bar{x}-\mu}{S/\sqrt{n}}$$

where S is sample standard deviation $\sqrt{\Sigma(x-\bar{x})^2/(n-1)}$. This value has to be larger than the critical value given in the table for a given level of significance and $v=n-1$, the degrees of freedom.

In testing for differences in means, we use the null hypothesis that the means are not different. The t-variate is given by

$$t=\frac{(\bar{x}_1-\bar{x}_2)}{S\sqrt{\frac{1}{n_1}+\frac{1}{n_2}}}$$

with (n_1+n_2-2) degress of freedom.

Here the value of S is obtained as

$$S=\sqrt{\frac{\Sigma(x_1-\bar{x}_1)^2+\Sigma(x_2-\bar{x}_2)^2}{n_1+n_2-2}}$$

The following illustrate the use of student-t distribution.

Illustration 20.1. Nine individuals are chosen at random from a population and their weights are found to be, in pounds, 110, 115, 118, 120, 122, 125, 128, 130, 139. In the light of these data, discuss the suggestion that the mean weight of population is 120 lbs.

Solution. Calculation of the average weight and its standard deviation:

No. of individuals	*Weights (in lbs)* (X)	*Deviations from the average (123)* (x)	x^2
1	110	−13	169
2	115	− 8	64
3	118	− 5	25
4	120	− 3	9
5	122	− 1	1
6	125	2	4
7	128	5	25
8	130	7	49
9	139	16	256
$n=9$	$\Sigma X=1,107$		$\Sigma x^2=602$

Average weight of the sample, or

$$\bar{X}=\frac{\Sigma X}{n}=\frac{1,107}{9}=123$$

Standard deviation of the sample,

$$S=\sqrt{\frac{\Sigma x^2}{n-1}}=\sqrt{\frac{602}{8}}=8.7$$

Substituting these values in the formula:

$$t=\frac{(\bar{X}-\mu)\sqrt{n}}{S}$$

$$=\frac{(123-120)\sqrt{9}}{8.7}=1.03$$

No. of degrees of freedom $=9-1=8$.

For 8 degrees of freedom at 5% level of significance, the critical value of $t=2.31$, the calculated value of t is less than this and hence we can conclude that the sufficient reason is not there to doubt that the mean weight in the universe is 120 lbs.

Illustration 20.2. Ten individuals are chosen at random from a population and their heights are found to be in inches, 63, 63, 66, 67, 68, 69, 70, 70, 71 and 71. In the light of the data discuss the suggestion that the mean height in the population is 66 inches.

Solution. The hypothesis to be tested is that the mean height in the population is 66 inches.

Calculation of the average height and its standard deviation:

No. of individuals	*Height (in inches) (X)*	*Deviation from the mean (67.8) (x)*	x^2
1	63	−4.8	23.04
2	63	−4.8	23 04
3	66	−1.8	3.24
4	67	−0.8	.64
5	68	0.2	.04
6	69	1.2	1.44
7	70	2.2	4 84
8	70	2.2	4.84
9	71	3.2	10.24
10	71	3.2	10.24
$n=10$	$\Sigma x=678$		$\Sigma x^2=81.6$

$$\bar{X}=\frac{\Sigma X}{n}=\frac{678}{10}=67.8$$

$$S=\sqrt{\frac{\Sigma x^2}{n-1}}=\sqrt{\frac{81.6}{9}}=\sqrt{9.07}=3.01.$$

$$t=\frac{(\bar{X}-\mu)\sqrt{n}}{S}=\frac{(67.8-66)\sqrt{10}}{3.01}=1.9.$$

No. of degrees of freedom$=10-1=9$.

For 9 degrees of freedom at 5% level of significance, the critical value of $t=2.262$. The calculated value of t is less than this value, hence our hypothesis holds good., *i.e.*, the mean height in the population can be 66 inches.

Illustration 20.3. The mean weekly sale of the Yum Yum chocolate bar in a chain of candy stores was 146.3 bars per store. After an advertising campaign the mean weekly sales in 22 stores for a typical week increased to 153.7 and showed a standard deviation of 17.2. Is the evidence conclusive that the advertising was successful? You are given that for 21 degrees of freedom, the value of t is 2.08 at 5% level of significance.

Solution. Let us take the hypothesis that advertising is not successful. Here in usual notations, we are given

$$\bar{X}=153.7,\ \mu=146.3,\ n=22 \text{ and } \sigma=17.2$$

$$t=\frac{\bar{X}-\mu}{\sigma}\sqrt{n-1}$$

$$=\frac{153.7-146.3}{17.2}\sqrt{21}$$

$$=1.92.$$

Number of degrees of freedom$=22-1=21$.

Since the calculated value of t is less than the critical value at 5% level of significance *i.e.*, the difference is not significant at 5 per cent level. Hence we cannot say that the advertising is successful.

Illustration 20.4 Strength tests carried out on samples of two yarns spun to the same count gave the following results:

	Number in sample	Sample mean	Sample variance
Yarn *A*	4	50	42
Yarn *B*	9	42	56

The strengths are expressed in pounds. Does the difference in mean strengths indicate a real difference in the mean strengths of the sour

Solution. We have

$$t=\frac{\bar{X}-\bar{Y}}{S\sqrt{\frac{1}{N_1}+\frac{1}{N_2}}}$$

Here $\bar{X}=50,\ \bar{Y}=42$

$\therefore\ \frac{1}{(N_1-1)}\Sigma(X_i-\bar{X})^2=42$ *i.e.*, $\Sigma(X_i-\bar{X})^2=3\times 42=126$

and $\Sigma(Y_i-\bar{Y})^2=8\times 56=448$

Thus $S^2=\frac{3\times 42+8\times 56}{9+4-2}=\frac{574}{11}=52.2$

or $S=7.2$

$$\therefore\quad t=\frac{8}{7.2\sqrt{\frac{1}{4}+\frac{1}{9}}}=\frac{8}{4.32}=1.85$$

Degrees of freedom$=4+9-2=11$

Referring to tables of t-distribution with 11 degrees of freedom, the calculated value is not at all significant. Thus test provides no evidence against the hypothesis.

Illustration 20.5. Two laboratories *A* and *B* carry out independent estimates of fat-content in ice-cream made by a firm. A sample is taken from each batch, halved, and the separate halves sent to the two laboratories. The fat-content obtained by the laboratories are recorded below:

Batch No.	1	2	3	4	5	6	7	8	9	10
Lab. *A*	7	8	7	3	8	6	9	4	7	8
Lab. *B*	9	8	8	4	7	7	9	6	6	6

(The fat-contents are given in grammes)

Is there a significant difference between the mean fat-content obtained by the two laboratories *A* and *B*?

Solution.

Laboratory A			*Laboratory B*		
(X_1)	*Deviation from 8* $x=(X_1-8)$	x^2	(X_2)	*Deviation from 8* $x'=X_2-8$	x'^2
7	−1	1	9	1	1
8	0	0	8	0	0
7	−1	1	8	0	0
3	−5	25	4	−4	16
8	0	0	7	−1	1
6	−2	4	7	−1	1
9	1	1	9	1	1
4	−4	16	6	−2	4
7	−1	1	6	−2	4
8	0	0	6	−2	4
67	−13	49	70	−10	32

Mean fat content obtained by lab^n^ *A*

$$\bar{X}_1=\frac{\Sigma X_1}{N_1}=\frac{67}{10}=6.7$$

Mean fat-content obtained by lab. *B*.

$$\bar{X}_2=\frac{\Sigma X_2}{N_2}=\frac{70}{10}=7$$

$$S=\sqrt{\frac{\Sigma x^2-\frac{(\Sigma x)^2}{N_1}+\Sigma x'^2-\frac{(\Sigma x')^2}{N_2}}{N_1+N_2-2}}$$

$$=\sqrt{\frac{49-\frac{(-13)^2}{10}+32-\frac{(-10)^2}{10}}{10+10-2}}$$

$$=\sqrt{\frac{81-16.9-10}{18}}=\sqrt{\frac{54.1}{18}}=\sqrt{3.009}$$
$$=1.732$$

Let us take the null hypothesis that the estimates of mean fat-content in ice-cream obtained by the laboratories A and B are the same

Applying t-test

$$t=\frac{\bar{X}_1-\bar{X}_2}{S\sqrt{\frac{1}{N_1}+\frac{1}{N_2}}}=\frac{6.7-7}{1.732\sqrt{\frac{1}{10}+\frac{1}{10}}}$$

$$|t|=\frac{0.3}{1.732\times0.14}=1.2$$

Number of degrees of freedom $=10+10-2=18$.

For 18 degrees of freedom at 5% level of significance the tabulated value of $t=2.10$. The calculated value of t is less than the tabulated value and hence our hypothesis holds good. In other words, estimate of fat content obtained by laboratory A does not differ significantly from estimate of fat content obtained by laboratory B.

Paired t-test for difference of means. Let us now consider the case when (i) the sample sizes are equal i.e. $N_1 = N_2 = N$ say and (ii) the two samples are not independent but the same observations are paired together *i.e.*, the pair of observations (X_{1i}, X_{2i}). $(i=1, 2..., n)$ corresponds to the same (ith sample) unit. The problem is to test if the sample means differ significantly or not.

For Example if we want to study the effect of training imparted to salesmen, say, for increasing the sales of a particular product. Let X_{1i} and X_{2i} $(i=1, 2,...,n)$ be the amount of sales by the ith individual, before and after the traning is given respectively. Here instead of applying the difference of the means test, we apply the paired t-test given below.

Here we consider the changes $x_i=X_{1i}-X_{2i}$ $(i=1, 2,..., n)$. Under the null hypothesis changes in the sales are due to fluctuations of sampling *i.e.*, training is not responsible for these increases in sales, the statistic

$$t=\frac{\bar{x}}{S/\sqrt{n}}$$

where $$\bar{x}=\frac{1}{n}\Sigma x$$

and $$S^2=\frac{1}{n-1}\Sigma(x_i-\bar{x})^2$$

$$=\frac{1}{n-1}\left[\Sigma x^2-\frac{(\Sigma x)^2}{n}\right]$$

follows student's t-distribution with $(n-1)$ $d.f.$

Illustration 20.6. The following data show weekely production for ten employees before change (X_1) and after change (X_2) in the production technique.

Employee	Weekly production	
	Before change (X_1)	After change (X_2)
A	24	26
B	26	26
C	20	22
D	21	22
E	23	24
F	30	30
G	32	32
H	25	26
I	23	24
J	23	25

Test whether there is any significant change in average production due to the changes in the production technique.

Solution. Let us take the null hypothesis that there is no significant change in average production due to changes in the production technique.

Employee	Weekly production before change (X_1)	Weekly produceion after change (X_2)	Difference X	Deviation from mean (x)	x^2
A	24	26	2	1	1
B	26	26	0	−1	1
C	20	22	2	1	1
D	21	22	1	0	0
E	23	24	1	0	0
F	30	30	0	−1	1
G	32	32	0	−1	1
H	25	26	1	0	0
I	23	24	1	0	0
J	23	25	2	1	1
			10		6

$$\bar{X}=\frac{\Sigma X}{N}=\frac{10}{10}=1$$

$$S=\sqrt{\frac{\Sigma x^2}{N-1}}=\sqrt{\frac{6}{10-1}}=\sqrt{\frac{2}{3}}=0.816$$

Applying *t*-test

$$t=\frac{\bar{X}\sqrt{N}}{S}$$

$$=\frac{1\times\sqrt{10}}{0.816}=\frac{3.162}{0.816}=3.87$$

No. of degrees of freedom $=10-1=9$.

For 9 degrees of freedom at 5% level of significance, the tabulated value of $t=2.26$. The calculated value of t is greater than this tabulated value. Hence the null hypothesis stands rejected *i.e.*, there is change in average production due to the change in production technique.

Illustration 20.7. To verify whether a course in accounting improved performance, a similar test was given to 12 participants both before and after the course. The original grades—recorded in alphabetical order of the participants—were 44, 40, 61, 52, 32, 44, 70, 41, 67, 72, 53 and 72. After the course the grades were, in the same order, 53, 38, 69, 57, 46, 39, 73, 48, 73, 74, 60, and 78.

(*a*) Was the course useful, as measured by performance on the test? Consider these 12 participants as a sample from a population.

(*b*) Would the same conclusion be reached if tests were not considered paired?

Solution. (*a*)

Grades before the course	*Grades after the course*	*Difference X*
44	53	9
40	38	—2
61	69	8
52	57	5
32	46	14
44	39	—5
70	73	3
41	48	7
67	73	6
72	74	2
53	60	7
72	78	6

The null hypothesis is that these changes are drawn from a normal distribution with mean 0 *i.e.*, the course has no beneficial effect.

$$\bar{X}=\frac{\Sigma x}{N}=\frac{60}{12}=5.$$

$$S^2=\frac{\Sigma(X-\bar{X})^2}{N-1}$$

$$=\frac{(9-5)^2+(-2-5)^2+...+(7-5)^2+(6-5)^2}{11}$$

$$=\frac{278}{11}=25.3.$$

Thus $\quad S=\sqrt{25.3}.$

Applying *t*-test

$$t=\frac{(\bar{X}-\mu)\sqrt{N}}{S}=\frac{\bar{X}\sqrt{N}}{S}$$

$$=\frac{5\times\sqrt{12}}{\sqrt{25.3}}=\frac{5\times 3.46}{5.03}=3.29.$$

Number of degrees of freedom$=12-1=11$.

For 11 degrees of freedom at 5% level of significance, the tabulated value of $t=2.20$. The calculated value of t is greater than the tabulated value of t. Hence we reject the null hypothesis *i.e.*, the course has a beneficial effect.

(*b*)

Before course			*After course*		
x_1	$(X_1-\bar{X}_1)$	$(X_1-\bar{x}_1)^2$	X_2	$(X_2-\bar{x}_2)$	$(X_2-\bar{x}_2)^2$
44	−10	100	53	− 6	36
40	−14	196	38	−21	441
61	7	49	69	10	100
52	− 2	4	57	− 2	4
32	−22	484	46	−13	169
44	−10	100	39	−20	400
70	16	256	73	14	196
41	−13	169	48	−11	121
67	13	169	73	14	196
72	18	324	74	15	225
53	− 1	1	60	1	1
72	18	324	78	19	361
		2176			2250

$$\bar{X}_1=\frac{648}{12}=54$$

$$\bar{X}_2=\frac{\Sigma X_2}{N_2}=\frac{708}{12}=59.$$

$$S=\sqrt{\frac{\Sigma(X_1-\bar{X}_1)^2+\Sigma(X_2-\bar{X}_2)^2}{N_1+N_2-2}}$$

$$=\sqrt{\frac{2176+2250}{12+12-2}}=\sqrt{\frac{4426}{22}}=14.18$$

Let us take the hypothesis that the course has no beneficial effect. Applying the formula

$$t=\frac{\bar{X}_1-\bar{X}_2}{S\sqrt{\frac{1}{N_1}+\frac{1}{N_2}}}=\frac{\bar{X}_1-\bar{X}_2}{S}\times\sqrt{\frac{N_1N_2}{N_1+N_2}}$$

$$|t|=\frac{|54-59|}{14.18}\times\sqrt{\frac{12\times 12}{12+12}}=\frac{5}{14.18}\times 2.45<0.86$$

Number of degrees of feedom$=12+12-2=22$.

For 22 degrees of freedom at 5% level of significance the tabulated value of $t=2.07$. Thus the calculated value of t is less than the tabulated value and hence we accept the hypothesis *i.e.*, the course has no beneficial effect.

Hence using the wrong procedure leads to wrong result.

***t*-Test of the significance of an observed correlation coefficient**. When we want to test whether r is or is not significantly greater than zero, that is to say, if we are interested in testing the 'null hypothesis' that the variables in the population are uncorrelated, the following statistics is to be used:

$$t=\frac{r}{\sqrt{1-r^2}}\times\sqrt{n-2}$$

t with $(n-2)$ degrees of freedom. s of freedom.

The following example will illustrate the use of this formula:

Illustration 20.8. A random sample of 18 pairs of observation from a normal population gives a correlation coefficient of 0.52. Is it likely that the variables in the population are uncorrelated?

Solution. Let us take the hypothesis that the variables in the population are uncorrelated.

Applying t-test,

$$t=\frac{r}{\sqrt{1-r^2}}\times\sqrt{n-2}=\frac{.52\times\sqrt{16}}{\sqrt{1-(.52)^2}}$$

$$=\frac{.52\times 4}{\sqrt{1-.27}}=\frac{2.08}{.85}=2.4.$$

No. of degrees of freedom $= 18-2=16$.

For 16 degrees of freedom at 5% level of significance, the table value of $t=2.12$. The calculated value of t is greater than the table value. Hence for 5% level of significance there is reason to doubt the hypothesis that the variables in the population are uncorrelated.

Illustration. 20.9. How many pairs of observations must be included in a sample in order that on observed correlation coefficient of value .42 shall have a calculated value of t greater than 2.72?

Solution.

$$t=\frac{r}{\sqrt{1-r^2}}\times\sqrt{N-2}$$

i.e.,
$$2.72=\frac{.42}{\sqrt{(1-(.42)^2}}\sqrt{N-2}$$

$$\frac{.42}{.907}\sqrt{N-2}=2.72$$

$$.463\sqrt{N-2}=2.72$$

$$\sqrt{N-2}=\frac{2.72}{.463}=5.9$$

$$N=2+(5.9)^2=36.81$$

So, $N=36.81$ or 37 approx.

Hence 37 pairs of observations must be included in a sample in order that an observed correlation coefficient of value .42 shall have the calculated value of t greater than 2.72.

20.3. THE Z-TEST FOR CORRELATION

The t-test for correlation r discussed above is applicable only for determining whether the computed r is significantly different from zero. When we want to test the sample correlation against any other theoretical value of r (other than zero), or if it is desired to test whether the two given samples have come from same population or not, t-test cannot be used.

Professor Fisher has shown that if r is changed to another statistic z by a suitable transformation then this testing is possible. He has derived that if z is calculated by $z=\frac{1}{2}\log_e\frac{1+r}{1-r}$, where e is base for natural logarithm, *i.e.*, $1.1513\log\frac{1+r}{1-r}$, the distribution of z will be approximately normal. The standard deviation of z distribution will be $\sqrt{\frac{1}{n-3}}$ (where n is the size of sample) and the arithmetic mean of this distribution will be the z corresponding to the population of r.

The following illustrations will explain the use of the test.

Illustration 20.10. A random sample of 30 pairs of obeservations from a normal population shows a correlation coefficient of .75. Is this consistent with the assumption that the correlation in the population is .55?

Solution. Here sample correlation is 0.75 and we have to test whether this sample can come from a population with correlation 0.55.

Now, $z=1.1513 \log \frac{1+r}{1-r}$

Sample $z=1.1513 \log \frac{1+0.75}{1-0.75}$

$=1.1513\times\log\frac{1.75}{.25}$

$=1.1513 \log 7$

$=1.1513\times0.8451=0.972.$

Population $z=1.1513 \log \frac{1+.55}{1-.55}$

$=1.1513 \log \frac{1.55}{0.45}=1.1513\times.5378=.619.$

According to z-test devised by Fisher, the distribution of z is a normal distribution with mean 0.619 and standard deviation $\frac{1}{\sqrt{N-3}}$, *i.e.* $\sqrt{\frac{1}{30-3}}$ or .192. The sample value to be tested is 0.972

Now, $z=\frac{X-\mu}{\sigma}=\frac{.972-.619}{.192}=1.84.$

As this value is less than 1.96 (*i.e.*, 5% level of significance value from normal distribution) the difference between sample value and population value is not significant. In other words, this sample can come from a population with correlation 0.55.

Illustration 20.11. Two independent samples have 18 and 19 pairs of observations with correlation coefficient .55 and .75 respectively. Are these values of r consistent with the hypothesis that the

samples are drawn from the same population?

Solution. Our hypothesis that the samples are drawn from the same population,

$$z_1 = \tfrac{1}{2} \log_e \frac{1+r}{1-r}$$

$$z_2 = \tfrac{1}{2} \log_e \frac{1+r}{1-r}$$

According to Fisher, $z_1 - z_2$ is distributed normally with mean 0 and standard deviation $\sqrt{\dfrac{1}{n_1-3} + \dfrac{1}{n_2-3}}$

$$z_1 = \tfrac{1}{2} \log_e \frac{1+.55}{1-.55}$$

$$= 1.1513 \log 3{:}44$$

$$= 1.1513 \times .5378 = .619.$$

$$z_2 = \tfrac{1}{2} \log_e \frac{1+.75}{1-.75}$$

$$= 1.1513 \log 7$$

$$= 1.1513 \times .8451$$

$$= .972$$

$$z_1 - z_2 = .972 - .619 = .353.$$

Standard error of distribution of $(z_1 - z_2)$

$$= \sqrt{\frac{1}{N_1-3} + \frac{1}{N_2-3}} = \sqrt{\frac{1}{15} + \frac{1}{16}}$$

$$= \sqrt{\frac{41}{400}} = .32$$

$$\therefore \qquad z = \frac{|z_1 - z_2|}{\text{S.E. of } z_1 - z_2} = \frac{.353}{.32} = 1.107$$

This is less than 1.96 (5% level of significance). There is, therefore no reason to doubt the hypothesis that the samples are drawn from the same normal population.

20.4. THE F-DISTRIBUTION AND VARIANCE RATIO TEST

Suppose we take a sample of n_1 items from a normal population (with n_1 not large), and calculate the sample variance

$$S_1^2 = \frac{\Sigma(x - \bar{x}_1)^2}{n_1 - 1}$$

and then take a sample of size n_2 and calculate its variance

$$S_2^2=\frac{\Sigma(x-\bar{x}_2)^2}{n_2-1}$$

the ratio of the two variances S_1^2/S_2^2 has a sampling distribution which is known as F-distribution. F-distribution is a theoretical sampling distribution with two parameters ν_1 and ν_2. These parameters are identified as the degrees of freedom in the two samples, namely (n_1-1) and (n_2-1) respectively.

The expected value of F under the null hypothesis that the two samples have same variance is one and it extends from zero to infinity. Clearly, it is not a symmetrical distribution. A table in appendix gives some important values of ordinates for 5% and 1% level of significance for various values of ν_1 and ν_2. The values in the table represent that value of F beyond which the upper tail area is 5% and 1%. Since the variance of the two samples may be different in either direction, a two-tail test is appropriate, and we reject the null hypothesis if the value of F obtained is larger than the critical value from the tables. Note that by properly labling the two samples as sample 1 and sample 2, we can always test only at the upper tail, by getting F to be larger than 1.

Illustration 20.12. In one sample of 8 observations the sum of the squares of deviation of sample values from the sample mean was 84.4 and in the other sample of 10 observations it was 102.6. Test whether the difference is significant at 5 per cent level, given that the 4 per cent point of F for $n_1=7$ and $n_2=9$ degrees of freedom is 3.29.

Solution. Here $N_1=8$, $N_2=10$

and $\Sigma(X-\bar{X})^2=84.4$, $\Sigma(Y-\bar{Y})^2=102.6$

$$\therefore \quad S_1^2=\frac{1}{N_1-1}\Sigma(X-\bar{X})^2=\frac{84.4}{7}=12.057$$

$$S_2^2=\frac{1}{N_2-1}\Sigma(Y-\bar{Y})^2=\frac{102.6}{9}=11.4$$

$$\therefore \quad F=\frac{S_1^2}{S_2^2}=\frac{12.057}{11.4}=1.057$$

Tabulated value of F for (7, 9) d. f. at 5% level of significance is 3.29.

Since the observed value of F is less than the tabulated value of F we conclude that the difference between variance is insignificant.

20.5. ANALYSIS OF VARIANCE

Suppose we want to assess the performance of students of various schools in a common examination. The mean scores of each school will show a variation, as also the scores of individual students within each school. It is at times difficult to tell at a glance whether the variations between schools are significant compared to variations within schools. It is for this reason that techniques of *analysis of variance* have been developed.

The 'null' hypothesis used in the analysis of variance is that there is no variation in the means of the populations from which various samples come, or in other words, all samples are drawn from the same population. We estimate the mean μ of this common population by the pooled mean of all the samples.

We then determine the sum of the squares of individual deviations from this population mean. This is denoted by SST (or sum of squares, total). Next, we calculate the sum of squares of deviations *between* samples (SSB). This is obtained by taking $(\bar{x}-\mu)^2$ for each sample, multiplying by number n of each sample, and adding.

Thus,
$$\text{SST} = \sum_{\substack{\text{all}\\ \text{samples}}} \left[\sum_{\substack{\text{all}\\ \text{items}}} (x-\mu)^2 \right]$$

and

$$\text{SSB} = n \sum_{\substack{\text{all}\\ \text{samples}}} (\bar{x}-\mu)^2$$

We next assert that the difference between SST and SSB is due to the variations *within* samples, also termed the *error*, since this is the true random variation.

So
$$\text{SSE} = \text{SST} - \text{SSB}$$

After we know the sum of squares of 'between samples' and 'within sample' variations, we find the *means* of these. Thus,

$$\text{MSB} = \frac{\text{SSB}}{k-1}$$

Where k is the number of samples. This is like the *variance* of sample means, and has $(k-1)$ degrees of freedom. Similary,

$$\text{MSE} = \frac{\text{SSE}}{k(n-1)}$$

where n is the number of samples and MSE is like variance of individual values with subtotal constraints and, therefore, with $k(n-1)$ degrees of freedom.

The ratio of MSB to MSE is then the F statistics

$$F = \frac{\text{MSB}}{\text{MSE}}$$

with $(k-1)$ and $k(n-1)$ degrees of freedom. We test on this F for the required level of significance. If F is less than one, clearly between sample variations are negligible. When F is greater than one, but still less than the critical F for $(k-1)$ and $k(n-1)$ degrees of freedom, the null hypothesis cannot be rejected, and the samples may be assumed to be coming from the same population.

Usually the calculation of SST and SSB as ontlined above is quite cumbersome. The following shortcut method may, therefore, be employed to great advantage:

We take deviations about any convenient value to reduce the bulk of figures. Then, if x is the deviation

So obtained.

$$\text{SST} = \sum_{\substack{\text{all}\\ \text{samples}}} \left[\sum_{\substack{\text{all}\\ \text{items}}} x^2 \right] - \frac{1}{k.n} \left\{ \sum_{\substack{\text{all}\\ \text{samples}}} \left[\sum_{\substack{\text{all}\\ \text{items}}} x \right] \right\}^2$$

$$\text{and SSB} = \frac{1}{n} \sum_{\substack{\text{all}\\ \text{samples}}} \left[\sum_{\substack{\text{all}\\ \text{items}}} x \right]^2 - \frac{1}{k.n} \left\{ \sum_{\substack{\text{all}\\ \text{samples}}} \left[\sum_{\substack{\text{all}\\ \text{items}}} x \right] \right\}^2$$

Here the last term may be treated as correction term, CT.

Illustration 20.13. To assess the significance of possible varitions in performance between the grammar schools of a city a common test was given to a sample of students taken at random from the fifth form of each of the four schools concerned. The results are given as follows. Make an analysis of the variance of the data.

Schools			
A	*B*	*C*	*D*
8	7	5	10
7	5	3	5
4	5	4	6
5	4	4	4
5	3	3	8
5	4	5	7
6	6	4	8
6	4	4	-

Solution. To simplify calculations, let us take 5 as the working origin:

A	*B*	*C*	*D*
3	2	0	5
2	0	—2	0
—1	0	—1	1
0	—1	—1	—1
0	—2	—2	3
0	—1	0	2
1	1	—1	3
1	—1	—1	—1
Total 6	—2	—8	12

$$CT=\frac{1}{8\times 4}[6-2-8+12]^2=\frac{64}{32}=2$$

Squaring the items,

A	*B*	*C*	*D*
9	4	0	25
4	0	4	0
1	0	1	1
0	1	1	1
0	4	4	9
0	1	0	4
1	1	1	9
1	1	1	1
Total 16	12	12	50=90

Total sum of squares=SST=90—2=88

Sum of squares between samples

$$\text{SSB}=\frac{1}{8}[6^2+2^2+8^2+12^2]-CT$$

$$=\frac{248}{8}-2=29.$$

Sum of squares within samples

SSE=Total sum of squares—Sum of squares between samples

=88—29=59.

Analysis of variance table:

Source of Variation	*S.S.*	*d.f.*	*M.S.*	*F.*
Between Samples, B	29	3	9.7	4.6
Within Samples, E	59	28	2.1	
Total	88	31		

$$F=\frac{9.7}{2.1}=4.6;\ \nu_1=3;\ \nu_2=28$$

For $\nu_1=3$ and $\nu_2=28, F_{0.05}=2.95$. The calculated value of F is greater than the table value and hence the difference between the school means is significant.

Illustration 20.14. In a feeding experiment on swines, three rations R_1, R_2, R_3 were tried. The animals were put into three classes of three each according to Litter and initial body weight. The following table gives the gains in body weight in kg in a certain period.

	Class I	*Class II*	*Class III*
R_1	4	16	10
R_2	14	18	19
R_3	3	14	7

Analyse the data and state your conclusion. Has division into classes proved effective?

Solution. Let us take the hypothesis that the division into classes has no effect on gain in body weight.

Taking 10 as the working origin to simplify calculation.

	Class I	*Class II*	*Class III*	*Total*
R_1	−6	6	0	0
R_2	4	8	9	21
R_3	−7	4	−3	−6
Totals	−9	18	6	15

$$\text{Correction factor}=CT=\frac{(15)^2}{9}=25.$$

Taking the squares of the items to find out total sum of squares,

	Class I	*Class II*	*Class III*	Total
R_1	36	36	0	72
R_2	16	64	81	161
R_3	49	16	9	74
Total	101	116	90	307

Total sum of squares$=307-25=282$.

Sum of squares between classes

$$=\frac{(-9)^2+(18)^2+(6)^2}{3}-25$$

$$=\frac{441}{3}-25=122.$$

Sum of squares between rations

$$=\frac{0^2+(21)^2+(-6)^2}{3}-25$$

$$=159-25=134.$$

Analysis of variance table:

Source of Variation	*S.S.*	*d.f.*	*M.S.*	*Variance ratio*
Between Rations	134	2	67	$\frac{67}{6.5}=10.3$
Between Classes	122	2	61	$\frac{61}{6.5}=9.38$
Residual	26	4	6.5	
Total	282	8		

$\nu_1=2;\ \nu_2=4;\ F_{0.05}=6.94.$

Calculated values of F_1 and F_2 are 9.38 and 10.3 respectively. Since these values are greater than the table value at 5% level of significance, our hypothesis stands rejected. Hence we conclude that division into classes has proved effective.

20 6. CHI SQUARE DISTRIBUTION

Another very useful sampling distribution is the so called chi-square (pronounced ki-square) distribution. It looks much like a normal distribution, except that it starts with zero, and is skewed with a long tail to the right. As for the t and F distribution, the χ^2-distribution, is not a single distribution, but a family with one parameter ν, the degrees of freedom. As before, the degrees of freedom measures the number of variables that can be freely or arbitrarily chosen under some overall constraints. Some critical values of χ^2 have been given in a table in appendix. Note that as degrees of freedom ν increase, the value of χ^2 for a given level of significance increases.

The χ^2-distribution serves as a sampling distribution for various statistics, and is widely used. The following are the three most important situations where χ^2-test is used:

(1) *To test the discrepancies between observed and expected frequencies.* χ^2-test can be applied in those cases where we want to know whether the difference between the observed and expected frequencies arises due to chance or whether it results from inadequacy of the theory to fit the observed facts.

(2) *To test the goodness of fit.* χ^2-test is the most important of all tests used to find out closeness of fit. When we fit an ideal frequency curve whether normal or some other type to the data, we are interested in finding out as to how will this curve fit with the observed facts. This is a problem of goodness of fit. χ^2-test can be applied to all forms of curve fitting.

(3) *To determine association between two or more attributes.* χ^2-test is widely used to find out whether or not there is any association between two or more attributes. For example, by applying χ^2-test we can find out whether there is any association between the colour of father's eye and son's eye. In such cases we procced on the 'null hypothesis' that there is no association between the attributes. If the calculated value of χ^2 (at a certain lavel of significance) is less than the table value the hypothesis holds good, otherwise we reject the hypothesis and draw the conclusion that there is association between the attributed.

20.7. TEST FOR GOODNES OF FIT

Suppose under some null hypothesis, we expect the frequencies corresponding to certain values of a discrete random variable to be $E_1, E_2, E_3, \ldots, E_n$ etc.

If the observed or given frequencies are $O_1, O_2, O_3 \ldots, O_n$ etc. it can be shown that the statistics

$$\Sigma \frac{(O-E)^2}{E}$$

follows the χ^2-distribution with the appropriate degrees of freedom. The term 'degrees of freedom' refers to the number of classes the frequencies of which could be filled in arbitrarily without violating any of the totals, sub-totals etc. For example, 5 coins are thrown 700 times and the following results are obtained:

No. of heads	Actual frequencies
0	17
1	132
2	230
3	204
4	103
5	14
Total	700

Now, if we write the expected frequencies we have the liberty to write any five figures but the sixth figure must be equal to 700 minus the total of five figures we have written for the simple reason that the total of expected frequencies must be the same as that of actual frequencies. Hence there are five degress of freedom in this case. In such cases the number of degrees of freedom are one less than the number of cases, *i.e.*, $n-1$.

In case of contingency tables, the degrees of freedom are obtained by applying the formula:

$$v=(r-1)(c-1)$$

where v refers to the degrees of freedom.

r refers to the number of rows, and

c refers to the number of columns.

Thus, in a 2×2 table, the degrees of freedom are:

$$v=(2-1)(2-1)=1$$

(because there are two rows and two columns in 2×2 table) and in a 3×3 contingency table, the degrees of freedom are:

$$v=(3-1)(3-1)=4.$$

It should be noted that table values of χ^2 are ordinarily available upto 30 degrees of freedom and for higher values we make use of the fact that the distribution of χ^2 becomes nearly normal.

Conditions for the application of χ^2-test. The following precautions should be observed while applying the χ^2-test:

(1) N, *i.e.*, the number of observations must be sufficiently large otherwise the differences between the actual and observed frequencies would not be normally distributed. It is difficult to say what constitutes largeness or smallness but as an arbitrary figure we may say that N should not be less than 30.

(2) No theoretical frequency should be very small. Again, it is difficult to decide what constitutes smallness. In the opinion of Yule and Kendall 5 should be regarded as the very minimum and 10 i

better. When the theoretical frequencies are less than 10 the adjoining classes should be merged together so that the frequency exceeds or in case of 2×2 contingency table the use of Yate's correction should be made.

Yate's correction. One of the conditions for the applicability of χ^2-test is that no theoretical or expected frequency shold be less than 5 in any case, 10 is still better. When the theoretical frequencies are smaller than 10 and especially when smaller than 5, the ordinary table values of χ^2 are inaccurate. This is especially true when there is only one degree of freedom. It is true to a lesser extent for two or three degrees of freedom. However, the error is negligible when the degrees of freedom exceed three.

When there is only one degree of freedom, a simple variation in the formula of χ^2 will adjust the 'calculated' χ^2 so that it is comparable with table values of χ^2. In a 2×2 table the adjustment consists of adding .5 to those observed frequencies which are less than 10 and subtracting .5 to those frequencies which are more than 10. This is known as Yate's correction. Yate's correction should not be applied when degree of freedom is greater than one. It should always be made when degrees of freedom is only one and N is small.

The following will illustrate the test for goodness of fit.

Illustration 20.15. A die is thrown 132 times with the following results:

No. turned up	1	2	3	4	5	6
Frequency	16	20	25	14	29	28

Test the hypothesis that the die is unbiased.

Solution. On the basis of the hypothesis that the die is unbiased, we should expect each number to turn up $\frac{132}{6}=22$ times.

Applying χ^2-test:

O	E	$(O-E)^2$	$(O-E)^2/E$
16	22	36	1.64
20	22	4	.18
25	22	9	.41
14	22	64	2.91
29	22	49	2.23
28	22	36	1.64
			9.01

$$\chi^2 = \frac{\Sigma(O-E)^2}{E} = 9.01.$$

No. of degrees of freedom $= n-1 = 6-1 = 5$.

For 5 degrees of freedom at 5% level of significance, the table value of $\chi^2 = 11.07$. The calculated value of χ^2 is less than the table value and hence there is no evidence against the hypothesis that the die is unbiased.

Illustration 20.16. The figures given below are (*a*) the observed frequencies of a distribution, and (*b*) the frequencies of the Poisson distribution having the same mean and total frequency as in (*a*). Apply the χ^2-test of goodness of fit.

(*a*)	305	365	210	80	28	9	3
(*b*)	301	361	217	88	26	6	1

Solution.

O	E	$(O-E)^2$	$(O-E)^2/E$
305	301	16	.05
365	361	16	.04
210	217	49	.23
80	88	64	.73
28	26	4	.15
9	6]	25	3.60
3	1]		
			4.80

$$\chi^2 = \frac{\Sigma(O-E)^2}{E} = 4.8.$$

No. of degress of freedom $= 7-3 = 4$.

(The number of degrees of freedom is one for each class less one for each 'restraint'. The original 7 classes have been reduced to 6 by grouping, thus reducing the degrees of freedom by 1. In addition, the mean and the total frequency of the original distribution have been used in calculating the theoretical frequencies, thus introducing two restraints. The number of degrees of freedom is accordingly 4.)

For 4 degrees of freedom at 5% level of significance, the table value of $\chi^2 = 9.49$. The calculated value of χ^2 is less than the table value and hence the fit is good.

Illustration 20.17. The figures given below are (*a*) the observed frequencies of a distribution, and (*b*) the frequencies of the normal distribution having the same mean, standard deviation and total frequency as in (*a*). Apply the χ^2-test of goodness of fit.

(*a*) 1 12 66 220 495 792 924 792 495 222 66 12 1

(*b*) 2 15 66 210 484 799 943 799 484 210 66 15 1

Solution. We are given (*a*) the observed frequencies, and (*b*) the expected frequencies of the normal distribution. We can apply the χ^2-test of goodness of fit.

O	*E*	$(O-E)^2$	$(O-E)^2/E$
1 12	2 15	16	.94
66	66	0	.00
220	210	100	.48
495	484	121	.25
792	799	49	.06
924	943	361	.38
792	799	49	.06
495	484	121	.25
222	210	144	.70
66	66	0	.00
12 1	15 1	9	.56
			3.68

$$\chi^2 = \frac{\Sigma(O-E)^2}{E} = 3.68.$$

The number of degrees of freedom = 13 − 5 = 8.

(The number of degrees of freedom is one for each class, less one for each 'restraint'. The 13 original classes have been reduced to 11 by grouping, thus reducing the degrees of freedom by 2. In addition, the mean, the standard deviation and the total frequency of the original distribution have been used in calculating theoretical frequencies, thus introducing three restraints. The number of degrees of freedom is accordingly 8.)

For 8 degrees of freedom at 5% level of significance, the table value of $\chi^2 = 15.50$. The calculated value of χ^2 is less than the table value and hence the fit is good.

Illustration 20.18. Investigate the association between the darkness of eye colour in father and son from the following data:

	Frequency
Fathers with dark eyes and sons with dark eyes	50
Fathers with dark eyes and sons with not dark eyes	79
Fathers with not dark eyes and sons with dark eyes	89
Fathers with not dark eyes and sons with not dark eyes	782
Total	1,000

Solution. The above information can be arranged in the form of a 2×2 table as follows:

Colour of fathers' eyes

Colour of son's eyes	Dark	Not dark	Total
Dark	50	89	139
Not dark	79	782	861
Total	129	871	1,000

Let us take the hypothesis that there is no association between the colour of the fathers' and that of the sons' eyes. On the basis of this hypothesis, the expected frequencies are:

Colour of fathers' eye

Colour of son's eyes	Dark	Not dark	Total
Dark	$\frac{139}{1000}\times 129 = 18$ approx	121	139
Not dark	111	750	861
Total	129	871	1,000

O	E	$(O-E)^2$	$(O-E^2)/E$
50	18	1,024	57.0
79	111	1,024	9.2
89	121	1,024	8.5
782	750	1,024	1.4
			76.1

$$\chi^2=\frac{\Sigma(O-E)^2}{E}=76.1.$$

No. of degrees of freedom$=(r-1)(c-1)=1$.

For one degree of freedom at 5% level of significance, the table value of $\chi^2=3.84$. The calculated value of χ^2 is much greater than the table value and hence our hypothesis stands rejected. We, therefore, conclude that there is association between darkness of eye colour in father and son.

Illustration 20.19. In an experiment on immunisation of cattle from tuberculosis the following results were obtained:

	Affected	Not affected
Inoculated	12	26
Not Inoculated	16	6

Discuss the effect of vaccine in controlling susceptibility to tuberculosis.

Solution. Let us take the hypothesis that vaccine has no affect in controlling susceptibility to tuberculosis. On the basis of this hypothesis the expected frequencies are:

	Affected	*Not Affected*	*Total*
Inoculated	$\frac{38}{60}\times 28=18$ approx.	20	38
Not Inoculated	10	12	22
Total	28	32	60

O	E	$(O-E)^2$	$(O-E)^2/E$
12.5	18	30.25	1 68
15.5	10	30.25	3.02
25.5	20	30.25	1.51
6.5	12	30.25	2.52
			8.73

$$\chi^2=\frac{\Sigma(O-E)^2}{E}=8.73.$$

$$d.f.=(r-1)(c-1)=(2-1)(2-1)=1.$$

For one degree of freedom at 5% level of significance, the table value of $\chi^2 = 3.84$. The calculated value of χ^2 is greater than the table value and hence our hypothesis stands rejected. We, therefore conculude that vaccine is effective in controlling susceptibility to tuberculosis.

Illustration 20.20. In a recent diet survey the following results were found in an Indian city:

No. of families	*Hindus*	*Muslims*	*Total*
Taking tea	1,236	164	1,400
Not taking tea	564	36	600
Total	1,800	200	2,000

Discuss whether there is any significant difference between the two communities in the matter of tea-taking.

Solution. Let us take the hypothesis that there is no difference between the two communities in the matter of tea-taking. On the basis of this hypothesis the expected frequencies would be:

No. of families	*Hindus*	*Muslims*	*Total*
Taking tea	$\frac{1,400}{2,000} \times 1,800 = 1,260$	140	1,400
Not taking tea	540	60	600
Total	1,800	200	2,000

O	*E*	$(O-E)^2$	$(O-E)^2/E$
1,236	1,260	576	0.457
564	540	576	1.070
164	140	576	4.110
36	60	576	9.600
			15.237

$$\chi^2 = \frac{\Sigma(O-E)^2}{E} = 15.237.$$

$$d.f. = (r-1)(c-1) = (2-1)(2-1) = 1.$$

For one degree of freedom at 5% level of significance, the table value of $\chi^2 = 3.84$. The calculated value of χ^2 is greater than the table value. Our hypothesis, therefore, stands rejected and we conclude that there is significant difference between the twocommunities in the matter of tea-taking

Illustration 20.21. Genetic theory states that children having one parent of blood type M and the other of blood type N will always be of one of the three types M, MN, N and that proportions of these types will on average be 1 : 2 : 1. A report states that out of 300 children having one M parent and one N parent, 30% were found to be of type M, 45% type MN and the remainder of type N. Test the hypothesis by χ^2-test.

Solution. If the genetic theory be true, 300 children should be distributed as follows :

$$\text{Children of type } M = \frac{300 \times 1}{1+2+1} = 75$$

Children of type $MN = 75 \times 2$. *i.e.*, 150

and Children of type $N = 75$.

The observed frequencies are :

$$\text{Type } M = \frac{300 \times 30}{100} = 90, \text{ Type } MN = \frac{300 \times 45}{100} = 135$$

and type $N = 75$.

$$\chi^2 = \Sigma \frac{(o-e)^2}{e} = \frac{(90-75)^2}{75} + \frac{(135-150)^2}{150} + \frac{(75-75)^2}{75}$$

$$= \frac{225}{75} + \frac{225}{150} + 0 = 4.5.$$

No. of degrees of freedom $= n - 1 = 3 - 1 = 2$.

Value of χ^2 for 2 degrees of freedom from tables is 5.99. As the calculated value is less than the tabulated value the difference between observed and calculated values is not significant.

Hence there is no positive evidence against the hypothesis $M : MN : N :: 1 : 2 : 1$ as far as χ^2-test is concerned.

Illustration 20.22. A certain drug is claimed to be effective in curing colds. In an experiment on 164 people with colds, half of them were given the drug and half of them given sugar pills. The patients reactions to the treatment are recorded in the following table. Test the hypothesis that the drug is no better than sugar pills for curing colds.

	Helped	Harmed	No Effect
Drug	52	10	20
Sugar Pills	44	12	26

So ution. Our hypothesis is that the drug is no better than sugar pills for curing colds. On the basis of this hypothesis expected frequencies are

	Helped	Harmed	No Effect	Total
Drugs	$\frac{82}{164}\times 96=48$	$\frac{82}{164}\times 22=11$	$\frac{82}{164}\times 46=23$	82
Sugar Pills	$\frac{82}{164}\times 96=48$	$\frac{82}{164}\times 22=11$	$\frac{82}{164}\times 46=23$	82
Total	96	22	46	164

Therefore

O	E	$(O-E)^2$	$(O-E)^2/E$
52	48	16	0.333
44	48	16	0.333
10	11	1	0.091
12	11	1	0.091
20	23	9	0.391
26	23	9	0.391
			1.630

$$\chi^2=\Sigma\frac{(O-E)^2}{E}=1.63$$

No. of degrees of freedom$=(r-1)(c-1)=(2-1)(3-1)=2$.

For two degrees of freedom at 5% level of significance, the table value of χ^2 is 5.991. The calculated value of χ^2 is less than the table value and, therefore, the hypothesis may be accepted at 5% level of significance. Hence, drug is no better than sugar pills in curing colds.

Illustration 20.23. 1600 families were selected at random in a city to test the belief that high income families usually send their childeren to public schools and low income families often send their children to Governments schools. The following results were obtained.

Income \ School	Public	Govt.	Total
Low	494	506	1000
High	162	438	600
Total	656	944	1600

Test whether income and type of school are independent.

Solution. Let us take the hypothesis that income and type of schools are independent. On the basis of this hypothesis the expected frequencies are:

Income \ School	Public	Govt.	Total
Low	$\frac{1000}{1600}\times 656=410$	590	1000
High	246	354	600
	656	944	1600

Therefore,

O	E	(O—E)	$(O-E)^2$	$(O-E)^2/E$
494	410	84	7056	17.21
162	246	—84	7056	28.68
506	590	—84	7056	11.96
438	354	84	7056	19.93
				77.78

$$x^2=\Sigma\frac{(O-E)^2}{E}=77.78$$

No. of degrees of freedom$=(r-1)(c-1)=1$

For one degree of freedom at 5% level of significance, the table value of $\chi^2=3.84$. The calculated value of χ^2 is much greater than the table value and hence our hypothesis stands rejected. We, therefore, conclde that there is association between income and type of school.

Illustration 20.24. A die is tossed 120 times with the following results:

No. turned up:	1	2	3	4	5	6	Total
Frequency:	30	25	18	10	22	15	120

Test the hypothesis that the die is unbiased.

Solution. On the basis of the hypothesis that the die is unbiased, we should expect each no. to turn up $\frac{120}{6}=20$ times. Applying χ^2-test

O	E	$(O-E)^2$	$(O-E)^2/E$
30	20	100	5.00
25	20	25	1.25
18	20	4	0.20
10	20	100	5.00
22	20	4	0.20
15	20	25	1.25
			$\Sigma \frac{(O-E)^2}{E}=12.9$

$$\chi^2=\Sigma\,\frac{(O-E)^2}{E}=12.9$$

No. of degrees of freedom $=n-1=6-1=5$.

For 5 degrees of freedom at 5% level of significance, table value of $\chi^2=11.07$. The calculated value of χ^2 is greater than the table value and hence the hypothesis that the die is unbiased is rejected at 5% level of significance.

Illustration. 20.25. Two investigators study the income of a group of persons by the method of sampling. Following results were obtained by them:

Investigator	Poor	Middle class	Well-to do	Total
A	160	30	10	200
B	140	120	40	300
Total	300	150	50	500

Show that sampling technique of at least one of the investigator is suspect.

Solution. Let us suppose that the sampling techniques of both the investigators are the same. If that is the case the results obtained by them must be on the same lines *i.e.*, the division of persons into various groups should be independent of investigators. Calculating cell frequencies on this basis we have

Investigator	Poor	Middle class	Well-to-do	Total
A	$200\times\frac{300}{500}=120$	$200\times\frac{150}{500}=60$	20	200
B	180	90	30	300
Total	300	150	50	500

Calculation of χ^2

O	E	$(O-E)$	$(O-E)^2$	$(O-E)^2/E$
160	120	40	1600	13.33
30	60	−30	900	15.00
10	20	−10	100	5.00

140	180	—40	1600	8.89
120	90	30	900	10.00
40	30	10	100	2.33
				55,55

Degrees of freedom$=(r-1)(c-1)=(2-1)(3-1)=2$

$$\chi^2=\Sigma\ \frac{(O-E)^2}{E}=55.55$$

χ^2 value for 2 degrees of freedom at 1% level of significance is 9.21. As the calculated value of χ^2 is much greater than table value, the difference between observed and calculated frequencies is highly significant. Hence our hypothesis stands rejected. So division of persons into three income groups depend on investigators *i.e.*, different for two investigators. Hence if the method of one of them in correct the method of the other must be wrong.

Illustration 20.26. The following table represents the number of boys and girls who chose the five possible answers to an item of an attitude scale.

	Strongly approve	Approve	Indifferent	Disapprove	Strongly disapprove
Boys	50	60	15	50	25
Girls	20	30	10	30	30

Do this data indicate a significant sex difference in the attitude towards the item.

Solution. Let there be no sex difference in the attitude. Calculating cell frequencies on this basis we have

	Strongly Approve	*Approve*	*Indiffe-rent*	*Dis-approve*	*Strongly Dis-approve*	*Total*
Boys	$\frac{70\times200}{320}=44$	$\frac{90\times200}{320}=56$	$\frac{25\times200}{320}=16$	$\frac{80\times200}{320}=50$	34	200
Girls	26	34	9	30	21	120
Total	70	90	25	80	55	320

Calculation of χ^2

O	E	$(O-E)$	$(O-E)^2$	$(O-E)^2/E$
50	44	6	36	0.82
60	56	4	16	0.29
15	16	—1	1	0.06
50	50	0	0	0.00
25	34	—9	81	2.38
20	26	—6	36	1.38
30	34	—4	16	0.47
10	9	1	1	0.11
30	30	0	0	0.00
30	21	9	81	3.86
				9.37

The table value for (5—1)(2—1) *i.e.,* 4 degrees of freedom is 9.5 at 5% level of significance.

As calculated value is less than table value, the evidence against the hypothesis that there is no difference in sex attitudes is insufficient.

20.8. MISCELLANEOUS ILLUSTRATIONS

Illustration 20.27. A group of 7 week-old chickens reared on a high protein diet weigh 12, 15, 11, 16, 14, 14 and 16 ounce, a second group of 5 chickens reared on a low protein diet weight 8, 10, 14, 10 and 13 ounce. Test whether there is a significant evidence that additional protein has increased the weight of chickens.

Solution. Let the null hypothesis be that there is no change in weight. As these samples are small, we work with *t*-Statistics. We first calculate *S*.

X_1	$(X_1-\bar{X}_1)$	$(X_1-\bar{X}_1)^2$	X_2	$(X_2-\bar{X}_2)$	$(X_2-\bar{X}_2)^2$
12	—2	4	8	—3	9
15	+1	1	10	—1	1
11	—3	9	14	+3	9
16	+2	4	10	—1	1
14	0	0	13	+2	4
14	0	0			
16	+2	4			
98	0	22	55		24

$$\bar{X}_1=\frac{98}{7}=14;\ \bar{X}_2=\frac{55}{5}=11$$

$$S=\sqrt{\frac{\Sigma(x_1-\bar{x}_1)^2+\Sigma(x_2-\bar{x}_2)^2}{n_1+n_2-2}}=\sqrt{\frac{22+24}{7+5-2}}$$

$$=2.14$$

$$\therefore \quad t=\frac{\bar{X}_1-\bar{X}_2}{S}\sqrt{\frac{n_1n_2}{n_1+n_2}}=\frac{14-11}{2.14}\sqrt{\frac{7\times 5}{7+5}}$$

$$=2.397$$

Degrees of freedom $=n_1+n_2-2=10$

This is greater than the critical value at 5% level and 10 d.f. Hence there is sufficient evidence to condude that the additional protein increases body weight.

Illustration 20.28. The mean life of a sample of 10 electric bulbs was found to be 1456 hrs. With S.D.=423 hrs. A second sample of 17 bulbs chosen from a different batch showed a mean life of 1280 hrs with S.D=398 hrs. Is there a significant difference in the means of the two samples

Solution. Let the null hypothesis be that there is no difference. As samples are small. we work with *t*-statistics.

$$S=\sqrt{\frac{(N_1-1)S_1^2+(N_2-1)S_2^2}{N_1+N_2-2}}$$

$$=\sqrt{\frac{9(423)^2+16(398)^2}{10+17-2}}=407.18.$$

$$\therefore \quad t=\frac{\bar{X}_1-\bar{X}_2}{S}\sqrt{\frac{n_1n_2}{n_1+n_2}}=\frac{1456-1280}{407.18}\sqrt{\frac{10\times 17}{10+17}}$$

$$=1.085$$

Degrees of freedom $=10+17-2=25$

The critical value at 5% level for $v=25$ is 2.06. There, the difference is not significant.

Illustration 20.29. A sample of 28 pairs gave a correlation coefficient of 0.40. If 'ρ' denotes the population correlation, test the hypothesis that (*i*) $\rho=0$ and (*ii*) $\rho=0.30$

Solution.

We work with Z-transformation.

$$Z=1.1513 \log\frac{1+r}{1-r}$$

For $r=0.4$: $Z=1.1513 \log\frac{1.4}{0.6}=0.42$

For $r=0.3$: $Z=1.1513 \log\frac{1.3}{0.7}=0.31$

For $r=0$: $Z=1.1583 \log \frac{1}{1}=0.0$

Under the hypothesis $\rho=0$,

$$\text{Normal variate}=\frac{0.42-0}{\sqrt{\frac{1}{28-3}}}=2.1$$

Thus under this hypothesis the difference is significant at 5% level.
Under the hypothesis $\rho=0.30$

$$\text{Normal variate}=\frac{0.42-0.31}{\sqrt{\frac{1}{28-3}}}=0.55$$

This is not significant, so the second hypothesis may be correct.

PROBLEMS

1. A sample of 16 students gives the arithmetic mean of marks as 53.8 and the estimate of population standard deviation as 5.2. Can you say that population mean of marks is 50?
(Ans. 't'=2.92)
2. A sample of size 10 has arithmetic mean as 56.8 and standard deviation 18.0, Can it come from a population with mean 50?
(Ans 't' value in the present case=1.13)
3. In an examination in Psychology 12 students in one class had a mean grade of 78 with a standard deviation 6 while 15 students in another class had a mean grade of 74 with standard deviation 8. Is there a significant difference between the means of the two groups?
(Ans. 't' value for testing difference of mean=1.43)
4. Two independent different samples of size 21 each give the following set of observation. Do the mean of two differ significantly?
(Ans. 't' value in the present case=1.6.)
5. A certain stimulus administered to each of 12 patients resulted in the following increase of blood pressure.

   5, 2, 8, —1, 3, 0, —2, 1, 5, 0, 4 and 6.

   Can it be concluded that the stimulus will in general be accompanied by an increase in blood pressure.
(Ans. 't' value corresponding to observed values of increase = 2.89)
6. The weight of 10 tins of 'Dalda' Vanspati taken from a filling machine are found to be in kgs as 4.2, 4.1, 4.0, 3.8, 3.7, 3.8, 3.6, 3.9, 4.2, and 3.6. The machine is expected to fill only 4, 00 kg of vanaspati in a tin on an average. Does the machine require resetting.
(Ans. 't' value =—6 approx.)
7. The incomes from a random sample of engineers in industry I are Rs 630, 650, 680, 690 710 and 720 p.m. The incomes of a similar sample from industry II are Rs 610, 620, 650, 660, 620, 690, 700, 710, 720 and 730 p.m.

Discuss the validity of the suggestion that industry I pays its engineers much better than industry II.
(Ans. 't' value=0.1)

8. Increase in weights in kgs in a particular period of 10 students of a certain age group of High School fed with nourishing 'Complan' were observed as 5,2,6. −1,0,4,3,—2,1,4. Twelve students of the same age group but of another High School were fed with another nourishing food 'Astra' and the increase in weights in kgs in the same period observed where 2,8, —1,5,3,0,6,1. —2,0,4,5. Test whether the two foods 'Complan' and 'Astra' differ significantly as regards the effect on the increase in weight.
(Ans. t value of the difference = 0.3 approx.)

9. Two horses, A and B, were tested according to the time (in seconds) taken to run a particular track with the following results.

   Horse A : 28.30,32,33,33,29 and 34

   House B : 29,30.30,24,27, and 29.

   Analyse the above data and report whether or not you can discriminate between the running time of two horses.
   (Ans. t=2.4)

10. Rainfall at two places A and B for 10 years are as below:

Years	1	2	3	4	5	6	7	8	9	10
Place A	40	30	34	39	43	25	49	40	45	55
Place B	39	28	34	35	41	23	45	37	43	55

Is there any significant difference in the rainfalls for two places taking data as (a) two independent samples (b) paired up values?
(Ans. 't'=0 502. 't'=4.47)

11. Two random samples drawn from two normal populations are sample I : 30,26,36,37,33.32,28,34,35 and 29. Sample II : 37,43,52,45,42,44,48,38,51 53,40 and 47. Test whether the two populations have the same variance.
(Ans. F=3.1)

12. Fstimate the range within which the population mean and standard deviation will lie, if the sample of size 14 gives mean as 34.7 and staudard deviatien as 8.7.
(Ans. t =13)

13. A sample of 15 students give the coefficient of correlation between height and weight as 0.8. A sample of 12 office employees gives the similar coefficient as 0.6. Is the difference significant?
(Ans. Z for office employees=0.692
Z for students =1.098)

14. Find the least value of r in a sample of 27 pairs significant at 5% level.
(Ans. r=0.308)

15. 30 persons in the income group of Rs 20,000 to Rs 25,000 were askeb to supply before a specified date returns of their annual income for some purposes connected with taxation. But when the date arrived only 20 returns of total income as noted below had been received in the following order.
20620, 21720, 24010, 20090, 20940, 21510, 20340, 22420, 22180, 22600, 21940, 23080, 22840, 23510, 24260, 23740, 24720, 21310, 21300 ane 24530. Do the data confirm the belief that persons with bigger income delay submission of returns more than others? You may assume that the square

of the test criterion t is equal to $(n-2)\,\rho^2/(1-\rho^2)$, where ρ is the coeffici- of rank correlation and n is the number of observations on which correlation is used.

(Ans. $t=3.38$)

16. The following table gives the number of aircraft accidents that occurred during the various days of the week. Test whether the accidents are uniformaly distributed over the week.

Days:	Mon.	Tue.	Wed.	Thur.	Fri.	Sat.
No. of accidents:	14	18	12	11	15	14.

[Ans. $(O-E)^2/E=2.14$]

17. In a radio listener's preference survey 120 persons were interviewed and their opinions were as follows:

Type of music	*Language A*	*Language B*
I	13	45
II	39	23

Examine whether the preference for music type is dependent on language. Also test the hypothesis that half of the radio listeners prefer language *A* to *B*.

(Ans. $\chi^2=2.13$)

18. The following table shows price increases and decreases in markets where credit squeeze is in operation and where it is not in operation.

Credit squeeze	*Price-decreases*	*Price-increase*	*Total*
In operation	862	10	872
Not in operatton	582	18	600
Total	1444	28	1472

Find whether the credit squeeze has been effective in checking price increase.

(Ans. $\chi^2=6.54$)

19. An enquiry was conducted to determine how the financial conditions of students who earn while they learn affect their academic success and the following results were obtained.

	Successful	*Not Successful*	*Total*
Self supporting	7	3	10
Not self supporting	15	5	20
Total	22	8	30

What conclusion can be drawn from the above data?

(Ans. $\chi^2=0.5$)

20. In a breeding experiment, it was expected that ducks would be hatched in the ratio of 1 duck with a white bib to every 3 ducks without bibs. Of 84 ducks hatched 29 had white bibs. Are these data compatible with expectation?

(Ans. $\chi^2=.03$)

Chapter 21

ATTRIBUTES AND THEIR ASSOCIATION

21.1. INTRODUCTION

In the chapter on 'Classification' it has been pointed out that the characteristics possessed by the individual items of a data may be classified into (*i*) numerical, and (*ii*) descriptive.

Characteristics which are capable of quantitative measurements, *e.g.*, age, weight, wages, length, etc., are put in the first category, and those which cannot be quantitatively measured, *e.g.*, sex, religion, complexion, etc., are put in the second category.

In the former case the observation themselves are quantitative in character while in the latter case the quantitative character arises solely in the process of counting. If, for example, we measure the heights of the students of a class such measures are themselves quantitative in character. On the other hand, if we record the number of students who are tall and the number of students who are not tall, then the quantitative character arises only in the process of counting.

Observations based on numerical characteristics are termed 'statistics of variables,' while observations based on descriptive characteristics are termed as 'statistics of attributes.' So far we have dealt with the classification, summarisation, interpretation and correlation of the statistics of variables. In this chapter we will discuss the method of determining if there exists any relation between different attributes.

21.2. DICHOTOMY AND NOTATION

Descriptive characteristics may be classified according to the presence or absence of an attribute. In this way two distinct and mutually exclusive classes are formed. Such a classification is termed as classification by dichotomy (cutting into two). For developing a theory it is essential to assign simple notation for the classes and for their observed results.

Attributes are generally denoted by letters, *viz.*, *A*, *B*, *C*, etc. An individual possessing the attribute *A* will be called '*A*'. A group or a class in which all the individuals possess attribute *A*, will be called 'the class A.' Likewise, we use small letters *a*, *b*, *c*...or greek letters α, β, γ...for the absence of attributes. An individual who does not possess attribute *A* will be called 'α'. A class in which all the members do not possess attribute *A* will be called 'α'.

Thus, if *A* stands for the attribute 'male' 'α' would represent female; if *B* stands for 'married', 'β' would stand for unmarried. Combinations of several attributes will be denoted by writing collectively the letters representing several attributes. As an illustration, if '*A*' stands for 'male' and 'B' for 'married', then ,*AB*, means married male and '*A*β' means unmarried male. If we add a third attribute, literacy, denoted by '*C*', then '*ABC*' will denote a literate married male, 'αβγ' will denote an unmarried woman who is illiterate, and so on.

Class frequencies are represented by enclosing the corresponding class symbols in brackets. Hence, (*A*) stands for the number of *A*'s, *i.e.*, the number of individuals who possess attribute '*A*'; (*ABC*) stands for the number of those persons who possess all the three attributes, *A*, *B* and *C*. Positive attributes are those which are represented by capital letters *A*, *B*, *C*...; whereas negative attributes are represented by lower case letters *a*, *b*, *c*,...or greek letters α, β, γ. Similarly, *A*, *AB*, *ABC* are positive classes and α, αβ, αβγ are negative classes

21.3. ORDER OF CLASSES

The class in which only one attribute is considered is called the class of the first order. The class in which two attributes are considered is called the class of the second order. The class in which three attributes are considered is called the class of the third order; and so on.

Thus,

A, *B*, *C*, α, β, γ are classes of 1st order.

AB, *A*β, α*B*, αβ, *AC*, *A*γ, α*C*, αγ, *BC*, *B*γ, β*C*, βγ } are classes of 2nd order.

ABC, *AB*γ, *A*βγ, *A*β*C*, α*BC*, α*B*γ, αβ*C*, αβγ } are classes of 3rd order.

N which represents the total number of items is taken as a class of '0' order as no attribute is specified for it.

We notice that total number of classes of '0', 1st, 2nd and 3rd order for a three attribute case (as enumerated above) are 27 or 3^3. In case of two attributes all possible classes will be $N, A, B, \alpha, \beta, AB, A\beta, \alpha\beta$, *i.e.*, 9 or 3^2. In general, we can say that for n attributes total number of classes possible will be 3^n.

These classes are not all independent. Number of items belonging to different classes (*i.e.*, class frequencies) are connected with one another by simple algebraic relations. In fact, any class frequency can be expressed as a sum of class frequencies of higher order. Let us consider an illustration.

Illustration 21.1. Let A stand for male, α stand for female, B stand for married and β stand for unmarried, C stand for literate and γ, stand for illiterate, and N stand for total population. Obtain the relations between classes of different order.

Solution. It is clear that the total population will be equal to the total number of men plus the total number of women. *i.e.*,

$$N=(A)+(\alpha) \quad \text{...(i)}$$

Similarly,

$$N=(B)+(\beta) \quad \text{...(ii)}$$

and

$$N=(C)+(\gamma) \quad \text{...(iii)}$$

Now men may be married and unmarried and as such total number of men will be equal to the total number of married men plus the unmarried men.

Thus,

$$(A)=(AB)+(A\beta) \quad \text{...(iv)}$$

and similarly,

$$(\alpha)=(\alpha B)+(\alpha\beta)$$

$$(B)=(AB)+(\alpha B)$$

and

$$(\beta)=(A\beta)+(\alpha\beta) \quad \text{...(v)}$$

The above relation shows a first order class frequency in terms of 2nd order. Substituting (*iv*) and (*v*) in (*i*), we can show the '0' order class frequency as sum of second order class frequencies, *i.e.*,

$$N=(AB)+(A\beta)+(\alpha B)+(\alpha\beta) \quad \text{...(vi)}$$

Now married men (AB) may be literate or illiterate and therefore the number of married men (AB) will always be equal to the sum of literate married mean (ABC) and illiterate married men ($AB\gamma$).

Thus,

$$\left.\begin{aligned}(AB)&=(ABC)+AB\gamma)\\(A\beta)&=(A\beta C)+(A\beta\gamma)\\(\alpha B)&=(\alpha BC)+(\alpha B\gamma)\\(\alpha\beta)&=(\alpha\beta C)+(\alpha\beta\gamma)\end{aligned}\right\} \text{showing a 2nd order in terms of 3rd order class frequency}$$

Substituting these values in (*iv*) and (*vi*), we can show a first order and 0 order in terms of 3rd order class frequency.

Thus,

$$A=(ABC)+(AB\gamma)+(A\beta C)+(A\beta\gamma)$$

and

$$N=(ABC)+(AB\gamma)+(A\beta C)+(AB\gamma)+(\alpha BC)+\alpha B\gamma) \\ +(\alpha\beta C)+(\alpha\beta\gamma) \quad ...(vii)$$

Now, (ABC), $(AB\gamma)$ etc., cannot be divided further unless there is a fourth characteristic also. If there is a fourth characteristic then each of the eight third order class frequency can be divided into two each.

The illustration points out two more things:

(*i*) The process of sub-division will stop when all the attributes are taken into consideration.

(*ii*) All the class frequencies can be expressed as an aggregate of some of the class frequencies of highest order.

In a given set of attributes, all the classes specified by the highest order class frequency are called a set of ultimate class frequency. In two attribute cases, there are (AB), $(A\beta)$, (αB) and $(\alpha\beta)$, or 2^2 class frequencies in ultimate set. In case of three attributes as stated in relation (*vii*) above, there are 8 or 2^3 class frequencies. In general, for n attributes there are 2^n class frequencies in a set of ultimate class frequency.

21.4. A FUNDAMENTAL SET

We have stated that all the class frequencies which may be possible in a study can be expressed as a total of some or all the ultimate class frequencies. Such a set of class frequencies which can specify all the remaining class frequencies is called a fundamental set. A set of ultimate class frequencies is not the only fundamental set. A set consisting of N and all combinations of positive attributes can also specify the remaining class frequencies. This set is called a positive set. For a two attributes case, a positive set consists of N (A), (B) and (AB) class frequencies. In case of three attributes positive set will have N, (A), (B), (C), (AB), (BC), (AC), and (ABC) class frequencies in it. An illustration will show how a positive set can give rise to all the remaining class frequences.

Let $N=100$, $(A)=60$, $(B)=30$ and $(AB)=20$.

The remaining class frequencies, are α, β, $(\alpha\beta)$, $(A\beta)$ and (αB), we have,

$$N=(A)+(\alpha) \quad \text{or} \quad (\alpha)=N-(A)=100-60=40$$
$$N=(B)+(\beta) \quad \text{or} \quad (\beta)=N-(B)=100-30=70$$
$$(A)=(AB)+(A\beta) \quad \text{or} \quad (A\beta)=A-(AB)=60-20=40$$
$$(B)=(AB)+(\alpha B) \quad \text{or} \quad (\alpha B)=B-(AB)=30-20=10$$
$$(\beta)=(A\beta)+(\alpha\beta) \quad \text{or} \quad (\alpha\beta)=\beta-(A\beta)=70-40=30$$

Out of the given set N, (A), (B) and (AB) no one class frequency can be expressed as the sum or the difference of the two or more of the remaining class frequencies of this set. Such a group of class frequencies is called a set of independent class frequencies. To make a set, a fundamental set we need 4, *i.e.*., 2^2 independent class frequencies in case of two attributes. Similarly, in case of three attributes we need 8 or 2^3 independent class frequencies. In general. for n-attributes we need 2^n independent class frequencies to form a fundamental set. In addition to ultimate set and a positive set there can be several other fundamental sets.

21.5. METHOD FOR DETERMINING CLASS FREQUENCIES

From a set of given class frequencies to find the remaining class frequencies one has to first write the inter-relation between them and then substitute the known values. Writing inter-relations is not always very simple, specially when there are three or more attributes. If ultimate set or a positive set of class frequencies are given, inter-relations can be derived easily by the method given below:

(A) is the class frequency obtained by attribute A dividing items N. Let us denote it as $A.N$.

Similarly, $(\alpha)=\alpha.\ N$ and $(AB)=AB.N$ etc.

Now, $N=(A)+(\alpha)=A.N+\alpha.N$.

Symbolically, this can be stated as $(A+\alpha)\ N$.

or $\quad N=(A+\alpha)\ N$, *i.e.*, we can take as if $A+\alpha=1$

Symbolically, thus, $A=1-\alpha$, or $\alpha=1-A$

Similarly, $B+\beta=C+\gamma=1$, *etc.*.

If a set of a positive class frequencies is given, the remaining class frequencies can be determined by converting every negative class into a positive class with the help of these relations. To take an illustration, take the case of three attributes. A positive set consists of N, (A), (B), (C) (AB), (BC) and (ABC). Let us determine the class frequency $(\alpha\beta\gamma)$. From, the relations given above,

$$(\alpha\beta\gamma)=(\alpha\beta\gamma)N=(1-A)\,(1-B)\,(1-C)\,N$$
$$=(1-A-B+AB)\,(1-C)N$$
$$=(1-A-B-C+AB+BC+AC-ABC)N$$
$$=N-(A)-(B)-(C)+(AB)+(BC)+(AC)-(ABC)$$

Similarly, all the other class frequencies can be determined.

When ultimate set is given to determine other frequencies, express them in terms of highest order by introducing sub-division according to the attributes not included in the class frequency to be determined To illustrate, let us take three attributes and find (α).

$$(\alpha)=\alpha.\ N=\alpha(B+\beta).N=(\alpha B+\alpha\beta).N$$
$$=(\alpha B+\alpha\beta)\ (C+\gamma).N=(\alpha BC)+(\alpha B\gamma)+(\alpha\beta\gamma)+(\alpha\beta C)$$

Similarly, other class frequencies can be determined when there are only two attributes. Calculations of the class frequencies can be done easily by a two-way table formed by taking one attribute on horizontal side and the other attribute on vertical side. This table is:

		Characteristic A		
		A	α	*Total*
Characteristic B	*B*	(*AB*	(α*B*)	(*B*)
	β	(*A*β)	(αβ)	(β)
	Total	(*A*)	(α)	*N*

In this table fill the values of the given class frequencies and determine the rest by addition or substraction as indicated by the table.

Illustration 21.2. Find the remaining class frequency, when $N=100$ $(\alpha)=40$, $(AB)=10$, $(\alpha\beta)=20$.

Solution. The remaining class frequencies are (*A*), (*A*β), (αβ), (*B*) and (β). We have the table as below:

		Characteristic A		
		A	α	Total
Characteristic B	*B*	(*AB*) 10	(α*B*)	(*B*)
	β	(*A*β)	(αβ) 20	(β)
	Total	(*A*)	(α) 40	*N* 100

The table suggests $(A)=N-(\alpha)=100-40=60$ ...(*i*)

$$(\alpha B)=\alpha-(\alpha\beta)=40-20=20 \qquad \text{...(ii)}$$

Then, with the help of (*i*)

$$(A\beta)=(A)-(AB)=60-10=50 \qquad \text{...(iii)}$$

and with the help of (*iii*)

$$(\beta)=(A\beta)+(\alpha\beta)=50+20=70$$

and with the help of (*ii*)

$$(B) = (AB)+(\alpha B) = 10+20 = 30.$$

21.6. CONSISTENCY OF DATA

A set of given figures is said to be consistent if no class frequency calculated from it is negative. As a class frequency of any order can be expressed as a sum of two or more ultimate class frequencies, the condition for consistency of data becomes that no ultimate class frequency is negative.

Illustration 21.3. Test for consistency, given $N=100$, $(A)=70$, $(B)=60$, $(AB)=15$.

Solution. This is a case of two attributes. The remaining ultimate class frequencies are $(A\beta)$, (αB) and $(\alpha\beta)$.

$(A\beta)=(A)-(AB)=70-15=55$

$(\alpha B)=(B)-(AB)=60-15=45$

$(\alpha\beta)=N-(A)-(B)+(AB)=100-70-60+15=-15.$

As $(\alpha\beta)$ is negative, the given data is inconsistent.

Illustration 21.4. Is the data given below consistent?

$N=1{,}000$, $(A)=500$, $(B)=550$, $(C)=450$

$(AB)=200$, $(BC)=250$, $(AC)=150$, $(ABC)=120$.

Solution. There are three attributes. So the remaining ultimate class frequencies will be $(AB\gamma)$, $(A\beta C)$, $(A\beta\gamma)$, (αBC), $(\alpha B\gamma)$, $(\alpha\beta C)$ and $(\alpha\beta\gamma)$.

$$\begin{aligned}\text{Now, } (\alpha\beta\gamma)&=N-(A)-(B)-(C)+(AB)+(BC)+(AC)-(ABC)\\&=1{,}000-500-550-450+200+250+150-120\\&=1{,}600-1{,}620=-20\end{aligned}$$

As $(\alpha\beta\gamma)$ is negative, it is not necessary to calculate the other ultimate class frequencies. As soon as any one of the ultimate class frequency comes out to be negative we can infer that the given data is inconsistent.

In problems involving three or more variables this procedure for testing consistency will become very lengthy. The condition that no ultimate class frequencies be negative can be written in terms of mathematical inequalities which are easy to apply. Sometimes incomplete data (*i.e.*, data from which all the ultimate class frequencies canot be calculated) may be given. In such case this procedure of testing consistence will not be applicable. But the inequalities referred to above give rise to a new set of inequalities which may be applicable in these cases.

For three attributes A, B, C, the inequalities are:

(1) $(ABC) \nless 0$ otherwise (ABC) will be negative
(2) $(ABC) \ngtr (AB)$,, $(AB\gamma)$,, ,, ,,
(3) $(ABC) \ngtr (BC)$,, (αBC) ,, ,, ,,
(4) $(ABC) \ngtr (AC)$,, $A\beta C)$,, ,, ,,
(5) $(ABC) \nless (AB)+(AC)-(A)$,, $(A\beta\gamma)$,, ,, ,,
(6) $(ABC) \nless (BC)+(AC)-(B)$,, (αBC) ,, ,, ,,
(7) $(ABC) \nless (BC)+(AC)-(C)$
(8) $(ABC) \ngtr (AB)+(AC)+(BC)$,, $(\alpha\beta\gamma)$,, ,, ,,
$-(A)-(B)-(C)+N$

These relations give four new relations for testing inconsistency :

(9) $(A)+(AC)+(BC) \nless (A)+(B)+(C)-N$, from Nos. (1) and (8)
(10) $(AC)+(AB)-(BC) \ngtr (A)$, from Nos. (3) and (5)
(11) $(AB)+(BC)-(AC) \ngtr (B)$, from Nos. (4) and (6)
(12) $(AC)+(BC)-(AB) \ngtr (C)$, from Nos. (2) and (7)

Illustration 21.5. Can there be inconsistency in the data given below:

$N=100$, $(A)=40$, $(AB)=38$, $(AC)=35$, $(B)=39$ and $(BC)=37$.

Solution. As (ABC) and (C) are not given, so relation Nos. 1 to 9 and 12 are not applicable. Here only relation Nos. 10 and 11 can test the inconsistency. Applying No. 10 relation, we get

$$(AC)+(AB)-(BC) \ngtr (A)$$

or $35+38-37 \ngtr 40$,......satisfied

Applying relation No. 11, we get

$$(AB)+(BC)-(AC) \ngtr (B)$$

or $38+37-35 \ngtr 39$,...not statisfied

Hence there is inconsistency in data.

Illustration 21.6. Apply inequality method for testing inconsistency for the Illustration 21.4.

Solution. Here $N=1000$, $(A)=500$, $(B)=550$, $(C)=450$.
$(AB)=200$, $(BC)=250$, $(AC)=150$, $(ABC)=120$.

Here all the 12 relations can be applied. By inspection relation Nose. 1 to 4 are satisfied.

No. (5) gives $120 \nless 200+150-500$...satisfied.
No. (6) gives $120 \nless 200+250-500$...satisfied.
No. (7) gives $120 \nless 250+150-450$...satisfied

No. (8) gives $120 \ngtr 200+250+150-500-550-450+1,000$

or $120 \ngtr 1,600-1,500$......not satisfied.

Hence data is inconsistent. Relation No. 8 not satisfied means $(\alpha\beta\gamma)$ will be negative and the same result was obtained earlier.

21.7. ASSOCIATION

The attributes A and B will be said to be associated only if they appear together in a large number of cases than is to be expected if they were independent. The mere fact that A and B are found together in a fairly high proportion will not be enough to warrant any kind of association between them. What is necessary is that this proportion should be higher than what can occur by chance if the two were only independent.

If there is no relationship of any kind between two attributes A and B, *i.e.*, they are independent, we expect that the proportion of A amongst B's is same as proportion of A amongst not B's.

or
$$\frac{(AB)}{(B)} = \frac{(A\beta)}{(\beta)}$$

This, in other words, means that the presence or absence of B has nothing to do with the occurrence of A. If proportion of A amongst B is greater than proportion of A in 'not B's' then we expect that A has a greater chance to occur with B than with β, *i.e.*, A and B have positive association or simply association. If the proportion of A amongst B is less than the proportion amongst β, than A has a tendency to go with β and not with B. This is called negative association or disassociation between A and B. A negative association between A and B means that A has a tendency to occur with β, *i.e.*, there is a positive association between A and β. A positive association between any two characters A and B can similarly be translated as the negative association between any one positive character and the other negative character, *i.e.*, A and β or α and B.

There are three methods of studying association:

1. Comparison of proportions,
2. Comparison of expected frequencies and
3. Coefficient of association.

Comparison of proportions. As stated above in the definitions the first method for studying association is:

If $\frac{(AB)}{(B)} = \frac{(A\beta)}{(\beta)}$, then A and B are independent of one another.

$$\text{If } \frac{(AB)}{B} > \frac{(A\beta)}{(\beta)}, \text{ then } A \text{ and } B \text{ are positively associated.}$$

$$\text{If } \frac{(AB)}{(B)} < \frac{(A\beta)}{(\beta)}, \text{ then } A \text{ and } B \text{ a[re n]egatively associated.}$$

Just as we have studied the proportions of *A*'s amongst *B* and not *B*'s, similarly we can study the proportions of *B*'s amongst *A*'s and not *A*'s, *i.e.*,

A and B are independent, *i.e.*, $\frac{(AB)}{(A)} = \frac{(\alpha B)}{(\alpha)}$,

A and *B* are positively associated, if $\frac{(AB)}{(A)} > \frac{(\alpha B)}{(\alpha)}$,

A and *B* are negatively associated, if $\frac{(AB)}{(A)} < \frac{(\alpha B)}{(\alpha)}$.

Comparison of proportions of *A*'s in *B* with *A*'s in total items or comparison of proportions of *B*'s in *A*'s with *B*'s in total items can also point out the association between *A* and *B*. If proportion of *A*'s in *B* is higher than the proportion of *A*'s in total number of items then it means *A* and *B* has a tendency to be present together, *i.e.*, they have a positive association. The three relations on these comparisons are, thus,

A and *B* are independent of one another, if

$$\frac{(AB)}{(B)} = \frac{(A)}{N} \text{ or } \frac{(AB)}{(A)} = \frac{(B)}{N}$$

A and *B* are possitively associated, if

$$\frac{(AB)}{(B)} > \frac{(A)}{N} \text{ or } \frac{(AB)}{(A)} > \frac{(B)}{N}$$

Out of these four comparisons of proportions the first two tests give more striking differences in proportions. In the last two cases the right hand side includes all items, *i.e.*, it includes those also which have been considered on the left hand side. In the first two cases of comparisons the items of both sides are mutually exclusive. It may be pointed out that these tests for association are less than satisfactory. They do not point out if the association is significant, or arises merely because of chance.

Comparison of expected frequencies. One of the methods under comparison of proportions is that *A* and *B* are independent of one another, if $\frac{(AB)}{(A)} = \frac{(B)}{N}$. This can be taken as $(AB) = \frac{(A)(B)}{N}$

The value on the right hand side is called frequency on the assumption that *A* and *B* are independent, which can also be deduced from

the principle of probability:

Probability of getting $A=(A)/N$

Probability of getting $B=(B)/N$.

From the theorem on multiplication of probabilities we have that if A and B are independent of one another, probability of getting A and B together is $P(A) \times P(B)$ or $\frac{(A) \times (B)}{N \times N}$

So the expected number of items out of N, in which A and B should come together, is $\frac{(A)}{N} \times \frac{(B)}{N} \times N = \frac{(A)(B)}{N}$.

Thus, even if A and B are independent of on another we expect them to occur together in as many as $\frac{(A)(B)}{N}$ itemes. If they come together in more cases than this then they have a tendency to occur together, *i.e.*, they have positive association. If they come in less number then $\frac{(A)(B)}{N}$, then they have negative association. So, to state briefly, A and B are associated positively when $(AB) > \frac{(A)(B)}{N}$, associated negatively when $(AB) < \frac{(A)(B)}{N}$, and independent of one another when

$$(AB) = \frac{(A)(B)}{N}.$$

Coefficient of association. Just as we have obtained the expected frequency of (AB) when A and B are independent of one another, we can also find the expected frequency of (Ab), (aB) and (ab), when A and B are independent.

$$(AB) = \frac{(A)(B)}{N} \qquad \ldots(i)$$

$$(\alpha\beta) = \frac{(\alpha)(\beta)}{N} \qquad \ldots(ii)$$

$$(A\beta) = \frac{(A)(\beta)}{N} \qquad \ldots(iii)$$

$$(\alpha B) = \frac{(\alpha)(B)}{N} \qquad \ldots(iv)$$

(i) and (ii) gives $(AB)(\alpha\beta) = \frac{(A)(B)(\alpha)(\beta)}{N}$ ...(v)

(iii) and (iv) gives $(A\beta)(\alpha B) = \frac{(A)(\beta)(\alpha)(B)}{N}$ ...(vi)

As the right hand side of (*v*) and (*vi*) are the same,

$(AB)(\alpha\beta)=(A\beta)(\alpha B)$ or $(AB)(\alpha B)-(AB)(\alpha B)=0$ ...(*vii*)

The expression [as given in (*vii*)] can also be used as a method for testing association. If it is zero there is no association, if it is positive there is positive association, and if it is negative association is negative.

Based on this, Yule has suggested that

$$Q=\frac{(AB)(\alpha\beta)-(A\beta)(\alpha B)}{(AB)(\alpha\beta)+(A\beta)(\alpha B)}$$

can be used as the coefficient of association. This coefficient (Q) gives a value zero when A and B are independent. When they are perfectly positively associated, *i.e.*, all A's, occur with B or to say that no A comes with B, *i.e.*, $(A\beta)=0$, the value of Q will be [by putting $(A\beta)=0$ in its value] $+1$. Similarly, when there is a perfect negative association, *i.e.*, no A comes with B or $(AB)=0$, the value of Q will be -1. For associations of intensity less than perfect, the value of Q will be in between $+1$ and -1. Thus, Q is also a useful measure of intensity of association. Hence it can be used in comparing intensity of association.

Illustration 21.7. Out of 70,000 literates in a particular district of India the number of criminals was 500. Out of 930,000 literates in the same district the number of criminals was 15,000. On the basis of these figures, do you find any association between illiteracy and criminality.

Solution. Denoting illiteracy by A, literacy by α, criminals by B, and noncriminals by b, the data can be written as $(A)=930,000$, $(\alpha)=70,000$, $(AB)=15,000$ and $(\alpha B)=500$.

Now,

$$\frac{(AB)}{(A)}=\frac{15,000}{930,000}=.016$$

and

$$\frac{(\alpha B)}{(\alpha)}=\frac{500}{70,000}=.007$$

As $\frac{(AB)}{(A)}>\frac{(\alpha B)}{(\alpha)}$, there is a positive association between A and B, *i.e.*, illiteracy and criminality.

Illustration 21.8. Of 1,600 persons in a locality exposed to small-pox, 300 in all were attacked. Of 1,600 persons 400 had been vaccinated and of these only 30 were attacked. Can vaccination be regarded as a preventive measure for small-pox?

Solution. Let attacked be denoted by A and vaccinated by B. The given data then is:

$N=1,600$, $(A)=300$, (B) 400 and $(AB)=30$.

Let there be no association between vaccination and attack of small-pox.

Then expected frequency of

$$(AB)=\frac{(A)(B)}{N}$$

or

$$(AB)=\frac{300\times 400}{1600}=75$$

As the actual number of (AB), *i.e.*, 30 is less than the expected number of (AB) which is 75, there is negative association between A and B *i.e.*, attack and vaccination. In other words, it means that there is positive association between vaccination (B) and not attacked from small-pox (α) *i.e.*, vaccination is effective in preventing small-pox.

Illustration 21.9. Compare the association between literacy and unemployment in urban and rural areas from the following observations:

	Urban	Rural
	(in lakhs)	
Total adult males	25	200
Literate males	10	40
Unemployed males	5	12
Literate and Unemployed males	3	4

Solution. Let literacy be A and employment be B, we have

	N	(A)	(B)	(AB)
Urban	25	10	5	3
Rural	200	40	12	4

The association in two groups is to be compared, coefficient of association is the best method. For this we need (AB), $(A\beta)$, (αB) and $(\alpha\beta)$.

For urban areas:

$(AB)=3$,

$(A\beta)=(A)-(AB)=10-3=7$.

$(\alpha B)=(B)-(AB)=5-3=2$,

$(\alpha\beta)=N-(A)-(\alpha B)=25-10-2=13$.

For Rural areas:

$(AB)=4$,

$(\alpha B)=(B)-(AB)=12-4=8$,

$(A\beta)=(A)-(AB)=40-4=36$,

$(\alpha\beta)=N-(A)-(\alpha B)=200-40-8=152$.

∴ **Coefficient of association for urban areas:**

$$= \frac{(AB)(\alpha\beta)(\alpha B)(A\beta)}{(AB)(\alpha\beta) + (\alpha B)(A\beta)}$$

$$= \frac{3\times 13 - 2\times 7}{3\times 13 + 2\times 7} = .47.$$

Coefficient of association for rural areas

$$= \frac{4\times 152 - 36\times 8}{4\times 152 + 36\times 8} = \frac{320}{896} = .356.$$

Thus, there is a positive association between literacy and unemployment in both rural and urban areas. It is however, more in the case of urban areas than in the rural.

21.8. PARTIAL AND ILLUSORY ASSOCIATION

Let A denote vaccination, B exemption from attack and C good hygienic conditions. Suppose we find in a group of persons that association between A and B is positive, we may conclude from this that association is due to some causal relation between vaccination and exemption from attack. But this conclusion may not be correct. It may happen that vaccination was taken by well-to-do people, who live under good hygienic conditions where chance of attack is very small and vaccination was not taken by poor people who looked at it with suspicion and were living under unhygienic conditions where the chance of attack is very high. Thus, exemption from attack may not necessarily be on account of effective vaccination but may be due to good living conditions. This means that in drawing conclusions about associations between attributes A and B one has to make sure about the effect of other attributes also which may influence both A and B. If there exists an attribute, say C, besides A and B, then association between A and B should be studied in two different groups of items one in which C is present, *i.e.*, sub-population C, and other in which C is not present *i.e.*, sup-population γ. If the association in the two sub-populations are similar then one can be more sure about the conclusion of causal relationship between A and B. The association in sub-populations are called partial associations. The coefficient of partial association between A and B in sub-population C may be denoted as $Q_{AB.C}$ and in sub-population γ by $Q_{AB.\gamma}$. To calculate them, the method would be the same as used for computation of Q_{AB}.

$$Q_{AB.C}=\frac{(ABC)(\alpha\beta C)-(A\beta C)(\alpha BC)}{(ABC)(\alpha\beta C)+(A\beta C)(\alpha BC)}$$

$$Q_{AB.\gamma}=\frac{(AB\gamma)(\alpha\beta\gamma)-(A\beta\gamma)(aB\gamma)}{(AB\gamma)(\alpha\beta\gamma)+(A\beta\gamma)(\alpha B\gamma)}$$

Illustration 21.10. A doctor claims that a vaccine prepared by him is effective for cholera. He took 86 persons and gave vaccination to 55 persons. 41 persons did not have attack of cholera and out of this 32 were those who were vaccinated. On further examining the records it was found that 35 persons were poor and living in locations with very bad hygienic conditions and 51 persons were well-to-do and were living in localities with good hygienic conditions. 34 persons amongst the second group had no attack of cholera and 45 amongst them were vaccinated. The number of persons living in good localities, vaccinated and who had no attack of cholera was 30. Discuss the claim of the doctor.

Solution. Let us denote exemption from cholera as A, vaccination as B and good hygienic conditions as C. The first part of the data is then $N=86$, $(A)=41$, $(B)=55$ and $(AB)=32$. Taking all the 86 persons together association between A and B can be studied by Q_{AB}, *i.e.*,

$$\frac{(AB)(\alpha\beta)-(A\beta)(\alpha B)}{(AB)(\alpha\beta)+(A\beta)(\alpha B)}.$$

Now,

$$(A\beta)=(A)-(AB)=41-32=9.$$
$$(\alpha B)=(B)-(AB)=55-32=23,$$
$$\alpha=N-A=86-41=45,$$

and

$$(\alpha\beta)=\alpha-(\alpha B)=45-23=22$$

$$\therefore \quad Q_{AB}=\frac{32\times 22-9\times 23}{32\times 22+9\times 23}=\frac{497}{911}=.54$$

So there is a positive association between A and B, *i.e.*, vaccination and exemption from cholera. So it seems that the claim of doctor is correct.

But if we utilize the supplementary information let us see what happens. The additional information in terms of symbols is $(C)=51$, $(AC)=34$, $(BC)=45$ and $(ABC)=30$. To see that the association between vaccination (B) and exemption from cholera (A) is real or is due to the effect of good hygienic conditions (C); we have to examine the partial associations between A and B in two groups separately; one who live in good hygienic conditions, *i.e.*, population (C) and the other who do not live in good localities, *i.e.*, sub-population γ.

Now,
$$Q_{(AB.C)}=\frac{(ABC)(\alpha\beta C)-(\alpha BC)(A\beta C)}{(ABC)(\alpha\beta C)+(\alpha BC)(A\beta C)}$$

and
$$Q_{AB.\gamma}=\frac{(AB\gamma)(\alpha\beta\gamma)-(\alpha B\gamma)(A\beta\gamma)}{(AB\gamma)(\alpha\beta\gamma)+(\alpha B\gamma)(A\beta\gamma)}$$

So we require the eight class frequencies of third order:

$$(AB\gamma)=(AB)-(ABC)=32-30=2,$$
$$(A\beta C)=(AC)-(ABC)=34-30=4$$
$$(\alpha BC)=(BC)-(ABC)=45-30=15,$$
$$(\alpha B\gamma)=(\alpha B)-(\alpha BC)=23-15=8,$$
$$(\alpha C)=(C)-(AC)=51-34=17,$$
$$(\alpha\beta C)=(\alpha C)-(\alpha BC)=17-15=2,$$
$$(A\gamma)=(A)-(AC)=41-34=7,$$
$$(A\beta\gamma)=(A\gamma)-(AB\gamma)=7-2=5,$$
$$(\alpha\beta\gamma)=(\alpha\beta)-(\alpha\beta C)=22-2=20.$$

$\therefore$
$$Q_{AB.C}=\frac{30\times2-15\times4}{30\times2+15\times4}=0$$

and
$$Q_{AB.\gamma}=\frac{2\times20-8\times5}{2\times20+8\times5}=0$$

This means there is no association between vaccination and exemption of cholera for those persons who live in good localities and also those who do not live in good localities. So, really speaking there is no association between vaccination and exemption from cholera. The value of Q_{AB} may be positive on account of the third character C and hence the claim of the doctor is not correct.

As in the above illustration it may happen that the partial association between A and B, both in sub-population C and sub-population γ, are zero but still there may be some association between A and B in the total population, this may be due to the fact that both A and B are associated with C.

To see this let us find out the association between A and C and B and C in the foregoing illustration:

$$Q_{AC}=\frac{(AC)(\alpha\gamma)-(A\gamma)(\alpha C)}{(AC)(\alpha\gamma)+(A\gamma)(\alpha C)}$$

and

$$Q_{BC}=\frac{(BC)(\beta\gamma)-(B\gamma)(\beta C)}{(BC)(\beta\gamma)+(B\gamma)(\beta C)}$$

Now

$$(\alpha\gamma)=(\alpha B\gamma)+(\alpha\beta\gamma)=8+20=28$$
$$(\beta\gamma)=(A\beta\gamma)+(\alpha\beta\gamma)=5+20=25$$
$$(B\gamma)=(AB\gamma)+(\alpha B\gamma)=2+8=10$$
$$(\beta C)=(A\beta C)+(\alpha\beta C)=4+2=6$$

$$\therefore \quad Q_{AC}=\frac{34\times 28-7\times 17}{34\times 28+7\times 17}=\frac{7\times 17\times 7}{7\times 17\times 9}=.78.$$

$$Q_{AC.B}=\frac{(ABC)(\alpha\beta\gamma)-(AB\gamma)(\alpha BC)}{(ABC)(\alpha B\gamma)+(AB\gamma)(\alpha BC)}=\frac{30\times 8-2\times 15}{30\times 8+2\times 15}=\frac{210}{270}=.0.78$$

$$Q_{AC.\beta}=\frac{(A\beta C)(\alpha\beta\gamma)-(A\beta\gamma)(\alpha\beta C)}{(A\beta C)(\alpha\beta\gamma)+(A\beta\gamma)(\alpha\beta C)}=\frac{4\times 20-5\times 2}{4\times 20+5\times 2}=\frac{70}{90}=.78$$

and

$$Q_{BC}=\frac{45\times 25-10\times 6}{45\times 25+10\times 6}=\frac{1065}{1185}=.9$$

$$Q_{BC.A}=\frac{(ABC)(A\beta\gamma)-(AB\gamma)(A\beta C)}{(ABC)(A\beta\gamma)+(AB\gamma)(A\beta C)}=\frac{30\times 5-2\times 4}{30\times 5+2\times 4}=\frac{142}{158}=.9$$

$$Q_{BC.\alpha}=\frac{(\alpha BC)(\alpha\beta\gamma)-(\alpha B\gamma)(\alpha\beta C)}{(\alpha BC)(\alpha\beta\gamma)+(\alpha B\gamma)(\alpha\beta C)}=\frac{15\times 20-8\times 2}{15\times 20+8\times 2}=\frac{284}{316}=.9$$

$\therefore\ Q_{AC}=Q_{AC.B}=Q_{AC.\beta}$ and $Q_{BC}=Q_{BC.A}=Q_{BC.\alpha}$.

Here in both the cases association in the whole population and the association in the two sub-populations are same. So there is high real association between good hygienic conditions and vacci-- nation and also between good hygienic conditions and exemption from attack. These two associations mean that persons under hygienic conditions have a tendency to be vaccinated and also be exempted from attack of cholera, and persons living under un-hygienic conditions have a tendency not to take vaccination and get more attack of cholera. So when all persons are taken together it appears as if vaccination and exemption from attack go together, *i.e.*, there is a positive association between $Q_{AB}=48$ in this illustration as calculated earlier).

In fact, this is not true as suggested by two partial associations, $Q_{AB.C}=Q_{AB.\gamma}=0$. Such a misleading association between A and B in the whole population is called illusory ssociation. If two partial associations between A and B in sub-populations C and γ are zero and association between A and C and B and C are of same sign, the illusory association between A and B will be positive. If association between A and C and B and C are of opposite sign, the illusory association between A and B will be negative. Such a misleading association may arise on account of mixing up records.

21.9. MANIFOLD CLASSIFICATION AND CONTINGENCY TABLES

Instead of dividing population under consideration into two parts we may divide it into more parts according to each attribute observed. For example, students may be classified according to proficiency in games as—good at games, average at games and poor in games and according to second character proficiency in studies as —1st class, 2nd class, 3rd class and who failed in annual examination. In general, we may divide population into S parts A_1, A_2, A_3,...A_s, according to first characteristic then each of these into 't' parts B_1, B_2, B_3,...B_t according to second characteristic. Further, each of these may be divided into u parts C_1, C_2...C_u according to a third character C and so on. This kind of classification is called manifold classification.

If there are two characters only one having 's' alternatives and other 't' all the class frequencies formed may be presented as below.

Atribute	A_1	A_2	A_3	....	A_s	*Total*
B_1	(A_1B_1)	(A_2B_1)	(A_3B_1)	....	(A_sB_1)	(B_1)
B_2	(A_1B_2)	(A_2B_2)	(A_3B_2)	....	(A_sB_2)	(B_2)
B_3	(A_1B_3)	$A_2B_3)$	(A_3B_3)	....	(A_sB_3)	(B_3)
.	.	.	.		.	.
.	.	.	.		.	.
B_t	(A_1B_t)	(A_2B_t)	(A_3B_t)		(A_sB_t)	(B_t)
Total	A_1	(A_2)	(A_3)	....	(A_s)	N

Such a table is called contingency table. It is generalized form of 2×2 fold table. In its main body it has t rows and s columns. So it is denoted $t\times s$ fold table. This table has $t\times s$ ultimate class frequencies and in all $(t+1)$ $(s+1)$ class frequency.

21.10. ASSOCIATION IN CONTINGENCY TABLES

To study association between A and B in such a case the table may be divided in 2×2 table in several ways and association in each part can be studied by methods discussed earlier. This meth--od when carried out fully will he very laborious though it may give information on all possible combinations of A and B. Generally, we may be interested in knowing if A's on the whole depend on B's or not. For this a coefficient which gaves over all effect will be required.

To obtain such a coefficient and to test whether that coefficient is significant, *i.e.*, that the observed value is not purely because of sampling variations, we calculate the expected frequencies under a 'null hypothesis' that A's and B's are completely independent of one another.

If A's and B's are independent, then expected frequency of the cell A_m—B_n is $\frac{(A_m)(B_n)}{N}$. If (A_mB_n) is equal to this expected frequency for all cells (all values of m and n), the variables will be truly independent. On the other hand, if they are not independent, then this difference will not be zero. So a measure based on this difference gives a measure of association. Denoting (A_mB_n) as O (for observed), and $(A_m)(B_n)/N$ as E (for expected under null hypothesis) we, obtain the sum

$$\Sigma\frac{(O-E)^2}{E}$$

It can be shown that this measure in a truly independent population follows the χ^2-distribution (chi-square) outlined in chapter 20. For this reason this is denoted as χ^2 (chi-square) and can be taken as a measure of association.

This χ^2 is called *square contingeney*. Sometimes a *mean contingency* ϕ^2 (phi-square) is also used, which is defined as χ^2/N. These two are measures of association between A and B. They being sum of squares cannot be negative, but can be zero.

21.11. PEARSON'S CO-EFFICIENT OF MEAN SQUARE CONTINGENCY

The quantity ϕ^2 or χ^2 is not suitable in itself to form a coefficient because its limits vary in different cases. Karl Pearson, therefore, proposed the coefficient C defined as

$$C=\sqrt{\frac{\phi^2}{1+\phi^2}}=\sqrt{\frac{\chi^2}{N+\chi^2}}$$

This is called *coefficient of mean square contingency* or *coefficient of contingency*. In general, no sign is attached to this. This only shows the degree of association. If there is no association its value will be zero and its value cannot exceed 1. In fact, the maximum value of C for perfect association is never 1 and it depends on the number of subdivisions 's' and 't'. It reaches 1 only when there are infinite divisions. Yule has shown that for a $t\times t$ table the value of C cannot exceed the following limits:

When $t=2$, C cannot exceed 0.707
" $t=3$ " " " 0.816
" $t=10$ " " " 0.949, etc.

Hence this coefficient from different systems of classification are not strictly comparable. C can only be used with advantage to compare association in two tables of the same fold.

To remedy the defect stated above, Tschuprow proposed the coefficient T which is defined as

$$T^2 = \frac{\phi^2}{\sqrt{(s-1)(t-1)}} \quad \text{or} \quad \frac{C^2}{(1-C^2)\sqrt{(s-1)(t-1)}}$$

As Yule has stated this varies between 0 and 1 in the desired manner when $s=t$.

These coefficients are of great use in many analysis but the importance of detailed analysis still remains. Every table should be carefully examined for any special way of distribution of frequencies.

21.12. SIGNIFICANCE OF ASSOCIATION

When a sample is taken from a population, the observed association between *A* and *B* in the selected group may or may not indicate an association in the whole population. To study it, χ^2 as defined earlier can be used. The χ^2 reflect the difference between the given frequencies and those obtained on the assumption of independence. Tables are available for the χ^2 value that may rise due to chance at a specified level of siginificance. If calculated value of χ^2 is greater than the value given in the table, the difference in observed and the expected frequency is more than what is permissible on chance basis (at the given level of significance) alone. This means that the assumption of independence is not true, or in other words there is an assoication. If χ^2 value calculated is equal to, or less than the value in the table, the difference in observed and expected frequency is within the chance limits. It means there is not enough evidence of real difference between the observed and expected frequencies. In other words, assumption of independence may be taken as correct for the population from which sample has been taken.

Generally 5 per cent level of significance is considered as reasonable level of chance for such testing. For more strict test 1 per cent level can also be used. The table gives values of χ^2 for different levels of significance and different degrees of freedom. In broad terms degrees of freedom refers to the number of classes the frequencies of which could be filled in without violating any of the totals or subtotals. In the $t \times s$ table first $(s-1)$ class frequesncy in any row can be arbitrarily assigned, but the *S*th frequency has to be adjusted to justify the sub-total. Similarly, we can arbitrarily chose only $(t-1)$ rows because the *t*th row will be fixed to justify column totals. Thus, a $t \times s$ contingency table has $(t-1)(s-1)$ degrees of freedom. For a

2×2 table, there is only one degree of freedom.

Illustration 21.11. In a sample study about the coffee habits in two towns following data was observed:

Town *X*: 50 per cent were male, 30 per cent were coffee drinkers and 18 per cent were male coffee drinkers.

Town *Y*: 45 per cent were male, 25 per cent were coffee drinkers and 16 per cent were male **coffee drinkers.**

Is there any association between sex and coffee habit? If so in which town is it greater?

Solution. Let males be denoted by *A* and Coffee drinkers by *B*.

Putting the data in contingency table from, and calculating the unknown class frequencies, we get:

Atribute	*Town X* Male *A*	Female α	*Total*	*Town Y* Male *A*	Female α	*Total*
Coffee drinkers (*B*)	18	12	30	16	9	25
Non-drinkers (β)	32	38	70	29	46	75
Total	50	50	100	45	55	100

Let there be no association between sex and coffee habits. Calculating expected frequencies on this assumption, we get:

Items	*Town X* Observed value (O)	Expected value (E)	$\frac{(O-E)^2}{E}$	*Town Y* Observed value (O)	Expected value (E)	$\frac{(O-E)^2}{E}$
(*AB*)	18	$\frac{30\times50}{100}=15$	0.600	16	$\frac{45\times25}{100}=11.25$	2.005
($A\beta$)	32	50—15=35	0.257	29	45–11.25=33.75	0.669
(αB)	12	30—15=15	0.600	9	25—11.25=13.75	1.641
($\alpha\beta$)	38	50—15=35	0.257	46	55—11.75 = 43.25	0.547
Total	100	100	1.714	100	100	4.862

$$\therefore \quad \chi^2=\Sigma\frac{(O-E)^2}{E} \text{ for town } X=1.714 \text{ and town } Y=4.862.$$

Degrees of freedom=(2—1)(2—1)=1.

Table value of χ^2 at 5 per cent level of significance for 1 degree of freedom=3.84.

As for town *X* the calculated value of χ^2 is less than 3.84, there is no positive evidence against the hypothesis of independence. Or, in broad terms, it may be taken that there is no association. For town *Y* the calculated value is more than the table value. Hence hypothesis of independence is rejected. In other words, there is sufficient evidence to believe that there is an association between sex and coffee habits. As given $(AB) = 16$ is greater than its expected value 11.25 there is a positive association between A (male) and B (drinking coffee).

Illustration 21.12. Given $N=1500$, $(A)=383$, $(B)=350$, $(AB)=35$. Prepare 2×2 fold contingency Table. Compute Yule's coefficient of Association. and interpret the result.

Solution. Putting the given values in table form, we have

Atributes		*Chac. A*		
		A	*α*	*Total*
Char.	*B*	(AB) 35	(αB)	(B) **350**
B	β	$(A\beta)$	$(\alpha\beta)$	(β)
Total		(A) 383	(α)	N 1500

The table suggests
$$(\alpha)=N-(A)=1500-383=1117$$
$$(\beta)=N-(B)=1500-360=1140$$
$$(\alpha B)=(B)-(AB)=\overline{350}-35=315$$
$$(A\beta)=(A)-(AB)=383-35=348$$
$$(\alpha\beta)=(\alpha)-(\alpha B)=1117-315=802$$

Hence the final table is:

2×2 Contingency table

Attributes		*Char. A*		
		A	*α*	*Total*
Char.	*B*	35	315	350
B	β	348	802	11[illegible]
Total		383	1117	1500

$$\text{Yule's coefficient of association}=\frac{(AB)(\alpha\beta)-(A\beta)(\alpha B)}{(AB)(\alpha\beta)+(A\beta)+(\alpha B)}$$
$$=\frac{35\times 802-348\times 315}{35\times 802+348\times 315}$$
$$=\frac{28070-109620}{28070+109620}=-.59 \text{ Approx.}$$

Illustration 21.13. Define independence in probability sense. X and Y are two variables—X taking the values 1, 2, 3, 4, and Y—1, 2, 3. The frequency for different values of x and y are as follows:

		x 1	2	3	4
y	1	14	26	22	18
	2	21	39	33	27
	3	35	65	55	45

Examine from definition or otherwise whether x and y are independent.

Solution. Two quantities A and B are said to be independent of each other if the probability of A multiplied by probability of B is equal to the probability of joint occurrence of A and B. The given table is:

		x 1	2	3	4	*Total*
	1	14	26	22	18	80
y	2	21	39	33	27	120
	3	35	65	55	45	200
	Total	70	130	110	90	400

Let us assume that two variables are independent. Then Prob. of $x \times$ Prob. of y = Prob. joint occurrence of x and y.

or $$N \times P(x) \times P(y) = N \times P(x, y)$$
$$= \text{expected no. of } x \text{ and } y \text{ out of } N$$

From the above table: $P(x_1) = \frac{70}{400}$, $P(y_1) = \frac{80}{400}$.

$\therefore$ Expected no. of x_1 and y_1 out of 400

$$= \frac{70}{400} \times \frac{80}{400} \times 400 = \frac{70 \times 80}{400} = 14.$$

Calculating in this manner the expected values for all combinations of x and y values, we have:

Expected values on the assumption of x and y independent.

		x 1	2	3	4
	1	14	26	22	18
y	2	21	39	33	27
	3	35	65	55	45

As the expected values are the same as the given values, the probability of joint occurrence of x and y is the same as obtained by assumption of independence.

Hence x and y are independent or there is no association between x and y.

Illustration 21.14. Calculate the coefficient of contingency from the following data relating social status and intelligence:

Social Status	*Dull*	*Average*	*Brilliant*
Lower Middle	22	35	23
Middle	38	70	32
Upper Middle	60	20	20

Solution. Given table is:

Social Status	*Dull* (I_1)	*Average* (I_2)	*Briliant* (I_3)	*Total*
Lower Middle (S_1)	22	35	23	80
Middle (S_2)	38	70	32	140
Upper Middle (S_3)	60	20	20	100
Total	120	125	75	320

Let the two attributes, social status and intelligence be independent of each other. On this assumption, the expected cell frequency corresponding to any cell is the product of row and column totals in which that cell occurs divided by the total number of items. So we have

CALCULATION OF χ^2

Cell	freq Observed O	Expected freq. E	$O-E$	$\frac{(O-E)^2}{E}$
S_1I_1	22	(80×120)/320=30	−8.0	2.133
S_1I_2	35	(80×125)/320=31.25	3.75	0.450
S_1I_3	23	(80×75)/320=18.75	4.25	0.963
S_2I_1	38	(140×120)/320=52.5	−14.5	4.004
S_2I_2	70	(140×125)/320=54.7	15.3	4.279
S_2I_3	32	(140×75)/320=32.8	−0.8	0.019
S_3I_1	60	(100×120)/320=37.5	22.4	13.500
S_3I_2	20	(100×125)/320=39.1	−19.1	9.330
S_3I_3	20	(100×75)/320=23.4	−3.4	0.494
Total	320	320	0	35.172

$$\therefore \quad \chi^2=\Sigma\frac{(O-E)^2}{E.}=35.172$$

Coefficient of contingency

$$C=\sqrt{\frac{\chi^2}{N+\chi^2}}=\sqrt{\frac{35.172}{320+35.172}}$$

$$=\sqrt{\frac{35.172}{355.172}}=\sqrt{0.903}=0.95 \text{ approx.}$$

Illustration 21.15. Is there any association between social status and intelligence in the data of illustration 21.14 above.

Solution. Degreess of freedom in a contingency table=$(r-1)(c-1)$, where r=no. of rows and c=no. of columns. So in the present case degree of freedom=$(3-1)(3-1)=4$.

For 4 degrees of freedom, value of χ^2 at 5% level of significance from tables is 9.49.

As the calculated value (35.172) is greater than the table value the difference between the observed and expected values is significant. Hence there is a significant association between social status and intelligence.

Illustration 21.16. Examine, by any suitable method, whether the nature of area is related to voting preference in the election for which the data are tabulated below:

Votes for	*A*	*B*	*Total*
Rural Area	620	480	1100
Urban Area	380	520	900
Total	1000	1000	2000

Solution. Let voting for *A* or *B* be independent of area:

CALCULATION OF χ^2

Cell	*Observed freq.* *O*	*Expected freq.* *E*	*O—E*	$\frac{(O-E)^2}{E}$
A—Rural	620	550	70	8.9
A—Urban	380	450	—70	10.9
B—Rural	480	550	—70	8.9
B—Urban	520	450	70	10,9
Total	2000	2000	0	39.6

$$\chi^2 = \Sigma \frac{(O-E)^2}{E} = 39.6 \text{ with one degree of freedom.}$$

As this value is much greater then the χ^2 value for one degree of freedom at 5% level of significance, hypothesis of independence is rejected. Hence voting for *A* and *B* is affected by area. As the votes for *A* in rural area is more than expected, so rural area prefers *A* and urban area prefers *B*.

Illustration 21.17. In a sample study of 100 items, two characteristics A and B were noted. Following were the various class frequencies.

$$N=100,\ (A)=80,\ (B)=70,\ (AB)=62$$

Is there any association between A and B?

Solution. The remaining class frequencies are

$$(\alpha)=N-(A)=100-80=20,\ (\beta)=N-(B)=100-70=30$$

$$(A\beta)=(A)-(AB)=80-62=18,\ (\alpha B)=(B)-(AB)=70-62=8$$

and $(\alpha\beta)=(\alpha)-(\alpha B)=20-8=12$. Putting in the tabular form:

	A	α	*Total*
B	62	8	70
β	18	12	30
Total	80	20	100

Let A and B be independent. Calculating the expected values we get:

CALCULATION OF χ^2

Items	*Observed freq.* O	*Expected freq.* E	$O-E$	$\frac{(O-E)^2}{E}$
AB	62	(70×80)/100=56	6	0.643
$A\beta$	18	(30×80)/100=24	−6	1.500
αB	8	(20×70)/100=14	−6	2.571
$\alpha\beta$	12	(20×30)/100= 6	6	6 000
Total	100	100	0	10.714

$$\therefore \qquad \chi^2=\Sigma\frac{(O-E)^2}{E}=10.714$$

Degree of freedom $=(2-1)(2-1)=1$.

From tables, χ^2 value for one degree of freedom at 1% level of significance is 6.64. The calculated value is much more than this. Therefore hypothesis of independence is rejected.

Hence A and B are associated. They are associated positively because the given value of (AB) is more than the expected value on the assumption of independence.

Illustration 21.18. Following two tables give the condition of home and condition of child in a sample study in two towns. Is there any association between the two? If so, in which place is it more?

Town A:

Condition of child	Condition of home: Clean	Not clean
Clean	70	30
Not clean	20	60

Town *B*:

Condition of home	Condition of child: Clean	Fairly Clean	Dirty
Clean	76	40	24
Not Clean	44	20	46

Solution. As it is sample study for finding association, χ^2-test will have to be applied. Further as the two contingency tables are of different order for comparing intensity of association Tschuprow's coefficient will be suitable measure.

The given data is

Condition of home	Town A Condition of child: Clean (Y_1)	Not clean (Y_2)	Total	Town B Condition of child: Clean (Y_1)	Fairly clean (Y_2)	Dirty (Y_3)	Total
Clean (X_1)	70	20	90	76	40	24	140
Not clean (X_2)	30	60	90	44	20	46	110
Total	100	80	180	120	60	70	250

Let there be no association between conditions of home and conditions of child. Calculating expected values on this assumption we have:

CALCULATION OF χ^2

Items	Town A: Observed values O	Expected values E	$\frac{(O-E)^2}{E}$	Town B: Observed values O	Expected values E	$\frac{(O-E)^2}{E}$
Y_1X_1	70	50	8.00	76	67.20	1.153
Y_1X_2	30	50	8.00	44	52.80	1.467
Y_2X_1	20	40	10.00	40	33.60	1.219
Y_2X_2	60	40	10.00	20	26.40	1.552
Y_3X_1	—	—	—	24	39.20	5.892
Y_3X_2	—	—	—	46	30.80	7.501
Total	180	180	36.00	250	250.00	18.784

Now $\chi^2 = \Sigma \frac{(O-E)^2}{E}$

$\therefore$ For town A, $\chi^2 = 36.00$, with degree of freedom as $(2-1)(2-1)$ *i.e.*, 1

For town B, $\chi^2 = 18.784$, with degrees of freedom as $(3-1)(2-1)$ *i.e.*, 2.

From Table χ^2 value for 1 degree of freedom at 1% level of significance is 6.35 and for 2 degrees of freedom 9.210. As in both the cases calculated value of χ^2 is greater than the table value, hypothesis of independence stands rejected. Or in other words there is highly significant association between home conditions and child conditions.

Tschuprow's coefficient T is given by

$$T^2 = \frac{C^2}{(1-C)\sqrt{(s-1)(t-1)}}$$

where $C^2 = \frac{\chi^2}{N+\chi^2}$ and $(s-1)(t-1)$ = degrees of freedom.

$$C^2 \text{ for town } A = \frac{36}{180+36} = \frac{1}{6} = 0.1667$$

$$C^2 \text{ for town } B = \frac{18.784}{250+18.784} = \frac{18.784}{268.784} = .0699$$

$$\text{For town } A,\ T^2 = \frac{\frac{1}{6}}{(1-\frac{1}{6})\sqrt{1}} = \frac{1}{5} = 0.2$$

or $$T = \sqrt{0.2} = 0.447$$

$$\text{For town } B,\ T^2 = \frac{0.0699}{(1-0.0669)\sqrt{2}} = \frac{0.0699}{.9301 \times 1.414} = .0531$$

$\therefore$ $$T = \sqrt{0.0531} = .230$$

As Tschuprow's coefficient for town A is greater, there is greater association in town A between condition of child and condition of home.

Illustration 21.19. A list of 1000 items classified according to three attributes is given below. A, B and C denote the presence and a, b, c denote absence of the attributes. Show if there is any association between A and B and A and C.

$ABC = 300$, $ABc = 200$, $AbC = 20$, $Abc = 80$,

$aBC = 150$, $aBc = 50$, $abC = 120$, $abc = 80$

Solution. Coefficient of association gives the indication of association.

Coeff. of association for A and $B(Q_{AB})=\dfrac{(AB)(ab)-(Ab)(aB)}{(AB)(ab)+(Ab)(aB)}$

and Coeff. of association for A and C $(Q_{AC})=\dfrac{(AC)(ac)-(aC)(Ac)}{(AC)(ac)+(aC)(Ac)}$

Now $(AB)=(ABC)+(ABc)=300+200=500$

$(ab)=(abC)+(abc)=120+80=200$

$(Ab)=(AbC)+(Abc)=20+80=100$

$(aB)=(aBC)+(aBc)=150+50=200$

$(AC)=(ABC)+(AbC)=300+20=320$

$(ac)=(aBc)+(abc)=50+80=130$

$(aC)=(aBC)+(abC)=150+120=270$

$(Ac)=(ABc)+(Abc)=200+80=280$

$$\therefore \quad Q_{AB}=\frac{500\times 200-100\times 200}{500\times 200+100\times 200}=\frac{80000}{120000}=0.67$$

and $$Q_{AC}=\frac{320\times 130-270\times 280}{320\times 130+270\times 280}=\frac{-34000}{117200}=-0.29$$

Hence there is positive association between A and B, and a negative association of lesser degree between A and C.

Note. A still better conclusion can be drawn by studying partial association.

Illustration 21.20 Calculate the partial and total association between A and B, and A and C from the following data. Give your comments.

$ABC=100$, $AB\gamma=110$ $\quad A\beta C=105$, $A\beta\gamma=95$.
$\alpha BC=107$, $\alpha B\gamma=55$. $\quad \alpha\beta C=86$ $\quad \alpha\beta\gamma=89$.

Solution. Coefficient of partial associations:

(i) Between A and B for sub-population C

$$\left(Q_{AB.C}\right)=\frac{(ABC)(\alpha\beta C)-(A\beta C)(\alpha BC)}{(ABC)(\alpha\beta C)+(A\beta C)(\alpha BC)}$$

$$=\frac{100\times 86-105\times 107}{100\times 86+105\times 107}=\frac{-2635}{19835}=-0.1328.$$

(ii) Between A and B for sub-population γ

$$\left(Q_{AB.\gamma}\right)=\frac{(AB\gamma)(\alpha\beta\gamma)-(A\beta\gamma)(\alpha B\gamma)}{(AB\gamma)(\alpha\beta\gamma)+(A\beta\gamma)(\alpha B\gamma)}$$

$$=\frac{110\times 89-95\times 55}{110\times 89+95\times 55}=\frac{4565}{15015}=0.3040.$$

(iii) Between A and C for sub-population B

$$\left(Q_{AC.B}\right)=\frac{(ABC)(\alpha B\gamma)-(AB\gamma)(\alpha BC)}{(ABC)(\alpha B\gamma)+(AB\gamma)(\alpha BC)}$$

$$= \frac{100 \times 55 - 110 \times 107}{100 \times 55 + 110 \times 107} = \frac{-6270}{17270} = -0.3631$$

(iv) Between A and C for sub-population β

$$\left(Q_{AC.\beta}\right) = \frac{(A\beta C)(\alpha\beta\gamma) - (A\beta\gamma)(\alpha\beta C)}{(A\beta C)(\alpha\beta\gamma) + (A\beta\gamma)(\alpha\beta C)}$$

$$= \frac{105 \times 89 - 95 \times 86}{105 \times 89 + 95 \times 86} = \frac{1175}{17615} = 0.0671$$

Total association between A and B

$$\left(Q_{AB}\right) = \frac{(AB)(\alpha\beta) - (A\beta)(\alpha B)}{(AB)(\alpha\beta) + (A\beta)(\alpha B)}$$

And total association between A and C

$$\left(Q_{AC}\right) = \frac{(AC)(\alpha\gamma) - (A\gamma)(\alpha C)}{(AC)(\alpha\gamma) + (A\gamma)(\alpha C)}$$

Now $(AB) = (ABC) + (AB\gamma) = 100 + 110 = 210$

$(\alpha\beta) = (\alpha\beta C) + (\alpha\beta\gamma) = 86 + 89 = 175$

$(A\beta) = (A\beta C) + (A\beta\gamma) = 105 + 95 = 200$

$(\alpha B) = (\alpha BC) + (\alpha B\gamma) = 107 + 55 = 162$

$(AC) = (ABC) + (A\beta C) = 100 + 105 = 205$

$(\alpha\gamma) = (\alpha B\gamma) + (\alpha\beta\gamma) = 55 + 89 = 144$

$(A\gamma) = AB\gamma + (A\beta\gamma) = 110 + 95 = 205$

$(\alpha C) = (\alpha BC) + (\alpha\beta C) = 107 + 86 = 193$

$$\therefore \quad Q_{AB} = \frac{210 \times 175 - 200 \times 162}{210 \times 175 + 200 \times 162} = \frac{4350}{69150} = 0.0629$$

$$Q_{AC} = \frac{205 \times 144 - 205 \times 193}{205 \times 144 + 205 \times 193} = \frac{-10045}{69085} = 0.1454$$

The above calculation shows than A and B have small positive association in the whole population. But the association between A and B in the items which have attribute C is negative (—0.1328), and in the terms which do not have attribute C the association is positive, (0.3040). Thus the conclusion only on the basis of total association is illusory. For correct picture of association between A and B all the three associations between A and B must be taken together. Similarly negative total association between A and C is illusory. There is a high negative association between A and C for those items which have attribute B and small positive association between A and C for the items which do not have attribute B. Thus there is a tendency for A and B to occur together when C is absent and A and C to come together when B is absent.

Illustration 21.21. Comment critically on the following statements.

(*a*) More surtax payers die in a year than the general death rate would indicate, thus it is unhealthy to be rich.

(*b*) There should be no discrimination in regard to the rates of dearness allowance payable to bachelor and married gazetted officers because enquiries show that 80% of bachelor government employees have aged parents or other dependents to support.

(*c*) 99% of the people who drank beer die before reaching 100 years of age. Therefore drinking beer is bad for longevity.

Solution. (*a*) Surtax payers are usually aged persons. More aged persons die in a year than the general death rate would indicate. This is why more surtax payers die in a year then the general death rate would indicate. In fact there may be negative association between richness and unhealthiness. So the result drawn is illusory. The whole thing can be put in other manner. Take a group of persons which have all conditions similar to that of surtax payers except that they are not surtax payers. If in this group the death rate is less than the surtax payers' group then only we can say, it is unhealthy to be rich.

(*b*) 80% of bachelor goverment servants have aged parents or other dependents to support. Married people also have parents. The conclusion drawn is one-sided. We should find out wether married people have also to support parents and other dependents in addition to their wife and children. If so, married people have more dependents and hence there is no justification in the given argument.

(*c*) The conclusion, drinking beer is bad for longevity can only be drawn, if we compare two groups. One who drinks beer with the one who does not drink beer. If the percentage of people in the group who drink beer is larger than the other, then only given conclusion is justified otherwise not. It may happen that more than 99% people who do not take beer may die before reaching 100, in such a case not taking beer may prove to be bad for longevity.

Illustration. 21.22. Comment critically on the following statements:

(*a*) Road accidents resulted in 4513 deaths in 1958 and 5250 in 1967 while the number of women drivers increased in the period. Hence women make bad drivers.

(*b*) Of 300 persons inoculated against influenza during the epidemic only 12% were affected. This shows a marked improvement comparing with the remainder of the population for which the equivalent figure was 28%.

(c) Non-cultivating land-owners should be deprived of their ownership rights without payment of compensation because 90% of the land owned by them is inherited property for which they have paid nothing.

Solution. (*a*) Women can be proved bad drivers only if number of accidents per man driver. The increase in number of accidents may be due to increase in population, more congestion on roads etc. In fact for such a conclusion as given here sufficient data is not given.

(*b*) There are two attributes here. Inoculation and occurrence of influenza. We are given that out of inoculated only 12% were affected by influenza and out of not inoculated 28% were affected. So the conclusion that inoculation is effective is correct, provided the two groups of persons, inoculated and not inoculated are identical in respect of other things such as health conditions, standard of living, sanitation of localities etc. If it is not so, the positive association between inoculation and not affected by influenza may be due to the positive association of inoculation with some third attribute like good health or good standard of living or good sanitation of localities etc.

(*c*) There are two main attributes here. First, owners of land with two alternatives: (*a*) Cultivating owners and (*b*) non-cultivating owners. Second, property with two alternatives (*a*) inherted and (*b*) self-acquired. According to the given statement compensation is not to be paid if there is a positive association between non-cultivating owners and inherited property. But cultivating owners can also have inherited property. It is just possible that more than 90% of their land owned may be inherited. In that case if inheritance is the only critera for deprivation of ownership without payment of compensation, cultivating owners should also be deprived of ownership without payment. In the problem we are not given the percentage of the land inherited by the cultivating owners. So statistically there is no justification in drawing the conclusion as given in the problem. Moreover there are many other considerations also such as Government policy, political and social considerations which would determine the payment of compensation, in the absence of which nothing definite can be said.

21.13 MISCELLANEOUS ILLUSTRATIONS

Illustration. 21.23. Do you find any association between the tempers of brothers and sisters from the following data:

Good natured brothers, Good natured sisters = 1040

Good natured brothers, Sullen sisters = 160

Sullen brothers, Good natured sisters = 180

Sullen brothers, Sullen sisters = 120

Solution. Let A and B denote good natured brother and sister respectively,

and α and β the sullen natured brother and sister respectively.

Thus, $(AB)=1040$; $(A\beta)=160$; $\alpha B=180$

and $(\alpha\beta)=120$

Yules coefficient of association

$$\phi=\frac{(AB)(\alpha\beta)-(A\beta)(\alpha B)}{(AB)(\alpha\beta)+(A\beta)(\alpha B)}$$

$$=\frac{1040\times 120-160\times 180}{1040\times 120+160\times 180}$$

$$=0.625.$$

a positive association

Illustration. 21.24. Do you find any inconsistency in the data

$N=1000$, $(A\beta)=483$, $(A\gamma)=378$, $(B\gamma)=226$

$(A)=525$, $(B)=312$, $(C)=470$, and $(ABC)=25$

Solution. As there are 3 attributes, there will be 8 class frequencies: (ABC), $(AB\gamma)$, $(A\beta C)$, $(A\beta\gamma)$, (αBC), $(\alpha B\gamma)$, $(\alpha\beta C)$, and $(\alpha\beta\gamma)$.

$(AB)=(A)-(A\beta)=525-483=42$

$(AC)=(A)-(A\gamma)=525-378=147$

$(BC)=(B)-(B\gamma)=312-226=86$

$(AB\gamma)=(AB)-(ABC)=42-25=17$

$(A\beta C)=(AC)-(ABC)=147-25=122$

$(A\beta\gamma)=(A\beta)-(A\beta C)=483-122=361$

$(\alpha BC)=(BC)-(ABC)=86-25=61$

$(\alpha B\gamma)=(\alpha B)-(\alpha BC)=(B)-(AB)-(\alpha BC)$

$=312-42-61=209$

$(\alpha\beta C)=(\alpha C)-(\alpha BC)=(C)-(AC)-(\alpha BC)$

$=470-147-61=262$

$(\alpha\beta\gamma)=(\alpha\beta)-(\alpha\beta C)=(\beta)-(A\beta)-(\alpha\beta C)$

$=N-(B)-(A\beta)-(\alpha\beta C)$

$=1000-312-483-262=-57$

As one class frequency is negative, the data is inconsistent.

Illustration 21.26. In a co-educational institution, out of 200

students, 150 were boys. They took an examination and it was found that 120 boys passed and 10 girls failed. Is there any association between sex and success in examination?

Solution. Let A denote a boy, and B denote a success. The contingency table is:

	Passed B	Failed β	Total
Boys (A)	120	30	150
Girl (α)	40	10	50
Total	160	40	200

Based on the null hypothesis that there is no association, the expected frequencies are

$$E_{AB}=\frac{150\times 160}{200}=120$$

$$E_{A\beta}=150-120=30$$

$$E_{\alpha B}=160-120=40$$

and $E\alpha\beta = 50-40 = 10$

Since all the expected frequencies are the same as the observed, we may safely say that sex and sucess in exammation are not associated.

PROBLEMS

1. From the following data calculate the class frequencies not given: $N=100$, $(A)=50$, $(B)=70$ and $(AB)=30$.
2. From the following set of ultimate class frequency, calculate the set of postives class frequencies: $(AB) = 10$, $(A\beta) = 20$, $(\alpha B) = 15$ and $(\alpha\beta) = 25$ (Ans. $(A)=30$, $(B)=25$ and $N=70$)
3. Of 598 men in a locality exposed to cholera 147 in all were attacked, of 598 men 137 were inoculated and of these only 14 were attacked. Find the number of persons not inoculated not attacked, inoculated not attacked and not inoculated attacked.
4. In certain class it was found that 60% of students passed in terminal examination. 30% students passed in terminal and annual examination while 25% were such who passed in annual but failed in terminal examination. Find the percentage of students who passed in annual examination, passed in terminal but failed in annual examination, and failed

both examinations.

(**Ans. Percentage of students who passed in annual = 55%, percentage of students who failed in annual but passed in terminal = 30% and who failed in both examinations = 15%**)

5. Find all the remaining class frequencies from the following data: $N=800$, $(A)=244$, $(B)=301$, $(C)=150$, $(AB)=125$, $(AC)=72$, $(BC)=60$ and $(ABC)=32$.
6. Given the following set of class frequencies, find the remaining class frequencies: $(\alpha\beta\gamma)=45$, $(\alpha BC)=37$, $(\alpha B\gamma)=28$, $(ABC)=20$. $(AB\gamma)=27$, $(A\beta C)=39$, $(A\beta\gamma)=46$, $(\alpha BC)=33$.
7. From the following set of class frequencies, find the remaining class frequencies: $N=1{,}000$, $(A)=205$, $(B)=115$, $(C)=100$, $(AB\gamma)=21$, $(A\beta C)=18$, $(\alpha BC)=17$ $(ABC)=12$.
8. In a certain college it was found that 380 students passed in annual examination, 100 students passed in all the three, first terminal, second terminal and annual examination, 120 students passed in two terminals but failed in annual, while 170 passed in annual but failed in both the terminals. Find the number of students who passed at least two examinations.

   (Ans. 330 students passed at least two examinations).
9. Obtain all the class frequencies in the following example: At an examination at which 600 candidates appeared, boys outnumbered girls by 16%. Also those passing the examination exceed in number those failing by 310. The number of successful boys choosing science subject was 300, while among the girls offering arts subjects there were 25 failures, altogether only 135 offered arts and 33 among them failed. Boys failing in the examination numbered 18.
10. In a report the following was given: "50% of items have characters A and B, 35% have A but not B, 25% have B but not A." Show there must be misprint in the data.
11. In a report on consumers' preference it was given that out of 500 persons surveyed 410 preferred variety A, 380 preferred variety B and 270 persons were such who gave their likings for both the varieties. Is there any inconsistency in the data?

    (Ans. The data are inconsistent.)
12. Prove that the following data are inconsistent: $N=1{,}000$, $(A)=525$, $(B)=312$, $(C)=470$, $(AB)=483$, $(AC)=378$, $(BC)=226$, $(ABC)=25$.
13. A study was made about the studying habits of the students of a certain university. Following summary was given at one place in the report: "Of the students surveyed 75% were from well-to-do families, 55% were boys and 60% were irregular in their studies. Out of the irregular ones 50% were boys and 2/3 from well-to-do families. The percentage of irregular boys from well-to-do families was 8% of all the students." Show that there is some inconsistency in the data.
14. **A universe consists of three attributes each of which is divisible to two parts. What are the different class frequencies obtainable?**

    A market investigator returns the following:

Of 1000 people consulted 811 liked chocolates, 752 liked toffee and 418 liked boiled sweets; 570 liked chocolates and toffee, 356 liked chocolates and boiled sweets and 348 liked toffee and boiled sweets, 297 liked all three. Show that this information as it stands must be incorrect.

15. The following is a summary of the statistical features of a census of ration cards:

Item No.	*Categories*	*Total No. of cards*
1.	The whole of census	1,000
2.	Permanent residents	510
3.	Male	490
4.	Consumers of rice	427
5.	Permanent male residents	189
6.	Consumers of rice among permanent residents	140
7.	Males consuming rice	97

Show that the entry against item No. 7 is inconsistent with the entries against all the previous items 1, 2, 3, 4, 5 and 6 taken together.

16. 700 person of London were asked by a B.B.C. investigator to give the nationality of the music they liked. They returned the following data: 570 liked English, 650 liked French, 480 liked German, 400 liked English and French, 360 liked French and German, 240 liked English and German and 225 liked all the three.
Show that the information as it stands must be incorrect.

17. The following table gives the number of persons suffering from certain infirmities in UP in 1971:

Sex	*Total No.*	*Insane*	*Deaf mutes*	*Deaf Mutes and insane*
Males	260 lakhs	12,650	21,301	545
Females	241 lakhs	9,055	14,136	317

Trace the association between insanity and deaf muteness for males and females of UP separately.

18. Calculate the coefficient of association between intelligence in father and son from the following data:

Intelligent fathers with intelligent sons	248
Intelligent fathers with dull sons	81
Dull fathers with intelligent sons	92
Dull fathers with dull sons	57

19. From the table below, compare the intensity of association between literacy and unemployment among males in urban areas with that in the rural areas:

	Total (N)	*Literate* (A)	*Unemployment* (B)	*Literate unemployed* (AB)
Urban (Lakhs)	25	10	5	3
Rural (Lakhs)	200	40	12	4

(**Ans.** Urban Areas = .472, Rural Areas = .375)

20. Calculate the coefficient of association between extravagance in fathers and sons from the following data:

Extravagant fathers with extravagant sons	327
Extravagant fathers with miserly sons	545
Miserly fathers with extravagant sons	741
Miserly fathers with miserly sons	234

21. Define independence in probability. X and Y are two variables X taking the values 1, 2, 3, 4, and Y 1, 2, 3; the frequency for different values of x and y are as follows:

		x		
	1	2	3	4
y 1	14	25	22	18
2	21	39	33	27
3	35	65	55	45

Examine from definition or otherwise, whether x and y are independent.

22. Given that 50% of the inmates of a work house are 'men', 60% are 'aged', 80% 'nonable bodied', 35% 'aged mean'. 45% 'nonable bodied men and 42% 'nonable bodied' and aged; find the greatest and least possible percentage of nonable bodied aged men.

(Ans. The percentage of nonable bodied aged men must be in between 30% to 32%).

Chapter 22

VITAL STATISTICS

22.1. INTRODUCTION

A statistical study of human population has two aspects, *viz.*, (*i*) a study of the composition of population at a point of time, and (*ii*) a study of the changes that occur during a given period *i.e.*, growth or decline of the population. Change in the population is the outcome of events like births, deaths, migration, marriages, divorces, etc., called "vital events".

These two aspects of study have given rise to two methods for the collection of population data: (*i*) census taking, and (*ii*) registration of vital events. Vital statistics is the application of statistical methods to the study of these facts, and has been defined as "the registration, separation, transcription, collection, compilation and preservation of data pertaining to the dynamics of the population, in particular pertaining to births, deaths, marital status and the data and facts incidental thereto". The distinction between these two methods of collecting demographic data is that the former is a record of persons while the latter is a record of events.

The population of a given geographic area at any point of time may be expressed as

$$P(t)=P(o)+B(t)-D(t)+I(t)-E(t)$$

where $P(t)$ represents total population at a given point of time,

$P(o)$ total population at the point of time taken as base,

$B(t)$ total number of births during the given period,

$D(t)$ total number of deaths during the given period,

$I(t)$ total numbrer of immigrant, and

$E(t)$ total number of emigrants.

There are thus four factors (or constants) which affect the size of population, *viz.*, births, deaths, immigration and emigration. Of these deaths and births are ordinarily the chief determinants of the chang in population and hence a great emphasis on the study of Mortality and Fertility.

22.2. MEASURES OF MORTALITY

Mortality is not a single factor to be expressed as single number of index. The risk of death has to be measured in several aspects and as such various kinds of death rates are employed.

Total crude death rate. A very common measure of decreases in population due to death is the crude death rate. It represents deaths per thousand of the population. The formula for calculating this rate is

$$\text{C.D.R.} = \frac{\text{Number of deaths in the population of a given geographic area during a given year}}{\text{Mid-year total population of the given geographic area during the given year}} \times 1{,}000$$

Within broad limits, it gives the probability of dying of a person in the population.

Crude death rate is widely used as an index of mortality. It is easy to compute and also to understand. It requires only the total number of deaths and the total mid-year population in a given period of time. It gives preliminary indication of the level of mortality. But crude death rate has its limitations too—particularly in inter-area comparison. Mortality normally varies with age. If the age structures of the two populations are different, the comparison of crude death rates may be misleading. A population having a large number of old persons will show a higher crude death rate than the population having a large number of young persons.

Crude death rate can be easily adopted for making comparisons for the same area from year to year, because the composition of the population is not likely to change much within a year. But for fitting a long term trend the effect of such changes should be allowed for.

Age specific death rates. Crude death rate simply reveals the average number of deaths per 1,000 persons including infants. children, adolescents, young and old persons. It does not give an exact idea about the death rate in a particular section of the population. In order to study the mortality conditions for a particular section of the population, say, death rate among infants under one year of age, we have to study the deaths occurring among the infants. Such a specific study is known as age specific death rates. In practice, insurance companies are interested in death rates in various age-groups of the population. Particularly, age specific death rates afford a sound basis for comparison between two populations. Here we shall study some of the type of age specific death rates.

Infant mortality rate. In infant mortality rate, we study the death rates under one year of age of the newly born babies in a given period of time. Generally, the risk of death is greater during the first year of life then afterwards, and as such infants mortality rate is higher than the crude death rate. It is, therefore, regarded as one of the most sensitive index of health conditions of the general population. The formula of its calculation is:

$$\text{Annual infant mortality rate} = \frac{\text{Total number of deaths under one year of age which occurred among the population in a given year.}}{\text{Total number of live births which occurred among the population of the given geographic area during the same year}} \times 1{,}000$$

But the calculation of infant mortality rate may suffer from one limitation, *viz*., babies who die soon after their births, may not be registered at all, as births or deaths. But even such, it has its uses for comparison purposes. When such a rate is computed separately for rural and urban areas, it will reveal clearly the uneven distribution of health facilities. The infant mortality rates may be computed separately for various geographical sub-divisions of specific population groups in order to study the effect of factors as overcrowding, unemployment, social habits and customs, dietary habits etc.

Mortality during childhood. Just as we calculate the infant mortality rate, *i.e.*, death rate among infants under one year of age, similarly we can calculate the death rate for children between 1-4 years of age and between 5-15 years of age. Usually mortality rate during childhood is lower than the infant mortality rate.

Maternal mortality rate or mortality of reproductive ages. Maternal mortality rates measures the risk of dying from causes associated with child-birth in the various age-groups in the reproductive span of life 15-49 years of age. At these ages, death rates among women is generally higher that those of men. The formula is:

$$\text{Annual maternal mortality rate} = \frac{\text{Total number of deaths due to child birth among the female population of given geographic area during a given year}}{\text{Total number of the birth which occurred among the population of the given geographic area during the same year}} \times 1000$$

Mortality at advanced ages. Death rate for higher age-group, for 60 years and above, is quite high, perhaps greater than infant mortality rate. Generally, for advanced ages only one death rate is calculated, 60 years or above, or it may be 65 years or above.

For purposes of comparison and to find out the trend or mortality conditions, it is better to consider age specific death rates rather than the crude death rate of the population at all ages. Taking the death rate of a particular year as the base, we can find out the general trend of the death rate in various sectors of the population and then we can make comparison of the various regions or the various countries. Further, for comparison of age specific death rates in the population of two regions, the age intervals should be uniform.

22.3. STANDARD DEATH RATE

We have seen that crude death rates cannot be used for comparisons of two populations because the two populations may have different compositions as regards age and sex. Age specific and sex specific death rates give a large bulk of data which does not facilitate comparison. As always in statistics, we like to obtain an index representing the overall death rate. For purpose of comparing the death rates in two geographic areas, it is therefore, essential that age and sex differences in the compositions of two populations should be eliminated in construction of such an index. There are two methods of doing it; *viz.*, (1) Direct, and (2) Indirect, and the rate obtained by each one of these methods is called the standardised Death Rate.

Direct method. Under this method the mortality rates at each age-group in the two geographical areas are applied to some common standard population. Thus, we would get a total death rate in that standard population if it were exposed to mortality rates of one area and another total death rate when it is exposed to mortality rates of another area. These total rates, called standerdised rates, show what would be the mortality rates in each one of the two areas if they had populations which were similar in their age and sex distributions. These rates have only one purpose, *i.e.*, comparison. By themselves they are fictitious.

If the age specific mortality rates of Town A and B are as given in Table 22.1, one can easily see the shortcoming of the crude death rate. A comparison of the two districts shows that in every age-

TABLE 22.1. CALCULATION OF CRUDE DEATH RATE

Age-group	District A			District B		
	Population	*Death*	*Rate per thousand*	*Population*	*Death*	*Rate per thousand*
0-10	4,000	36	9	3,000	30	10
10-25	12,000	48	4	20,000	100	5
25-60	6,000	66	11	4,000	48	12
60 and over	8,000	158	19.75	3,000	60	20
	30,000	308	10.26	30,000	238	7.93

group *A* has a lower death rate than *B*. Yet its crude death rate is higher than that of *B*. The fallacy is due to the fact that different age groups having varying death rates have unequal importance in the two populations.

However, if we apply these rates to a standard population as shown in Table 22.2 we get a truer picture of mortality rate in the districts.

TABLE 22.2. CALCULATION OF STANDARD DEATH RATE (DIRECT METHOD

Age-group	*Standara Population* *Population*	*Town A* *Mortality rate*	*Total deaths*	*Town B* *Mortality rate*	*Total death*
0-10	1,000	9	9	10	10
10-25	4,000	4	16	5	20
25-60	3,000	11	33	12	36
60 and over	2,000	19.75	39.5	20	40
All ages	10,000		97.75		106
Standardised death rate per 1,000			9.75		10.6

Indirect method. The direct method of standardising the death rate requires a knowledge of the mortality rates of each age-group in the population for which standardised death rate is needed (*see* Table 22.2). Sometimes this information is not available. In such instances indirect method of standardisation is applied. The first step in this method is the selection of a series of standard death rates for each age-group (*see* col. 2. Table 22.3.). These rates are then applied to the population at various age-groups, in the areas, the standardised rate of which is sought, to determine the number of deaths that would have occurred in each area if it had the standard mortality rates (*see* cols. 4 and 6 of Table 22.3). It is thus found that if standard mortality rates are applied to the population of Town *A* and Town *B* respectively there would be 1,072 deaths in Town *A*, and 1,124 deaths in Town *B*, or rates of mortality at all ages would be 16.75 and 18.73 per thousand respectively. These rates are called index rates for their level is an index of the type of population from which they have been derived. Thus, if a population has a large proportion of old persons and infants its index rate would be higher than that of a population composed of young persons. It the two rates are different (as they are in the illustration) it shows that the two populations are not identical so far as their age structure is concerned and hence if their crude death rates are to be compared some adjustment of them (crude rates) is necessary to allow for the difference in population type. These adjustments (called standardising factors) are determined by dividing the standard mortality rate (for all ages (15 in this case) by the index rates of each one of the two towns, *viz.* 15/16.75, *i.e.*, 0.896 for Town *A* and 15/18.73, *i.e.*, 0.801 for Town *B*. Now if the given crude rate for Town *A* is 20 and for Town *B* is 22 the standardised death rates would be as in Table 22.3

TABLE 22.3. CALCULATION OF STANDARD DEATH RATE (INDIRECT METHOD)

Age-group	*Standard Mortality rate*	*Town A*		*Town B*	
		Population	*Number of deaths that would occur at standaed rate*	*Population*	*Number of deaths that would occur at standard rate*
Under 2	64	3,000	192	5,000	320
2-10	7	10,000	70	12,000	84
10-20	4	10,000	40	10,000	40
20-60	8	32,500	260	25,000	200
60 and above	60	8,500	510	8,000	480
	15	64,000	1,072	60,000	1,124

$$\text{Index For Town } A = 15/\left(\frac{1{,}072}{64{,}000} \times 1{,}000\right) = 0.896$$

$$\text{For Town } B = 15/\left(\frac{1{,}124}{60{,}000} \times 1{,}000\right) = 0.801$$

Crude death rates given

Town A : 20

Town B : 22

Standard death rates

Town A : $20 \times 0.896 = 17.92$

Town B : $22 \times 0.801 = 17.622$

According to this, Town B is healthier than Town A.

22.4. MEASURES OF FERTILITY

To study the growth of the population of a given geographic area in a given period of time we have to take into consideration the number of live births that occur. As Barclay has put it, 'fertility is an actual level of performance, based on the numbers of live births that occur' It is ascertained from the data collected by registration of births, and is measured as the frequency of births in a population. To calculate the frequency or speed by which population is increasing we use some fertility rates. Some of the important fertility rates are as follows:

Crude birth rate. It gives the average number of births per 1,000 persons in the population of a given area during a given period of time. It is calculated as follows:

$$\text{Annual crude birth rate} = \frac{\text{Total number of live births which occurred among the population of a given geographic area during a given year}}{\text{Mid-year total population of the given geographic area during the same year}} \times 1{,}000$$

Crude birth rate is similar to the crude death rate and suffers from the same limitations as that of the crude edath rate. It is affected by several factors like age and sex structure of the population, marriage rate, migration etc. In order to make comparison of the crude birth rate between the two populations, allowance should be made for differences in age and sex distribution of the population.

General fertility rate. Crudé birth rate suffers from a great limitation in as much as it relates the total number of live births to the total mid-year population. But, in fact, the total number of live births depends upon the population of woman of the child-bearing age. To study the fertility rate as such we should relate the total number of live births to the total female population of the child-bearing age. The rate so calculated is known as general fertility rate. It is computed as follows:

$$\text{General fertility rate} = \frac{\text{Total number of live births in a given population in a particular year}}{\text{Total number of females in reproductive span of life (say, 15-49 years) at the mid-year}} \times 1{,}000$$

Age specific fertility rate. Although general fertility rate is an improvement over crude birth rate, it gives only a general view of the fertility rate of the child-bearing age-group (15 to 49 years) as a whole. For a detailed study we may calculate what are known as age specific fertility rates for different child-bearing age- groups. It is calculated as follows:

$$\text{Annual age specific fertility rate} = \frac{\text{Number of live births which occurred to females of a specified age-group of the population of a given geographic area during a given year}}{\text{Mid-year female population of the specified age-group in the given geographic area during the same year}} \times 1{,}000$$

Age specific fertility rates are similar to age specific death rates. It reveals the distripution of the frequencies of birth among the female population according to age. These rates afford a detailed analysis of fertility in a given population of a given period.

Total fertility rate. If we find out the sum of age specific fertility rates at each age-group interval from 15 to 49 years of age it will give us the total fertility rate. This rate indicates, with fertility as it is, how many children will be born per thousand woman, arriving at child-bearing age provided none of these women dies before having passed the child-bearing age.

Sex ratio at birth. If we find out the ratio of total live births of males divided by total live births of females in the population of a given area in a particular period per 100, it is sex ratio at birth. It is computed as follows:

$$\text{Sex ratio at birth} = \frac{\text{Total number of live births of males in the population of a given area during a given year}}{\text{Total number of live births of females of the given area during the same year}} \times 100$$

Gross reproduction rate. 'Gross reproduction rate is the age specific fertility rates calculated from *female births* for each single year of child bearing age'. In order to calculate the age specific rate, we have to relate the females born to mothers of each specific year of age or age-groups to the total number of women of that age or the age-group of a given geographic area.

$$\text{Age specific reproduction rate} = \frac{\text{Number of } \textit{female} \text{ live births to a specified age-group of mothers of a given geographic area during a given years}}{\text{Ttoal mid year female population in that specific age-group in the given geographic area during the same year}} \times 1{,}00$$

Summing up these age specific reproduction rates for all ages in the reproductive span of life, a measure of population growth called gross reproduction rate (G.R.R.) is obtained. It provides an upper limit of the rate of population growth indicating the average number of daughters that would be born to each group of 1,000 woman beginning life together, if none died before reaching the end of the child-bearing period, and if they experienced the current rate of fertility. If G.R.R. is 1, it indicates that the current generation of females of child-bearing age will maintain itself on the basis of current fertility rate but without mortality. But if it is less than one, then no amount of reduction of deaths will enable it to escape decline sooner or later.

The computation of G.R.R. depends on the availability of the data, *i.e.*, classification of births according to age of mother at the time of birth, and according to sex. But this two way classification may not always be available. In the absence of such data, we can approximately find out the value of G.R.R. by an alternative method —by multiplying total fertility rate by the proportion of births that were female on the assumption that sex ratio at birth, *i.e.*, the ratio of the number of male births to the number of female births remained constant over all ages of mothers.

$$\text{G.R.R.} = \text{Total Fertility Rate} \times \frac{\text{Number of female births}}{\text{Total number of births}}$$

Here total fertility rate is the total number of children that would ever be born to a given group of women, if the group pased through its reproductive span of life with these birth rates at each year of age and if none died before reaching the end of the reproductive period of life.

G.R.R. provides the hypothetical upper limit of the rate of population growth. It recognises the current rate of fertility. Its drawback is that it ignores the current mortality. Some of the females who begin life together may die before reaching the upper limit of the child-bearing age. But G.R.R. does not take this aspect of mortality into consideration. This drawback is removed in the computation of net repreduction rate

Net reproduction rate. Net reproduction rate indicates the average number of daughters that would be born to a group of women beginning their life together if they are subject to the fertility and mortality rates throughout their reproductive span of life. It is computed by multiplying the age specific fertility rates of each age, by the survival factor of that age or age-group. The sum of these specific fertility rates will be N.R.R. Survival factor, *i.e.*, the proportion of female survivors to that age is available from the life table. N.R.R. uses the same age specific fertility rates, as the G.R.R. but it takes into consideration the survival factor taken from a life table.

$$NRR = \sum_{15}^{49} b_x L_x$$

where b_x represents female births at each age x,

L_x the number of years lived at each age per woman born to the original group of females and, $\sum_{15}^{49}$ represents the sum of these rates for the reproductive span of life taken from 15 to 49 years of age.

N.R.R. cannot exceed G.R.R. because it takes the mortality factor into consideration. If it is one, it indicates that on the basis of current fertility and mortality rates, a group of newely born females will exactly replace itself in the next generation, *i.e.*, the tendency of the population to remain constant. It will show a tendency of increase or decrease in population if it is greater than one or less than one, as the case may be.

However, both the gross production rate and the net reproduction rate should not be used for forecasting future population changes. Firstly, because they do not take into consideration the factor of migration. Secondly, the rates of fertility and mortality are unlikely to be the same as at present.

Illustration 22.1. The following data gives the number of women child bearing ages and yearly births by quinquinnial age groups for a city. Calculate the general fertility rate and total fertility rate. If the ratio of male to female children is 13 : 12, what is the gross reproduction rate?

Age group	15–19	20–24	25–29	30–34	35–39	40–44	45–49
Female pop. in (000)	16	15	14	13	12	11	9
Births	400	1710	2100	1430	960	330	36

Solution

Age group	*Female Pop. in 000's (P)*	*Births (B)*	*Specific fertility rate (B/P)*
15–19	16	400	25
20–24	15	1710	114
25–29	14	2100	150
30–34	13	1430	110
35–39	12	960	80
40–44	11	330	30
45–49	9	36	4
Total	90	6966	513

General fertility rate

$$= \frac{\text{Total births in one year} \times 1000}{\text{Female pop. of child bearing ages}}$$

$$= \frac{6966 \times 1000}{90{,}000} = 77.4 \text{ per thousand}$$

Total fertility rate *i.e.*, number of children born to one woman as she passes through child bearing ages

$$= \frac{\Sigma \text{ Specific fertility rates of each child bearing year}}{1000}$$

When data are given in 5 yearly groups it become,

(Sum of specific fertility rates of all groups of 5 years × 5)/1000.

So here total fertility

$$= \frac{513 \times 5}{1000} = 2.565.$$

Gross reproduction rate *i.e.*, number of female children born to one woman as she passes through child bearing ages

$$= \frac{\text{Total fertility} \times \text{Female births}}{\text{Total Births}}$$

Here ratio of male to female is 13 : 12. So we can say, out of 13+12 there are 12 female births

$$\therefore \text{ Gross reproduction rate} = \frac{2.565 \times 12}{25} = 1.231.$$

Illustration 22.2. Compute the gross and net reproduction rates from the data given below:

Age groups	15—19	20—24	25—29	30—34	35—39	40—44	45—49
Femal pop. (000)	1558	1112	1595	1629	1627	1522	1401
Female births	18900	71100	96900	64200	34900	10800	800
Survival rate	0.914	0.899	0.884	0.868	0.852	0.834	0.813

Solution. Gross reproduction rate *i.e.*, number of female children born to one woman as sne passes through child bearing ages, is sum of specific fertility rate (female births) per woman for all child bareing ages.

When data is in groups of 5 years, it is equal to the sum of specific fertility rate per woman (female reproductive rate) for all groups ×5.

Net reproduction rate is the number of female children, surviving till their reproductive ages, born to one woman as she passes through child bearing ages. So it is sum of the specific fertility rate per woman for all ages × survival rates. For data expressed in 5 yearly age groups, it is sum of the specific fertility rate per woman for various groups × survival rates × 5.

CALCULATION OF GROSS AND NET REPRODUCTION RATES*

Age	*Female pop. (000)*	*Femal births*	*Specific fertility* rate per woman (F)*	*Survival rate (S)*	*Female offspring of Survivors F×S*
15—19	1558	18900	0.012	0.914	0.011
20—24	1112	71100	0.064	0.899	0.057
25—29	1595	96900	0.061	0.884	0.054
30—34	1629	64200	0.039	0.868	0.034
35—39	1627	34900	0.021	0.852	0.018
40—44	1522	10800	0.007	0.834	0.006
45—49	1401	800	0.001	0.813	0.001
Total			0.205		0.181

*Specific fertility rate per woman = Female births/Female pop. For group 15—49 it 18900/1558000=0.012 and so on.

Gross reproduction rate $= \Sigma F \times 5 = 0.205 \times 5 = 1.025$.

Net reproduction rate $= \Sigma F \times S \times 5 = 0.181 \times 5 = 0.905$.

Illustration 22.3. From the following data calculate the gross and net reproduction rates of females. Can you also calculate similar rates for males?

Ages	Population in 000's Male (P_m)	Femate (P_f)	Children born to females Male (B_m)	Female (B_f)	Suvival rate to the middle of age group Male (S_m)	Female (S_f)
15–20	10.3	10	312	300	0.902	0.90
20–25	9.4	9	692	630	0.888	0.89
25–30	8.2	8	477	480	0.879	0.88
30–35	7,1	7	293	280	0.871	0.87
35–40	5.9	6	160	150	0.853	0.85
40–45	4.9	5	32	35	0.831	0.83

Solution. Calculation for Gross and net female reproduction rates.

Age group	15–20	20–25	25–30	30–35	35–40	40–45	Total
Annual av. birth rate $B_f/P_f = F_f$	30	70	60	40	25	7	232
$F_f \times S_f$	27	62.3	52.8	34.8	21.3	5.8	204.0

∴ Gross reproduction rate

$$\frac{\Sigma F_f \times 5}{1000} = \frac{232 \times 5}{1000} = 1.16$$

Net reproduction rate = children born to a woman surviving till the end of reproductive age $= 5 \times \Sigma F_f S_f / 100 = 204 \times 5/1000 = 1.02$.

So the gross reproduction rate is 1.16 and net reproduction rate is 1.02.

From this data male gross and net reproduction rates cannot be calculated. Reproduction rate would mean the number of male babies replaced by one man in his life-time. Here we are given male children born to females in various age groups. This number for a particular group may not be due to male population of only that group. In fact what we require is that against each age group what is the number of children born due to the male population of that age group only?. Those children may be born to women of that age group or even other groups.

Illustration 22.4. With the help of an example show that while calculating total fertility rates if number of women at each age is decreasing and fertility rate is increasing, the result obtained from data expressed in groups of five years is lower than the actual value and if fertility rate is decreasing the result is higher than actual one.

Solution. Consider any one age group of five years. Let the figures for population and fertility rates at various ages be as given below:

Age	*Pop. in (000)*	*1st case*		*2nd case*	
		Fertility rates	*No. of births*	*Fertility rates*	*No. of births*
1st Year	10	5	50	20	200
2nd "	8	8	64	18	144
3rd "	6	12	72	15	90
4th "	5	17	85	10	50
5th "	4	21	84	7	28
Total	33	63	355	70	512

Case I. Average fertility for the whole group

$$=\frac{\text{Total births}}{\text{Female pop. in (000)}}=\frac{355}{33}=10.8.$$

If we find the number of children born to 1000 ladies as they pass through various child bearing ages, when data in individual years is given it will be the sum of specific fertility rates for individual years *i.e.*, 63. But if we use the average fertility rate for the group, it will be the specific fertility rate of the group × 5 *i.e.*, 10.8 × 5 = 54.

So the result obtained by 5--yearly groups is lower than the actual result *i.e.*, when figures for individual years are known.

Case II. Average fertility rate for the group $=\frac{512}{33}=15.5$

No. of children born to 1000 females as they pass through the group, when data for individual year is known = 70.

No. of children born, when group totals only known = 15.5 × 5 = 77.5. So in this case the result obtained by 5 -- yearly groups is higher than the actual results.

Note. This is also true for gross reproduction rate and net reproduction rate. The difference in the two results is due to the fact that for calculating rate by 5 yearly average we assume that the population and rates for each of the 5 years are equal.

22.5. LIFE TABLES

A life table is just another and effective way of expressing the death rates experienced by a population during a chosen period of time. It contains eight columns as shown below:

LIFE TABLE*

Age	Living at age X	Dying between ages X and X+1	Mortality rate	Survivance rate	Living between age X and X+1	Living above age X	Mean after life time at age X
x	l_x	d_x	q_x	p_x	L_x	T_x	e°_x
0	100,000	1,710	.01710	.98290	99,145	4,086,420	40.86
1	98,290	1,592	.01620	.98380	97,494	3,987,275	40.57
2	96,698	1,483	.01534	.98466	95,957	3,889,781	40.23
3	95,215	1,383	.01452	.98548	94,523	3,793,824	39.84
4	93,832	1,293	.01378	.98622	93,186	3,699,301	39.52
—	—	—	—	—	—	—	—
—	—	—	—	—	—	—	—
—	—	—	—	—	—	—	—
95	21	9	.40957	.59043	16	32	1.52
96	12	6	.42933	.57068	9	16	1.34
97	6	3	.44964	.55036	5	7	1.17
98	3	2	.47046	.52954	2	2	0.67
99	1	1	.49176	.50824	...	...	...

*Adapted from Life Table (1941-50) from Paper No. 2 of Census of India, 1951.

The construction of life table. The basis of the table is the value commonly known as q_x which is the probability of dying between any age X and age $X+1$, where X can have any value between zero and the longest duration of life. For example, q (10) is the probability that a person who has reached his tenth birthday would die before completing his eleventh year. These probabilities, one for each year of age, are calculated from the death rates experienced by the population in a particular year. The probabilities are stated in the column headed q_x in the table given above. Once these values are known, the construction of the life table is a simple process. The next step is to assume an arbitrary number say 100,000 at age zero. By relating the probability of dying before the attainment of first birthday we find the number who die in the first year of life. This number of total deaths in this group of 100,000 infants before they attain their first birthday would be written in the d_x column. By subtracting these deaths from the initial group of 100,000 we have the number of survivors at age one. For these survivors at age one we know the probability of dying between age one and age two. By relating this probability of those who have attained the age one, the number of deaths between age one and age two can be calculated. Subtracting these deaths from the group which survived first year we shall get the survivors at the end of second year. This process would be repeated till all are dead.*

This process has clearly indicated that if q_x values are known to us, the l_x and d_x columns can be easily constructed.

On the basis of the above discussion we are in a position to explain more fully the contents of the first four columns of the table.

The first column (called 'x') gives years of age.

The second column (l_x) gives the number of persons surviving at each successive age starting out together at birth. Thus, the number 100,000 in the l_x column against '0' year age indicates the number that began their life together and are running the first year of their life. The figure 98,290 against one year age indicates the number who have completed first year of their age and are running the second. Likewise, 96,698 shows the number that completed second year and running the third.

The third column (d_x) indicates the number of deaths in the year x. Thus, 1,710 are the number of deaths in the group before it attains one year of age ; 1,592 deaths among those who have attained one year of age but have not completed their second year.

The fourth column (q_x) gives the mortality rates to which the population group would be exposed throughout their lives.

The fifth column (p_x) gives the probability of living from one age to the next, *i.e.*, from age x te age $x+1$. Since the individuals must either live or die in a particular year of life, $q_x+p_x=1$. p_x, therefore, equals $1-q_x$.

The sixth column (L_x) gives years of life lived by the group between the ages x and $x+1$. This means that if a group of 100,000 infants began life togther and if 1,710 die during the first year of their lives, the L_0 would be $100{,}000-\frac{1}{2}\times 1{,}710=99{,}145$. L_1 would be $98{,}290-\frac{1}{2}\times 1{,}592=97{,}494$. This is based on the assumption that deaths at each year of age are evenly spread throughout the year.

The seventh column (T_x) gives the total number of years lived by the group from the age x until all of them die.

$$T_x=L_x+L_{x+1}+L_{x+2}\ldots\ldots L_n$$

or
$$T_0=L_0+L_1+L_2\ldots\ldots L_{99}$$

i.e.,
$$T_0=99{,}145+97{,}494+95{,}957\ldots\ldots 9+5+2$$
$$=4086{,}420$$

Illustration 22.5. Fill in the blanks which are marked with a

*It may, however, be stated here that in the absence of adequate mortality data, the d_x column may be obtained if the number of persons living at successive ages are known.

query in the following skeleton life table and explain the meanings of the symbols at the heads of the columns:

Age x...	l_x	d_x	p_x	q_x	L_x	T_x	e_x	m_x
30	762227	?	?	?	?	27296732	?	?
31	758580	—	—	—	—	?	—	—

Solution. Here

d_x=number of persons dying between ages x and $x+1$

$=l_x-l_{x+1}=762227-758580=3647$

p_x=probability of a persons living between ages x and $x+1$

$$=\frac{l_{x+1}}{l_x}=\frac{758580}{762227}=0.9952$$

$$q_x=1-p_x=0.00478$$

L_x=number of persons of a stationary population living between ages x and $x+1$.

$$=\tfrac{1}{2}(l_x+l_{x+1})=\tfrac{1}{2}(762227+758580)=760404$$

T_x=number of persons of age x and above in a stationary population at any moment.

$$T_{x+1}=T_x-L_x=27296732-760404=26536328.$$

$$e_x^0=\text{expectation of life}=\frac{T_x}{l_x}$$

$$=\frac{27296732}{762227}=36 \text{ approx.}$$

m_x=central death rate at age x

$$=\frac{d_x}{L_x}=\frac{3647}{760404}=0.00479$$

Illustration 22.6. Following figures gives the mortality rate of the life table (q_x) at the ages 0, 1, 2, 3, 4, 5 taken from the Life table of Males (All India) for 1931. Taking the starting l as 10.000 calculate the l_x, d_x, p_x, L_x values for the various ages.

0.2487, 0.0918, 0.0564, 0.0392, 0.0274, and 0.0193.

Solution. The inter relations between the various columns of life table are:

$$l_x \times p_x=d_x,\ l_{x+1}=l_x-d_x,\ p_x=1-q_x,$$

and $$L_x=(l_x+l_{x+1})/2.$$

We are given that l_0 = 10,000.

$$p_0=1-q_0=1-0.2487=0.7513.$$

$$l_1=l_0-d_0=10{,}000-2487=7513$$

So $$L_0=(10{,}000+7513)/2=8757$$

Calculating similarly other values, the various columns of life-table are as below:

Age	l_x	d_x	q_x	p_x	L_x
0	10000	2487	0.2487	0.7513	8757
1	7513	690	0.0918	0.9082	7168
2	6823	385	0.0564	0.9436	6631
3	6438	252	0.0392	0.9608	6312
4	6186	169	0.0274	0.9726	6102
5	6017	116	0.0193	0.9807	5959
6	5901	—	—	—	—

Illustration 22.7. Fill in the blanks of the following skeleton table which are marked with question marks.

Age x	l	d	p	q	L	T	e°
20	693435	2762	?	?	?	35081126	?
21	620673	—	—	—	—	?	?

Solution. The relations connecting the various columns, required her, are:

$$q_x = d_x/l_x, \qquad p_x = 1 - q_x, \qquad L_x = (l_x + l_{x+1})/2$$

$$T_x = T_{x+1} + L_x, \qquad e^\circ_x = T_x/l_x.$$

Taking age 20 as x, *i.e.*, age 21 as $x+1$, the various required values are:

$$q_{20} = d_{20}/l_{20} \qquad = 2762/693435 = 0.00398$$

$$p_{20} = 1 - q_{20} \qquad = 1 - 0.00398 = 0.99602$$

$$L_{20} = (l_{20} + l_{21})/2 \qquad = (693435 + 690673)/2 = 692054$$

$$T_{21} = T_{20} - L_{20} \qquad = 35081126 - 692054 = 34389072$$

$$e^\circ_{20} = T_{20}/l_{20} \qquad = 35081126/693435 = 50.59$$

$$e^\circ_{21} = T_{21}/i_{21} \qquad = 34389072/690673 = 49.79$$

Illustration. 22.8. The l_x column of a certain life table for ages 0, 1, 2, 3, 4, 5,...is given by a series 1000, 983, 967, 952, 938, 925,... Calculate d_x, q_x, p_x, L_x columns for ages 0, 1, 2, 3, 4.

Solution. We know $d_x = l_x - l_{x+1}$, $\quad q_x = d_x/l_x$

$$p_x = 1 - q_x \text{ and } L_x = (l_x + l_{x+1})/2$$

Calculating in this way, the required values are below

Ages	l_x	$d_x = l_x - l_{x+1}$	$q_x = d_x/l_x$	$p_x = 1 - q_x$	$L_x = (l_x + l_{x+1}/2)$
0	1000	17	.0170	.9830	991.5
1	983	16	.0163	.9837	975.0
2	967	15	.0155	.9845	959.5
3	952	14	.0147	.9853	945.0
4	938	13	.0139	.9861	931.5
5	925	⋮	⋮	⋮	⋮
	—	⋮	⋮	⋮	⋮

Illustration 22.9. The l_x column of a certain life-table for ages 0, 1, 2, 3. . .100 is given by series 100, 99, 98, 97. . .0. Calculate the following values:

$$e^{\circ}_0,\ T_5,\ L_7,\ q_{10},\ p_{20},\ e_{25},\ \text{and } d_{70}.$$

Solution. The given data is: $l_0=100$, $l_1=99$, $l_2=98$, $l_3=97$. . This means in general $l_x=100-x$.

According to the relations between various columns

$d_{70}=l_{70}-l_{71}=(100-70)-(100-71)=30-29=1$

$q_{10}=d_{10}/l_{10}=(l_{10}-l_{11})/l_{10}=[100-10)-(100-11)]/(100-10)$
$=1/90=0.0111$

$p_{20}=1-q_{20}=1-[(l_{20}-l_{21})/l_{20}]$
$=1-[(100-20)-(100-21)]/(100-20)=1-1/80=0.9875$

$L_7=(l_7+l_8)/2=[(100-7)+(100-8)]/2=(93+92)/2=92.5$

T column is the cumulative of L column, when cumulation is done from the highest ages possible side to the age zero.

So the general value of $L_x=(l_x+l_{x+1})/2$
$=(100-x)+(100-x-1)/2$
$=(200-2x-1)/2=(199-2x)/2=99.5-x.$

The highest possible age is 100. At this $L_{100}=99.5-100=-0.5$.

But L value can never be negative. It can at the most be zero. So at this age it is taken as zero.

$L_{99}=99.5-99=0.5$, $L_{98}=99.5-98=1.5$, $L_{97}=99.5-97=2.5$...etc.
So L value for various ages 100, 99, 98, 97, 96......0 are 0, 0.5, 1.5, 2.5, 3.5,......99.5.

Now T_5 is ΣL_x where summation is taken over the x values 100, 99, etc. upto 5. So $T_5=L_{100}+L_{99}+L_{98}+\ldots+L_5$.

i.a., $T_5=0+0.5+1.5+2.5+\ldots\ldots+94.5$ (as $L_5=99.5-5=94.5$).

This is an A.P. with number of terms $(n)=95$, 1st term $(a)=0.5$, and last term $(l)=94.5$. So its sum$=\frac{n}{2}(a+l)$ *i.e.*,

$$T_5=\frac{95}{2}(0.5+94.5)=\frac{95\times 95}{2}=\frac{9025}{2}=4512.5.$$

$e^0_0=T_0/l_0$. As above.

$T_0=0.5+1.5+2.5+\ldots\ldots 99.5$ as $(L_0=99.5)$
$=\frac{100}{2}(0.5+99.5)=5000.$

Now l_0 = 100. So e_0^0 = 5000/100 = 50 years.

$e_{25}=(T_{25}/l_{25})-\frac{1}{2}$. As above $L_{25}=99.5-25=74.5$

$\therefore\ T_{25}=0.5+1.5-\ldots\ldots 74.5=\frac{75}{2}(0.5+74.5)=2812.5.$

$\therefore\ e_{25}=(2812.5/75)-\frac{1}{2}=37.5-0.5=37$ years.

Illustration 22.10. Given that the complete expectation of life at ages 30 and 31 for a particular group are respectively 21.39 and 20.91 years and that the number living at age 30 is 41, 176, find the number that attains the age 31.

Solution. We have $\overset{\circ}{e}_{30}=21.39$, $\overset{\circ}{e}_{31}=20.91$

$$\therefore \quad e_{30} = 21.39 - 0.5 = 20.89 \text{ and } e_{31} = 20.91 - 0.5 = 20.41$$

$$p_{31}=\frac{e_{30}}{1+e_{31}}=\frac{20.89}{1+20.41}=\frac{20.89}{21.41}$$

$$p_{31}=\frac{l_{31}}{l_{30}} \quad \therefore \quad l_{31}=l_{30}\times p_{31}=41176\times\frac{20.89}{21.41}$$

$$=40176 \text{ approx.}$$

Illustration 22.11. Given the following table for l_x the number of rabbits living at age x, complete the life table for rabbits.

x	0	1	2	3	4	5	6
l_x	100	90	80	75	60	30	0

Solution. $d_0=l_0-l_1=100-90=10,\ d_1=l_1-l_2=90-80=10,$...etc.

$$p_0=\frac{l_1}{l_0}=\frac{90}{100}=.9,\ p_1=\frac{l_2}{l_1}=\frac{80}{90}=.88,\text{......etc.}$$

$$q_0=1-p_0=1-.9=.1,\ q_1=1-p_1=1-.88=.12,\text{......etc.}$$

$$L_0=\frac{l_0+l_1}{2}=\frac{100+90}{2}=95,\ L_1=\frac{l_1+l_2}{2}=\frac{90+80}{2}$$

$$=85,\text{......etc}$$

T_x column is the cumulative values of L_x starting from highest age *i.e.* 6. $e^0_x=T_x/l_x$, so it can be calculated after completing the T_x column. Completing in this way the table is

Age (x)	l_x	d_x	p_x	q_x	L_x	T_x	l_0^0
0	100	10	.90	.10	95	385	3.85
1	90	10	.89	.11	85	290	3.22
2	80	5	.94	.06	775	20.5	2.56
3	75	15	.80	.20	67.5	127.5	1.70
4	60	30	.50	.50	45	60	1.0
5	30	30	.00	1.00	15	16	0.5
6	0	—	—	—	—		

PROBLEMS

1. What do you understand by 'Vital statistics'? Write note on 'Life table' and its importance.

2. Explain the meaning of 'Net Reproduction Rate.' How is it calculated?
3. Distinguish between crude and standardised death rates, and gross and net reproduction rates. Why is this distinction necessary?
4. Calculate the crude and standardised death rates of the local population from the following data and compare them with the crude death rate of the standard population. What inference do you draw from the comparison?

Age-group	*Standard Population*	*Population Deaths*	*Local Population*	*Population Deaths*
0-10	600	18	400	16
10-20	1,000	5	1,500	6
20-60	3,000	24	2,400	24
60-100	400	20	700	21
Total	5,000	67	5,000	67

5. Compute the crude and standardised rates of unemployment from the following data, taking the standard population as the population of two districts together.

	District A		*District B*	
Age-group	*Population in thousands*	*Unemployment Rate %*	*Population in thousands*	*Unemployment Rate %*
15—29	200	9	100	5
30—44	450	12	600	15
45—59	300	20	150	10

6. Construct an example to show how an incorrect picture can be presented by crude death rates when we employ them for comparing the salubriousness of two places. Show how the standardised death rates can be used to avoid such pitfalls.
7. The following data give the number of women of child-bearing ages and yearly birth by quinquennial age-groups for a city. Calculate the general fertility rate and total fertility rate. If the ratio of male to female children is 13 : 12, what is the gross reproduction rate?

Age-group	15-19	20-24	25-29	30-34	35-39	50-44	45-49
Female pop. ('000)	16	15	14	13	12	11	9
Births	400	1,710	2,100	1,430	900	330	36

8. Compute the gross and net reproduction rates from the data given below:

Age-group	15-19	20-24	25-29	30-34	35-39	40-44	45-49
Female pop. ('000)	1,558	1,112	1,595	1,629	1,627	1,522	1,401
Female births	18,900	71,100	96,900	64,200	34.900	10,800	800
Survival rate	0.914	0.899	0.884	0.868	0.852	0.834	0.813

9. From the information given below, calculate the gross reproduction rate per female, the net reproduction rate per female. The female sex ratio is 0.49148.

Age-group	*Number of females*	*Number of births during year of enquiry*	*Survival rate per person from life table*
13-17	1701	100	0.59825
18-22	1797	431	0.57582
23-27	1491	383	0.54785
28-32	1414	140	0.50780
33-37	1001	67	0.45670
38-42	1071	67	0.40301
43-47	818	20	0.38897
48-52	893	1	0.29527

Chapter 23

STATISTICAL QUALITY CONTROL

23.1. INTRODUCTION

Statistical quality control is one of the more useful and economically important applications of the theory of sampling in the industrial field. An important feature of modern industry is repetitive work turning out a large number of presumably "identical products". But no two pieces off the same machine are identical in measurable characteristics in spite of all the precision of modern engineering. The variation may be infinitesimal but it does exist. Because of the inevitability of the occurrence of variations the users of the products (assembly plants) set standards of quality to which products must conform if they are to be considered satisfactory for use. These standards specify not only a basic norm but also the upper and lower limits within which a products will be considered satisfactory. These limits are called tolerances and represent the allowable variation in the measurable characteristic of the products so far as its use is concerned. If the size of the product falls outside the range of these specifications it is considered unfit for use. This being so, a manufacturer is faced with the problem of ensuring that his products are of requisite quality and the size of the characteristics in terms of which their quality is measured does not fall outside the maximum and minimum tolerances stipulated by the assembly plants. One method usually employed to ensure that defective products are not passed into stock from the factory is to have a 100 per cent inspection system that is to say that each unit of product is inspected to assess its quality. But this system has some serious defects. Human nature being what it is, even a hundred per cent inspection is no guarantee about the quality of the product passed into stock. Besides, the cost of such inspection may often be formidable.

A suitable system of inspection would be that which (*i*) detects the defect as it occurs, *i.e.*, at its origin, and (*ii*) introduces a continuous sample inspection in place of 100 per cent inspection.

The technique of quality control provides such a system. Thus, statistical quality control is a system which consists of (*i*) sampling inspection of manufactured products at each stage of its production, and (*ii*) statistical inference regarding the variability of its quality with the help of simple devices like charts etc. This technique was originated in the work of Walter A. Shewhart and it was during World War II that its greatest development took place.

There are two aspects of statistical quality control, *viz*., (*i*) process control, and (*ii*) products or lot control, also called acceptance inspection. The aim of the first is to evaluate the performance of each individual process and thus, to foresee the variability in the quality of output of each process in the immediate future. The aim of the second is to see that a lot put in the market does not contain a large number of unsatisfactory units.

23.2. PROCESS CONTROL

The variations that occur in a production process may be attributed to two main causes:

(1) *Random or chance causes*. These are very many in number, each one exercising only a trivial effect on the quality of the product. These causes are inherent in a production process in the sense that they will continue to operate and connot be removed completely. Small variations in the skill of manual operators or in the quality of raw materials are causes which come under this category. Since these varitions are caused by innumerable forces any effort to eliminate then is likely to be uneconomic and may even prove a failure.

(2) *Assignable causes*. These are causes which can be identified and are responsible for important and larger variation in the quality of the product. These causes normally interfere with the economic working in the plant and it is economically necessary to eliminate them as soon as as they are discovered. Excessive wear on the cutting tool, mechanical fault in plant, bad handling of the machine by the operator are a few examples of such causes

If the variations in the quality of particular product are such that they can in entirely be attributed to chance causes alone (that is to say if the variation is such as would occur in random sampling from stable population) the process is said to be in a state of statistical control. If this is the case, the variability of the quality of product cannot be altered unless the production process itself is altered and as such prediction about the future behaviour of the data can be made. If, however, it is said that a process is *out of statistical control* it implies that certain assignable causes are present affecting the variability of quality. The object of statistical quality control is to detect such assignable causes as soon as they occur in the production process.

From what has been said before, it follows that the technique of process control implies (*i*) determination of the way in which variations in quality would be distributed when the process is under control, and (*ii*) checking on a continuous basis whether the variability in quality conforms to this distribution or not. If the variations do not fall within this distribution, it is a warning that the process is going out of control and remedial measures should be adopted.

Now if production is influenced by random causes only, the various units produced constitute a single homogenous population and the variations in its size can be described by sampling distributions. Thus, if the mean value and standard deviation of a certain measurable quality characteristic X (say the length of a screw) are $\bar{X}$ and σ, respectively, when the process is in control, then if samples of size N are taken, the sample mean $\bar{x}$ will be approximately normally distributed about a mean $\bar{X}$ with a standard error $\frac{\sigma}{\sqrt{N}}$. This means that approximately 1 out of 20 samples drawn from a population will have means lying outside $\bar{X} \pm 1.96\sigma/\sqrt{N}$ and 1 out of 100 will have them lying outside the limits $\bar{X} \pm 2.576\sigma/\sqrt{N}$. Very rarely a sample mean would lie farther away from $\bar{X}$ than by $3\frac{\sigma}{\sqrt{N}}$ just due to chance. This means that if we get a sample mean farther from X than $\pm 3\sigma/\sqrt{N}$ we should suspect that there is some assignable cause present which accounts for it. This is the basis of all statistical quality control procedures.

This checking up of quality characteristic is done with the use of control charts.

23.3. CONTROL CHART

As stated above the main tool of process control is the control chart. There are different types of control charts, depending on different ways of assessing the quality of a product. Many characteristics are measurable, such as length of a screw, tensile strength of a yarn, resistance of wire, life of an electric bulb, etc. Such variables are continuous. For controlling such qualities two types of charts are are plotted—one for mean of the measurements ($\bar{X}$-chart) and another for the range of the measurements (R-chart).

Sometimes the characteristic representing the quality of a product is discrete, such as thread count of a piece of cloth, number of surface defects on a polished surface etc. In such cases the number of defects on an item may be nil, one, two or more. The total number of defects may be very large, but it will always be an integral value and never a faction like 27.1 defects, etc. Under such a circumstance control charts for average number of defects per item or count chart (C-chart) is plotted.

There is still a third way in which quality may be measured. Many times it may be possible only to classify a produced item as good or bad, *e.g.*, a container produced may have a leak. Whether the leak is at one place or at more than one place, the item produced is useless. In such a case to assess quality it is only possible to determine the proportion of bad or defective items in a sample. In such a case control chart for proportion of defectives (p-Chart) or number of defectives in a sample of fixed size is prepared.

All types of control charts (*viz* $\bar{X}$, R, C or p charts) are similalr in composition and construction. All of them represent how quality characteristic is changing its value from one sample to another sample. The various stages in their constructions are:

(1) Take sample number against X-axis and the quality characteristic *i.e.*, $\bar{X}$, R, p or C along Y-axis.

(2) Mark the central line corresponding to the average value or the get value of the quality characteristic, *i.e.*, what the process is capable of achieving in items of quality characteristic.

(3) Plot the upper and lower tolerance limits. These limits point out the values of qualities beyond which products will not be accepted. This step is optional.

(4) Calculate the control limits for variable under consideration and mark them. These limits define the maximum possible variations due to chance. Control limits should always be within tolerance limits. Calculation of these limits is explained in a later section.

(5) Select samples at fixed intervals of time from the items produced. Assess the characteristic representing the quality of the product. Plot their values against the various sample numbers. If the value plotted is $\bar{X}$, the chart is called $\bar{X}$-Chart. If the value plotted is p (proportion of defectives in the sample), the chart is called

p-chart, etc.

A process is considered out of control and an action to check and correct the process is taken when –

(1) a plotted point falls outside the control limits;

(2) several points lie close to the control limits; and

(3) there is an unusual non-random arrangements of points.

These control charts have the following advantages. (1) They provide visual aids, (2) are easy to prepare, (3) are simple device of checking whether variations conform to chance variations or are more than that, (4) give early warning of a trouble, and (5) they are flexible. From the unusual arrangement of points an alert person might suspect the incoming trouble at quite an early stage.

Calculations of control limits. In drawing conclusions with the help of control charts, two types of errors can be made: (*i*) concluding that the process is out of control when actually it is in control, and (*ii*) concluding that process is in control when it is actually out of control. The control limits are so set that there is some kind of economic balance between these two errors. They are so set that the engineer is in a position to detect serious troubles and may not waste time on minor ones. It has been stated by the central limit theorem that even if the distribution of items produced is not very close to normal distribution, the distribution of variation in the arithmetic means of the measurements taken from random samples drawn from various lots of production, when process is in control, follows normal distribution very closely. In a normal distribution it is only 3 in 1,000 chance that a value lie outside the range A.M. $\pm$ 3 St. Deviation. With these limits as control limits it will be highly improbable that a plotted point may be outside the control limit provided the process is in control. Thus, if a plotted point is outside these limits (called 3 sigma limits), it is a warning that variations are perhaps not due to chance but due to some serious troubles. So control limits in all the charts are usually plotted at 3 sigma limits.

In $\bar{X}$ charts these limits will be generally found at the start by taking the arithmetic mean of the mean values of the first 20 or 25 samples when the process is in control. '*N*' in the formula is size of sample selected and 'σ' is the standard deviation of the whole production (population). In usual problems of testing of hypothesis σ is estimated from the standard deviations of the sample. But in plotting control chart σ is usually estimated from the average range of the first 20 or 25 samples. This estimation is usually done with the help of standard tables called A_2 table (See Table 23.1) which directly give the value of $3/\sigma\sqrt{n}$ as the ratio to range for different sizes of samples. Similarly, in case of *R*-chart, control limits are set at average range$\pm 3\times$St. Error of Range. These values are estimated from standard tables called D_3 and D_4 tables. These given the values of lower and upper control limits as a ratio to range for different sizes of samples.

TABLE 23.1. A_2, D_3 AND D_4 TABLES

Size of sample *n*	A_2	D_3	D_4
2	1.881	0	3.267
3	1.023	0	2.575
4	0.7285	0	2.282
5	0.5768	0	2.115
6	0.4833	0	2.004
7	0.4193	0.076	1.924
8	0.3726	0.136	1.864
9	0.3367	0.184	1.816
10	0.3082	0.223	1.777
11	0.2851	0.256	1.744
12	0.2658	0.284	1.716

Note. When plotting $\bar{X}$-chart and *R*-chart usually small samples of size up to 12 are selected out of lots of approximately 100 items produced. This is only because of practical convenience.

For estimating control limits for *p*-chart (proportion of defectives), first average value of *p*, *i.e.*, $\bar{p}$ is estimated from first 20 to 25 samples. This $\bar{p}$ gives the central line. From binomial distribution we know standard error of $\bar{p}$ is $\sqrt{\bar{p}(1-\bar{p})/n}$ where *n* is the size or sample. So the control limits for *p*-chart are $\bar{p}\pm\sqrt{\bar{p}(1-\bar{p})/n}$. If the control chart of number of defectives is plotted, standard error would be $\sqrt{n\bar{p}\,(1-\bar{p})}$ and hence the control limits would be number defectives$\pm 3\sqrt{n\bar{p}\,(1-\bar{p})}$.

Similarly for *C*-chart, first average number of defects per item ($\bar{C}$) are estimated. Theory of Poisson Distribution gives standard error of $\bar{C}$ as $\sqrt{\bar{C}}$. So the control limits are $\bar{C}\pm 3\sqrt{\bar{C}}$.

Examples of Different Charts

Illustration 23.1. *X and R charts.* The nominal dimension of a component is 0.1350″ with a tolerance of $\pm$0.0032″. The data given below are in 0.0001 units above and below the nominal dimension. Plot the control chart for average and range:

Observation No.	No. of samples									
	1	2	3	4	5	6	7	8	9	10
1	5	0	−15	−15	−5	10	10	−5	5	20
2	25	10	−10	0	5	−15	15	−5	5	−10
3	−25	0	0	−25	5	−10	5	5	−5	0
4	0	5	−10	−5	0	−10	−10	−15	−5	−5
5	15	0	0	5	−10	5	−15	10	10	5

Solution. Mean and range of 5 observations for various samples in units of 0.0001 inch are given as:

Sample No.	1	2	3	4	5	6	7	8	9	10
Mean ($\bar{X}$)	4	3	−7	−8	−1	−4	1	−2	2	2
Range (R)	50	10	15	30	15	25	30	25	15	30

Tolerance limits are 0.135″±.0032″, *i.e.*, 0.1382″ to 0.1318″.

Process average $(\bar{\bar{X}}) = \frac{\Sigma \bar{X}}{N} = \frac{-10}{10} = -1,$

i.e., $0.1350 - 1 \times .0001 = 0.1349$ inch.

Mean Range $(\bar{R}) = \frac{\Sigma R}{N} = \frac{245}{10} = 24.5,$

i.e., $24.5 \times .0001 = .00245$ inch.

A_2 value for a sample of size 5 from the tables $= 0.5768$

$$\therefore \qquad A_2 = \frac{3\sigma/\sqrt{n}}{\bar{R}}$$

or

$$\frac{3\sigma}{\sqrt{n}} = \bar{R} \times A_2$$

$$= .00245 \times .5768$$

$$= .0014$$

$\therefore$ Control limits for $\bar{X}$ chart $= \bar{\bar{X}} \pm \frac{3\sigma}{\sqrt{n}} = 0.1349 \pm .0014.$

or 0.1363″ to 0.1335″.

D_3 and D_4 values for sample of size 5 are 0 and 2.115.

$\therefore$ Lower control limit for R chart $= D_3 \times \bar{R} = 0$

and upper control limit for $\bar{R}$ chart $= D_4 \times \bar{R}$

$$= 2.115 \times .00245 = .0052''$$

Now to plot the control chart taken the values of $\bar{\bar{X}}$ and $\bar{R}$ along Y-axis and sample number along X-axis. Here the target value 0.1350″ and the process average 0.1349 are different. This means machine has been set for 0.1350″ but actually it is capable of producing average size of 0.1349″. So central line will be marked at

0.1349". If actual target value 1350" is to be achieved, some correction must be made in the process. Plotting these values we obtain control charts as in Fig. 23.1.

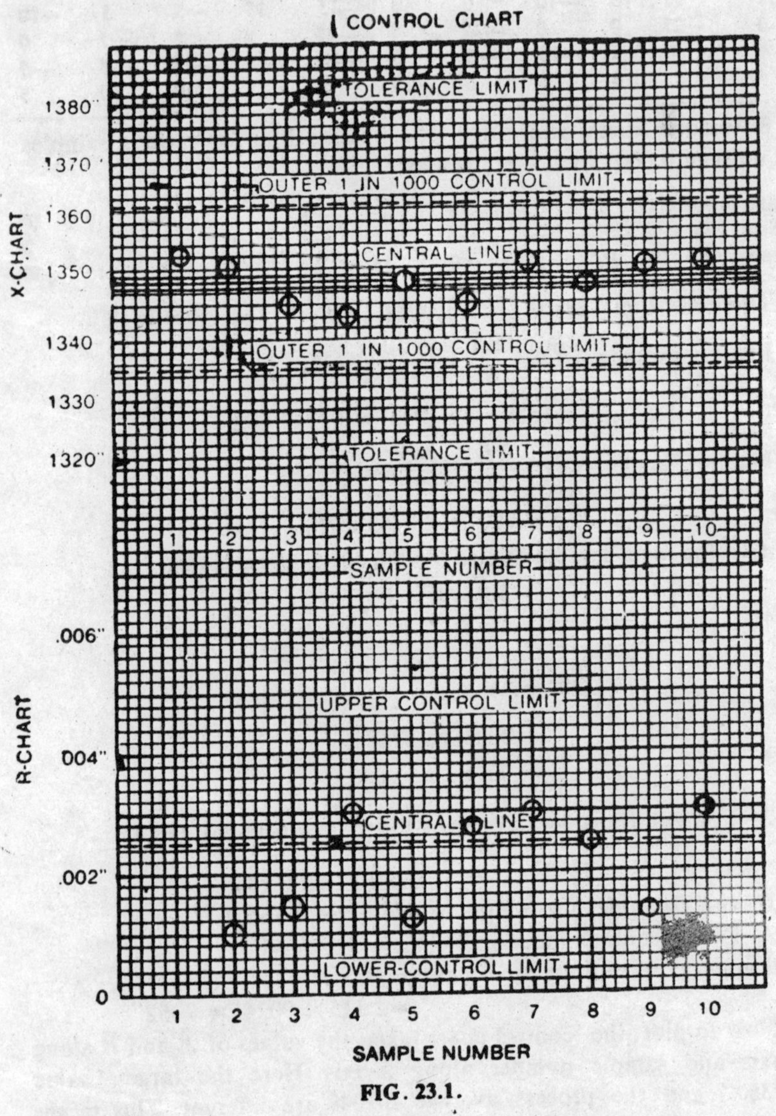

FIG. 23.1.

Illustration 23.2. *Chart for number of defectives.* In the manufacture of certain tanks following data were observed.

Sample No.	1	2	3	4	5	6	7	8	9	10
Size	200	200	225	75	190	210	500	212	200	188
No. of defectives	20	19	23	8	18	22	51	21	19	19

Draw the control chart for number of defectives.

Solution. In the present case size of sample is not the same. Control limits depend on size of sample (n). So this means for each sample there would be separate control limits. But if variation in size of sample is not large, the average size of sample is usually taken for calculating the control limits. In the present example except for sample Nos. 4 and 7 all other samples are approximately of the same size. So separate control limits may be calculated for samples 4 and 7 and for others average size of the sample may be used.

Average size of sample (except samples Nos. 4 and 7)$=\frac{1,625}{8}$ $=203.1$. The fraction of defectives is more or less same in all cases.

So $\bar{p}=\frac{\text{Total defectives in all the ten samples}}{\text{Total items in all the ten samples}}=\frac{230}{2,200}=0.1.$

Standard error of number of defectives for average size

$$=\sqrt{203.1\times.1\times.9}=4.28.$$

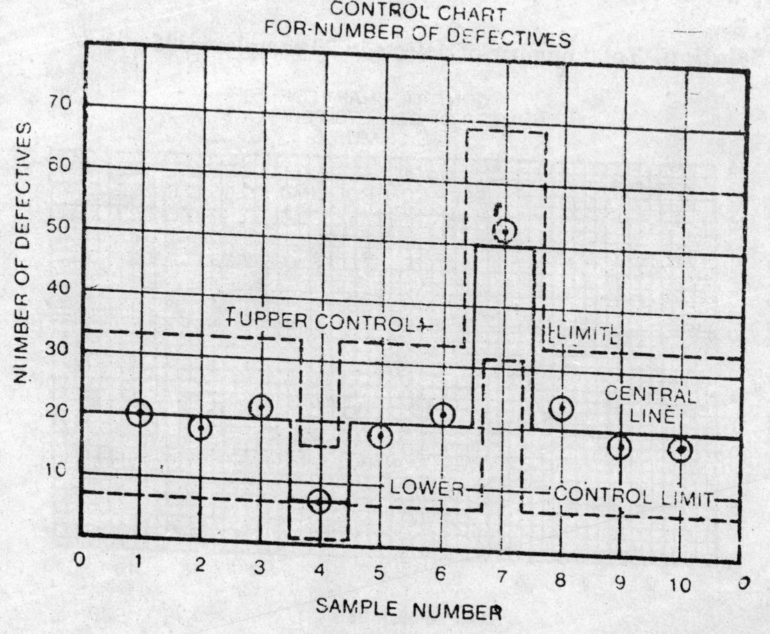

FIG. 23.2.

The average number of defectives for 8 samples leaving 4 and 7

$= \frac{220-59}{8} = 20.1$. This gives the cental line.

$\therefore$ 3 sigma control limits for average size of sample

$$= 20.1 \pm 3 \times 4.28 = 7.3 \text{ to } 33.9.$$

For sample Nos. 4 and 7 due to difference in size, the average number of defectives. *i.e.*, central line, cannot be at 20.1. The position of their central lines can be found by multiplying the size of the sample and proportion of defectives, *i.e.*, $\bar{p}$. So for number 4 central line will be at $75 \times .1$ *i.e.*, 7.5 and for number 7 it will be at $500 \times .1$, *i.e.*, 50.

Similarly, standard error for number $4 = \sqrt{75 \times .1 \times .9} = 2.598$ and No. 7 will be $\sqrt{500 \times .1 \times .9} = 6.32$.

Therefore, control limit for No. 4 sample $= 7.5 \pm 3 \times 2.598$, *i.e.*, 15.3 and 0, and control limits for No. 7 sample $= 50 \pm 3 \times 6.32$, *i.e.*, 69 and 31.

Plotting these values, control chart is as shown in Fig. 23.2.

Illustration: 23.3. *C*-Chart. Following 20 figures correspond to the number of defects (spots) in 20 sheets of $1' \times 2'$ selected as 20 samples during a process of manufacturing metal sheets. Plot a suitable control chart.

27, 26, 20, 19, 12, 16, 22, 29, 13, 17, 22, 26, 25, 18, 17, 13, 19, 15, 18, 22.

Solution. Total number of defects in 20 samples $= 396$

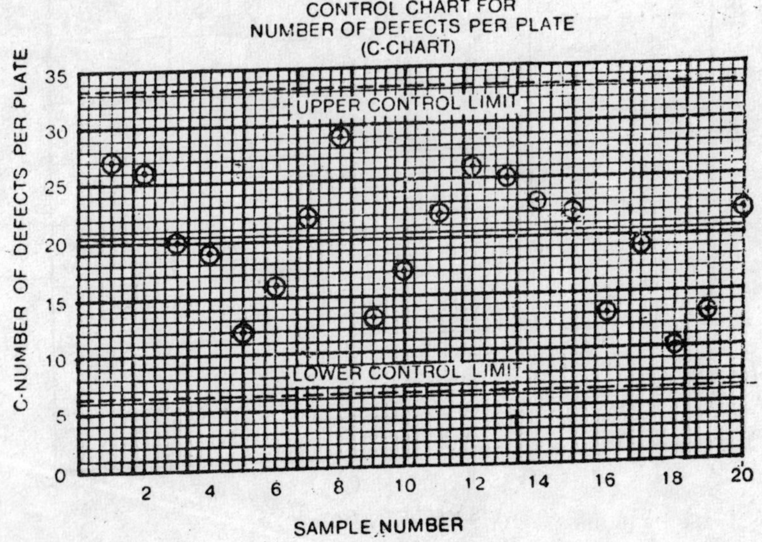

FIG. 23.3.

$\therefore$ Average number of defects per sheet $(\bar{C}) = \frac{396}{20}$

$= 19.8$

Standard error of number of defects $= \sqrt{(\bar{C})}$

$= \sqrt{19.8} = 4.45.$

$\therefore$ 3 sigma limits for C-chart $= \bar{C} \pm 3\sqrt{(\bar{C})}$

$= 19.8 \pm 3 \times 4.45 = 33.10$ to $6.45.$

To plot the control chart plot the central line at 19.8 defects and control limits at 33.15 to 6.45. The chart is as shown in Fig. 23.3.

23.4. PRODUCT CONTROL

Another aspect of the statistical quality control is the product control. The object of this is to decide whether to accept or reject a lot. This is also done by sampling inspection. First a sample of small size is taken from the lot. If the quality of the product as assessed by this sample is very good, the lot is accepted for marketing. If it is very bad the lot is rejected. If no clear cut conclusion can be drawn a second sample is selected from the lot and the lot is accepted or rejected on the basis of two samples combined. Sometimes more samples may have to be taken for a clear cut decision. The plans for selecting sample depends on the following objects:

(1) The producer's risk, *i.e.*, chance of rejecting a good lot is small (a specified value).

(2) The consumer's risk, *i.e.*, chance of accepting a bad lot is small (another specified value).

(3) The average quality of goods sent out of the factory may not be worse than some specification.

(4) The amount of inspection be minimum.

If the process is in control, the variability in quality will be small and hence the extent of inspection for accepting or rejecting the lot will be very small. But if the process is out of control, *i.e.*, the variability in quality is large, inspection would be more frequently needed as some items may lie outside the tolerance limits.

Advantage of statistical quality control. In conclusion, the advantages may be summarised as below:

(1) It is a technique which provides a continuous inspection of the product at various stages of the manufacturing process.

(2) It eliminates the need of 100 per cent inspection of the fini-

shed product and is usually more efficient and less costly than 100 per cent inspection.

(3) It reduces waste of time and material to absolute minimum by giving early warning about the occurrence of defects.

(4) The technique is quite simple and can be operated by semi-skilled operators.

(5) Rajections by buyers are almost reduced to nil.

(6) Savings in terms of the factors stated above mean less cost of production and hence may ultimately lead to more profit.

Illustration 23.4. Quality control is maintained in a factory with the help of mean ($\bar{X}$) and standard deviation (σ) charts. Ten items are chosen in every sample. 18 samples in all were chosen when $\Sigma\bar{X}$ was 595.8 and $\Sigma\sigma$ was 8.28. Determine 3σ limits for $\bar{X}$ and σ charts You may use the following factors for finding the 3σ limits.

For $n=10$ ∴ $A_1=1.03$, $B_3=0.28$, $B_4=1.72$.

Solution. Mean of means $(\bar{\bar{X}})=\frac{\Sigma\bar{X}}{18}=\frac{595.8}{18}=33.1$ (central line for $\bar{X}$ chart).

Mean of σ $(\bar{\sigma})$ $=\frac{\Sigma\sigma}{18}=\frac{8.2}{18}=.46$ (central line for σ chart

3σ limits of $\bar{X}$ chart is given by $\bar{\bar{X}}\pm A_1\bar{\sigma}$

i.e., $33.1\pm1.03\times.46$ or $33.1\pm.47$.

∴ Lower control limit for $\bar{X}$ chart is 32.63 and upper control limit is 33.57.

Lower 3σ limit for σ chart is given by $B_3\bar{\sigma}$ *i.e.*, $0.28\times.46=.13$

Upper 3σ " " " $B_4\bar{\sigma}$ *i.e.*, $1.72\times.46=.79$

Illustration 23.5. The following data gives readings for 10 samples of size 6 in the production of certain component.

Sample	1	2	3	4	5	6	7	8	9	10
Mean $\bar{X}$	383	508	508	582	557	337	514	618	707	753
St. Dev. (s)	30.5	41.6	39.5	32.2	27.4	24.2	48.7	8.9	13.1	33.9
Range (R)	95	128	100	91	68	65	148	28	37	80

Draw the control charts for $\bar{X}$, s and R calculating the limits for $\bar{X}$ in two ways. Can the within -- group variability be regarded as; homogeneous? Can one assume that all groups are from homogeneous lot?

Solution. Mean of $R=\Sigma R/N=\frac{840}{10}=84.0$ *i.e.*, central line of R.

Mean of $s=\frac{\Sigma s}{N}=\frac{300}{10}=30.0$ *i.e.*, central line of s.

$$\text{Mean of } \bar{X} = \frac{5460}{10} = 546, \textit{ i.e., } \text{central line of } \bar{X}.$$

Estimate of population standard deviation can now be obtained

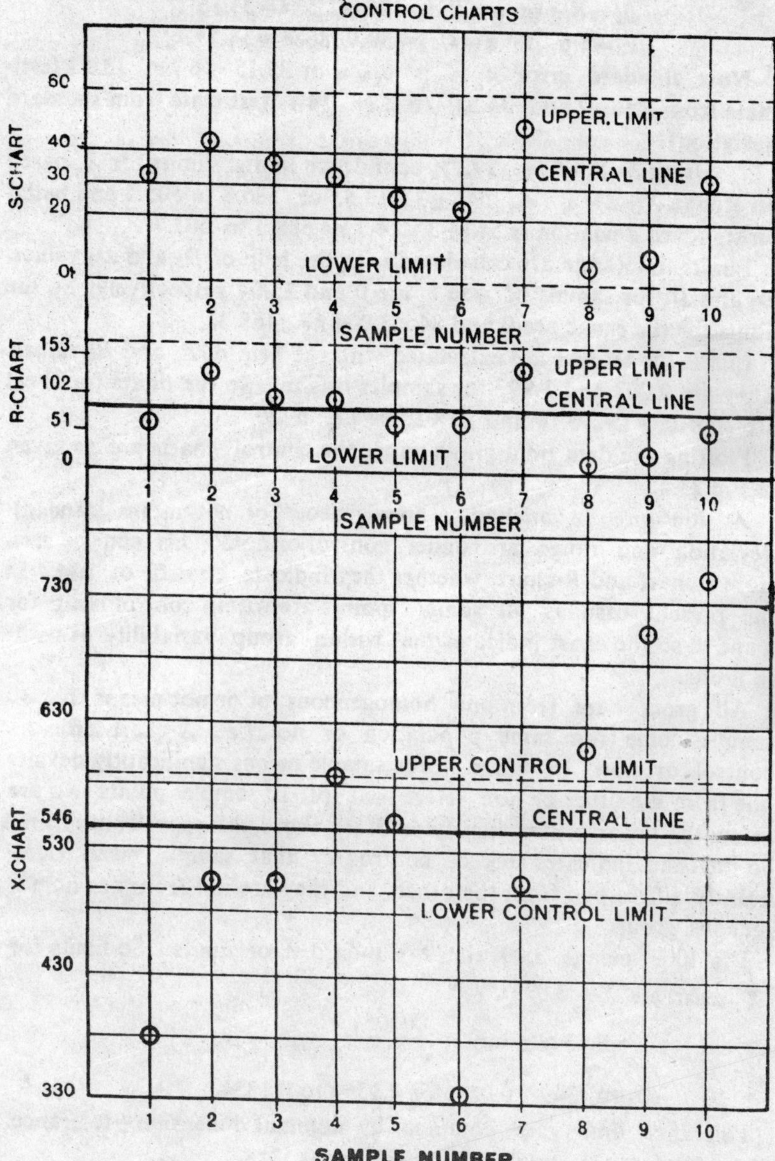

FIG 23.4

in two ways, one with the help of R and other with the help of s. From tables d_2 and c_2 values for sample of size 6 are 2.534 and 0.8686 respectively.

$$d_2=\bar{R}/\sigma \text{ or } \sigma=\bar{R}/d_2=84.0/2.534=33.15$$

$$c_2=\bar{s}/\sigma \text{ or } \sigma=\bar{s}/c_2=30.0/.8686=34.53$$

Now standard error of $\bar{X}$ is $\sigma/\sqrt{n}$ or $33.15/\sqrt{6}$ *i.e.*, 13.53 (estimate from Range) and $34.53/\sqrt{6}$ *i.e.*, 14.1 (estimate from standard deviation)

∴ Using 3σ limit *i.e.*, 99.7% confidence limits, limits for $\bar{X}$, based on Range $\bar{X}\pm3\times\sigma$ *i.e.*, $546\pm3\times13.53$ or 586.6 to 505.4 and based on standard deviation is $546\pm3\times14.1$ or 588.3 to 503.7.

Limits for Range are calculated with the help of D_3 and D_4 values. D_3 and D_4 for sample of size 6 are 0 and 2.004 respectively. So the limits for the range are 0 and 84×2.004 *i.e.*, 168.3.

Limits for S-chart are calculated with the help of B_3 and B_4 values. They are 0.003 and 1.997 for samples of size. So the limits for chart are $30\times.003$ *i.e.*, 0.10 and 30×1.997 *i.e.*, 59.9.

Plotting the data on a graph paper the control charts are as given in Fig 23.4.

Within--group variability homogeneous or not means standard deviation and range are under control or not. This can be seen from s-chart and R-chart, whether they indicate control or not. In the present case as all sample points are within control limits for s and R so the chart indicates that within group variability is within limits.

All groups are from one homogeneous lot or not means that all samples come from same population or not, *i.e.*, $\bar{X}$-chart indicates control or not. In other words sample means significantly deviate one from the other or not. Here out of 10 sample points two are below the lower control limits and 3 are above the upper control limit So the chart indicates lack of control or that sample mean significantly differs one from the other and they are not from one homogeneous group.

1 in 1000 means 3.09 times standard error limits. So limits for $\bar{X}$ - chart are $\bar{X}\pm\ 3.09\ \sigma/\sqrt{N}$

$$=0.1349\pm3.09\times\frac{.00105}{\sqrt{5}}$$

$$=0.1349\pm0.0015 \text{ or } 0.1364 \text{ to } 0.1334.$$

Tolerance limits are obtained by nominal dimension ± tolerance allowed *i.e.*, $0.1350\pm.0032$ or 0.1382 and 0.1318.

Limits for range can be obtained by the standard tables. For 1 in 1000 *i.e.*, .001 probability R/σ values for lower and upper limits for sample of size 5 are 0.37 and 5.48 respectively. Substituting the calculated value σ *i.e.*, 0.0005 this directly gives lower and upper limits R.

Lower limit for $R = 0.37 \times 0.00105 = 0.0004$

Upper limit for $R = 5.48 \times 0.00105 = 0.0058$

Illustration 23.6. Average proportion of defectives in first 10 samples of size 150 each was observed to be 0.04. What are the 1 in 1000 control limits? If later on it is noticed that machine is only producing 2% detective items what are the revised control limits on percent--defectives?

Solution. Here $p=0.04$, $n=150$

$\therefore$ Standard error of proportions

$$=\sqrt{\frac{p\times(1-p)}{n}}=\sqrt{\frac{.04\times.96}{150}}=0.016$$

1 in 1000 control limits for proportions (p) are given by $p\pm3.09$ S.E. *i.e.*, here by $0.04\pm3.06\times0.016$ or $0.04\pm.05$ or. 09 and $-.01$. But proportions cannot be negative. So the lower limit is 0 and not $-.01$ Hence the required limits are 0 to 0.09.

In the second case $p=2\%$

So the standard error of percentages

$$=\sqrt{\frac{p\times(1-p)}{n}}=\sqrt{\frac{2\times98}{150}}=1.14$$

$\therefore$ 1 in 1000 control limits for percentages (p)

$$=p\pm3.09 \text{ S.E. } i.e., 2\pm3.09\times1.14$$

or 2 ± 3.5 or 5.5 to 0% (as negative percentages are not possible).

Illustration 23.7. It was found that when a manufacturing process is under control, the average number of defectives per sample batch of 10 is 1.2. What limits would you set in a quality control chart based on the proportion of defectives in sample batches of 10?

Solution. Proportion of defectives (p)

$$=\frac{1.2}{10}=0.12$$

$\therefore$ Standard error of number defectives

$$=\sqrt{np(1-p)}=\sqrt{10\times0.12\times0.88}$$

$$=\sqrt{1.056}=1.03$$

Usually 3 sigma limits are marked on the control chart. So in the present case control limits for number of defectives=average number of defectives$\pm 3\times$St. Error

$$=1.2\pm 3\times 1.03=1.2\pm 3.09 \text{ or } 4.29 \text{ and } 0$$

Illustration 23.8. After the chart was set up the process was observed to be well under control for a long period with an average of 4% defective and it was decided to reduce inspection for control. A sample of size 25 was decided upon. Obtain the control limits for the revised chart. At a later date, the examination of the conttol chart gave the following number of defectives per sample over the last 20 samples : 0, 0, 2, 3, 0, 2, 1, 1, 0, 0, 1, 2, 1, 1, 1, 2, 4, 4, 3, 3. The quality control incharge decided on the basis of the data that the process has to be checked. Is the decision justified?

Solution. Average percentage of defective is 4%

$\therefore$ Average number of defectives in a sample of $25=\dfrac{25\times 4}{100}=1.$

As number of defective items are noted, control limits may be set in terms of it.

Now $p=.04$, $\qquad n=25.$

$\therefore$ Standard error of number defectives$=\sqrt{25\times .04\times .96}=.98.$

$\therefore$ 3σ limits are $1\pm 3\times .98=-1.94$ to 3.94.

As number of defective cannot be negative, the limits are 0 to 3.94 and 2σ limits are $1\pm 2\times .98=-.96$ to 2.96 or 0 to 2.96.

Plotting these limits and the given number of defectives in last 20 samples the control chart is as in Fig 23.5

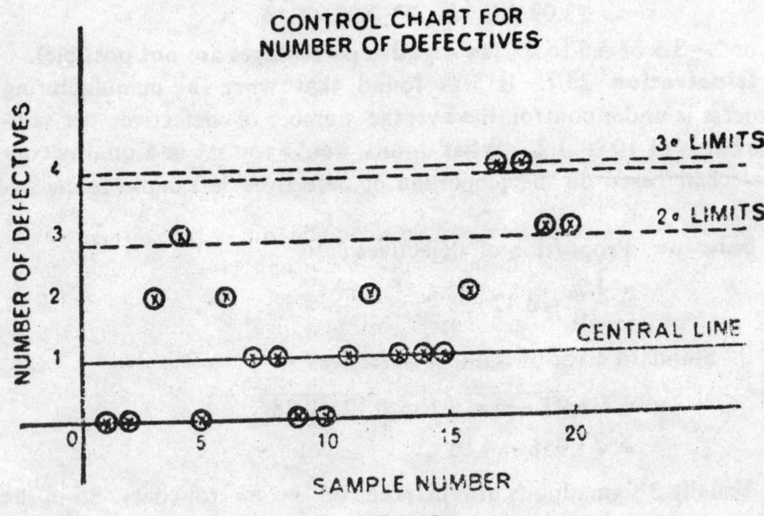

FIG 23.5

We notice from the chart that last 4 points are outside the 2σ limits and 2 of them outside even 3σ limits. For a process in control only 1 in 20 samples should be outside the 2σ limits and not outside the 3σ limits. So it is justified to assume that process is not under control.

Illustration 23.9. The following figures give the number of defects found in items of cotton piece goods inspected every day in a certain month: 1, 1, 3, 7, 8, 1, 2, 6, 1, 1, 10, 5, 0, 19, 16, 20, 1, 6, 12, 4, 5, 1, 8, 7, 9, 2, 3, 14, 6, 8. Do these data come from a controlled process?

Solution. Average number of defects $=\frac{\Sigma \text{ defects}}{n}=\frac{187}{30}=6.2=c$

Standard error of number of defects $=\sqrt{c}=\sqrt{6.2}=2.49$

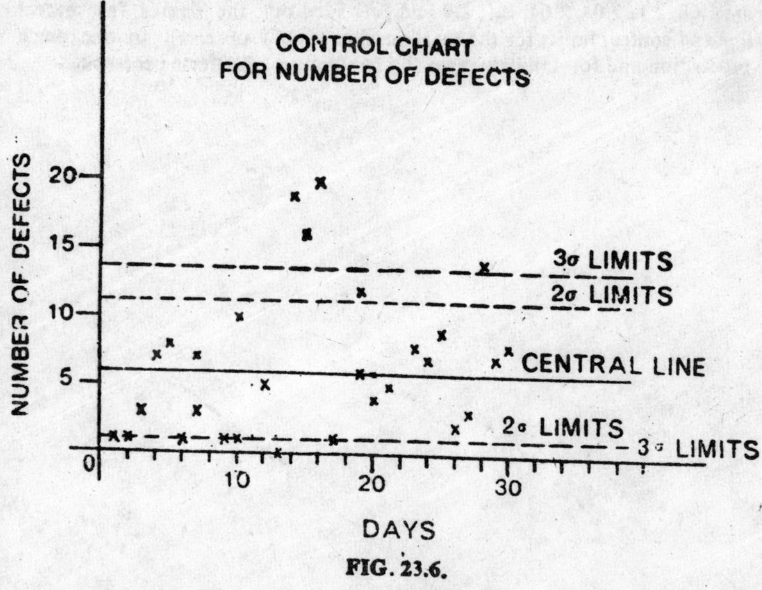

FIG. 23.6.

∴ 2σ limits for the control chart of c will be $c\pm 2\sqrt{c}$ or $6.2\pm$ (2.49) or 1.22 to 11.18.

The 3σ limits are $c\pm 3\sqrt{c}$ or $6.2\pm 3(2.49)$ or -1.27 to 13.67. As number of defects cannot be negative they are 0 to 13.67.

The central line is at 6.2. Plotting these the control chart is as shown in Fig. 23.6.

From the chart we notice that out of 30 points 4 points are outside 3σ limits. In fact none should have been. Also we notice a large number of points are above control line. So we can say process is not under control and the tendency is to have more defects than expected.

PROBLEMS

1. The following data give readings for 10 samples of size 6 in the production of a certain component:

Sample	1	2	3	4	5	6	7	8	9	10
Mean ($\bar{X}$)	383	508	505	582	557	337	514	614	707	753
S.D. (σ)	30.5	41.6	39.5	32.2	27.4	24.2	48.7	8.9	13.1	33.9
Range (R)	95	128	100	91	68	65	148	28	37	80

Draw the control charts for ($\bar{X}$), σ and R, calculating the limits for $\bar{X}$ in two ways. Can the within group variability be regarded as homogeneous? Can one assume that all groups are from homogeneous lot?

2. In a plant making about 95 trucks per day number of defects at the end of assembly line is noted. Average number of defects per truck for 7 weeks are 1.68, 2.1, 2.03, 2.04, 2.1, 2.0 and 1.7. Find out the figures for central line and control limits for the total number of (C) observed in one day's production and for daily averages (C) for number of defects per truck.

Chapter 24

STATISTICS FOR BUSINESS DECISIONS

24.1 INTRODUCTION

Throughout this text so far, we have shown how statistics helps make business decisions. The selection of the length of the handle of a spade was made by calculating the mode of the requirements of the various workers. Selection of the optimum strategy for advertising a product was made on the basis of statistical market tests.The future prediction of the demand of a product was based on trend values corrected for seasonal influences. The efficacy of a new strategy was tested using the various statistical tests of significance. In fact all the statistical techniques introduced so far have been used for making some or the other decision relevant to a businessman.

In this chapter we introduce yet another technique of business decision-making suitable to the situation where a businessman has to chose one from amongst several strategies of action in the absence of exact knowledge about what the future holds for him.

24.2 THE DECISION PROBLEM UNDER UNCERTAINLY

Consider the case of a baker who bakes cakes every night for sale on the next date. Any cake left unsold at the end of the day must be discarded for reasons of health regulations, and therefore,he can't bake too many cakes. At the same time he cannot afford to bake too few since if the demand is more than what he baked, it represents a loss of potential income. He wishes he could predict the exact demand so that he could bake exactly that many cakes and optimize his profit. But that is not to be. The closest he can get to predicting the demand is to project the probable demand on the basis of his past experience. The task of statistics is to help him arrive at a rational decision about the number he should bake depending on the meagre data available.

Consider another problem of a similar nature. To service demand at a rent-a-car agency, the manager needs to decide on the number of cars he should have, If the number of cars is more than that demanded, it represents a loss on investment, while if the demand is more than what he has set up to meet, it results in a loss of income opportunity.

The elements common to most decision problems of this type are:

1. *A desired objective.* The objective in both the above problems is to minimise losses, the actual loss due to unutilized capacity on one hand, and the opportunity loss on the other.

2. *Multiple courses of action.* The decision involves chosing from the various courses of action available. The numbers of cake the baker bakes or the number of cars to be stocked constitute the multiple courses of action available to the decision-maker.

3. *Multiple outcomes of the state of nature.* The actual demand for cakes or for cars is outside the control of the decision maker. Each one of these has multiple possible outcomes. What the demand will be on any given day cannot be controlled or ever predicted. The actual demand on any given day will be a given fact, after that day. The unpredictable, uncontrollable factors on which the profit depends are quaintly termed as the *state-of-nature* variables, and the actual state that obtains on a given day is termed as the *outcome* of the state of nature.

4. *Uncertainity.* The absence of exact knowledge about the outcome of the *state-of-nature* on any given period is an important characterising feature of this class of problems. Though we do not know exactly what the demand will be tomorrow, we may be able to assign probabilities to the various levels of demands (i.e., the outcomes) based on our past experience, market-survey or on hunch.

Consider our baker. Let us assume that the past experience suggests the following distribution for the probability of the occurance of the demand at various levels :

TABLE 24.1 DEMAND PROBABILITIES

Demand (No. of cakes)	*Probability*
11	0.1
12	0.15
13	0.4
14	0.25
15	0.1
	1.0

Thus, the various possible *outcomes* are demands of 11,12, 13, 14 or 15 cakes with the probabilities of occurence as shown in Table 24.1.

5. *A calculable measure of benefits under each course-of-action and outcome combination*

Let us assume that each cake costs Rs. 25 to bake, and that the baker sells it for Rs. 40. Let us also assume that any unsold cake is worthless.

Clearly the various 'Courses of action' or the options available to the baker are to bake 11, 12,13, 14 or 15 cakes. Now, he cannot be sure of the exact outcome of the state of nature. If he bakes 12, and the demand is 12 it is a very happy situation. But if the demand is only 11, he would have an unsold cake left at the end of the day representing a loss of Rs 25. On the other hand if the demand was actually 14 he would have two unsatisfied customers and a loss of potential profits.

Let us construct a 2-fold table to show the profits he makes under various combinations of 'actions' and 'outcomes'

TABLE 24.2 PROFIT TABLE

Action	*Profit under outcome (Rs.)*				
	11(01)	12(02)	13(03)	14(04)	15(05)
11 (A1)	165	165	165	165	165
12 (A2)	140	180	180	180	180
13 (A3)	115	155	195	195	195
14 (A4)	90	130	160	210	210
15 (A5)	65	105	135	185	225

The entry under action 14 and outcome 12 represents profit on the sale of 12 as demanded (Rs 15 x 12) *minus* loss on 2 discarded cakes (Rs 25 x 2) which comes to Rs 130. The entry under action 12 and outcome 14 represents the profit on sale of 12 (Rs 15 x 12) which is Rs. 180, with the demand of 13th and 14th cake remaining unmet.

If our baker was meek he might select the 'action' of baking only 11 cakes. He would be assured of a profit of Rs. 165 in all eventualities. But this, clearly is not the best strategy for he could earn as much as Rs. 225 if he bakes 15 and the demand is for 15. There is a risk involved though, since if he bakes 15 and the demand is only 11, earnings dip to Rs 65 only.

This table can be termed as a *conditional* profit table since it tabulates the profits under specified conditions of outcome and course-of-action.

24.3 EXPECTED PROFITS

The next step is to use the probabilities of the various demands and calculate the expected profit P for each of the actions. Using the probabilities of Table 24.1, the expected profits can be calculated by weighting profits against each demand by the probability of that demand.

E (Profit when A = 11) $= 0.1 \times 165 + 0.15 \times 165 + 0.4 \times 165 + 0.25 \times 165 + 0.1 \times 165 =$ Rs 165

E (Profit when A = 12) $= 0.1 \times 140 + 0.15 \times 180 + 0.4 \times 180 + 0.25 \times 180 + 0.1 \times 180 =$ Rs 176

E (Profit when A = 13) $= 0.1 \times 115 + 0.15 \times 155 + 0.4 \times 195 + 0.25 \times 195 + 0.1 \times 195 =$ Rs 181

E (Profit when A = 14) $= 0.1 \times 90 + 0.15 \times 130 + 0.4 \times 160 + 0.25 \times 210 + 0.1 \times 210 =$ Rs 166

E (Profit when A = 15) $= 0.1 \times 65 + 0.15 \times 105 + 0.4 \times 135 + 0.25 \times 185 + 0.1 \times 225 =$ Rs 145

It is clear that the expected profit peaks when the baker bakes 13 cakes. With 15 cakes baked, his profit could go upto Rs 225, but it could be as low as Rs 65 giving an expected value of only Rs 145.

Notice that our baker still does not know what the demand is going to be 'tomorrow', but he has used his past experience (in the form of a probability distribution) to determine the best baking action. There is no guarantee that he would make Rs 181. In fact he would not make Rs 181 ever for it would be 115, 155, or 195 depending on what the demand is. But if he sticks to this course of action for a long period of time, and provided the pattern of demand does not change, he will make Rs.181 on the average. This is the best he can do with the limited information about future demands that he has.

Illustration 24.1. A fruit wholeseller buys cases of strawberries for Rs. 200 each and sells them for Rs 500 each. Any case left would at the end of the day have a salvage value of only Rs 50. An analyis of past sales record reveals the following probability distribution for the daily number of cases sold.

Daily sale	*Probability*
10	0.15
11	0.25
12	0.40
13	0.20
	1.00

What is the optimum stock action for the fruit-seller?

Solution. The following profit table may be constructed for the various combinations of 'actions' and 'outcomes'. Note that a case sold results in a profit of Rs.300 while a case unsold results in a loss of Rs 150. The expected profit for each course of action is now calculated by weighting the various profits by the corresponding demand probability. This gives the following profit table.

PROFIT TABLE (RUPEES)

Action	*Outcome (demand)*			
	10	11	12	13
10	3000	3000	3000	3000
11	2850	3300	3300	3300
12	2700	3150	3600	3600
13	2550	3000	3450	3900

We next calculate the expected profit for each course of action by weighting the profit for each outcome (for the corresponding course of action) with the probability of occurence of that outcome of the state of nature, as below :

E (Profit when 10 stocked) $= 0.15 \times 3000 + 0.25 \times 3000 + 0.4 \times 3000 + 0.2 \times 3000 =$ Rs 3000

E (Profit when 11 stocked) $= 0.15 \times 2850 + 0.25 \times 3300 + 0.4 \times 3300 + 0.2 \times 3300 =$ Rs 3232.50

E (Profit when 12 stocked) $= 0.15 \times 2700 + 0.25 \times 3150 + 0.4 \times 3600 + 0.2 \times 3600 =$ Rs 3352.50

E (Profit when 13 stocked) $= 0.15 \times 2550 + 0.25 \times 3000 + 0.4 \times 3450 + 0.2 \times 3900 =$ Rs 3292.50

This example clearly brings out the nature of the decision problems under uncertainity. Thus a 'safe' action is the stocking of only 10 cases which would give a profit of Rs 3000 each day. Adding another case to the stock may result in set-backs where profit declines to Rs 2850 only, but it is expected to occur only on 15 percent of the days. Stocking more may give more days with profit declining, but 'on the average' the profits will be better. The optimal stock-strategy is to have12 cases.

24.4 EXPECTED VALUE OF PERFECT INFORMATION

Suppose the baker of the previous section could do a 'survey' and determine 'tomorrow's demand 'today', what is going to be his profit? Ofcourse the demand will still vary, but can be predicted. If he knows the demand is going to be 11 then he bakes 11 cakes and earns Rs 165. Table 24.3 gives the profits under perfect informations about the demand.

TABLE 24.3 PROFITS UNDER PERFECT INFORMATION

Action	Outcome 11	12	13	14	15
11	165	-	-	-	-
12	-	180	-	-	-
13	-	-	195	-	-
14	-	-	-	210	-
15	-	-	-	-	225

In this table there are entries only when *Action* and *Outcome* match, since there is perfect information.

The expected profit in this case is again obtained by weighting each figure by the corresponding probability.

Thus:

E (Profit under perfect information) = 0.1×165+0.15×180 +0.4×195+0.25×210+0.1×225=Rs 196.50

This value of Rs 196.50 for the expected profit is significant in that it represents the maximum possible profit 'in the long run' from the market, there being no loss due to unpredicted demand.

Thus, by investing extra effort and money (in conducting the survey), the baker can increase his profit from Rs 181 (the profit he makes when he uses optimal strategy under uncertainity) to Rs 196.50. Thus, his profit can increase by a maximum of Rs 15.50 per day (on an average) if he has the perfect information. This, therefore, represents the *expected value of perfect information.* If conducting the survey costs him an amount less than this, the information is useful, but if collecting the information costs more than Rs 15.50 per day, he may as well not worry about it.

Illustration 24.2 What is the expected value of perfect information about the market in Illustration 24.1?

Solution : If the fruit seller knew in advance what his sale is going to be, his profit table is going to look like this.

PROFIT TABLE UNDER PERFECT INFORMATION

Demand	*Action (Stock)*	*Profit*	*Probability*
10	10	3000	0.15
11	11	3300	0.25
12	12	3600	0.40
13	13	3900	0.20

His expected profit would then be

$0.15 \times 3000 + 0.25 \times 3300 + 0.40 \times 3600 + 0.20 \times 3900 = $ Rs 3495

This is the maximum average profit he can expect under the perfect information. Since the optimal expected profit under uncertainity is Rs 3352.50, the expected value of perfect information is Rs (3495 – 3352.50) = Rs 242.50 per day.

24.5 MARGINAL ANALYSIS

In many problems the number of possible actions or outcomes is very large and the amount of effort required in calculating profit tables and expected values of various actions becomes phenomenal. Under such circumstances it becomes advantageous to use what is termed as *marginal analysis*. We introduce marginal analysis first with the example of the baker, and then generalize it to apply in other cases.

Marginal analysis looks at the *increase* in profit as a single unit is added to the stocked items. It is based on the fact that as we add one additional item to a *given* stock level, only two distinct fates await it. It will be either sold or not sold. The probability that it is sold is the probability that the demand is *atleast* one unit more than the base level of stock, and the probability that it is not sold is obviously equal to (1–probability that it is sold).

Let us start with the baker's action of storing the minimum number of cakes at 11. His expected profit has been calculated as Rs 165.

Now if he bakes the twelfth cake, the condition that it does not sell is that the demand is less than 12, or is 11. The probability of *not* selling, therefore, is 0.1. The probability that the sale is made is, therefore, (1-0.1) = 0.9.

If the twelfth cake is sold, the *marginal profit, MP*, is Rs 15, while the *marginal loss, ML*, in case it is not sold, is Rs 25.

We can next calculate the *expected marginal profit, EMP*, as

$$\begin{aligned} EMP\,(12) &= EP \times \text{(probability of sale)} - EL \times \text{(probability of not selling)} \\ &= 15 \times 0.9 - 25 \times 0.1 \\ &= \text{Rs. } 11 \end{aligned}$$

Thus , the baker expects to earn an average of Rs 11 more from this twelfth cake and should, therefore, bake it everyday.

Let us next consider the thirteenth cake. The probability p that it sells is the probability that the demand is 13 or more, i.e., (0.4 + 0.25 + 0.1) = 0.75, and the probability that it does not sell is (1–0.75) = 0.25, and the expected marginal profit for this thirteenth cake is :

$$\begin{aligned} EMP_{(13)} &= EP \times p - EL \times (1 - p) \\ &= 15 \times 0.75 - 25 \times 0.25 = \text{Rs } 5 \end{aligned}$$

This is also positive, and hence it is worthwhile for the baker to bake the thirteenth cake as well.

Similarly, the probability that the fourteenth cake will be sold is (0.25 + 0.1) = 0.35 and that it will not be sold is (1–0.35) = 0.65, and

$$EMP_{(14)} = 15 \times 0.35 - 25 \times 0.65$$
$$= \text{Rs. } -7.5$$

Since this is negative, one may not like to stock 14 cakes. Thus, the optimal strategy for the baker is found to be a stock level of 13, as determined earlier.

We generalise the above procedure by introducing x^* as the optimal value of the *action* variable x. A stock level x is desirable if EMP (x) > 0, and not, if $EMP\ (x) < 0$. Thus, the optimal value x^* is obtained by the condition

$$EMP\ (x^*) \geq 0$$

If $p\ (x^*)$ represents the probability that x* th item is sold, then

$$EMP\ (x^*) = MP \times p^* - ML\ (1-p^*) \geq 0$$

Simplifying, we get

$$p(x^*) \geq \frac{ML}{MP + ML} \equiv p^*$$

The expression on the right of the inequality denotes the *minimum desired* probability of selling *at least* one additional item. The value of x^* upto which $p\ (x^*)$ remains greater than p^* is, thus, the optimal level of inventory.

In our case,

$$p^* = \frac{ML}{MP + ML} = \frac{25}{15+25} = 0.62$$

The probability $p\ (x)$ of selling the x_{th} item when the inventory level is (x^*-1) is the probability of the demand being x or more. Thus,

$p\ (11) = 1$

$p\ (12)$ = probability of the demand being

12, 13, 14, or 15 = $0.15 + 0.4 + 0.25 + 0.1 = 0.9$

Similarly,

$p\ (13) = 0.4 + 0.25 + 0.1 = 0.75$,

$p\ (14) = 0.25 + 0.1 = 0.35$, and

$p\ (15) = 0.1$

Sure $p\ (14) < p^* = 0.625$, it is not profitable, in the long run, to stock the 14th cake. Thus, 13 cakes is the optimum stocking action.

Illustration 24.3. Determine the optimal stock action of the fruit wholeseller of Illustration 24.1 by marginal analysis.

Solution. The marginal profit, MP, is Rs 300, while the marginal loss ML is Rs 150. The minimum required probability of selling an additional case is, therefore.

$p^* = 150/(300 + 150) = 0.33$
Probability of selling 11th case = 0.85
Probability of selling 12th case = 0.60
Probability of selling 13th case = 0.20

Since the probability of selling the 13th case is less than the minimum required probability calculated above, the 13th case should not be stocked or the optimal action is storing 12 cases only, the same as calculated before.

24.6 USING NORMAL DISTRIBUTION IN MARGINAL ANALYSIS

In many decision problems, the number of outcomes (and hence possible actions) is so large that it is difficult to assign a probability to each outcome. It is possible in many such situations to model the probability distribution of the outcomes by a normal curve with a specified mean and standard deviation. We can then use the marginal analysis outlined above together with the normal probability tables to calculate the optimum action. This is illustrated below :

Our baker also sells bread which has a mean daily demand of 100 loaves with a standard deviation of 15. If the cost of each loaf is Rs 4, and it sells for Rs 6, but at the end of the day its value reduces to only Rs 3, what is the optimal lot he should bake everyday?

We first calculate, the minimum required probability p^* of the marginal sale. Marginal profit is (Rs. 6–Rs. 4) = Rs. 2. Marginal loss is Rs. (4–1) = Rs. 3.

So $p^* = 3/(3 + 2) = 0.6$

Thus, the optimal baker's action is where the probability of selling an additional loaf is 0.6. Since we have assumed the demand to have a normal distribution with mean = 100 and standard deviation = 15, the probability that the demand is atleast x is the area to the right of x. We want this area to be 0.6. The corresponding standard normal variate should be negative, so that the area from it to the mean is 0.1.

From tables this is obtained as Z = –0.253. Thus,

$$-0.253 = \frac{x - 100}{15}, \text{ which gives}$$

$$x = 100 - 0.253 \times 15 = 96.2 \text{ or } 96$$

Thus, the baker can expect marginal profit only if his stock is 96 loaves, beyond that he expects a loss for each additional loaf baked and stocked.

Illustration 24.4: A newspaper vendor buys newspaper at Rs 1 each and sells for Rs 2. At the end of the day, the scrap value of the paper is only Rs 0.20. If the demand of newspaper is estimated as a normal distribution with mean 210 and standard deviation of 15, what should be the number of copies purchased by the vendor?

Solution.

The marginal profit MP = Rs (2-1) = Rs 1.00

The marginal loss ML = Rs (1-0.20) = Rs 0.80

The minimum acceptable probability of selling an additional copy should be

$$p^* = 0.80 \,/\, (100 + 0.80) = 0.4444$$

Since this is less than 0.5, the standard normal variate for optimal stock action should be the one, the area to the right of which is 0.4444. From the table of normal curves it is seen as 0.14.

$$\text{Therefore,} \quad 0.14 = \frac{x - 210}{15}$$

or $x = 210 + 0.14 \times 15 = 212.1$

Thus 212 copies of the newspaper should be stocked each day for maximum average profit in the long run.

Illustration 24.5 Assume that a stores manager sells a perishable article whose daily demand can be approximated by a normal distribution having a mean of 50 units and a standard deviation of 15 units. If the cost price of the item is Rs 40 and it sells for Rs 90 per unit, and if the salvage value at the end of the day is zero, calculate the optimum opening inventory level each day.

Solution :

Here, MP = Rs 90–Rs 40 = Rs 50, and the marginal loss ML = Rs 40. Thus, the cut-off probability p^* is

$$p^* = \frac{ML}{MP + ML} = \frac{40}{40+50} = 0.44$$

This implies that the manager should stock at the level where the marginal probability of sale, i.e., the probability of the demand at this level or exceeding it should by 0.444

Since 0.444 is less than 0.50, the standard ordinate is to the right of the mean. From the normal curve tables, we see that the corresponding value of Z, the area to the right of which is 0.444 is 0.14. Thus

$$0.14 = \frac{x^* - 50}{15}$$

or $x^* = 50 + 0.14 \times 15 = 50 + 2.1 = 52.1$ units.

The stores manager should buy 52 units each day, since the 53rd unit will result in a marginal loss on the average.

24.7 DECISION TREES

A decision tree is a graphic model of the decision process used in complex situations where there are many levels of actions and many states of nature to be considered. As before, the 'outcomes' of the states of nature are uncertain, only their probabilities are known from past experience. A decision tree is so named because of its similarity in appearance to a tree with a stem branching off into a number of branches, each of which again branches into still another 'generation' of branches. Decision-tree analysis is useful in making decisions concerning acquisitions and disposable of capital assets, investments, project management and new-product strategies.

A decision tree is like the branching probability tree introduced in chapter 16. But a decision tree contains not only the probabilities of outcomes but also the conditional payoffs attached to these outcomes. Because of this, a decision tree provides a very graphic tool for calculating the expected pay-off for the various strategies.

We introduce the concept of decision trees with the following example : A movie distributor is considering the strategy for the release of a movie. Given the uncertainity of the acceptance of any movie by the public he is debating whether he should go for a low publicity release or a high-impact expensive release. From his past experience with movies of a similar kind, he draws the following table for expected pay-offs.

TABLE 24.4 PROFIT TABLE

Level of Success	*Probability*	*Profit in Rs lakh for two levels of publicity*	
		Low	*High*
Super hit	0.1	100	80
Hit	0.3	20	18
Commission-earner	0.4	5	5
Flop	0.2	(–5)	(–2)

He wants to decide on what course of action to take. Here, using the nomenclature introduced earlier, level of success is what cannot be predicted with certainty, and hence can be termed as a 'state of nature', with four different outcomes. There are two courses of action available.

We next draw the decision tree which clarifies the structure of the problem graphically. Fig. 24.1 shows the relevant decision tree. We start at the decision point and draw two branches for the two courses of action. For each course of action branch, we branch off for the states of nature, writing

their probabilities and the resultant pay-offs.

Course of Action | State of Nature | Pay-Off Rs. | Expected Pay-Off
Low Publicity
Superhit (0.1) 100
Hit (0.3) 20
Comm. Earner (0.4) 5
Flop (0.2) (-5)
17
High Publicity
Superhit (0.1) 80
Hit (0.3) 18
Comm. Earner (0.4) 5
Flop (0.2) (-2)
15

Fig. 24.1 Decision Tree

Then we calculate the expected profits by weighting the payoffs for each state of nature within each course of action by the respective probabilities of occurrence of the states of nature. A comparison of the expected profits completes the decision-tree analysis. In this case, the distributor expects to earn more if he goes in for a low publicity release. Of course, he risks more if the picture flops.

Illustration 24.6. A businessman has option to buy 3 plants, the choice depending upon the expected sales volume. The payoff table is organized as below.

PROFIT TABLE (RUPEES IN LAKHS)

	Demand		
Plant	*low*	*Moderate*	*High*
A	–5	5	20
B	2	8	8
C	4	5	5
Probabilities	0.3	0.4	0.3

Construct a decision tree and recommend which plant should be bought.

Solution: The figure below shows the decision tree, and the calculations of the expected profits for the three courses of action. Since the expected profit for plant A is highest the optimal course of action appears to be the purchase of plant A.

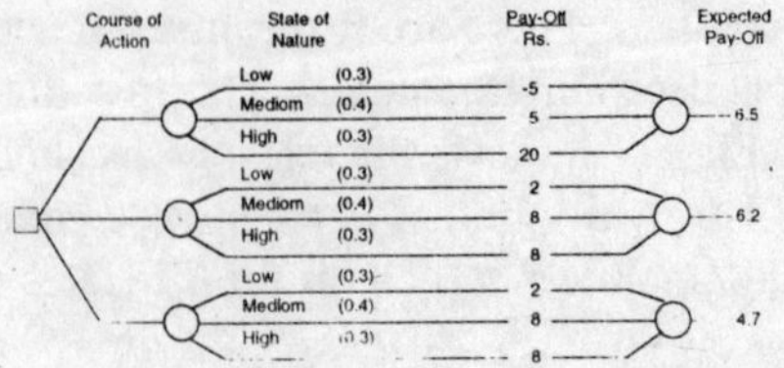

Fig. 24.2 Decision Tree For Illustration 24.6

Illustration 24.7. Him Fut operates a small ski-resort at Kufri. His profit depends upon the amount of snow-fall in the region. From his past experience he draws up the following table

Amount of Snow-fall (state of nature)	*Probability of occurrence*	*Seasonal profit, Rs.*
> 50 cm (A)	0.4	1,20,000
20 - 50 cm (B)	0.2	40,000
less than 20 cm (C)	0.4	(-) 40,000

He is considering two proposals. One : leasing snow-making equipment which will render his resort free from the vagaries of nature and he will make the full profit of Rs 1.2 lakh minus the leasing cost of Rs 1.2,000 and the operating costs. The operating cost, however, varies with the state-of-nature, being Rs 10,000 if the natural snow-fall is 50 cm or above, Rs 50,000 if between 20-50 cm, and Rs 90.000 if less than 20 cm.

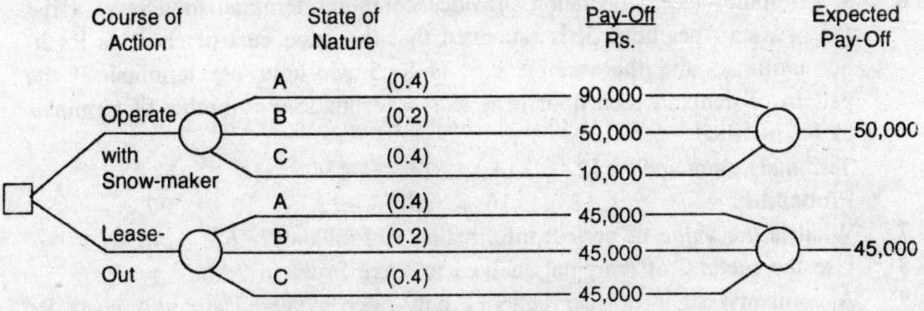

Fig. 24.3 Decision-tree for Illustration 24.7

The second proposal he has is from an international chain of resorts which proposes to lease his resort for a fixed fee of Rs 45,000 for the season. In that case, all operating expenses will be borne by the chain of resorts.

Draw a decision-tree and help Mr. Him Fut make a rational decision.

Solution: The decision tree for this problem is drawn and shown as Fig, 24.3. The payoffs are conditional payoffs for the state-of-nature and course-of-action combinations. The expected payoffs are the probability weighted payoff for each course of action. Thus, the expected payoff for 'operate with snow-maker' is 0.4 x 90,000 + 0.2 x 58,000 + 0.4 x 10,000 = Rs 58,000. We notice the maximum expected profit and is, therefore, the recommended strategy.

Problems

24.1 A contractor likes to work with permanant staff only. He pays each worker Rs.50 a day and charges his customers Rs.120 a manday. How many workers should he hire if the demand of mandays per month (20 working days) has the following probability distribution?

Hours	10,000	12,000	14,000	16,000
Probability	0.2	0.3	0.4	0.1

24.2 Determine how much is the worth of perfect information in Problem 24.1.

24.3 A car leasing company plans to buy a fleet of cars, the daily acquisition cost of which turns out to be Rs.150. If the rent per day is Rs.300 and if the demand distribution is as below, calculate the optimal fleet size the company should have.

Demand	8	9	10	11	12	13	14
Probability	0.1	0.15	0.15	0.2	0.2	0.1	0.1

24.4 Solve Problem 24.1 again using marginal analysis.

24.5 Solve Problem 24.3 again using marginal analysis.

24.6 A Computer-access operation provides computer terminal to user at a flat fee of Rs.50 per hour. It is estimated that the fixed cost per hour is Rs.20 per terminal, and the variable cost is Rs.5 per hour per terminal. If the pattern of demand each hour is as shown, estimate the number of terminals to be installed.

Terminals, demand	12	13	14	15	16	17
Probability	.12	.16	.20	.29	.16	.07

24.7 What is the value of perfect information in Problem 24.6?

24.8 Use the method of marginal analysis to solve Problem 24.6.

24.9 A company can order car batteries only once a year. Battery A costs Rs 1,000 and sells for Rs 1,600, but after one year the change of model forces a discount of Rs.800. Battery B costs Rs 1,400 and sells for Rs 2,400 but can be sold after a year only on a discount of Rs 1,200. How many batteries of each type should be stocked, given the following demand pattern.

	Mean sale	*Standard deviation*
A	1400	120
B	700	60

24.10 An ex-maharaja is approached by a hotel chain A with an offer for converting his palace into a hotel and paying him a rental of Rs 50,000 every month. This sets the ex-maharaja thinking and he calculates that if he converts and runs the hotel himself he can expect the following profits depending on the three level of demands.

Demand level	*Probability*	*Profit (loss)/month*
low	0.3	(10,000)
moderate	0.5	60,000
high	0.2	1,00,000

He makes further enquiries and gets another offer from a chain B which quotes a rental on the average occupancy rate.

Occupancy rate	*Probability*	*Rental*
below 50	0.3	10,000
50 - 100	0.5	50,000
above 100	0.2	80,000

Carry out a decision-tree analysis and recommend the course of action.

24.11 A car dealership is planning to open a car repair garage. It has estimated the following demand for mechanic hours (per year) :

Hours:	10,000	12,000	14,000	16,000
Probability:	0.2	0.3	0.4	0.1

The dealership plans to pay each mechanic Rs 18 per hour and to charge the customers at Rs 32/hour. Each mechanic will work 40 hours per week and will get a 2-week vacation. Determine how many mechanics must be hired, and how much should the dealership pay to get perfect information about the number of mechanics that they need?

24.12 A small-time hawker hits upon the idea of selling souvenir T-shirts at the world cup. Each printed shirt will cost him Rs 25, and he estimates that he can sell it for Rs 60. After the event, the unsold shirts can be remaindered for Rs 5 each only. A consultant advices him that at similar events, the demand of T-shirts has a normal distribution with an average of 500 and a standard deviation of 60. How many T-shirts should he order?

24.13 Sahni Tyre Company must order its requirements of tyres at the beginning of each year. There are two types of tyres. Tyre A costs Rs 1,500, sells for Rs 2,200 and has a demand estimated at an average of 600 per year with a standard deviation of 40. Tyre B has the corresponding figures as Rs 1,800, Rs 2,300, 800 and 35. If at the end of the year the unsold tyres are to be discounted by 50 percent, estimate the stocks of A and B that should be ordered.

24.14 Mohan has inherited Rs 20 lakh and is considering 3 possible investment avenues.

1. Invest in a new amusement park with a potential of making a Rs 20 lakh profit (probability 0.4), but with a high risk of losing all (p = 0.6).
2. Invest in real estate with a assured profit of Rs 1. lakh but with a possibility of yielding Rs 1.5 lakh, Rs 2.0 lakh, Rs 2.5 lakh, or even Rs 3.0 lakh, with probabilities estimated as 0.3, 0.25, 0.2 and 0.05 respectively.
3. Invest in government securities yielding 8.25 percent.

Construct a decision tree and select the investment options to optimize yield.

Chapter 25

Linear Programming

25.1 Introduction

Linear programming is a mathematical technique that helps organizations in finding optimal solutions to certain class of operational problems. The technique was first suggested in 1939 by L.V. Kantorovich and then independently in 1947 by G.B. Dantzig. Linear programming methods are probably the most successful of the methods for optimizing the use of resources in manufacturing and service industries and in military operations.

Several techniques are available within the linear programming methods. We shall begin with a graphical method and then introduce a technique known as Simplex method.

25.2 Linear Programming Problems

Given below are two of the various classes of problems for which solutions can be obtained us ing linear programming techniques:

Allocation Problem: A manufacturer has a plant where he can make various kinds of products. The resources required for the manufacturing have limited availability. For example, he may have only a certain number of machine hours available per month, or the foundry may be able to cast only a given weight of metal every month. The available skilled man-hours or other resources such as power, raw materials, etc., may also restrict him. Each of the various products uses different amounts of resources per unit and gives different amount of profit. The problem is one of the optimized allocation of resources to arrive at a product mix which will maximize the profits.

In algebraic terms, let $x_1, x_2, x_3, ..., x_n$ denote the number of units of product $1, 2, 3, \ldots, n$. Let p_i^j represent the amount of jth input needed per unit of product i. Thus, the total requirement of resource j is given by $p_1{}^j x_1 + p^{2\,j} x_2 + p_3{}^j x_3$ $+ ... + p_n{}^j x_n$ or simply $\sum_{i=1}^{n} p_i^j x_i$ with summation carried out over i from 1 to n. Let the total availability of the resource j be b^j. Let the profit per unit of product i be a_i. Then the allocation problem can be stated mathematically as follows:

Maximize the criterion function (profit here)

$$U = \sum_{i=1}^{n} a_i\, x_i \text{, i.e., } a_1 x_1 + a_2 x_2 + a_3 x_3 + \dots + a_n\, x_n,$$

subject to the constraints

$\sum_{i=1}^{n} p_i^{\,j} x_i \le b^j$ for all j's (a total of m such constraints, m being the number of resources with limited availability)

The complete definition of the mathematical problem will also need the obvious constraints that the numbers x_i must not be negative for all i's (a total of n such constraints). This problem and its variants are known as the *allocation problem.*

Transportation Problem

Consider next the problem of a wholesale distributor who wishes to supply a number of items from m warehouses to n stores. He can supply each store from each of the warehouses, but the cost of transportation to each store varies with the warehouses. Let x_{ij} represent the quantity that is shipped to store i from the warehouse j. Let c_{ij} represent the unit cost of transporting one item to store i from the warehouse j. The transportation problem then consists of determining x_{ij} such that the total transportation cost is minimized. In algebraic terms the problem can be written as:

Minimize the criterion function (the total transportation cost here)

$U = \sum_{i=1}^{n}\sum_{j=1}^{m} c_{ij}\, x_{ij}$, summed, in turn, over all warehouses (values of j from 1 to m), and all stores (values of i from 1 to n) is minimum, subject to the constraints:

x_{ij} is not negative, for all values i from 1 to n and all values of j from 1 to m (a total of constraints),

$\sum_{j=1}^{m} x_{ij} = a_i$, (where summation is over *all* j's from 1 to m) for all values i from 1 to (a total of n constraints),

$$\sum_{i=1}^{n} x_{ij} \leq b_j,$$ (where summation is over *all* i's from 1 to n) for all values j from 1 to m (a total of m constraints).

Here, a_i represents the total requirement of store i, and b_j represents the total availability at warehouse j. This problem and its variants are known as the *transportation problem.*

It may be noticed that the criterion functions U, as well as all the constraints in the two problems outlined above are *linear* functions of the variables x_i's or x_{ij}'s. This is an essential condition for the linear programming methods discussed below.

25.3 Graphical Method

For those problems of linear programming where only two variables (two x's, x_1 and x_2) are involved, one can use a graphical method. Though the real life problems seldom involve as few as two variables, it is instructive to build the graphical methods for the insight is helps develop. We introduce here the graphical technique through two illustrations.

Illustration 25.1: Let us take as our first example a furniture shop manufacturing two products: desks and chairs. Let us say that in the manufacture of these items four inputs or resources are required, all of them having a limited availability: carpentry man-hours, upholstery man-hours, good quality wood, and laminated topping for desks. Table 25.1 gives the constraints on the availability of the four resources and the requirements per desk or chair. The profit per desk is Rs 500 and per chair is Rs 120.

Table 25.1

Resources Required for Desks and Chairs

Resource	*Availability per week*	*Requirement*	
		Per Desk	*Per Chair*
Carpentry man-hours	800	32	16
Upholstery man-hours	480	–	12
Wood (m^3)	90	3.3	2
Laminate (m^2)	60	3	–

Solution: We first cast the optimization problem in the mathematical terms. If x_1 denotes the number of desks made per week and x^2 the number of chairs made per week, the linear programming problem consists of maximizing the profit function

$$U(x_1, x_2) = 500x_1 + 120x_2$$

subject to the constraints

$$x_1 \geq 0$$
$$x_2 \geq 0$$
$$32x_1 + 16x_2 \leq 800$$
$$12x_2 \leq 480$$
$$3.3x^1 + 2x_2 \leq 90$$
$$3x_1 \leq 60$$

The first two constraints are the trivial ones that the number of desks or chairs made cannot be negative. The third is the constraint on the available carpentry hours, fourth on the upholstery hours, and the last two are constraints on materials.

If we plot the equation $32x_1 + 16x_2 = 800$ on an x_1, x_2 plot, we get a straight line as shown in Figure 25.1. This line represents the various mixes of products x_1 and x_2 that will consume the total available carpentry man-hours. Thus, if we make only desks, we can make 25 of them in the available carpentry time, but if we make only chairs, we can make 50 of them. Clearly, the shaded area represents those combinations of x_1 and x_2 that satisfy the carpentry man-hours constraints (with x_1 and x_2 being greater than zero). Any point above this line will violate this constraint.

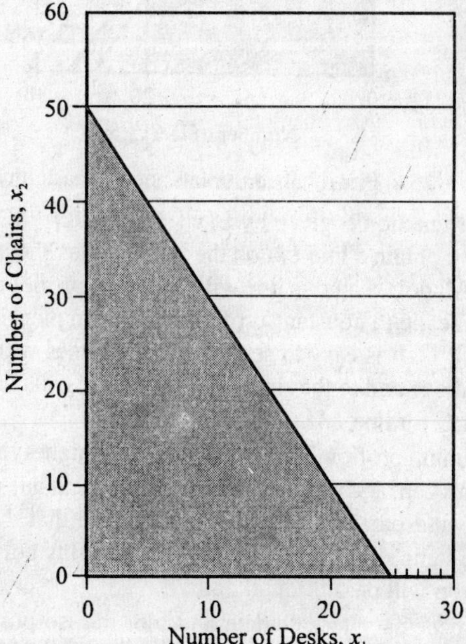

Figure 25.1. Plot of the carpentry man-hour constraint

Similarly we can plot all other constraints as shown in Figure 25.2. The shaded area then represents those combinations of the numbers of desks and chairs that can be made in the shop meeting *all* constraints. Any combination outside this area is not feasible.

All the points within the shaded polygon do not contribute equally to the profit function U. The problem consists of finding within the shaded area that combination of x_1 and x_2 that will maximize the profit function U.

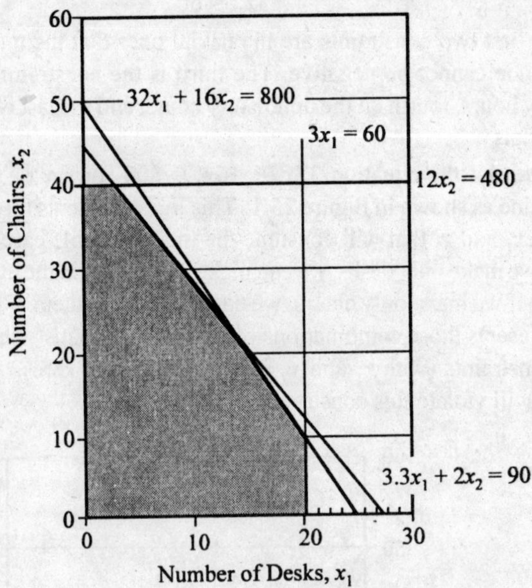

Figure 25.2. Plot of all constraints and the feasibility polygon,

The profit function is given by $U(x_1, x_2) = 500x_1 + 120x_2$. For a constant given profit, we obtain a line LM on the x_1, x_2 plot as shown for $U = 60{,}000$ in Figure 25.3. All points on this line will result in a profit of Rs. 60,000. This is an iso-profit line, i.e., a constant-profit line. We can plot similar lines for other values of profit U. It is easy to see that all such lines will be *parallel* to LM. They will be above and to the right for values of U greater than 60,000, which is the value of U for line LM.

The maximum profit will be obtained on the highest iso-profit line within the feasible polygon, and that line would pass through one of the vertices of the polygon. Drawing parallel lines, it is easy to see that the line of maximum profit with the same slope as LM, within the feasibility polygon, and farthest to the right and top will pass through point C.

It is not necessary to calculate and plot the iso-profit lines. Since the maximum profit will occur at one of the vertices towards the top and right of

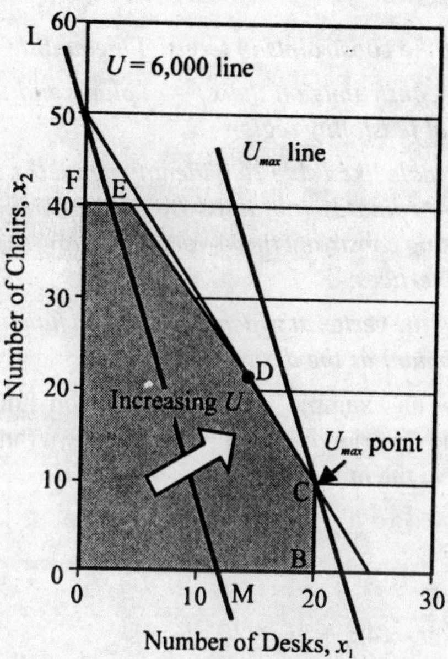

Figure 25.3 The plot of feasible area with iso-profit lines.

the feasibility polygon, it is enough to calculate the profits at each of these. We first calculate the locations of the candidate vertices *C, D* and *E*. Point *C* is the intersection of lines $3x_1 = 60$ and $32x_1 + 16x_2 = 800$, and can be obtained by solving the two simultaneously. Thus, point *C* corresponds to $x_1 = 20$ and $x_2 = 10$. Similarly, point *D*, which is the intersection of $32x_1 + 16x_2 = 800$ and $3.3x_1 + 2x_2 = 90$ and can be located by solving these two equations simultaneously. This gives the product mix *D* as $x_1 = 14.3$ and $x_2 = 21.4$. Table 25.1 shows profit at each of these vertices.

Table 25.2

Profits at the Vertices of the Feasibility Polygon

Point	x_1	x_2	*Profit, Rs.*
C	20	10	11,200
D	14.3	21.4	9,718
E	3	40	6,300

Thus, the product mix represented by point *C* is the mix that maximizes profit for the shop.

The graphical method of solving linear programming problems of this kind can be summarized as below:

1. *Identify the variables* x_1 *and* x_2.
2. *Write the criterion function or the profit function in terms of variables* x_1 *and* x_2.

3. *Write all the constraints in terms of inequalities (or equations).*
4. *Plot the constraints on the $x_1 - x_2$ plane and identify the convex polygonal feasibility region.*
5. *Set up a table like Table 25.2 identifying all the candidate vertices, their co-ordinates (obtained by solving the equations for the intersecting constraint lines), and evaluating the criterion function at these vertices.*
6. *Selecting the vertex at which the criterion function is the maximum (or minimum) as the desired solution.*

We give below an example where the criterion function needs to be minimized. Here the criterion function is pushed down (and to the left) as far as possible to achieve the minimum.

Illustration 25.2:

Minimize $\quad U = 2x_1 + 4x_2$

Subject to

$$4x_1 + 2x_2 \geq 12$$
$$3x_1 + 5x_2 \geq 15$$
$$5x_1 + 4x_2 \geq 20$$
$$x_1 \geq 1$$
$$x_2 \geq 0$$

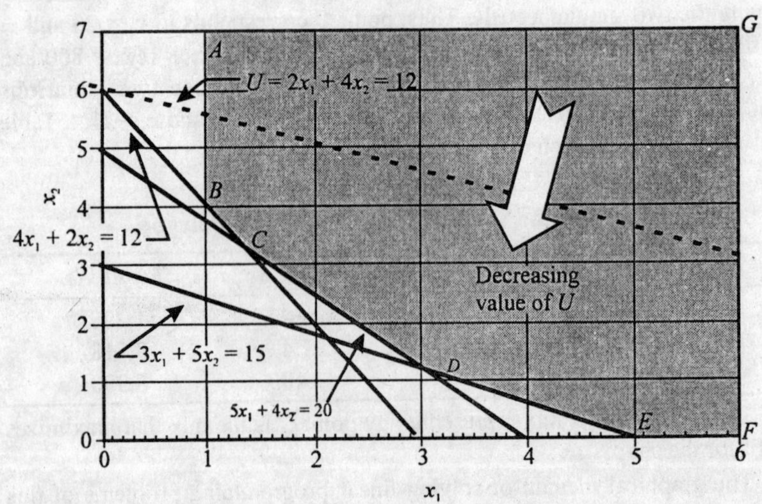

Figure 25.4. Feasibility polygon for the minimization problem.

Solution: Figure 25.4 shows the graph for this linear programming problem. The shaded region *ABCDEFG* represents the feasibility region in this case. The dotted line represents a line for constant value 12 of the criterion function. As has been clarified earlier, since the slopes of all constant *U* lines are the same as of this dotted line, and since the value of *U* decreases as we move down (and to the left), the value of *U* is minimized at one of the lower vertices of the feasibility polygon. The candidate vertices are *B, C, D, E* and *F*. Instead of graphically determining the values, we construct Table 25.3 where the location of the vertices has been determined by solving the equations of the appropriate lines simultaneously. Thus, vertex *D* is the intersection of lines $5x_1 + 4x_2 = 20$ and . Solving these two simultaneously, we get the coordinates of point *D* as (3.08, 1.15).

TABLE 25.3

***U* at the Vertices of the Feasibility Polygon**

Point	x_1	x_2	*U*
B	1	4	18
C	1.33	3.33	16
D	3.08	1.15	10.76
E	5	0	10
F	6	0	12

Clearly, point *E* represents the optimum mix where *U* has the minimum value.

25.4 The Simplex Method

The Simplex method is a very efficient technique for handling linear programming problems with any number of variables and any number of constraints. It has exceptional power and generality and has been the first technique that revolutionized operations planning in the 1960's. It is applicable to a wide variety of problems and has provided management with a remarkable tool to solve those problems that seemed insolvable until the 1960's.

We will introduce the Simplex method through a simple illustration in two variables only.

Illustration 25.3: A shop manufactures two products *A* and *B* using two machines. Each unit of product *A* requires two hours of time on each machine, while each unit of product *B* requires three hours on the first machine (which has a total of 12 hours available) but only one hour on the second machine (which has a total of 8 hours available). If the profit per unit of *A* is Rs. 6 and per unit of *B* is Rs. 7, find the product mix that maximizes the profit.

Solution: Let x_1 and x_2 represent the numbers of units of *A* and *B* in the product mix. The mathematical statement of the problem can be cast in the following form:

Minimize $U = 6x_1 + 7x_2$

Subject to

$$2x_1 + 3x_2 \le 12$$
$$2x_1 + x_2 \le 8$$
$$x_1 \ge 0;\ x_2 \ge 0$$

Figure 25.5 shows the graphical determination of the solution.

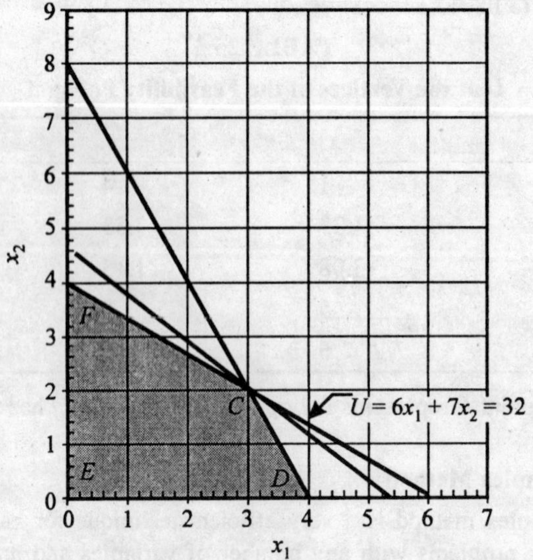

Figure 25.5. Graphical solution of Illustration 25.3

We next outline the Simplex method for obtaining the solution to this problem.

In the Simplex method, we deal with equations rather than inequalities as they appear in the two constraints above. Therefore, the first step consists of changing these inequalities in to equations.

Step 1: Introduce slack variables x_3 and x_4 to convert the constraint inequalities into equations. Assume that to fill in the slack time available, if any, on the machines, the shop makes two more products. Let it make x_3 and x_4 units of these two products, each unit taking one hour (for the purpose of simplification only) on one of the machines. Thus, product 3 takes one hour on the first machine (and none on the second), and product 4 takes one hour on the second machine (and none on the first). Let the two slack products contribute nothing to the profit function U. The modified mathematical statement of the problem then reads:

Solve the system of equations:

$$U - 6x_1 - 7x_2 - 0x_3 - 0x_4 = 0$$

$$2x_1 + 3x_2 + x_3 + 0x_4 = 12$$

$$2x_1 + x_2 + 0x_3 + x_4 = 8$$

$$x_1 \geq 0;\ x_2 \geq 0;\ x_3 \geq 0;\ x_4 \geq 0$$

We now have five variables and only three equations to determine them. Thus, there are many solutions available, and the desired solution is one that makes the value of U the maximum.

The Simplex method starts with assuming that two of the variables have values of zero and the values of the other two variables can be obtained from the two constraint equations.

We introduce some new terminology here: The variables that are taken to be non-zero, i.e., those variables whose value is obtained from the constraint equations are said to be the *basic* variables.

Step 2: Construct the Simplex table as shown in Table 25.4 with the assumption that we start with setting x_1 and x_2 as zero, and treating x_3 and x_4 as basic.

Table 25.4

Starting Simplex Table for Illustration 25.3

Basic Variables	*U*	x_1	x_2 ↓	x_3	x_4	*Right-hand side*
⟶ *U*	1	–6	–7	0	0	0
x_3	0	2	<u>3</u>	1	0	12
x_4	0	2	1	0	1	8

The numbers in various columns are the coefficients of the various terms in the system of simultaneous equations given above. The first row below the heading gives the criterion function in the new format. What do the various numbers in the table represent? One very insightful interpretation suggests that the number at the intersection of x_i-row and x_j-column tells us how much reduction in x_i must occur if we increase the value of x_j by one (that always being the entry at $x_j - x_j$ intersection).

The basic variables to begin with are x_3 and x_4. Clearly, the values of x_3 and x_4 obtained from the two constraint equations with x_1 and x_2 set as zero are 12 and 8 respectively. The profit is clearly zero with these. Graphically, this starting solution corresponds to the corner *E* of the feasibility polygon of Figure 25.5.

Just like the graphical solution, the Simplex method is also based on exploring the vertices of the feasibility polygon. We now attempt to move to the other vertices with better or higher value of *U*. This is done by making one of the currently *non-basic* variables (x_1 or x_2) basic, by making its value non-zero. In the process, one of the currently *basic* variables (x_3 or x_4) is made *non-basic*, by trading it in to make its value zero.

Let us next decide which vertex, *D* or *F*, we should explore. If we increase x_1 by 1, the profit will increase by 6, while increasing x_2 by 1 increases the profit by 7. It is therefore prudent to first make x_2 as basic. This can be translated into a thumb rule:

• *Rule 1: Make the variable that contributes most to the profit (as indicated by the largest negative entry in the U row of the Simplex table) a basic variable.*

We next determine which of the currently basic variable to make non-basic when x_2 is made basic. If we make x_3 non-basic, we can trade in its current value of 12 for 12/3 = 4 x_2's, but if we make x_4 non-basic, we can trade in its current value of 8 for 8/1 = 8 x_2's. But a value of 8 for x_2 is not possible, since any value larger than 4 is going to violate the first constraint. Thus, x_3 needs to be made non-basic. This again can be translated into a thumb rule:

• *Rule 2: Make the variable that has the lowest trade-in value the non-basic variable.*

The Simplex table is now modified as in Table 25.5. Here old rows have been set in lighter type and the new replacing rows are added near the bottom (in bold type) using the following rules. The entry at the intersection of the row of leaving basic-variable and the column of the entering basic-variable plays a crucial role in the revision and is termed as the *pivot* or the *pivot element.* Thus, **3** (bold and underlined) is the pivot element at this stage.

• *Rule 3: The entry at the intersection of the row of leaving basic-variable and the column of the entering basic variable is the* pivot *for the next stage.*

The new x_2 row is obtained by dividing each entry of the leaving x_3 row by the value of the pivot element, as shown. Remembering that the rows in our table represent algebraic equations, we can modify the other rows representing other equations by adding or subtracting appropriate multiples of this new entering equation from the old equations.

Table 25.5

Modified Simplex Table for Illustration 25.3

	Basic Variables	*U*	x_1	x_2	x_3	x_4	*Right-hand side*
First Stage	*U*	1	–6	–7	0	0	0
	x_3	0	2	<u>3</u>	1	0	12
	x_4	0	2	1	0	1	8
	U	**1**	**–4/3**	**0**	**7/3**	**0**	**28**
	$\mathbf{x_2}$	**0**	**2/3**	**1**	**1/3**	**0**	**4**
	→ $\mathbf{x_4}$	**0**	<u>**4/3**</u> ↑	**0**	**–1/3**	**1**	**4**

Since we now require in the new x_4-row a 0 in the x_2-column, we subtract the new x_2-row from the old x_4-row. The result is shown in the Table 25.5. The new *U*-row is obtained similarly by requiring that entry in the x_2-column is zero. (Entry under basic variables should be zero).

This new table now represents the solution point $x_1 = 0, x_2 = 4, x_3 = 0$, and $x_4 = 4$. The profit now is 28 units.

Since the coefficient of x_1 in the modified *U* equations is still negative, it means that increasing the value of x_1 can yet increase the value of *U*. Therefore, we try one more modification and pivoting. Using thumb rules 1 and 2 given above, we see that x_1 is the new entering basic variable and x_3 is the leaving basic variable. Using Rule 3 above, the pivot is now 4/3.

Table 25.6

Next Simplex Table for Illustration 25.3

	Basic Variables	*U*	x_1	x_2	x_3	x_4	*Right-hand side*
First Stage	*U*	1	–6	–7	0	0	0
	x_3	0	2	<u>3</u>	1	0	12
	x_4	0	2	1	0	1	8
Second Stage	*U*	1	–4/3	0	7/3	0	28
	x_2	0	2/3	1	1/3	0	4
	x_4	0	<u>4/3</u>	0	-1/3	1	4
	U	**1**	**0**	**0**	**2**	**1**	**32**
Final Stage	x_2	0	0	1	1/2	–1/2	2
	x_1	0	1	0	–1/4	3/4	3

New x_1 row is obtained scaling the old x_4 row by 3/4, the reciprocal of the value of the pivot. The new x_2 row is obtained by subtracting 2/3 times the new x_1-row from the old x_2-row to make entry in x_1-column as zero, and the new U row is obtained by adding 4/3 times the x_1 row from the old U row to make entry under x_1 column as zero. This is the solution point $x_1 = 3, x_2 = 2, x_3 = 0$, and $x_4 = 0$. The profit now is 32 units. This is the optimal solution. This is indicated by the fact that there is no negative entry in the new U row.

Let us summarize the Simplex technique:

1. Set up the problem mathematically introducing slack variables, one in each of the inequalities to convert them into equations.
2. Construct the Simplex table by choosing the slack variables as the initial basic variables, and setting the values of the other variables as zero.
3. Select the entering and the leaving variables by using Rules 1 and 2:
   Rule 1: *Make the variable that contributes most to the profit (as indicated by the largest negative entry in the U row of the Simplex table) a basic variable.*
   Rule 2: *Make the variable that has the lowest trade-in value the non-basic variable.*
4. Do the pivoting using Rule 3:
   Rule 3: *The entry at the intersection of the row of leaving basic-variable and the column of the entering basic variable is the* pivot *for the next stage.*
   The new *entering* row is obtained by dividing each entry of the leaving row by the value of the pivot element, so that 1 replaces the value of the pivot. Remembering that the rows in our table represent algebraic equations, we can modify the other rows representing other equations by adding or subtracting appropriate multiples of this new entering equation from the old equations. This is done so that all values in the column of the entering basic variables are zero (except for the 1 in the entering row).
5. Check the criterion function row (i.e., the U row) to see if there is any negative entry in it. Absence of any negative entry will indicate that the optimal solution has been obtained. If not, go through the cycle again.

We illustrate the use of the Simplex method through one more example.

Illustration 25.4: A company sells two different products A and B. The incremental profits on the two are Rs. 60 and 40, respectively. The two are produced in a common production process, which has a capacity of 30,000 man-hours per year. It takes 3 hours to produce A and one hour to produce B. A maximum of 8,000 units of A and 12,000 units of B can be sold in a year. Obtain the best mix of the product.

Solution: Let x_1 represent the number of units of A and x_2 the number of units of B made per year. The mathematical statement of the problem then becomes:

Maximize $U = 60x_1 + 40x_2$

Subject to constraints

$$3x_1 + x_2 \leq 30,000$$
$$x_1 \leq 8,000$$
$$x_2 \leq 12,000$$
$$x_1 \geq 0; x_2 \geq 0$$

Figure 25.6 shows the graphical solution of this problem. The maximum profit occurs at point D whose coordinates are $x_1 = 6,000$ and $x_2 = 12,000$. The maximum profit itself is Rs. 8,40,000.

To use the Simplex technique, we first cast the problems in terms of equalities by introducing three slack variables, one for each inequality.

$$U - 60x_1 - 40x_2 - 0x_3 + 0x_4 + 0x_5 = 0$$

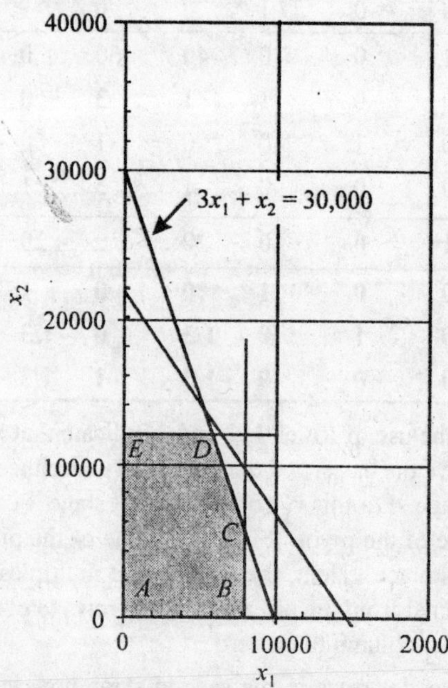

Figure 25.6 Graphical solution of Illustration 25.4

$$3x_1 + x_2 + x_3 = 36,000$$
$$x_1 + x_4 = 8,000$$
$$x_2 + x_5 = 12,000$$
$$x_1 \geq 0; x_2 \geq 0; x_3 \geq 0; x_4 \geq 0; x_5 \geq 0$$

The number of basic variables is 5 – 2 = 3. We choose x_3, x_4 and x_5 as the initial basic variables. Table 25.7 is the Simplex table:

Table 25.7
Simplex Table for Illustration 25.4

	Basic Variables	*U*	x_1	x_2	x_3	x_4	x_5	*Right-hand side*
First Stage	*U*	1	–60	–40	0	0	0	0
	x_3	0	3	1	1	0	0	30,000
	x_4	0	**<u>1</u>**	0	0	1	0	8,000
	x_5	0	0	1	0	0	1	12,000
Second Stage	*U*	1	0	–40	0	60	0	4,80,000
	x_3	0	0	**<u>1</u>**	1	–3	0	6,000
	x_1	0	1	0	0	1	0	8,000
	x_5	0	0	1	0	0	1	12,000
Third Stage	*U*	1	0	0	40	–60	0	7,20,000
	x_2	0	0	1	1	–3	0	6,000
	x_1	0	1	0	0	1	0	8,000
	x_5	0	0	0	–1	**<u>3</u>**	1	6,000
	U	**1**	**0**	**0**	**20**	**0**	**20**	**8,40,000**
Final Stage	x_2	**0**	**0**	**1**	**0**	**0**	**1**	**12,000**
	x_1	**0**	**1**	**0**	**1/3**	**0**	**–1/3**	**6,000**
	x_4	**0**	**0**	**0**	**–1/3**	**1**	**1/3**	**2,000**

In the first stage, the use of Rules 1, 2 and 3 indicate that x_1 should enter the basis, x_4 should leave the basis, and the bold (and underlined) 1 is the *pivot* to obtain the second stage. For this x_4 row of the first stage is copied as x_1 row dividing it by the value of the pivot. Since the value of the pivot is 1 in this case, no division takes place. Next, the appropriate multiples of the new x_1 row is added to/subtracted from the old x_3, x_5 and *U* rows to obtain the entries in the new rows under x_1 column to be zero.

The use of the relevant rules in the second stage now indicates that x_2 should enter the basis, x_3 should leave the basis, and that the bold and underlined 1 is the next *pivot* to obtain the third stage. The procedure of pivoting is similar to the one described in the previous paragraph.

The use of the relevant rules in the third stage now indicates that x_4 should enter the basis, x_5 should leave the basis, and that the bold and underlined 3 is the next *pivot* to obtain the fourth stage.

Since there now is no negative entry in the *U* column, the solution can now be taken as optimum, giving the maximum profit. Thus, the maximum profit occurs at $x_1 = 6{,}000$ and $x_2 = 12{,}000$. The maximum profit itself is Rs. 8,40,000.

25.4 Simplex Method for the Minimization Problems

The Simplex method described in the previous section is applicable to problems in linear programming with any number of variables and with any number of constraints. But the method described above can be applied as it is only to those problems in which the constraints are of the type 'less than or equal to'. For other types of constraints some changes in strategy are necessary. The illustration below gives the procedural changes required to handle such problems properly.

Illustration 25.5 *(A minimization problem with 'more than or equal to' type constraints inequalities and/or with 'equal to' type constraints)*: A ballast must weigh 150 kg. It can be made from two raw materials, *A* (with a cost of Rs. 20 per unit) and *B* (with a cost of Rs. 80 per unit). At least 14 units of *B* and no more than 20 units of *A* must be used. Each unit of *A* weighs 5 kg and each unit of *B* weighs 10 kg. How many units of each type of raw material must be used for a product to minimize cost?

Solution: Let x_1 and x_2 represent the numbers of units of *A* and *B* used in a ballast. The mathematical statement of the problem is:

Minimize $U = 20x_1 + 80x_2$

Subject to constraints:

$$5x_1 + 10x_2 = 150$$

$$x_2 \geq 14$$

$$x_1 \leq 20$$

$$x_1 \geq 0$$

These constraints are plotted in Figure 25.7. Since one of the constraints is an equation and *NOT* an inequality, the feasibility region reduces to merely a feasibility line, which extends from $(x_1 = 0, x_2 = 15)$ to $(x_1 = 2, x_2 = 14)$. But since we can use only whole units, the feasible domain consists of just these two points, which have been marked with dark circles in the figure. It is clear that the minimum cost occurs at point $(x_1 = 2, x_2 = 14)$. This cost is Rs. 1,160.

While formulating the problem for Simplex method, we immediately realize that the formulation introduced in the last section will not do. The Simplex method requires that all the slack variables be positive, so that the slack in the inequalities of the type *'more than or equal to'* cannot be handled without a trick. The trick consists of introducing what are termed as *artificial variables* as described below.

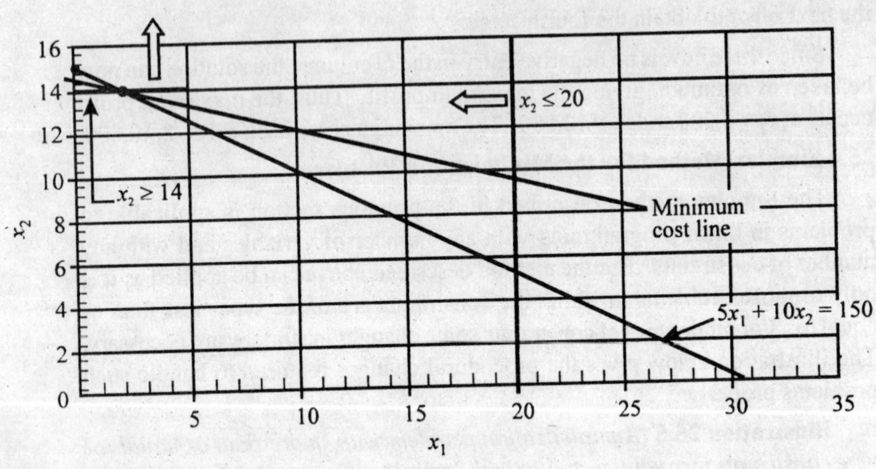

Figure 25.7. Graphical solution of Illustration 25.5.

Further, since the rules of the Simplex methods have been developed for a maximization problem, it will cause least confusion if this problem is recast as a maximization problem. Thus, instead of minimizing $20x_1 + 80x_2$, we may as well maximize $U = -20x_1 - 80x_2$. The problem of ballast is, therefore cast as:

Solve the system of equation

$$U + 20x_1 + 80x_2 + Ma_1 + 0x_3 + Ma_2 + 0x_4 = 0 \quad (1)$$

$$5x_1 + 10x_2 + a_1 = 150 \quad (2)$$

$$x_2 - x_3 + a_2 = 14 \quad (3)$$

$$x_1 = x_4 = 20 \quad (4)$$

$$x_1 \ge 0; x_2 \ge 0; \ x_3 \ge 0; \ x_4 \ge 0; a_2 \ge 0$$

Here x_3 and x_4 are the slack variables. However, since x_3 does not have +1 as coefficient, it cannot be used directly as a basic variable. Then again, the equality constraint cannot accommodate a slack variable, there being no slack in it. It is for this purpose we introduce two artificial variables a_1 and a_2. These are artificial in the sense that we know to begin with that their final value should be zero. To ensure this the weights associated with these variables in the cost function is set as M, a very large number. Clearly, M being very large, the minimization process will ensure that the value of a_1 and a_2 will come out to be zero. The weights of the artificial variables a_1 and a_2 in the criterion function (which is negative of cost) is accordingly set as $-M$.

We start with x_4, a_1 and a_2 as the initial choice of basic variables, this being the only choice possible among the variables introduced (coefficients in the constraint equations required to be non-negative). However, the criterion function equation is not in its desirable form as the coefficients of x_4, a_1 and a_2 are not all zero. To make the coefficients of x_4, a_1 and a_2 equal to zero, we subtract from equation (1) above, the equations (2) and (3), each multiplied by M. The resulting equation then is:

$$U + (-5M + 20)\,x_1 + (-11M + 80)\,x_2 + Mx_3 + 0x_4 + 0a_1 + 0a_2 = -164\,M$$

We use this equation in the formulation of the Simplex table.

Table 25.8

Simplex Table for Illustration 25.5

	Basic Variables	*U*	x_1	x_2	x_3	x_4	a_1	a_2	*Right-hand side*
First Stage	U	1	$-5M+20$	$-11M+80$	M	0	0	0	$-164M$
	x_4	0	1	0	0	1	0	0	20
	a_1	0	5	10	0	0	1	0	150
	a_2	0	0	$\underline{1}$	-1	0	0	1	14
Second Stage	U	1	$-5M+20$	0	$-10M+80$	0	0	$11M-80$	$-10M$ $-1{,}120$
	x_4	0	1	0	0	1	0	0	20
	a_1	0	5	0	$\underline{10}$	0	1	-10	10
	x_2	0	0	1	-1	0	0	1	14
Third Stage	U	1	-20	0	0	0	$M-8$	M	$-1{,}200$
	x_4	0	1	0	0	1	0	0	20
	x_3	0	$\underline{1/2}$	0	1	0	1/10	-1	1
	x_2	0	1/2	1	0	0	1/10	0	15
Final Stage	U	**1**	**0**	**0**	**40**	**0**	$M-4$	$M-8$	**−1,160**
	x_4	**0**	**0**	**0**	**−2**	**1**	**−1/5**	**−2**	**18**
	x_1	**0**	**1**	**0**	**2**	**0**	**1/5**	**−2**	**2**
	x_2	**0**	**0**	**1**	**−1**	**0**	**0**	**−1**	**14**

Thus, the optimal solution is that the minimum cost occurs at $x_1 = 2$ and $x_2 = 14$. The minimum cost is Rs. 1,160. Note that the negative sign obtained for U is because we had changed the sign of the cost and thus, U is interpreted as *negative* of cost.

Summary of key steps in handling minimization problem with equality constraints and *greater than or equal to* constraints:

1. Minimization problem is handled by converting it into a maximization problem by taking the negative of the cost funcion as the criterion function.

2. Equality constraints are handled by introducing an artificial variable in each one of them. The artificial variables are given a large negative *weight in the U function* (so that it is finally driven out of the solution in the maximization problem).

3. 'Greater than or equal to constraints are first converted into equalities by introducing slack variables with negative signs. But since these slack variables with the negative signs cannot be used as the initial basic variables, artificial variables (with positive signs) are introduced. These too are used as initial basic variable and are given large negative weights in the U function (so that they are finally driven out of the solution).

25.5 Problems with Multiple Optimal Solutions

Recall that in the graphical method of solving the linear programming problem, the constant U lines are all parallel to one another. The maximum value of the criterion function is obtained at a vertex of the feasibility polygon through which the highest constant U line passes. It is easy to see that if a bounding line of the polygon, i.e., a constraint line has the same slope as the iso-U lines, the largest U line will lie all along this edge of the polygon, rather than pass through a vertex. In such a case, instead of a single optimal solution all points on that line will represent optimal solutions. This is illustrated by the following example.

Illustration 25.6:

Maximize $U = 4x_1 + 3x_2$

Subject to the constraints $-x_1 + 2x_2 \leq 4$

$$4x_1 + 3x_2 \leq 16$$

$$x_1 - x_2 \leq 3$$

$$x_1 \geq 0;\ x_2 \geq 0$$

Solution: Figure 25.8 shows the plot of various constraints and the feasibility polygon. One constant U line is also shown. All other constant U lines are parallel to this line. Clearly, the value of U along the whole line CD is the same, since the line CD has the same slope as the all constant-U lines.

For Simplex method the relevant equations are:

$$U - 4x_1 - 3x_1 = 0$$

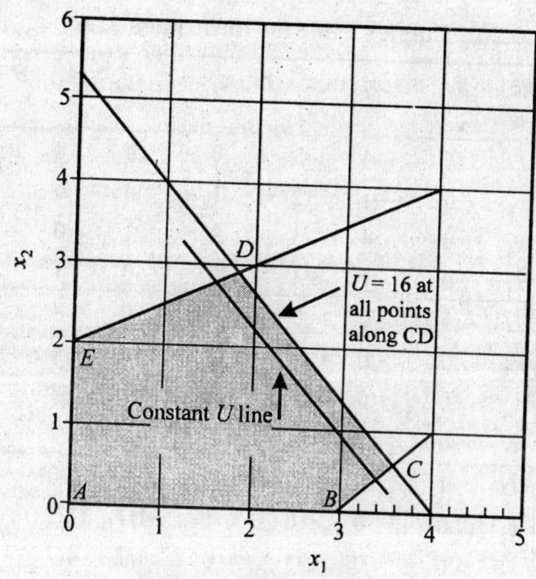

Figure 25.8 Graphical Solution of Illustration 25.6.

$$-x_1 + 2x_2 + x_3 = 4$$

$$4x_1 + 3x_2 + x_4 = 16$$

$$x_1 - x_2 + x_5 = 3$$

$$x_1 \geq 0;\ x_2 \geq 0;\ x_3 \geq 0;\ x_4 \geq 0;\ x_5 \geq 0$$

We have introduced here slack variables x_3, x_4 and x_5, one for converting each inequality (of less than or equal to type) into an equation. Table 25.9 shows the Simplex table for the problem. We start with x_3, x_4 and x_5 as the basic variables and write the first stage table. Using Rules 1 & 2 above, we see that variable x_1 must enter the basis and variable x_5 must leave it. The bold underlined entry 1 in x_1-column and x_5-row is the pivot element. Doing the pivoting operation such that all other entries in x_1 column become zero, we obtain the second-stage table. The application of the relevant rules now suggests that variable x_2 must enter the basis and variable x_4 must leave it. The bold underlined entry 7 in x_2-column and x_4-row is the next pivot element. Doing the pivoting operation, we obtain the third-stage table. Since the entries in the x_1 and x_2 columns of U row are zero, the solution at this stage is optimal. Thus, at $x_1 = 25/7$ and $x_2 = 4/7$ we get an optimum solution with $U = 16$. This corresponds to point C in Figure 25.8.

Table 25.9

Simplex Table for Illustration 25.6

	Basic Variables	U	x_1	x_2	x_3	x_4	x_5	Right-hand side
First Stage	U	1	–4	–3	0	0	0	0
	x_3	0	–1	2	1	0	0	4
	x_4	0	4	3	0	1	0	16
	x_5	0	1	–1	0	0	1	3
Second Stage	U	1	0	–7	0	0	4	12
	x_3	0	0	1	1	0	1	7
	x_4	0	0	7	0	1	–4	4
	x_1	0	1	–1	0	0	1	3
Third Stage	U	1	0	0	0	1	0	16
	x_3	0	0	0	1	–1/7	**11/7**	45/7
	x_2	0	0	1	0	1/7	–4/7	4/7
	x_1	0	1	0	0	1/7	3/7	25/7
	U	1	0	0	0	1	0	16
Final Stage	x_5	0	0	0	7/11	–1/11	1	45/11
	x_2	0	0	1	4/11	1/11	0	32/11
	x_1	0	1	0	–3/11	10/77	0	20/11

But the presence of 0 in the x_5-column of the U-row suggests that we can proceed further by letting x_5 enter as a basic variable and letting x_3 leave the basic variables without changing the value of the criterion function U. The new values $x_1 = 20/11$ and $x_2 = 32/11$ correspond to point D in Figure 25.8. This is also an optimum solution with $U = 16$, as was to be expected.

Summary Rule: *Multiple optimal solutions are indicated whenever a non-basic variable has a zero entry in the U row of the optimal table.*

26.6 Unbounded Solution

There are certain situations where the linear programming problem does not have a bounded solution. This is indicated when we can determine which of the non-basic variables enters the solution but cannot find any that may leave the solution. This is explained through the following illustration.

Illustration 25.7: Consider the following problem:

Maximize $U = 3x_1 + 2x_2$

Subject to the constraints

$$x_1 - x_2 \leq 2$$
$$-3x_1 + x \leq 4$$
$$x_1 \geq 0;\ x_2 \geq 0$$

Solution: We introduce appropriate slack variables and recast the equations

$$U - 3x_1 - 2x_2 - 0x_3 - 0x_4 = 0$$
$$x_1 - x_2 + x_3 = 2$$
$$-3x_1 + x_2 + x_4 = 4$$

Table 25.10 shows the development of the solution by Simplex method.

Table 25.10

Next Simplex Table for Illustration 25.7

	Basic Variables	*U*	x_1	x_2	x_3	x_4	*Right-hand side*
First Stage	*U*	1	–3	–2	0	0	0
	x_3	0	<u>1</u>	–1	1	0	2
	x_4	0	–3	1	0	1	4
Second Stage	*U*	1	0	–5	3	0	6
	x_1	0	1	–1	1	0	2
	x_4	0	0	–2	3	1	10

The first stage table was obtained directly from the equations. Application of Rules 1 and 2 indicates that x_1 must enter as a basic variable and x_3 must leave. Pivoting about the indicated entry we get $x_1 = 2$ and $U = 6$.

Applying Rule 1 to second stage table suggests that x_2 should enter as a basic variable, but Rule 2 does not give any variable as a leaving variable. This is because coefficients of all the entries in x_2-column are negative, and therefore can form no positive ratios. This means that x_1 can be increased indefinitely without violating any constraint and there is no upper limit on the value of the criterion function *U*.

This linear programming problem is said to have *unbounded* solution.

The reason why the solution is unbounded is clear if we look at the graph of this problem as shown in Figure 25.9. The feasibility polygon is unbounded and so iso-*U* line for howsoever a large value of *U* always has points within the feasible region.

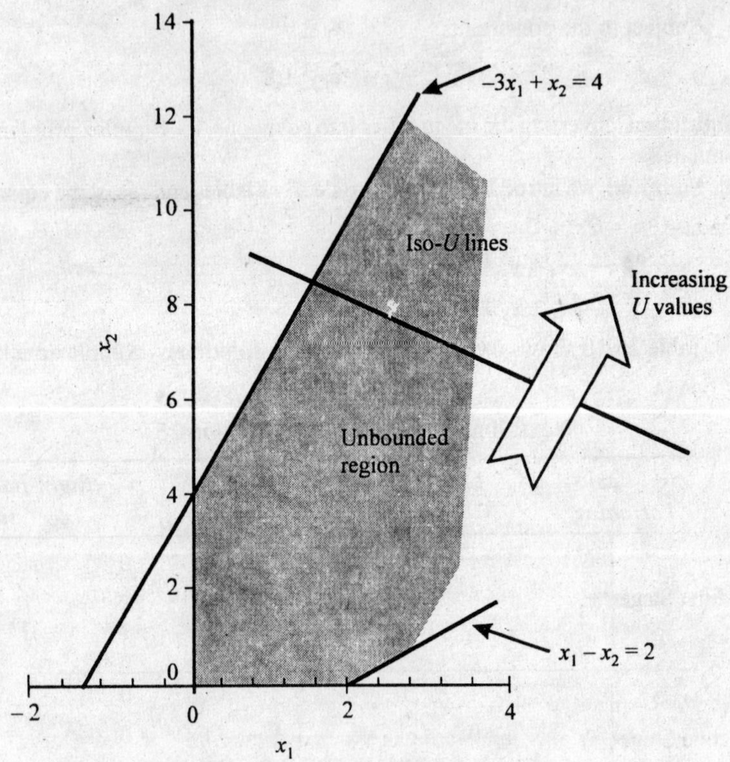

Figure 25.9 Graphical Solution for Illustration 25.7.

Summary Rule: *Whenever Rule 1 can identify a variable to be brought in as a basic variable, but Rule 2 cannot be applied because all the entries in the subject column are negative, the solution of the problem is unbounded.*

26.7 Degeneracy

A solution is known as a degenerate solution if one of the constraints is not necessary for the solution. As a trivial example consider two constraints $x_1 \leq 30$ and $x_1 \leq 45$. Clearly, the second constraint is redundant for it is automatically satisfied if the first one is. Most redundant constraints will not be so easily recognizable, and not all redundant constraints will lead to degeneracy. The following illustration shows how to recognize degeneracy and suggests a method of handling it.

Illustration 25.8: Maximize $U = 4x_1 + 3x_2$

Subject to constraints $4x_1 + 2x_2 \leq 10$

$$2x_1 + \frac{8}{3}x_2 \leq 8$$

$$x_1 \geq 0$$

$$x_2 \geq 1.8$$

Solution: Converting the inequalities into equations, the Simplex problem is formulated:

$$U - 4x_1 - 3x_2 - 0x_3 - 0x_4 - 0x_5 + Ma_1 = 0$$

$$4x_1 + 2x_2 + x_3 = 10$$

$$2x_1 + \frac{8}{3}x_2 + x_4 = 8$$

$$x_2 - x_5 + a_1 = 1.8$$

Here x_3, x_4 and x_5 are the slack variables. However, since x_5 does not have +1 as coefficient, it cannot be used directly as a basic variable. It is for this purpose we introduce an artificial variable a_1. This is artificial in the sense that we know to begin with that its final value should be zero. To ensure this the weight associated with this variable in the criterion function is set as $-M$, a very large number. Clearly, this will ensure that the value of a_1 will comes out to be zero.

We start with x_3, x_4 and a_1 as the initial choice of basic variables, this being the only choice possible among the variables introduced (coefficients in the constraint equations being non-negative). However, the criterion function equation is not in its desirable form as the coefficient of a_1 is not zero. To make the coefficient of a_1 equal to zero, we subtract from the *U*-equation above, the last equation multiplied by *M* to get:

$$U - 4x_1 - (3 + M)x_2 - 0x_3 - 0x_4 + Mx_5 + 0a_1 = -1.8\,M$$

We use this as the *U*-equation in the Simplex Table. 25.11. shown on the next page.

Here the choice of the pivot in the first stage is simple. But in the second stage, while x_1 is a clear choice of variable to go in as a basic variable, variables x_3 and x_4 tie for the outgoing one on the basis of Rule 2, the trade-in value for each of them being 1.6. This indicates a *degeneracy*. Fortunately, the Simplex procedure will give the correct result if we make an arbitrary choice (with the option to change the choice if the first one does not give a result). Let us replace x_4 by x_1, and obtain the third-stage table. Since there are no negative entries in *U*-row, this table is optimal and the solution is $x_1 = 1.6$, $x_2 = 1.8$, with $U = 11.8$.

Table 25.11

Simplex Table for Illustration 25.8

	Basic Variables	*U*	x_1	x_2	x_3	x_4	x_5	a_1	*Right-hand side*
	U	1	–4	–3–*M*	0	0	+*M*	0	–1.8*M*
First Stage	x_3	0	4	2	1	0	0	0	10
	x_4	0	2	8/3	0	1	0	0	8
	a_1	0	0	<u>**1**</u>	0	0	–1	1	1.8
Second Stage	*U*	1	–4	0	0	0	–3	3+*M*	5.4
	x_3	0	4	0	1	0	2	–2	6.4
	x_4	0	<u>**2**</u>	0	0	1	+8/3	–8/3	3.2
	x_2	0	0	1	0	0	–1	1	1.8
	U	**1**	**0**	**0**	**0**	**2**	**7/3**	**–7/3+*M***	**11.8**
Third Stage	x_3	**0**	**0**	**0**	**1**	**–2**	**–10/3**	**–10/3**	**0**
	x_1	**0**	**1**	**0**	**0**	**1/2**	**+4/3**	**–4/3**	**1.6**
	x_2	**0**	**0**	**1**	**0**	**0**	**–1**	**1**	**1.8**

If we had made x_3 as our choice for the in-going variable after the second stage, the third stage table would have been as in Table 25.12. Note here that this is not final because there is a negative entry in the *U*-row. Variable x_5 must enter as a basic variable. To decide on which variable leaves, note that x_4 has the least trade-in value, and the pivot element is the x_1-row, x_5-column entry 5/3 shown bold and underlined. Carrying out the indicated pivoting operation, we recover the same optimized solution as before.

A look at the graphical construction reveals the reason why the solution is degenerate. As can be seen very clearly, the feasible region on this chart is quite peculiarly situated with respect to the line of the first constraint. This line is entirely outside the feasibility polygon. This implies that whenever the second and the third constraints are met, the first one is automatically met. Hence it is a redundant constraint. Whenever this happens, degeneracy in Simplex solution results.

Table 25.12
Alternate Development in Illustration 25.8

	Basic Variables	*U*	x_1	x_2	x_3	x_4	x_5	a_1	*Right-hand side*
New Third Stage	*U*	1	0	0	1	0	–1	1+*M*	11.8
	x_1	0	1	0	1/4	0	1/2	–1/2	1.6
	x_4	0	0	0	–1/2	1	<u>5/3</u>	–5/3	0
	x_2	0	0	1	0	0	–1	1	1.8
	U	**1**	**0**	**0**	**7/10**	**3/5**	**0**	***M***	**11.8**
New Final Stage	x_1	**0**	**1**	**0**	**2/5**	**–3/10**	**0**	**0**	**1.6**
	x_5	**0**	**0**	**0**	**–3/10**	**3/5**	**1**	**–1**	**0**
	x_2	**0**	**0**	**1**	**–3/10**	**3/5**	**0**	**0**	**1.8**

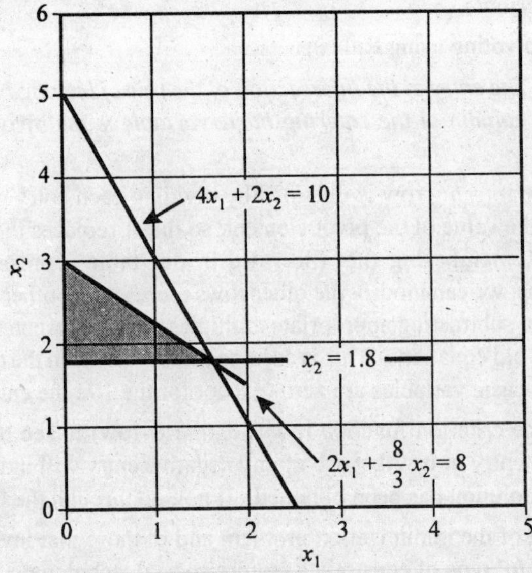

Figure 25.10. Graph for Illustration 25.8

25.8 Summary

In this chapter we have considered a class of problems that involve maximization or minimization with certain constraints. The constraints and the criterion functions are all linear in the variables of interest. We first introduced the graphical method wherein the constraint equations are plotted on an $x_1 - x_2$ graph to arrive at the feasibility polygon. The extremum occurs at one of the vertices of the feasibility polygon. The method consists of determining these vertices and evaluating the criterion function at each of these to locate the extremum.

The Simplex method of solving the linear programming problems uses a tabular approach. The principal steps in the process have been summarized before as:

- Set up the problem mathematically introducing slack variables, one in each of the inequalities to convert them into equations.
- Construct the Simples table by choosing the slack variables as the initial basic variables, and setting the value of the other variables as zero.
- Select the entering and the leaving variables by using Rules 1 and 2:

  Rule 1: *Make the variable that contributes most to the profit (as indicated by the largest negative entry in the U-row of the Simplex table) a basic variable.*

  Rule 2: *Make the variable that has the lower trade-in value the non-basic variable.*

- Do the pivoting using Rule 3:

  Rule 3: *The entry at the intersection of the row of leaving basic-variable and the column of the entering basic variable is the* pivot *for the next stage.*

  The new *entering* row is obtained by dividing each entry of the leaving row by the value of the pivot element, so that 1 replaces the value of the pivot. Remembering that the rows in our table represent algebraic equations, we can modify the other rows representing other equations by adding or subtracting appropriate multiples of this new entering equation from the old equations. This is done so that all values in the column of the entering basic variables are zero (except for the 1 in the entering row).

- Check the criterion function row (i.e., the *U*-row) to see if there is any negative entry in it. Absence of any negative entry will indicate that the optimal solution has been obtained. If not, go through the cycle again.

Handling of the minimization problem and of those that involve 'greater than on equal to' type of constraints require special technique:

- Minimization problem is handled by converting it into a maximization problem by taking the negative of the cost function as the criterion function.
- Equality constraints are handled by introducing an artificial variable in each one of them. The artificial variables are given a large negative *weight in the U function* (so that it is finally driven out of the solution in the maximization problem).

- 'Greater than or equal to constraints are first converted into equalities by introducing slack variables with negative signs. But since these slack variables with the negative signs cannot be used as the initial basic variables, artificial variables (with positive signs) are introduced. These too are used as initial basic variable and are given large negative weights in the *U* function (so that they are finally driven out of the solution).

Multiple optimal solutions are indicated whenever a non-basic variable has a zero entry in the *U* row of the optimal table.

Whenever Rule 1 above can identify a variable to be brought in as a basic variable, but Rule 2 cannot be applied because all the entries in the subject column are negative, the solution of the problem is unbounded.

Though we have restricted in all our illustrations to problems in two variables only, the Simplex method is directly extendable to more than two variables. The method is amenable to automation with simple digital computer programmes and is used extensively in operations planning in industry and in military operations for which it was first conceived.

Problems

25.1. Two grades of gasoline are obtained by blending together two types of blending gasolines whose characteristics are shown below:

Characteristics of Blending Gasolines

	Octane Rating	*Vapour pressure*	*Amount Available*
Type I	104	4	30,000 barrels
Type II	94	9	70,000 barrels

The characteristics of the final gasolines are given below:

Characteristics of Product Gasolines

	Minimum Octane Rating	*Maximum Vapour pressure*	*Maximum Sales*	*Selling Price/Barrel*
Aviation gasoline	102	5	20,000 barrels	Rs. 18,000
Motor fuel	96	8	Any amount	Rs. 14,000

When gasoline is mixed, both the octane rating and vapour pressure of the resulting mixture is the weighted average of the constituents where the weights are proportional to the mixing volumes.

Formulate the mathematical statement of the linear programming problem when the company wishes to maximize its sales revenue.

25.2. A manufacturer of toothpaste has three plants located in Amritsar, Bhopal, and Jalgaon. Major distribution warehouses are located at Noida, Kanpur, Raipur, Patna and Thane. Sales requirement from each of the warehouses for the next year are given in the table below:

Sales Requirements

Warehouse	Amount of sales ('000 cases)
Noida	50
Kanpur	10
Raipur	60
Patna	30
Thane	20
Total	170

The following table gives the relative cost of shipping from each factory to each warehouse:

Cost of shipping 1,000 cases

To →		*Noida*	*Kanpur*	*Raipur*	*Patna*	*Thane*
From ↓	*Capacity*					
Amritsar	100,000	100	150	180	180	200
Bhopal	50,000	80	100	70	120	100
Jalgaon	30,000	180	170	160	190	40

Formulate the mathematical statement of the linear programming problem to determine the shipping schedule that will minimize the overall transportation cost.

25.3. A blender of whiskey imports three grades A, B and C, and mixes them according to the following recipe:

Recipe of the Two brands

Blend	*Composition*	*Price per bottle, Rs.*
Supreme	Not less than 50% A Not more than 30% C	225
Excellent	Not less than 15% A Not more than 65% C	150

He can secure the following *maximum* supplies:

Availability and Price of the Inputs

Grade	*Quantities available, bottles*	*Price per bottle, Rs.*
A	2,000	200
B	3,000	150
C	100	100

Set up the linear programming problem to maximize the profit.

25.4. A small shop makes two types of locks, having the following requirements:

Requirements of Locks

Lock type	*Turning min.*	*Grinding min.*	*Assembly min*	*Cost of materials Rs.*	*Sales Price Rs*
A	5	5	5	9	45
B	4	8	6	11	100

The availability of turning is 40 hours per day at Rs. 50 per hour, of grinding is 50 hours at Rs. 75 per hour, and of assembly is 50 hours at Rs. 30 per hour. Set up the linear programming problem and determine how many of each type should be made each day.

25.5. Solve the following linear programming problem by graphical method and interpret the results.

Maximize $U = -2x_1 + x_2$

Subject to $5x_1 - x_2 \geq 20$

$x_1 \geq 5$

$x_2 \geq 0$

25.6. Consider the following problem:

Maximize $8x_1 + 4x_2$

Subject to $x_1 \geq 14$

$-x_2 \geq -20$

$10x_1 + 5x_2 = 150$

$x_1 > 0;\ x_2 > 0$

(a) Find an optimal solution using graphical method.

(b) Find an optimal solution using Simplex method.

(c) Is the solution unique? Interpret.

25.7. Maximize $2x_1 + x_2$

Subject to $2x_1 + x_2 \leq 14$

$2x_2 - x_2 \leq 8$

$x_1 - x_2 \leq 3$

$x_1 \geq 0; x_2 \geq 0$

25.8. Maximize $2x_1 + 4x_2$

Subject to $x_1 \leq 10$

$x_1 + 2x_2 = 30$

$x_1 \geq 0; x_2 \geq 14$

25.9. Maximize $4x_1 + 3x_2 + 2x_3$

Subject to $x_1 + x_2 - x_3 \leq 10$

$4x_1 + x_2 - x_3 \leq 20$

$2x_2 + x_3 = 10$

$x_1 \geq 0; x_2 > 0; x_3 \geq 0$

25.10. The Electrodyne Company makes two kinds of test equipment: products *AX*-1 and *AX*-21. The physical plant is organized into three main shops: Motherboard department (*MB*), Chassis department (*C*), and Final-assembly and Testing department (*AT*). The following tables give the relevant data:

Time Requirements and Monthly Capacities

	AX-1	*AX-21*	*Hours Available per month*
Motherboard department (*MB*)	4	2	2,000
Chassis department (*C*)	5	2	2,500
Final-assembly and Testing department (*AT*)	6	3	4,200

Costs and Prices

Costs and Prices *Rs. per unit*	*AX-1*	*AX-21*
Sale Price	20,000	5,500
Cost		
Variable Labour	2,000	500
Material	5,000	1,000
Overheads	4,000	1,000

Current market forecast for the two products is for 400 units of *AX*-1 and 600 units of *AX*-21. Formulate the linear programming problem, and solve by graphical and Simplex method.

Chapter 26

THEORY OF GAMES

26.1. INTRODUCTION

Theory of games is concerned with the development of strategies where two or more competitors, or opponents, are engaged in a situation of conflicting interests, and when the outcomes depend on the decisions of all the parties involved in competition. In the earlier chapter on *Business Decisions* only those processes were considered where an agent (or a *player* as termed herein) made decisions in response to conditions which are pre-existing (but are unknown to the player) and which do not depend on the actions or decisions of the player (that is, they are *dead variables*). In contrast, the theory of games deals with those situations where a *player's* decisions are strategic reactions to the actions (*live variables*) of the other agents involved in the competitive *game*. Each player is assumed to develop a strategy in a rational manner to resolve the situation of conflict in its favour after considering the strategy options of all the players involved in the *game*.

The theory of games was first proposed by John von Neumann and Oskar Morgenstern in their famous book *Theory of Games and Economic Behaviour* (Princeton University Press, 1944), wherein they provided a new approach to many problems involving conflict situations. This approach now forms a bedrock of decision making under situations of conflict in many areas in economics, business administration, sociology, political science, as well in military strategy.

26.2. THE GAME SITUATION

Within the context of the theory of games, a competitive situation is termed as a *game* if it has the following properties:

(*i*) The number of players is finite (We will, however for the purpose of illustrations here take the number of players as two to reduce the complexity),

(*ii*) There is a conflict of interest between the players,

(*iii*) Each of the player has available to it a finite number of strategic options, (*iv*) Each player knows the strategy options available to each of the players,

(*v*) The outcome (i.e., the *pay-off*) of the games depends on the strategic choices made by each of the players.

(*vi*) Each player has complete knowledge the pay-off for each combination of the various options available to all the players,

(*vii*) Each player plays the game rationally, i.e., uses the rules to maximise its profit (or minimise its loss).

(*viii*) The following table illustrates a simple two-player zero-sum game in which there are three options available to each player:

TABLE 26.1. A SIMPLE TWO-PLAYER THREE-OPTIONS-EACH GAME

		Player *Y*		
		Press Button *D*	Press Button *E*	Press Button *F*
Player *X*	Press Button *A*	*X* wins 10 points	*Y* wins 8 points	*Y* wins 2 points
	Press Button *B*	*X* wins 6 points	*X* wins 7 points	*X* wins 3 points
	Press Button *C*	*Y* wins 9 points	*Y* wins 6 points	*X* wins 2 points

There are two players, *X* and *Y*. Each player has three strategies available to it. Player *X* can press either button *A*, button *B*, or button *C*. Player *Y* can choose between buttons *D*, *E* and *F*. The game table above is essentially a *rules-of-pay-off matrix*. This is a zero sum game, that is, what *X* wins *Y* looses, and vice versa.

This table is, conventionally, abbreviated as follows.

TABLE 26.2. PAY-OFF TABLE FROM THE POINT OF VIEW OF PLAYER X

		Y's strategies		
		y_1	y_2	y_3
X's strategies	x_1	+ 10	− 8	− 2
	x_2	+ 6	+ 7	+ 3
	x_3	− 9	− 6	− 2

The strategies available to X are x_1, x_2 and x_3 and that to Y are y_1, y_2 and y_3. Since the above pay-offs are shown from the point of view of the player X, a positive pay-off is a winning for X, while a negative pay-off is a winning for Y.

26.3. THE MINIMAX AND MAXIMIN DECISION RULES

Under the conditions, there are 9 (= 3 × 3) possible combinations of actions of X and Y. Pursuant to the assumptions of the theory of games, each of the two players is perfectly aware of these moves and the consequent pay-offs. Each of the player attempts to maximise its profit or minimize its loss.

The conservative approach is to *assume the worst*. Thus, if X chooses the strategy x_1, it would expect the worst, that is, expect that Y will choose y_2 resulting in the minimum profit of –8, or a maximum loss of 8 units. Using the same logic, strategy x_2 will be expected to invite strategy y_3 resulting in a minimum profit of 3 units, and the strategy x_3 will be expected to invite strategy y_1 resulting in a minimum profit of –9 units (that is, a loss of 9 units). The row minima are shown conveniently in the last column of Table 26.3.

TABLE 26.3. DETERMINING MAXIMUM OF ROW MINIMA AND MINIMUM OF COLUMN MAXIMA

		Y's strategies			
		y_1	y_2	y_3	Row minima
X's strategies	x_1	+10	– 8	– 2	– 8
	x_2	+ 6	+ 7	+ 3	+ 3
	x_3	– 9	– 6	– 2	– 9
Column maxima		+10	+ 7	+ 3	

Under the conservative principle, then, X should choose the strategy that should *maximize* its *minimum* profit. Under this *maximin* decision rule, then, X should choose the strategy x_2.

Similarly, from the point of view of player Y, choice of strategy y_1 would be expected to be responded by strategy x_1 resulting in the minimum loss of 10. Note that since the table is constructed from the point of view of X, positive values represent loss for Y. Using the same logic, strategy y_2 will be expected to invite strategy x_2 resulting in a maximum loss of 7 units, and the strategy y_3 will be expected to invite strategy x_2 resulting in a maximum loss of 3 units. These column maxima are shown in the last row of Table 26.3. The conservative decision rule for Y is the *minimax* rule, that is, Y should choose the strategy that should *minimize* its *maximum* loss. Under this minimax decision rule, Y should choose the strategy y_3.

One may argue that X should adopt the strategy x_1 which may possible give it a maximum pay-off of +12. But that is a false promise. Under the rules of the game with perfect information to all, Y will be at a liberty to choose y_3, resulting in a loss of 2 units to X (and a profit of 2 to Y). Thus, the *maximin* strategy for X, and the *minimax* strategy for Y is the correct solution. The difference in the recommended strategies for the two players arises because the pay-off table has been drawn from X's perspective, which is exactly the *negative* from the Y's perspective.

Under the maximin strategy, X ensures that his profit does not dip below three units, and by adopting minimax strategy Y ensures that X does not take away more than 3 units as profit.

26.4, SADDLE POINT, EQUILIBRIUM AND THE VALUE OF THE GAME

It is interesting to note that the maximum of the row's minimum (maximin for X), and the minimum of the column's maximum (minimax for Y), happen to be the same element in the pay-off table, shown circled. Such a point is aptly termed as a *saddle point* after the centre point of a horse saddle, which truly is the maximum of the minimum-height points looking side ways, and minimum of the maximum-height points looking along the length of the saddle. When a game situation has a saddle point, it is said to be a case of *equilibrium*. In such a game there is only one strategy each for both players, which they play every time. Thus, X always plays x_2, while Y always plays y_3.

In each game, thus, X gets 3 units pay-off while Y pays out that amount. This is termed as the *value* of the game. Clearly, this is a biased game, biased towards X and against Y. An unbiased, or a fair game is one in which the value is zero.

Not all game situations have a saddle point. Such games are not in equilibrium. We give next an example of a game in which no saddle point exists.

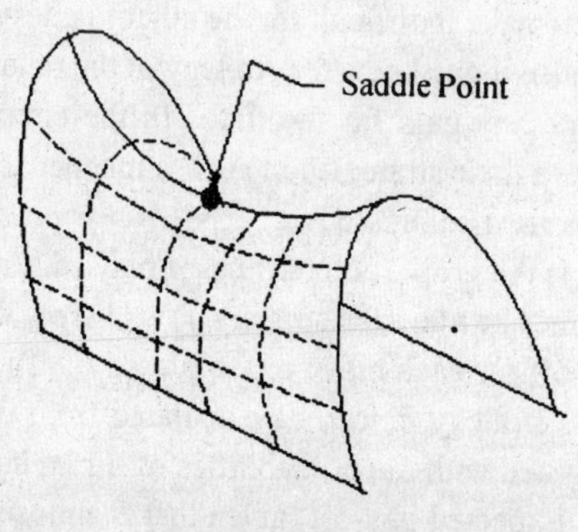

FIG. 26.1 Saddle Point

26.5. GAMES WITHOUT SADDLE POINT

The illustration below is an example of a game without saddle point. In such a situation, equilibrium is not obtained and there is no clear cut solution. In such a situation we resort to what is termed as a *mixed strategy*.

Illustration 26.1. The following matrix represents the pay-off in a game being played by two players, *A* and *B*. Determine the optimal strategy from the view point of the two players and determine the value of the game.

TABLE 26.4. DETERMINING MAXIMUM OF ROW MINIMA AND MINIMUM OF COLUMN CAXIMA

		B's strategies		
		b_1	b_2	Row minima
A's strategies	a_1	+10	− 8	− 8
	a_2	− 6	+ 7	− 6
Column maxima		+10	+ 7	

Solution. The row maxima and the column minima have been obtained and shown in the above table itself. It is clear that this game does not have a saddle point. From the viewpoint of player *A*, the strategy a_2 is the best, where his maximin is –6, or a loss of 6 units. From the minimax principle for *B*, the optimum strategy is b_2, with a loss (or profit for *A*) of 7 units. This clearly in not an equilibrium case. For if *B* knew that *A* has chosen a_2, he will immediately choose b_1 resulting in loss to *A* of 6 units. And if *A* knew that *B* has chosen b_1, he will choose a_1 resulting in his profit of 10 units. In this situation, both *A* and *B* will make it difficult for the other to guess what strategy he or she is playing, because if one knew the strategy of the other, he/she will choose its strategy to negate any gain for the first. In the circumstances, both players randomly choose their strategies in such a manner that their expected gain (or loss) is maximised (minimised).

If the proportion of times *A* plays the strategy a_1 is p, then the proportion of times he plays the strategy a_2 is $(1-p)$. We can then determine his expected pay-off for each move of the player *B*. The expected pay-off (for *A*) when *B* plays strategy b_1 can be calculated from the first column values of pay-offs weighted with the probabilities of *A* playing the two strategies as:

Expected pay-off (given that *B* chooses b_1)

$$= 10 \times p + (-6)(1-p) = 16p - 6 \qquad \text{(a)}$$

Similarly, the expected pay-off for A when B plays strategy b_2 can be calculated from the second column values as:

Expected pay-off (given that B chooses b_2)

$$= (-8) \times p + (+7)(1-p) = 7 - p) \quad \text{(b)}$$

A little thought would convince you that the optimal strategy for A is when the pay-off becomes independent of the strategy adopted by B, or in other words, the two expected pay-offs calculated above should be the same. The value of p, the proportion of times A plays strategy a_1 can, thus, be determined by equating the two pay-offs calculated above.

$$16p - 6 = 7 - p$$

or $$16p = 13$$

or $$p = 13/17 = 0.765$$

Thus, A would be best off playing the two strategies randomly (so that B cannot guess which one is being played any single time), in such a manner that the strategy a_1 is played 0.765 proportion of times (that is, a little over three fourth of times), and the strategy a_2 is played the remaining proportion of times. This is what is termed as a *mixed strategy*.

What about the strategy of player B? We calculate it in the same fashion as above. Let q be the proportion of times the player B plays strategy b_1, and $(1 - q)$ the proportion he plays strategy b_2. Then the expected pay-off (for A) when A plays strategy a_1 can be calculated from the first row values of pay-offs weighted with the probabilities of B playing the two strategies as:—

Expected pay-off (given that A chooses a_1)

$$= 10 \times q + (-8)(1-q) = 18q - 8 \quad \text{(c)}$$

Similarly, the expected pay-off (for A) when A plays strategy a_2 can be calculated from the second row values as:

Expected pay-off (given that A chooses a_2)

$$= (-6) \times q + 7(1-q) = 7 - 13q \quad \text{(d)}$$

Again, the optimal strategy for B is when the pay-off becomes independent of the strategy A adopts, or in other words, the two expected pay-offs calculated above should be the same. The value of q, the proportion of times B plays strategy b_1 can be determined by equating the two pay-offs calculated above.

$$18q - 8 = 7 - 13q$$

or $$31q = 15$$

or $$q = 15/31 = 0.484$$

Thus, B would be best off playing the two strategies randomly (so that A cannot guess which one is being played any single time), in such a manner that the strategy b_1 is played 0.484 proportion of times (that is, a little less than 1 times out of 2), and the strategy b_2 is played the remaining proportion of times.

Calculation of the value of the game is simple. Since the pay-off to A remains the same whether he plays strategy a_1 or a_2, we could use either of the two equations, (a) or (b) to calculate the pay off. Inserting the value of $p = 13/17$, we obtain a pay-off (for A) of $16 \times (13/17) - 6$ or 6.235 units from equation (a). We obtain the same pay-off from equation (b). By using the value of p in (b), we get the pay-off to A as $7 - (13/17)$ or 6.235, confirming our earlier assertion.

Note further, that since this is a constant- or zero-sum game, we should get the same value for the game from equations (c) and (d). We leave it to the reader to verify. Table 26.5 summarises the results:

TABLE 26.5. SUMMARY OF THE MIXED STRATEGY FOR ILLUSTRATION 26.1

	Strategy	*Probability*
For A	a_1	0.765
	a_2	0.235
For B	b_1	0.484
	b_2	0.516

The above procedure for calculating the mixed strategies and the value of two-person constant-sum game can be easily cast in terms of formulae. Let a_{ij} represents the pay-off to player A when A plays strategy i and B plays the strategy j. The subscripts i and j will take values 1 and 2 in the case when each player has only two strategies. The pay-off table then looks like Table 26.6 shown below:

TABLE 26.6 SYMBOLS FOR PAY-OFFS FOR PLAYER A

		B's strategies	
		b_1	b_2
A's strategies	a_1	a_{11}	a_{12}
	a_2	a_{21}	a_{22}

The proportion of times the player A should play strategy a_1 is given by p, where

$$p = \frac{a_{22} - a_{21}}{(a_{11} + a_{22}) - (a_{12} + a_{21})} \tag{26.1}$$

and the proportion of times the player A should play strategy a_2 is given by $(1 - p)$.

Similarly, the proportion of times the player b should play strategy b_1 is given by q, where

$$q = \frac{a_{22} - a_{12}}{(a_{11} + a_{22}) - (a_{12} + a_{21})} \tag{26.2}$$

and the proportion of times the player B should play strategy b_2 is given by $(1 - q)$.

The value V of the game (from the perspective of the player A) is given by

$$V = \frac{a_{11}a_{22} - a_{12}a_{21}}{(a_{11} + a_{22}) - (a_{12} + a_{21})} \tag{26.3}$$

These formulae can easily be verified from the first principles enunciated above.

26.6. THE DOMINANCE RULE

If in the pay-off table of a game, each one of a row, say ith element is larger (or equal) to the corresponding element of another row, say jth, it is obvious that from the point of view of the player A, the strategy being represented by the ith row is preferable over the strategy represented by the jth row, no matter what strategy the player B chooses. In this situation the ith row is said to *dominate* the jth row, and the *dominance rule* permits us to simply delete the jth row and thus simplify the table. A similar rule holds for the columns. Here the column whose each element is *less* than the corresponding element of another column is said to dominate the later column, and that later column is deleted reducing the number of columns in the table. The following illustration shows the operation of this rule.

Illustration 26.2. Company A has three strategies available to it, while its competitor B also has three available strategies. The pay-offs (from the point of view of A are given in Table 26.7 below:

TABLE 26.7. PAY-OFF MATRIX FOR ILLUSTRATION 26.2

		B's strategies		
		b_1	b_2	b_3
	a_1	10	–7	–2
A's strategies	a_2	6	6	3
	a_3	–10	–6	3

Solution. Compare the pay-offs for strategies a_2 and a_3 and notice that pay-offs for strategy a_2 are better than (or equal to) the pay-offs for strategy a_3, irrespective of what strategy the player B chooses. This clearly means that there can be no advantage to A in choosing strategy a_3. Thus, strategy a_2 *dominates* strategy a_3, and we might as well drop this whole row from the table:

TABLE 26.8. SIMPLIFIED PAY-OFF MATRIX FOR ILLUSTRATION 26.2

		B's strategies		
		b_1	b_2	b_3
A's strategies	a_1	10	−7	−2
	a_2	6	6	3

Then again, compare columns pertaining to strategies b_1 and b_3. Every single entry in column for b_1 is larger than the corresponding value in the column for b_3. This signifies that player B will payout *more* following the strategy b_1 than following the strategy b_3. Thus, the strategy b_3 *dominates* strategy b_1 as far as player B is concerned, and the whole column may as well be dropped. The simplified table then looks as:

TABLE 26.9. FURTHER SIMPLIFICATION TO THE PAY-OFF MATRIX FOR ILLUSTRATION 26.2

		B's strategies		
		b_2	b_3	Row minima
A's strategies	a_1	−7	−2	−7
	a_2	6	(3)	3
Column maxima		6	3	

This table clearly has a saddle point, and therefore, a_2 and b_3 are the strategies of choice for players A and B respectively. The value of the game is clearly 3 units.

One would notice that in the table above, the row for a_2 dominates the row for a_1, and therefore the table is further simplified leaving only one real choice of strategy for A.

TABLE 26.10. THE SIMPLEST PAY-OFF MATRIX FOR ILLUSTRATION 26.2

		B's strategies	
		b_2	b_3
A's strategies	a_2	6	3

This, too, gives the same answer as before.

Illustration 26.3. Determine the optimum strategies and the value of the 2×5 game whose pay-off table is given below.

TABLE 26.11. PAY-OFF TABLE FOR A 2 × 5 GAME

		Strategies for *Y*					
		y_1	y_2	y_3	y_4	y_5	Row minima
Strategies for *X*	x_1	3	6	– 3	0	– 1	– 3
	x_2	2	3	– 1	2	– 4	– 4
Column maxima		3	6	– 1	2	– 1	

Solution. There clearly is no saddle point in this game. This implies that a mixed strategy will be called for in this game. Further, it is clear that strategies y_3 and y_5 dominate the other strategies so that the player *Y* can discard the other strategies right away. The simplified table then is

TABLE 26.12. SIMPLIFIED PAY-OFF TABLE FOR THE ABOVE

		Strategies for *Y*	
		y_3	y_5
Strategies for *X*	x_1	– 3	– 1
	x_2	– 1	– 4

Using the notations of Table 26.6 and formula (26.1), we straight away get the value of *p*, the proportion of times the player *X* should play the strategy x_1, as

$$q = \frac{a_{25} - a_{15}}{(a_{13} + a_{25}) - (a_{15} + a_{23})} = \frac{(-4) - (-1)}{[(-3) + (-4)] - [(-1) + (-1)]} = \frac{-3}{-5} = 0.6$$

Similarly, from formula (26.2), the value of q, the proportion of times the player Y should play the strategy y_3, is obtained as

$$q = \frac{a_{25} - a_{15}}{(a_{13} + a_{25}) - (a_{15} + a_{23})} = \frac{(-4) - (-1)}{[(-3) + (-4)] - [(-1) + (-1)]} = \frac{-3}{-5} = 0.6$$

and from formula (26.3) the value V of the game is obtained as:

$$V = \frac{a_{13}a_{25} - a_{15}a_{23}}{(a_{13} + a_{25}) - (a_{15} + a_{23})} = \frac{(-3)(-4) - (-1)(-1)}{[(-3) + (-4)] - [(-1) + (-1)]} = \frac{11}{-5} = -2.2$$

26.7. GRAPHICAL SOLUTION OF 2 × *n* OR *m* × 2 GAMES

When one of the players has only two strategies available, one can solve the problems through a very ingenious graphical scheme. The problem should either be $2 \times n$ or $m \times 2$ game, or reducible to it by one or more application of the dominance rule. We shall introduce this graphical scheme through the following illustration.

Illustration 26.4. Find the optimum strategies and the value of the game whose pay-off table is given below:

TABLE 26.13. PAY-OFF TABLE FOR A 2× 4 GAME

		Strategies for Y			
		y_1	y_2	y_3	y_4
Strategies for X	x_1	–2	4	6	–6
	x_2	9	–7	5	8

Solution: Figure 26.2 shows a graphical construct of the pay-offs from the point of view of the player X. When the player X plays strategy x_1 a proportion of p times (and consequently, the strategy x_2 a portion of $(1 - p)$ times, the expected value of his pay-off when the player Y plays the strategy y_1 is $(-2)p + 9(1 - p)$ or $-11p + 9$. This is represented by a straight line as shown in Figure 26.2.

This figure has been drawn by constructing two vertical *pay-off* scales. In the horizontal separation between the two, the scale for p is laid out, running from 0 to 1. The line for strategy y_1 is laid out by connecting 9 on the left pay-off scale (representing pay-off when the value of p is 0, that is, only the strategy x_2 is played) to – 2 on the right pay-off scale (representing pay-off when the value of p is 1, that is, only the strategy x_1 is played). The line labelled y_1, thus, represents the various pay-offs as p varies from 0 to 1 and the player Y plays the strategy y_1. Similarly, the lines for the other strategies of Y are drawn.

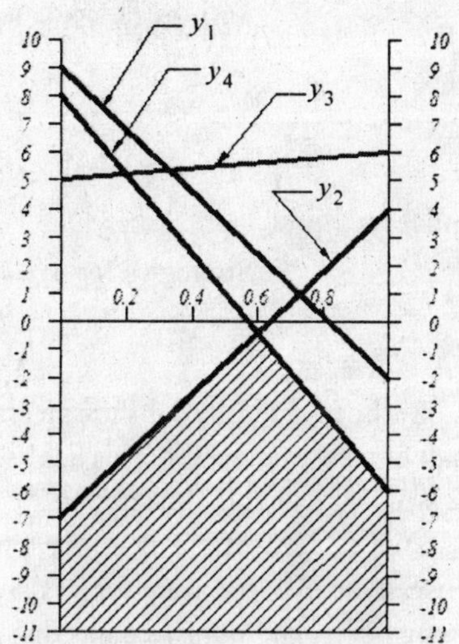

FIG. 26.2 Plot of the pay-offs in Illustration 26.4

A player is concerned with maximizing the worst-case pay-off. Clearly, the upper boundary of the hatched region represents the worst-case pay-off for any given value of p. The highest point on this boundary is, thus, the optimum strategy. This is the point of intersection of the pay-off lines when player Y plays strategies y_2 and y_4. Line labelled y_2 is given by $4p+(-7)(1-p)=11p-7$, since the return from playing x_1 a p proportion of times is $4p$, and playing x_2 a $(1-p)$ proportion of times is $(-7)(1-p)$. Similarly the line labelled y_4 is given by $-6p+8(1-p) = -14p+8$. The pay-off at the point of intersection is independent of the strategy of Y. Thus, equating the two pay-offs, we get

$$11p-7=-14p+8$$

or $$25p=15,$$

or $$p=0.6$$

The value of the game can be evaluated by substituting for the value of p in either of the two expressions to obtain -0.4 as the value of the game.

A fresh look at Figure 26.2 reveals that from the pint of view of player Y, the strategy y_1 is dominated by y_4, and strategy y_3 is dominated by y_2. Thus, it should be clear from the beginning that Y should never play y_1 or y_3. We could have discarded these with the result that the problem simplifies to a 2×2 game (without a saddle point) which could easily be solved by the procedure outlined in the previous section.

As the next illustration we take an example where this simplification is not available.

Illustration 26.5. Obtain the optimal strategy and the value of the game whose pay-off is given in Table 26.14.

TABLE 26.14. PAY-OFF TABLE FOR A 2×3 GAME

		Strategies for B		
		b_1	b_2	b_3
Strategies for A	a_1	3	4	8
	a_2	8	6	2

Solution. Figure 26.3 shows a graphical construct of the pay-offs from the point of view of player A. This has been drawn using the procedure outlined in the previous example (Illustration 26.4). Two vertical pay-off scales (both from the point of view of player A) have been drawn. In the horizontal separation between the two, the scale for p is laid out, running from 0 to 1. The line for strategy b_1 is laid out by connecting 8 on the left pay-off scale (representing pay-off when the value of p is 0, that is, only the strategy a_2 is played) to 3 on the right pay-off scale (representing pay-off when the value of p is 1, that is, only the strategy a_1 is played). The lines labelled b_1, thus, represents the various pay-offs as p varies from 0 to 1 and the player B plays the strategy b_1. Similarly, the lines for the other strategies of B are drawn.

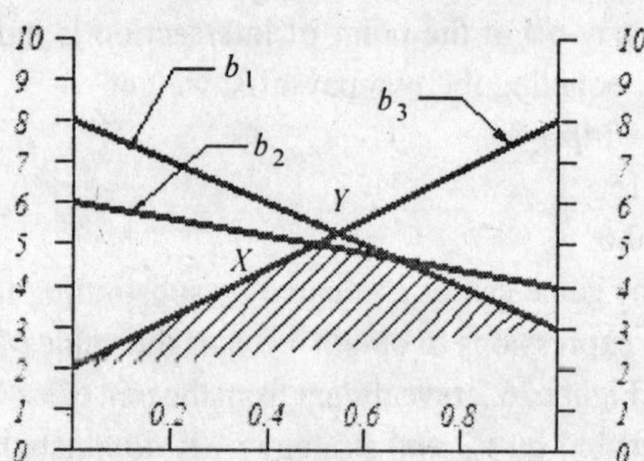

FIG. 26.3 Plot of the pay-offs in Illustration 26.5

A player is concerned with maximizing the worst-case pay-off. Clearly, the upper boundary of the hatched region represents the worst-case pay-off for any given value of p. The highest point on this boundary is thus, the optimum strategy. Thus, the point X, which is the intersection of the lines representing strategies b_2 and b_3, is the solution point.

To obtain that point, the pay-offs for the two written in terms of p are equated. The pay-off for b_2 is $4p + 6(1 - p) = 6 - 2p$. The pay-off for b_3 is $8p + 2(1 - p) = 6p + 2$. Equating the two, we get $p = 0.5$. Thus, A should play the two strategies with equal frequencies. The value of the game for A is evaluated by using this value of p in the pay-off expression giving $V = 5$. The value of q, the proportion of times player B should play strategy b_2 can then be obtained by solving the 2×2 game with strategies a_1, a_2, b_2 and b_3, and using formula 26.2.

26.8. FORMULATION OF $m \times n$ GAMES AS LINEAR PROGRAMMING PROBLEMS

If each of the players has *more* than two strategies available to him even after the application of the rules of dominance and if there is no saddle point, neither of the two methods outlined above will be able to determine the optimum strategy. In such cases the technique of linear programming offers a viable method of solution. We illustrate the formulation of a game situation as a linear programming problem through an example.

Illustration 26.6. Determine the optimum strategies and the value for the following game:

TABLE 26.15. PAY-OFF TABLE FOR A 3 × 3 GAME WITH NO SADDLE POINT

		B's strategies		
		b_1	b_2	b_3
A's strategies	a_1	8	9	3
	a_2	2	5	6
	a_3	4	1	7

Solution. This problem does not have a saddle point and has no obvious dominance and can not be resolved to either a 2×2 game or a 2×3 or 3×2 game that can solved by graphical methods. Such a problem can be formulated and solved as a linear programming problem. The problem can be viewed as a maximization problem from the point of view of player A who is interested in maximizing his minimum expected gain.

Let us introduce the variables x_1, x_2 and x_3 as the proportions of times player A plays the strategies a_1, a_2 and a_3. Similarly, let the variables y_1, y_2 and y_3 denote the proportion of times player B plays the strategies b_1, b_2 and b_3. Let E_1, E_2 and E_3 be the returns to player A when player B plays strategies b_1, b_2 and b_3, respectively. Clearly,

$$E_1 = 8x_1 + 2x_2 + 4x_3$$

$$E_2 = 9x_1 + 5x_2 + x_3, \text{ and}$$

$E_3 = 3x_1 + 6x_2 + 7x_3$

The value of the game V is the maximum of the minimum returns he obtains for each of the strategies of player B. In other words, regardless of the strategy player B chooses, the return for A is at least equal to (or greater than or equal to) V. Thus,

$E_1 = 8x_1 + 2x_2 + 4x_3 \geq V$
$E_2 = 9x_1 + 5x_2 + x_3 \geq V$ and
$E_3 = 3x_1 + 6x_2 + 7x_3 \geq V$

operate as constraints on the selection of the values of x_1, x_2 and x_3.

We are also given that

$x_1 + x_2 + x_2 = 1$, since x_1, x_2 and x_3 are the proportions and their sum should be equal to 1. Also, x_1, x_2 and x_3 are individually positive.

The values of x_1, x_2 and x_3 are to be chosen such that V is a maximum.

The presence of V, the utility function, in the constraints can not be handled by routine simplex method. A very ingenious trick can get rid of V.

If V is assumed as positive, we can divide across by V in every constraint to obtain

$8X_1 + 2X_2 + 4X_3 \geq 1$
$9X_1 + 5X_2 + X_3 \geq 1$
$3X_1 + 6X_2 + 7X_3 \geq 1$
$X_1 + X_2 + X_3 = 1 / V$, and
X_1, X_2 and $X_3 \geq 3$

where X_1, X_2 and X_3 are new variables x_1/V, etc.

Since V is to be maximized, we may take the fourth condition above as the utility function to write the linear programming problem as:

Choose X_1, X_2 and X_3 such that $X_1 + X_2 + X_3$ is a minimum (V maximum implies that $1/V$ is a minimum), subject to

$8X_1 + 2X_2 + 4X_3 \geq 1$
$9X_1 + 5X_2 + X_3 \geq 1$
$3X_1 + 6X_2 + 7X_3 \geq 1$, and
$X_1, X_2, X_3 \geq 0$

If we had formulated the problem from the point of view of player B, we would need to select appropriate values of y_1, y_2 and y_3 to minimise his pay-out V. Remember that Table 26.15 has been written in terms of the pay-off to player A. Let E'_1 E'_2 and E'_3 be the returns to player B when player A plays strategies a_1, a_2 and a_3, respectively. Clearly,

$E'_1 = 8y_1 + 9y_2 + 3y_3$
$E'_2 = 2y_1 + 5y_2 + 6y_3$, and
$E'_3 = 4y_1 + y_2 + 7y_3$

The value of the game V is the minimum of the maximum pay-out of player B for each of the strategies of player A. In other words, regardless of the strategy player A chooses, the pay-out for B is at most equal to (or lesser than or equal to) V. Thus,

$$E_1' = 8y_1 + 9y_2 + 3y_3 \le V$$
$$E_2' = 2y_1 + 5y_2 + 6y_3 \le V, \text{ and}$$
$$E_3' = 4y_1 + y_2 + 7y_3 \le V$$

operate as constraints on the selection of the values of y_1, y_2 and y_3.

We are also given that $y_1 + y_2 + y_3 = 1$, since y_1, y_2 and y_3 are the proportions and their sum should be equal to 1. Also, y_1, y_2 and y_3 are individually positive.

The values of y_1, y_2 and y_3 are to be chosen such that V is a minimum. As before, we divide across by V in every constraint to obtain

$$8Y_1 + 9Y_2 + 3Y_3 \le 1$$
$$2Y_1 + 5Y_2 + 6Y_3 \le 1, \text{ and}$$
$$4Y_1 + Y_2 + 7Y_3 \le 1$$
$$Y_1 + Y_2 + Y_3 = 1/V, \text{ and}$$
$$Y_1, Y_2 \text{ and } Y_3 \ge 0$$

where Y_1, Y_2 and Y_3 are new variables y_1/V, etc.

Since V is to be minimized, we may take the fourth condition above as the utility function to write the linear programming problem as:

Choose Y_1, Y_2 and Y_3 such that $Y_1 + Y_2 + Y_3$ is a maximum (V minimum implies that $1/V$ is a maximum), subject to

$$8Y_1 + 9Y_2 + 3Y_3 \le 1$$
$$2Y_1 + 5Y_2 + 6Y_3 \le 1$$
$$4Y_1 + Y_2 + 7Y_3 \le 1, \text{ and}$$
$$Y_1, Y_2 \text{ and } Y_3 \ge 0$$

These, then, are the two alternate formulations of the game situation as linear programming problem.

We had seen in Chapter 25 that simplex algorithm is a lot easier to operate when the constraints are *less than or equal to* type than when they are *more than or equal to* type. It is for this reason that it is easier to work with the second formulation, that is, the minimization formulation from the point of view of player B. We shall illustrate the solution of this second formulation.

As outlined in Section 25.4, we introduce slack variables Y_4, Y_5 and Y_6 to convert the inequalities in constraints to equalities. The optimization problem from the point of view of player B, then is:

Maximise $Y_1 + Y_2 + Y_3 + 0Y_4 + 0Y_5 + 0Y_6$ (a)

subject to $8Y_1 + 9Y_2 + 3Y_3 + Y_4 = 1$ (b)

$$2Y_1 + 5Y_2 + 6Y_3 + Y_5 = 1$$
$$4Y_1 + Y_2 + 7Y_3 + Y_6 = 1, \text{ and}$$
$$Y_1, Y_2, Y_3, Y_4, Y_5 \text{ and } Y_6 \geq 0$$

For the purpose of Simplex table, we convert expression (a) into an equation:

$$U - Y_1 - Y_2 - Y_3 - 0Y_4 - 0Y_5 - OY_6 = 0$$

and construct the Simples table shown as Table 26.16.

Here, for the initial table we have selected the slack variables Y_4, Y_5 and Y_6 as the basic initial variables. Application of *Rule 1* of Simplex procedure (given in Section 25.4 of the last chapter) identifies Y_1 as the new basic variable, and Y_4 as the variable to go out. The entry 8 at the intersection of Y_1 column and Y_4 row is the pivot element.

In the second stage table Y_3 as the new basic variable, and Y_6 as the variable to go out. The entry 11/2 at the intersection of Y_3 column and Y_6 row is the pivot element.

TABLE 26.16. SIMPLEX TABLE FOR ILLUSTRATION 26.5

		U	Y_1	Y_2	Y_3	Y_4	Y_5	Y_6	*Right-hand side*
First stage	U	1	-1	-1	-1	0	0	0	0
	Y_4	0	**8**	9	3	1	0	0	1
	Y_5	0	2	5	6	0	1	0	1
	Y_6	0	4	1	7	0	0	1	1
Second stage	U	1	0	1/8	-5/8	1/8	0	0	1/8
	Y_1	0	1	9/8	3/8	1/8	0	0	1/8
	Y_5	0	0	11/4	21/4	-1/4	1	0	3/4
	Y_6	0	0	-7/2	**11/2**	-1/2	0	1	1/2
Third stage	U	1	0	-3/11	0	3/44	0	5/44	2/11
	Y_1	0	1	15/11	0	7/44	0	-3/44	1/11
	Y_5	0	0	**67/11**	0	5/22	1	-21/22	3/11
	Y_3	0	0	-7/11	1	-1/11	0	2/11	1/11
Final stage	U	1	0	0	0	21/268	12/268	19/268	13/67
	Y_1	0	1	0	0	29/268	-15/67	39/268	2/67
	Y_2	0	0	1	0	5/134	11/67	-21/134	3/67
	Y_3	0	0	0	1	-9/134	7/67	11/134	8/67

We proceed in a similar fashion till there are no negative entry in the U row. This gives us the optimal solution. Thus, the optimal solution is $Y_1 = 2/67$, $Y_2 = 3/67$, $Y_3 = 8/67$, and $U = 13/67$.

We know that $Y_4 + Y_2 + Y_3 = 1/V$, so that $1/V = 2/67 + 3/67 + 8/67 = 13/67 = 0.194$, and, therefore, V, the value of the game is 5.15.

$$Y_1 = y_1 / V, \text{ therefore, } y_1 = Y_1 \times V = \frac{2}{67} \times 5.15 = 0.154$$

Similarly, $y_2 = 0.231$ and $y_3 = 0.615$.

Please note that this formulation is valid only when the pay-off for A is positive. This is ensured if all the entries in the pay-off table are positive.

Consider the pay-off in Table 26.17. Here some of the values are negative. One simple way to take care of this problem is to add a constant to each term so that each entry becomes positive. Since the minimum value is – 4, we add 5 to each term to get the pay-off matrix as in Table 26.18.

TABLE 26.17

		B's strategies		
		b_1	b_2	b_3
A's strategies	a_1	3	4	– 12
	a_2	– 3	0	1
	a_3	– 1	– 4	2

TABLE 26.18

		B's strategies		
		b_1	b_2	b_3
A's strategies	a_1	8	9	3
	a_2	2	5	6
	a_3	4	1	7

Table 26.18 has exactly the same entries as Table 26.15, and therefore, the solution will be same. However, we will have to subtract 5 from the value of the previous game, since we had earlier added 5 to each of the table entries. Thus, the value of this game is just 5.15 – 5 or 0.15.

PROBLEMS

1 Find the saddle point, if it exists, for the game with the following pay-off table:

		B's strategies	
		b_1	b_2
A's strategies	a_1	3	7
	a_2	− 5	5

2 Find the saddle point, if it exists, for the game with the following pay-off table:

		Strategies for Y			
		y_1	y_2	y_3	y_4
Strategies for X	x_1	4	2	8	5
	x_2	3	1	6	2

3. Find the saddle point, if it exists, for the game with the following pay-off table:

		Strategies for Y			
		y_1	y_2	y_3	y_4
Strategies for X	x_1	2	3	4	15
	x_2	9	11	8	13
	x_3	5	13	6	4

4. Find the saddle point, if it exists, for the game with the following pay-off table:

		Strategies for Y		
		y_1	y_2	y_3
Strategies for X	x_1	30.000	−21.000	1.000
	x_2	18.000	14.000	12.000
	x_3	−6.000	28.000	4.000
	x_4	18.000	6.000	2.000

5. There are two stores A and B in a small town. Both stores sell the same types of goods and with similar reputations. Both are planning a new-year sale in their stores. Each store has a limited budget for publicity so that it can chose only one medium for advertisement, newspaper, radio or television. Their media consultant has prepared the following table which shows the net swing of customers from Store B to Store A if the two stores use a given combination of the publicity media. Thus, if Store A uses newspaper but B uses television, A stands to lose 80 customers to store B. Determine the optimum strategy foe this game situation and the value of the game.

		B's strategies		
		Newspaper	*Radio*	*Television*
A's strategies	*Newspaper*	30	40	– 80
	Radio	0	15	– 20
	Television	90	20	50

6. Reduce the following game to a 2 × 2 game and determine the optimum strategy and the value of the game.

		Strategies for Y			
		y_1	y_2	y_3	y_4
Strategies for X	x_1	3	2	4	0
	x_2	3	4	2	4
	x_3	4	2	4	0
	x_4	0	4	0	8

7. Use the graphic method to solve the following game:

		Strategies for Y		
		y_1	y_2	y_3
Strategies for X	x_1	5	– 1	6
	x_2	1	3	– 1

8. Use the graphic method to solve the following game:

		B's strategies	
		b_1	b_2
A's strategies	a_1	2	5
	a_2	4	1

9. Use the graphic method to solve the following game:

		Strategies for Y				
		y_1	y_2	y_3	y_4	y_5
Strategies for X	x_1	2	–4	6	–3	5
	x_2	–3	4	–4	1	0

10. Use the graphic method to solve the following game:

		Strategies for Y		
		y_1	y_2	y_3
Strategies for X	x_1	3	4	12
	x_2	8	6	3

11. In a two-player zero-sum game the pay-off table is given as:

		Strategies for Y		
		y_1	y_2	y_3
Strategies for X	x_1	5	–2	–2
	x_2	–2	–2	2
	x_3	–2	1	–1

(a) Show that the game has no saddle point.
(b) Show that the game cannot be reduced by the application of dominance rule.
(c) Formulate the game as a linear programming problem.
(d) Obtain the value of the game.

12. Formulate the following as a linear programming problem and find the value of the game:

Strategies for Y

Strategies for X		y_1	y_2	y_3	y_4
	x_1	4	−8	20	16
	x_2	7	12	15	−18
	x_3	21	−5	−12	10

APPENDIX

LOGARITHMS

Logarithm of a given number to a certain base is the power to which that base must be raised in order to obtain the given number. Thus, the logarithm (or simply log) of 124 to base 4 is 5. This means that if 4 (base) is raised to the power 5 it will give us the required number.

The system of logarithms that is commonly used is on the base 10, so that when we speak of the logarithm of a number what we really mean is the power to which 10 must be raised in order to obtain that number. Thus, the logarithm of 100 is 2, for $10^2=100$, logarithm of 1,000 is 3 for $10^3=1000$, and so on....

Natural Number	**Power to which** *10 is raised*	*The Logarithm*
1,000	10^3	3.000
100	12^2	2.000
10	10^1	1.000
1	10^0	0.000
.1	10^{-1}	—1 or $\bar{1}$
.01	10^{-2}	—1 or 2

It can be seen from the above that the log of 10 is 1 and log of 100 is 2. Hence the log of any number in between 10 and 100 would be in between 1 and 2 which means that the log for any number greater than 10 and less than 100 will be 1+a fraction. Thus, a logarithm may consist of two parts: (*i*) an intergral part, and (*ii*) decimal part. The 'integral part' is called the *Characteristic* and 'decimal part' the *Mantissa*.

Finding the Logarithm. To find the logarithm of a number, we will have to find out the characteristic as also the mantissa

Finding the Characteristic. To determine the characteristic of given number the following rules are applied:

(*i*) The characteristic of any number greater than 1 is positive and is one less than the number of digits to the left of the decimal point.

(*ii*) The characteristic of any number less than 1 is negative and is greater by one than the number of zeroes between the decimal point and the first significant figure of the given number.

Following the above rules characteristic of a few numbers are as below:

Number	*Characteristic*
3,758	3
375·8	2
37·58	1
3·758	0
0·3758	—1 or $\bar{1}$
0·03758	—2 or $\bar{2}$
0·003758	—3 or $\bar{3}$

The minus sign of the characteristic is written at the top of the figure and is usually designated as bar 1, bar 2, bar 3, bar 4, etc.

Finding the Mantissa. Mantissa of a given number can be found in the logarithmic table. The first step in the process of finding the mantissa is to obtain significant figure of the given number. To get the significant figures omit all zeroes before the first and after the last non-zero digit. Thus, 375 is the significant figure of 37, 500, as well as ·00375, and ·375.

The mantissa of all numbers having the same significant figures is the same. Thus, the mantissa of 8, 80, 800, 8000, or ·008 is the same.

The first two significant figures of the number are then found at the extreme left of the logarithmic table. Thus, if we have to find the mantissa of 375 we look for 37 at the extreme left of the logarithmic table, and then move along the horizontal line to the number in the vertical column headed by the third figure 5, we obtain the mantissa 5.740

$$\log 375 = 2{\cdot}5740$$

If, however, the significant figure of a number comprises of only one digit, two zeroes may be added to its right (if it consists of two digits, one zero may be added) so that the method of finding the mantissa explained above may be adopted. Thus, the mantissa of 8, 80, or 800 is 9,031.

If a number comprises of four significant digits the mantissa is found by using the mean difference columns at extreme right. Thus, to find log of 3,758 we procced as:

Mantissa of	log 375 = 5740	
Mean difference for	8 =	9
or	log of 3,758 =	3·5749
Similarly log of	37·58 =	1·5749

Mantissa is always a positive number.

ANTILOGARITHMS

To find the natural number of a given logarithm we have to make use of the table of antilogarithms. The mantissa will enable us to obtain the different digits of the given number and the characteristic will determine the location of the decimal point.

Illustration

Find the number whose log is 2·4725

Solution

From antilog tables, entry for 472	= 2,965
Mean difference for 5	= 3
Entry for 4,725	= 2,968
Hence the number whose log is 2·4725	= 296·8
Similarly, the number whose log is $\bar{2}$·4725	= ·02968

The Use of Logarithms

(1) Multiplication. If two or more numbers are to be multiplied find the sum of their logarithms. The number corresponding to this logarithm, obtained from the table of antilogarithms is the product required.

Illustration

Multiply 32·87 by 0·00238

Solution

log 32·87	= 1·5168
log 0·00238	= $\bar{3}$·3766
Sum	= $\bar{2}$·8934 = log ·07823
Hence 32·87 × 0·00238	= 0·07823.

(2) Division. If a number is to be divided by another, subtract the logarithm of the latter from the logarithm of the former. The difference so obtained is logarithm of the quotient – the number corresponding to which can be found from the table of antilogs.

Illustration

Multiply and divide 0·342 by 0·0902.

Solution

log 0·342	$=\bar{1}\cdot5340$
log 0·0902	$=\bar{2}\cdot9552$
Sum	$=\bar{2}\cdot4892=\log 0\cdot03084$
Difference	$=0\cdot5788=\log 3\cdot791$
Hence 0·342×0·0902	=0·03084
and 0·342÷0·0902	=3·791.

(3) Power of a Number. The square, cube or other power of a number is obtained by multiplying the logarithm of the number by the exponent of the power, the product is the logarithm of the required value.

Illustration

Calculate the value of $(0\cdot07)^3$.

Solution

log (0·07)	$=\bar{2}\cdot8451$
log $(0\cdot07)^3$	$=3\times\bar{2}\cdot8451$
	$=\bar{4}\cdot5353$
	=log ·000343
∴ $(\cdot07)^3$	=·000343.

(4) Root of a Number. If it is desired to extract any root of a given number, divide the logarithm of the number by the number which indicates the root.

The division of a logarithm presents some difficulty when the characteristic is negative and is not exactly divisible by the given value of the root. (It should be remembered that mantissa is always positive). The best way to overcome this difficulty is to make the characteristic exactly divisible and add compensating figure to mantissa.

Illustration 1.

Solve $\sqrt{144}$.

Solution

log 144=2·1584

Dividing 2·1584 by 2, we get

1·0792=log 12

∴ $\sqrt{144}=12$.

Illustration 2.

Solve $\sqrt{0{\cdot}625}$.

Solution

$$
\begin{aligned}
\log \sqrt{0{\cdot}625} &= \tfrac{1}{2}(\bar{1}{\cdot}7959) \\
&= \tfrac{1}{2}(\bar{2}+1{\cdot}7959) \\
&= \bar{1}{\cdot}8979 \\
&= \log 0{\cdot}7905 \\
\text{or} \quad \sqrt{0{\cdot}625} &= 0{\cdot}7905.
\end{aligned}
$$

STATISTICAL TABLES

LOGARITHMS

	0	1	2	3	4	5	6	7	8	9	1	2	3	4	5	6	7	8	9
10	0000	0043	0086	0128	0170						5	9	13	17	21	26	30	34	38
						0212	0253	0294	0334	0374	4	8	12	16	20	24	28	32	36
11	0414	0453	0492	0531	0569						4	8	12	16	20	23	27	31	35
						0607	0645	0682	0719	0755	4	7	11	15	18	22	26	29	33
12	0792	0828	0864	0899	0934						3	7	11	14	18	21	25	28	32
						0969	1004	1038	1072	1106	3	7	10	14	17	20	24	27	31
13	1139	1173	1206	1239	1271						3	6	10	13	16	19	23	26	29
						1303	1335	1367	1399	1430	3	7	10	13	16	19	22	25	29
14	1461	1492	1523	1553	1584						3	6	9	12	15	19	22	25	28
						1614	1644	1673	1703	1732	3	6	9	12	14	17	20	23	26
15	1761	1790	1818	1847	1875						3	6	9	11	14	17	20	23	26
						1903	1931	1959	1987	2014	3	6	8	11	14	17	19	22	25
16	2041	2068	2095	2122	2148						3	6	8	11	14	16	19	22	24
						2175	2201	2227	2253	2279	3	5	8	10	13	16	18	21	23
17	2304	2330	2355	2380	2405						3	5	8	10	13	15	18	20	23
						2430	2455	2480	2504	2529	3	5	8	10	12	15	17	20	22
18	2553	2577	2601	2625	2648						2	5	7	9	12	14	17	19	21
						2672	2695	2718	2742	2765	2	4	7	9	11	14	16	18	21
19	2788	2810	2833	2856	2878						2	4	7	9	11	13	16	18	20
						2900	2923	2945	2967	2989	2	4	6	8	11	13	15	17	19
20	3010	3032	3054	3075	3096	3118	3139	3160	3181	3201	2	4	6	8	11	13	15	17	19
21	3222	3243	3263	3284	3304	3324	3345	3365	3385	3404	2	4	6	8	10	12	14	16	18
22	3424	3444	3464	3483	3502	3522	3541	3560	3579	3598	2	4	6	8	10	12	14	15	17
23	3617	3636	3655	3674	3692	3711	3729	3747	3766	3784	2	4	6	7	9	11	13	15	17
24	3802	3820	3838	3856	3874	3892	3909	3927	3945	3962	2	4	5	7	9	11	12	14	16
25	3979	3997	4014	4031	4048	4065	4082	4099	4116	4133	2	3	5	7	9	10	12	14	15
26	4150	4166	4183	4200	4216	4232	4249	4265	4281	4298	2	3	5	7	8	10	11	13	15
27	4314	4330	4346	4362	4378	4393	4409	4425	4440	4456	2	3	5	6	8	9	11	13	14
28	4472	4487	4502	4518	4533	4548	4564	4579	4594	4609	2	3	5	6	8	9	11	12	14
29	4624	4639	4654	4669	4683	4698	4713	4728	4742	4757	1	3	4	6	7	9	10	12	13
30	4771	4786	4800	4814	4829	4843	4857	4871	4886	4900	1	3	4	6	7	9	10	11	13
31	4914	4928	4942	4955	4969	4983	4997	5011	5024	5038	1	3	4	6	7	8	10	11	12
32	5051	5065	5079	5092	5105	5119	5132	5145	5159	5172	1	3	4	5	7	8	9	11	12
33	5185	5198	5211	5224	5237	5250	5263	5276	5289	5302	1	3	4	5	6	8	9	10	12
34	5315	5328	5340	5353	5366	5378	5391	5403	5416	5428	1	3	4	5	6	8	9	10	11
35	5441	5453	5465	5478	5490	5502	5514	5527	5539	5551	1	2	4	5	6	7	9	10	11
36	5563	5575	5587	5599	5611	5623	5635	5647	5658	5670	1	2	4	5	6	7	8	10	11
37	5682	5694	5705	5717	5729	5740	5752	5763	5775	5786	1	2	3	5	6	7	8	9	10
38	5798	5809	5821	5832	5843	5855	5866	5877	5888	5899	1	2	3	5	6	7	8	9	10
39	5911	5922	5933	5944	5955	5966	5977	5988	5999	6010	1	2	3	4	5	7	8	9	10
40	6021	6031	6042	6053	6064	6075	6085	6096	6107	6117	1	2	3	4	5	6	8	9	10
41	6128	6138	6149	6160	6170	6180	6191	6201	6212	6222	1	2	3	4	5	6	7	8	9
42	6232	6243	6253	6263	6274	6284	6294	6304	6314	6325	1	2	3	4	5	6	7	8	9
43	6335	6345	6355	6365	6375	6385	6395	6405	6415	6425	1	2	3	4	5	6	7	8	9
44	6435	6444	6454	6464	6474	6484	6493	6503	6513	6522	1	2	3	4	5	6	7	8	9
45	6532	6542	6551	6561	6571	6580	6590	6599	6609	6618	1	2	3	4	5	6	7	8	9
46	6628	6637	6646	6656	6665	6675	6684	6693	6702	6712	1	2	3	4	5	6	7	7	8
47	6721	6730	6739	6749	6758	6767	6776	6785	6794	6803	1	2	3	4	5	5	6	7	8
48	6812	6821	6830	6839	6848	6857	6866	6875	6884	6893	1	2	3	4	4	5	6	7	8
49	6902	6911	6920	6928	6937	6946	6955	6964	6972	6981	1	2	3	4	4	5	6	7	8

LOGARITHMS

	0	1	2	3	4	5	6	7	8	9	123	456	789
50	6990	6998	7007	7016	7024	7033	7042	7050	7059	7067	1 2 3	3 4 5	6 7 8
51	7076	7084	7093	7101	7110	7118	7126	7135	7143	7152	1 2 3	3 4 5	6 7 8
52	7160	7168	7177	7185	7193	7202	7210	7218	7226	7235	1 2 2	3 4 5	6 7 7
53	7243	7251	7259	7267	7275	7284	7292	7300	7308	7316	1 2 2	3 4 5	6 6 7
54	7324	7332	7340	7348	7356	7364	7372	7380	7388	7396	1 2 2	3 4 5	6 6 7
55	7404	7412	7419	7427	7435	7443	7461	7459	7466	7474	1 2 2	3 4 5	5 6 7
56	7482	7490	7497	7505	7513	7520	7528	7536	7543	7551	1 2 2	3 4 5	5 6 7
57	7559	7566	7574	7582	7589	7597	7604	7612	7619	7627	1 2 2	3 4 5	5 6 7
58	7634	7642	7649	7657	7664	7672	7679	7686	7694	7701	1 1 2	3 4 4	5 6 7
59	7709	7716	7723	7731	7738	7745	7752	7760	7767	7774	1 1 2	3 4 4	5 6 7
60	7782	7789	7796	7803	7810	7818	7825	7832	7839	7846	1 1 2	3 4 4	5 6 6
61	7853	7860	7868	7875	7882	7889	7896	7903	7910	7917	1 1 2	3 4 4	5 6 6
62	7924	7931	7938	7945	7952	7959	7966	7973	7980	7987	1 1 2	3 3 4	5 6 6
63	7993	8000	8007	8014	8021	8028	8035	8041	8048	8055	1 1 2	3 3 4	5 5 6
64	8062	8069	8075	8082	8089	8096	8102	8109	8116	8122	1 1 2	3 3 4	5 5 6
65	8129	8136	8142	8149	8156	8162	8169	8176	8182	8189	1 1 2	3 3 4	5 5 6
66	8195	8202	8209	8215	8222	8228	8235	8241	8248	8254	1 1 2	3 3 4	5 5 6
67	8261	8267	8274	8280	8287	8293	8299	8306	8312	8319	1 1 2	3 3 4	5 5 6
68	8325	8331	8338	8344	8351	8357	8363	8370	8376	8382	1 1 2	3 3 4	4 5 6
69	8388	8395	8401	8407	8414	8420	8426	8432	8439	8445	1 1 2	2 3 4	4 5 6
70	8451	8457	8463	8470	8476	8482	8488	8494	8500	8506	1 1 2	2 3 4	4 5 6
71	8513	8519	8525	8531	8537	8543	8549	8555	8561	8567	1 1 2	2 3 4	4 5 5
72	8573	8579	8585	8591	8597	8603	8609	8615	8621	8627	1 1 2	2 3 4	4 5 5
73	8633	8639	8645	8651	8657	8663	8669	8675	8681	8686	1 1 2	2 3 4	4 5 5
74	8692	8698	8704	8710	8716	8722	8727	8733	8739	8745	1 1 2	2 3 4	4 5 5
75	8751	8756	8762	8768	8774	8779	8785	8791	8797	8802	1 1 2	2 3 3	4 5 5
76	8808	8814	8820	8825	8831	8837	8842	8848	8854	8859	1 1 2	2 3 3	4 5 5
77	8865	8871	8876	8882	8887	8893	8899	8904	8910	8915	1 1 2	2 3 3	4 4 5
78	8921	8927	8932	8938	8943	8949	8954	8960	8965	8971	1 1 2	2 3 3	4 4 5
79	8976	8982	8987	8993	8998	9004	9009	9015	9020	9025	1 1 2	2 3 3	4 4 5
80	9031	9036	9042	9047	9053	9058	9063	9069	9074	9079	1 1 2	2 3 3	4 4 5
81	9085	9090	9096	9101	9106	9112	9117	9122	9128	9133	1 1 2	2 3 3	4 4 5
82	9138	9143	9149	9154	9159	9165	9170	9175	9180	9186	1 1 2	2 3 3	4 4 5
83	9191	9196	9201	9206	9212	9217	9222	9227	9232	9238	1 1 2	2 3 3	4 4 5
84	9243	9248	9253	9258	9263	9269	9274	9279	9284	9289	1 1 2	2 3 3	4 4 5
85	9294	9299	9304	9309	9315	9320	9325	9330	9335	9340	1 1 2	2 3 3	4 4 5
86	9345	9350	9355	9360	9365	9370	9375	9380	9385	9390	1 1 2	2 3 3	4 4 5
87	9395	9400	9405	9410	9415	9420	9425	9430	9435	9440	0 1 1	2 2 3	3 4 4
88	9445	9450	9455	9460	9465	9469	9474	9479	9484	9489	0 1 1	2 2 3	3 4 4
89	9494	9499	9504	9509	9513	9518	9523	9528	9533	9538	0 1 1	2 2 3	3 4 4
90	9542	9547	9552	9557	9562	9566	9571	9576	9581	9586	0 1 1	2 2 3	3 4 4
91	9590	9595	9600	9605	9609	9614	9619	9624	9628	9633	0 1 1	2 2 3	3 4 4
92	9638	9643	9647	9652	9657	9661	9666	9671	9675	9680	0 1 1	2 2 3	3 4 4
93	9685	9689	9694	9699	9703	9708	9713	9717	9722	9727	0 1 1	2 2 3	3 4 4
94	9731	9736	9741	9745	9750	9754	9759	9763	9768	9773	0 1 1	2 2 3	3 4 4
95	9777	9782	9786	9791	9795	9800	9805	9809	9814	9818	0 1 1	2 2 3	3 4 4
96	9823	9827	9832	9836	9841	9845	9850	9854	9859	9863	0 1 1	2 2 3	3 4 4
97	9868	9872	9877	9881	9886	9890	9894	9899	9903	9908	0 1 1	2 2 3	3 4 4
98	9912	9917	9921	9926	9930	9934	9939	9943	9948	9952	0 1 1	2 2 3	3 4 4
99	9956	9961	9965	9969	9974	9978	9983	9987	9991	9996	0 1 1	2 2 3	3 3 4

ANTILOGARITHMS

	0	1	2	3	4	5	6	7	8	9	123	456	789
·00	1000	1002	1005	1007	1009	1012	1014	1016	1019	1021	0 0 1	1 1 1	2 2 2
·01	1023	1026	1028	1030	1033	1035	1038	1040	1042	1045	0 0 1	1 1 1	2 2 2
·02	1047	1050	1052	1054	1057	1059	1062	1064	1067	1069	0 0 1	1 1 1	2 2 2
·03	1072	1074	1076	1079	1081	1084	1086	1089	1091	1094	0 0 1	1 1 1	2 2 2
·04	1096	1099	1102	1104	1107	1109	1112	1114	1117	1119	0 1 1	1 1 2	2 2 2
·05	1122	1125	1127	1130	1132	1135	1138	1140	1143	1146	0 1 1	1 1 2	2 2 2
·06	1148	1151	1153	1156	1159	1161	1164	1167	1169	1172	0 1 1	1 1 2	2 2 2
·07	1175	1178	1180	1183	1186	1189	1191	1194	1197	1199	0 1 1	1 1 2	2 2 2
·08	1202	1205	1208	1211	1213	1216	1219	1222	1225	1227	0 1 1	1 1 2	2 2 3
·09	1230	1233	1236	1239	1242	1245	1247	1250	1253	1256	0 1 1	1 1 2	2 2 3
·10	1259	1262	1265	1268	1271	1274	1276	1279	1282	1285	0 1 1	1 1 2	2 2 3
·11	1288	1291	1294	1297	1300	1303	1306	1309	1312	1315	0 1 1	1 2 2	2 2 3
·12	1318	1321	1324	1327	1330	1334	1337	1340	1343	1346	0 1 1	1 2 2	2 2 3
·13	1349	1352	1355	1358	1361	1365	1368	1371	1374	1377	0 1 1	1 2 2	2 3 3
·14	1380	1384	1387	1390	1393	1396	1400	1403	1406	1409	0 1 1	1 2 2	2 3 3
·15	1413	1416	1419	1422	1426	1429	1432	1435	1439	1442	0 1 1	1 2 2	2 3 3
·16	1445	1449	1452	1455	1459	1462	1466	1469	1472	1476	0 1 1	1 2 2	2 3 3
·17	1479	1483	1486	1489	1493	1496	1500	1503	1507	1510	0 1 1	1 2 2	2 3 3
·18	1514	1517	1521	1524	1528	1531	1535	1538	1542	1545	0 1 1	1 2 2	2 3 3
·19	1549	1552	1556	1560	1563	1567	1570	1574	1578	1581	0 1 1	1 2 2	3 3 3
·20	1585	1589	1592	1596	1600	1603	1607	1611	1614	1618	0 1 1	1 2 2	3 3 3
·21	1622	1626	1629	1633	1637	1641	1644	1648	1652	1656	0 1 1	2 2 2	3 3 3
·22	1660	1663	1667	1671	1675	1679	1683	1687	1690	1694	0 1 1	2 2 2	3 3 3
·23	1698	1702	1706	1710	1714	1718	1722	1726	1730	1734	0 1 1	2 2 2	3 3 4
·24	1738	1742	1746	1750	1754	1758	1762	1766	1770	1774	0 1 1	2 2 2	3 3 4
·25	1778	1782	1786	1791	1795	1799	1803	1807	1811	1816	0 1 1	2 2 2	3 3 4
·26	1820	1824	1828	1832	1837	1841	1845	1849	1854	1858	0 1 1	2 2 3	3 3 4
·27	1862	1866	1871	1875	1879	1884	1888	1892	1897	1901	0 1 1	2 2 3	3 3 4
·28	1905	1910	1914	1919	1923	1928	1932	1936	1941	1945	0 1 1	2 2 3	3 4 4
·29	1950	1954	1959	1963	1968	1972	1977	1982	1986	1991	0 1 1	2 2 3	3 4 4
·30	1995	2000	2004	2009	2014	2018	2023	2028	2032	2037	0 1 1	2 2 3	3 4 4
·31	2042	2046	2051	2056	2061	2065	2070	2075	2080	2084	0 1 1	2 2 3	3 4 4
·32	2089	2094	2099	2104	2109	2113	2118	2123	2128	2133	0 1 1	2 2 3	3 4 4
·33	2138	2143	2148	2153	2158	2163	2168	2173	2178	2183	0 1 1	2 2 3	3 4 4
·34	2188	2193	2198	2203	2208	2213	2218	2223	2228	2234	1 1 2	2 3 3	4 4 5
·35	2239	2244	2249	2254	2259	2265	2270	2275	2280	2286	1 1 2	2 3 3	4 4 5
·36	2291	2296	2301	2307	2312	2317	2323	2328	2333	2339	1 1 2	2 3 3	4 4 5
·37	2344	2350	2355	2360	2366	2371	2377	2382	2388	2393	1 1 2	2 3 3	4 4 5
·38	2399	2404	2410	2415	2421	2427	2432	2438	2443	2449	1 1 2	2 3 3	4 4 5
·39	2455	2460	2466	2472	2477	2483	2489	2495	2500	2506	1 1 2	2 3 3	4 5 5
·40	2512	2518	2523	2529	2535	2541	2547	2553	2559	2564	1 1 2	2 3 4	4 5 5
·41	2570	2576	2582	2588	2594	2600	2606	2612	2618	2624	1 1 2	3 3 4	4 5 5
·42	2630	2636	2642	2649	2655	2661	2667	2673	2679	2685	1 1 2	2 3 4	4 5 6
·43	2692	2698	2704	2710	2716	2723	2729	2735	2742	2748	1 1 2	3 3 4	4 5 6
·44	2754	2761	2767	2773	2780	2786	2793	2799	2805	2812	1 1 2	3 3 4	4 5 6
·45	2818	2825	2831	2838	2844	2851	2858	2864	2871	2877	1 1 2	3 3 4	5 5 6
·46	2884	2891	2897	2904	2911	2917	2924	2931	2938	2944	1 1 2	3 3 4	5 5 6
·47	2951	2958	2965	2972	2979	2985	2992	2999	3006	3013	1 1 2	3 3 4	5 5 6
·48	3020	3027	3034	3041	3048	3055	3062	3069	3076	3083	1 1 2	3 4 4	5 6 6
·49	3090	3097	3105	3112	3119	3126	3133	3141	3148	3155	1 1 2	3 4 4	5 6 6

ANTILOGARITHMS

	0	1	2	3	4	5	6	7	8	9	1	2	3	4	5	6	7	8	9
·50	3162	3170	3177	3184	3192	3199	3206	3214	3221	3228	1	1	2	3	4	4	5	6	7
·51	3236	3243	3251	3258	3266	3273	3281	3289	3296	3304	1	2	2	3	4	5	5	6	7
·52	3311	3319	3327	3334	3342	3350	3357	3365	3373	3381	1	2	2	3	4	5	6	6	7
·53	3388	3396	3404	3412	3420	3428	3436	3443	3451	3459	1	2	2	3	4	5	6	6	7
·54	3467	3475	3483	3491	3499	3508	3516	3524	3532	3540	1	2	2	3	4	5	6	6	7
·55	3548	3556	3565	3573	3581	3589	3597	3606	3614	3622	1	2	2	3	4	5	6	7	7
·56	3631	3639	3648	3656	3664	3673	3681	3690	3698	3707	1	2	3	3	4	5	6	7	8
·57	3715	3724	3733	3741	3750	3758	3767	3776	3784	3793	1	2	3	3	4	5	6	7	8
·58	3802	3811	3819	3828	3837	3846	3855	3864	3873	3882	1	2	3	4	4	5	6	7	8
·59	3890	3899	3908	3917	3926	3936	3945	3954	3963	3972	1	2	3	4	5	5	6	7	8
·60	3981	3990	3999	4009	4018	4027	4036	4046	4055	4064	1	2	3	4	5	6	6	7	8
·61	4074	4083	4093	4102	4111	4121	4130	4140	4150	4159	1	2	3	4	5	6	7	8	9
·62	4169	4178	4188	4198	4207	4217	4227	4236	4246	4256	1	2	3	4	5	6	7	8	9
·63	4266	4276	4285	4295	4305	4315	4325	4335	4345	4355	1	2	3	4	5	6	7	8	9
·64	4365	4375	4385	4395	4406	4416	4426	4436	4446	4457	1	2	3	4	5	6	7	8	9
·65	4467	4477	4487	4498	4508	4519	4529	4539	4550	4560	1	2	3	4	5	6	7	8	9
·66	4571	4581	4592	4603	4613	4624	4634	4645	4656	4667	1	2	3	4	5	6	7	9	10
·67	4677	4688	4699	4710	4721	4732	4742	4753	4764	4775	1	2	3	4	5	7	8	9	10
·68	4786	4797	4808	4819	4831	4842	4853	4864	4875	4887	1	2	3	4	6	7	8	9	10
·69	4898	4909	4920	4932	4943	4955	4966	4977	4989	5000	1	2	3	5	6	7	8	9	10
·70	5012	5023	5035	5047	5058	5070	5082	5093	5105	5117	1	2	4	5	6	7	8	9	11
·71	5129	5140	5152	5164	5176	5188	5200	5212	5224	5236	1	2	4	5	6	7	8	10	11
·72	5248	5260	5272	5284	5297	5309	5321	5333	5346	5358	1	2	4	5	6	7	9	10	11
·73	5370	5383	5395	5408	5420	5433	5445	5458	5470	5483	1	3	4	5	6	8	9	10	11
·74	5495	5508	5521	5534	5546	5559	5572	5585	5598	5610	1	3	4	5	6	8	9	10	12
·75	5623	5636	5649	5662	5675	5689	5702	5715	5728	5741	1	3	4	5	7	8	9	10	12
·76	5754	5768	5781	5794	5808	5821	5834	5848	5861	5875	1	3	4	5	7	8	9	11	12
·77	5888	5902	5916	5929	5943	5957	5970	5984	5998	6012	1	3	4	5	7	8	10	11	12
·78	6026	6039	6053	6067	6081	6095	6109	6124	6138	6152	1	3	4	6	7	8	10	11	13
·79	6166	6180	6194	6209	6223	6237	6252	6266	6281	6295	1	3	4	6	7	9	10	11	13
·80	6310	6324	6339	6353	6368	6383	6397	6412	6427	6442	1	3	4	6	7	9	10	12	13
·81	6457	6471	6486	6501	6516	6531	6546	6561	6577	6592	2	3	5	6	8	9	11	12	14
·82	6607	6622	6637	6653	6668	6683	6699	6714	6730	6745	2	3	5	6	8	9	11	12	14
·83	6761	6776	6792	6808	6823	6839	6855	6871	6887	6902	2	3	5	6	8	9	11	13	14
·84	6918	6934	6950	6966	6982	6998	7015	7031	7047	7063	2	3	5	6	8	10	11	13	15
·85	7079	7096	7112	7129	7145	7161	7178	7194	7211	7228	2	3	5	7	8	10	12	13	15
·86	7244	7261	7278	7295	7311	7328	7345	7362	7379	7396	2	3	5	7	8	10	12	13	15
·87	7413	7430	7447	7464	7482	7499	7516	7534	7551	7568	2	3	5	7	9	10	12	14	16
·88	7586	7603	7621	7638	7656	7674	7691	7709	7727	7745	2	4	5	7	9	11	12	14	16
·89	7762	7780	7798	7816	7834	7852	7870	7889	7907	7925	2	4	5	7	9	11	13	14	16
·90	7943	7962	7980	7998	8017	8035	8054	8072	8091	8110	2	4	6	7	9	11	13	15	17
·91	8128	8147	8166	8185	8204	8222	8241	8260	8279	8299	2	4	6	8	9	11	13	15	17
·92	8318	8337	8356	8375	8395	8414	8433	8453	8472	8492	2	4	6	8	10	12	14	15	17
·93	8511	8531	8551	8570	8590	8610	8630	8650	8670	8690	2	4	6	8	10	12	14	16	18
·94	8710	8730	8750	8770	8790	8810	8831	8851	8872	8892	2	4	6	8	10	12	14	16	18
·95	8913	8933	8954	8974	8995	9016	9036	9057	9078	9099	2	4	6	8	10	12	15	17	19
·96	9120	9141	9162	9183	9204	9226	9247	9268	9290	9311	2	4	6	8	11	13	15	17	19
·97	9333	9354	9376	9397	9419	9441	9462	9484	9506	9528	2	4	7	9	11	13	15	17	20
·98	9550	9572	9594	9616	9638	9661	9683	9705	9727	9750	2	4	7	9	11	13	16	18	20
·99	9772	9795	9817	9840	9863	9886	9908	9931	9954	9977	2	5	7	9	11	14	16	18	20

RECIPROCALS

[Numbers in Mean Difference Columns to be Subtracted.]

	0	1	2	3	4	5	6	7	8	9	Mean Differences 1	2	3	4	5	6	7	8	9
1·0	1·000	9901	9804	9709	9615	9524	9434	9346	9259	9174									
1·1	·9091	9009	8929	8850	8772	8696	8621	8547	8475	8403									
1·2	·8333	8264	8197	8130	8065	8000	7937	7874	7813	7752									
1·3	·7692	7634	7576	7519	7463	7407	7353	7299	7246	7194									
1·4	·7143	7092	7042	6993	6944	6897	6849	6803	6757	6711	5	10	14	19	24	29	33	38	43
1·5	·6667	6623	6579	6536	6494	6452	6410	6369	6329	6289	4	8	13	17	21	25	29	33	38
1·6	·6250	6211	6173	6135	6098	6061	6024	5988	5952	5917	4	7	11	15	18	22	26	29	33
1·7	·5882	5848	5814	5780	5747	5714	5682	5650	5618	5587	3	6	10	13	16	20	23	26	29
1·8	·5556	5525	5495	5464	5435	5405	5376	5348	5319	5291	3	6	9	12	15	17	20	23	26
1·9	·5263	5236	5208	5181	5155	5128	5102	5076	5051	5025	3	5	8	11	13	16	18	21	24
2·0	·5000	4975	4950	4926	4902	4878	4854	4831	4808	4785	2	5	7	10	12	14	17	19	21
2·1	·4762	4739	4717	4695	4673	4651	4630	4608	4587	4566	2	4	7	9	11	13	15	17	20
2·2	·4545	4525	4505	4484	4404	4444	4425	4405	4386	4367	2	4	6	8	10	12	14	16	18
2·3	·4348	4329	4310	4292	4274	4255	4237	4219	4202	4184	2	4	5	7	9	11	13	14	16
2·4	·4167	4149	4132	4115	4098	4082	4065	4049	4032	4016	2	3	5	7	8	10	12	13	15
2·5	·4000	3984	3968	3953	3937	3922	3906	3891	3875	3861	2	3	5	6	8	9	11	12	14
2·6	·3846	3831	3817	3802	3788	3774	3759	3745	3731	3717	1	3	4	6	7	8	10	11	13
2·7	·3704	3690	3676	3663	3650	3636	3623	3610	3597	3584	1	3	4	5	7	8	9	11	12
2·8	·3571	3559	3546	3534	3521	3509	3497	3484	3472	3460	1	2	4	5	6	7	9	10	11
2·9	·3448	3436	3425	3413	3401	3390	3378	3367	3356	3344	1	2	3	5	6	7	8	9	10
3·0	·3333	3322	3311	3300	3289	3279	3268	3257	3247	3236	1	2	3	4	5	6	7	9	10
3·1	·3226	3215	3205	3195	3185	3175	3165	3155	3145	3135	1	2	3	4	5	6	7	8	9
3·2	·3125	3115	3106	3096	3036	3077	3067	3058	3049	3040	1	2	3	4	5	6	7	8	9
3·3	·3030	3021	3012	3003	2994	2985	2976	2967	2959	2950	1	2	3	4	4	5	6	7	8
3·4	·2941	2933	2924	2915	2907	2899	2890	2882	2874	2865	1	2	3	3	4	5	6	7	8
3·5	·2857	2849	2841	2833	2825	2817	2809	2801	2793	2786	1	2	2	3	4	5	6	6	7
3·6	·2778	2770	2762	2755	2747	2740	2732	2725	2717	2710	1	2	2	3	4	5	5	6	7
3·7	·2703	2695	2688	2681	2674	2667	2660	2653	2646	2639	1	1	2	3	4	4	5	6	6
3·8	·2632	2625	2618	2611	2604	2597	2591	2584	2577	2571	1	1	2	3	3	4	5	5	6
3·9	·2564	2558	2551	2545	2538	2532	2525	2519	2513	2506	1	1	2	3	3	4	4	5	6
4·0	·2500	2494	2488	2481	2475	2469	2463	2457	2451	2485	1	1	2	2	3	4	4	5	5
4·1	·2439	2433	2427	2421	2415	2410	2404	2398	2392	2387	1	1	2	2	3	3	4	5	5
4·2	·2381	2375	2370	2364	2358	2353	2347	2342	2336	2331	1	1	2	2	3	3	4	4	5
4·3	·2326	2320	2315	2309	2304	2299	2294	2288	2283	2278	1	1	2	2	3	3	4	4	5
4·4	·2273	2268	2262	2257	2252	2247	2242	2237	2232	2227	1	1	2	2	3	3	4	4	5
4·5	·2222	2217	2212	2208	2203	2198	2193	2188	2183	2179	0	1	1	2	2	3	3	4	4
4·6	·2174	2169	2165	2100	2155	2151	2146	2141	2137	2132	0	1	1	2	2	3	3	4	4
4·7	·2128	2123	2119	2114	2110	2105	2101	2096	2092	2088	0	1	1	2	2	3	3	4	4
4·8	·2083	2079	2075	2070	2066	2062	2058	2053	2049	2045	0	1	1	2	2	3	3	3	4
4·9	·2041	2037	2033	2028	2024	2020	2016	2012	2008	2004	0	1	1	2	2	2	3	3	4
5·0	·2000	1996	1992	1588	1984	1980	1976	1972	1969	1965	0	1	1	2	2	2	3	3	4
5·1	·1961	1957	1953	1949	1946	1942	1938	1934	1931	1927	0	1	1	2	2	2	3	3	3
5·2	·1923	1919	1916	1912	1908	1905	1901	1898	1894	1890	0	1	1	1	2	2	3	3	3
5·3	·1887	1883	1880	1876	1873	1869	1866	1862	1859	1855	0	1	1	1	2	2	2	3	3
5·4	·1852	1848	1845	1842	1838	1835	1832	1828	1825	1821	0	1	1	1	1	2	2	3	3

RECIPROCALS

[Numbers in Mean Difference Columns to be Subtracted.]

	0	1	2	3	4	5	6	7	8	9	Mean Differences 1 2 3	4 5 6	7 8 9
5·5	·1818	1815	1812	1808	1805	1802	1799	1795	1792	1789	0 1 1	1 2 2	2 3 3
5·6	·1786	1783	1779	1776	1773	1770	1767	1764	1761	1757	0 1 1	1 2 2	2 3 3
5·7	·1754	1751	1748	1745	1742	1739	1736	1733	1730	1727	0 1 1	1 1 2	2 2 3
5·8	·1724	1721	1718	1715	1712	1709	1706	1704	1701	1698	0 1 1	1 1 2	2 2 3
5·9	·1695	1692	1689	1686	1684	1681	1678	1675	1672	1669	0 1 1	1 1 2	2 2 3
6·0	·1667	1664	1661	1658	1656	1653	1650	1647	1645	1642	0 1 1	1 1 2	2 2 3
6·1	·1639	1637	1634	1631	1629	1626	1623	1621	1618	1616	0 1 1	1 1 2	2 2 2
6·2	·1613	1610	1608	1605	1603	1600	1597	1595	1592	1590	0 1 1	1 1 2	2 2 2
6·3	·1587	1585	1582	1580	1577	1575	1572	1570	1567	1565	0 0 1	1 1 1	2 2 2
6·4	·1562	1560	1558	1555	1553	1550	1548	1546	1543	1541	0 0 1	1 1 1	2 2 2
6·5	·1538	1536	1534	1531	1529	1527	1524	1522	1520	1517	0 0 1	1 1 1	2 2 2
6·6	·1515	1513	1511	1508	1506	1504	1502	1499	1497	1495	0 0 1	1 1 1	2 2 2
6·7	·1493	1490	1488	1486	1484	1481	1479	1477	1475	1473	0 0 1	1 1 1	2 2 2
6·8	·1471	1468	1466	1464	1462	1460	1458	1456	1453	1451	0 0 1	1 1 1	2 2 2
6·9	·1449	1447	1445	1443	1441	1439	1437	1435	1433	1431	0 0 1	1 1 1	2 2 2
7·0	·1429	1427	1425	1422	1420	1418	1416	1414	1412	1410	0 0 1	1 1 1	1 2 2
7·1	·1408	1406	1404	1403	1401	1399	1397	1395	1393	1391	0 0 1	1 1 1	1 2 2
7·2	·1389	1387	1385	1383	1381	1379	1377	1376	1374	1372	0 0 1	1 1 1	1 2 2
7·3	·1370	1368	1366	1364	1362	1361	1359	1357	1355	1353	0 0 1	1 1 1	1 2 2
7·4	·1351	1350	1348	1346	1344	1342	1340	1339	1337	1335	0 0 1	1 1 1	1 1 2
7·5	·1333	1332	1330	1328	1326	1325	1323	1321	1319	1318	0 0 1	1 1 1	1 1 2
7·6	·1316	1314	1312	1311	1309	1307	1305	1304	1302	1300	0 0 1	1 1 1	1 1 2
7·7	·1299	1297	1295	1294	1292	1290	1289	1287	1285	1284	0 0 0	1 1 1	1 1 1
7·8	·1282	1280	1279	1277	1276	1274	1272	1271	1269	1267	0 0 0	1 1 1	1 1 1
7·9	·1266	1264	1263	1261	1259	1258	1256	1255	1253	1252	0 0 0	1 1 1	1 1 1
8·0	·1250	1248	1247	1245	1244	1242	1241	1239	1238	1236	0 0 0	1 1 1	1 1 1
8·1	·1235	1233	1232	1230	1229	1227	1225	1224	1222	1221	0 0 0	1 1 1	1 1 1
8·2	·1220	1218	1217	1215	1214	1212	1211	1209	1208	1206	0 0 0	1 1 1	1 1 1
8·3	·1205	1203	1202	1200	1199	1198	1196	1195	1193	1192	0 0 0	1 1 1	1 1 1
8·4	·1190	1189	1188	1186	1185	1183	1182	1181	1179	1178	0 0 0	1 1 1	1 1 1
8·5	·1176	1175	1174	1172	1171	1170	1168	1167	1166	1164	0 0 0	1 1 1	1 1 1
8·6	·1163	1161	1160	1159	1157	1156	1155	1153	1152	1151	0 0 0	1 1 1	1 1 1
8·7	·1149	1148	1147	1145	1144	1143	1142	1140	1139	1138	0 0 0	1 1 1	1 1 1
8·8	·1136	1135	1134	1133	1131	1130	1129	1127	1126	1125	0 0 0	1 1 1	1 1 1
8·9	·1124	1122	1121	1120	1119	1117	1116	1115	1114	1112	0 0 0	1 1 1	1 1 1
9·0	·1111	1110	1109	1107	1106	1105	1104	1103	1101	1100	0 0 0	1 1 1	1 1 1
9·1	·1099	1098	1096	1095	1094	1093	1092	1090	1089	1088	0 0 0	0 1 1	1 1 1
9·2	·1087	1086	1085	1083	1082	1081	1080	1079	1078	1076	0 0 0	0 1 1	1 1 1
9·3	·1075	1074	1073	1072	1071	1070	1068	1067	1066	1065	0 0 0	0 1 1	1 1 1
9·4	·1064	1063	1062	1060	1059	1058	1057	1056	1055	1054	0 0 0	0 1 1	1 1 1
9·5	·1053	1052	1050	1049	1048	1047	1046	1045	1044	1043	0 0 0	0 1 1	1 1 1
9·6	·1042	1041	1039	1038	1037	1036	1035	1034	1033	1032	0 0 0	0 1 1	1 1 1
9·7	·1031	1030	1029	1028	1027	1026	1025	1024	1022	1021	0 0 0	0 1 1	1 1 1
9·8	·1020	1019	1018	1017	1016	1015	1014	1013	1012	1011	0 0 0	0 1 1	1 1 1
9·9	·1010	1009	1008	1007	1006	1005	1004	1003	1002	1001	0 0 0	0 0 1	1 1 1

SQUARE ROOTS
[From 1 to 10]

	0	1	2	3	4	5	6	7	8	9	Mean Differences		
											1 2 3	4 5 6	7 8 9
1·0	1·000	1·005	1·010	1·015	1·020	1·025	1·030	1·034	1·039	1·044	0 1 1	2 2 3	3 4 4
1·1	1·049	1·054	1·058	1·063	1·068	1·072	1·077	1·082	1·086	1·091	0 1 1	2 2 3	3 4 4
1·2	1·095	1·100	1·105	1·109	1·114	1·118	1·122	1·127	1·131	1·136	0 1 1	2 2 3	3 4 4
1·3	1·140	1·145	1·149	1·153	1·158	1·162	1·166	1·170	1·175	1·179	0 1 1	2 2 3	3 3 4
1·4	1·183	1·187	1·192	1·196	1·200	1·204	1·208	1·212	1·217	1·221	0 1 1	2 2 2	3 3 4
1·5	1·225	1·229	1·233	1·237	1·241	1·245	1·249	1·253	1·257	1·261	0 1 1	2 2 2	3 3 4
1·6	1·265	1·269	1·273	1·277	1·281	1·285	1·288	1·292	1·296	1·300	0 1 1	2 2 2	3 3 3
1·7	1·304	1·308	1·311	1·315	1·319	1·323	1·327	1·330	1·334	1·338	0 1 1	2 2 2	3 3 3
1·8	1·342	1·345	1·349	1·353	1·356	1·360	1·364	1·367	1·371	1·375	0 1 1	1 2 2	3 3 3
1·9	1·378	1·382	1·386	1·389	1·393	1·396	1·400	1·404	1·407	1·411	0 1 1	1 2 2	3 3 3
2·0	1·414	1·418	1·421	1·425	1·428	1·432	1·435	1·439	1·442	1·446	0 1 1	1 2 2	2 3 3
2·1	1·449	1·453	1·456	1·459	1·463	1·466	1·470	1·473	1·476	1·480	0 1 1	1 2 2	2 3 3
2·2	1·483	1·487	1·490	1·493	1·497	1·500	1·503	1·507	1·510	1·513	0 1 1	1 2 2	2 3 3
2·3	1·517	1·520	1·523	1·526	1·530	1·533	1·536	1·539	1·543	1·546	0 1 1	1 2 2	2 3 3
2·4	1·549	1·552	1·556	1·559	1·562	1·565	1·568	1·572	1·575	1·578	0 1 1	1 2 2	2 3 3
2·5	1·581	1·584	1·587	1·591	1·594	1·597	1·600	1·603	1·606	1·609	0 1 1	1 2 2	2 3 3
2·6	1·612	1·616	1·619	1·622	1·625	1·628	1·631	1·634	1·637	1·640	0 1 1	1 2 2	2 2 3
2·7	1·643	1·646	1·649	1·652	1·655	1·658	1·661	1·664	1·667	1·670	0 1 1	1 2 2	2 2 3
2·8	1·673	1·676	1·679	1·682	1·685	1·688	1·691	1·694	1·697	1·700	0 1 1	1 1 2	2 2 3
2·9	1·703	1·706	1·709	1·712	1·715	1·718	1·720	1·723	1·726	1·729	0 1 1	1 1 2	2 2 3
3·0	1·732	1·735	1·738	1·741	1·744	1·746	1·749	1·752	1·755	1·758	0 1 1	1 1 2	2 2 3
3·1	1·761	1·764	1·766	1·769	1·772	1·775	1·778	1·780	1·783	1·786	0 1 1	1 1 2	2 2 3
3·2	1·789	1·792	1·794	1·797	1·800	1·803	1·806	1·808	1·811	1·814	0 1 1	1 1 2	2 2
3·3	1·817	1·819	1·822	1·825	1·828	1·830	1·833	1·836	1·838	1·841	0 1 1	1 1 2	2 2 2
3·4	1·844	1·847	1·849	1·852	1·855	1·857	1·860	1·863	1·865	1·868	0 1 1	1 1 2	2 2 2
3·5	1·871	1·873	1·876	1·879	1·881	1·884	1·887	1·889	1·892	1·895	0 1 1	1 1 2	2 2 2
3·6	1·897	1·900	1·903	1·905	1·908	1·910	1·913	1·916	1·918	1·921	0 1 1	1 1 2	2 2 2
3·7	1·924	1·926	1·929	1·931	1·934	1·936	1·939	1·942	1·944	1·947	0 1 1	1 1 2	2 2 2
3·8	1·949	1·952	1·954	1·957	1·960	1·962	1·965	1·967	1·970	1·972	0 1 1	1 1 2	2 2 2
3·9	1·975	1·977	1·980	1·982	1·985	1·987	1·990	1·992	1·995	1·997	0 1 1	1 1 2	2 2 2
4·0	2·000	2·002	2·005	2·007	2·010	2·012	2·015	2·017	2·020	2·022	0 0 1	1 1 1	2 2 2
4·1	2·025	2·027	2·030	2·032	2·035	2·037	2·040	2·042	2·045	2·047	0 0 1	1 1 1	2 2 2
4·2	2·049	2·052	2·054	2·057	2·059	2·062	2·064	2·066	2·069	2·071	0 0 1	1 1 1	2 2 2
4·3	2·074	2·076	2·078	2·081	2·083	2·086	2·088	2·090	2·093	2·095	0 0 1	1 1 1	2 2 2
4·4	2·098	2·100	2·102	2·105	2·107	2·110	2·112	2·114	2·117	2·119	0 0 1	1 1 1	2 2 2
4·5	2·121	2·124	2·126	2·128	2·131	2·133	2·135	2·138	2·140	2·142	0 0 1	1 1 1	2 2 2
4·6	2·145	2·147	2·149	2·152	2·154	2·156	2·159	2·161	2·163	2·166	0 0 1	1 1 1	2 2 2
4·7	2·168	2·170	2·173	2·175	2·177	2·179	2·182	2·184	2·186	2·189	0 0 1	1 1 1	2 2 2
4·8	2·191	2·193	2·195	2·198	2·200	2·202	2·205	2·207	2·209	2·211	0 0 1	1 1 1	2 2 2
4·9	2·214	2·216	2·218	2·220	2·223	2·225	2·227	2·229	2·232	2·234	0 0 1	1 1 1	2 2 2
5·0	2·236	2·238	2·241	2·243	2·245	2·247	2·249	2·252	2·254	2·256	0 0 1	1 1 1	2 2 2
5·1	2·258	2·261	2·263	2·265	2·267	2·269	2·272	2·274	2·276	2·278	0 0 1	1 1 1	2 2 2
5·2	2·280	2·283	2·285	2·287	2·289	2·291	2·293	2·296	2·298	2·300	0 0 1	1 1 1	2 2 2
5·3	2·302	2·304	2·307	2·309	2·311	2·313	2·315	2·317	2·319	2·322	0 0 1	1 1 1	2 2 2
5·4	2·324	2·326	2·328	2·330	2·332	2·335	2·337	2·339	2·341	2·343	0 0 1	1 1 1	1 2 2

	0	1	2	3	4	5	6	7	8	9	Mean Differences 1 2 3	4 5 6	7 8 9
5·5	2·345	2·347	2·349	2·352	2·354	2·356	2·358	2·360	2·362	2·364	0 0 1	1 1 1	1 2 2
5·6	2·366	2·369	2·371	2·373	2·375	2·377	2·379	2·381	2·383	2·385	0 0 1	1 1 1	1 2 2
5·7	2·387	2·390	2·392	2·394	2·396	2·398	2·400	2·402	2·404	2·406	0 0 1	1 1 1	1 2 2
5·8	2·408	2·410	2·412	2·415	2·417	2·419	2·421	2·423	2·425	2·427	0 0 1	1 1 1	1 2 2
5·9	2·429	2·431	2·433	2·435	2·437	2·439	2·441	2·443	2·445	2·447	0 0 1	1 1 1	1 2 2
6·0	2·449	2·452	2·454	2·456	2·458	2·460	2·462	2·464	2·466	2·468	0 0 1	1 1 1	1 2 2
6·1	2·470	2·472	2·474	2·476	2·478	2·480	2·482	2·484	2·486	2·488	0 0 1	1 1 1	1 2 2
6·2	2·490	2·492	2·494	2·496	2·498	2·500	2·502	2·504	2·506	2·508	0 0 1	1 1 1	1 2 2
6·3	2·510	2·512	2·514	2·516	2·518	2·520	2·522	2·524	2·526	2·528	0 0 1	1 1 1	1 2 2
6·4	2·530	2·532	2·534	2·536	2·538	2·540	2·542	2·544	2·546	2·548	0 0 1	1 1 1	1 2 2
6·5	2·550	2·551	2·553	2·555	2·557	2·559	2·561	2·563	2·565	2·567	0 0 1	1 1 1	1 2 2
6·6	2·569	2·571	2·573	2·575	2·577	2·579	2·581	2·583	2·585	2·587	0 0 1	1 1 1	1 2 2
6·7	2·588	2·590	2·592	2·594	2·596	2·598	2·600	2·602	2·604	2·606	0 0 1	1 1 1	1 2 2
6·8	2·608	2·610	2·612	2·613	2·615	2·617	2·619	2·621	2·623	2·625	0 0 1	1 1 1	1 2 2
6·9	2·627	2·629	2·631	2·632	2·634	2·636	2·638	2·640	2·642	2·644	0 0 1	1 1 1	1 2 2
7·0	2·646	2·648	2·650	2·651	2·653	2·655	2·657	2·659	2·661	2·663	0 0 1	1 1 1	1 2 2
7·1	2·665	2·666	2·668	2·670	2·672	2·674	2·676	2·678	2·680	2·681	0 0 1	1 1 1	1 1 2
7·2	2·683	2·685	2·687	2·689	2·691	2·693	2·694	2·696	2·698	2·700	0 0 1	1 1 1	1 1 2
7·3	2·702	2·704	2·706	2·707	2·709	2·711	2·713	2·715	2·717	2·718	0 0 1	1 1 1	1 1 2
7·4	2·720	2·722	2·724	2·726	2·728	2·729	2·731	2·733	2·735	2·737	0 0 1	1 1 1	1 1 2
7·5	2·739	2·740	2·742	2·744	2·746	2·748	2·750	2·751	2·753	2·755	0 0 1	1 1 1	1 1 2
7·6	2·757	2·759	2·760	2·762	2·764	2·766	2·768	2·769	2·771	2·773	0 0 1	1 1 1	1 1 2
7·7	2·775	2·777	2·778	2·780	2·782	2·784	2·786	2·787	2·789	2·791	0 0 1	1 1 1	1 1 2
7·8	2·793	2·795	2·796	2·798	2·800	2·802	2·804	2·805	2·807	2·809	0 0 1	1 1 1	1 1 2
7·9	2·811	2·812	2·814	2·816	2·818	2·820	2·821	2·823	2·825	2·827	0 0 1	1 1 1	1 1 2
8·0	2·828	2·830	2·832	2·834	2·835	2·837	2·839	2·841	2·843	2·844	0 0 1	1 1 1	1 1 2
8·1	2·846	2·848	2·850	2·851	2·853	2·855	2·857	2·858	2·860	2·862	0 0 1	1 1 1	1 1 2
8·2	2·864	2·865	2·867	2·869	2·871	2·872	2·874	2·876	2·877	2·879	0 0 1	1 1 1	1 1 2
8·3	2·881	2·883	2·884	2·886	2·888	2·890	2·891	2·893	2·895	2·897	0 0 1	1 1 1	1 1 2
8·4	2·898	2·900	2·902	2·903	2·905	2·907	2·909	2·910	2·912	2·914	0 0 1	1 1 1	1 1 2
8·5	2·915	2·917	2·919	2·921	2·922	2·924	2·926	2·927	2·929	2·931	0 0 1	1 1 1	1 1 2
8·6	2·933	2·934	2·936	2·938	2·939	2·941	2·943	2·944	2·946	2·948	0 0 1	1 1 1	1 1 2
8·7	2·950	2·951	2·953	2·955	2·956	2·958	2·960	2·961	2·963	2·965	0 0 1	1 1 1	1 1 2
8·8	2·966	2·968	2·970	2·972	2·973	2·975	2·977	2·978	2·980	2·982	0 0 1	1 1 1	1 1 2
8·9	2·983	2·985	2·987	2·988	2·990	2·992	2·993	2·995	2·997	2·998	0 0 1	1 1 1	1 1 2
9·0	3·000	3·002	3·003	3·005	3·007	3·008	3·010	3·012	3·013	3·015	0 0 0	1 1 1	1 1 1
9·1	3·017	3·018	3·020	3·022	3·023	3·025	3·027	3·028	3·030	3·032	0 0 0	1 1 1	1 1 1
9·2	3·033	3·035	3·036	3·038	3·040	3·041	3·043	3·045	3·046	3·048	0 0 0	1 1 1	1 1 1
9·3	3·050	3·051	3·053	3·055	3·056	3·058	3·059	3·061	3·063	3·064	0 0 0	1 1 1	1 1 1
9·4	3·066	3·068	3·069	3·071	3·072	3·074	3·076	3·077	3·079	3·081	0 0 0	1 1 1	1 1 1
9·5	3·082	3·084	3·085	3·087	3·089	3·090	3·092	3·094	3·095	3·097	0 0 0	1 1 1	1 1 1
9·6	3·098	3·100	3·102	3·103	3·105	3·106	3·108	3·110	3·111	3·113	0 0 0	1 1 1	1 1 1
9·7	3·114	3·116	3·118	3·119	3·121	3·122	3·124	3·126	3·127	3·129	0 0 0	1 1 1	1 1 1
9·8	3·130	3·132	3·134	3·135	3·137	3·138	3·140	3·142	3·143	3·145	0 0 0	1 1 1	1 1 1
9·9	3·146	3·148	3·150	3·151	3·153	3·154	3·156	3·158	3·159	3·161	0 0 0	1 1 1	1 1 1

SQUARE ROOTS
[From 10 to 100]

	0	1	2	3	4	5	6	7	8	9	Mean Differences								
											1	2	3	4	5	6	7	8	9
10	3·162	3·178	3·194	3·209	3·225	3·240	3·256	3·271	3·286	3·302	2	3	5	6	8	9	11	12	14
11	3·317	3·332	3·347	3·362	3·376	3·391	3·406	3·421	3·435	3·450	1	3	4	6	7	9	10	12	13
12	3·464	3·479	3·493	3·507	3·521	3·536	3·550	3·564	3·578	3·592	1	3	4	6	7	8	10	11	13
13	3·606	3·619	3·633	3·647	3·661	3·674	3·688	3·701	3·715	3·728	1	3	4	5	7	8	10	11	12
14	3·742	3·755	3·768	3·782	3·795	3·808	3·821	3·834	3·847	3·860	1	3	4	5	7	8	9	11	12
15	3·873	3·886	3·899	3·912	3·924	3·937	3·950	3·962	3·975	3·987	1	3	4	5	6	8	9	10	11
16	4·000	4·012	4·025	4·037	4·050	4·062	4·074	4·087	4·099	4·111	1	2	4	5	6	7	9	10	11
17	4·123	4·135	4·147	4·159	4·171	4·183	4·195	4·207	4·219	4·231	1	2	4	5	6	7	8	10	11
18	4·243	4·254	4·266	4·278	4·290	4·301	4·313	4·324	4·336	4·347	1	2	3	5	6	7	8	9	10
19	4·359	4·370	4·382	4·393	4·405	4·416	4·427	4·438	4·450	4·461	1	2	3	5	6	7	8	9	10
20	4·472	4·483	4·494	4·506	4·517	4·528	4·539	4·550	4·561	4·572	1	2	3	4	6	7	8	9	10
21	4·583	4·593	4·604	4·615	4·626	4·637	4·648	4·658	4·669	4·680	1	2	3	4	5	6	8	9	10
22	4·690	4·701	4·712	4·722	4·733	4·743	4·754	4·764	4·775	4·785	1	2	3	4	5	6	7	8	9
23	4·796	4·806	4·817	4·827	4·837	4·848	4·858	4·868	4·879	4·889	1	2	3	4	5	6	7	8	9
24	4·899	4·909	4·919	4·930	4·940	4·950	4·960	4·970	4·980	4·990	1	2	3	4	5	6	7	8	9
25	5·000	5·010	5·020	5·030	5·040	5·050	5·060	5·070	5·079	5·089	1	2	3	4	5	6	7	8	9
26	5·099	5·109	5·119	5·128	5·138	5·148	5·158	5·167	5·177	5·187	1	2	3	4	5	6	7	8	9
27	5·196	5·206	5·215	5·225	5·235	5·244	5·254	5·263	5·273	5·282	1	2	3	4	5	6	7	8	9
28	5·292	5·301	5·310	5·320	5·329	5·339	5·348	5·357	5·367	5·376	1	2	3	4	5	6	7	7	8
29	5·385	5·394	5·404	5·413	5·422	5·431	5·441	5·450	5·459	5·468	1	2	3	4	5	5	6	7	8
30	5·477	5·486	5·495	5·505	5·514	5·523	5·532	5·541	5·550	5·559	1	2	3	4	4	5	6	7	8
31	5·568	5·577	5·586	5·595	5·604	5·612	5·621	5·630	5·639	5·648	1	2	3	3	4	5	6	7	8
32	5·657	5·666	5·675	5·683	5·692	5·701	5·710	5·718	5·727	5·736	1	2	3	3	4	5	6	7	8
33	5·745	5·753	5·762	5·771	5·779	5·788	5·797	5·805	5·814	5·822	1	2	3	3	4	5	6	7	8
34	5·831	5·840	5·848	5·857	5·865	5·874	5·882	5·891	5·899	5·908	1	2	3	3	4	5	6	7	8
35	5·916	5·925	5·933	5·941	5·950	5·958	5·967	5·975	5·983	5·992	1	2	2	3	4	5	6	7	8
36	6·000	6·008	6·017	6·025	6·033	6·042	6·050	6·058	6·066	6·075	1	2	2	3	4	5	6	7	7
37	6·083	6·091	6·099	6·107	6·116	6·124	6·132	6·140	6·148	6·156	1	2	2	3	4	5	6	7	7
38	6·164	6·173	6·181	6·189	6·197	6·205	6·213	6·221	6·229	6·237	1	2	2	3	4	5	6	6	7
39	6·245	6·253	6·261	6·269	6·277	6·285	6·293	6·301	6·309	6·317	1	2	2	3	4	5	6	6	7
40	6·325	6·332	6·340	6·348	6·356	6·364	6·372	6·380	6·387	6·395	1	2	2	3	4	5	6	6	7
41	6·403	6·411	6·419	6·427	6·434	6·442	6·450	6·458	6·465	6·473	1	2	2	3	4	5	5	6	7
42	6·481	6·488	6·496	6·504	6·512	6·519	6·527	6·535	6·542	6·550	1	2	2	3	4	5	5	6	7
43	6·557	6·565	6·573	6·580	6·588	6·595	6·603	6·611	6·618	6·626	1	2	2	3	4	5	5	6	7
44	6·633	6·641	6·648	6·656	6·663	6·671	6·678	6·686	6·693	6·701	1	2	2	3	4	5	5	6	7
45	6·708	6·716	6·723	6·731	6·738	6·745	6·753	6·760	6·768	6·775	1	1	2	3	4	4	5	6	7
46	6·782	6·790	6·797	6·804	6·812	6·819	6·826	6·834	6·841	6·848	1	1	2	3	4	4	5	6	7
47	6·856	6·863	6·870	6·877	6·885	6·892	6·899	6·907	6·914	6·921	1	1	2	3	4	4	5	6	7
48	6·928	6·935	6·943	6·950	6·957	6·964	6·971	6·979	6·986	6·993	1	1	2	3	4	4	5	6	6
49	7·000	7·007	7·014	7·021	7·029	7·036	7·043	7·050	7·057	7·064	1	1	2	3	4	4	5	6	6
50	7·071	7·078	7·085	7·092	7·099	7·106	7·113	7·120	7·127	7·134	1	1	2	3	4	4	5	6	6
51	7·141	7·148	7·155	7·162	7·169	7·176	7·183	7·190	7·197	7·204	1	1	2	3	4	4	5	6	6
52	7·211	7·218	7·225	7·232	7·239	7·246	7·253	7·259	7·266	7·273	1	1	2	3	3	4	5	6	6
53	7·280	7·287	7·294	7·301	7·308	7·314	7·321	7·328	7·335	7·342	1	1	2	3	3	4	5	5	6
54	7·348	7·355	7·362	7·369	7·376	7·382	7·389	7·396	7·403	7·409	1	1	2	3	3	4	5	5	6

SQUARE ROOTS
[From 10 to 100]

	0	1	2	3	4	5	6	7	8	9	Mean Differences 1 2 3	4 5 6	7 8 9
55	7·416	7·423	7·430	7·436	7·443	7·450	7·457	7·463	7·470	7·477	1 1 2	3 3 4	5 5 6
56	7·483	7·490	7·497	7·503	7·510	7·517	7·523	7·530	7·537	7·543	1 1 2	3 3 4	5 5 6
57	7·550	7·556	7·563	7·570	7·576	7·583	7·589	7·596	7·603	7·609	1 1 2	3 3 4	5 5 6
58	7·616	7·622	7·629	7·635	7·642	7·649	7·655	7·662	7·668	7·675	1 1 2	3 3 4	5 5 6
59	7·681	7·688	7·694	7·701	7·707	7·714	7·720	7·727	7·733	7·740	1 1 2	3 3 4	4 5 6
60	7·746	7·752	7·759	7·765	7·772	7·778	7·785	7·791	7·797	7·804	1 1 2	3 3 4	4 5 6
61	7·810	7·817	7·823	7·829	7·836	7·842	7·849	7·855	7·861	7·868	1 1 2	3 3 4	4 5 6
62	7·874	7·880	7·887	7·893	7·899	7·906	7·912	7·918	7·925	7·931	1 1 2	3 3 4	4 5 6
63	7·937	7·944	7·950	7·956	7·962	7·969	7·975	7·981	7·987	7·994	1 1 2	3 3 4	4 5 6
64	8·000	8·006	8·012	8·019	8·025	8·031	8·037	8·044	8·050	8·056	1 1 2	2 3 4	4 5 6
65	8·062	8·068	8·075	8·081	8·087	8·093	8·099	8·106	8·112	8·118	1 1 2	2 3 4	4 5 6
66	8·124	8·130	8·136	8·142	8·149	8·155	8·161	8·167	8·173	8·179	1 1 2	2 3 4	4 5 5
67	8·185	8·191	8·198	8·204	8·210	8·216	8·222	8·228	8·234	8·240	1 1 2	2 3 4	4 5 5
68	8·246	8·252	8·258	8·264	8·270	8·276	8·283	8·289	8·295	8·301	1 1 2	2 3 4	4 5 5
69	8·307	8·313	8·319	8·325	8·331	8·337	8·343	8·349	8·355	8·361	1 1 2	2 3 4	4 5 5
70	8·367	8·373	8·379	8·385	8·390	8·396	8·402	8·408	8·414	8·420	1 1 2	2 3 4	4 5 5
71	8·426	8·432	8·438	8·444	8·450	8·456	8·462	8·468	8·473	8·479	1 1 2	2 3 4	4 5 5
72	8·485	8·491	8·497	8·503	8·509	8·515	8·521	8·526	8·532	8·538	1 1 2	2 3 3	4 5 5
73	8·544	[illegible]	8·556	8·562	8·567	8·573	8·579	8·585	8·591	8·597	1 1 2	2 3 3	4 5 5
74	8·602	[illegible]	8·614	8·620	8·626	8·631	8·637	8·643	8·649	8·654	1 1 2	2 3 3	4 5 5
75	8·660	[illegible]	8·672	8·678	8·683	8·689	8·695	8·701	8·706	8·712	1 1 2	2 3 3	4 5 5
76	8·718	[illegible]	8·729	8·735	8·741	8·746	8·752	8·758	8·764	8·769	1 1 2	2 3 3	4 5 5
77	8·775	[illegible]	8·786	8·792	8·798	8·803	8·809	8·815	8·820	8·826	1 1 2	2 3 3	4 4 5
78	8·832	[illegible]	8·843	8·849	8·854	8·860	8·866	8·871	8·877	8·883	1 1 2	2 3 3	4 4 5
79	8·888	[illegible]	8·899	8·905	8·911	8·916	8·922	8·927	8·933	8·939	1 1 2	2 3 3	4 4 5
80	8·944	8·950	8·955	8·961	8·967	8·972	8·978	8·983	8·989	8·994	1 1 2	2 3 3	4 4 5
81	9·000	9·006	[illegible]	9·017	9·023	9·028	9·033	9·039	9·044	9·050	1 1 2	2 3 3	4 4 5
82	9·055	9·061	9·066	9·072	9·077	9·083	9·088	9·094	9·099	9·105	1 1 2	2 3 3	4 4 5
83	9·110	9·116	9·121	9·127	9·132	9·138	9·143	9·149	9·154	9·160	1 1 2	2 3 3	4 4 5
84	9·165	9·171	9·176	9·182	9·187	9·192	9·198	9·203	9·209	9·214	1 1 2	2 3 3	4 4 5
85	9·220	9·225	9·230	9·236	9·241	9·247	9·252	9·257	9·263	9·268	1 1 2	2 3 3	4 4 5
86	9·274	9·279	9·284	9·290	9·295	9·301	9·306	9·311	9·317	9·322	1 1 2	2 3 3	4 4 5
87	9·327	9·333	9·338	9·343	9·349	9·354	9·359	9·365	9·370	9·375	1 1 2	2 3 3	4 4 5
88	9·381	9·386	9·391	9·397	9·402	9·407	9·413	9·418	9·423	9·429	1 1 2	2 3 3	4 4 5
89	9·434	9·439	9·445	9·450	9·455	9·460	9·466	9·471	9·476	9·482	1 1 2	2 3 3	4 4 5
90	9·487	9·492	9·497	9·503	9·508	9·513	9·518	9·524	9·529	9·534	1 1 2	2 3 3	4 4 5
91	9·539	9·545	9·550	9·555	9·560	9·566	9·571	9·576	9·581	9·586	1 1 2	2 3 3	4 4 5
92	9·592	9·597	9·602	9·607	9·612	9·618	9·623	9·628	9·633	9·638	1 1 2	2 3 3	4 4 5
93	9·644	9·649	9·654	9·659	9·664	9·670	9·675	9·680	9·685	9·690	1 1 2	2 3 3	4 4 5
94	9·695	9·701	9·706	9·711	9·716	9·721	9·726	9·731	9·737	9·742	1 1 2	2 3 3	4 4 5
95	9·747	9·752	9·757	9·762	9·767	9·772	9·778	9·783	9·788	9·793	1 1 2	2 3 3	4 4 5
96	9·798	9·803	9·808	9·813	9·818	9·823	9·829	9·834	9·839	9·844	1 1 2	2 3 3	4 4 5
97	9·849	9·854	9·859	9·864	9·869	9·874	9·879	9·884	9·889	9·894	1 1 1	2 3 3	4 4 5
98	9·899	9·905	9·910	9·015	9·920	9·925	9·930	9·935	9·940	9·945	0 1 1	2 2 3	3 4 4
99	9·950	9·955	9·960	9·965	9·970	9·975	9·980	9·985	9·990	9·995	0 1 1	2 2 3	3 4 4

AREAS UNDER THE NORMAL CURVE

The table gives area from mean (0) to distance x/σ from mean (shaded area) expressed as decimal fractions of the total area 1·0000.

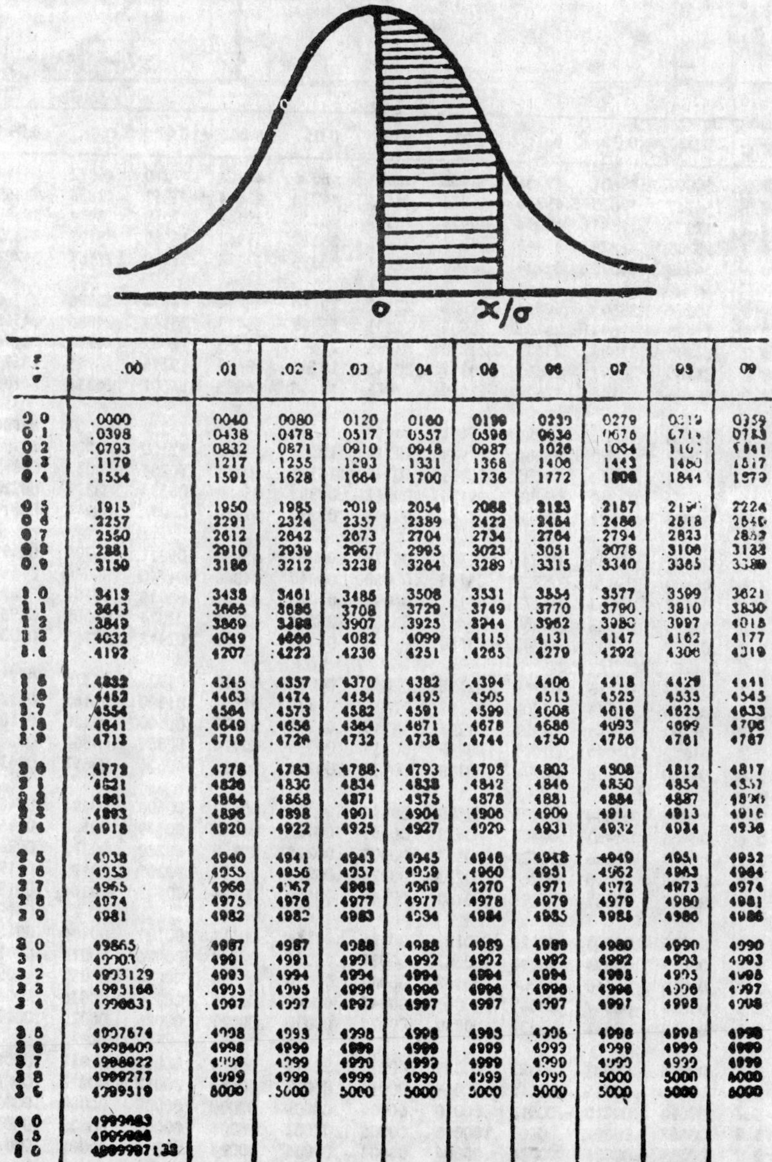

x/σ	.00	.01	.02	.03	.04	.05	.06	.07	.08	.09
0.0	.0000	0040	.0080	0120	0160	0199	.0239	0279	0319	0359
0.1	.0398	.0438	.0478	.0517	0557	0596	0636	0675	0714	0753
0.2	.0793	.0832	.0871	.0910	.0948	0987	1026	1064	1103	1141
0.3	.1179	.1217	.1255	.1293	.1331	1368	1406	1443	1480	1517
0.4	.1554	.1591	.1628	.1664	.1700	.1736	.1772	1808	.1844	1879
0.5	.1915	.1950	.1985	2019	.2054	.2088	.2123	.2157	2190	2224
0.6	.2257	.2291	.2324	2357	.2389	2422	.2454	.2486	2518	2549
0.7	2580	.2612	.2642	.2673	.2704	.2734	.2764	.2794	2823	2852
0.8	.2881	.2910	.2939	.2967	.2995	.3023	3051	3078	.3106	3133
0.9	.3159	.3186	.3212	.3238	.3264	.3289	.3315	.3340	3365	3389
1.0	.3413	.3438	3461	.3485	.3508	.3531	3554	.3577	3599	3621
1.1	.3643	.3665	.3686	.3708	.3729	3749	.3770	3790	.3810	3830
1.2	.3849	.3869	3888	.3907	.3925	.3944	.3962	.3980	3997	4015
1.3	.4032	4049	.4066	4082	.4099	.4115	4131	.4147	4162	4177
1.4	.4192	4207	.4222	.4236	.4251	.4265	.4279	4292	4306	4319
1.5	.4332	.4345	.4357	.4370	4382	.4394	4406	4418	4429	4441
1.6	.4452	4463	4474	4484	4495	4505	4515	4525	4535	.4545
1.7	.4554	.4564	.4573	4582	.4591	4599	4608	4616	.4625	4633
1.8	.4641	.4649	.4656	4664	.4671	.4678	4686	4693	4699	4706
1.9	.4713	.4719	.4726	4732	.4738	.4744	.4750	.4756	4761	4767
2.0	.4772	.4778	.4783	.4788	.4793	.4798	4803	4808	4812	4817
2.1	.4821	.4826	4830	4834	.4838	.4842	.4846	4850	4854	4857
2.2	.4861	.4864	4868	.4871	.4875	4878	4881	.4884	4887	4890
2.3	.4893	.4896	.4898	4901	.4904	.4906	.4909	4911	4913	4916
2.4	.4918	.4920	.4922	.4925	.4927	.4929	.4931	.4932	4934	4936
2.5	.4938	.4940	.4941	4943	4945	4946	.4948	4949	4951	4952
2.6	.4953	.4955	.4956	4957	4959	.4960	.4961	.4962	4963	4964
2.7	4965	4966	.4967	4968	4969	4970	4971	4972	4973	4974
2.8	4974	.4975	4976	.4977	4977	.4978	4979	4979	4980	4981
2.9	4981	.4982	.4982	4983	4984	4984	.4985	.4985	.4986	4986
3.0	49865	4987	4987	4988	.4988	4989	4989	4989	4990	4990
3.1	49903	4991	.4991	4991	4992	4992	.4992	4992	4993	4993
3.2	4993129	4993	4994	4994	4994	4994	.4994	4995	4995	4995
3.3	4995166	.4995	4995	4996	4996	4996	.4996	4996	4996	4997
3.4	4996631	.4997	4997	.4997	4997	4997	4997	4997	4998	4998
3.5	4997674	.4998	4998	4998	4998	4998	4998	4998	4998	4998
3.6	4998409	4998	4999	4999	4999	.4999	4999	4999	4999	4999
3.7	4998922	4999	4999	4999	4999	4999	4999	4999	4999	4999
3.8	4999277	.4999	4999	4999	4999	.4999	4999	5000	5000	5000
3.9	4999519	5000	.5000	5000	5000	5000	.5000	5000	5000	5000
4.0	4999683									
4.5	4999966									
5.0	4999997133									

AREAS UNDER THE NORMAL CURVE

The table gives the shaded area.

x/σ	0.00	0.01	0.02	0.03	0.04	0.05	0.06	0.07	0.08	0.09
0.0	.50000	49601	49202	48803	48405	48006	47608	47210	46812	46414
0.1	.46017	45620	.45224	.44828	.44433	44038	43644	43251	42858	42465
0.2	42074	.41683	41294	.40905	40517	40129	39743	39358	38974	38591
0.3	.38209	37828	37448	.37070	.36693	.36317	.35942	35569	35197	34827
0.4	34458	.34090	33724	.33360	32997	32636	32276	31918	31561	31207
0.5	.30854	30503	.30153	29806	.29460	29116	28774	28434	28096	27760
0.6	.27425	27093	.26763	.26435	.26109	.25785	.25463	.25143	24825	24510
0.7	24196	.23885	.23576	.23270	.22965	.22663	22363	.22065	21770	.21476
0.8	.21186	.20897	20611	.20327	.20045	.19766	19489	.19215	18943	.18673
0.9	.18406	.18141	.17879	.17619	.17361	17106	.16853	.16602	16354	.16109
1.0	15866	15625	.15386	15151	.14917	14686	.14457	14231	.14007	.13786
1.1	.13567	.13350	.13136	.12924	.12714	12507	.12302	12100	11900	11702
1.2	.11507	.11314	.11123	.10935	10749	.10565	.10383	.10204	.10027	.09853
1.3	.09680	.09510	.09342	.09176	.09012	.08851	.08691	.08534	.08379	.08226
1.4	.08076	.07927	.07780	.07636	.07493	.07353	.07215	.07078	06944	.06811
1.5	.06681	.06552	.06426	.06301	.06178	.06057	.05938	.05821	05705	05592
1.6	.05480	.05370	.05262	.05155	05050	.04947	.04846	.04746	.04648	04551
1.7	.04457	.04363	.04272	.04182	.04093	.04006	.03920	.03836	.03754	03673
1.8	.03593	.03515	.03438	.03362	.03288	.03216	.03144	.03074	.03005	.02938
1.9	.02872	.02807	.02743	.02680	.02619	.02559	.02500	.02442	.02385	.02330
2.0	.02275	.02216	.02169	.02118	.02068	.02018	.01970	.01923	.01876	01831
2.1	.01786	.01743	.01700	.01659	.01618	.01578	.01539	.01500	.01463	.01426
2.2	.01390	.01355	.01321	.01287	.01255	.01222	.01191	.01160	.01130	.01101
2.3	.01072	.01044	.01017	.00990	.00964	.00939	.00914	.00889	.00866	.00842
2.4	.00820	.00798	.00776	.00755	.00734	.00714	.00695	.00676	.00657	00639
2.5	.00621	.00604	.00587	.00570	.00554	.00539	.00523	00508	00494	.00480
2.6	.00466	.00453	.00440	.00427	00415	.00402	.00391	.00379	.00368	.00357
2.7	.00347	00336	.00326	.00317	.00307	.00298	.00289	00280	00272	.00264
2.8	.00256	00248	.00240	.00233	.00226	.00219	.00212	.00205	.00199	.00193
2.9	.00187	.00181	.00175	.00169	.00164	00159	.00154	.00149	.00144	.00139
3.0	.00135	.00131	.00126	.00122	00118	.00114	.00111	00107	.00104	.00100
3.1	.00097	.00094	.00090	.00087	.00084	.00082	.00079	.00076	.00074	.00071
3.2	.00069	.00066	00064	.00062	.00060	.00058	.00056	.00054	.00052	00050
3.3	00048	00047	.00045	.00043	.00042	.00040	.00039	.00038	00036	.00035
3.4	00034	00032	.00031	00030	.00029	.00028	.00027	.00026	.00025	.00024
3.5	.00023	00022	.00022	.00021	00020	.00019	.00019	.00018	.00017	.00017
3.6	00016	.00015	.00015	00014	.00014	.00013	.00013	.00012	.00012	.00011
3.7	.00011	.00010	.00010	.00010	.00009	.00009	.00008	.00008	.00008	.00008
3.8	.00007	.00007	.00007	.00006	.00006	.00006	.00006	00005	.00005	00005
3.9	00005	.00005	.00004	.00004	.00004	.00004	00004	.00004	.00003	00003

ORDINATES OF THE NORMAL CURVE

The table gives ordinates (y) erected at distance x from the mean for a standard normal curve, i.e., $y=(1/\sqrt{2\pi})\,e^{-x^2/2}$

z	00	01	02	03	04	05	06	07	08	09
.0	3989	3989	3989	3988	3986	3984	3982	3980	3977	3973
.1	3970	3965	3961	3956	3951	3945	3939	3932	3925	3918
.2	3910	3902	3894	3885	3876	3867	3857	3847	3836	3825
.3	3814	3802	3790	3778	3765	3752	3739	3725	3712	3697
.4	3683	3668	3653	3637	3621	3605	3589	3572	3555	3538
.5	3521	3503	3485	3467	3448	3429	3410	3391	3372	3352
.6	3332	3312	3292	3271	3251	3230	3209	3187	3166	3144
.7	3123	3101	3079	3056	3034	3011	2989	2966	2943	2920
.8	2897	2874	2850	2827	2803	2780	2756	2732	2709	2685
.9	2661	2637	2613	2589	2565	2541	2516	2492	2468	2444
1.0	2420	2396	2371	2347	2323	2299	2275	2251	2227	2203
1.1	2179	2155	2131	2107	2083	2059	2036	2012	1989	1965
1.2	1942	1919	1895	1872	1849	1826	1804	1781	1758	1736
1.3	1714	1691	1669	1647	1626	1604	1582	1561	1539	1518
1.4	1497	1476	1456	1435	1415	1394	1374	1354	1334	1315
1.5	1295	1276	1257	1238	1219	1200	1182	1163	1145	1127
1.6	1109	1092	1074	1057	1040	1023	1006	0989	0973	0957
1.7	0940	0925	0909	0893	0878	0863	0848	0833	0818	0804
1.8	0790	0775	0761	0748	0734	0721	0707	0694	0681	0669
1.9	0656	0644	0632	0620	0608	0596	0584	0573	0562	0551
2.0	0540	0529	0519	0508	0498	0488	0478	0468	0459	0449
2.1	0440	0431	0422	0413	0404	0396	0387	0379	0371	0363
2.2	0355	0347	0339	0332	0325	0317	0310	0303	0297	0290
2.3	0283	0277	0270	0264	0258	0252	0246	0241	0235	0229
2.4	0224	0219	0213	0208	0203	0198	0194	0189	0184	0180
2.5	0175	0171	0167	0163	0158	0154	0151	0147	0143	0139
2.6	0136	0132	0129	0126	0122	0119	0116	0113	0110	0107
2.7	0104	0101	0099	0096	0093	0091	0088	0086	0084	0081
2.8	0079	0077	0075	0073	0071	0069	0067	0065	0063	0061
2.9	0060	0058	0056	0055	0053	0051	0050	0048	0047	0046
3.0	0044	0043	0042	0040	0039	0038	0037	0036	0035	0034
3.1	0033	0032	0031	0030	0029	0028	0027	0026	0025	0025
3.2	0024	0023	0022	0022	0021	0020	0020	0019	0018	0018
3.3	0017	0017	0016	0016	0015	0015	0014	0014	0013	0013
3.4	0012	0012	0012	0011	0011	0010	0010	0010	0009	0009
3.5	0009	0008	0008	0008	0008	0007	0007	0007	0007	0006
3.6	0006	0006	0006	0005	0005	0005	0005	0005	0005	0004
3.7	0004	0004	0004	0004	0004	0004	0003	0003	0003	0003
3.8	0003	0003	0003	0003	0003	0002	0002	0002	0002	0002
3.9	0002	0002	0002	0002	0002	0002	0002	0002	0001	0001

ORDINATES OF THE NORMAL CURVE

The table gives ordinates erected at a distance x/σ from the mean expressed as a decimal fraction of maximum ordinate YO, i.e., ordinate at mean ($Yo = Ni/\sqrt{2\pi\sigma}$.)

x/σ	00	.01	.02	.03	.04	05	06	07	08	.09
0.0	.00000	.99995	.99980	.99955	.99920	.99875	.99820	99755	99685	99598
0.1	.99501	.99396	.99283	.99158	.99025	.98881	.98728	98565	98393	98211
0.2	.98020	.97819	.97609	.97390	.97161	.96923	.96676	96420	96156	95882
0.3	.95600	.95309	.95010	.94702	.94387	.94055	.93723	93382	93021	92677
0.4	.92312	.91939	.91558	.91169	.90774	.90371	.89961	89543	89119	[illegible]
0.5	.88250	.87805	.87253	.86896	.86482	.85962	.85488	.85006	.84519	84060
0.6	.83527	.83023	.82514	.82010	.81461	.80957	.80429	.79896	.79359	.78817
0.7	.78270	.77731	.77167	.76610	.76048	.75484	.74916	.74342	73769	73193
0.8	.72615	.72033	.71448	.70861	.70272	.69681	.69087	.68493	.67896	67298
0.9	.66689	.66097	.65494	.64801	.64287	.63683	.63077	.62472	61865	61259
1.0	.60653	.60047	.59440	.58834	.58228	.57623	.57017	.56414	.55810	55209
1.1	.54607	.54007	.53409	.52812	.52214	.51620	.51027	.50437	.49848	.49260
1.2	.48673	.48092	.47511	.46933	.46387	.45783	.45212	.44644	.44078	.43516
1.3	.42956	.42399	.41845	41294	.40747	.40202	.39661	.39123	.38569	38058
1.4	.37531	.37007	.36487	.35971	.35459	.34950	.34445	.33944	.33447	32954
1.5	.32465	.31980	.31500	.31023	.30580	.30082	.29618	29158	.28702	28251
1.6	.27804	.27361	.26923	.26489	.25059	.25634	.25213	.24797	.24385	23978
1.7	.23673	.23176	.22782	.22392	.22008	.21627	.21251	.20679	20511	20148
1.8	.19790	.19436	.19036	.18741	.18400	.18064	.17732	.17404	.17081	16762
1.9	.16448	.16137	.15831	.15530	.15232	.14939	.14650	.14364	.14083	[illegible]
2.0	.13534	.13265	.13000	.12740	.12483	.12230	.11981	.11737	.11496	[illegible]
2.1	.11025	.10795	.10570	.10347	.10129	.09914	.09702	.09495	.09290	[illegible]
2.2	.08892	.08698	.08507	.08320	.08136	.07956	.07778	.07604	.07433	07265
2.3	.07100	.06939	.06780	.06624	.06471	.06321	.06174	.06029	05888	05730
2.4	.05314	.05481	.05350	.05222	.05096	.04973	.04852	.04734	.04618	04505
2.5	.04394	.04285	.04179	.04074	.03972	.03873	03775	.03680	03586	03494
2.6	.03405	.03317	.03232	.03148	.03066	.02986	.02908	.02831	.02757	02681
2.7	.02612	.02542	.02474	.02408	.02343	.02280	.02218	.02157	.02098	02040
2.8	.01984	.01929	.01876	.01823	.01772	.01723	.01674	.01627	01581	01536
2.9	.01492	.01449	.01408	.01367	.01328	.01288	.01252	.01215	.01179	01145
3.0	.01111	.01078	.01046	.01015	.00984	.00955	.00926	.00898	.00871	00845
3.1	.00819	.00794	.00769	.00746	.00723	.00700	.00679	00658	00637	00617
3.2	.00598	00579	.00560	.00543	.00525	.00509	.00492	00477	00461	.00448
3.3	.00432	.00418	.00404	.00391	.00373	.00366	.00354	00342	.00331	00320
3.4	.00309	.00299	.00289	.00279	.00269	.00260	00251	.00243	.00235	00227
3.5	.00219	.00211	.00204	.00197	.00190	.00183	.00177	.00171	00165	00159
3.6	00153	.00148	.00143	.00138	.00133	.00128	00123	.00119	.00115	00110
3.7	00106	00103	.00099	.00095	.00092	.00088	.00085	00082	.00079	00076
3.8	00073	00070	.00068	.00065	.00063	.00060	.00058	00056	.00054	00057
3.9	00050	00048	.00046	.00044	.00043	.00041	00039	00038	00036	00035
4.0	00034									
4.5	00004									
5.0	00000									

VALUES OF STANDARD NORMAL VARIATE

(Explanation on Page xxiv)

P	·000	·001	·002	·003	·004	·005	·006	·007	·008	·009
·00	∞	3·0902	2·8782	2·7478	2·6521	2·5758	2·5121	2·4573	2·4089	2·3656
·01	2·3263	2·2904	2·2571	2·2262	2·1973	2·1701	2·1444	2·1201	2·0969	2·0749
·02	2·0537	2·0335	2·0141	1·9954	1·9774	1·9600	1·9431	1·9268	1·9110	1·8957
·03	1·8808	1·8663	1·8522	1·8384	1·8250	1·8119	1·7991	1·7866	1·7744	1·7624
·04	1·7507	1·7392	1·7279	1·7169	1·7060	1·6954	1·6849	1·6747	1·6646	1·6546
·05	1·6449	1·6352	1·6258	1·6164	1·6072	1·5982	1·5893	1·5805	1·5718	1·5632
·06	1·5548	1·5464	1·5382	1·5301	1·5220	1·5141	1·5063	1·4985	1·4909	1·4833
·07	1·4758	1·4684	1·4611	1·4538	1·4466	1·4395	1·4325	1·4255	1·4187	1·4118
·08	1·4051	1·3984	1·3917	1·3852	1·3787	1·3722	1·3658	1·3595	1·3532	1·3469
·09	1·3408	1·3346	1·3285	1·3225	1·3165	1·3106	1·3047	1·2988	1·2930	1·2873
·10	1·2816	1·2759	1·2702	1·2646	1·2591	1·2536	1·2481	1·2426	1·2372	1·2319
·11	1·2265	1·2212	1·2160	1·2107	1·2055	1·2004	1·1952	1·1901	1·1850	1·1800
·12	1·1750	1·1700	1·1650	1·1601	1·1552	1·1503	1·1455	1·1407	1·1359	1·1311
·13	1·1264	1·1217	1·1170	1·1123	1·1077	1·1031	1·0985	1·0939	1·0893	1·0848
·14	1·0803	1·0758	1·0714	1·0669	1·0625	1·0581	1·0537	1·0494	1·0450	1·0407
·15	1·0364	1·0322	1·0279	1·0237	1·0194	1·0152	1·0110	1·0069	1·0027	0·9986
·16	0·9945	0·9904	0·9863	0·9822	0·9782	0·9741	0·9701	0·9661	0·9621	0·9581
·17	0·9542	0·9502	0·9463	0·9424	0·9385	0·9346	0·9307	0·9269	0·9230	0·9192
·18	0·9154	0·9116	0·9078	0·9040	0·9002	0·8965	0·8927	0·8890	0·8853	0·8816
·19	0·8779	0·8742	0·8705	0·8669	0·8633	0·8596	0·8560	0·8524	0·8488	0·8452
·20	0·8416	0·8381	0·8345	0·8310	0·8274	0·8239	0·8204	0·8169	0·8134	0·8099
·21	0·8064	0·8030	0·7995	0·7961	0·7926	0·7892	0·7858	0·7824	0·7790	0·7756
·22	0·7722	0·7688	0·7655	0·7621	0·7588	0·7554	0·7521	0·7488	0·7454	0·7421
·23	0·7388	0·7356	0·7323	0·7290	0·7257	0·7225	0·7192	0·7160	0·7128	0·7095
·24	0·7063	0·7031	0·6999	0·6967	0·6935	0·6903	0·6871	0·6840	0·6808	0·6776
·25	0·6745	0·6713	0·6682	0·6651	0·6620	0·6588	0·6557	0·6526	0·6495	0·6464
·26	0·6433	0·6403	0·6372	0·6341	0·6311	0·6280	0·6250	0·6219	0·6189	0·6158
·27	0·6128	0·6098	0·6068	0·6038	0·6008	0·5978	0·5948	0·5918	0·5888	0·5858
·28	0·5828	0·5799	0·5769	0·5740	0·5710	0·5681	0·5651	0·5622	0·5592	0·5563
·29	0·5534	0·5505	0·5476	0·5446	0·5417	0·5388	0·5359	0·5330	0·5302	0·5273
·30	0·5244	0·5215	0·5187	0·5158	0·5129	0·5101	0·5072	0·5044	0·5015	0·4987
·31	0·4959	0·4930	0·4902	0·4874	0·4845	0·4817	0·4789	0·4761	0·4733	0·4705
·32	0·4677	0·4649	0·4621	0·4593	0·4565	0·4538	0·4510	0·4482	0·4454	0·4427
·33	0·4399	0·4372	0·4344	0·4316	0·4289	0·4261	0·4234	0·4207	0·4179	0·4152
·34	0·4125	0·4097	0·4070	0·4043	0·4016	0·3989	0·3961	0·3934	0·3907	0·3880
·35	0·3853	0·3826	0·3799	0·3772	0·3745	0·3719	0·3692	0·3665	0·3638	0·3611
·36	0·3585	0·3558	0·3531	0·3505	0·3478	0·3451	0·3425	0·3398	0·3372	0·3345
·37	0·3319	0·3292	0·3266	0·3239	0·3213	0·3186	0·3160	0·3134	0·3107	0·3081
·38	0·3055	0·3029	0·3002	0·2976	0·2950	0·2924	0·2898	0·2871	0·2845	0·2819
·39	0·2793	0·2767	0·2741	0·2715	0·2689	0·2663	0·2637	0·2611	0·2585	0·2559
·40	0·2533	0·2508	0·2482	0·2456	0·2430	0·2404	0·2378	0·2353	0·2327	0·2301
·41	0·2275	0·2250	0·2224	0·2198	0·2173	0·2147	0·2121	0·2096	0·2070	0·2045
·42	0·2019	0·1993	0·1968	0·1942	0·1917	0·1891	0·1866	0·1840	0·1815	0·1789
·43	0·1764	0·1738	0·1713	0·1687	0·1662	0·1637	0·1611	0·1586	0·1560	0·1535
·44	0·1510	0·1484	0·1459	0·1434	0·1408	0·1383	0·1358	0·1332	0·1307	0·1282
·45	0·1257	0·1231	0·1206	0·1181	0·1156	0·1130	0·1105	0·1080	0·1055	0·1030
·46	0·1004	0·0979	0·0954	0·0929	0·0904	0·0878	0·0853	0·0828	0·0803	0·0778
·47	0·0753	0·0728	0·0702	0·0677	0·0652	0·0627	0·0602	0·0577	0·0552	0·0527
·48	0·0502	0·0476	0·0451	0·0426	0·0401	0·0376	0·0351	0·0326	0·0301	0·0276
·49	0·0251	0·0226	0·0201	0·0175	0·0150	0·0125	0·0100	0·0075	0·0050	0·0025

VALUES OF '*t*'

The table gives points of *t*-distribution corresponding to degrees of freedom *y* and the upper tail area Q (suitable for use *n* one tail test).
For two tail test, area indicated by 2 Q should be used.

•	Q = 0·4 2Q = 0·8	0·25 0·5	0·1 0·2	0·05 0·1	0·025 0·05	0·01 0·02	0·005 0·01	0·0025 0·005	0·001 0·002	0·0005 0·001
1	0·325	1·000	3·078	6·314	12·706	31·821	63·657	127·32	318·31	636·62
2	·289	0·816	1·886	2·920	4·303	6·965	9·925	14·089	22·326	31·598
3	·277	·765	1·638	2·353	3·182	4·541	5·841	7·453	10·218	12·924
4	·271	·741	1·533	2·132	2·776	3·747	4·604	5·598	7·173	8·610
5	0·267	0·727	1·476	2·015	2·571	3·365	4·032	4·773	5·893	6·869
6	·265	·718	1·440	1·943	2·447	3·143	3·707	4·317	5·208	5·959
7	·263	·711	1·415	1·895	2·365	2·998	3·499	4·029	4·785	5·408
8	·262	·706	1·397	1·860	2·306	2·896	3·355	3·833	4·501	5·041
9	·261	·703	1·383	1·833	2·262	2·821	3·250	3·690	4·297	4·781
10	0·260	0·700	1·372	1·812	2·228	2·764	3·169	3·581	4·144	4·587
11	·260	·697	1·363	1·796	2·201	2·718	3·106	3·497	4·025	4·437
12	·259	·695	1·356	1·782	2·179	2·681	3·055	3·428	3·930	4·318
13	·259	·694	1·350	1·771	2·160	2·650	3·012	3·372	3·852	4·221
14	·258	·692	1·345	1·761	2·145	2·624	2·977	3·326	3·787	4·140
15	0·258	0·691	1·341	1·753	2·131	2·602	2·947	3·286	3·733	4·073
16	·258	·690	1·337	1·746	2·120	2·583	2·921	3·252	3·686	4·015
17	·257	·689	1·333	1·740	2·110	2·567	2·898	3·222	3·646	3·965
18	·257	·688	1·330	1·734	2·101	2·552	2·878	3·197	3·610	3·922
19	·257	·688	1·328	1·729	2·093	2·539	2·861	3·174	3·579	3·883
20	0·257	0·687	1·325	1·725	2·086	2·528	2·845	3·153	3·552	3·850
21	·257	·686	1·323	1·721	2·080	2·518	2·831	3·135	3·527	3·819
22	·256	·686	1·321	1·717	2·074	2·508	2·819	3·119	3·505	3·792
23	·256	·685	1·319	1·714	2·069	2·500	2·807	3·104	3·485	3·767
24	·256	·685	1·318	1·711	2·064	2·492	2·797	3·091	3·467	3·745
25	0·256	0·684	1·316	1·708	2·060	2·485	2·787	3·078	3·450	3·725
26	·256	·684	1·315	1·706	2·056	2·479	2·779	3·067	3·435	3·707
27	·256	·684	1·314	1·703	2·052	2·473	2·771	3·057	3·421	3·690
28	·256	·683	1·313	1·701	2·048	2·467	2·763	3·047	3·408	[illegible]
29	·256	·683	1·311	1·699	2·045	2·462	2·756	3·038	3·396	3·659
30	0·256	0·683	1·310	1·697	2·042	2·457	2·750	3·030	3·385	3·646
40	·255	·681	1·303	1·684	2·021	2·423	2·704	2·971	3·307	3·551
60	·254	·679	1·296	1·671	2·000	2·390	2·660	2·915	3·232	3·460
120	·254	·677	1·289	1·658	1·980	2·358	2·617	2·860	3·160	3·373
∞	·253	·674	1·282	1·645	1·960	2·326	2·576	2·807	3·090	3·291

Explanation of the table on left hand page, i.e., Value of Standard Normal Variate.

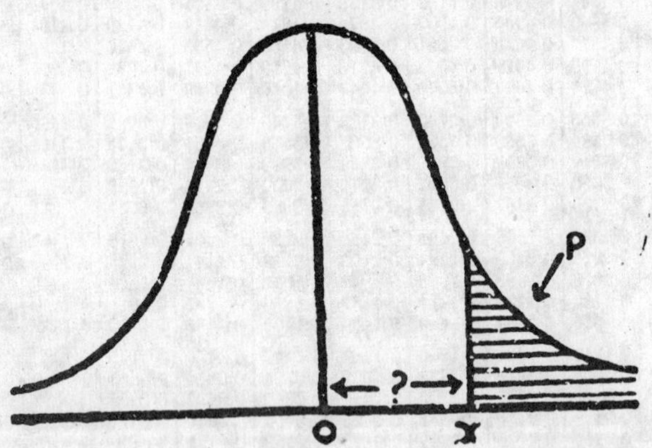

The table gives value of Standard Normal Variate *x* (marked by "?" in the diagram), corresponding to different values of area shown as *P* in the above diagram.

Upper 5% Point of F-distribution

ν_2 \ ν_1	1	2	3	4	5	6	7	8	9	10	12	15	20	24	30	40	60	120	∞
1	4052	4999·5	5403	5625	5764	5859	5928	5982	6022	6056	6106	6157	6209	6235	6261	6287	6313	6339	6366
2	98·50	99·00	99·17	99·25	99·30	99·33	99·36	99·37	99·39	99·40	99·42	99·43	99·45	99·46	99·47	99·47	99·48	99·49	99·50
3	34·12	30·82	29·46	28·71	28·24	27·91	27·67	27·49	27·35	27·23	27·05	26·87	26·69	26·60	26·50	26·41	26·32	26·22	26·13
4	21·20	18·00	16·69	15·98	15·52	15·21	14·98	14·80	14·66	14·55	14·37	14·20	14·02	13·93	13·84	13·75	13·65	13·56	13·46
5	16·26	13·27	12·06	11·39	10·97	10·67	10·46	10·29	10·16	10·05	9·89	9·72	9·55	9·47	9·38	9·29	9·20	9·11	9·02
6	13·75	10·92	9·78	9·15	8·75	8·47	8·26	8·10	7·98	7·87	7·72	7·56	7·40	7·31	7·23	7·14	7·06	6·97	6·88
7	12·25	9·55	8·45	7·85	7·46	7·19	6·99	6·84	6·72	6·62	6·47	6·31	6·16	6·07	5·99	5·91	5·82	5·74	5·65
8	11·26	8·65	7·59	7·01	6·63	6·37	6·18	6·03	5·91	5·81	5·67	5·52	5·36	5·28	5·20	5·12	5·03	4·95	4·86
9	10·56	8·02	6·99	6·42	6·06	5·80	5·61	5·47	5·35	5·26	5·11	4·96	4·81	4·73	4·65	4·57	4·48	4·40	4·31
10	10·04	7·56	6·55	5·99	5·64	5·39	5·20	5·06	4·94	4·85	4·71	4·56	4·41	4·33	4·25	4·17	4·08	4·00	3·91
11	9·65	7·21	6·22	5·67	5·32	5·07	4·89	4·74	4·63	4·54	4·40	4·25	4·10	4·02	3·94	3·36	3·78	3·69	3·60
12	9·33	6·93	5·95	5·41	5·06	4·82	4·64	4·50	4·39	4·30	4·16	4·01	3·86	3·78	3·70	3·62	3·54	3·45	3·36
13	9·07	6·70	5·74	5·21	4·86	4·62	4·44	4·30	4·19	4·10	3·96	3·82	3·66	3·59	3·51	3·43	3·34	3·25	3·17
14	8·86	6·51	5·56	5·04	4·69	4·46	4·28	4·14	4·03	3·94	3·80	3·66	3·51	3·43	3·35	3·27	3·18	3·09	3·00
15	8·68	6·36	5·42	4·89	4·56	4·32	4·14	4·00	3·89	3·80	3·67	3·52	3·37	3·29	3·91	3·13	3·05	2·96	2·87
16	8·53	6·23	5·29	4·77	4·44	4·20	4·03	3·89	3·78	3·69	3·55	3·41	3·26	3·18	3·10	3·02	2·03	2·84	2·75
17	8·40	6·11	5·18	4·67	4·34	4·10	3·93	3·79	3·63	3·59	3·46	3·31	3·16	3·08	3·00	2·92	2·83	2·75	2·65
18	8·29	6·01	5·09	4·58	4·25	4·01	3·84	3·71	3·60	3·51	3·37	3·23	3·08	3·00	2·92	2·84	2·75	2·66	2·57
19	8·18	5·93	5·01	4·50	4·17	3·94	3·77	3·63	3·52	3·43	3·30	3·15	3·00	2·92	2·84	2·76	2·67	2·58	2·49
20	8·10	5·85	4·94	4·43	4·10	3·87	3·70	3·56	3·46	3·37	3·23	3·09	2·94	2·86	2·78	2·69	2·61	2·52	2·42
21	8·02	5·78	4·87	4·37	4·04	3·81	3·64	3·51	3·40	3·31	3·17	3·03	2·88	2·80	2·72	2·64	2·55	2·46	2·36
22	7·95	5·72	4·82	4·31	3·99	3·76	3·59	3·45	3·35	3·26	3·12	2·98	2·83	2·75	2·67	2·58	2·50	2·40	2·31
23	7·88	5·66	4·76	4·26	3·94	3·71	3·54	3·41	3·30	3·21	3·07	2·93	2·78	2·70	2·62	2·54	2·45	2·35	2·25
24	7·82	5·61	4·72	4·22	3·90	3·67	3·50	3·36	3·26	3·17	3·03	2·89	2·74	2·66	2·58	2·49	2·40	2·31	2·21
25	7·77	5·57	4·68	4·18	3·85	3·63	3·46	3·32	3·22	3·13	2·99	2·85	2·70	2·62	2·54	2·45	2·36	2·27	2·17
26	7·72	5·53	4·64	4·14	3·82	3·59	3·42	3·29	3·18	3·09	2·93	2·81	2·66	2·58	2·50	2·42	2·33	2·23	2·13
27	7·68	5·49	4·60	4·11	3·78	3·56	3·39	3·26	3·15	3·06	2·93	2·78	2·63	2·55	2·47	2·38	2·29	2·20	2·10
28	7·64	5·45	4·57	4·07	3·75	3·53	3·36	3·23	3·12	3·03	2·90	2·75	2·60	2·52	2·44	2·35	2·26	2·17	2·06
29	7·60	5·42	4·54	4·04	3·73	3·50	3·33	3·20	3·09	3·00	2·87	2·73	2·57	2·49	2·41	2·33	2·23	2·14	2·03
30	7·56	5·39	4·51	4·02	3·70	3·47	3·30	3·17	3·07	2·98	2·84	2·70	2·55	2·47	2·39	2·30	2·21	2·11	2·01
40	7·31	5·18	4·31	3·83	3·51	3·29	3·12	2·99	2·89	2·80	2·66	2·52	2·37	2·29	2·20	2·11	2·02	1·92	1·80
60	7·08	4·98	4·13	3·65	3·34	3·12	2·95	2·82	2·72	2·63	2·50	2·35	2·20	2·12	2·03	1·94	1·84	1·73	1·60
120	6·85	4·79	3·95	3·48	3·17	2·96	2·79	2·66	2·56	2·47	2·34	2·19	2·03	1·95	1·86	1·76	1·66	1·53	1·38
∞	6·63	4·61	3·78	3·32	3·02	2·80	2·64	2·51	2·41	2·32	2·18	2·04	1·88	1·79	1·70	1·59	1·47	1·32	1·00

Upper 1% Points of F-distribution

v_2 \ v_1	1	2	3	4	5	6	7	8	9	10	12	15	20	24	30	40	60	120	∞
1	161·4	199·5	215·7	224·6	230·2	234·0	236·8	238·9	240·5	241·9	243·9	245·9	248·0	249·1	250·1	251·1	252·2	253·3	254·3
2	18·51	19·00	19·16	19·25	19·30	19·33	19·35	19·37	19·38	19·40	19·41	19·43	19·45	19·45	19·46	19·47	19·48	19·49	19·50
3	10·13	9·55	9·28	9·12	9·01	8·94	8·89	8·85	8·81	8·79	8·74	8·70	8·66	8·64	8·62	8·59	8·57	8·55	8·53
4	7·71	6·94	6·59	6·39	6·26	6·16	6·09	6·04	6·00	5·96	5·91	5·86	5·80	5·77	5·75	5·72	5·69	5·66	5·63
5	6·61	5·79	5·41	5·19	5·05	4·95	4·88	4·82	4·77	4·74	4·68	4·62	4·56	4·53	4·50	4·46	4·43	4·40	4·36
6	5·99	5·14	4·76	4·53	4·39	4·28	4·21	4·15	4·10	4·06	4·00	3·94	3·87	3·84	3·81	3·77	3·74	3·70	3·67
7	5·59	4·74	4·35	4·12	3·97	3·87	3·79	3·73	3·68	3·64	3·57	3·51	3·44	3·41	3·38	3·34	3·30	3·27	3·23
8	5·32	4·46	4·07	3·84	3·69	3·58	3·50	3·44	3·39	3·35	3·28	3·22	3·15	3·12	3·08	3·04	3·01	2·97	2·93
9	5·12	4·26	3·86	3·63	3·48	3·37	3·29	3·23	3·18	3·14	3·07	3·01	2·94	2·90	2·86	2·83	2·79	2·75	2·71
10	4·96	4·10	3·71	3·48	3·33	3·22	3·14	3·07	3·02	2·98	2·91	2·85	2·77	2·74	2·70	2·66	2·62	2·58	2·54
11	4·84	3·98	3·59	3·36	3·20	3·09	3·01	2·95	2·90	2·85	2·79	2·72	2·65	2·61	2·57	2·53	2·49	2·45	2·40
12	4·75	3·89	3·49	3·26	3·11	3·00	2·91	2·85	2·80	2·75	2·69	2·62	2·54	2·51	2·47	2·43	2·38	2·34	2·30
13	4·67	3·81	3·41	3·18	3·03	2·92	2·83	2·77	2·71	2·67	2·60	2·53	2·46	2·42	2·38	2·34	2·30	2·25	2·21
14	4·60	3·74	3·34	3·11	2·96	2·85	2·76	2·70	2·65	2·60	2·53	2·46	2·39	2·35	2·31	2·27	2·22	2·18	2·13
15	4·54	3·68	3·29	3·06	2·90	2·79	2·71	2·64	2·59	2·54	2·48	2·40	2·33	2·29	2·25	2·20	2·16	2·11	2·07
16	4·49	3·63	3·24	3·01	2·85	2·74	2·66	2·59	2·54	2·49	2·42	2·35	2·28	2·24	2·19	2·15	2·11	2·06	2·01
17	4·45	3·59	3·20	2·96	2·81	2·70	2·61	2·55	2·49	2·45	2·38	2·31	2·23	2·19	2·15	2·10	2·06	2·01	1·96
18	4·41	3·55	3·16	2·93	2·77	2·66	2·58	2·51	2·46	2·41	2·34	2·27	2·19	2·15	2·11	2·06	2·02	1·97	1·92
19	4·38	3·52	3·13	2·90	2·74	2·63	2·54	2·48	2·42	2·38	2·31	2·23	2·16	2·11	2·07	2·03	1·98	1·93	1·88
20	4·35	3·49	3·10	2·87	2·71	2·60	2·51	2·45	2·39	2·35	2·28	2·20	2·12	2·08	2·04	1·99	1·95	1·90	1·84
21	4·32	3·47	3·07	2·84	2·68	2·57	2·49	2·42	2·37	2·32	2·25	2·18	2·10	2·05	2·01	1·96	1·92	1·87	1·81
22	4·30	3·44	3·05	2·82	2·66	2·55	2·46	2·40	2·34	2·30	2·23	2·15	2·07	2·03	1·98	1·94	1·89	1·84	1·78
23	4·28	3·42	3·03	2·80	2·64	2·53	2·44	2·37	2·32	2·27	2·20	2·13	2·05	2·01	1·96	1·91	1·86	1·81	1·76
24	4·26	3·40	3·01	2·78	2·62	2·51	2·42	2·36	2·30	2·25	2·18	2·11	2·03	1·98	1·94	1·89	1·84	1·79	1·73
25	4·24	3·39	2·99	2·76	2·60	2·49	2·40	2·34	2·28	2·24	2·16	2·09	2·01	1·96	1·92	1·87	1·82	1·77	1·71
26	4·23	3·37	2·98	2·74	2·59	2·47	2·39	2·32	2·27	2·22	2·15	2·07	1·99	1·95	1·90	1·85	1·80	1·75	1·69
27	4·21	3·35	2·96	2·73	2·57	2·46	2·37	2·31	2·25	2·20	2·13	2·06	1·97	1·93	1·88	1·84	1·79	1·73	1·67
28	4·20	3·34	2·95	2·71	2·56	2·45	2·36	2·29	2·24	2·19	2·12	2·04	1·96	1·91	1·87	1·82	1·77	1·71	1·65
29	4·18	3·33	2·93	2·70	2·55	2·43	2·35	2·28	2·22	2·18	2·10	2·03	1·94	1·90	1·85	1·81	1·75	1·70	1·64
30	4·17	3·32	2·92	2·69	2·53	2·42	2·33	2·27	2·21	2·16	2·09	2·01	1·93	1·89	1·84	1·79	1·74	1·68	1·62
40	4·08	3·23	2·84	2·61	2·45	2·34	2·25	2·18	2·12	2·08	2·00	1·92	1·84	1·79	1·74	1·69	1·64	1·58	1·51
60	4·00	3·15	2·76	2·53	2·37	2·25	2·17	2·10	2·04	1·99	1·92	1·84	1·75	1·70	1·65	1·59	1·53	1·47	1·39
120	3·92	3·07	2·68	2·45	2·29	2·17	2·09	2·02	1·96	1·91	1·83	1·75	1·66	1·61	1·55	1·50	1·43	1·35	1·25
∞	3·84	3·00	2·60	2·37	2·21	2·10	2·01	1·94	1·88	1·83	1·75	1·67	1·57	1·52	1·46	1·39	1·32	1·22	1·00

VALUES of χ^2

The table gives values of χ^2 at probability leaves ·99, ·98, . . . ·01, ·001 for various degrees of freedom '*n*'

n	.99	.98	.95	.90	.30	.70	.50	.30	.20	.10	.05	.02	.01	.001
1	000157	000628	00393	.0158	0642	148	455	1 074	1 642	2 706	3 841	5 412	6 635	10 827
2	0201	0404	103	211	446	713	1 386	2 408	3 219	4 605	5 991	7 824	9 210	13 815
3	115	185	.352	584	1 005	1 424	2 366	3 665	4 642	6 251	7 815	9 837	11 345	16 268
4	297	429	711	1 064	1 649	2 195	3 357	4 878	5 989	7 779	9 488	11 668	13 277	18 465
5	554	.752	1.145	1.610	2 343	3 000	4 351	6 064	7 289	9 236	11 070	13 388	15 086	20 517
6	872	1 134	1.635	2 204	3 070	3 828	5 348	7 231	8 558	10 645	12 592	15 033	16.812	22 457
7	1.239	1.564	2 167	2 833	3.822	4 671	6 346	8 383	9 803	12 017	14 067	16.622	18 475	24 322
8	1 646	2 032	2 733	3 490	4 594	5.527	7 344	9.524	11 030	13.362	15 507	18 168	20 090	26 125
9	2 088	2 532	3.325	4.168	5 380	6 393	8.343	10.656	12.242	14 684	16 919	19 679	21.666	27 877
10	2 558	3.059	3.940	4 865	6.179	7 267	9 342	11.781	13.442	15.987	18 307	21.161	23 209	29 588
11	3 053	3.609	4 575	5 578	6 989	8 148	10 341	12.899	14.631	17.275	19 675	22.618	24 725	31 264
12	3 571	4 178	5 226	6 304	7 807	9 034	11 340	14.011	15 812	18.549	21 026	24 054	26 217	32 909
13	4 107	4 765	5 892	7 042	8.634	9 926	12 340	15.119	16·985	19 812	22.362	25 472	27 688	34 528
14	4 660	5 368	6 571	7 790	9 467	10 821	13.339	16.222	18 151	21 064	23 685	26 873	29.141	36 123
15	5 229	5 985	7.261	8.547	10 307	11 721	14 339	17.322	19·311	22.307	24 996	28 259	30 578	37.697
16	5 812	6.614	7.962	9 312	11.152	12 624	15.338	18.418	20.465	23.542	26 296	29 633	32 000	39 252
17	6 408	7.255	8.672	10.085	12.002	13 531	16.338	19.511	21 615	24.769	27 587	30 995	33 409	40 790
18	7 015	7.906	9 390	10.865	12 857	14 440	17.338	20.601	22.760	25.989	28 869	32 346	34 805	42 312
19	7.633	8 567	10.117	11 651	13 716	15 352	18.338	21.689	23 900	27 204	30 144	33 687	36 191	43 820
20	8.260	9 237	10.851	12.443	14.578	16 266	19 337	22.775	25 038	28 412	31 410	35 020	37 566	45 315
21	8.897	9 915	11.591	13.240	15.445	17 182	20.337	23.858	26 171	29 615	32 671	36 343	38 932	46 797
22	9 542	10 600	12 338	14 041	16 314	18 101	21 337	24.939	27 301	30 813	33 924	37 659	40 289	48 268
23	10 196	11 293	13 091	14.848	17 187	19.021	22 337	26 018	28 429	32 007	35 172	38 968	41 638	49 728
24	10 856	11 992	13 848	15 659	18 062	19 943	23 337	27 096	29 553	33 196	36 415	40 270	42 980	51 179
25	11 524	12 697	14 611	16.473	18 940	20 867	24 337	28.172	30 675	34 382	37 652	41 566	44 314	52 620
26	12 198	13 409	15 379	17 292	19 820	21 792	25 336	29 246	31 795	35 563	38 885	42 856	45 642	54 052
27	12 879	14 125	16 151	18.114	20·703	22 719	26 336	30 319	32 912	36 741	40 113	44 140	46 963	55 476
28	13 565	14 847	16 928	18.939	21 588	23 647	27 336	31 391	34 027	37 916	41 337	45 419	48 278	56 893
29	14 256	15.574	17 708	19 768	22 475	24 577	28 336	32 461	35 139	39 087	42 557	46 693	49 588	58 302
30	14 953	16 306	18 493	20 599	23 364	25 508	29 336	33 530	36 250	40 256	43 773	47 962	50 892	59 703